Current Topics in Behavioral Neurosciences

Volume 49

Current Topics in Behavioral Neurosciences provides critical and comprehensive discussions of the most significant areas of behavioral neuroscience research, written by leading international authorities. Each volume in the series represents the most informative and contemporary account of its subject available, making it an unrivalled reference source. Each volume will be made available in both print and electronic form.

With the development of new methodologies for brain imaging, genetic and genomic analyses, molecular engineering of mutant animals, novel routes for drug delivery, and sophisticated cross-species behavioral assessments, it is now possible to study behavior relevant to psychiatric and neurological diseases and disorders on the physiological level. The Behavioral Neurosciences series focuses on translational medicine and cutting-edge technologies. Preclinical and clinical trials for the development of new diagnostics and therapeutics as well as prevention efforts are covered whenever possible. Special attention is also drawn on epigenetical aspects, especially in psychiatric disorders.

CTBN series is indexed in PubMed and Scopus.

More information about this series at http://www.springer.com/series/7854

Naomi A. Fineberg • Trevor W. Robbins
Editors

The Neurobiology and Treatment of OCD: Accelerating Progress

Springer

Editors
Naomi A. Fineberg
Centre for Health Services and Clinical Research, School of Life and Medical Sciences
University of Hertfordshire
Hatfield, Hertfordshire, UK

Trevor W. Robbins
Department of Psychology and Behavioural and Clinical Neuroscience Institute
University of Cambridge
Cambridge, UK

ISSN 1866-3370 ISSN 1866-3389 (electronic)
Current Topics in Behavioral Neurosciences
ISBN 978-3-030-75392-4 ISBN 978-3-030-75393-1 (eBook)
https://doi.org/10.1007/978-3-030-75393-1

This Springer imprint is published by the registered company Springer Nature Switzerland AG.
The registered company address is: Gewerbestrasse 11, 6330 Cham, Switzerland

Preface

Obsessive-compulsive disorder (OCD) retains pole position in obsessive compulsive and related disorders (OCRDs) research. The OCRDs represent a broad spectrum of common, disabling and on the whole poorly understood brain disorders, sharing obsessive and compulsive thoughts and behaviours as their defining characteristics (Hollander et al. 2011). In recent years, following an exhaustive process of expert consultation and research review (Phillips et al. 2010), the Diagnostic and Statistical Manual of Mental Disorders (5th ed) (DSM-5; American Psychiatric Association 2013) and subsequently the International Classification of Diseases (11th Revision) (ICD-11; World Health Organization 2018) established the OCRDs as a brand new 'family' of disorders. In the DSM-5, OCD was re-positioned alongside diverse disorders including hoarding disorder, body dysmorphic disorder, skin picking disorder, trichotillomania and in the ICD-11 this grouping was subsequently extended to include hypochondriasis and olfactory reference disorder.

The landmark creation of the OCRDS by the DSM-5 and ICD-11, viewed by some as a rather bold move (Stein et al. 2020), was inspired not only by the evident overlap between constituent disorders in terms of specific clinical factors of nosological relevance, such as phenomenology, age at onset, clinical course, comorbidity profile, family history, pharmacotherapy and psychotherapy response (Fineberg et al. 2011a) but also by the convergence of findings from a number of seminal neurobiological studies, including brain imaging and neurocognitive paradigms (e.g. Baxter 1994; Rauch et al. 1994, 1997; Brody et al. 1998; Chamberlain et al. 2008, 2010) supported by findings from animal modelling (reviewed in Fineberg et al. 2010, 2011b; d'Angelo et al. 2014). Albeit few in number and mostly derived from relatively small human study samples, this emerging evidence persuasively suggested that alteration of the normal neurocognitive mechanisms mediating, *inter alia,* behavioural inhibition (motor inhibition, cognitive inflexibility), reversal learning and habit formation (shift from goal-directed to habitual responding) and the underpinning fronto-striatal neural circuitry contributed toward obsessive-compulsive symptoms both in OCD and also across the broader spectrum of the OCRDs.

The OCRDs, it was thought, may therefore derive from disruption of the normal functional inhibition of thoughts and behaviours naturally prone to excess, for example, cleaning, grooming, eating, purging, gambling and checking (Fineberg et al. 2010, 2018a).

The new classification was therefore intended to actively encourage scientists and clinicians to capitalize on these early findings, strengthen and generate new research heuristics, such as computational modelling approaches and develop more systematic and incisive methods for investigating these clinically important but relatively underexplored disorders from the perspective of behavioural neuroscience. One important implication of this change was a move away from traditional 'anxiety-based' conceptualizations of OCD, for example, as a disorder driven by pathological sensitivity to anxiety-inducing cues, toward a greater focus on the role of 'compulsivity'. Compulsions can be defined as repetitive, stereotyped thoughts and behaviours designed to reduce harm. However, they also lead to a profound experience of 'lack of control' (Fineberg et al. 2016). These neurocognitive mechanisms can now be reliably probed using a variety of computer-based tests and questionnaires, alongside electrophysiological and brain imaging paradigms. An improved understanding of the neural processes underpinning compulsivity across the OCRDs and the associated causal factors would be expected to inform the development of new preventative and therapeutic interventions (Fineberg et al. 2018a).

Remarkably, to a large extent, that promise has been realized. Acknowledging this, the field was exhaustively reviewed in a major volume edited by Pittenger (2017). However, as this volume shows, even since then, research into the OCRDs, and OCD in particular, has accelerated apace, bringing fresh insights into their etiologies, brain-based mechanisms and identifying promising new clinical interventions. Moreover, research success in the field of the OCRDs is paving the way for the study of other related areas of psychiatry and public health, such as the new and exciting field of behavioural addictions, including the increasing number of recognized disorders falling under the umbrella of problematic usage of the Internet (Fineberg et al. 2018b).

Contemporary research into OCRDs encompasses a broad and diverse range of disciplines and models, ranging from theoretical to therapeutic and from preclinical to clinical. This volume mirrors the broad scope of the research landscape and brings together a series of 'up to the minute' chapters, written by the leading experts in their field, succinctly reviewing those areas where most progress has been made. The chapters are written in accessible language aimed at the interested scientist, clinician or informed lay reader and wherever appropriate illustrated for clarity.

The aetiology of OCD is still somewhat of a mystery, even in this age of functional genomics and this issue is addressed, from different perspectives by four chapters in the present volume. Grunblatt (Genetics) reveals what progress has been made in the area which suggests that we are beginning to match what has been achieved for schizophrenia and depression, but we are still some way off from compiling a definitive genetic risk for OCD. Since the original discovery of the PANDAS (Pediatric Autoimmune Neuropsychiatric Disorders Associated with

Streptococcal Infections) syndrome, immunological factors have been considered to play an important role in OCD. Meyer's timely chapter takes account of the recent upsurge of interest in immunological factors in psychiatry, including central and peripheral markers of intra-cerebral gliosis. It is now generally recognized that most psychiatric disorders have their origins before the age of 24 years and OCD is no exception to this principle. Chapter "On the Development of OCD" therefore reviews developmental aspects of OCD and argues cogently for a longitudinal study approach. Animal models also have much to provide concerning the analysis of possible factors in the aetiology and neural circuitry of OCD—an area that has been enhanced by the use of modern neurobiological methods such as optogenetics (chapter "Animal Models for OCD Research").

There is no doubt that our enhanced methods of imaging the human brain have led to a much greater understanding of the neural substrates and circuitries of OCD, building on the original seminal findings described above. These advances are covered by chapters on structural neuroimaging (Veltman), molecular neuroimaging (PET and MRS) (Banca et al.), functional neuroimaging (Soriano-Mas) and electrophysiology (De Souza). The combination of enhanced imaging methods and more sophisticated theoretical psychological models of OCD is highlighted by the chapters on the goal-directed/habit learning hypothesis (Gillan), cognitive inflexibility (SR Chamberlain et al.) and cognitive neuroendophenotypes (Vaghi).

Finally, OCD is perhaps unusual in psychiatry in having so many approaches to treatment, psychological, pharmacological and surgical. Exciting advance in these domains is covered for psychotherapy (Patel et al.) and pharmacotherapy (Pittenger), counterpoised with the chapter "Pharmacogenetics of Obsessive-Compulsive Disorder: An Evidence-Update" (Zai), in which the known pharmacogenomic risk factors for treatment response are carefully reviewed. Non-invasive neurostimulation, in its various forms, targeting relatively superficial nodes and tracts within the candidate cortico-striatal circuitry, as described in the chapter "Invasive and Non-invasive Neurostimulation for OCD", is gaining traction as an efficacious and relatively well-tolerated treatment for those responding poorly to conventional approaches with drugs or cognitive behavioural therapy. However, for the purposes of neurosurgical interventions for OCD (neurostimulation and lesion surgery), 'treatment refractoriness' implies a biological substrate that is totally unresponsive to *all* available evidence-based psychological and pharmacological therapies and thus an underlying neurobiology that may be qualitatively different. Procedural refinements in deep brain stimulation (Bergfeld et al.) and ablative neurosurgery (Etherington et al.) targeting the deeper subcortical tracts, including evolving methods of irradiation, e.g. using the gamma knife or MRI guided focused ultrasound as lesion-creation methods not requiring craniotomy, provide exciting opportunities for future controlled studies to test these assumptions.

Through this comprehensive collection of papers, we not only present the current state of the art but also, as our understanding grows, highlight those new directions of discovery we believe will turn out most profitable for clinicians, researchers and patients in the years ahead (Vats et al.). These include the study of patient centred outcomes, developmental brain trajectories in spectrum conditions, pharmacogenomics, problematic use of the internet and digital interventions. As an example, digital medicine may aid the early identification of patients before their illness takes hold, provide digital biomarkers to subtype patients, predict the likelihood of treatment response or even act as a more proximal outcome measure of treatment response than do existing clinical ratings and provide more easily accessible or less costly forms of care. In order to address the unmet clinical needs of patients with OCRD and the substantial cost and burden falling on them and their families, it is essential to take an interdisciplinary approach. The work described in this volume is therefore expected to be invaluable in shaping the future of research in the field.

Acknowledgments

Much of the work of TWR on OCRDs has been supported by the Wellcome Trust and of NAF by the National Institute of Health Research and COST (European Cooperation in Science and Technology), linked to the Project 'European Research Network into Problematic Usage of the Internet' (EU-PUI) and funded by the Horizon 2020 Framework Programme of the European Union.

Hatfield, UK N. A. Fineberg
Cambridge, UK T. W. Robbins

References

American Psychiatric Association (2013) Diagnostic and statistical manual of mental disorders, 5th edn. Washington, DC

Baxter LR Jr (1994) Positron emission tomography studies of cerebral glucose metabolism in obsessive compulsive disorder. J Clin Psychiatry 55(Suppl):54–59

Brody AL, Saxena S, Schwartz JM, Stoessel PW, Maidment K, Phelps ME, Baxter LR Jr (1998) FDG-PET predictors of response to behavioral therapy and pharmacotherapy in obsessive compulsive disorder. Psychiatry Res 84(1):1–6. https://doi.org/10.1016/s0925-4927(98)00041-9

Chamberlain SR, Menzies L, Hampshire A, Suckling J, Fineberg NA, del Campo N, Aitken M, Craig K, Owen AM, Bullmore ET, Robbins TW, Sahakian BJ (2008) Orbitofrontal dysfunction in patients with obsessive-compulsive disorder and their unaffected relatives. Science 321 (5887):421–422

Chamberlain SR, Hampshire A, Menzies LA, Garyfallidis E, Grant JE, Odlaug BL, Craig K, Fineberg N, Sahakian BJ (2010) Reduced brain white matter integrity in trichotillomania: a diffusion tensor imaging study. Arch Gen Psychiatry 67(9):965–971. https://doi.org/10.1001/archgenpsychiatry.2010.109

d'Angelo CLS, Eagle DM, Grant JE, Fineberg NA, Robbins TW, Chamberlain SR (2014) Animal models of obsessive-compulsive spectrum disorders. CNS Spectrums 19(1):28–49

Fineberg NA, Robbins TW, Bullmore E, Potenza M, Menzies L, Chamberlain S, Sahakian B, Bechara A, Hollander E (2010) Probing compulsive and impulsive behaviors, from animal models to endophenotypes: a narrative review. Neuropsychopharmacology 35(3):591–604. Epub 2009 Nov 25

Fineberg NA, Saxena S, Zohar J, Craig K (2011a) Obsessive-compulsive disorder boundary issues. In: Hollander E, Zohar J, Sirovatka PJ, Regier DA (eds) Obsessive-compulsive spectrum disorders: refining the research agenda for DSM-V. American Psychiatric Publishing Inc., Arlington, pp 1–32

Fineberg NA, Chamberlain S, Hollander E, Boulougouris V, Robbins T (2011b) Translational approaches to obsessive-compulsive disorder: from animal models to clinical treatment. Br J Pharmacol 164(4):1044–1061

Fineberg NA, Menchon JM, Zohar J, Veltman DJ (2016) Compulsivity-A new trans-diagnostic research domain for the Roadmap for Mental Health Research in Europe (ROAMER) and Research Domain Criteria (RDoC) initiatives. Eur Neuropsychopharmacol 26(5):797–799

Fineberg NA, Apergis-Schoute AM, Vaghi MM, Banca P, Gillan CM, Voon V, Chamberlain SR, Cinosi E, Reid J, Shahper S, Bullmore ET, Sahakian BJ, Robbins TW (2018a) Mapping compulsivity in the DSM-5 obsessive compulsive and related disorders: cognitive domains, neural circuitry, and treatment. Int J Neuropsychopharmacol 21(1):42–58

Fineberg NA, Demetrovics Z, Stein DJ, Ioannidis K, Potenza MN, Grünblatt E, Brand M, Billieux J, Carmi L, King DL, Grant JE, Yücel M, Dell'Osso B, Rumpf HJ, Hall N, Hollander E, Goudriaan AE, Menchon J, Zohar J, Burkauskas J, Martinotti G, Van Ameringen M, Corazza O, Pallanti S (2018b) COST Action Network and Chamberlain SR. Manifesto for a European Research Network into problematic usage of the internet. Eur Neuropsychopharmacol 28 (11):1232–1246

Hollander E, Kim S, Braun A, Simeon D, Zohar Z (2011) Cross cutting issues and future directions for the obsessive-compulsive spectrum of disorders. In: Hollander E, Zohar J, Sirovatka PJ, Regier DA (eds) Obsessive-compulsive spectrum disorders; refining the research agenda for DSM-V. American Psychiatric Publishing Inc., Arlington, pp xix–xxiv

Phillips KA, Stein DJ, Rauch S, Hollander E, Fallon B, Barsky A, Fineberg NA, Mataix-Cols D, Ferrão YA, Saxena S, Wilhelm S, Kelly MM, Clark LA, Pinto A, Bienvenu OJ, Farrow J, Leckman J (2010) Should an obsessive-compulsive spectrum grouping of disorders be included in DSM-V? Depress Anxiety 27(6):528–555

Pittenger C (2017) Obsessive-compulsive disorder; phenomenology, pathophysiology and treatment. Oxford University Press, New York

Rauch SL, Jenike MA, Alpert NM, Baer L, Breiter HC, Savage CR, Fischman AJ (1994) Regional cerebral blood flow measured during symptom provocation in obsessive-compulsive disorder using oxygen 15-labeled carbon dioxide and positron emission tomography. Arch Gen Psychiatry 51(1):62–70

Rauch SL, Savage CR, Alpert NM, Dougherty D, Kendrick A, Curran T, Brown HD, Manzo P, Fischman AJ, Jenike MA (1997) Probing striatal function in obsessive-compulsive disorder: a PET study of implicit sequence learning. J Neuropsychiatry Clin Neurosci 9(4):568–573

Stein DJ, Szatmari P, Gaebel W, Berk M, Vieta E, Maj M, de Vries YA, Roest AM, de Jonge P, Maercker A, Brewin CR, Pike KM, Grilo CM, Fineberg NA, Briken P, Cohen-Kettenis PT, Reed GM (2020) Mental, behavioural and neurodevelopmental disorders in the ICD-11: an international perspective on key changes and controversies. BMC Med 18(1):21. https://doi.org/10.1186/s12916-020-1495-2

World Health Organization (2018) International classification of diseases for mortality and morbidity statistics (11th Revision)

Contents

Genetics of OCD and Related Disorders; Searching for Shared Factors

Edna Grünblatt

Contents

Abstract Obsessive compulsive disorder (OCD) and several other obsessive-compulsive related disorders (OCRDs) including hoarding disorder, body dysmorphic disorder (BDD), skin picking disorder, trichotillomania and the newly arising public health conditions of online gaming and gambling disorders, under the umbrella of Problematic Usage of the Internet (PUI), not only share some common phenotypes, but there is evidence to suggest share some genetic risk factors. The simple fact that these disorders segregate within families points to this notion. However, the current data are still scarce. This chapter focuses on identifying the shared genetic factors. To address this question, a systematic review of the literature investigating genetic findings in OCD and OCRDs including PUI was conducted, with a focus on their genetic similarities. Greater knowledge of the specific genetic risks shared among OCRDs would be expected to open new avenues in the understanding of the biological mechanisms causing the development of these phenotypes, as well as provide opportunities to develop new animal and cellular models testing new therapy avenues.

E. Grünblatt (✉)
Department of Child and Adolescent Psychiatry and Psychotherapy, Psychiatric University Hospital Zurich, University of Zurich, Zurich, Switzerland

Neuroscience Center Zurich, University of Zurich and ETH Zurich, Zurich, Switzerland

Zurich Center for Integrative Human Physiology, University of Zurich, Zurich, Switzerland
e-mail: edna.gruenblatt@kjpd.uzh.ch

Curr Topics Behav Neurosci (2021) 49: 1–16
https://doi.org/10.1007/7854_2020_194
Published Online: 6 February 2021

Keywords Body dysmorphic disorder · Genetic · Genome-wide association study · GWAS · Heritability · Obsessive compulsive disorder · Obsessive-compulsive related disorder · Polygenic risk score · Polymorphisms · Problematic use of internet · PRS · Skin picking · SNP · Trichotillomania

1 Introduction

According to DSM-5 classification (American Psychiatric Association 2013), the diagnosis of obsessive compulsive disorder (OCD) now heads a chapter of various different obsessive-compulsive and related disorders (OCRDs) including hoarding disorder, body dysmorphic disorder (BDD), skin picking disorder and trichotillomania. All these disorders have shown some evidence of overlap in their genetic risk factors (Browne et al. 2014). In addition, the newly arising public health conditions of online gaming and gambling disorders, included under the umbrella of Problematic Usage of the Internet (PUI), and considered as newly emerging obsessive-compulsive (OC)-related or addictive disorders (Fineberg et al. 2018), also share some of the same risk genes found to associate with OCD and other OCRDs, however the data are still rather scarce. In this chapter, the current knowledge on their heritability, shared candidate genes, and findings originating from genome-wide association studies (GWAS) will be presented. Since several reviews already exist relating to several of these disorders (Browne et al. 2014; Fineberg et al. 2018; Zai et al. 2019; Chattopadhyay 2012), this chapter focuses in particular on the search for shared genetic factors. Such shared genetic factors may explain the high familial occurrence of OCRDs, which, in the case of OCD, involves a risk of occurring among first-degree relatives of affected probands approaching 55% (Browne et al. 2014). Indeed, the risk is likely to be even higher, if the spectrum of OC disorders is taken into account, thought specifically due to the presence of shared genetic risk factors. Greater knowledge about the specific shared genetic risks will also open new avenues in the understanding of the biological mechanisms causing the development of these phenotypes, and provide opportunities to develop new animal and cellular models testing new therapy avenues.

To address this question, a systematic review of the published studies investigating current knowledge of the genetics of OCD and other OCRDs, including PUI, was conducted, with a focus on searching for their genetic similarities.

2 Search Strategy

A review of the literature was conducted searching articles that included studies describing genetics and heritability in OCD, OCRDs and PUI. Articles were identified in PubMed and Google Scholar. The final search was launched on December 2nd, 2019. Literature was searched using the keywords: (American Psychiatric Association 2013) (Internet OR "computer gaming" OR "computer game" OR "text messaging" OR "online buying" OR "online pornography" OR "online shopping" OR "video gaming" OR "online gambling" OR "social media" OR cybersex OR "cell phone" OR digital*) AND (problem* OR behav* OR addict*) AND ("mental health" OR psychiatr*) AND (gen* OR SNP* OR "genome wide" OR GWA* OR polymorphism*); (Browne et al. 2014) (Internet OR "computer gaming" OR "computer game" OR "text messaging" OR "online buying" OR "online pornography" OR "online shopping" OR "video gaming" OR "online gambling" OR "social media" OR cybersex OR "cell phone" OR digital*) AND (problem* OR behav* OR addict* OR "mental health" OR psychiatr*) AND (heredit* OR heritab*); (Fineberg et al. 2018) (Internet OR "computer gaming" OR "computer game" OR "text messaging" OR "online buying" OR "online pornography" OR "online shopping" OR "video gaming" OR "online gambling" OR "social media" OR cybersex OR "cell phone" OR digital*) AND (problem* OR behav* OR addict* OR "mental health" OR psychiatr*) AND (gen* OR SNP* OR "genome wide" OR GWA* OR polymorphism*); (Zai et al. 2019) (substance OR SUD OR addict*) AND (OCD OR "obsessive compulsive") AND (gene* OR SNP* OR polymorphism* OR "genome wide" OR GWA*); (Chattopadhyay 2012) (gen* OR SNP* OR "genome wide" OR GWA* OR polymorphism*) AND (Skin picking" OR Trichotillomania OR excoriation); (Burton et al. 2018) (gen* OR SNP* OR "genome wide" OR GWA* OR polymorphism*) AND hoarding; (Monzani et al. 2014) (gen* OR SNP* OR "genome wide" OR GWA* OR polymorphism*) AND "body dysmorphic".

3 Prevalence and Heritability of OCD, OC-Related Disorders and PUI

Family studies and twin studies have shown that OCD has a higher prevalence in individuals with relatives affected with OCD or OCRDs, and shows heritability ranging from 27 to 65%, see review by Zai et al. (2019) (Box 1, Table 1).

Box 1 Definitions

Heritability – is an estimate of the proportion of variation in a phenotypic trait in the population that is due to genetic variation.

Prevalence – is the proportion of a population affected by a specific trait or disorder.

SNPs – Single nucleotide polymorphisms, represent a single substitute of one nucleotide in the sequence of the genome. It may be a substitute of one of the nucleotides: A, Adenine; T, Thymine; C, Cytosine; G, Guanine.

GWAS – Genome-wide association study assessing the associations of large sets of genetic variants, usually SNPs, and disorders or traits.

OR – odds ratio describing the association between a specific gene allele and the occurrence of a disorder. OR $= 1$ implies no association between the genotype and disease, OR > 1 means the allele demonstrates increased risk of the disorder, OR < 1 means the allele demonstrates protective effects.

Table 1 Comparison of familial prevalence and heritability of OCD, OC-related disorders and PUI

Disorder	Prevalence in first-degree relatives of affected probands [%]	Reference	Heritability Twin studies	Reference	Heritability GWAS	Reference
OCD	10–12%	Nestadt et al. (2000), Pauls et al. (2014), Pauls (2010)	42–65%	Pauls et al. (2014)	37%	Davis et al. (2013)
Adult onset OCD			27–47%	van Grootheest et al. (2005), Zilhão et al. (2019)		
Childhood onset OCD			45–65%	Walitza et al. (2010), Burton et al. (2020)		
OC dimensions			74% twins and family studies	Burton et al. (2018)		
Hoarding	1.5%	Nordsletten et al. (2013)	45–60% 71% in family study	Burton et al. (2018), Monzani et al. (2014), Nordsletten et al. (2013)		
Hoarding symptoms			65–71%	Mathews et al. (2007), Ivanov et al. (2017)		
Body dysmorphic disorder	6%	Bienvenu et al. (2012)	43%	Monzani et al. (2014)		
Body dysmorphic symptoms			37–49%	Enander et al. (2018)		
Skin picking disorder	17%	Bienvenu et al. (2012)	40–47%	Monzani et al. (2012b, 2014)		
Trichotillomania	4%	Bienvenu et al. (2012)	32–76%	Monzani et al. (2012b), Novak et al. (2009)		
PUI	0.3–36.7% 0.6–2.2% Female adolescents	Cao and Su (2007), Fu et al. (2010), Park et al.	41–59% 48–58%	Li et al. (2014), Deryakulu and Ursavas		

(continued)

Table 1 (continued)

Disorder	Prevalence in first-degree relatives of affected probands [%]	Reference	Heritability Twin studies	Reference	Heritability GWAS	Reference
	4.3–9.7% Male adolescents	(2008), Shaw and Black (2008), Siomos et al. (2008), Yen et al. (2008), Fam (2018), Li et al. (2018), Karhulahti and Koskimaa (2019), Gomez et al. (2020), Kumar et al. (2019), Sayyah and Khanafereh (2019)	Female 48–66% Male	OF (2014), Vink et al. (2016), Long et al. (2016), Hahn et al. (2017)		

GWAS genome-wide association study, *OC* obsessive compulsive, *OCD* obsessive-compulsive disorder, *PUI* problematic usage of the internet

Interestingly, both in twin and family studies, OC trait dimensions as a whole showed heritability of 74% (CI 60–83%) (Burton et al. 2018), while for hoarding and hoarding symptoms slightly lower heritability was found (Table 1) (Burton et al. 2018; Monzani et al. 2014; Nordsletten et al. 2013; Mathews et al. 2007; Ivanov et al. 2017). When comparing the aforementioned prevalence within family members and heritability of OCD, OC dimensions and hoarding, with BDD and body dysmorphic symptoms, a lower familial prevalence as well as lower heritability was reported for BDD (Table 1) (Bienvenu et al. 2012; Monzani et al. 2012a; Enander et al. 2018). Skin picking disorders have shown a slightly higher familial prevalence of around 17% while heritability was found to be around 40–47% (Table 1) (Monzani et al. 2012b, 2014; Bienvenu et al. 2012). Trichotillomania shows a low familial prevalence of around 4%, however its heritability reaches up to 76% according to some reports (Table 1) (Bienvenu et al. 2012; Monzani et al. 2012b; Novak et al. 2009). Considering PUI in this context, depending on the type of PUI, sex, education, ethnicity and culture, the current prevalence is estimated at between 1% up to nearly 40% (Cao and Su 2007; Fu et al. 2010; Park et al. 2008; Shaw and Black 2008; Siomos et al. 2008; Yen et al. 2008; Fam 2018; Li et al. 2018; Karhulahti and Koskimaa 2019; Gomez et al. 2020; Kumar et al. 2019; Sayyah and Khanafereh 2019). Males seem to show a higher prevalence of PUI than females

(Table 1). Furthermore, the heritability of PUI is estimated to range from around 40% up to nearly 70% (Table 1), once again with higher heritability in males (Li et al. 2014; Deryakulu and Ursavas OF 2014; Vink et al. 2016; Long et al. 2016; Hahn et al. 2017).

Summarising the above, the prevalence and heritability of OCD, OCRDs and PUI show some similarities. However, in some of the OCRDs, due to a rather broader spectrum of symptoms, the heritability seems to be slightly lower.

4 Candidate Gene Studies

Following the compelling evidence of heritability, candidate gene studies have been conducted looking for risk genes for OCD, various OCRDs and PUI. A summary of the current findings for candidate gene associations with hoarding, BDD, trichotillomania and PUI compared to OCD (excluding risk genes that are exclusive for OCD only) is presented in Table 2.

Some genetic overlaps could be found between OCRDs and/or PUI and OCD, however, not always with a full coverage. For example, one of the most frequent gene variants studied in various psychiatric disorders, the serotonin transporter gene, coded by the *SLC6A4* gene, has consistently shown a significant association with OCD following meta-analysis (Taylor 2013; Walitza et al. 2014; Grünblatt et al. 2018). The serotonin-transporter-linked polymorphic region (5-HTTLPR) of the *SLC6A4* gene, combined with the point mutation rs25531, has resulted in significantly increased risk of OCD for carriers of the L_A-allele with an odds ratio (OR) = 1.21 (Taylor 2013; Walitza et al. 2014; Grünblatt et al. 2018). On the other hand, hoarding, BDD, trichotillomania and PUI showed either significant risk for carriers of the $L_G + S$ allele, or carriers of the S-allele, or no association at all (see Table 2) (Sinopoli et al. 2019; Wang et al. 2014; Hemmings et al. 2006; Lee and Ham 2008). However, for each of the other OCRDs, only one study was reported, with limited power of detection, in contrast to the power that meta-analyses conducted in OCD represent. Therefore, further genetic studies urgently need to be conducted for these phenotypes to verify the reported findings.

The dopamine metabolising enzyme catechol-O-methyltransferase (*COMT*) rs4680 (Val158Met) demonstrated a significant association with OCD in male patients in whom Met-allele carriers showed a high risk according to meta-analysis results (Taylor 2013). Similarly, in one study, female carriers of the Met/Met genotype showed significant risk for hoarding (Melo-Felippe et al. 2016). However, in a study of adolescents with excessive Internet video game playing, male carriers of the Val-allele showed only a trend for a risk, but due to very small sample size no conclusion could be made (Han et al. 2007). Similarly, the dopamine D2 receptor *DRD2* gene variant (rs1800497/Taq1A1) was found to have some tendency for association with OCD, with A2-allele carriers having an OR = 1.25 (Taylor 2013), while two studies produced similar findings for PUI, with an OR ranging from 1.08 to 1.55 (Han et al. 2007; Paik et al. 2017).

Table 2 Comparison of candidate gene associations between obsessive-compulsive disorder (OCD), other OC-related disorders and problematic usage of the Internet (PUI)

Gene	OCD*	Hoarding	Body dysmorphic disorder	Trichotillomania	PUI
Serotonergic					
SLC6A4 (5-HTTLPR)	+ rs25531 M all L_A OR = 1.21 (Taylor 2013; Walitza et al. 2014; Grünblatt et al. 2018)	+ rs25531 N = 1 GE male; L_G + S OR = 1.35 (Sinopoli et al. 2019)	N = 1 GE all; S OR = 0.99[b] (Wang et al. 2014)	N = 1 CC all; S OR = 0.9[b] (Hemmings et al. 2006)	N = 1 CC male; S OR = 1.51[a] (Lee and Ham 2008)
HTR2A	(rs6311) M all A OR = 1.219 (Taylor 2013)	n.a.	n.a.	(rs6313) N = 1 CC all; T OR = 2.00 (Hemmings et al. 2006)	n.a.
MAOA (EcoRV)	M male; T OR = 2.87 (Taylor 2013)	n.a.	n.a.	n.a.	n.a.
Glutamatergic					
SLC1A1	(rs301443) M all; C OR = 1.17 (rs12682807) M male; C OR = 1.65 (Stewart et al. 2013b)	(rs301434-rs3780412) N = 1 CC all; G-A OR = 2.12 (rs3780412) Female; G OR = 2.13 (de Salles Andrade et al. 2019)	n.a.	n.a.	n.a.
SAPAP/ DLGAP	(rs9952159 DLGAP1) M-GWAS all; T OR = 1.20 (International Obsessive Compulsive Disorder Foundation Genetics C, Studies OCDCGA 2018)	n.a.	n.a.	(rs6662980 SAPAP3) N = 1 FAM all; G OR = 1.56 (Bienvenu et al. 2009) (rs11583978 SAPAP3) N = 1 CC all; C OR = 1.14 (Boardman et al. 2011)	n.a.
GRID2	(rs1030757) M-GWAS all; C OR = 1.18 (International	n.a.	n.a.	n.a.	n.a.

(continued)

Table 2 (continued)

Gene	OCD*	Hoarding	Body dysmorphic disorder	Trichotillomania	PUI
	Obsessive Compulsive Disorder Foundation Genetics C, Studies OCDCGA 2018)				
Dopaminergic					
COMT (Val158Met / rs4680)	M male; Met OR = 1.54 (Taylor 2013)	N = 1 CC female; Met/Met OR = 3.76 (Melo-Felippe et al. 2016)	n.a.	n.a.	N = 1 CC male; Val OR = 1.61[a] (Han et al. 2007)
DRD2 (Taq1A1, rs1800497)	M all; A2 OR = 1.25[b] (Taylor 2013)	n.a.	n.a.	n.a.	N = 2 CC male; A2 OR = 1.08–1.55[a] (Han et al. 2007; Paik et al. 2017)
Neurotrophins					
BDNF (rs6265, Val66Met)	M all; Met OR = 1.013[b] (Taylor 2013)	N = 1 CC all; Val OR = 1.68 (Timpano et al. 2011)	n.a.	n.a.	n.a.
NTRK3	(rs1017412) M all; T OR = 1.053[b] (rs7176429) M all; G OR = 0.983[b] (Taylor 2013)	(rs1017412) N = 1 CC all; T OR = 2.16 (rs7176429) T OR = 2.78 (Alonso et al. 2008)	n.a.	n.a.	(rs2223310) N = 1 CC male; C OR = 0.154 (Kim et al. 2016)
Others					
CHRNA4	n.a.	n.a.	n.a.	n.a.	(rs1044396) N = 3 CC male; T OR = 0.21–0.85 (Jeong et al. 2017; Montag et al.

(continued)

Table 2 (continued)

Gene	OCD*	Hoarding	Body dysmorphic disorder	Trichotillomania	PUI
					2012; Park et al. 2018)
CRHR1	n.a.	n.a.	n.a.	n.a.	(rs28364027) N = 1 CC male; A OR = 2.63 (Park et al. 2018)

A Adenine, *BDNF* brain-derived neurotrophic factor, *C* cytosine, *CC* case-control, *CHRNA4* choline receptor nicotinic alpha-4 subunit, *COMT* catechol-O-methyltransferase, *CRHR1* corticotropin-releasing hormone receptor 1, *DLGAP/SAPAP* disks large-associated protein/ SAP90/PSD-95-associated protein, *DRD2* dopamine D2 receptor, *FAM* family study, *G* guanine, *GE* general population, *GRID2* glutamate ionotropic receptor delta type 2 subunit, *GWAS* genome-wide association studies, *HTR2A* serotonin receptor 2A, L_a, long-allele variant including the A-allele of the rs25531, L_g long-allele variant including the G-allele of the rs25531, *M* meta-analysis, *N* number of studies, *NTRK3* neurotrophic receptor tyrosine kinase 3, *MAOA* monoamine oxidase A, *Met* methionine, *n.a.* not available, *OC* obsessive compulsive, *OCD* obsessive compulsive disorder, *OR* odds ratio, *PUI* problematic usage of the internet, *rs#* SNP ID number according to the dbSNP hosted by the NCBI, *S* short-allele variant, *SLC1A1* glutamate transporter, *SLC6A4 (5-HTTLPR)* serotonin transporter gene, *T* thymine, *Val* valine

*Only significant results were listed for OCD- for detailed review see (Zai et al. 2019)

[a]trend with none significant results

[b]none significant results

Neurotrophins, such as brain-derived neurotrophic factor (*BDNF*) and neurotrophic tyrosine kinase receptor type 3 (*NTRK3*) genes, were hypothesised to associate with OCD, OCRDs and PUI (Table 2). However, *BDNF* rs6265/Val66Met could not be proven to associate with OCD following meta-analysis (Taylor 2013), while in only one study of hoarding, an association was found with Val-allele carriers (Timpano et al. 2011). Similarly, *NTRK3* rs1017412 and rs7176429 were not proven to associate with OCD (Taylor 2013), while one study found an association between T-allele carriers and hoarding (Alonso et al. 2008). On the other hand, in a pilot study of Internet gaming disorder, male carriers of the T-allele for rs2223310 showed a significant association with an OR = 0.154 (Kim et al. 2016).

The gene coding for the nicotinic acetylcholine receptor subunit alpha 4 (*CHRNA4*), also known to interact with several dopaminergic genes (Markett et al. 2013; Quan et al. 2017; Breckel et al. 2015), and linked with smoking addiction, has been studied for its association with PUI in three independent samples, in which rs1044396 T-allele carriers were found to show a protective effect with OR = 0.21–0.85 (Jeong et al. 2017; Montag et al. 2012; Park et al. 2018). Although nicotinic acetylcholine receptors have been shown to modulate dopamine activity and release (Wen et al. 2016; Gozen et al. 2016), this gene has not been studied in OCD or other OCRDs, so no further comparison is available.

Stress and stress-related genes, such as the corticotropin-releasing hormone receptor 1 (*CRHR1*) (Ray et al. 2013; Muller and Wurst 2004; Blomeyer et al. 2008; Chen et al. 2010), have been the subject of investigation in several psychiatric and addictive disorders. In one study investigating its association with Internet gaming addiction, a significant association was observed in male A-allele carriers (rs28364027) with OR = 2.63. However, again, this gene has not been studied in OCD or other OCRDs.

5 Genome-Wide Association Studies

Unexpectedly, few genome-wide association studies (GWAS) have been conducted in this field to date. Two GWAS for OCD have been performed, followed by a meta-analysis (Mattheisen et al. 2015; Stewart et al. 2013a; International Obsessive Compulsive Disorder Foundation Genetics C, Studies OCDCGA 2018). No genome-wide significance was found due to a persistently small sample size for such a study. However, some interesting candidates genes with low p-values were reported. For example, one was the gene for the glutamate ionotropic receptor delta type subunit 2 (*GRID2*), with the rs1030757 C-allele associating with OCD (OR = 1.18) (International Obsessive Compulsive Disorder Foundation Genetics C, Studies OCDCGA 2018). However, no current study exists investigating the association of this gene with other OCRDs or PUI. The *SAPAP1/DLGAP1* gene involved in the postsynaptic scaffolding of the glutamatergic neuronal system was found to associate with OCD with the single nucleotide polymorphism (SNP) rs9952159 with an OR = 1.20 (International Obsessive Compulsive Disorder Foundation Genetics C, Studies OCDCGA 2018). Another member of the *SAPAP* gene group, *SAPAP3*, was found in one study to associate with trichotillomania (Boardman et al. 2011), but this was not confirmed in any further independent study (Table 2).

For hoarding traits, one GWAS study has been reported (Perroud et al. 2011). The study used a large sample of Caucasian twins with 3,410 participants, to associate genome-wide genes for hoarding traits. No genome-wide significance was found, however, two loci on chromosome 5 and 6 showed a trend for a significant association with hoarding traits (Perroud et al. 2011). The most significant SNP was rs3747767 on the *SH3BGRL2* gene that encodes the SH3 domain binding glutamate rich protein like 2. Therefore, similarly to OCD (which disorder shows some indication of association with two genes involved in the glutamatergic system: *GRID2* and *DLGAP1),* hoarding traits also seem to be linked to the glutamatergic system. However, in order to draw any meaningful conclusions, larger studies with independent verification cohorts will be needed.

To date, no genome-wide studies exist for BDD, skin picking disorders, trichotillomania or PUI. Such studies are still pending, in particular those combining the GWAS findings with polygenic risk scores (PRS) for various relevant traits, such as anxiety, depression, extraversion, neuroticism, loneliness, chronotype (morningness), education, intelligence, or well-being (Howard et al. 2019; Otowa et al. 2016; van den Berg et al. 2016; Nagel et al. 2018; Gao et al. 2017; Okbay et al. 2016a, 2016b; Jansen et al. 2019a; Savage et al. 2018). Such an approach has fruitfully been used in the general population as well as in various psychiatric disorders such as schizophrenia, ADHD, or cocaine addiction (Cabana-Dominguez et al. 2019; Jansen et al. 2019b; Polimanti et al. 2020; Salvatore et al. 2020; Hermosillo et al. 2020). An attempt to correlate such traits with genetic risk loads might help either to elucidate the aetiology and pathology of the disorder, or to predict the likelihood of developing the disorder, as well as finding their common molecular pathways (Cabana-Dominguez et al. 2019; Jansen et al. 2019b; Richardson et al. 2019; Ni et al. 2019).

6 Conclusion

The research field investigating the genetics of OCD and other OCRDs is still lagging behind many other psychiatric disorders, pointing to the urgent need for further intensive studies assessing the similar genetic risk factors as well as their unique genetic signatures. From the current, rather scarce, data available, it is difficult to draw any conclusion regarding any specific gene variant or a system pathway (e.g. glutamatergic, dopaminergic) to play a role in OCD, OCRD or PUI. However, some indications suggest that there is combined involvement of the serotonergic, dopaminergic and glutamatergic systems in OCD and other OCRDs, as well as to some extent in PUI, in which some addiction-related systems such as the nicotinic and stress-related genes may also play some role. Nevertheless, the data rather point to the notion that we are dealing with overlapping spectrum of traits that most likely are polygenic, with a collection of hundreds to thousands of gene variants with very small effect sizes when taken separately. The use of GWAS in a great number of individuals, introducing polygenic risk score (PRS) analysis may overcome this obstacle. A multi-dimensional approach to phenotyping of individuals that not only includes a diagnosis of OCD, but also includes all the other OCRDs and PUI, as well as their latent traits, together with bigger sample sizes, is desirable. With the help of GWAS, PRS and consortia such a goal could certainly be achieved by mobilising worldwide research groups, as well as fund-raising to support such studies.

References

Alonso P, Gratacos M, Menchon JM, Segalas C, Gonzalez JR, Labad J et al (2008) Genetic susceptibility to obsessive-compulsive hoarding: the contribution of neurotrophic tyrosine kinase receptor type 3 gene. Genes Brain Behav 7(7):778–785

American Psychiatric Association (2013) Diagnostic and statistical manual of mental disorders, fifth edition (DSM-5), 5th edn. American Psychiatric Association, Washington

Bienvenu OJ, Wang Y, Shugart YY, Welch JM, Grados MA, Fyer AJ et al (2009) Sapap3 and pathological grooming in humans: results from the OCD collaborative genetics study. Am J Med Genet B Neuropsychiatr Genet 150B(5):710–720

Bienvenu OJ, Samuels JF, Wuyek LA, Liang KY, Wang Y, Grados MA et al (2012) Is obsessive-compulsive disorder an anxiety disorder, and what, if any, are spectrum conditions? A family study perspective. Psychol Med 42(1):1–13

Blomeyer D, Treutlein J, Esser G, Schmidt MH, Schumann G, Laucht M (2008) Interaction between CRHR1 gene and stressful life events predicts adolescent heavy alcohol use. Biol Psychiatry 63(2):146–151

Boardman L, van der Merwe L, Lochner C, Kinnear CJ, Seedat S, Stein DJ et al (2011) Investigating SAPAP3 variants in the etiology of obsessive-compulsive disorder and trichotillomania in the South African white population. Compr Psychiatry 52(2):181–187

Breckel TP, Giessing C, Gieseler A, Querbach S, Reuter M, Thiel CM (2015) Nicotinergic modulation of attention-related neural activity differentiates polymorphisms of DRD2 and CHRNA4 receptor genes. PLoS One 10(6):e0126460

Browne HA, Gair SL, Scharf JM, Grice DE (2014) Genetics of obsessive-compulsive disorder and related disorders. Psychiatr Clin North Am 37(3):319–335

Burton CL, Park LS, Corfield EC, Forget-Dubois N, Dupuis A, Sinopoli VM et al (2018) Heritability of obsessive-compulsive trait dimensions in youth from the general population. Transl Psychiatry 8(1):191

Burton C, Barta C, Cath D, Geller D, Van den Heuvel OA, Yao Y et al (2020) Genetics of obsessive-compulsive disorder and Tourette disorder. In: Baune B (ed) Psychiatry. Academic Press, San Diego, pp 239–252

Cabana-Dominguez J, Shivalikanjli A, Fernandez-Castillo N, Cormand B (2019) Genome-wide association meta-analysis of cocaine dependence: shared genetics with comorbid conditions. Prog Neuro-Psychopharmacol Biol Psychiatry 94:109667

Cao F, Su L (2007) Internet addiction among Chinese adolescents: prevalence and psychological features. Child Care Health Dev 33(3):275–281

Chattopadhyay K (2012) The genetic factors influencing the development of trichotillomania. J Genet 91(2):259–262

Chen AC, Manz N, Tang Y, Rangaswamy M, Almasy L, Kuperman S et al (2010) Single-nucleotide polymorphisms in corticotropin releasing hormone receptor 1 gene (CRHR1) are associated with quantitative trait of event-related potential and alcohol dependence. Alcohol Clin Exp Res 34(6):988–996

Davis LK, Yu D, Keenan CL, Gamazon ER, Konkashbaev AI, Derks EM et al (2013) Partitioning the heritability of Tourette syndrome and obsessive compulsive disorder reveals differences in genetic architecture. PLoS Genet 9(10):e1003864

de Salles Andrade JB, Giori IG, Melo-Felippe FB, Vieira-Fonseca T, Fontenelle LF, Kohlrausch FB (2019) Glutamate transporter gene polymorphisms and obsessive-compulsive disorder: a case-control association study. J Clin Neurosci 62:53–59

Deryakulu D, Ursavas OF (2014) Genetic and environmental influences on problematic internet use: a twin study. Comput Hum Behav 39:331–338

Enander J, Ivanov VZ, Mataix-Cols D, Kuja-Halkola R, Ljotsson B, Lundstrom S et al (2018) Prevalence and heritability of body dysmorphic symptoms in adolescents and young adults: a population-based nationwide twin study. Psychol Med 48(16):2740–2747

Fam JY (2018) Prevalence of internet gaming disorder in adolescents: a meta-analysis across three decades. Scand J Psychol 59(5):524–531

Fineberg NA, Demetrovics Z, Stein DJ, Ioannidis K, Potenza MN, Grunblatt E et al (2018) Manifesto for a European research network into problematic usage of the internet. Eur Neuropsychopharmacol 28(11):1232–1246

Fu KW, Chan WS, Wong PW, Yip PS (2010) Internet addiction: prevalence, discriminant validity and correlates among adolescents in Hong Kong. Br J Psychiatry 196(6):486–492

Gao J, Davis LK, Hart AB, Sanchez-Roige S, Han L, Cacioppo JT et al (2017) Genome-wide association study of loneliness demonstrates a role for common variation. Neuropsychopharmacology 42(4):811–821

Gomez P, Feijoo S, Brana T, Varela J, Rial A (2020) Minors and online gambling: prevalence and related variables. J Gambl Stud 36:735–745

Gozen O, Nesil T, Kanit L, Koylu EO, Pogun S (2016) Nicotinic cholinergic and dopaminergic receptor mRNA expression in male and female rats with high or low preference for nicotine. Am J Drug Alcohol Abuse 42(5):556–566

Grünblatt E, Marinova Z, Roth A, Gardini E, Ball J, Geissler J et al (2018) Combining genetic and epigenetic parameters of the serotonin transporter gene in obsessive-compulsive disorder. J Psychiatr Res 96:209–217

Hahn E, Reuter M, Spinath FM, Montag C (2017) Internet addiction and its facets: the role of genetics and the relation to self-directedness. Addict Behav 65:137–146

Han DH, Lee YS, Yang KC, Kim EY, Lyoo IK, Renshaw PF (2007) Dopamine genes and reward dependence in adolescents with excessive internet video game play. J Addict Med 1(3):133–138

Hemmings SM, Kinnear CJ, Lochner C, Seedat S, Corfield VA, Moolman-Smook JC et al (2006) Genetic correlates in trichotillomania – a case-control association study in the South African Caucasian population. Isr J Psychiatry Relat Sci 43(2):93–101

Hermosillo RJM, Mooney MA, Fezcko E, Earl E, Marr M, Sturgeon D et al (2020) Polygenic risk score-derived subcortical connectivity mediates attention-deficit/hyperactivity disorder diagnosis. Biol Psychiatry Cogn Neurosci Neuroimaging 5(3):330–341

Howard DM, Adams MJ, Clarke TK, Hafferty JD, Gibson J, Shirali M et al (2019) Genome-wide meta-analysis of depression identifies 102 independent variants and highlights the importance of the prefrontal brain regions. Nat Neurosci 22(3):343–352

International Obsessive Compulsive Disorder Foundation Genetics C, Studies OCDCGA (2018) Revealing the complex genetic architecture of obsessive-compulsive disorder using meta-analysis. Mol Psychiatry 23(5):1181–1188

Ivanov VZ, Nordsletten A, Mataix-Cols D, Serlachius E, Lichtenstein P, Lundstrom S et al (2017) Heritability of hoarding symptoms across adolescence and young adulthood: a longitudinal twin study. PLoS One 12(6):e0179541

Jansen PR, Watanabe K, Stringer S, Skene N, Bryois J, Hammerschlag AR et al (2019a) Genome-wide analysis of insomnia in 1,331,010 individuals identifies new risk loci and functional pathways. Nat Genet 51(3):394–403

Jansen AG, Dieleman GC, Jansen PR, Verhulst FC, Posthuma D, Polderman TJC (2019b) Psychiatric polygenic risk scores as predictor for attention deficit/hyperactivity disorder and autism spectrum disorder in a clinical child and adolescent sample. Behav Genet

Jeong JE, Rhee JK, Kim TM, Kwak SM, Bang SH, Cho H et al (2017) The association between the nicotinic acetylcholine receptor alpha4 subunit gene (CHRNA4) rs1044396 and internet gaming disorder in Korean male adults. PLoS One 12(12):e0188358

Karhulahti VM, Koskimaa R (2019) On the prevalence of addicted or problematic gaming in Finland. Addict Behav Rep 10:100225

Kim JY, Jeong JE, Rhee JK, Cho H, Chun JW, Kim TM et al (2016) Targeted exome sequencing for the identification of a protective variant against internet gaming disorder at rs2229910 of neurotrophic tyrosine kinase receptor, type 3 (NTRK3): a pilot study. J Behav Addict 5 (4):631–638

Kumar S, Singh S, Singh K, Rajkumar S, Balhara YPS (2019) Prevalence and pattern of problematic internet use among engineering students from different colleges in India. Indian J Psychiatry 61(6):578–583

Lee BT, Ham BJ (2008) Serotonergic genes and amygdala activity in response to negative affective facial stimuli in Korean women. Genes Brain Behav 7(8):899–905

Li M, Chen J, Li N, Li X (2014) A twin study of problematic internet use: its heritability and genetic association with effortful control. Twin Res Hum Genet 17(4):279–287

Li L, Xu DD, Chai JX, Wang D, Li L, Zhang L et al (2018) Prevalence of internet addiction disorder in Chinese university students: a comprehensive meta-analysis of observational studies. J Behav Addict 7(3):610–623
Long EC, Verhulst B, Neale MC, Lind PA, Hickie IB, Martin NG et al (2016) The genetic and environmental contributions to internet use and associations with psychopathology: a twin study. Twin Res Hum Genet 19(1):1–9
Markett S, Reuter M, Montag C, Weber B (2013) The dopamine D2 receptor gene DRD2 and the nicotinic acetylcholine receptor gene CHRNA4 interact on striatal gray matter volume: evidence from a genetic imaging study. NeuroImage 64:167–172
Mathews CA, Nievergelt CM, Azzam A, Garrido H, Chavira DA, Wessel J et al (2007) Heritability and clinical features of multigenerational families with obsessive-compulsive disorder and hoarding. Am J Med Genet B Neuropsychiatr Genet 144B(2):174–182
Mattheisen M, Samuels JF, Wang Y, Greenberg BD, Fyer AJ, McCracken JT et al (2015) Genome-wide association study in obsessive-compulsive disorder: results from the OCGAS. Mol Psychiatry 20(3):337–344
Melo-Felippe FB, de Salles Andrade JB, Giori IG, Vieira-Fonseca T, Fontenelle LF, Kohlrausch FB (2016) Catechol-O-methyltransferase gene polymorphisms in specific obsessive-compulsive disorder Patients' subgroups. J Mol Neurosci 58(1):129–136
Montag C, Kirsch P, Sauer C, Markett S, Reuter M (2012) The role of the CHRNA4 gene in internet addiction: a case-control study. J Addict Med 6(3):191–195
Monzani B, Rijsdijk F, Anson M, Iervolino AC, Cherkas L, Spector T et al (2012a) A twin study of body dysmorphic concerns. Psychol Med 42(9):1949–1955
Monzani B, Rijsdijk F, Iervolino AC, Anson M, Cherkas L, Mataix-Cols D (2012b) Evidence for a genetic overlap between body dysmorphic concerns and obsessive-compulsive symptoms in an adult female community twin sample. Am J Med Genet B Neuropsychiatr Genet 159B (4):376–382
Monzani B, Rijsdijk F, Harris J, Mataix-Cols D (2014) The structure of genetic and environmental risk factors for dimensional representations of DSM-5 obsessive-compulsive spectrum disorders. JAMA Psychiat 71(2):182–189
Muller MB, Wurst W (2004) Getting closer to affective disorders: the role of CRH receptor systems. Trends Mol Med 10(8):409–415
Nagel M, Jansen PR, Stringer S, Watanabe K, de Leeuw CA, Bryois J et al (2018) Meta-analysis of genome-wide association studies for neuroticism in 449,484 individuals identifies novel genetic loci and pathways. Nat Genet 50(7):920–927
Nestadt G, Samuels J, Riddle M, Bienvenu OJ 3rd, Liang KY, LaBuda M et al (2000) A family study of obsessive-compulsive disorder. Arch Gen Psychiatry 57(4):358–363
Ni G, Amare AT, Zhou X, Mills N, Gratten J, Lee SH (2019) The genetic relationship between female reproductive traits and six psychiatric disorders. Sci Rep 9(1):12041
Nordsletten AE, Reichenberg A, Hatch SL, Fernandez de la Cruz L, Pertusa A, Hotopf M et al (2013) Epidemiology of hoarding disorder. Br J Psychiatry 203(6):445–452
Novak CE, Keuthen NJ, Stewart SE, Pauls DL (2009) A twin concordance study of trichotillomania. Am J Med Genet B Neuropsychiatr Genet 150B(7):944–949
Okbay A, Beauchamp JP, Fontana MA, Lee JJ, Pers TH, Rietveld CA et al (2016a) Genome-wide association study identifies 74 loci associated with educational attainment. Nature 533 (7604):539–542
Okbay A, Baselmans BM, De Neve JE, Turley P, Nivard MG, Fontana MA et al (2016b) Genetic variants associated with subjective well-being, depressive symptoms, and neuroticism identified through genome-wide analyses. Nat Genet 48(6):624–633
Otowa T, Hek K, Lee M, Byrne EM, Mirza SS, Nivard MG et al (2016) Meta-analysis of genome-wide association studies of anxiety disorders. Mol Psychiatry 21(10):1391–1399
Paik SH, Choi MR, Kwak SM, Bang SH, Chun JW, Kim JY et al (2017) An association study of Taq1A ANKK1 and C957T and − 141C DRD2 polymorphisms in adults with internet gaming disorder: a pilot study. Ann General Psychiatry 16:45

Park SK, Kim JY, Cho CB (2008) Prevalence of internet addiction and correlations with family factors among South Korean adolescents. Adolescence 43(172):895–909
Park J, Sung JY, Kim DK, Kong ID, Hughes TL, Kim N (2018) Genetic association of human Corticotropin-releasing hormone receptor 1 (CRHR1) with internet gaming addiction in Korean male adolescents. BMC Psychiatry 18(1):396
Pauls DL (2010) The genetics of obsessive-compulsive disorder: a review. Dialogues Clin Neurosci 12(2):149–163
Pauls DL, Abramovitch A, Rauch SL, Geller DA (2014) Obsessive-compulsive disorder: an integrative genetic and neurobiological perspective. Nat Rev Neurosci 15(6):410–424
Perroud N, Guipponi M, Pertusa A, Fullana MA, Iervolino AC, Cherkas L et al (2011) Genome-wide association study of hoarding traits. Am J Med Genet B Neuropsychiatr Genet 156 (2):240–242
Polimanti R, Walters RK, Johnson EC, McClintick JN, Adkins AE, Adkins DE et al (2020) Leveraging genome-wide data to investigate differences between opioid use vs. opioid dependence in 41,176 individuals from the Psychiatric Genomics Consortium. Mol Psychiatry
Quan J, Ong ML, Bureau JF, Sim LW, Sanmugam S, Abdul Malik AB et al (2017) The influence of CHRNA4, COMT, and maternal sensitivity on orienting and executive attention in 6-month-old infants. Brain Cogn 116:17–28
Ray LA, Sehl M, Bujarski S, Hutchison K, Blaine S, Enoch MA (2013) The CRHR1 gene, trauma exposure, and alcoholism risk: a test of G x E effects. Genes Brain Behav 12(4):361–369
Richardson TG, Harrison S, Hemani G, Davey SG (2019) An atlas of polygenic risk score associations to highlight putative causal relationships across the human phenome. eLife 8
Salvatore JE, Barr PB, Stephenson M, Aliev F, Kuo SI, Su J et al (2020) Sibling comparisons elucidate the associations between educational attainment polygenic scores and alcohol, nicotine and cannabis. Addiction 115(2):337–346
Savage JE, Jansen PR, Stringer S, Watanabe K, Bryois J, de Leeuw CA et al (2018) Genome-wide association meta-analysis in 269,867 individuals identifies new genetic and functional links to intelligence. Nat Genet 50(7):912–919
Sayyah M, Khanafereh S (2019) Prevalence of internet addiction among medical students: a study from southwestern Iran. Cent Eur J Public Health 27(4):326–329
Shaw M, Black DW (2008) Internet addiction: definition, assessment, epidemiology and clinical management. CNS Drugs 22(5):353–365
Sinopoli VM, Erdman L, Burton CL, Park LS, Dupuis A, Shan J et al (2019) Serotonin system genes and obsessive-compulsive trait dimensions in a population-based, pediatric sample: a genetic association study. J Child Psychol Psychiatry 60(12):1289–1299
Siomos KE, Dafouli ED, Braimiotis DA, Mouzas OD, Angelopoulos NV (2008) Internet addiction among Greek adolescent students. Cyberpsychol Behav 11(6):653–657
Stewart SE, Yu D, Scharf JM, Neale BM, Fagerness JA, Mathews CA et al (2013a) Genome-wide association study of obsessive-compulsive disorder. Mol Psychiatry 18(7):788–798
Stewart SE, Mayerfeld C, Arnold PD, Crane JR, O'Dushlaine C, Fagerness JA et al (2013b) Meta-analysis of association between obsessive-compulsive disorder and the 3′ region of neuronal glutamate transporter gene SLC1A1. Am J Med Genet B Neuropsychiatr Genet 162(4):367–379
Taylor S (2013) Molecular genetics of obsessive-compulsive disorder: a comprehensive meta-analysis of genetic association studies. Mol Psychiatry 18(7):799–805
Timpano KR, Schmidt NB, Wheaton MG, Wendland JR, Murphy DL (2011) Consideration of the BDNF gene in relation to two phenotypes: hoarding and obesity. J Abnorm Psychol 120 (3):700–707
van den Berg SM, de Moor MH, Verweij KJ, Krueger RF, Luciano M, Arias Vasquez A et al (2016) Meta-analysis of genome-wide association studies for extraversion: findings from the genetics of personality consortium. Behav Genet 46(2):170–182
van Grootheest DS, Cath DC, Beekman AT, Boomsma DI (2005) Twin studies on obsessive-compulsive disorder: a review. Twin Res Hum Genet 8(5):450–458

Vink JM, van Beijsterveldt TC, Huppertz C, Bartels M, Boomsma DI (2016) Heritability of compulsive internet use in adolescents. Addict Biol 21(2):460–468

Walitza S, Wendland JR, Gruenblatt E, Warnke A, Sontag TA, Tucha O et al (2010) Genetics of early-onset obsessive-compulsive disorder. Eur Child Adolesc Psychiatry 19(3):227–235

Walitza S, Marinova Z, Grünblatt E, Lazic SE, Remschmidt H, Vloet TD et al (2014) Trio study and meta-analysis support the association of genetic variation at the serotonin transporter with early-onset obsessive-compulsive disorder. Neurosci Lett 580:100–103

Wang SK, Lee YH, Kim JL, Chee IS (2014) No effect on body dissatisfaction of an interaction between 5-HTTLPR genotype and neuroticism in a young adult Korean population. Clin Psychopharmacol Neurosci 12(3):229–234

Wen L, Yang Z, Cui W, Li MD (2016) Crucial roles of the CHRNB3-CHRNA6 gene cluster on chromosome 8 in nicotine dependence: update and subjects for future research. Transl Psychiatry 6(6):e843

Yen JY, Ko CH, Yen CF, Chen SH, Chung WL, Chen CC (2008) Psychiatric symptoms in adolescents with internet addiction: comparison with substance use. Psychiatry Clin Neurosci 62(1):9–16

Zai G, Barta C, Cath D, Eapen V, Geller D, Grunblatt E (2019) New insights and perspectives on the genetics of obsessive-compulsive disorder. Psychiatr Genet 29(5):142–151

Zilhão N, Boomsma D, Smit D, Cath D (2019) Genetics of obsessive – compulsive disorder and Tourette's syndrome. In: Miu A, Homberg J, Lesch K (eds) Genes, brain, and emotions: interdisciplinary and translational perspectives. Oxford University Press, New York

On the Development of OCD

T. U. Hauser

Contents

Abstract OCD most often arises before adulthood with adolescence being a particularly vulnerable period. This is also a time when both brain and cognition undergo fundamental developmental change and reorganisation. However, the neurocognitive mechanisms that drive the emergence of OCD during development are still largely unknown. In this chapter, I review the relatively sparse literature on the developmental aspects of OCD and I discuss the symptomatic, cognitive and neural patterns in OCD and the developing mind. I highlight how we need to understand the emergence of cognitive impairments and neural alteration in a developmental context if we want to understand more about the mechanisms that give rise to OCD. Moreover, I outline how we best approach this challenge to overcome the current limitations in research.

Keywords Developmental computational psychiatry · Juvenile OCD · Longitudinal studies

---

T. U. Hauser (✉)
Max Planck UCL Centre for Computational Psychiatry and Ageing Research, London, UK

Wellcome Centre for Human Neuroimaging, University College London, London, UK
e-mail: t.hauser@ucl.ac.uk

Curr Topics Behav Neurosci (2021) 49: 17–30
https://doi.org/10.1007/7854_2020_195
Published Online: 13 February 2021

1 Introduction

Obsessive-compulsive disorder (OCD) is a neglected disorder. The public acknowledgment of OCD is still low, the stigmatisation is high, and a misconception of the disorder is widespread with a colloquial notation of 'being a bit OCD' playing down the suffering of those affect by OCD. Such ignorance not only contributes to patients suffering 7–10 years before receiving their diagnosis (Benatti et al. 2016), but also seems to substantially affect research funding for this common disorder. Despite a prevalence of 2–3% (Kessler et al. 2005), less than 1% of the mental health funding is directed towards OCD (Woelbert et al. 2019), a number substantially lower than other more prominent but less common mental health disorders. This figure gets even worse when we look at OCD in children and adolescents, as only 26% of total mental health funding is dedicated to research in youths (Woelbert et al. 2019). It is thus not surprising that we still know little about the developmental aspects of OCD and that our knowledge lacks much behind other developmental disorders, such as attention-deficit/hyperactivity disorder (ADHD). This book chapter will thus not only discuss what we know about developmental aspects in OCD, but also outline the methods and questions that should help us to better understand the importance of development in OCD.

2 Why Development?

OCD is often regarded as an adulthood disorder and most of the experimental and neurocognitive studies have been conducted in adult samples (cf. all remaining book chapters). This might, at least in part, be driven by the early epidemiological studies that concluded that the prevalence of juvenile OCD is only 0.2% to 1.2%, thus much lower than adult OCD (Shafran 2003). And although the prevalence estimates are still relatively heterogeneous, a commonly accepted prevalence of approx. 2% means that OCD during childhood and adolescence is no less common than in adulthood (Zohar 1999; Shafran 2003). In fact, a juvenile manifestation of OCD is a common feature and the majority of all adult OCD patients report a disorder onset before their coming of age (Rasmussen and Eisen 1992; Kessler et al. 2005; Fineberg et al. 2013c).

To better understand the role of development in OCD, it is important to understand at what age OCD is typically emerging. Epidemiological research suggests that there are two periods of heightened incidence: an early-onset (EO) period during late childhood around 11 years, and a late-onset (LO) in early adulthood around 22 years (Geller et al. 1998; Taylor 2011; Fineberg et al. 2013c; Walitza et al. 2019). Relatively little is known about OCD in infants and during early childhood. This is in part because some of the behaviours characteristic for OCD – such as ordering or symmetry – are part of the normal childhood development and should not be misinterpreted as OCD symptoms. Moreover, many features of adult OCD, such as insight or ego-dystonia, are rarely present in young children as they require cognitive skills that do not develop until later in life. However, there are several case studies and first on-going larger patient studies report OCD in children as young as 4 years old although such cases are relatively rare (Fineberg et al. 2013c, 2019; Walitza et al. 2019).

The high prevalence of OCD in childhood and adolescence as well as the bimodal age-of-onset not only shows that juvenile OCD is a relevant field of study, but the finding of 'sensitive periods' with heightened incidence of OCD strongly suggests that neurocognitive development is a critical contributing factor to OCD. If development would be of no importance, then prevalence would be steadily accumulating throughout all ages, which is clearly not the case (Fineberg et al. 2013c). The bimodal age-of-onset, however, means that there are (at least) two periods during development when one is particularly vulnerable for developing OCD.

3 How to Study Development of a Disorder?

Understanding how and when OCD emerges during development is critical for two reasons. Firstly, it will allow us to better understand the aetiology of OCD, the factors and triggers that give rise or protect from developing OCD symptoms. Secondly, if we know the course of OCD development, then we might be able to improve treatment predictions or even intervene before the disorder has fully developed. For example, if we were able to predict that a certain set of symptoms and cognitive markers is likely to lead to clinical OCD within a few years, then we could target these children with interventions and prevent a disorder from manifesting fully (Fineberg et al. 2019). Although such a 'prodromal OCD' (Fineberg et al. 2019) state is being discussed, our knowledge is far from sufficient to make any predictions about the course of OCD symptoms.

The key challenge in developmental research is to gather the right data that allows adequate inference. To date, most of our data is cross-sectional (Fig. 1). This means that we compare two different groups of participants in a between-subjects design. Such studies can compare patients of the same age, but with different age-of-onset (e.g. early vs late onset). Alternatively, one can compare children and adults with OCD and assess differences. The former approach provides a control for current age, but these groups will inherently differ in their chronicity, the EO group probably has had more treatment and medication, and possibly even forms its own treatment-resistant subgroup within EO patients. Any differences in symptoms, brain or behaviour are thus difficult to interpret and to draw any conclusions about underlying aetiological mechanisms. The shortcoming of the latter approach (children vs adults), however, is that it does not control for participant age and neurocognitive development. Even when comparing each patient group to age-matched controls, finding an impairment in one but not the other patient group does not necessarily have any meaningful implications, because each (absence of a) group difference has to be analysed in the context of the current developmental ability. For example, if a

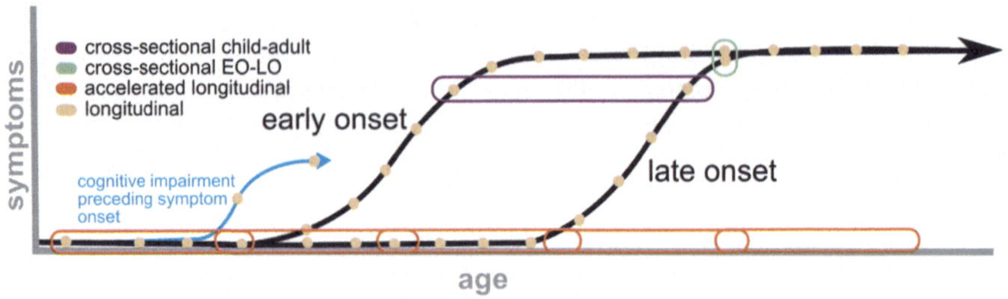

Fig. 1 A developmental perspective on OCD. Disorders never emerge out of nothing and the timepoint of their emergence can provide critical insights into the mechanisms that cause the disorder. The common dichotomy between early and late onset OCD highlights the relevance of (adolescent) development and the necessity of studying OCD in the context of juvenile development. Studying the symptomatic, cognitive and neural development can be pursued using a variety of study designs that differ in terms of complexity, costs and analytic sophistication. Simple cross-sectional comparisons allow to compare subjects of different age-of-onset (green) or different age groups (purple). Fully longitudinal studies (beige) allow tracing individual developmental trajectories and reveal the neurocognitive mechanisms that may precede disorder onset. Accelerated longitudinal designs (orange) strike a balance between wealth of data and practicability and promise relevant new insights in the upcoming years

certain cognitive function is not yet or only minimally developed in the young sample, then an absence of a difference in that group does not mean that their function is intact, but more likely that is not yet developed at the level that differences would be detectable. Lastly, all cross-sectional studies have the common weakness that they are unable to determine whether any (neurocognitive) impairment is a cause or a consequence. This is of particular worry for many cognitive and neuroscience studies that aim at understanding the aetiological mechanisms underlying OCD. A change in a brain circuitry in OCD patients does not have to be the driver that caused OCD. It can equally likely be a consequence of the disorder or just a coincidental alteration driven by factors that are linked but not relevant to the disorder (e.g. altered social interactions due to illness).

The much slower and more laborious alternative is longitudinal studies. These studies assess the patients not only once, but repeatedly over time following them throughout their lives. Such design has multiple advantages, most prominently that one can characterise changes of symptoms and cognition on a within-subject level, meaning that one is able to trace each participant's individual developmental trajectory. One has thus the possibility to determine when OCD symptoms start to emerge, and how this emergence is linked to other personal and environmental factors. With sufficient temporal granularity, it is even possible to determine the directionality of effects, and thus determine whether neural and cognitive deficits precede clinical symptoms, or vice versa. For example, if a cognitive impairment precedes an increase in OCD symptom, then it is possible that given cognitive feature is causally driving the emergence of symptoms. If, however, the symptoms emerge first, then the cognitive feature is unlikely to be relevant for the aetiology of OCD, but rather a consequence or an irrelevant by-product of the disorder. Knowing the directionality is absolutely essential if one wants to develop novel interventions that target cognitive or neural processes (e.g. cognitive bias modification, neurofeedback). The chances that such training will be successful are considerably higher in the case when a cognitive impairment is shown to be driving the symptom emergence.

However, running longitudinal studies is challenging. Firstly, they evidently take a long time. This means that it may take decades before first results can be published – a pace that is at odds with the current academic research and funding situation. Moreover, if one wants to study the emergence of OCD, then such studies have to include participants prior to their disorder onset. This poses the challenge of how to select the sample so to have enough participants that develop a disorder, without assessing a vast majority of subjects that will never develop any symptoms. Risk-enhanced samples in which participants are included because of a known vulnerability (e.g. familial history of OCD) could help increase sensitivity. Another challenge is to determine the right frequency of assessments. Thereby a balance between practicalities and statistical sophistication needs to be found. As many assessments as possible (maybe even experience sampling methods) allows to maximise sensitivity for detecting changes and determining directionality but is very costly and is likely to lead to increased attrition rates. This is why longitudinal data in OCD are still rare (Fineberg et al. 2013a, b, c).

A viable alternative to pure longitudinal studies are accelerated longitudinal study designs. These designs have found much application lately (e.g. Kiddle et al. 2017) as they are more feasible but still allow conducting longitudinal analyses. Key to this design it that one does not follow a single participant group along their entire childhood and adolescence, but to follow multiple cohorts of different ages for a shorter duration. This allows to trace how symptoms emerge, but also to determine at what age such an emergence is most likely.

4 OCD Symptoms During Development

In recent years, there was a lot of interest in whether juvenile OCD is different from adult OCD in its form and expression, and whether it might represent an entirely different disorder with distinct genetic causes. Multiple studies indeed revealed that there are some key differences between EO and LO patients. Firstly, the gender proportions are different. Whilst males and females are about equally often affected by adult OCD, youths with OCD are predominantly male, with a preponderance of roughly 2/3 males (Geller et al. 1998). Relatedly, juvenile OCD patients have a different comorbidity profile. They have many more neurological motor- and impulsivity-related disorders, such as Tic-disorders or ADHD (Geller et al. 1998; Walitza et al. 2019). Moreover, juvenile OCD is assumed to be more heritable than adult OCD (Walitza et al. 2010). Early genetic studies indeed suggested distinct genetic profiles, but such distinctions have become less clear in more recent years (Grünblatt et al. 2014).

However, recent evidence also suggests that both EO and LO OCD are not so different in many aspects. When investigating the actual symptom profiles, then juvenile and adult OCD patients are strikingly similar. When analysing symptom expressions, then both EO and LO OCD show highly similar profiles. Whilst some studies report slightly different obsession but not compulsion profiles (EO having more aggressive and less sexual obsessions) (Geller et al. 2001), factor analyses show a similar symptom structure in children, adolescents and adults suggesting that the OCD symptoms only differ minimally between these age groups (Stewart et al. 2008). This thus suggests that the clinical expression of the disorder itself is highly similar, and that the many differences can be well explained by developmental differences and altered relevance of certain symptom categories at different developmental stages. Altered cognitive abilities most probably also underly the reduced illness insight and a less prominent ego-dystonia (Farrell et al. 2006) with youths having less metacognitive insight (Weil et al. 2013; Moses-Payne et al. 2020).

Taken together, these findings suggest that OCD in childhood and adulthood disorder expression are more similar than in other mental health disorders, such as ADHD, which show substantially different symptom profiles across development. The comorbidity profiles reflect the developmental preponderance of psychiatric disorders, with a dominance of motor-related comorbidities during youth. These different comorbidity profiles may not indicate a distinct aetiology, but rather a heightened vulnerability to mental health problems per se and to an expression of disorder that is most likely for given age. The heightened heritability in juvenile OCD can also be understood as increased (additive) vulnerability factors that thus drive an early emergence of the disorder. It is thus possible that OCD in youth and adulthood are very much the same disorder, and that their characteristic profiles are a mere reflection of the developmental context.

5 Cognitive Markers in Development

Understanding the cognitive and neural mechanisms impaired in a psychiatric disorder has been believed to be of high importance, as these features may shed light on the aberrant causes that underlie the clinically observed symptoms (Montague et al. 2012; Hauser et al. 2016, 2018). The premise of this computational psychiatry approach is that the illness symptoms that are observed in clinical practice are expressions of underlying, aberrant brain processes. These processes are assumed to lead to cognitive biases or deficits that in turn facilitate the emergence of clinical symptoms (cf Hauser et al. 2018).

In the last two decades, a variety of cognitive deficits have been observed in adult OCD, many of which are discussed in more detail in adjacent chapters (e.g. chapters “Cognitive Inflexibility in OCD and Related Disorders” and “Recent Developments in the Habit Hypothesis of OCD and Compulsive Disorders”). However, much less is known about cognitive impairments in youths with OCD. Only a small corpus of studies reports findings in juvenile OCD patients, and first review articles on cognitive deficits are being published at the time this author writes this chapter (Loosen and Hauser 2020; Marzuki et al. 2019).

The paucity of developmental studies stands in stark contrast to the importance of such studies. Many of the adult neurocognitive studies investigate impairments in the belief that cognitive impairments are aetiologically involved in the causation of OCD. However, when studying cognition in adult patients, the cause and effect question cannot be answered and many potentially confounding factors cannot or only unsatisfactorily be addressed using traditional approaches (Pingault et al. 2018). Of particular importance is hereby the impact of illness and the adverse consequences (e.g. reduced social interactions, unemployment, medication effects) and how this impacts on cognition. Moreover, disorder-related distress and distraction (e.g. on-going intrusive obsessions that withdraw cognitive resources) may also alter behaviour in cognitive tests, even though the cognitive ability itself may not affected (i.e. observed performance does not necessarily need to reflect ability).

An area that has drawn much attention in youth and adult OCD research is decision making and learning. Based on the premise that youths with OCD may suffer from some fundamental decision making deficits, several studies investigated basic decision making, in which subjects have to make decisions based on explicitly stated choice options (e.g. a more risky vs safe choice option). In most of these tasks, no differences were found between juvenile patients and controls. OCD patients were not more risk-taking (Hybel et al. 2017), did not gamble more often (Drechsler et al. 2015) or discount delayed rewards more (Vloet et al. 2010; Carlisi et al. 2017; Marzuki et al. 2019). A common feature of all of these tasks is that they explicitly state the outcome sizes and possibilities. In tasks that do not state these determinants explicitly, youths with OCD indeed seem to show some impairments compared to age-matched controls. They perform worse when making decisions under ambiguity (Hauser et al. 2017a; Norman et al. 2018), when having to adapt to sudden changes in reward-contingencies (Gottwald et al. 2018) or when they can gather further information (Hauser et al. 2017b) (for detailed review cf. Loosen and Hauser 2020; Marzuki et al. 2019).

Although multiple studies found impairments in these ambiguity-related tasks, the underlying mechanisms are unclear and likely to be heterogeneous. Several of these tasks are based on learning stimulus-outcome associations, but on top have additional cognitive features that change (e.g. sudden reversals). Simple behavioural outcome measures (such as the number of errors) thus often fall short of identifying the exact mechanisms that is impaired. More recent methods of computational modelling (Montague et al. 2012; Hauser et al. 2018) help to overcome this limitation. By using algorithms from artificial intelligence, these methods allow to tease apart these different mechanisms and to identify the exact cognitive process that is altered. Whilst several studies have already used such methods in adults OCD, only the first handful of studies in juvenile OCD are surfacing (Loosen and Hauser 2020). It is thus critical that more studies use these computational methods to determine the precise impairments in juvenile OCD.

From the relatively sparse data that are available today, it does not seem as if paediatric OCD patients are suffering from a learning impairment per se as they are mostly unimpaired when learning pure stimulus-outcome associations (Loosen and Hauser 2020; Marzuki et al. 2019). However, youths with OCD struggle when having to adjust to complex unexpected changes in stimulus-outcome contingencies, often termed as cognitive flexibility (Loosen and Hauser 2020). How and why these impairments arise and how they relate to similar symptoms in adulthood remains unclear and requires not only longitudinal studies, but also novel computational models that capture these tasks.

A further area in which youths with OCD seem to differ from same-aged controls is information gathering. A clinical symptom common in both juvenile and adult OCD is that patients often report an increased indecisiveness, even when the decision is entirely unrelated to their OCD. Studies using different tasks (serial information gathering, perceptual decision making) showed that paediatric OCD patients show more cautious decision making with an increased information gathering behaviour before committing to a decision (Erhan et al. 2017; Hauser et al. 2017b). These tasks do not involve any learning, which means that this impairment could be a distinct feature from the above-mentioned deficit in complex learning tasks. Interestingly, the findings in adults OCD are somewhat more heterogeneous (cf Loosen and Hauser 2020), which could mean that this excessive information gathering is a feature that only exists in patients with early-onset OCD, or that this deficit is critical early in the disorder but vanishes once the disorder becomes more chronic.

Another domain that has found much attention in adult OCD is the formation of habits (cf chapter "Recent Developments in the Habit Hypothesis of OCD and Compulsive Disorders"). It is believed that one of the driving factors underlying OCD is that patients with OCD have a strong tendency to form habits, i.e. automatic stimulus-response patterns in which responses are automatically triggered when being presented with an associated stimulus. A computational task commonly used for probing such habit-like behaviour assesses how much a subject is model-based, i.e. how strongly she exploits her knowledge of the task structure to guide her behaviour (Daw et al. 2011). In adult OCD patients, and in sub-clinical adults with high OCD scores, a reduction in model-based reasoning has been found repeatedly (Gillan and Robbins 2014; Gillan et al. 2016). To my knowledge, no single study has investigated model-based reasoning in juvenile OCD. However, knowing whether a similar deficit is already present in juvenile OCD could help understand better whether such reduced model-basedness is indeed a driving factor underlying OCD. Adolescence is therefore a critical age, because preliminary studies have shown that model-based reasoning is an ability that is only fully developed once we reach adulthood (Smid et al. n.d.; Decker et al. 2016). A study that sheds important insight did not investigate paediatric OCD patients, but subjects with high OCD scores in an accelerated longitudinal study during adolescence (Vaghi et al. 2020). The authors show that model-based reasoning is still diminished during adolescence, but is fully matured in early adulthood. The authors also demonstrate that a reduced model-basedness is linked to heightened OCD symptoms. Critically, because the authors had multiple measurements of the same subjects across several years, they were able to investigate whether and how OCD symptoms and model-basedness influenced each other over time. Vaghi et al. (2020) found that the OCD symptoms preceded the reduction in model-based reasoning. This means that adolescents with high OCD scores will later on develop an impairment in model-basedness, but not the other way around. This suggests that the reduced model-basedness found in adult OCD is more likely a consequence of OCD symptoms, rather than a cause thereof. However, whether this is only true for model-based reasoning, or also for other forms of goal-directed control (Gottwald et al. 2018) remains to be determined. It is thus critical to further investigate when and how OCD symptoms drive a reduction in model-basedness, and whether this also holds in adolescent OCD patients (cf Gottwald et al. 2018).

In summary, the current corpus of research in youths with OCD is rather limited and it is difficult to draw definite conclusions about cognitive impairments, or even compare it to adult OCD (Loosen and Hauser 2020; Marzuki et al. 2019). Several aspects seem to be impaired with indecisiveness and complex reasoning being the most promising. It is of highest importance to understand whether the cognitive profile in adolescence differs from adult OCD and how these deficits emerge and interact with OCD symptoms during development.

6 Neural Features in Juvenile OCD

The number of studies that studied functional and structural brain differences in juvenile OCD seems substantially larger than the number of studies on cognition. This is particularly so when looking at structural studies, but there are also multiple reviews of functional neuroimaging deficits in juvenile OCD (Huyser et al. 2009; Brem et al. 2012). Importantly, recent heroic efforts in pooling structural neuroimaging studies of juvenile and adult OCD have allowed to make more definite answers about structural brain differences and to overcome the generally small sample sizes of individual studies (Boedhoe et al. 2016, 2018).

Similar to adult OCD studies, most studies in youths with OCD have focused on fronto-striatal deficits. This is primarily based on the belief that OCD arises from an imbalance of direct and indirect pathways of fronto-striatal loops. Such fronto-striatal loops are well described in basic neuroscience and we have by now good models that allow us to understand how individual impairments may arise from an imbalance between these excitatory and inhibitory loops (Haber and Knutson 2010; Frank 2011; Maia and Frank 2011; Hauser et al. 2016). Several different models of how these fronto-striatal loops are impaired have been put forward to explain OCD symptoms (e.g. Maia et al. 2008; van den Heuvel et al. 2010; Brem et al. 2012; Robbins et al. 2012; Grünblatt et al. 2014).

Many of the individual functional and structural neuroimaging studies have revealed impairments in these fronto-striatal loops. In juvenile OCD patients, these alterations were primarily found in prefrontal areas such as the orbitofrontal cortex, anterior cingulate and dorsolateral prefrontal cortex (Brem et al. 2012; Walitza et al. 2014; Hauser et al. 2017a). These are also regions that are known to be relevant for computing in the complex cognitive behaviours that are found impaired in OCD (cf Loosen and Hauser 2020; Robbins et al. 2012 for more detailed discussion). In fact, the error-related negativity, an EEG component arising from the anterior cingulate cortex reflecting error processing (Holroyd and Coles 2002; Debener et al. 2005), has consistently been found to be increased in youths with OCD (Marzuki et al. 2019).

Despite the relatively consistent findings of these smaller studies, the recent mega-analyses of pooled structural neuroimaging studies draw a somewhat different picture. First, these studies reveal that adult OCD patients and juvenile OCD have different patterns of brain differences, meaning that adult OCD patients show deficits in areas that are different from youths with OCD (Boedhoe et al. 2016, 2018). Second, these mega-analyses suggest that juvenile OCD patients mainly exhibit differences in the thalamus and in parietal areas (Boedhoe et al. 2016, 2018). Whilst the thalamus is known to be an important hub in fronto-striatal loops, parietal areas have previously received little attention in juvenile OCD. Lastly, the reported effects are of a relatively small effect size. Although this is in part driven by the heterogeneity of protocols used across sites, this also suggests that there might be a substantial neural heterogeneity, even within the juvenile OCD samples.

The reported heterogeneity between juvenile and adult neuroimaging findings highlights the importance of longitudinal studies also in neuroimaging. It is well known that the brain shows on-going developmental changes well into adulthood (Gogtay et al. 2004; Paus 2005; Whitaker et al. 2016; Ziegler et al. 2019), and it is thus well possible that neuroimaging differences in juvenile OCD can only be found in areas that are already fully matured, because the more subtle impairments in late-maturing areas are only fully visible once maturation has finished in adulthood.

Such a developmental embedding of brain changes can be found when using longitudinal approaches. In an early longitudinal study with paediatric OCD patients, Huyser et al. (2014) found a time-by-disorder interaction in the orbitofrontal cortex. Whilst the control youths showed a steady decrease in volume in this area, the OCD patients showed an on-going increase. This directly speaks to distinct developmental trajectories between youths with and without OCD and highlights the shortcoming of above cross-sectional studies that would not be able to detect such differences in on-going maturation.

A more recent study in a large population-based sample has shown similar effects (Ziegler et al. 2019). In this sample of late adolescents, the authors did not report any difference between subjects with high and low OCD symptoms when looking at brain myelin from a purely cross-sectional perspective. However, by exploiting the longitudinal data and looking at individual brain maturation trajectories, the authors found that OCD symptoms were linked to a reduced on-going maturation in the striatum as well as cingulate and dorsolateral prefrontal cortex.

These studies thus exemplify the importance of longitudinal studies that are not only able to reveal altered processes that would not be detectable using traditional approaches, but may also help identify aberrant brain development that precedes the emergence of OCD during adolescence.

7 Outlook

Understanding how OCD emerges during development is crucial for understanding the environmental, psychological, cognitive and neural mechanisms that are critically involved in the pathogenesis of this debilitating disorder. Despite the importance, the field itself is still in its infancy and requires substantial efforts to shed light onto the processes underlying OCD. In this chapter, I have highlighted the need for sophisticated study designs and methods that allow drawing conclusions about directionality and causality of effects. I have discussed first studies that venture into this direction and revealed interesting and novel insights into juvenile OCD. In the years to come, we need to characterise the developmental trajectories of OCD symptoms and how they are related to associated changes in brain and cognition. Our approaches should thereby be guided by other, more developed fields that have already provided substantial novel insights into the heterogeneity of symptom and developmental trajectories (Pingault et al. 2011, 2015) and that developed novel methods that allow tracing the causal structure of cross-modal impairments (Pingault et al. 2018). All these efforts will help us better understand the factors that give rise to OCD and to develop new criteria for youths at high risk of OCD, and to even design novel preventive interventions.

Acknowledgements TUH is supported by a Wellcome Sir Henry Dale Fellowship (211155/Z/18/Z), a grant from the Jacobs Foundation (2017-1261-04), the Medical Research Foundation, and a 2018 NARSAD Young Investigator grant (27023) from the Brain & Behavior Research Foundation. The Max Planck UCL Centre is a joint initiative supported by UCL and the Max Planck Society. The Wellcome Centre for Human Neuroimaging is supported by core funding from the Wellcome Trust (203147/Z/16/Z).

Conflict of Interest The author declares no competing financial interests.

References

Benatti B, Camuri G, Dell'Osso B, Cremaschi L, Sembira E, Palazzo C, Oldani L, Dobrea C, Arici C, Primavera D, Carpiniello B, Castellano F, Carrà G, Clerici M, Baldwin DS, Altamura AC (2016) Which factors influence onset and latency to treatment in generalized anxiety disorder, panic disorder, and obsessive-compulsive disorder? Int Clin Psychopharmacol 31:347–352

Boedhoe PSW et al (2016) Distinct subcortical volume alterations in pediatric and adult OCD: a worldwide meta- and mega-analysis. Am J Psychiatry 174:60–69. https://doi.org/10.1176/appi.ajp.2016.16020201

Boedhoe PSW et al (2018) Cortical abnormalities associated with pediatric and adult obsessive-compulsive disorder: findings from the ENIGMA obsessive-compulsive disorder working group. Am J Psychiatry 175:453–462

Brem S, Hauser TU, Iannaccone R, Brandeis D, Drechsler R, Walitza S (2012) Neuroimaging of cognitive brain function in paediatric obsessive compulsive disorder: a review of literature and preliminary meta-analysis. J Neural Transm Vienna Austria 119:1425–1448

Carlisi CO, Norman L, Murphy CM, Christakou A, Chantiluke K, Giampietro V, Simmons A, Brammer M, Murphy DG, Mataix-Cols D, Rubia K (2017) Comparison of neural substrates of temporal discounting between youth with autism spectrum disorder and with obsessive-compulsive disorder. Psychol Med 47:2513–2527

Daw ND, Gershman SJ, Seymour B, Dayan P, Dolan RJ (2011) Model-based influences on humans' choices and striatal prediction errors. Neuron 69:1204–1215

Debener S, Ullsperger M, Siegel M, Fiehler K, von Cramon DY, Engel AK (2005) Trial-by-trial coupling of concurrent electroencephalogram and functional magnetic resonance imaging identifies the dynamics of performance monitoring. J Neurosci 25:11730–11737

Decker JH, Otto AR, Daw ND, Hartley CA (2016) From creatures of habit to goal-directed learners: tracking the developmental emergence of model-based reinforcement learning. Psychol Sci 27:848–858

Drechsler R, Hauser TU, Iannaccone R, Brandeis D, Walitza S, Brem S (2015) Dealing with uncertainty – decision-making, executive function and cognitive style in adolescents with ADHD compaired to adolescents with OCD

Erhan C, Bulut GÇ, Gökçe S, Ozbas D, Turkakin E, Dursun OB, Yazgan Y, Balcı F (2017) Disrupted latent decision processes in medication-free pediatric OCD patients. J Affect Disord 207:32–37

Farrell L, Barrett P, Piacentini J (2006) Obsessive–compulsive disorder across the developmental trajectory: clinical correlates in children, adolescents and adults. Behav Change 23:103–120

Fineberg NA, Hengartner MP, Bergbaum C, Gale T, Rössler W, Angst J (2013a) Lifetime comorbidity of obsessive-compulsive disorder and sub-threshold obsessive-compulsive symptomatology in the community: impact, prevalence, socio-demographic and clinical characteristics. Int J Psychiatry Clin Pract 17:188–196

Fineberg NA, Hengartner MP, Bergbaum C, Gale T, Rössler W, Angst J (2013b) Remission of obsessive-compulsive disorders and syndromes; evidence from a prospective community cohort study over 30 years. Int J Psychiatry Clin Pract 17:179–187

Fineberg NA, Hengartner MP, Bergbaum CE, Gale TM, Gamma A, Ajdacic-Gross V, Rössler W, Angst J (2013c) A prospective population-based cohort study of the prevalence, incidence and impact of obsessive-compulsive symptomatology. Int J Psychiatry Clin Pract 17:170–178

Fineberg NA et al (2019) Early intervention for obsessive compulsive disorder: an expert consensus statement. Eur Neuropsychopharmacol J Eur Coll Neuropsychopharmacol 29:549–565

Frank MJ (2011) Computational models of motivated action selection in corticostriatal circuits. Curr Opin Neurobiol 21:381–386

Geller D, Biederman J, Jones J, Park K, Schwartz S, Shapiro S, Coffey B (1998) Is juvenile obsessive-compulsive disorder a developmental subtype of the disorder? A review of the pediatric literature. J Am Acad Child Adolesc Psychiatry 37:420–427

Geller DA, Biederman J, Faraone S, Agranat A, Cradock K, Hagermoser L, Kim G, Frazier J, Coffey BJ (2001) Developmental aspects of obsessive compulsive disorder: findings in children, adolescents, and adults. J Nerv Ment Dis 189:471–477

Gillan CM, Robbins TW (2014) Goal-directed learning and obsessive-compulsive disorder. Philos Trans R Soc Lond B Biol Sci 369

Gillan CM, Kosinski M, Whelan R, Phelps EA, Daw ND (2016) Characterizing a psychiatric symptom dimension related to deficits in goal-directed control. eLife 5

Gogtay N, Giedd JN, Lusk L, Hayashi KM, Greenstein D, Vaituzis AC, Nugent TF, Herman DH, Clasen LS, Toga AW, Rapoport JL, Thompson PM (2004) Dynamic mapping of human cortical development during childhood through early adulthood. Proc Natl Acad Sci U S A 101:8174–8179

Gottwald J, de Wit S, Apergis-Schoute AM, Morein-Zamir S, Kaser M, Cormack F, Sule A, Limmer W, Morris AC, Robbins TW, Sahakian BJ (2018) Impaired cognitive plasticity and goal-directed control in adolescent obsessive–compulsive disorder. Psychol Med 48:1900–1908

Grünblatt E, Hauser TU, Walitza S (2014) Imaging genetics in obsessive-compulsive disorder: linking genetic variations to alterations in neuroimaging. Prog Neurobiol 121:114–124

Haber SN, Knutson B (2010) The reward circuit: linking primate anatomy and human imaging. Neuropsychopharmacol Off Publ Am Coll Neuropsychopharmacol 35:4–26
Hauser TU, Fiore VG, Moutoussis M, Dolan RJ (2016) Computational psychiatry of ADHD: neural gain impairments across Marrian levels of analysis. Trends Neurosci
Hauser TU, Iannaccone R, Dolan RJ, Ball J, Hättenschwiler J, Drechsler R, Rufer M, Brandeis D, Walitza S, Brem S (2017a) Increased fronto-striatal reward prediction errors moderate decision making in obsessive–compulsive disorder. Psychol Med:1–13
Hauser TU, Moutoussis M, Iannaccone R, Brem S, Walitza S, Drechsler R, Dayan P, Dolan RJ (2017b) Increased decision thresholds enhance information gathering performance in juvenile obsessive-compulsive disorder (OCD). PLoS Comput Biol 13:e1005440
Hauser TU, Will G-J, Dubois M, Dolan RJ (2018) Annual research review: developmental computational psychiatry. J Child Psychol Psychiatry. https://onlinelibrary.wiley.com/doi/abs/10.1111/jcpp.12964. Accessed 22 Oct 2018
Holroyd CB, Coles MGH (2002) The neural basis of human error processing: reinforcement learning, dopamine, and the error-related negativity. Psychol Rev 109:679–709
Huyser C, Veltman DJ, de Haan E, Boer F (2009) Paediatric obsessive-compulsive disorder, a neurodevelopmental disorder? Evidence from neuroimaging. Neurosci Biobehav Rev 33:818–830
Huyser C, van den Heuvel OA, Wolters L, de Haan E, Lindauer R, Veltman DJ (2014) A longitudinal VBM study in paediatric obsessive–compulsive disorder at 2-year follow-up after cognitive behavioural therapy. World J Biol Psychiatry 15:443–452
Hybel KA, Mortensen EL, Lambek R, Thastum M, Thomsen PH (2017) Cool and hot aspects of executive function in childhood obsessive-compulsive disorder. J Abnorm Child Psychol 45:1195–1205
Kessler RC, Berglund P, Demler O, Jin R, Merikangas KR, Walters EE (2005) Lifetime prevalence and age-of-onset distributions of DSM-IV disorders in the National Comorbidity Survey Replication. Arch Gen Psychiatry 62:593–602
Kiddle B, Inkster B, Prabhu G, Moutoussis M, Whitaker KJ, Bullmore ET, Dolan RJ, Fonagy P, Goodyer IM, Jones PB (2017) Cohort profile: the NSPN 2400 cohort: a developmental sample supporting the Wellcome Trust NeuroScience in psychiatry network. Int J Epidemiol. https://academic.oup.com/ije/advance-article/doi/10.1093/ije/dyx117/4644452. Accessed 22 Nov 2017
Loosen AM, Hauser TU (2020) Towards a computational psychiatry of juvenile obsessive-compulsive disorder. Neurosci Biobehav Rev 118:631–642
Maia TV, Frank MJ (2011) From reinforcement learning models to psychiatric and neurological disorders. Nat Neurosci 14:154–162
Maia TV, Cooney RE, Peterson BS (2008) The neural bases of obsessive-compulsive disorder in children and adults. Dev Psychopathol 20:1251–1283
Marzuki AA, Pereira de Souza AMFL, Sahakian BJ, Robbins TW (2019) Are candidate neurocognitive endophenotypes of OCD present in paediatric patients? A systematic review. Neurosci Biobehav Rev. http://www.sciencedirect.com/science/article/pii/S014976341930908X. Accessed 13 Dec 2019
Montague PR, Dolan RJ, Friston KJ, Dayan P (2012) Computational psychiatry. Trends Cogn Sci 16:72–80
Moses-Payne ME, Habicht J, Bowler A, Steinbeis N, Hauser TU (2020) I know better! Emerging metacognition allows adolescents to ignore false advice. PsyArXiv. https://doi.org/10.31234/osf.io/gb9f4
Norman LJ, Carlisi CO, Christakou A, Murphy CM, Chantiluke K, Giampietro V, Simmons A, Brammer M, Mataix-Cols D, Rubia K (2018) Frontostriatal dysfunction during decision making in attention-deficit/hyperactivity disorder and obsessive-compulsive disorder. Biol Psychiatry Cogn Neurosci Neuroimaging 3:694–703
Paus T (2005) Mapping brain maturation and cognitive development during adolescence. Trends Cogn Sci 9:60–68

Pingault J-B, Tremblay RE, Vitaro F, Carbonneau R, Genolini C, Falissard B, Côté SM (2011) Childhood trajectories of inattention and hyperactivity and prediction of educational attainment in early adulthood: a 16-year longitudinal population-based study. Am J Psychiatry 168:1164–1170

Pingault J-B, Viding E, Galéra C, Greven CU, Zheng Y, Plomin R, Rijsdijk F (2015) Genetic and environmental influences on the developmental course of attention-deficit/hyperactivity disorder symptoms from childhood to adolescence. JAMA Psychiat 72:651–658

Pingault J-B, O'Reilly PF, Schoeler T, Ploubidis GB, Rijsdijk F, Dudbridge F (2018) Using genetic data to strengthen causal inference in observational research. Nat Rev Genet 19:566–580

Rasmussen SA, Eisen JL (1992) The epidemiology and clinical features of obsessive compulsive disorder. Psychiatr Clin North Am 15:743–758

Robbins TW, Gillan CM, Smith DG, de Wit S, Ersche KD (2012) Neurocognitive endophenotypes of impulsivity and compulsivity: towards dimensional psychiatry. Trends Cogn Sci 16:81–91

Shafran R (2003) Obsessive-compulsive disorder in children and adolescents. In: obsessive-compulsive disorder: theory, research and treatment. Wiley, Hoboken

Smid C, Kool W, Hauser TU, Steinbeis N (n.d.) Model-based decision-making and its cost/benefit arbitration in childhood. (In prep)

Stewart SE, Rosario MC, Baer L, Carter AS, Brown TA, Scharf JM, Illmann C, Leckman JF, Sukhodolsky D, Katsovich L, Rasmussen S, Goodman W, Delorme R, Leboyer M, Chabane N, Jenike MA, Geller DA, Pauls DL (2008) Four-factor structure of obsessive-compulsive disorder symptoms in children, adolescents, and adults. J Am Acad Child Adolesc Psychiatry 47:763–772

Taylor S (2011) Early versus late onset obsessive-compulsive disorder: evidence for distinct subtypes. Clin Psychol Rev 31:1083–1100

Vaghi MM, Vasa F, Moutoussis MM, Kievit R, Hauser TU, Vertes PE, Shahar N, Romero-Garcia R, Kitzbichler MG, Bullmore ET, NSPN Consortium, Dolan RJ (2020) Compulsivity disrupts adolescent development of goal-directed control and fronto-striatal functional connectivity. Proc Natl Acad Sci USA

van den Heuvel OA, van der Werf YD, Verhoef KMW, de Wit S, Berendse HW, Wolters EC, Veltman DJ, Groenewegen HJ (2010) Frontal-striatal abnormalities underlying behaviours in the compulsive-impulsive spectrum. J Neurol Sci 289:55–59

Vloet TD, Marx I, Kahraman-Lanzerath B, Zepf FD, Herpertz-Dahlmann B, Konrad K (2010) Neurocognitive performance in children with ADHD and OCD. J Abnorm Child Psychol 38:961–969

Walitza S, Wendland JR, Gruenblatt E, Warnke A, Sontag TA, Tucha O, Lange KW (2010) Genetics of early-onset obsessive-compulsive disorder. Eur Child Adolesc Psychiatry 19:227–235

Walitza S, Brem S, Hauser TU, Grünblatt E (2014) Wie biologisch sind Zwangsstörungen? Kindh Entwickl 23:75–85

Walitza S, Van Ameringen M, Geller D (2019) Early detection and intervention for obsessive-compulsive disorder in childhood and adolescence. Lancet Child Adolesc Health

Weil LG, Fleming SM, Dumontheil I, Kilford EJ, Weil RS, Rees G, Dolan RJ, Blakemore S-J (2013) The development of metacognitive ability in adolescence. Conscious Cogn 22:264–271

Whitaker KJ et al (2016) Adolescence is associated with genomically patterned consolidation of the hubs of the human brain connectome. Proc Natl Acad Sci 113:9105–9110

Woelbert E, Kirtley A, Balmer N, Dix S (2019) How much is spent on mental health research: developing a system for categorising grant funding in the UK. Lancet Psychiatry 6:445–452

Ziegler G, Hauser TU, Moutoussis M, Bullmore ET, Goodyer IM, Fonagy P, Jones PB, Lindenberger U, Dolan RJ, NSPN Consortium (2019) Compulsivity and impulsivity traits linked to attenuated developmental frontostriatal myelination trajectories. Nat Neurosci 22:992–999

Zohar AH (1999) The epidemiology of obsessive-compulsive disorder in children and adolescents. Child Adolesc Psychiatr Clin N Am 8:445–460

Inflammation, Obsessive-Compulsive Disorder, and Related Disorders

Jeffrey Meyer

Contents

Abstract Initial reports supporting the possibility of inflammation in the brain in obsessive-compulsive disorder (OCD) evolved from the models of Sydenham's Chorea, and Pediatric Autoimmune Neuropsychiatric Disorder Associated with Streptococcus (PANDAS), which implicated excessive autoimmune responses following exposure to group A B-hemolytic streptococcal infections. Subsequently, this model was expanded to Pediatric Autoimmune Neuropsychiatric Syndrome (PANS) which applied the same concept but included other infections. A critical shortcoming of this model was that it was attributable to a small minority of OCD cases. The relationship between inflammation and OCD was more broadly demonstrated through translocator protein (TSPO) positron emission tomography imaging, a method that detects gliosis, an important component of brain inflammation, in neuropsychiatric diseases, including morphological activation and proliferation of microglia and to some extent astroglia. This method identified greater TSPO binding in the cortico-striatal-thalamo-cortical circuit in OCD, providing a direct brain measure of an important component of inflammation. To identify OCD cases with

J. Meyer (✉)
Campbell Family Mental Health Research Institute, CAMH, Toronto, ON, Canada

Department of Psychiatry, University of Toronto, Toronto, ON, Canada
e-mail: jeffrey.meyer@utoronto.ca

Curr Topics Behav Neurosci (2021) 49: 31–53
https://doi.org/10.1007/7854_2020_210
Published Online: 24 February 2021

prominent elevations in TSPO binding in clinical research settings with lower cost peripheral markers, a promising approach is to apply blood serum biomarkers of inflammatory molecules produced by activated microglia and astroglia (gliosis). Such measures may aid stratification in future clinical trials. Several inflammatory-modifying interventions, including celecoxib, minocycline, and n-acetylcysteine, have been tested as treatments in randomized double-blind placebo controlled clinical trials and there is a tendency toward positive results, although these medications are not optimized for brain penetration and sample sizes for most trials were small. Future clinical trials of medications that target gliosis in OCD should apply larger sample sizes, ideally incorporating stratification approaches to enrich samples for the presence of gliosis.

Keywords Autoimmune · Gliosis · Neuroinflammation · Obsessive-compulsive disorder

1 Introduction

Optimal evidence for neuroinflammation in the pathophysiology of obsessive-compulsive disorder (OCD) should include evidence of causality, presence of brain inflammation as demonstrated by either neuroimaging or postmortem markers of inflammatory cells and reductions of symptoms with targeting of inflammatory cells in the disease. An advantage of considering this framework is that elements of it are testable; however, it has some limitations since OCD may be a heterogeneous disorder; and pathological inflammatory alterations in neuropsychiatric illness may be complex with similarly complex downstream cellular changes that may not be straightforwardly reversible. The evidence in support of this fundamental model is presented, prioritizing investigations of OCD, or subtypes of OCD. First, causal models of Sydenham's Chorea, pediatric autoimmune neuropsychiatric disorder associated with group A Beta-Hemolytic Streptococcus (GABHS) (PANDAS) and pediatric acute neuropsychiatric syndrome (PANS) are presented. Second evidence from a neuroimaging marker of gliosis is shown, followed by postmortem investigation. Finally, a summary of clinical trials repurposing inflammatory modulating medications and their effect on OCD symptoms is presented.

2 Sydenham's Chorea, PANDAS, and PANS

Sydenham's chorea is a syndrome of either a full body or half body expression of jerky involuntary movements that may be associated with reduced muscle tone, motor impersistence, dysarthria, weakness, and tics (Dean and Singer 2017; Walker and Wilmshurst 2010). Associated psychiatric symptoms may include obsessive-compulsive behaviors, which is the rationale for its relevance to OCD (Swedo 1994).

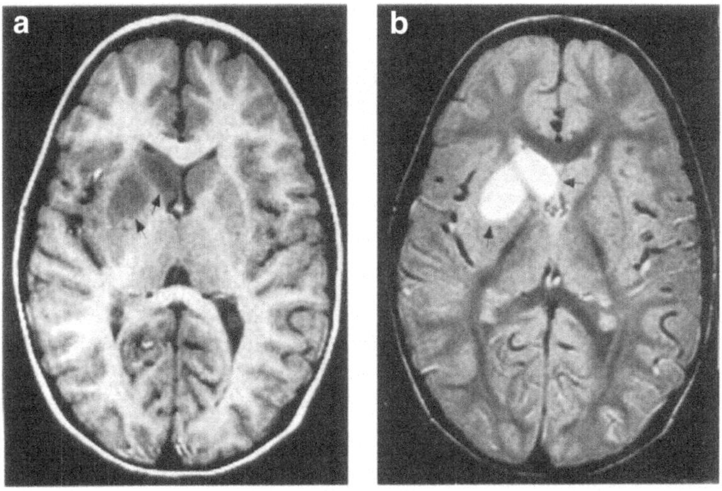

Fig. 1 MRI of Sydenham Chorea Case MRI of the brain of a 3-year-old boy performed 3 days after the onset of hemichorea (Sydenham Chorea) revealed abnormal signal and enlargement of the contralateral caudate and putamen. MRI of patient at presentation. (**a**) Axial TI-weighted image at the level of the basal ganglia shows swelling of the right caudate nucleus and putamen (arrows). (**b**) Axial proton density image, at same level as A, displays diffuse increased signal involving the same structures (arrows) (Emery and Vieco 1997)

Other psychiatric symptoms may include anxiety, lability, distractibility, and anorexia (Dean and Singer 2017; Walker and Wilmshurst 2010). Sydenham's chorea is part of a wider multisystem response to GABHS, which may also include lesions to cardiac valves, carditis, arthritis, arthralgias; as well as dermatological manifestations of subcutaneous nodules and erythema marginatum (Lee et al. 2009). The usual age of onset is between 5 and 15 years and it is more likely to occur in girls. The chorea occurs 4–8 weeks after exposure to GABHS pharyngitis but the psychiatric symptoms may occur earlier. The neurological symptoms usually resolve in 1–6 months; although one prospective study found that symptoms persisted for more than 2 years in 50% of cases studied (Cardoso et al. 1999).

The pathophysiological mechanism implicated in Sydenham's chorea is cross-reactivity between antigens on cells in the basal ganglia and thalamus with gangliosides or other antigens on the cell wall of GABHS (see Figs. 1 and 2) (Husby et al. 1976). Injuries involving inflammation to the basal ganglia and/or thalamus are associated with dysregulated movement control in neuropsychiatric diseases and there are case reports of abnormal structure on MRI proximal to the onset of symptoms (Emery and Vieco 1997). More specific evidence in support includes immunoglobulin reactivity in the blood of affected people to caudate and thalamic tissue (Husby et al. 1976). Interestingly, serum and monoclonal antibodies in affected patients were shown to activate calmodulin-dependent protein kinase II activity in a human neuronal cell line, demonstrating that antibody binding may affect neuronal signal transduction (Kirvan et al. 2003a, 2006). Moreover, evidence of similar abnormal immune responses are demonstrated more broadly in Sydenham's chorea as evidenced by the rheumatic heart valve lesions consequent

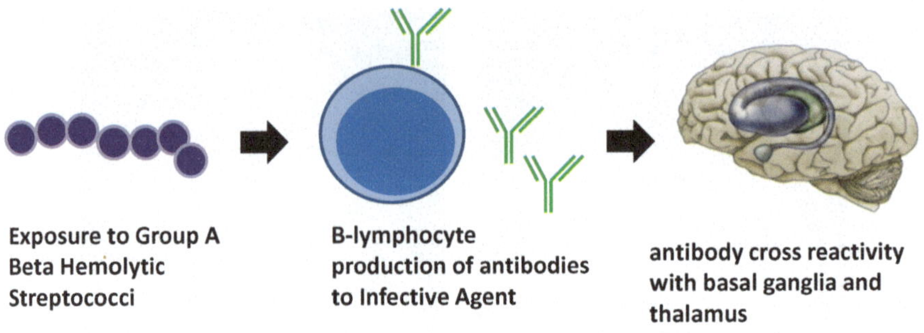

Fig. 2 Mechanism of PANDAS This involves exposure to Group A Beta-Hemolytic Streptococci which has antigens that provoke B-lymphocyte mediated production of antibodies that cross-react with similar antigens in the basal ganglia and thalamus

to anti-streptococcal antibodies from B lymphocytes that cross-react with antigens on heart valves.

Several case series of OCD characterized by prepubertal acute onset, episodic course, and concurrent neurological abnormalities such as choreiform movements occurring or exacerbating after exposure to infection have been documented (Swedo et al. 1998, 2012a). This syndrome, representing in part a subset of Sydenham's chorea, has been termed PANDAS. Subsequent cases series of the same syndrome occurring after different initial infections has been termed PANS (Swedo et al. 2012b). Consistent with reports in Sydenham's chorea, there are some reports of anti-neuronal antibodies and anti-basal ganglia antibodies in blood serum at a higher frequency in PANDAS as compared to controls (Singer et al. 2004; Dale et al. 2005; Kiessling et al. 1994). Translocator protein (TSPO) is mostly found on outer mitochondrial membranes and it is overexpressed during inflammatory changes, such as the activation that occurs in microglia and to a lesser extent, astroglia. Hence, TSPO imaging is a brain measure of activated microglia and activated astroglia, which may be termed gliosis (discussed in more detail in this chapter under, "Translocator Protein Imaging in Obsessive Compulsive Disorder"). A TSPO PET imaging study applying [^{11}C]-(*R*)-PK11195 reported increased binding in the bilateral caudate and bilateral lentiform nucleus in 17 cases of PANDAS compared to 15 healthy adults. This result is supportive of the autoimmune-neuroinflammatory model although a limitation is that [^{11}C]-(*R*)-PK11195 is a first generation TSPO PET radiotracer and has low sensitivity (Kobayashi et al. 2018). Overall, these collective findings argue that PANDAS is a subset of Syndenham's chorea that leads to OCD and PANS is considered to have a similar pathophysiological mechanism.

Several studies, albeit some preliminary, suggest that interventions targeting the mechanism of Sydenham's chorea reduce PANDAS severity. Antibiotic prophylaxis with azithromycin or penicillin was associated with both reduced recurrences of streptococcal infection and neuropsychiatric symptoms in a small double-blind randomized controlled trial (Snider et al. 2005). In a retrospective review of

35 cases of PANDAS in which therapeutic plasma apheresis was applied to lower plasma antibody level, a high proportion of cases reported sustained benefit (Latimer et al. 2015). A study in 95 participants found that early intervention with nonsteroidal anti-inflammatory treatment modestly reduced the duration of flare of symptoms in PANS/PANDAS and that maintenance NSAID use was associated with shorter flare duration (Brown et al. 2017).

3 Neuroanatomy Implicated in OCD

Studies of neuroinflammation are guided by the preexisting literature regarding the brain structures implicated in OCD. Arguably, the strongest regional convergence across investigation of neurochemical abnormalities in OCD, such as 5-HT_{2A}, 5-HTT, 5-HT_{1B}, and mGluR5 receptor binding (Adams et al. 2005; Pittenger et al. 2016; Hesse et al. 2011; Akkus et al. 2014; Wong et al. 2008; Matsumoto et al. 2010) and fluorodeoxyglucose uptake (Saxena et al. 1998) occurs in the dorsal caudate and the orbitofrontal cortex. However, the symptom-neuropathological theory of OCD also implicates a broader group of structures in the CSTC circuit such as the thalamus, ventral striatum, dorsal putamen, and anterior cingulate cortex (Matsumoto et al. 2010; Saxena et al. 1998; Bloch et al. 2006; Pittenger et al. 2011) (see Fig. 3). There is an accumulation of empirical support for neuroinflammatory pathologies of the cortico-striato-thalamo-cortical (CSTC) circuit leading to obsessive-compulsive symptoms, since illnesses affecting the CSTC including vascular disease, tumors, Huntington's disease, Tourette's Disorder and Sydenham's Chorea, are associated with disturbances of complex motor behavior.

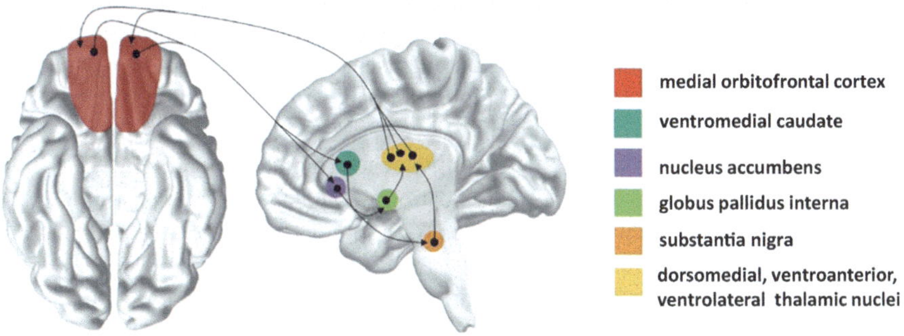

Fig. 3 Corticostriatal Circuitry of Obsessive-Compulsive Disorder. The caudate, orbitofrontal cortex, basal ganglia, ventral striatum, and thalamus highly implicated in the corticostriatal circuitry of OCD across neurochemical, metabolic, and functional imaging studies and are hence candidate structures for investigations of neuroinflammation in OCD (Fettes et al. 2017)

4 Translocator Protein Imaging in Obsessive-Compulsive Disorder

Gliosis is a brain response to injury or infection in the brain that involves both proliferation of glia and morphological changes in glia. Morphological changes during microglial and astroglial activation include increased cell body size, shorter thicker dendrites; or, in the case of microglia it is additionally possible to change to ameboid shape. During gliosis, glial cells participate in certain sets of functions, some of which influence the immune response. The range of functions for activated microglia and astroglia may include producing cytokines, complement proteins, reactive oxygen species and proteinases, as well as removing debris and synaptic remodeling; Gliosis is often viewed as an inflammatory response, but it has a range of severity and may be used by the brain to respond to more modest intensity of injury such as neurodegeneration.

TSPO imaging in neuropsychiatric illness is a method to detect mainly activated microglia and to a lesser extent astroglia. There is considerable evidence that TSPO is a marker of microglial activation based on reports in rodents that lipopolysaccharide administration, toxin, and stroke were associated with measurements of greater activated microglia expressing TSPO, and that lipopolysaccharide administration in humans was also associated with greater TSPO binding (Banati et al. 1997; Martin et al. 2010; Sandiego et al. 2015). In rodent studies, in which additional markers of microglial activation and observation are possible, the temporal course and/or expression of TSPO is dominated by its expression in microglia. The possibility of some species differences has been raised (Owen et al. 2017) and in human postmortem studies of neuropsychiatric illnesses such as AD, HIV encephalitis, multiple sclerosis, amyotrophic lateral sclerosis, frontotemporal dementia, and stroke, TSPO is often expressed in both activated microglia and activated astroglia (Cosenza-Nashat et al. 2009; Venneti et al. 2008). TSPO is also expressed to a lesser extent in vascular endothelium (Beltazar et al. 2018). Overall, given these findings, TSPO binding in health is mostly attributed to its expression in endothelial cells and the additional elevation in TSPO expression in disease states is usually ascribed to gliosis from proliferation of activated microglia and astroglia (Beltazar et al. 2018; Meyer et al. 2020).

In the early 1990s the only PET radioligand for translocator protein was [^{11}C](R)PK11195 but formal modeling quantification was delayed in humans until 2006 which demonstrated much better properties for applying this radiotracer with a two tissue compartment and arterial sampling versus reference tissue approaches, presumably because no brain region with free and non-displaceable binding characteristics representative of gray matter have been shown to be substantially devoid of TSPO receptors (Meyer 2017). Over the mid-to-late 2000s, a new generation of TSPO binding radiotracers were discovered which showed a greatly improved ratio of specific binding relative to free and non-displaceable binding (Kobayashi et al. 2018). There are a number of such radiotracers including [^{11}C]PBR28, [^{18}F]FEPPA, [^{18}F]PBR111, [^{18}F]PBR111, [^{18}F]DPA714 and [^{11}C]ER176 (Meyer 2017). [^{11}C]PBR28 is probably the most widely applied technique but [^{18}F]FEPPA is also applied at a number of sites. They are similar although [^{11}C]PBR28 has slightly less stable binding (total distribution volume (V_T))values than [^{18}F]FEPPA over the PET scanning period in humans and there are some, although not completely consistent, reports that [^{11}C]PBR28 has radioactive, brain penetrant metabolites in rodent brain (Imaizumi et al. 2007; Wilson et al. 2008). The properties of [^{18}F]PBR111 sacrifice some of its ratio of specific binding to free and non-displaceable binding for excellent reversibility of its time activity curve, a useful asset for quantitating TSPO binding in white matter. It has a lesser number of quantifiable regions due to binding of radioactive metabolites to bone. [^{11}C]ER176 is notable insofar as its distribution volume measure is much less affected by the homozygous state of single nucleotide polymorphism rs6971 (Ikawa et al. 2017), found in 1–10% of subjects, depending on their ethnicity (for other second-generation radioligands, subjects who are homozygous for this genotype are typically excluded from applied neuroimaging studies). This polymorphism causes a single amino acid substitution that reduces the binding of TSPO to all second-generation radioligands. While reliability of TSPO V_T for most ligands has not been formally reported, it is common for such data to remain unpublished during the first several years of radiotracer application.

Prior to neuroimaging TSPO in OCD, autoimmune mechanisms were considered for several reasons, including a high prevalence rate of OCD in autoimmune disorders (e.g., systemic lupus erythematosus and multiple sclerosis (Marrie et al. 2015; Bachen et al. 2009)) as well as the previously discussed case series of OCD following certain types of infections in children, known as PANDAS or PANS) (Swedo et al. 2012c) (Kirvan et al. 2003b). While the autoimmune theory of OCD was largely restricted to the basal ganglia and thalamus in PANDAS and PANS, abnormalities of CSTC circuitry are implicated in OCD given a confluence of human neurochemical imaging studies previously discussed in this chapter. This circuit was implicated in a cardinal [^{18}F]FEPPA PET study in which, 30–36% greater TSPO V_T was found within the dorsal caudate, orbitofrontal cortex, thalamus, ventral striatum, and dorsal putamen of 20 participants with OCD compared to matched healthy controls, with lower TSPO elevations observed in other gray matter regions sampled (see Fig. 4 and Table 1) (Attwells et al. 2017).

Because findings of elevated TSPO in neuropsychiatric illness are best explained by gliosis, these data argue several points. First inflammatory processes are relevant to adulthood in a high proportion of OCD cases rather than only during childhood in a small subset of OCD cases; and second, that the aberrant inflammatory processes in OCD extend beyond the striatum. It also raises the clinical question of whether interventions that inhibit or modulate some of the downstream effects of gliosis might have therapeutic effects in a subpopulation of adult OCD with more prominent TSPO V_T.

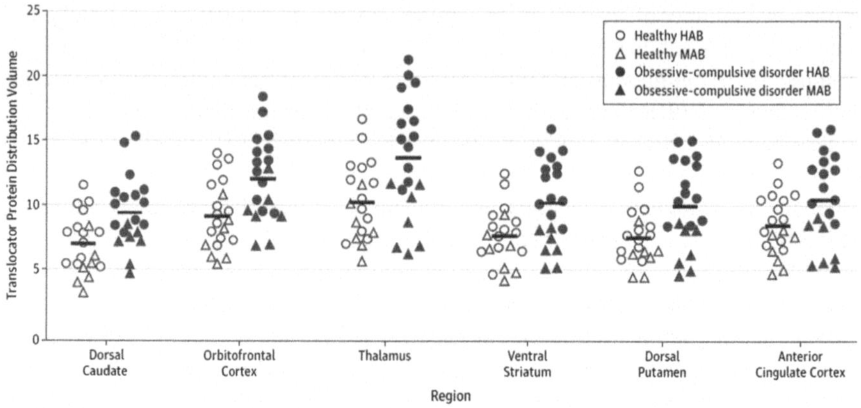

Fig. 4 Elevated Translocator Protein Distribution Volume in Obsessive-Compulsive Disorder (Attwells et al. 2017). Translocator protein distribution volume was significantly greater across brain regions assessed in participants with obsessive-compulsive disorder ($n = 20$) compared with health control individuals ($n = 20$). The single nucleotide polymorphism rs6971 of the TSPO gene, which influences binding of second-generation translocator protein positron emission tomography radiogliands, was included as a nuisance factor in the analyses of variance. Translocator protein distribution volume values represent raw values unadjusted for genotype. For this polymorphism, high-affinity homozygotes are denoted as HAB and mixed-affinity heterozygotes are denoted as MAB (mixed-affinity binding). The dark horizontal bars indicate the mean for each group (Attwells et al. 2017)

5 Postmortem Investigation

There have been few postmortem investigations in OCD, although one, albeit in a small sample of six OCD cases and eight controls completed a transcriptome analysis which implicated inflammation related pathways (Lisboa et al. 2019). In this study, the transcriptome was sampled from the caudate nucleus, nucleus accumbens, and putamen. Enrichment analysis of differential enrichment genes frequently showed changes related to immune response in the caudate nucleus as well as alterations associated with synapse transmission and ion transport. Strengths of the study included a short mean postmortem interval of 15 h in each group and availability of sample for the regions evaluated. It should be noted, however, that the mean age of the sample was 79 years in the OCD group and 75 in the control group, so these findings may reflect the state of OCD at a more advanced age.

6 Inflammation in Other Related Disorders

Major depressive disorder represents a common comorbidity for which there is substantial evidence of neuroinflammation. TSPO PET imaging studies of major depressive episodes (MDE) of major depressive disorder consistently identify

Table 1 Greater TSPO V_T associated with OCD diagnosis (Attwells et al. 2017)

	TSPO V_T, mean (SD)							Effect			
	Obsessive compulsive disorder			Healthy				Diagnosis		Genotype	
Region of interest	HAB (n = 13)	MAB (n = 7)	Total (n = 20)	HAB (n = 13)	MAB (n = 7)	Total (n = 20)	Percent difference	$F_{1,37}$	P	$F_{1,37}$	P
Dorsal caudate	10.7 (2.3)	6.9 (1.4)	9.4 (2.7)	7.8 (2.1)	5.2 (1.7)	6.9 (2.3)	35.6	15.5	<0.001	23.9	<0.001
OFC	13.4 (2.8)	9.3 (2.1)	12.0 (3.2)	10.1 (2.5)	7.5 (2.0)	9.2 (2.6)	30.9	13.5	0.001	17.5	<0.001
Thalamus	16.2 (3.2)	8.9 (2.4)	13.7 (4.6)	11.3 (3.0)	8.3 (2.0)	10.2 (3.0)	33.5	13.2	0.001	26.5	<0.001
Ventral striatum	12.1 (2.2)	6.8 (1.3)	10.2 (3.2)	8.3 (2.1)	6.4 (1.8)	7.6 (2.2)	33.8	14.4	0.001	25.6	<0.001
Dorsal putamen	11.8 (2.4)	6.6 (1.7)	10.0 (3.3)	8.3 (2.1)	6.1 (1.5)	7.5 (2.1)	32.6	12.8	0.001	26.2	<0.001
ACC	12.4 (2.3)	6.9 (1.7)	10.5 (3.4)	9.5 (2.0)	6.5 (1.4)	8.5 (2.3)	23.5	9.5	0.004	40.3	<0.001
MPFC	12.8 (2.4)	7.1 (1.7)	10.8 (3.5)	9.4 (2.1)	6.7 (1.3)	8.4 (2.2)	27.5	11.9	0.001	34.8	<0.001
DLPFC	13.0 (2.1)	7.3 (1.9)	11.0 (3.4)	10.0 (2.0)	7.1 (1.8)	9.0 (2.3)	22.0	9.1	0.005	38.5	<0.001
VLPFC	13.9 (2.6)	8.2 (1.8)	11.9 (3.6)	10.7 (2.2)	7.7 (2.0)	9.6 (2.5)	23.9	10.0	0.003	31.8	<0.001
INS	13.3 (2.6)	7.5 (1.8)	11.3 (3.6)	9.9 (2.3)	6.9 (1.5)	8.8 (2.5)	27.7	11.4	0.002	33.0	<0.001
Temporal cortex	13.5 (2.5)	7.8 (2.1)	11.5 (3.6)	10.4 (2.3)	7.2 (1.8)	9.3 (2.6)	24.3	9.5	0.004	33.5	<0.001
Inferior parietal cortex	14.1 (3.1)	8.1 (1.8)	12.0 (4.0)	10.9 (2.2)	7.9 (1.9)	9.8 (2.6)	22.2	7.5	0.009	29.8	<0.001
Occipital cortex	13.9 (2.6)	7.8 (2.1)	11.8 (3.8)	10.6 (2.4)	7.4 (1.9)	9.5 (2.7)	23.8	8.9	0.005	33.9	<0.001
Hippocampus	12.2 (2.3)	6.9 (2.0)	10.3 (3.3)	8.7 (2.5)	6.8 (2.0)	8.0 (2.5)	28.9	9.4	0.004	20.3	<0.001

ACC anterior cingulate cortex, *ANOVA* analysis of variance, *DLPFC* dorsolateral prefrontal cortex, *HAB* high-affinity binding, *INS* insula, *MAB* mixed-affinity binding, *MPFC* medial PFC, *OFC* orbitofrontal cortex, *SD* standard deviation, *TSPO* translocator protein, *TSPO* V_T translocator protein density measured by distribution volume, *VLPFC* ventrolateral PFC

[a]Indicates binding to the single-nucleotide polymorphism rs6971 of the TSPO gene known to influence binding of second-generation TSPO radioligands, including [^{18}F]FEPPA

greater TSPO total distribution volume (V_T), an index of TSPO density, in the prefrontal cortex (PFC) and anterior cingulate cortex (ACC), as demonstrated across six different studies at four different sites totalling 142 patients and 93 controls (Setiawan et al. 2015, 2018; Richards et al. 2018; Li et al. 2018; Holmes et al. 2018), with ACC values elevated 15–67% and PFC values elevated 25–35%. One additional [^{11}C]PBR28 study reported a negative result within a sample combining 5 MDE subjects and four recovered depressed subjects, however, the same group reports that the five subjects in a current MDE also had elevated TSPO V_T values in gray-matter regions (Cosgrove 2013). The TSPO PET findings in MDE are largely attributed to microglial activation and proliferation, since markers of astrogliosis are lower in the prefrontal and anterior cingulate cortex in MDE (Rajkowska and Stockmeier 2013) and rodent studies find a preferential expression of TSPO in activated microglia as compared to astroglia (Banati et al. 1997; Martin et al. 2010; Sandiego et al. 2015). There are a few postmortem studies investigating markers of microglial activation which vary in regards to sample size, regions sampled and methodology (reviewed (Enache et al. 2019)), and this influences the subtype of inflammatory cell identified. Most studies focused on suicide rather than MDE and had small sample sizes of 12 or less MDE cases (Enache et al. 2019). There was one study (Clark et al. 2016) which used a protocol more comparable to the TSPO PET imaging studies and had a sample size greater than 12 participants. This postmortem study (Clark et al. 2016) was by far the largest study representing almost half of MDE cases across postmortem studies focusing on microglial activation in MDE, having investigated 45 MDE cases as compared to 36 controls. In addition, this study sampled a similar region, the ventrolateral prefrontal cortex and used a marker representing microglial proliferation and activation, Iba-1, a similar label in some respects to TSPO as their binding has been correlated in postmortem tissue (Van Weehaeghe et al. 2020). Greater Iba-1 level was identified, consistent with the neuroimaging studies of TSPO (Clark et al. 2016).

7 Biomarkers of Inflammation

Given evidence of central inflammation in OCD, based on the elevated TSPO V_T in the CSTC regions, the high prevalence of OCD in autoimmune disorder (Marrie et al. 2015; Bachen et al. 2009), and the pathophysiological mechanism Sydenham's chorea implicated in PANDAS and PANS (Swedo et al. 2012c), peripheral biomarkers should be investigated to determine their relationship to brain inflammation in OCD to aid stratification of patients in clinical settings. One model to consider is that states of elevated peripheral inflammation are communicated to the brain through mechanisms including cytokine stimulation of peripheral nerves, macrophage-like cells proximal to the ventricles secreting cytokines, and cytokine transporters at the blood–brain barrier (Dantzer et al. 2008), so it is theoretically possible that peripheral markers could be useful surrogates under such circumstances. However, there is a lack of evidence for heightened peripheral inflammation during most of the course of OCD illness as suggested by a meta-analysis of 538 OCD cases and 463 healthy controls, demonstrating a lack of consistent differences in a number of serum markers including tumor necrosis factor α (TNFα), IL-6, IL-1β, IL-4, and IL-10 (Cosco et al. 2019). The number of participants compared for each marker totaled between 106 and 292. So, although heightened peripheral inflammation may occur during predisposing events for Sydenham's Chorea, PANDAS or PANS, it is unlikely to be the most prominent change during much of the OCD illness. Thus, it could be argued that biomarkers representative of central inflammation should be developed based on central brain changes of gliosis as the main source of bodily inflammation.

Easily applicable, replicable, low cost measure indicative of gliosis that are acceptable to patients with OCD are challenging to develop. Although positron emission tomography (PET) imaging of the translocator protein (TSPO) replicates well in diseases such as major depressive disorder, Alzheimer's disease and schizophrenia, and other radiotracers targeting P2X7, P2Y12 and CSF-1 receptors are being advanced (Meyer et al. 2020; Narayanaswami et al. 2018; Horti et al. 2019), these methods are expensive, require scarce resources, and typically need arterial blood sampling for kinetic modeling because microglia are ubiquitous throughout the brain. Another approach is cerebrospinal fluid measurement of concentration of products consequent to greater indoleamine 2,3-dioxygenase activity and/or cytokines but the necessary lumbar puncture is difficult for many patients (Galasko and Montine 2010; Wang and Miller 2018). In regards to magnetic resonance imaging (MRI) methods with paramagnetic probes like Cd-bis-5-HT-DTPA that measure myeloperoxidase activity across activated microglia, neutrophils, and monocytes, their sensitivity to detect gliosis in neuropsychiatric disease is not yet established and it is possible that their application will be restricted to disease states with substantial blood–brain barrier breakdown (Pulli and Chen 2014; Attwells et al. 2020a).

A recent [^{18}F]FEPPA PET study of TSPO V_T evaluated the relationship of TSPO V_T to composite peripheral measures (Attwells et al. 2020a). The perspective taken was that activated microglia in brain were considered an important source of inflammatory markers. Thus, several such products that are also actively transported out of the central nervous system into blood were measured, including serum prostaglandin E_2 (PGE_2), prostaglandin F_2 alpha ($PGF_{2\alpha}$), and TNF_α (Minghetti et al. 1997; Akanuma et al. 2014; Tachikawa et al. 2014; Banks 2015; Hagen et al. 1977). Since PGE2, PGF2α, and TNF_α are also produced by the immune response of peripheral tissues (Kalinski 2012) and microglial activation in brain is influenced by peripheral inflammation (Block et al. 2007), a control measure of an established index of peripheral inflammation, serum C-reactive protein (CRP) concentration, was also applied. TSPO V_T were measured in 3 cohorts: dorsal caudate TSPO V_T of 20 subjects with obsessive-compulsive disorder, prefrontal cortex TSPO V_T of 20 subjects with major depressive episodes from major depressive disorder; and 56 subjects with treatment resistant major depressive episodes from major depressive disorder. $\ln(PGE_2/CRP)$ and $\ln(TNF_\alpha/CRP)$ consistently correlated with TSPO V_T ($R^2 = 0.36$–0.11, $p = 0.0030$ to $p = 0.0076$, see Fig. 5) (Attwells et al. 2020a). The ln is the natural logarithm with base ~2.7, so for example, with this function 2.7

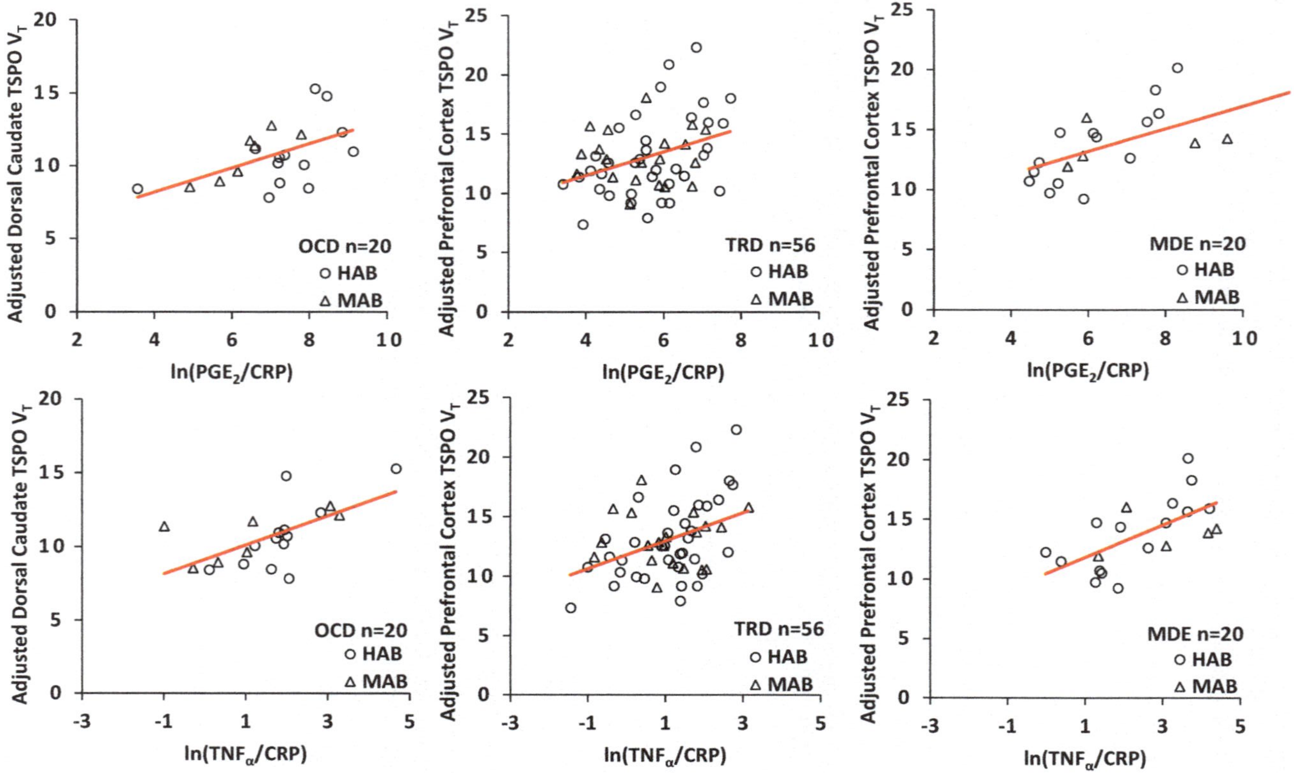

Fig. 5 Blood Serum Markers Significantly Correlated with Predictive of Translocator Protein Distribution Volume in all Three Cohorts (Attwells et al. 2020a). Ln(PGE_2/CRP) and ln(TNFα/CRP) were highly significant correlates of prefrontal cortex TSPO V_T in MDE and TRD cohorts, and dorsal caudate TSPO V_T in

OCD cohort. To address the effect of the rs6971 genotype on TSPO V_T, the differential effect of genotype in a linear regression was found for TSPO V_T. Then, the MAB TSPO V_T values were adjusted by adding the differential effect of genotype to the TSPO V_T values. (Adjusted prefrontal cortex TSPO V_T = unadjusted TSPO V_T + b1*genotype, where b1 = 4.939 in MDE cohort and b1 = 4.678 in TRD cohort. Adjusted dorsal caudate TSPO V_T = unadjusted TSPO V_T + b1*genotype, where b1 = 4.230 in OCD cohort). *Abbreviations*: *CRP* c-reactive protein, *HAB* high-affinity binders, *MAB* mixed-affinity binders, *MDE* medication free major depressive episodes secondary to major depressive disorder, *OCD* obsessive-compulsive disorder, PGE_2 prostaglandin E2, *TNFα* tumor necrosis factor alpha, *TRD* treatment resistant major depressive episodes secondary to major depressive disorder, *TSPO* V_T translocator protein distribution volume becomes 1, 2.7 squared becomes 2. Assessment of threshold serum values for ln (PGE_2/CRP) and ln (TNFalpha/CRP) to predict highly elevated TSPO V_T in the PFC of the former two groups demonstrated that a positive predictive value (PPV) of 80% was possible while retaining 40% of participant samples and that receiver operating curves (ROC) ranged from 75 to 81%. While these two sets of markers were ultimately the best replicated, demonstrating significant correlations to TSPO V_T in the primary region of interest within the three different samples, it was also notable that ln (PGF_2/CRP) predicted 56% of the variance of TSPO V_T in the OCD sample, so it will be important in future study to determine if this is an OCD-specific finding. These results suggest that markers such as the ratios of TNFα/CRP, PGE_2/CRP, which replicated across diagnoses, or PGF_2/CRP, which accounted for a very high proportion of variance in dorsal caudate TSPO V_T in OCD, may be applied in clinical trials to advance participant stratification to detect those with greater gliosis to match to new therapeutics targeting neuroinflammation for OCD.

Another direction of interest for developing peripheral biomarkers is to measure antibodies reflecting exposure to group A streptococcus and/or antibodies against neurons. For example, serum anti-streptolysin-O titres (ASOT) may reflect exposure to group A streptococcus and antibodies against neurons include enolase and aldolase C. In a study of 96 OCD and 50 healthy controls, western immunoblotting was conducted against ASOT, neuronal-specific enolase, aldolase C, and pyruvate kinase. (The latter marker may be found in red blood cells as well as on neuronal surfaces). Interestingly, 13 of the OCD cases and none of the control group were positive for neuronal-specific enolase, leading to an uncorrected significance of 0.004 (Nicholson et al. 2012). Five main comparisons were reported, with the remaining findings being negative, so arguably the most conservative corrected significance was 0.02 which would still be a significant result (Nicholson et al. 2012). Replication has been mixed with one group reporting no elevation in enolase titres in childhood OCD (Gause et al. 2009) and another group, reporting higher enolase titres in childhood OCD (Nicolini et al. 2015).

Given that there is some uncertainty as to whether antibodies in the central nervous system may transport to blood serum, a study by Bhattacharyya et al. (Bhattacharyya et al. 2009) evaluated antibody binding to basal ganglia and thalamus homogenates within the cerebrospinal fluid in 23 OCD cases and 23 controls with Western blot and found differential CSF antibody level at several molecular weights. This suggests that there are more antibodies against antigens of some molecular weights within homogenates of basal ganglia and thalamus in the CSF, which suggests greater autoimmunity to antigens within these weight ranges. However, the specific identity of the antigens isn't known. In the same study, a similar comparison of blood serum, after corrections for multiple comparisons, found a trend toward greater presence of an antibody against a 43Kd protein.

The conceptual underpinnings for measuring anti-neuronal antibodies are appealing but further studies are needed in this area to achieve consistent findings across different geographic regions and sites with the same methodologies and target antigens. A challenge in study design is that given the findings of Bhattacharyya et al. (Bhattacharyya et al. 2009), which identify greater antibodies in the CSF in OCD to a variety of antigens contained in the basal ganglia and thalamus homogenates (mixed tissue), it is plausible that there is substantial heterogeneity of neuronal antigenic exposures and responses in OCD, making this difficult. For example, it is possible that certain infectious and antigenic exposures predispose to OCD, but these might vary by geographic region and time so it is difficult then to obtain consistency across studies.

8 Investigations of Putative Inflammation Targeting Interventions in OCD

Several medications with inflammatory modulating properties such as celecoxib, minocycline, and n-acetylcysteine, have been investigated in randomized double-blind placebo controlled studies in OCD (see Table 2). Celecoxib is a cyclooxygenase-2 (COX-2) inhibitor approved for treating osteoarthritis in many countries, including the United States, Canada, and most of Europe. Minocycline is a second-generation tetracycline antibiotic with off target anti-inflammatory effects that includes lowering of lipopolysaccharide induced brain expression of major histocompatibility complex II, tumor necrosis factor α, inferferon γ, interleukin 6 and interleukin IL-1β mRNA and/or protein; and reduction of microglial and astroglial activation and proliferation in animal disease models of neurodegeneration, traumatic brain injury, and sickness behavior (Cheng et al. 2015; Garrido-Mesa et al. 2013; Homsi et al. 2010; Seabrook et al. 2006; Kobayashi et al. 2013). N-acetylcysteine (NAC) is a modified amino acid with a number of implicated properties including anti-inflammatory, antioxidant, and antiapoptotic effects, the former including reductions of gliosis in models of traumatic brain injury, stroke and lipopolysaccharide administration (Samuni et al. 2013; Berk et al. 2013).

A limitation of these interventions is that brain engagement in humans is not fully established. In regards to celecoxib (Attwells et al. 2020b) there are currently no PET COX-2 radioligands validated in humans with the most advanced radiotracer being [^{11}C]MC1 which did not demonstrate specific binding in healthy baboons (Kim et al. 2018). On the other hand, administration of intravenous [^{11}C] radiolabelled celecoxib in humans was associated with a peak standard uptake value of 1.2–2 in the brain (Kumar et al. 2018), which is comparable to some other radiolabelled drugs with high target occupancies like venlafaxine and citalopram (Schou et al. 2015). In addition, the ratio of cerebrospinal fluid to unbound plasma concentration of celecoxib is reported as 2:1 which is favorable. Minocycline does reduce gliosis in rodents but the dosing of minocycline in such studies, is 3–30 times larger relative to body mass (Martin et al. 2011; Wolf et al. 2020; Converse et al. 2011; Plane et al. 2010). There is one human study suggesting brain engagement of minocycline based on a 20% reduction in TSPO V_T reported across brain regions assessed in seven

Table 2 Randomized double-blind placebo controlled trials of putative inflammatory influencing interventions in OCD

Trial	Intervention	Participants	Result
Sayyah et al. (2011)	Celecoxib 200 mg bid add-on	Total: 52 Celecoxib: 27 Placebo: 25	Celecoxib > placebo on Y-BOCS
Shalbafan et al. (2015)	Celecoxib taper to 200 mg bid add-on	Total: 50 Celecoxib: 25 Placebo: 25	Celecoxib > placebo on Y-BOCS
Esalatmanesh et al. (2016)	Minocycline taper to 200 mg bid add-on	Total: 94 Minocycline: 47 Placebo: 47	Minocycline > placebo on Y-BOCS
Paydary et al. (2016)	NAC taper to 1,000 mg bid add-on	Total: 44 NAC: 22 Placebo: 22	No significant difference on final severity
Afshar et al. (2012)	NAC 2400 mg add-on	Total: 39 NAC: 19 Placebo: 20	NAC > placebo on Y-BOCS
Sarris et al. (2015)	NAC 1500 mg bid monotherapy or add-on	Total: 35 NAC: 20 Placebo: 15	No significant difference on final severity
Costa et al. (2017)	NAC 3000 mg daily monotherapy or add-on	Total: 35 NAC: 16 Placebo: 19	No significant difference On final severity
Ghanizadeh et al. (2017)	NAC 2400 mg daily add-on	Total: 29 NAC: 18 Placebo: 11	NAC > placebo on Y-BOCS
Li et al. (2020)	NAC 900 mg tid monotherapy or add-on	Total: 11 NAC: 5 Placebo: 6	NAC > placebo on CY-BOCS

cases treated with minocycline versus five receiving placebo, in traumatic brain injury cases within 6 months of the injury, but ideally the sample size should have been larger (Scott et al. 2018). To best support trials in OCD, ideally, brain engagement of minocycline with TSPO V_T as an endpoint marker of target engagement should be demonstrated in OCD rather than traumatic brain injury. In regards to NAC, effects of reducing gliosis are demonstrated in animal models at substantially higher dosing than in humans (Berk et al. 2013) but it would be best to demonstrate effects of NAC on TSPO V_T or another brain marker of gliosis, preferably in OCD in humans. Although such data is not available, there is some other evidence that NAC, at high doses, may affect human brain: Cysteine is an amino acid precursor to glutathione (GSH), a prominent antioxidant in the brain and modified n-acetyl-cysteine is believed to avoid first pass liver metabolism and deliver cysteine more effectively to the brain. NAC administration has been shown to increase GSH levels in the brain in humans, suggesting brain penetration, but at higher dosing than what is used in clinical trials, since the effect on GSH was shown after a single dose sevenfold greater given by infusion (Holmay et al. 2013). Thus evidence for brain engagement is a major limitation for these interventions to target inflammation in OCD, particularly for minocycline and n-acetylcysteine.

Even though brain penetration is not fully established, there is a tendency toward positive results in the clinical trials adding inflammatory modulating medications to serotonin reuptake inhibitors. While there are only a few trials, with most having smaller sample sizes as compared to industry funded studies that established SSRI as a treatment for OCD, it is notable that the two for celecoxib (Shalbafan et al. 2015; Sayyah et al. 2011), the one for minocycline (Esalatmanesh et al. 2016) and three of six NAC trials (Ghanizadeh et al. 2017; Costa et al. 2017; Paydary et al. 2016; Afshar et al. 2012; Sarris et al. 2015; Li et al. 2020) find greater effects for active treatment over placebo (see Table 2). Sample size of the clinical trials was small with the largest studies being the two evaluating celecoxib that each had approximately 50 participants (Shalbafan et al. 2015; Sayyah et al. 2011), and the minocycline trial that had 94 participants (Esalatmanesh et al. 2016). Thus, while the clinical interventions show some promise, ideally target engagement of the therapeutic should be clearly demonstrated and larger sized clinical trials should be completed.

9 Conclusions

There is promising and increasingly substantial evidence for neuroinflammation in the pathophysiology of OCD. First, as demonstrated by the examples of Sydenham's Chorea, PANDAS and PANS, there is an accumulation of observations that certain infections are associated with subsequent onset of OCD symptoms in children. A limitation of this evidence is that a small minority of OCD cases implicate this specific mechanism. However, there is a second key element of evidence, that TSPO V_T, a marker of gliosis in neuropsychiatric disorder, is elevated in the CTSC circuit in OCD. Since gliosis is a broad characterization of inflammation, future studies should endeavor to characterize this in more detail with additional markers of gliosis and neuroinflammation. A third element of evidence is the generally favorable findings of medications with neuroinflammatory properties repurposed in randomized double-blind placebo controlled clinical trials in OCD. The main limitation of these studies is that the brain penetration of the interventions should ideally be shown in the clinical setting, and most of the clinical trials are of modest sample size. So, there is important evidence for neuroinflammatory changes in OCD, but future advances should address the limitations in order to optimally translate novel inflammatory modulating interventions.

References

Adams KH, Hansen ES, Pinborg LH, Hasselbalch SG, Svarer C, Holm S, Bolwig TG, Knudsen GM (2005) Patients with obsessive-compulsive disorder have increased 5-HT2A receptor binding in the caudate nuclei. Int J Neuropsychopharmacol 8(3):391–401

Afshar H, Roohafza H, Mohammad-Beigi H, Haghighi M, Jahangard L, Shokouh P, Sadeghi M, Hafezian H (2012) N-acetylcysteine add-on treatment in refractory obsessive-compulsive disorder: a randomized, double-blind, placebo-controlled trial. J Clin Psychopharmacol 32 (6):797–803

Akanuma S, Higuchi T, Higashi H, Ozeki G, Tachikawa M, Kubo Y, Hosoya K (2014) Transporter-mediated prostaglandin E(2) elimination across the rat blood-brain barrier and its attenuation by the activation of N-methyl-D-aspartate receptors. Drug Metab Pharmacokinet 29(5):387–393

Akkus F, Terbeck S, Ametamey SM, Rufer M, Treyer V, Burger C, Johayem A, Mancilla BG, Sovago J, Buck A, Hasler G (2014) Metabotropic glutamate receptor 5 binding in patients with obsessive-compulsive disorder. Int J Neuropsychopharmacol 17(12):1915–1922

Attwells S, Setiawan E, Wilson AA, Rusjan P, Mizrahi R, Miler L, Xu C, Richter MA, Kahn A, Kish SJ, Houle S, Ravindran L, Meyer JH (2017) Inflammation in the neurocircuitry of obsessive-compulsive disorder. JAMA Psychiat 74:833–840

Attwells S, Setiawan E, Wilson AA, Rusjan PM, Miler L, Xu C, Hutton C, Husain MI, Kish S, Vasdev N, Houle S, Meyer JH (2020a) Replicating predictive serum correlates of greater translocator protein distribution volume in brain. Neuropsychopharmacology 45(6):925–931

Attwells S, Setiawan E, Rusjan PM, Xu C, Hutton C, Rafiei D, Varughese B, Kahn A, Kish SJ, Vasdev N, Houle S, Meyer JH (2020b) Translocator protein distribution volume predicts reduction of symptoms during open-label trial of celecoxib in major depressive disorder. Biol Psychiatry 88(8):649–656. https://doi.org/10.1016/j.biopsych.2020.03.007

Bachen EA, Chesney MA, Criswell LA (2009) Prevalence of mood and anxiety disorders in women with systemic lupus erythematosus. Arthritis Rheum 61:822–829

Banati R, Myers R, Kreutzberg GW (1997) PK ('peripheral benzodiazepine')--binding sites in the CNS indicate early and discrete brain lesions: microautoradiographic detection of [3H]PK11195 binding to activated microglia. J Neurocytol 26:77–82

Banks WA (2015) The blood-brain barrier in neuroimmunology: Tales of separation and assimilation. Brain Behav Immun 44:1–8

Beltazar C, Harrison-Brown M, Middleton RJ, Banati R, Liu GJ (2018) Cellular sources and regional variations in the expression of the neuroinflammatory marker translocator protein (TSPO) in the normal brain. Int J Mol Sci 19:E2707

Berk M, Malhi GS, Gray LJ, Dean OM (2013) The promise of N-acetylcysteine in neuropsychiatry. Trends Pharmacol Sci 34(3):167–177

Bhattacharyya S, Khanna S, Chakrabarty K, Mahadevan A, Christopher R, Shankar SK (2009) Anti-brain autoantibodies and altered excitatory neurotransmitters in obsessive-compulsive disorder. Neuropsychopharmacology 34(12):2489–2496

Bloch MH, Landeros-Weisenberger A, Kelmendi B, Coric V, Bracken MB, Leckman JF (2006) A systematic review: antipsychotic augmentation with treatment refractory obsessive-compulsive disorder. Mol Psychiatry 11(7):622–632

Block ML, Zecca L, Hong JS (2007) Microglia-mediated neurotoxicity: uncovering the molecular mechanisms. Nat Rev Neurosci 8(1):57–69

Brown KD, Farmer C, Freeman GM Jr, Spartz EJ, Farhadian B, Thienemann M, Frankovich J (2017) Effect of early and prophylactic nonsteroidal anti-inflammatory drugs on flare duration in pediatric acute-onset neuropsychiatric syndrome: an observational study of patients followed by an academic community-based pediatric acute-onset neuropsychiatric syndrome clinic. J Child Adolesc Psychopharmacol 27(7):619–628

Cardoso F, Vargas AP, Oliveira LD, Guerra AA, Amaral SV (1999) Persistent Sydenham's chorea. Mov Disord 14(5):805–807

Cheng S, Hou J, Zhang C, Xu C, Wang L, Zou X, Yu H, Shi Y, Yin Z, Chen G (2015) Minocycline reduces neuroinflammation but does not ameliorate neuron loss in a mouse model of neurodegeneration. Sci Rep 5:10535
Clark SM, Pocivavsek A, Nicholson JD, Notarangelo FM, Langenberg P, McMahon RP, Kleinman JE, Hyde TM, Stiller J, Postolache TT, Schwarcz R, Tonelli LH (2016) Reduced kynurenine pathway metabolism and cytokine expression in the prefrontal cortex of depressed individuals. J Psychiatry Neurosci 41(6):386–394
Converse AK, Larsen EC, Engle JW, Barnhart TE, Nickles RJ, Duncan ID (2011) 11C-(R)-PK11195 PET imaging of microglial activation and response to minocycline in zymosan-treated rats. J Nucl Med 52(2):257–262
Cosco TD, Pillinger T, Emam H, Solmi M, Budhdeo S, Matthew Prina A, Maes M, Stein DJ, Stubbs B, Carvalho AF (2019) Immune aberrations in obsessive-compulsive disorder: a systematic review and meta-analysis. Mol Neurobiol 56(7):4751–4759
Cosenza-Nashat M, Zhao ML, Suh HS, Morgan J, Natividad R, Morgello S, Lee SC (2009) Expression of the translocator protein of 18 kDa by microglia, macrophages and astrocytes based on immunohistochemical localization in abnormal human brain. Neuropathol Appl Neurobiol 35:306–328
Cosgrove KP (2013) Personal communication from Hannestad et al. The neuroinflamation marker translocator protein is not elevated in individuals with mild to moderate depression: a [11C] PBR28 PET study. Brain Behav Immun 33:131–138
Costa DLC, Diniz JB, Requena G, Joaquim MA, Pittenger C, Bloch MH, Miguel EC, Shavitt RG (2017) Randomized, double-blind, placebo-controlled trial of N-acetylcysteine augmentation for treatment-resistant obsessive-compulsive disorder. J Clin Psychiatry 78(7):e766–e773
Dale RC, Heyman I, Giovannoni G, Church AW (2005) Incidence of anti-brain antibodies in children with obsessive-compulsive disorder. Br J Psychiatry 187:314–319
Dantzer R, O'Connor JC, Freund GG, Johnson RW, Kelley KW (2008) From inflammation to sickness and depression: when the immune system subjugates the brain. Nat Rev Neurosci 9 (1):46–56
Dean SL, Singer HS (2017) Treatment of Sydenham's chorea: a review of the current evidence. Tremor Other Hyperkinet Mov 7:456
Emery ES, Vieco PT (1997) Sydenham chorea: magnetic resonance imaging reveals permanent basal ganglia injury. Neurology 48(2):531–533
Enache D, Pariante CM, Mondelli V (2019) Markers of central inflammation in major depressive disorder: a systematic review and meta-analysis of studies examining cerebrospinal fluid, positron emission tomography and post-mortem brain tissue. Brain Behav Immun 81:24–40
Esalatmanesh S, Abrishami Z, Zeinoddini A, Rahiminejad F, Sadeghi M, Najarzadegan MR, Shalbafan MR, Akhondzadeh S (2016) Minocycline combination therapy with fluvoxamine in moderate-to-severe obsessive-compulsive disorder: a placebo-controlled, double-blind, randomized trial. Psychiatry Clin Neurosci 70(11):517–526
Fettes P, Schulze L, Downar J (2017) Cortico-striatal-thalamic loop circuits of the orbitofrontal cortex: promising therapeutic targets in psychiatric illness. Front Syst Neurosci 11:25
Galasko D, Montine TJ (2010) Biomarkers of oxidative damage and inflammation in Alzheimer's disease. Biomark Med 4(1):27–36
Garrido-Mesa N, Zarzuelo A, Galvez J (2013) Minocycline: far beyond an antibiotic. Br J Pharmacol 169(2):337–352
Gause C, Morris C, Vernekar S, Pardo-Villamizar C, Grados MA, Singer HS (2009) Antineuronal antibodies in OCD: comparisons in children with OCD-only, OCD+chronic tics and OCD +PANDAS. J Neuroimmunol 214(1–2):118–124
Ghanizadeh A, Mohammadi MR, Bahraini S, Keshavarzi Z, Firoozabadi A, Alavi SA (2017) Efficacy of N-acetylcysteine augmentation on obsessive compulsive disorder: a multicenter randomized double blind placebo controlled clinical trial. Iran J Psychiatry 12(2):134–141

Hagen AA, Gerber JN, Sweeley CC, White RP, Robertson JT (1977) Levels and disappearance of prostaglandin F2alpha in cerebral spinal fluid: a clinical and experimental study. Stroke 8 (6):672–675

Hesse S, Stengler K, Regenthal R, Patt M, Becker GA, Franke A, Knupfer H, Meyer PM, Luthardt J, Jahn I, Lobsien D, Heinke W, Brust P, Hegerl U, Sabri O (2011) The serotonin transporter availability in untreated early-onset and late-onset patients with obsessive-compulsive disorder. Int J Neuropsychopharmacol 14(5):606–617

Holmay MJ, Terpstra M, Coles LD, Mishra U, Ahlskog M, Oz G, Cloyd JC, Tuite PJ (2013) N-acetylcysteine boosts brain and blood glutathione in Gaucher and Parkinson diseases. Clin Neuropharmacol 36(4):103–106

Holmes SE, Hinz R, Conen S, Gregory CJ, Matthews JC, Anton-Rodriguez JM, Gerhard A, Talbot PS (2018) Elevated Translocator protein in anterior cingulate in major depression and a role for inflammation in suicidal thinking: a positron emission tomography study. Biol Psychiatry 83 (1):61–69

Homsi S, Piaggio T, Croci N, Noble F, Plotkine M, Marchand-Leroux C, Jafarian-Tehrani M (2010) Blockade of acute microglial activation by minocycline promotes neuroprotection and reduces locomotor hyperactivity after closed head injury in mice: a twelve-week follow-up study. J Neurotrauma 27(5):911–921

Horti AG, Naik R, Foss CA, Minn I, Misheneva V, Du Y, Wang Y, Mathews WB, Wu Y, Hall A, LaCourse C, Ahn HH, Nam H, Lesniak WG, Valentine H, Pletnikova O, Troncoso JC, Smith MD, Calabresi PA, Savonenko AV, Dannals RF, Pletnikov MV, Pomper MG (2019) PET imaging of microglia by targeting macrophage colony-stimulating factor 1 receptor (CSF1R). Proc Natl Acad Sci U S A

Husby G, van de Rijn I, Zabriskie JB, Abdin ZH, Williams RC Jr (1976) Antibodies reacting with cytoplasm of subthalamic and caudate nuclei neurons in chorea and acute rheumatic fever. J Exp Med 144(4):1094–1110

Ikawa M, Lohith TG, Shrestha S, Telu S, Zoghbi SS, Castellano S, Taliani S, Da Settimo F, Fujita M, Pike VW, Innis RB (2017) Biomarkers consortium Radioligand project T. 11C-ER176, a Radioligand for 18-kDa Translocator protein, has adequate sensitivity to robustly image all three affinity genotypes in human brain. J Nucl Med 58(2):320–325

Imaizumi M, Kim HJ, Zoghbi SS, Briard E, Hong J, Musachio JL, Ruetzler C, Chuang DM, Pike VW, Innis RB, Fujita M (2007) PET imaging with [11C]PBR28 can localize and quantify upregulated peripheral benzodiazepine receptors associated with cerebral ischemia in rat. Neurosci Lett 411(3):200–205

Kalinski P (2012) Regulation of immune responses by prostaglandin E2. J Immunol 188(1):21–28

Kiessling LS, Marcotte AC, Culpepper L (1994) Antineuronal antibodies: tics and obsessive-compulsive symptoms. J Dev Behav Pediatr 15(6):421–425

Kim MJ, Shrestha SS, Cortes M, Singh P, Morse C, Liow JS, Gladding RL, Brouwer C, Henry K, Gallagher E, Tye GL, Zoghbi SS, Fujita M, Pike VW, Innis RB (2018) Evaluation of two potent and selective PET radioligands to image COX-1 and COX-2 in rhesus monkeys. J Nucl Med 59 (12):1907–1912

Kirvan CA, Swedo SE, Heuser JS, Cunningham MW (2003a) Mimicry and autoantibody-mediated neuronal cell signaling in Sydenham chorea. Nat Med 9(7):914–920

Kirvan CA, Swedo SE, Heuser JS, Cunningham MW (2003b) Mimicry and autoantibody-mediated neuronal cell signaling in Sydenham chorea. Nat Med 9:914–920

Kirvan CA, Swedo SE, Kurahara D, Cunningham MW (2006) Streptococcal mimicry and antibody-mediated cell signaling in the pathogenesis of Sydenham's chorea. Autoimmunity 39(1):21–29

Kobayashi K, Imagama S, Ohgomori T, Hirano K, Uchimura K, Sakamoto K, Hirakawa A, Takeuchi H, Suzumura A, Ishiguro N, Kadomatsu K (2013) Minocycline selectively inhibits M1 polarization of microglia. Cell Death Dis 4:e525

Kobayashi M, Jiang T, Telu S, Zoghbi SS, Gunn RN, Rabiner EA, Owen DR, Guo Q, Pike VW, Innis RB, Fujita M (2018) 11C-DPA-713 has much greater specific binding to translocator protein 18 kDa (TSPO) in human brain than 11C-(R)-PK11195. J Cereb Blood Flow Metab 38:393–403

Kumar JSD, Bai B, Zanderigo F, DeLorenzo C, Prabhakaran J, Parsey RV, Mann JJ (2018) In vivo brain imaging, biodistribution, and radiation dosimetry estimation of [(11)C]Celecoxib, a COX-2 PET ligand, in nonhuman primates. Molecules 23(8)

Latimer ME, L'Etoile N, Seidlitz J, Swedo SE (2015) Therapeutic plasma apheresis as a treatment for 35 severely ill children and adolescents with pediatric autoimmune neuropsychiatric disorders associated with streptococcal infections. J Child Adolesc Psychopharmacol 25(1):70–75

Lee JL, Naguwa SM, Cheema GS, Gershwin ME (2009) Acute rheumatic fever and its consequences: a persistent threat to developing nations in the 21st century. Autoimmun Rev 9 (2):117–123

Li H, Sagar AP, Keri S (2018) Translocator protein (18kDa TSPO) binding, a marker of microglia, is reduced in major depression during cognitive-behavioral therapy. Prog Neuro-Psychopharmacol Biol Psychiatry 83:1–7

Li F, Welling MC, Johnson JA, Coughlin C, Mulqueen J, Jakubovski E, Coury S, Landeros-Weisenberger A, Bloch MH (2020) N-acetylcysteine for pediatric obsessive-compulsive disorder: a small pilot study. J Child Adolesc Psychopharmacol 30(1):32–37

Lisboa BCG, Oliveira KC, Tahira AC, Barbosa AR, Feltrin AS, Gouveia G, Lima L, Feio Dos Santos AC, Martins DC Jr, Puga RD, Moretto AC, De Braganca Pereira CA, Lafer B, Leite REP, Ferretti-Rebustini REL, Farfel JM, Grinberg LT, Jacob-Filho W, Miguel EC, Hoexter MQ, Brentani H (2019) Initial findings of striatum tripartite model in OCD brain samples based on transcriptome analysis. Sci Rep 9(1):3086

Marrie RA, Fisk JD, Tremlett H, Wolfson C, Warren S, Tennakoon A, Leung S, Patten SB, CIHR Team in the Epidemiology and Impact of Comorbidity on Multiple Sclerosis (2015) Differences in the burden of psychiatric comorbidity in MS vs the general population. Neurology 85:1972–1979

Martin A, Boisgard R, Thézé B, van Camp N, Kuhnast B, Damont A, Kassiou M, Dolié F, Tavitian B (2010) Evaluation of the PBR/TSPO radioligand [(18)F]DPA-714 in a rat model of focal cerebral ischemia. J Cereb Blood Flow Metab 30:230–241

Martin A, Boisgard R, Kassiou M, Dolle F, Tavitian B (2011) Reduced PBR/TSPO expression after minocycline treatment in a rat model of focal cerebral ischemia: a PET study using [(18)F]DPA-714. Mol Imaging Biol 13(1):10–15

Matsumoto R, Ichise M, Ito H, Ando T, Takahashi H, Ikoma Y, Kosaka J, Arakawa R, Fujimura Y, Ota M, Takano A, Fukui K, Nakayama K, Suhara T (2010) Reduced serotonin transporter binding in the insular cortex in patients with obsessive-compulsive disorder: a [11C]DASB PET study. NeuroImage 49(1):121–126

Meyer J (2017) Novel phenotypes detectable with PET in mood disorders: elevated monoamine oxidase A and Translocator protein level. In: Vasdev N, Alavi A (eds) PET clinics: novel PET radiotracers with potential clinical applications, vol 12. Elsevier, pp 361–371

Meyer JH, Cervenka S, Kim M-J, Kreisl WC, Henter ID, Innis RB (2020) Neuroinflammation in psychiatric disorders: PET imaging and promising new targets. Lancet Psychiatry 7(12):1064–1074

Minghetti L, Polazzi E, Nicolini A, Creminon C, Levi G (1997) Up-regulation of cyclooxygenase-2 expression in cultured microglia by prostaglandin E2, cyclic AMP and non-steroidal anti-inflammatory drugs. Eur J Neurosci 9(5):934–940

Narayanaswami V, Dahl K, Bernard-Gauthier V, Josephson L, Cumming P, Vasdev N (2018) Emerging PET radiotracers and targets for imaging of Neuroinflammation in neurodegenerative diseases: outlook beyond TSPO. Mol Imaging 17:1536012118792317

Nicholson TR, Ferdinando S, Krishnaiah RB, Anhoury S, Lennox BR, Mataix-Cols D, Cleare A, Veale DM, Drummond LM, Fineberg NA, Church AJ, Giovannoni G, Heyman I (2012) Prevalence of anti-basal ganglia antibodies in adult obsessive-compulsive disorder: cross-sectional study. Br J Psychiatry 200(5):381–386

Nicolini H, Lopez Y, Genis-Mendoza AD, Manrique V, Lopez-Canovas L, Niubo E, Hernandez L, Bobes MA, Riveron AM, Lopez-Casamichana M, Flores J, Lanzagorta N, De la Fuente-Sandoval C, Santana D (2015) Detection of anti-streptococcal, antienolase, and anti-neural antibodies in subjects with early-onset psychiatric disorders. Actas Esp Psiquiatr 43 (2):35–41

Owen DR, Narayan N, Wells L, Healy L, Smyth E, Rabiner EA, Galloway D, Williams JB, Lehr J, Mandhair H, Peferoen LA, Taylor PC, Amor S, Antel JP, Matthews PM, Moore CS (2017) Pro-inflammatory activation of primary microglia and macrophages increases 18 kDa translocator protein expression in rodents but not humans. J Cereb Blood Flow Metab 37 (8):2679–2690

Paydary K, Akamaloo A, Ahmadipour A, Pishgar F, Emamzadehfard S, Akhondzadeh S (2016) N-acetylcysteine augmentation therapy for moderate-to-severe obsessive-compulsive disorder: randomized, double-blind, placebo-controlled trial. J Clin Pharm Ther 41(2):214–219

Pittenger C, Bloch MH, Williams K (2011) Glutamate abnormalities in obsessive compulsive disorder: neurobiology, pathophysiology, and treatment. Pharmacol Ther 132(3):314–332

Pittenger C, Adams TG Jr, Gallezot JD, Crowley MJ, Nabulsi N, James R, Gao H, Kichuk SA, Simpson R, Billingslea E, Hannestad J, Bloch M, Mayes L, Bhagwagar Z, Carson RE (2016) OCD is associated with an altered association between sensorimotor gating and cortical and subcortical 5-HT1b receptor binding. J Affect Disord 196:87–96

Plane JM, Shen Y, Pleasure DE, Deng W (2010) Prospects for minocycline neuroprotection. Arch Neurol 67(12):1442–1448

Pulli B, Chen JW (2014) Imaging Neuroinflammation - from bench to bedside. J Clin Cell Immunol 5

Rajkowska G, Stockmeier CA (2013) Astrocyte pathology in major depressive disorder: insights from human postmortem brain tissue. Curr Drug Targets 14(11):1225–1236

Richards EM, Zanotti-Fregonara P, Fujita M, Newman L, Farmer C, Ballard ED, Machado-Vieira-R, Yuan P, Niciu MJ, Lyoo CH, Henter ID, Salvadore G, Drevets WC, Kolb H, Innis RB, Zarate CA Jr (2018) PET radioligand binding to translocator protein (TSPO) is increased in unmedicated depressed subjects. EJNMMI Res 8(1):57

Samuni Y, Goldstein S, Dean OM, Berk M (2013) The chemistry and biological activities of N-acetylcysteine. Biochim Biophys Acta 1830(8):4117–4129

Sandiego CM, Gallezot JD, Pittman B, Nabulsi N, Lim K, Lin SF, Matuskey D, Lee JY, O'Connor KC, Huang Y, Carson RE, Hannestad J, Cosgrove KP (2015) Imaging robust microglial activation after lipopolysaccharide administration in humans with PET. Proc Natl Acad Sci U S A 112:12468–12473

Sarris J, Oliver G, Camfield DA, Dean OM, Dowling N, Smith DJ, Murphy J, Menon R, Berk M, Blair-West S, Ng CH (2015) N-acetyl cysteine (NAC) in the treatment of obsessive-compulsive disorder: a 16-week, double-blind, randomised, placebo-controlled study. CNS Drugs 29 (9):801–809

Saxena S, Brody AL, Schwartz JM, Baxter LR (1998) Neuroimaging and frontal-subcortical circuitry in obsessive-compulsive disorder. Br J Psychiatry Suppl 35:26–37

Sayyah M, Boostani H, Pakseresht S, Malayeri A (2011) A preliminary randomized double-blind clinical trial on the efficacy of celecoxib as an adjunct in the treatment of obsessive-compulsive disorder. Psychiatry Res 189(3):403–406

Schou M, Varnas K, Lundquist S, Nakao R, Amini N, Takano A, Finnema SJ, Halldin C, Farde L (2015) Large variation in brain exposure of reference CNS drugs: a PET study in nonhuman primates. Int J Neuropsychopharmacol 18(10):pyv036

Scott G, Zetterberg H, Jolly A, Cole JH, De Simoni S, Jenkins PO, Feeney C, Owen DR, Lingford-Hughes A, Howes O, Patel MC, Goldstone AP, Gunn RN, Blennow K, Matthews PM, Sharp DJ (2018) Minocycline reduces chronic microglial activation after brain trauma but increases neurodegeneration. Brain 141(2):459–471

Seabrook TJ, Jiang L, Maier M, Lemere CA (2006) Minocycline affects microglia activation, Abeta deposition, and behavior in APP-tg mice. Glia 53(7):776–782

Setiawan E, Wilson AA, Mizrahi R, Rusjan PM, Miler L, Rajkowska G, Suridjan I, Kennedy JL, Rekkas PV, Houle S, Meyer JH (2015) Role of translocator protein density, a marker of neuroinflammation, in the brain during major depressive episodes. JAMA Psychiat 72 (3):268–275

Setiawan E, Attwells S, Wilson AA, Mizrahi R, Rusjan PM, Miler L, Xu C, Sharma S, Kish S, Houle S, Meyer JH (2018) Association of translocator protein total distribution volume with duration of untreated major depressive disorder: a cross-sectional study. Lancet Psychiatry 5 (4):339–347

Shalbafan M, Mohammadinejad P, Shariat SV, Alavi K, Zeinoddini A, Salehi M, Askari N, Akhondzadeh S (2015) Celecoxib as an adjuvant to fluvoxamine in moderate to severe obsessive-compulsive disorder: a double-blind, placebo-controlled, randomized trial. Pharmacopsychiatry 48(4–5):136–140

Singer HS, Loiselle CR, Lee O, Minzer K, Swedo S, Grus FH (2004) Anti-basal ganglia antibodies in PANDAS. Mov Disord 19(4):406–415

Snider LA, Lougee L, Slattery M, Grant P, Swedo SE (2005) Antibiotic prophylaxis with azithromycin or penicillin for childhood-onset neuropsychiatric disorders. Biol Psychiatry 57 (7):788–792

Swedo SE (1994) Sydenham's chorea. A model for childhood autoimmune neuropsychiatric disorders. JAMA 272(22):1788–1791

Swedo SE, Leonard HL, Garvey M, Mittleman B, Allen AJ, Perlmutter S, Lougee L, Dow S, Zamkoff J, Dubbert BK (1998) Pediatric autoimmune neuropsychiatric disorders associated with streptococcal infections: clinical description of the first 50 cases. Am J Psychiatry 155 (2):264–271

Swedo SE, Leckman JF, Rose NR (2012a) From research subgroup to clinical syndrome: modifying the PANDAS criteria to describe PANS (pediatric acute-onset neuropsychiatric syndrome). Pediatr Ther 2(2):113

Swedo S, Leckman JF, Rose N (2012b) From research subgroup to clinical syndrome: modifying the PANDAS criteria to describe PANS (pediatric acute-onset neuropsychiatric syndrome). Pediatr Therapeut 2:1–8

Swedo SE, Leckman JF, Rose NR (2012c) From research subgroup to clinical syndrome: modifying the PANDAS criteria to describe PANS (pediatric acute-onset neuropsychiatric syndrome). Pediatr Therapeut 2:1–8

Tachikawa M, Hosoya K, Terasaki T (2014) Pharmacological significance of prostaglandin E2 and D2 transport at the brain barriers. Adv Pharmacol 71:337–360

Van Weehaeghe D, Van Schoor E, De Voch J, Koole M, Attili B, Celen S, Declercq L, Thal DR, Van Damme P, Bormans G, Van Laere K (2020) TSPO versus P2X7 as a target for neuroinflammation – an in vitro and in vivo study. J Nucl Med 61(4):604–607

Venneti S, Wang G, Nguyen J, Wiley CA (2008) The positron emission tomography ligand DAA1106 binds with high affinity to activated microglia in human neurological disorders. J Neuropathol Exp Neurol 67:1001–1010

Walker KG, Wilmshurst JM (2010) An update on the treatment of Sydenham's chorea: the evidence for established and evolving interventions. Ther Adv Neurol Disord 3(5):301–309

Wang AK, Miller BJ (2018) Meta-analysis of cerebrospinal fluid cytokine and tryptophan Catabolite alterations in psychiatric patients: comparisons between schizophrenia, bipolar disorder, and depression. Schizophr Bull 44(1):75–83

Wilson AA, Garcia A, Parkes J, McCormick P, Stephenson KA, Houle S, Vasdev N (2008) Radiosynthesis and initial evaluation of [18F]-FEPPA for PET imaging of peripheral benzodiazepine receptors. Nucl Med Biol 35(3):305–314

Wolf BJ, Brackhan M, Bascunana P, Leiter I, Langer BLN, Ross TL, Bankstahl JP, Bankstahl M (2020) TSPO PET identifies different anti-inflammatory minocycline treatment response in two rodent models of epileptogenesis. Neurotherapeutics 17(3):1228–1238

Wong DF, Brasic JR, Singer HS, Schretlen DJ, Kuwabara H, Zhou Y, Nandi A, Maris MA, Alexander M, Ye W, Rousset O, Kumar A, Szabo Z, Gjedde A, Grace AA (2008) Mechanisms of dopaminergic and serotonergic neurotransmission in Tourette syndrome: clues from an in vivo neurochemistry study with PET. Neuropsychopharmacology 33(6):1239–1251

Animal Models for OCD Research

Brittany L. Chamberlain and Susanne E. Ahmari

Contents

Abstract OCD has lagged behind other psychiatric illnesses in the identification of molecular treatment targets, due in part to a lack of significant findings in genome-wide association studies. However, while progress in this area is being made, OCD's symptoms of obsessions, compulsions, and anxiety can be deconstructed into distinct neural functions that can be dissected in animal models. Studies in rodents and non-human primates have highlighted the importance of cortico-basal ganglia-thalamic circuits in OCD pathophysiology, and emerging studies in human post-mortem brain tissue point to glutamatergic synapse abnormalities as a potential cellular substrate for observed dysfunctional behaviors. In addition, accumulated evidence points to a potential role for neuromodulators including serotonin and

B. L. Chamberlain and S. E. Ahmari (✉)
Department of Psychiatry, Translational Neuroscience Program, University of Pittsburgh, Pittsburgh, PA, USA

Center for Neuroscience Program and Center for the Neural Basis of Cognition, University of Pittsburgh, Pittsburgh, PA, USA
e-mail: ahmarise@upmc.edu

Curr Topics Behav Neurosci (2021) 49: 55–96
https://doi.org/10.1007/7854_2020_196
Published Online: 25 March 2021

dopamine in both OCD pathology and treatment. Here, we review current efforts to use animal models for the identification of molecules, cell types, and circuits relevant to OCD pathophysiology. We start by describing features of OCD that can be modeled in animals, including circuit abnormalities and genetic findings. We then review different strategies that have been used to study OCD using animal model systems, including transgenic models, circuit manipulations, and dissection of OCD-relevant neural constructs. Finally, we discuss how these findings may ultimately help to develop new treatment strategies for OCD and other related disorders.

Keywords Animal models · Anxiety · Basal ganglia · Chemogenetic · Cortico-striato-thalamo-cortical circuits · Decision-making · Fear learning · Obsessive compulsive disorder (OCD) · Optogenetic · Orbitofrontal cortex (OFC) · Prefrontal cortex · Striatum · Transgenic · Translatable tasks

1 Distinguishing Features of OCD That Can Be Modeled in Animals

1.1 Clinical Phenomenology of OCD

Diagnostic Criteria and Symptoms Though a detailed description of the clinical phenomenology of OCD is beyond the scope of this chapter, there are several key features that are important to highlight when considering the effective use of animal models to investigate this disorder.

First, according to DSM-5 criteria, OCD is diagnosed by the presence of obsessions and/or compulsions. Though it is not necessary to have both obsessions and compulsions to meet diagnostic criteria, these symptom elements appear to be linked together in most patients, which in turn may yield insight into potential underlying neurobiological substrates. In turn, animal models can be used to dissect the key circuits potentially involved in these processes and to test alternative theories for OCD etiology.

Second, although OCD was classified as an Anxiety Disorder in DSM-IV, it was shifted into a new category of Obsessive-Compulsive and Related Disorders in DSM-5 due to potentially distinct underlying circuit abnormalities. Specifically, as detailed further below, individuals with OCD and related disorders (e.g., skin-picking, pathological grooming) typically demonstrate abnormal neural activity in cortico-basal ganglia-thalamic loops, while abnormal neural activity tends to be more prominently localized to cortico-limbic circuits (e.g., connections between mPFC, amygdala, and hippocampus) in anxiety disorders such as generalized anxiety disorder (GAD), post-traumatic stress disorder (PTSD), and social anxiety disorder (Marras et al. 2016; Stein et al. 2010). However, despite this new DSM-5 classification schema, it is important to note that anxious affect is often linked with obsessions and compulsions, and anxiety is therefore a key symptom experienced by people with OCD. Although it is difficult to determine if symptoms of anxiety in OCD patients are the cause or the consequence of obsessions due to the lack of fine-grained longitudinal studies in humans, animal models can be used to investigate these two possibilities by directly testing the relationship between anxiety and compulsive behaviors using causal manipulations.

Third, OCD is a heterogeneous disorder. Over the years, many rubrics have been proposed to split OCD into meaningful subtypes. These initially centered on dividing OCD into different categories based on the content of obsessions and compulsions. This practice has gained some support in the nosology literature, with the identification of four subtypes including (1) contamination obsessions/ cleaning rituals; (2) taboo thoughts/ checking rituals; (3) symmetry obsessions/ arranging rituals, and (4) harm-related OCD (Pinto et al. 2009); hoarding is considered distinct from the other categories due to identification of distinct underlying circuits (Mataix-Cols et al. 2010; Saxena 2008). While there is evidence that different types of symptom content may have distinct underlying neural circuits (Mataix-Cols et al. 2004; Gilbert et al. 2008; van den Heuvel et al. 2009), especially with respect to hoarding, recent evidence calls this idea into question by demonstrating that individual patients (a) often have highly dynamic symptoms that fluctuate between different types of content throughout their lifetimes, and (b) often have co-existing symptoms at any one timepoint within multiple different symptom domains (Bloch et al. 2013). Thus, there is significant interest in identifying biologically meaningful subtypes that can be localized to distinct neural substrates. Strong candidates for such divisions using the existing literature include early- vs late-onset OCD (early = higher male prevalence) (Geller 2006; Stewart et al. 2004); and tic vs non-tic associated OCD (tics = more striatal involvement) (Leckman et al. 2010; Leckman et al. 2009; McDougle et al. 1994). However, large-scale neuroimaging studies, as have been performed in the field of depression (Lynch et al. 2020; Drysdale et al. 2017), are necessary to further optimize our definitions of neurobiologically-grounded symptom subtypes in order to identify metrics that aid in both individual diagnosis and prediction of treatment response.

1.2 Neural Circuit Abnormalities Associated with OCD

OCD neuroimaging studies serve as the basis for robust theories regarding neural circuit abnormalities in the disorder. While this rich literature is covered in detail elsewhere in this volume, here we provide an abbreviated overview to frame the discussion of probing OCD-relevant circuits in experimental animal systems.

Cortico-Basal Ganglia-Thalamic Circuits Mounting evidence indicates that aberrant activity in cortico-striatal-thalamo-cortical (CSTC) circuitry is associated with OCD symptoms (Piras et al. 2015; Pujol et al. 2004; Rotge et al. 2009; Rotge et al. 2010; Kim et al. 2001; Christian et al. 2008; Yun et al. 2020). Specifically, fMRI and

PET studies have repeatedly reported abnormal patterns of blood flow and glucose metabolism consistent with increased activity in orbitofrontal cortex (OFC), anterior cingulate cortex (ACC), caudate, and thalamus both at rest and with symptom provocation (Graybiel and Rauch 2000; Saxena et al. 1998; Menzies et al. 2008b; Breiter et al. 1996; Rauch et al. 1994; Hou et al. 2012; Kwon et al. 2003; Saxena et al. 1999). Effective treatment with either serotonin reuptake inhibitors (SRIs) and exposure therapy with response prevention (EXRP) normalizes these activity patterns (Saxena et al. 1999; Rauch et al. 1994; Menzies et al. 2008a; Breiter et al. 1996), further suggesting that abnormal activity in these circuits is associated with OCD symptoms. In addition, functional connectivity studies have predominantly demonstrated increased connectivity between both lateral and medial OFC and caudate/ ventral striatum, respectively (Harrison et al. 2009), though decreased functional connectivity has also been observed (Posner et al. 2014). Together, these data support a model in which excessive and/or abnormally-patterned activity in CSTC circuits leads to OCD symptoms due to impairments in key neural functions associated with these circuits (Graybiel and Rauch 2000).

Other Cortical Regions Additional cortical areas of potential importance to OCD pathophysiology have emerged in more recent neuroimaging studies. First, abnormal processing of feelings of disgust in contamination-related OCD has been localized to the insula (Stein et al. 2006; Husted et al. 2006; Shapira et al. 2003). Hyperactivation of anterior insula was also observed in OCD subjects during an emotional go/no-go task while inhibiting responses to either fear-evoking or disgusting images (Berlin et al. 2015). In addition, accumulating evidence suggests that abnormal signaling in supplementary and pre-supplementary motor areas, which play important roles in motor planning, may contribute to OCD symptoms (Rauch et al. 1997; Rotge et al. 2009; Saxena et al. 2001; Maia et al. 2008). Specifically, excessive baseline and symptom-associated activity in premotor cortical areas has been consistently observed in both OCD and Tourette Syndrome (TS) in functional imaging studies (Braun et al. 1993; Stern et al. 2000; Bohlhalter et al. 2006; Hampson et al. 2009; de Wit et al. 2012; Grutzmann et al. 2016; Yucel et al. 2007), and repetitive transcranial magnetic stimulation (rTMS) of SMA using protocols that decrease activity may be effective for OCD (Lee et al. 2017; Mantovani et al. 2013; Pallanti et al. 2016; Mantovani et al. 2010; Bloch et al. 2016). Below, we highlight efforts to test causality of these findings using rodent models.

Anxiety and Depression-Related Circuits While a majority of accumulated evidence highlights abnormalities in CSTC circuits in OCD, these findings do not exclude contributions of other brain networks to symptoms. For example, emerging evidence from targeted studies suggests abnormal task- or symptom-related activity in amygdala (Thorsen et al. 2019; Paul et al. 2019; Thorsen et al. 2018a; Rus et al. 2017; Simon et al. 2014; Stern et al. 2013; van den Heuvel et al. 2004) and mPFC (Paul et al. 2019; Apergis-Schoute et al. 2017; Stern et al. 2013) in OCD patients, consistent with the prominence of anxiety and comorbid depression in this disorder (see (Wood and Ahmari 2015) for review). These findings are also supported by neuroimaging datasets identifying resting state (Fullana et al. 2017; Thorsen et al.

2018b; Pico-Perez et al. 2019; Gottlich et al. 2015; Gottlich et al. 2014) and structural (Reess et al. 2016; Subira et al. 2016; Atmaca et al. 2008) abnormalities in limbic networks in OCD.

Conclusion While significant insights regarding dysfunctional circuits in OCD have been obtained via human neuroimaging studies, they are inherently limited due to their lack of cellular resolution and inability to directly test cause and effect. As described below, recent technological advances now allow direct tests of causality in animal models (Kim et al. 2017; Rajasethupathy et al. 2016; Lerner et al. 2016; Deisseroth 2015; Bernstein et al. 2012; Roth 2016; Sternson and Roth 2014).

1.3 Genetics of OCD

Though the genetics of OCD is reviewed extensively in chapter "Genetics of OCD and Related Disorders; Searching for Shared Factors", here we highlight factors relevant to translation of these findings into animal models. Strong evidence from twin and family studies indicates significant genetic contributions to OCD (Pauls et al. 2014), particularly in pediatric-onset cases (Pauls 2008). A meta-analysis of twin studies reported that additive genetic effects account for 40% of phenotypic variance and noted that shared environment did not display a significant contribution (Taylor 2011). Although subsequent genetic association studies have generated important advances and led to the development of new mechanistic hypotheses, two main strategies are currently being used to address limitations of these studies: (1) genome-wide association studies (GWAS); and (2) identification of de novo mutations of large effect size.

GWAS Studies in OCD To date, the 2 major GWAS in OCD (Stewart et al. 2013; Mattheisen et al. 2015) suggest that genes that encode for proteins found at and around excitatory synapses may be particularly important for OCD etiology, though no specific polymorphisms have yet reached genome-wide significance in a combined meta-analysis (International Obsessive Compulsive Disorder Foundation Genetics Collaborative and O. C. D. Collaborative Genetics Association Studies 2018). In the first study, disease-associated single nucleotide polymorphisms (SNPs) included *DLGAP1,* a gene that encodes an excitatory post-synaptic density scaffolding protein, and *ISM1*, a gene that is highly correlated with brain-wide expression of multiple glutamate-related genes (Stewart et al. 2013). A second GWAS reported a significant association near the *PTPRD* gene, which encodes the receptor protein tyrosine phosphatase Type D. PTPRD promotes the differentiation of glutamatergic synapses and interacts with SLITRK proteins which are critical for proper formation of excitatory synapses (Mattheisen et al. 2015).

Identification of De Novo Mutations To date, one study has identified de novo mutations in OCD by performing whole exome sequencing in 184 trios (i.e., 2 unaffected parents and one child with an OCD diagnosis) (Cappi et al. 2020). Two high confidence risk genes were reported: *CHD8 (Chromodomain Helicase*

DNA Binding Protein 8), which has previously been shown to be strongly associated with autism (Sanders et al. 2012; Neale et al. 2012; Bernier et al. 2014; O'Roak et al. 2012), and *SCUBE1 (Signal Peptide, CUB Domain and EGF Like Domain Containing 1)*, whose function in the brain is not yet well understood (Grimmond et al. 2000; Yang et al. 2002). If validated in subsequent studies, these genes would be useful candidates for generation of mouse models.

2 Strategies for Studying OCD Using Animal Model Systems

2.1 Transgenic Models for Studying OCD

Large-scale collaborative projects are beginning to make inroads into the genetic underpinnings of OCD. However, because completion of these studies will take significant time and resources, simultaneous efforts have been underway to translate the findings from the initial genetic association studies into mice. In addition, important discoveries with relevance to the mechanisms underlying OCD-relevant behaviors have been made using transgenic models often generated for other scientific purposes. Below, we will review several transgenic models that have recently revealed interesting insights regarding both anxiety-like behavior and abnormal repetitive and compulsive behaviors including compulsive grooming, perseverative responding during reversal learning, and perseverative hyperlocomotion (Table 1).

***Hoxb8*-Knockout Mouse** One of the first transgenic models to serendipitously find a potential mechanism underlying OCD-relevant compulsive grooming behavior was the *Hoxb8*-knockout (KO) mouse (Greer and Capecchi 2002). A loss-of-function mutation of the *Hoxb8* gene resulted in excessive grooming, a patterned sequenced behavior that is thought to have relevance for OCD pathology (Greer and Capecchi 2002). Notably, although *Hoxb8,* which encodes one of a complex of transcription factors that direct anteroposterior patterning in embryonic development (Trainor and Krumlauf 2000), is widely expressed in mouse CNS (olfactory bulb, brainstem, hippocampus, frontal cortex, caudate-putamen) (Capecchi 1997), cortical expression is strongest in critical brain regions implicated in the pathophysiology of OCD–orbitofrontal cortex and anterior cingulate cortex. As in several other mouse models of compulsive behavior (see below), *Hoxb8*-KOs initiated more grooming bouts than wild-type (WT) littermates and groomed for significantly longer periods of time, resulting in extensive hair loss and formation of skin lesions. Furthermore, excessive grooming was normalized by chronic (>7 days) fluoxetine treatment, supporting the potential clinical relevance of this model (Nagarajan et al. 2018). Since *Hoxb8* is also highly expressed within the dorsal spinal cord, early reports argued that excessive grooming could be a consequence of altered nociceptive responses (Holstege et al. 2008). However, while multiple studies have reported nociception abnormalities in *Hoxb8*-KOs (Greer and Capecchi 2002; Holstege et al.

Table 1 Animal models used for investigation of neural substrates underlying OCD

Model	OCD-relevant characteristics	Putative mechanisms (brain regions; circuits; cell types)
Genetic		
Hoxb8-KO mouse	Excessive self-grooming → lesions	Loss of bone marrow-derived microglia
Sapap3-KO mouse	– Excessive self-grooming → lesions – Increased anxiety – Impaired reversal learning – Impaired habit formation	– Loss of SAPAP3 in CS synapses→ striatal hyperactivity & excessive grooming – Reduced fEPSP amplitude – Increased NR2B subunits at cortico-striatal synapses – Decreased strength of LOFC→CS synapses – Increased strength of M2→CS synapses
Slitrk5-knockout mouse	Excessive self-grooming → lesions – Increased anxiety	– Increased OFC activity (measured via FosB) – Reduced complexity of striatal dendritic arbors – Reduced expression of AMPA/NMDA receptors
Spred2-KO mouse	Excessive self-grooming → lesions	– Increased EPSCs in DLS and LA – Putative release of inhibition of Ras/ERK-MAPK signaling pathway
Slc1a1 ST mouse	– Decreased AMPH response (decreased stereotypy, decreased hyperlocomotion)	Decreased AMPH-induced c-Fos, D1R density, and basal extracellular dopamine in DS
Slc1a1-OE mouse	– Increased AMPH response (increased stereotypy, hyperlocomotion) – grooming – anxiety-like behavior (OF, EPM)	Altered striatal NMDAR subunit composition and impaired NMDA-dependent LTD
DAT knockdown mouse	Stereotyped syntactic grooming chains	Elevated extracellular striatal dopamine levels
5-HT_{2C} KO mouse	Methodical screen chewing; repetitive head dipping	Not yet localized
Circuit manipulation		
D1-cholera toxin (CT) mice	Repetitive selection of actions	– Hyperactivity in D1+ neurons – Possible localization: basal ganglia, intercalated nucleus of the amygdala, piriform cortex, and somatosensory cortex
OFC-VMS optogenetic stimulation	Progressive increase in grooming	Progressive increase in evoked striatal activity
LOFC-CS optogenetic stim in *Sapap3*-KOs	Reduced self-grooming	Dampened striatal activity via boosting FSIs

(continued)

Table 1 (continued)

Model	OCD-relevant characteristics	Putative mechanisms (brain regions; circuits; cell types)
BLA-DLS optogenetic stimulation	Progressive increase in self-grooming & anxiety-like behavior (open field, EPM)	– Reduced paired pulse ratio in D1-SPNs – Increased NMDA/AMPA ratios – Blocked by DNA methyltransferase inhibitor RG108
ChAT interneuron ablation (DLS)	– Increased stress-induced grooming – AMPH-induced stereotypies	Impact on DLS activity (no effects with DMS ablation)
PV interneuron ablation (DS)	Increased stress-induced grooming	No effects on AMPH-induced stereotypies
Pharmacologic manipulation		
Quinpirole	Repetitive checking behavior	– D2/3-receptor agonism – Dysregulation of dopaminergic signaling & cortico-basal ganglia circuit function – Increased excitability of BLA-mPFC inputs & increased mPFC firing
RU24969	Perseverative hyperlocomotion, PPI deficits	5-HT1B receptor agonism localized to ventromedial OFC
mCPP	Increased directional persistence in T-maze rewarded alternation task	5-HT_{2C} agonism in OFC – No contribution from mPFC/NAc

BLA basolateral amygdala, *DLS* dorsolateral striatum, *DMS* dorsomedial striatum, *DS* dorsal striatum, *CS* central striatum, *LA* lateral amygdala, *OFC* orbitofrontal cortex, *mPFC* medial prefrontal cortex, *LOFC* lateral orbitofrontal cortex, *M2* supplementary motor cortex, *VMS* ventromedial striatum, *AMPH* amphetamine, *5-HT_{1B}* serotonin 1B, *5-HT_{2C}* serotonin 2C, *NR2B* NMDA receptor subtype B, *DAT* dopamine transporter, *ChAT* choline acetyltransferase, *PV* parvalbumin, *FSIs* fast-spiking interneurons, *SPN* spiny projection neuron, *fEPSP* field excitatory post-synaptic potential, *LTD* long-term depression, *KO* knockout, *OE* overexpresser, *OF* open field, *EPM* elevated plus maze

2008; Chen et al. 2010), these are dissociable from the excessive grooming phenotype (Chen et al. 2010; Trankner et al. 2019), suggesting that abnormal pain responses do not underlie this compulsive behavior.

Additional mechanistic investigation using this model system found that *Hoxb8* is expressed in a subset of bone marrow-derived microglia rather than neurons; as a result, *Hoxb8*-KOs have a 15% reduction in total microglia (Chen et al. 2010). Interestingly, bone marrow transplants from WT mice into irradiated *Hoxb8* mutants led both to repopulation of *Hoxb8*-KO brains with donor WT microglia and a significant decline in grooming, suggesting that loss of this microglia subpopulation plays a causal role in the excessive grooming phenotype (Chen et al. 2010). This model may therefore provide a useful platform for investigation of potential links between neuroimmune dysfunction and OCD, such as have been proposed in pediatric autoimmune neuropsychiatric disorders associated with *Streptococcal* infection (PANDAS) (Swedo et al. 1998).

***Sapap3*-Knockout Mouse** While it is difficult to investigate cellular and circuit mechanisms in human subjects, animal models of compulsive behavior that display altered activity in cortico-basal ganglia loops can be used for this purpose. One such model that is widely used is *Sapap3*-KO mice. *Sapap3* family members are scaffolding proteins found in the postsynaptic density that act to regulate trafficking of AMPA- and NMDA-type glutamate receptors to the postsynaptic membrane of excitatory synapses (Welch et al. 2007). Mice with global deletion of the *Sapap3* gene, which is the only family member that is specifically enriched in the striatum, display a compulsive grooming phenotype which often results in self-inflicted skin lesions; this phenotype is reversed by subchronic (6-days) treatment with fluoxetine (Welch et al. 2007) (Fig. 1). When striatal *Sapap3* expression is rescued via lentivirus injection, *Sapap3*-KO grooming is reduced to WT levels, emphasizing the importance of striatal post-synaptic signaling in development of this phenotype. *Sapap3*-KOs also display other OCD-relevant phenotypes including increased levels

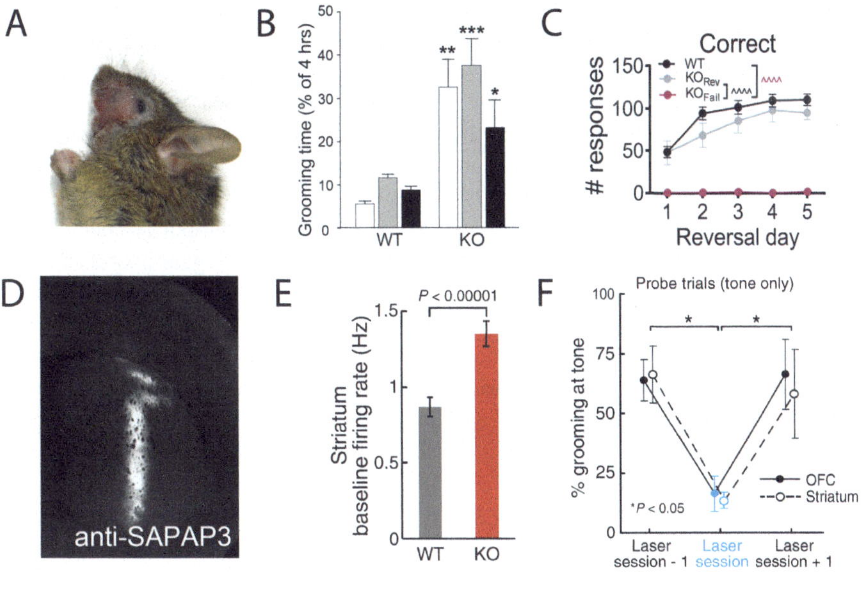

Fig. 1 Summary of initial findings from the *Sapap3*-KO model. (**a**) Homozygous constitutive knockout of *Sapap3* leads to the development of skin lesions resulting from (**b**) excessive grooming. (**c**) *Sapap3*-KOs also display heterogeneous impairments in deterministic reversal learning, with approximately 50% of KOs failing to learn a new rule. (**d**) Lentiviral rescue of SAPAP3 in the central striatum leads to reversal of excessive grooming behavior. (**e**) *Sapap3*-KOs show baseline increased firing rates in the central striatum. (**f**) Optogenetic channelrhodopsin-mediated stimulation of LOFC-central striatum terminals leads to reversal of excessive grooming in *Sapap3*-KOs. (**a**, **b**, **d**) reproduced with permission from Welch et al. (2007); (**c**) reproduced with permission from Manning et al. (2019); (**e**, **f**) reproduced with permission from Burguiere et al. (2013)

of anxiety, impaired cognitive flexibility, and abnormal habit formation (Welch et al. 2007; Manning et al. 2019; van den Boom et al. 2019; Hadjas et al. 2019), many of which are reversed by treatment with the serotonin-reuptake inhibitor fluoxetine providing additional support for the relevance of this animal model for mechanistic investigations.

In addition, *Sapap3*-KOs display altered cortico-striatal transmission (Welch et al. 2007; Burguiere et al. 2013). Ex vivo recordings from striatal slices of *Sapap3*-KO mice and WT littermates show that field excitatory postsynaptic potentials (fEPSPs) are significantly reduced in *Sapap3*-KOs relative to wild-type controls. However, axonal excitability and presynaptic functions are unchanged, indicating that the reduction in fEPSPs is the result of postsynaptic impairments (Welch et al. 2007). Later ex vivo studies also established that these abnormalities are localized to cortico-striatal, and not thalamo-striatal, synapses (Wan et al. 2014); and identified additional strengthened supplementary motor area (M2)-striatal and weakened lateral orbitofrontal cortex (lOFC) -striatal post-synaptic connections (Corbit et al. 2019). Furthermore, in vivo electrophysiological recordings in *Sapap3*-KOs and WTs show significant elevation in baseline firing rates of striatal spiny projection neurons (SPNs) in KOs (Burguiere et al. 2013). Optogenetic stimulation of lOFC terminals in the central striatum resulted in a reduction of striatal SPN firing rates (presumably through restoration of fast-spiking interneuron (FSI)-mediated feed-forward inhibition) and a subsequent significant reduction of conditioned grooming in *Sapap3*-KOs (Burguiere et al. 2013), suggesting a direct relationship between abnormal cortico-striatal signaling and compulsive behavior.

Interestingly, our recent post-mortem study of differential gene expression in OFC and striatum of subjects with OCD and matched comparison subjects also supported a potential role for *Sapap3* expression in OCD pathophysiology. Specifically, expression of the human homolog *DLGAP3* gene was significantly lower in medial and lateral OFC, nucleus accumbens core, and head of caudate in OCD subjects (Piantadosi et al. 2019). In addition, although DLGAP3/SAPAP3 has not reached genome-wide significance in association studies of OCD, multiple rare DLGAP3 missense variants have been observed in OCD and trichotillomania (Koizumi et al. 1991), and a family based association study found a link between DLGAP3 and grooming disorders (Bienvenu et al. 2009). Together, these findings provide further support for the hypothesis that decreases in *Sapap3* protein are involved in the generation of compulsive behaviors.

***Slitrk5*-Knockout Mouse** Shortly after initial characterization of the *Sapap3*-KO strain, *Slitrk5*, another synaptic protein, was implicated in OCD-relevant behaviors. *Slitrk5* is a member of the *Slitrk* family of transmembrane proteins that control neurite outgrowth and synaptogenesis through transsynaptic interactions with presynaptic binding partners (Aruga et al. 2003). While all *Slitrks* are localized to CNS synapses, *Slitrk5* is specifically localized to excitatory synapses and enriched in striatal neurons (Yim et al. 2013; Song et al. 2017), where it interacts with the presynaptic adhesion molecule protein tyrosine phosphatase delta (PTPdelta) (Song et al. 2015). Constitutive *Slitrk5*-KO leads to excessive grooming to the point of lesion development at 3 months of age; grooming is normalized by chronic (21 days) fluoxetine administration (Shmelkov et al. 2010). In addition to this compulsive behavior, *Slitrk5*-KOs showed increased anxiety-like behaviors (reduced time spent/ entries into open arms of elevated-plus maze; reduced center time in open field). Increased marble burying was also noted; however, the relevance of this test to OCD is unclear because it is most highly correlated with digging behavior, which can be increased due to many different motivational factors (e.g., anxiety, generalized hyperactivity, hunger) (de Brouwer et al. 2019; de Wolmarans et al. 2016). Investigation of the neural circuit abnormalities underlying these behavioral findings revealed that FosB expression was specifically upregulated in OFC of *Slitkr5*-KOs, potentially consistent with functional neuroimaging reports of OFC hyperactivity in OCD patients (Saxena et al. 1998; Beucke et al. 2013). In addition, *Slitrk5*-KOs had reduced striatal volume, complexity of dendritic arbors in striatal SPNs, and expression of both AMPA and NMDA glutamate receptors (NR2A, NR2B, GluR1, GluR2) in the striatum (Shmelkov et al. 2010). Together these results suggested that cortico-striatal dysfunction may be responsible for the observed behavioral abnormalities in *Slitkr5*-KOs.

Potential links have been identified between the *Slitrk5* gene and OCD and related disorders. In the second GWAS of OCD, while no associations met genome-wide significance, the strongest detected was a polymorphism upstream of the human PTPdelta gene (*PTPRD*) which interacts with a related member of the *Slitrk* protein family, *Slitrk3* (Mattheisen et al. 2015). In addition, resequencing the protein-coding sequence of the *Slitrk5* gene from 377 OCD patients showed that all identified rare non-synonymous mutations impaired activity in an in vitro assay of synaptogenesis, while none of the mutations in pseudo-matched controls had a significant impact on *Slitrk5* function (Song et al. 2017). Further dissection using the *Slitrk5*-KO model system may therefore help identify how alterations in cortico-striatal synaptic interactions lead to OCD pathophysiology.

***Spred2*-Knockout Mouse** In another recent study, investigators knocked out *Spred2*, one of the members of the *Sprouty/Spred* family that inhibits the Ras/ ERK-MAPK signaling pathway, which in turn plays an important role in the promotion of cellular growth and proliferation. Similar to other models discussed above, *Spred2*-KO mice had excessive grooming resulting in skin lesions (Ullrich et al. 2011; Ullrich et al. 2018). Grooming behavior was normalized by chronic (2 weeks) fluoxetine treatment, and lesions were improved by selumetinib (1 week), a mitogen-activated protein kinase (MEK)1/2 inhibitor that specifically blocks Ras/ ERK-MAPK activity and therefore may rescue inhibition of the pathway that is normally inhibited by *Spred2* (Ullrich et al. 2018). Compared to WT littermates, ex vivo electrophysiological recordings in *Spred2*-KOs showed both increased excitatory post-synaptic current (EPSC) magnitude in dorsolateral striatal neurons after stimulation of corpus callosum fibers and in lateral amygdala neurons following stimulation of thalamic fibers. Although increased cortico-striatal neurotransmission is consistent with a large body of literature suggesting that CSTC circuit dysfunction underlies OCD pathophysiology, these thalamo-amygdala projection findings also highlight the possible importance of other limbic circuits in OCD-relevant behaviors – for example, through the influence of affective state (e.g., anxiety/fear levels) on behavioral selection. No genetic link has yet been made between *Spred2* and OCD in humans. However, keeping the overarching caveats regarding the challenges with reproducibility of association studies in mind, these findings could potentially be consistent with observed associations of BDNF and TrkB sequence variants with OCD (Taylor 2013; Hemmings et al. 2008; Hall et al. 2003) since BDNF/TrkB interaction stimulates the Ras/ERK-MAPK pathway (Gudasheva et al. 2017).

Although an initial study found that *Spred2*-KOs have increased stress hormone production (ACTH and CRH) due to a lack of *Spred2* inhibition of Ras/ERK-MAPK signaling (Ullrich et al. 2011), formal testing of anxiety-related behavior showed that *Spred2*-KOs spent more time in the anxiety-provoking open arms of the elevated plus maze and light side of the light-dark box (Ullrich et al. 2018). While grooming data during elevated plus maze testing was not reported, an increased number of grooming bouts in *Spred2*-KOs vs. WTs throughout light-dark box testing was noted. The possibility that compulsive or perseverative behaviors may confound tests typically used to assess anxiety-like behaviors is therefore an important consideration when characterizing animal models that may have relevance to OCD.

***Slc1a1*-Knockout and Overexpression** Although the field is moving toward the use of transgenic models based on either GWAS hits or rare variants of large effect size, useful biological insights have been gained using models based on candidate genes. For example, the *Slc1a1/Eaac1* gene, which encodes the excitatory amino acid transporter 3 (EAAT3), has been consistently highlighted in OCD linkage and association studies (Shugart et al. 2009; Dickel et al. 2006; Arnold et al. 2006; Veenstra-VanderWeele et al. 2001). Importantly, *Slc1a1* is localized perisynaptically and is enriched in the CSTC circuits implicated in OCD (Burguiere et al. 2015; Holmseth et al. 2012; Bjorn-Yoshimoto and Underhill 2016; Wendland et al. 2009). It serves three primary functions: buffering glutamate concentrations at extra-synaptic NMDA and AMPA receptors, taking up glutamate as a precursor for GABA synthesis, and taking up cysteine to help protect against oxidative stress (Peghini et al. 1997; Scimemi et al. 2009). Multiple groups have therefore investigated the functional impact of under- or over-expression of *Slc1a1* on behaviors relevant to OCD pathophysiology.

Several studies have now used different approaches to downregulate EAAT3 expression and examine the impact on both behavior and ex vivo electrophysiologic measures. In the first study, targeted disruption of the *Eaac1* gene via insertion of a pgk-Neo cassette led to spontaneous hypolocomotion and dicarboxylic aminoaciduria due to renal dysfunction; no abnormalities were noted on the other behavioral tests examined (rotarod; Morris water maze), but no specific OCD-relevant behaviors were examined (Peghini et al. 1997). In a later study, we developed a flexible mouse model system (Tanaka et al. 2010) that permitted either up- or down-regulation of *Slc1a1* to examine the impact of expression of this gene on

OCD-relevant behaviors (Zike et al. 2017). Interruption of *Slc1a1* expression via insertion of a STOP-tetO element prior to the promoter led to functional knockout [*Slc1a1*-STOP (ST) mice]. While spontaneous behavior was unaltered in ST mice (grooming, locomotion, elevated plus maze), they exhibited reduced amphetamine-induced locomotion and stereotypy and reduced SKF-38393-induced grooming. Examination of amphetamine-induced activity in dorsal striatum showed an increase in cFos positive cells in WT mice that was absent in ST mice. D1 receptor density and basal levels of extracellular dopamine in dorsal striatum were also reduced in ST mice relative to WT littermates. Furthermore, viral-mediated rescue of EAAT3 in midbrain of ST mice normalized amphetamine-induced stereotypy and locomotion. Together, these findings indicate that reduction of EAAT3 is protective against pharmacologic induction of repetitive behaviors, suggesting a potential therapeutic treatment target. Later studies in EAAT3-heterozygous mice found no changes in anxiety-like or compulsive behaviors, suggesting that strong blockade of this transporter would be necessary for therapeutic effects (Gonzalez et al. 2017).

In a later study also using the pgk-Neo cassette insertion model (Peghini et al. 1997), Bellini and colleagues (Bellini et al. 2018) performed a thorough behavioral characterization in young mice from post-natal day 14 to post-natal day 35 to encompass the period at which EAAC1 reaches peak expression in the striatum. In addition to subtle motor deficits, *Eaac1*-KOs showed increased anxiety-like behavior as measured by more time spent in the closed arms of an elevated-plus maze and increased marble burying. In addition, *Eaac1*-KOs had significant increases in spontaneous grooming time and disrupted grooming syntax relative to WT littermates. Ex vivo electrophysiological studies in dorsolateral striatum demonstrated increased sensitivity to metabotropic glutamate receptor (mGluR) blockers in *Eaac1*-KOs, suggesting increased phasic activation of mGluR in the absence of EAAC1. In addition, mRNA and protein expression of D1Rs throughout striatum was reduced in *Eaac1*-KOs, but the density of D1- and D2-SPNs did not differ by genotype. This suggests that EAAC1 may promote striatal D1R expression by limiting local mGluR activation. Note, though these behavioral results differ from those seen in Zike et al. (2017), the two studies were performed at very different developmental stages (Bellini et al. (2018): early post-natal to weaning; Zike et al. (2017)): adult), which is likely to impact behavioral task performance.

Notably, the *Slc1a1* allele with the strongest association to OCD results in increased EAAT3 expression (Wendland et al. 2009). Consistent with this directionality and the initial findings from Zike et al. (2017), conditional EAAT3 overexpression in the forebrain (EAAT3^{glo}/CMKII mice: frontal cortex, hippocampus, and striatum) leads to several behavioral abnormalities relevant to OCD, including increased basal grooming and increased anxiety-like behaviors in the open field, light-dark chamber, and marble burying tests (Delgado-Acevedo et al. 2019). Similar to other transgenic models discussed above, these behaviors were normalized by chronic (21 days) fluoxetine treatment. In addition, although evaluation of reversal learning, fear conditioning, and extinction training showed no significant differences between genotypes, EAAT3^{glo}/CMKII mice showed increased freezing indicative of a fear respons when spontaneous recovery was assessed 2 weeks later. Finally, ex vivo electrophysiological recordings of cortico-striatal synapses from EAAT3^{glo}/CMKII mice found evidence for a change in NMDAR subunit composition and impaired NMDA-dependent long-term depression (LTD). These findings suggest that EAAT3 regulates NMDAR-dependent synaptic plasticity in striatum and that resulting alterations in striatal glutamatergic transmission may lead to anxiety-like and repetitive behaviors relevant to OCD.

Dopamine Transporter Gene (DAT) Knockdown Mice Transgenic approaches can also be used to generate neurotransmitter abnormalities proposed to play a role in disease pathophysiology. For example, it has been proposed that dysfunction in dopamine signaling may be involved in OCD pathophysiology. Berridge et al. (2005) therefore generated DAT-knockdown (KD) mice to test the behavioral impact of elevated extracellular dopamine levels in striatum. DAT-KD mice showed stereotyped and predictable syntactic grooming chains, consistently completing the entire complex sequence of grooming movements (elliptical, unilateral, and bilateral facial grooming followed by flank grooming) each time the sequence was initiated. This sequential super-stereotypy supports the idea that hyper-dopaminergic function in striatum could underlie behavioral rigidity relevant to OCD and other neuropsychiatric conditions.

Serotonin Receptor Modulation Evidence from genetic association studies (Sanders et al. 2002; Mundo et al. 2000; Camarena et al. 2004), neuroimaging studies (Perani et al. 2008; Hesse et al. 2005; Pogarell et al. 2003; Zitterl et al. 2007) (though see van der Wee et al. 2004; van der Wee et al. 2004), and pharmacological challenge studies (Barr 1992, #320} suggests that alterations in serotonergic signaling may contribute to OCD symptomology. Furthermore, serotonin reuptake inhibitors (SRIs) are the only proven pharmacological monotherapy for OCD (though note that full remission only occurs in 10–15% of patients (Hirschtritt et al. 2017; Romanelli et al. 2014; Simpson 2010). Though the relationship between these accumulated findings and pathophysiology is still unclear, this evidence has prompted the use of animal models with various serotonergic alterations to investigate the role of serotonergic signaling in OCD-relevant behaviors.

Clinical studies have repeatedly reported that administering meta-chlorophenylpiperazine (mCPP), a nonspecific serotonin agonist that largely acts at 5-HT$_{2C}$, 5-HT$_{1A}$, and 5-HT$_{1D}$ receptors, exacerbates symptoms in unmedicated OCD patients (Zohar and Insel 1987; Hollander et al. 1992; Broocks et al. 1998). In rodents, systemic administration of mCPP or 5-methoxy-N,N-dimethyl tryptamine (5-MeODMT), a different nonselective serotonin agonist, induces increased directional persistence in T-maze rewarded alternation tasks in a subset of animals with baseline tendencies to persist in one arm of the maze prior to drug delivery (Yadin et al. 1991; Tsaltas et al. 2005). Papakosta et al. (2013) expanded on these initial findings by assessing the relative contributions of 5-HT$_{2A}$ and 5-HT$_{2C}$ receptors to these mCPP effects). While neither 5-HT$_{2A}$ nor 5-HT$_{2C}$ receptor antagonists affected baseline directional persistence, 5-HT$_{2C}$ receptor antagonism provided partial protection against directional persistence induced by mCPP administration, implicating

$5\text{-HT}_{2C}R$ activation as the main contributor to mCPP effects in the spontaneous alternation task. In addition, mice with constitutive knockout of the 5-HT_{2C} receptor gene, which were originally described as obese and hyperphagic ((Tecott et al. 1995; Nonogaki et al. 1998; Vickers et al. 1999), engage in methodical screen chewing behavior and exhibit repetitive head dipping behavior (Chou-Green et al. 2003). $5\text{-HT}_{2C}R$ antagonism also improves initial reversal learning on a serial spatial reversal learning task (Boulougouris et al. 2008); this effect was specifically localized to OFC, and not medial prefrontal cortex (mPFC) or nucleus accumbens (NAc), via local infusion (Boulougouris and Robbins 2010). While these findings largely suggest that $5\text{-HT}_{2C}R$ activation contributes to the manifestation of perseverative behaviors, it is important to note that other studies have reported $5\text{-HT}_{2C}R$ agonist-induced decreases in perseverative behaviors such as schedule-induced polydipsia, nestlet shredding, pharmacologically-induced scratching, and marble burying (though see critique of this behavioral test in (Silverman et al. 2020; Jimenez-Gomez et al. 2011; de Brouwer et al. 2019) agonism (Martin et al. 1998; Rodriguez et al. 2017).

Findings from the rodent literature investigating the potential role of serotonin in OCD have been augmented by non-human primate studies that employ challenging cognitive tasks closely aligned with paradigms used in humans. In particular, extensive work from Clarke and colleagues has explored the role of 5-HT in cognitive inflexibility as it relates to OCD and other neuropsychiatric conditions. By introducing targeted 5-HT lesions throughout marmoset PFC, they found that selective depletion of 5-HT in PFC significantly impaired subjects' ability to reverse reward contingencies in a visual discrimination reversal task, while maintenance of the previously learned discrimination and acquisition of a new discrimination were spared (Clarke et al. 2004). Further investigation showed that this impairment arose from an inability to disengage from the previously rewarded discrimination rather than learned avoidance of the previously unrewarded discrimination (Clarke et al. 2007). Interestingly, selective PFC 5-HT depletion does not impair attentional set shifting (Clarke et al. 2005), suggesting that central serotonin modulates changes in motivational significance that impact behavior rather than higher-order processes of attentional control.

2.2 *Circuit Manipulation to Study OCD*

Several studies have recently capitalized on the development of precise circuit-manipulation technologies to translate the rich OCD neuroimaging literature highlighted in Sect. 1.2 into experimental animal systems. Here we discuss these efforts.

D1-CT Mice One of the earliest studies to investigate OCD-relevant circuit abnormalities in an animal model system was Campbell et al. (Campbell et al. 1999). The intent of this work was to hyperactivate cells expressing D1-receptors throughout the brain using novel transgenic technology that permitted expression of the cholera-toxin light chain under the control of the D1 promoter, with a goal of overactivating neurons without impacting cell health. Serendipitously the authors noted an interesting heterogeneous expression pattern in one of their transgenic lines that correlated with a perseverative behavior phenotype. Specifically, regardless of what actions the D1-CT mice initiated in their unconstrained home cage environment (e.g., eating, drinking, locomoting, licking, cage-top hanging), they performed significantly longer bouts of the behavior compared to their littermate control counterparts. In addition, subsequent studies using the D1-CT model have reported tic-like behaviors; dissection of circuits underlying tics and perseverative behaviors in this model system could therefore potentially yield insights regarding the biological underpinnings of high levels of comorbidity between OCD and Tourette Syndrome (de Vries et al. 2016; Grados and Mathews 2009; Bloch and Leckman 2009). However, gleaning mechanistic information from the D1-CT model is challenging because the cholera-toxin light chain is expressed in many different regions of the brain – including basal ganglia, intercalated nucleus of the amygdala, piriform cortex, and somatosensory cortex – making it difficult to pinpoint the exact circuits leading to the key abnormalities relevant to OCD and related disorders. Although the authors proposed that a final common net effect of hyperactivity in these individual regions would be hyperactivity of the OFC, this could not be directly tested using this system. Thus, although the D1-CT model has allowed the identification of several interesting targets for future investigation, the development of techniques for increasingly precise circuit manipulations is now enabling more refined tests of circuit involvement in specific OCD-relevant behaviors, as described below.

Optogenetic Investigation of OCD-Relevant Behaviors In the late 2000s, the advent of optogenetic technology began to allow precise investigation of the role of neural circuits in the generation of behavior. This technique takes advantage of restricted expression of light-activated ion channels in particular neural populations to allow spatially and temporally specific reversible stimulation in awake behaving animals (Tye and Deisseroth 2012). This technical advance was initially applied to the field of preclinical OCD research in a pair of complementary studies. First, building off of the findings from both the D1-CT mouse model and human neuroimaging findings described above, Ahmari et al. (2013) used optogenetics to directly generate hyperactivity in mOFC-ventral striatal projections, and determine the impact on behavior. This attempt to simulate human neuroimaging findings in an animal model system led to unexpected results. While acute optogenetic stimulation with channelrhodopsin (ChR2) did not lead to OCD-relevant behaviors, and instead led to hyperlocomotion, brief but repeated stimulation over the course of 5–7 days led to a progressive increase in grooming (Kalueff et al. 2016). Notably, this increase in grooming was observed an hour after stimulation occurred, and gradually increased in magnitude with each progressive day. This behavioral pattern was highly suggestive of cortico-striatal plasticity, which was confirmed by demonstrating an increase in evoked striatal activity over the course of multiple days using in vivo electrophysiology. Interestingly, both the behavioral phenotype of increased grooming and the increased evoked striatal activity were reversed by chronic (2 weeks), but not acute (1–7 days), treatment with fluoxetine, further suggesting potential relevance to mechanisms that may have importance for treatment of OCD.

In parallel, a study by Burguiere et al. (2013) used optogenetics to probe OFC-striatal circuits in a transgenic OCD-relevant model that displays compulsive grooming, the *Sapap3*-KO (Fig. 1; also see Sect. 2.1 for details). After training animals on a conditioned grooming task in which a tone predicted the timed release of a water drop onto the face, they found that *Sapap3*-KOs persistently groomed in response to the water-drop-predicting tone, whereas WT littermate controls adaptively timed their grooming response to the onset of the water drop. Interestingly, in vivo recordings showed decreased lateral OFC activity during persistent tone-associated grooming in *Sapap3*-KOs, and optogenetically boosting activity in either lOFC or lOFC-central striatal (CS) projections normalized *Sapap3*-KO behavior. In addition, the authors presented evidence that this behavioral change may be linked to dysfunctional feed-forward inhibition of lOFC projections onto CS fast-spiking interneurons (FSIs), resulting in CS hyperactivity. These results therefore support a model in which impaired lOFC-striatal inhibitory control can lead to online emergence of compulsive behaviors.

Several other studies have since used optogenetics to dissect circuits underlying OCD-related phenotypes in both disease models and healthy animals. Building on previous work, Corbit et al. (2019) determined that lOFC and ALM (anterior lateral motor area; also known as M2, the mouse equivalent of pre-supplementary/supplementary motor area) are the two main sources of cortical projections to CS, the striatal region associated with compulsive grooming in *Sapap3*-KOs (Burguiere et al. 2013; Welch et al. 2007). Projection-specific probing of these circuits using optogenetics and ex vivo electrophysiology demonstrated that *Sapap3*-KOs had decreased synaptic strength in projections from lOFC to CS SPNs compared to WTs, consistent with prior work (Burguiere et al. 2013). However, projections from ALM to CS SPNs were significantly strengthened, suggesting a potential role for the ALM-CS circuit in driving both CS hyperactivity and compulsive grooming behavior. A recent preprint further supported a role for ALM-CS projections in grooming behavior by demonstrating that in vivo optogenetic stimulation of ALM-CS projections can directly trigger grooming bouts (Corbit et al. 2020). Together, this work has interesting implications for OCD, since transcranial magnetic stimulation (TMS) of supplementary and pre-supplementary motor cortex has been suggested as a potential treatment target (Lee et al. 2017; Mantovani et al. 2010; Pallanti et al. 2016).

3 Chemogenetic Investigation of OCD- and TS-Relevant Behaviors

3.1 Striatal Histidine Signaling

While optogenetic techniques allow manipulation of distinct cell types and circuits with high spatial and temporal resolution, they are generally not conducive to sustained activation or inhibition of specific brain regions, cell types, or circuits that may be important for modeling of pathophysiologic processes (though see description of stabilized step function opsins (SSFOs) (Lerner et al. 2016)). The recent development of chemogenetic techniques has filled this gap (see (Sternson and Roth 2014; Roth 2016) for review). Using these approaches, mutant G-protein coupled receptors expressed in cell types of interest can be selectively engaged by designer synthetic molecules administered either systemically or locally via intracerebral infusion. One set of studies used this approach to demonstrate a role for histamine signaling in the generation of repetitive behaviors. After identification of histidine decarboxylase (Hdc) [the enzyme responsible for synthesizing histamine from histidine] as a rare cause of Tourette Syndrome (TS) with high-incidence of comorbid OCD (Fernandez et al. 2012; Ercan-Sencicek et al. 2010), parallel studies using histidine decarboxylase knockout mice (Hdc-KO) demonstrated increased stereotypic behavior following amphetamine administration (Baldan et al. 2014) and elevated grooming behavior following acute stress (Xu et al. 2015). Additional studies in Hdc-KOs showed dopaminergic abnormalities in striatum (Castellan Baldan et al. 2014) including elevated basal levels of dopamine (Baldan et al. 2014; Rapanelli et al. 2014) and significant reductions in extracellular dopamine levels following local striatal application of histamine in Hdc-KOs (Baldan et al. 2014). Further implicating striatal Hdc signaling through histamine receptor 3 (H3R) in TS/OCD-relevant behaviors (Rapanelli et al. 2017), activation of H3R with the specific agonist RAMH induced stereotypy in Hdc-KOs and led to increased cellular activation in dorsal striatum (DS). In addition, chemogenetic inhibition of RAMH-activated cells in DS following subcutaneous RAMH administration prevented RAMH-induced stereotypy in Hdc-KOs (Rapanelli et al. 2017), indicating that H3R activation in dorsal striatum can directly cause stereotypic behavior. Further dissection of the upstream circuits showed that chemogenetic inhibition of Hdc-expressing neurons in the tuberomammilary nucleus (TMN) of WT mice led to both increased grooming and stereotypy and increased cFos expression in dorsal striatum, prelimbic cortex, and infralimbic cortex. An activity-regulated ligand-gated Cre-recombinase was then used to tag dorsal striatal and mPFC cells that exhibited increased activity following TMN inactivation, followed by bidirectional chemogenetic manipulation. Specific chemogenetic activation of tagged dorsal striatal neurons (and not mPFC neurons) resulted in profound increases in grooming. Together, these studies suggest that alterations in histaminergic neuromodulation of the striatum can lead to pathologic repetitive behaviors of relevance for TS, OCD, and related neuropsychiatric illnesses.

3.2 Striatal Interneuron Manipulation

Another set of studies recently used chemogenetic approaches to translate post-mortem findings from TS, which is highly comorbid with OCD, into animal models. A significant reduction (50–60%) in interneurons is seen in post-mortem striatal tissue from TS patients (though note some studies were unable to dissociate reduced cell numbers from reduced transcript or protein expression) (Kataoka et al. 2010; Kalanithi et al. 2005; Lennington et al. 2016). Pittenger and colleagues attempted to recapitulate these pathologic changes by performing targeted ablations of striatal cholinergic (ChAT) (Xu et al. 2015) and fast-spiking parvalbumin (PV) interneurons (Xu et al. 2016) in mice.

First, Xu et al. (2015) used an immunotoxin to selectively kill ChAT interneurons in dorsolateral striatum (DLS) or dorsomedial striatum (DMS) in independent cohorts of mice and determine the impact of these focused ablations. Although DLS ChAT-ablated mice had no baseline behavioral changes, they exhibited increased grooming following unpredictable presentation of acoustic startle stimuli that persisted after the startle block. They also displayed elevated stereotypic behaviors 30 min after amphetamine administration that were not observed in controls. In contrast, DMS ChAT-ablated mice showed no changes in either acute stress-induced grooming or amphetamine-induced stereotypies. Together, these data suggest that DLS ChAT interneurons play a uniquely important role in modulating both stress-induced and pharmacologically-induced repetitive behaviors, supporting a potential link between these behaviors and circuits classically associated with habits.

Next, Xu et al. (2016) used the same approach to selectively ablate PV+ interneurons in dorsal striatum. Similar to DLS ChAT-ablated mice (Xu et al. 2015), PV-ablated mice had no behavioral changes at baseline, but showed increased grooming during and following acoustic startle testing. In addition, after acquisition of cued fear conditioning, PV-ablated mice exhibited increased grooming in the 30 min following conditioned stimulus (tone) presentation, suggesting induction of repetitive behavior by conditioned fear. Unlike ChAT-ablated mice, however, there were no differences in amphetamine-induced stereotypic behaviors compared to controls, suggesting that ChAT+ and PV+ interneurons may modulate behavior in striatal ensembles that are only partially overlapping. These data are consistent with the idea that stochastic input from PV+ interneurons onto SPNs creates feed-forward inhibition, enabling selection of favorable actions while suppressing incompatible alternative actions (Owen et al. 2018). In the context of OCD and related disorders, it is therefore possible that disruption of FSI-mediated feed-forward inhibition of maladaptive behaviors could lead to interrupted execution of adaptive, voluntary actions. Similarly, decreased ChAT-mediated modulation of SPNs after action initiation could theoretically enable the continuation of persistent, repetitive behaviors. Together, these studies demonstrate the utility of translating post-mortem human findings into animal models to enable mechanistic dissection.

Investigation of Circuits Underlying Symptom Domains that Cut Across Psychiatric Diagnoses Because this review is focused on the use of animal models for the study of OCD, this discussion is mostly limited to models that have been explicitly tied to OCD through either genetic findings, circuit-based human studies, or pharmacologic induction of repetitive behavior. However, it is important to note that abnormal repetitive behaviors are a symptom construct that cuts across many different neuropsychiatric disorders, so insights about these behaviors may be gleaned from models designed to investigate other highly comorbid illnesses. For example, the *Shank-3* gene, which encodes a post-synaptic scaffolding protein similar in function to *Sapap3*, has been consistently highlighted as an autism risk gene. Notably, in addition to displaying deficits in social interactions, the *Shank-3* knockout mouse model for autism spectrum disorder displays compulsive grooming and anxiety phenotypes similar to those observed in the *Sapap3*-KO model for OCD (Mei et al. 2016; Peca et al. 2011). Thus, evidence that striatal synaptic abnormalities in the *Shank3*-KO model are primarily localized to striatopallidal SPNs and that artificial enhancement of striatopallidal SPN activity using DREADDs (Designer Receptors Exclusively Activated by Designer Drugs)-based chemogenetics normalized repetitive grooming (Wang et al. 2017), may yield insight into circuit abnormalities that underlie symptoms in people with comorbid autism and OCD.

3.3 Dissection of Circuits Underlying Constructs Relevant to OCD

Much work has gone into identifying neural constructs that are reliably abnormal in people with OCD (see (Ahmari et al. 2014) for discussion). Several key constructs have collectively emerged as potential focuses for preclinical dissection using the techniques described above. These include reversal learning, balance between habit-learning/goal-directed behavior, sensorimotor gating, response inhibition, and avoidance learning. Though detailed discussion of preclinical studies investigating each of these constructs is beyond the scope of this review, and exhaustive reviews have been published elsewhere (reversal learning (Manning and Ahmari 2018); habit learning (Gillan et al. 2016); sensorimotor gating (Geyer and Dulawa 2003; Swerdlow et al. 2001); response inhibition (Bari and Robbins 2013); avoidance learning (Geramita et al. 2020; LeDoux and Daw 2018)), we will briefly discuss key studies that have used circuit-specific approaches to investigate the neural circuits underlying repetitive checking, sensorimotor gating, reversal learning, and habit learning in the context of OCD below.

In addition to these neural constructs that may be relatively disease-specific, there are also fundamental neural processes that are broadly impacted in many different neuropsychiatric disorders due to their centrality to human behavior. These include processes associated with key circuits implicated in OCD pathology (e.g., OFC-basal ganglia-thalamic loops) such as value-based decision-making, action initiation, reinforcement learning (including negative reinforcement), and action termination. However, because human clinical studies are still defining how these core processes are impaired in OCD, making preclinical translation difficult, here we focus instead on the preclinical literature investigating key theories that are specifically relevant to OCD.

3.4 *Dissecting Abnormal Neural Constructs in OCD*

Quinpirole-Induced Checking Repetitive checking behavior is a commonly observed compulsion in OCD. Though it is difficult to model the obsessions related to this behavior in animals, administration of the D2/3 receptor agonist quinpirole in rats leads to increased checking-like behavior in an unconstrained open field environment, facilitating mechanistic investigation (Szechtman et al. 1998). After injection, rats revisit particular objects and/or locations in a large open field in a stereotyped order that is specific to individual animals (Dvorkin et al. 2006). Studies using this model have observed dysregulation in dopaminergic signaling (Szechtman et al. 1998) and cortico-basal ganglia circuits (Dvorkin et al. 2010; Djodari-Irani et al. 2011), suggesting possible mechanisms underlying repetitive checking. More recently, this system was adapted in a new operant-based task (observing response task – ORT) designed to further explore cognitive processes underlying repetitive checking behavior (Eagle et al. 2014). In this task, rats learned to press an "observing lever" to acquire information about the location of an active lever that delivers food rewards, and administration of quinpirole after training led to an increase in both functional (i.e., information obtaining) and non-functional presses of the "observing lever." Finally, the quinpirole model was recently translated into mice in order to facilitate optogenetic and chemogenetic circuit dissection (Sun et al. 2019). After quinpirole administration, mice did not demonstrate increased checking in an open field, but did show increased time at a "home base" location, as well as increased frequency of visits to a sipper port containing 10% sucrose. This behavior was associated with enhanced excitability of basolateral amygdala (BLA) inputs to mPFC and increased firing rate of mPFC GABAergic interneurons measured using ex vivo electrophysiology. Both optogenetic and chemogenetic interventions supported a model in which repeated quinpirole administration activates a BLA → mPFC interneuron→mPFC pyramidal neuron circuit, resulting in net reduction of activity in the mPFC that leads to checking-like behavior. These behaviors were reversed by either optogenetic inhibition of BLA → mPFC terminals, chemogenetic activation of mPFC glutamatergic neurons, or chemogenetic inhibition of mPFC GABAergic neurons, which could have implications for designing neurostimulation-based treatments if these findings translate into higher-order primates.

Sensorimotor Gating Sensorimotor gating deficits, which prominently feature in many neuropsychiatric disorders including schizophrenia, have also been demonstrated in both medicated (Hoenig et al. 2005; Swerdlow et al. 1993) and unmedicated (Ahmari et al. 2012) patients with OCD. Interestingly, these deficits have been shown to be most prominent in patients with comorbid tics (Ahmari et al. 2012), as well as in female subjects (Steinman et al. 2016; Ahmari et al. 2016), which may yield insight into circuit and hormonal mechanisms in this disorder. To date, there has been limited dissection of this construct in OCD-relevant animal models. One exception is a series of studies by the Dulawa group, which demonstrated that repetitive systemic 5-HT1B agonist administration in WT mice led to thigmotaxis (i.e., perseverative locomotion) and prepulse inhibition (PPI) deficits that were resolved by chronic (4 weeks) (but not acute) fluoxetine (Shanahan et al. 2009; Dulawa and Geyer 2000; Sipes and Geyer 1994; Cheetham and Heal 1993; Chaouloff et al. 1999). Subsequent studies using autoradiography and genetic rescue have localized these effects to the ventromedial OFC (Shanahan et al. 2011), consistent with human OCD imaging studies demonstrating abnormal activity in this brain region. Future work is needed to further refine the underlying circuits. In addition, recent work from our lab using the *Sapap3*-KO transgenic model demonstrates that PPI disruption is significantly correlated with severity of compulsive grooming. Both PPI disruption and compulsive grooming severity were associated with reduced dopamine D1 and D2/3 receptor density in the nucleus accumbens core (NAcc), suggesting that elevation of NAcc dopamine may be a common mechanism leading to these OCD-relevant behaviors (Manning et al. 2020).

Imbalance Between Goal-Directed and Habitual Behavior Evidence from the clinical literature has lent support to the theory that OCD symptoms stem from an imbalance in circuits controlling goal-directed and habitual behavior – specifically, that dysfunction in goal-directed responding can lead to increased reliance on habits to guide actions (Gillan and Robbins 2014). For example, OCD patients had decreased ability to refrain from responding to devalued outcomes in an outcome-devaluation "slips of action" test, which was associated with a selective deficit in knowledge of the response-outcome contingencies acquired during training (i.e., goal-directed associations) (Gillan et al. 2011). Furthermore, people with OCD are more prone to developing habits in avoidance (i.e., persistently selecting actions to prevent administration of a shock despite explicit knowledge that the stimulus no longer predicts a shock) (Gillan et al. 2015; Gillan et al. 2014). It is still unclear how these abnormalities map onto OCD symptoms (e.g., increased performance of habits in avoidance was not correlated with severity of compulsive behavior on the clinical Y-BOCS scale), and alternative theories which feature a prominent role for anxiety in the disorder have not yet been tested in animal models (Piantadosi and Ahmari 2015). Nevertheless, identifying the neural circuit mechanisms underlying habit formation is likely to be important both for understanding OCD pathophysiology and developing new treatment strategies (Gillan and Robbins 2014). Here we therefore review key preclinical studies that have investigated the role of cortico-striatal circuits in forming habits (including habits in avoidance) and flexibly switching between goal-directed and habitual behaviors.

Early studies using lesion and chemical inactivation methods implicated the dorsolateral striatum (DLS) and infralimbic (IL) cortex in the formation and expression of habits (Kimchi et al. 2009; Tang et al. 2007). To assess IL's specific role in executive control over habit expression with temporal precision, (Smith et al. 2012) rats were first overtrained in a T-maze until they developed habitual performance (i.e., selected the devalued arm). Optogenetic inhibition of IL while rats performed a post-devaluation probe test led to a sharp decline in selection of the devalued arm, indicating that IL activity is necessary for habit expression and/or inhibition of IL can restore goal-directed strategies. Surprisingly, rats also appeared to develop a new habit following repeated optogenetic disruption of IL. Building on these findings, simultaneous chronic tetrode recordings from DLS and IL showed that DLS ensemble activity shifted from a pattern of heightened activity throughout the T-maze run to a "task-bracketing" pattern of ensemble activity as rats reached learning criterion in initial training. In contrast, IL activity was largely unchanged throughout initial training but switched to a similar task-bracketing pattern when rats were halfway through over-training, suggesting a role for IL in habit formation (Smith and Graybiel 2013). This research provides a template for further investigation of habit formation, crystallization, and cessation by precisely manipulating relevant neural circuitry and recording neuronal activity throughout longitudinal studies.

Impairments in flexible switching between goal-directed and habitual responding via differential engagement of cortico-striatal circuits may also contribute to OCD symptoms. Parallel cortico-striatal circuits are thought to modulate goal-directed and habitual actions, with the dorsomedial striatum guiding goal-directed behavior and dorsolateral striatum guiding habitual behavior (Yin et al. 2005; Yin et al. 2004, 2006). To probe the neural mechanisms governing the shift between these response strategies, Gremel and Costa (2013) designed a novel instrumental task that required mice to shift between habitual [Random Interval (RI) training schedule] or goal-directed [Random Response (RR) training schedule] responding. Selective excitotoxic lesions of DMS biased mice toward habitual responding, while mice with DLS lesions showed goal-directed action selection. These findings indicate that goal-directed and habitual strategies are both encoded in parallel, and that disruption of critical circuitry for one strategy leads to a bias toward using the remaining circuit for action execution. Subsequent simultaneous multielectrode recordings from DMS, DLS, and OFC neurons identified populations of neurons that were differentially responsive to lever-pressing in both training contexts (RR and RI). Interestingly, stronger modulation of either OFC or DMS neurons during RR vs. RI correlated with increased sensitivity to changes in outcome value, suggesting a shift toward goal-directed responding as OFC and DMS are engaged. In addition, chemogenetic inhibition of OFC led to impaired goal-directed responding in the RR context, suggesting that the OFC conveys information related to outcome value that is used to guide appropriate goal-directed responding.

Recent studies have also used the transgenic *Sapap3*-KO mouse model to investigate goal-directed and habitual behavior. Hadjas et al. (2019) found that *Sapap3*-

KOs formed habits more rapidly than WT littermates when trained on a random-interval schedule that is known to elicit habitual responding given sufficient training time (Hadjas et al. 2019). In contrast, Ehmer et al. (2020) found that KOs had delayed habit formation and preservation of goal-directed response strategies compared to WTs on a similar task. Notably, high-grooming KOs showed the strongest goal-directed behavior evidenced by reward devaluation in both RR and RI contexts, suggesting a dissociation between behavioral compulsivity and habitual responding. Future studies probing neural circuit changes that underlie maladaptive action selection in *Sapap3*-KOs will be necessary to resolve the discrepancies between these two studies.

Finally, preclinical investigations of the neural mechanisms of persistent avoidance, which could have relevance to habits in avoidance, are now emerging. Though a detailed discussion of the circuits underlying freezing (presumptive relationship to fear) and avoidance (presumptive relationship to anxiety) is beyond the scope of this review (see Geramita et al. 2020, LeDoux et al. 2017; Diehl et al. 2019, for details), there is evidence that a specific circuit connecting lateral amygdala (LA), basolateral amygdala (BLA), and nucleus accumbens (NAc) is required for avoidance behavior in multiple paradigms (LeDoux et al. 2017). Moreover, a series of studies from the laboratory of Greg Quirk has highlighted circuit mechanisms leading to pathologic persistent avoidance, which may be specifically relevant to behaviors that are observed in OCD. First, they developed a novel task in which rats learned to avoid a shock associated with a tone conditioned stimulus (CS) by mounting a platform within an operant chamber (platform avoidance task). Inactivating either PL, VS, or BLA via muscimol led to impaired expression of avoidance, while inactivation of IL impaired avoidance extinction while leaving avoidance expression intact (Bravo-Rivera et al. 2014). In follow-up studies, a subset of rats did not reduce avoidance behaviors after extinction, which should have rendered the CS non-threatening (Overmier and Brackbill 1977; Bravo-Rivera et al. 2015). This lack of extinction was correlated with increased activity in PL, VS, and BLA and decreased activity in IL as measured by c-fos. Additional studies indicated that inhibitory neuronal responses in PL were associated with "avoidability" (i.e., whether a particular threat being signaled was avoidable), and that optogenetic activation of rostral PL neurons to neutralize this inhibitory response impaired avoidance (Diehl et al. 2018).

The platform avoidance paradigm has also been modified to model aspects of exposure therapy with response prevention (EXRP), a first-line therapy for OCD (Rauch et al. 1994). In this variant of the task, a plexiglass barrier was introduced to physically prevent animals from mounting the platform while the tone CS was extinguished. Though most rats showed decreased avoidance after this training experience, a subset continued to show avoidance by mounting the platform in the presence of the tone (Rodriguez-Romaguera et al. 2016); this percentage was increased by extending the initial conditioning training from 8 to 20 days (Martínez-Rivera 2018). In the 2016 study (Rodriguez-Romaguera et al. 2016), this persistent avoidance was normalized via muscimol inactivation of the lateral OFC, consistent with clinical OCD studies relating OFC hyperactivity to symptoms

(Breiter et al. 1996; Rauch et al. 1994; Hou et al. 2012; Kwon et al. 2003; Saxena et al. 1999). In addition, ventral striatal DBS reduced persistent avoidance, potentially yielding insight into mechanisms underlying DBS efficacy. Ongoing studies are continuing to dissect mechanisms underlying both adaptive and maladaptive avoidance behaviors, in the hopes of identifying new treatment targets (Diehl et al. 2019).

Reversal Learning Clinical studies demonstrate that people with OCD have impaired cognitive flexibility – i.e., the ability to appropriately adjust behavior in a changing environment to achieve optimal outcomes (Dajani and Uddin 2015). In particular, abnormalities in reversal learning performance and/or associated neural activity have been demonstrated in OCD using both behavioral and functional neuroimaging studies (Lagemann et al. 2012; Remijnse et al. 2006; Chamberlain et al. 2008; Gottwald et al. 2018; Tezcan et al. 2017; Remijnse et al. 2009; Valerius et al. 2008). However, heterogeneity of these findings indicates the need for mechanistic studies to further dissect the neural correlates of OCD-relevant cognitive inflexibility. While this review will not exhaustively appraise the neural circuits underlying reversal learning, we will highlight key studies emphasizing dissection of OFC (Nahum et al. 2011; O'Doherty et al. 2001; Nagahama et al. 2001; Kringelbach and Rolls 2003; Remijnse et al. 2005; Ghahremani et al. 2010)) and ACC circuits (Kringelbach and Rolls 2003; Ghahremani et al. 2010; D'Cruz et al. 2011; Meder et al. 2017)), which have both been implicated in OCD pathophysiology and are recruited during reversal learning tasks in healthy participants.

Although OFC hyperactivity has been reported in OCD patients at baseline and during symptom provocation (Breiter et al. 1996; Rauch et al. 1994; Hou et al. 2012; Kwon et al. 2003; Saxena et al. 1999), decreased OFC activity has been reported in OCD patients during reversal learning paradigms (Remijnse et al. 2006; Chamberlain et al. 2008; Apergis-Schoute et al. 2017; Gu et al. 2008; Vaghi et al. 2017), and individuals with OFC lesions have significant reversal learning impairments (Rolls et al. 1994; Fellows and Farah 2003; Hornak et al. 2004). Similarly, extensive rodent lesion and inactivation studies support a necessary role for OFC in reversal learning (Bohn et al. 2003; Brigman and Rothblat 2008; Burke et al. 2009; Ghods-Sharifi et al. 2008; Boulougouris et al. 2007; Schoenbaum et al. 2002; McAlonan and Brown 2003). In addition, extensive lesion and inactivation studies in non-human primates (Rygula et al. 2010; Rudebeck and Murray 2008) support a necessary role for OFC in reversal learning (Bohn et al. 2003; Brigman and Rothblat 2008; Burke et al. 2009; Ghods-Sharifi et al. 2008; Boulougouris et al. 2007; Schoenbaum et al. 2002; McAlonan and Brown 2003; Dias et al. 1996; Jones and Mishkin 1972). Furthermore, targeted inactivation in rodents has attempted to disentangle the role of medial vs. lateral OFC in probabilistic (Dalton et al. 2016) and serial (Hervig et al. 2020) reversal learning. Though there are some discrepancies in findings, mOFC disruption impacted perseverative errors [(Dalton et al. 2016): increased perseverative errors; (Hervig et al. 2020): decreased perseverative errors], while lOFC inactivation increased perseverative errors (Hervig et al. 2020) and decreased lose-shift probabilities only following incorrect trials (Dalton et al.

2016), indicating that lOFC may integrate negative feedback signals to update response selection. Further support for this idea is provided by Groman et al. (2019), in which targeted ablation of OFC-striatal projections using a retro-Cre virus (NAcc) and a floxed diphtheria toxin receptor (OFC) led to perseverative responses specifically following incorrect unrewarded choices on a probabilistic reversal learning task. In contrast, ablating terminals projecting from the OFC to amygdala increased correct responding during reversal and lose-shift probability, suggesting encoding of action value over time.

The ACC, which is thought to integrate information about reward and alert subjects to unexpected outcomes in order to direct reward-guided decision- making (Amiez et al. 2006; Brown and Braver 2005; Noonan et al. 2011; Chudasama et al. 2013), is likewise engaged during human reversal learning and is hyperactive in OCD patients at baseline and during symptom provocation (Breiter et al. 1996; Rauch et al. 1994; Hou et al. 2012; Kwon et al. 2003; Saxena et al. 1999). Consistent with these findings, single-trial optogenetic inhibition during a probabilistic reversal learning task in rodents reduces the influence of the previous outcome or state transition on subsequent choice, which could lead to perseverative errors by preventing lose-shift responses (Akam et al. 2017). Future work to dissociate value-based decision-making and value representations in both OFC and ACC using more nuanced tasks such as the two-step task (Miller et al. 2018) will further delineate how flexible decision-making is encoded in prefrontal cortical circuits and how disruptions to this encoding may manifest as cognitive impairments in OCD and related disorders.

4 Conclusion: How Do We Get from Animal Models to Targeted Treatments?

Because of the recent explosion in neuroscience tool development and relatively low cost of exome sequencing, we now have an unprecedented ability to monitor activity of specific cell types and circuits during complex animal behaviors and subsequently manipulate these neural populations to test causality. In turn, these mechanistic investigations can yield information regarding potential novel treatment targets for OCD. Though circuit dissection during behavioral tasks is important in its own right, such investigations are most likely to be useful for treatment development if they serve as a key node in an integrated set of translational studies.

Acknowledgements This work was supported by NIMH R01MH104255, NIMH R01MH119837, NIMH R21 MH116330, Burroughs Wellcome Career Award for Medical Scientists, Klingenstein-Simons Fellowship Award in the Neurosciences, One Mind Rising Star Award, and the FFOR Consortium (SEA); and NIGMS T32GM008208 (BLC). Figures created with BioRender.com.

Disclosure/Conflicts of Interest No conflicts of interest to disclose.

References

Ahmari SE, Risbrough VB, Geyer MA, Simpson HB (2012) Impaired sensorimotor gating in unmedicated adults with obsessive-compulsive disorder. Neuropsychopharmacology 37:1216–1223

Ahmari SE, Spellman T, Douglass NL, Kheirbek MA, Simpson HB, Deisseroth K, Gordon JA, Hen R (2013) Repeated cortico-striatal stimulation generates persistent OCD-like behavior. Science 340:1234–1239

Ahmari SE, Eich T, Cebenoyan D, Smith EE, Blair Simpson H (2014) Assessing neurocognitive function in psychiatric disorders: a roadmap for enhancing consensus. Neurobiol Learn Mem 115:10–20

Ahmari SE, Risbrough VB, Geyer MA, Simpson HB (2016) Prepulse inhibition deficits in obsessive-compulsive disorder are more pronounced in females. Neuropsychopharmacology 41:2963–2964

Akam T, Rodrigues-Vaz I, Marcelo I, Zhang X, Pereira M, Oliveira RF, Dayan P, Costa RM (2017) Anterior cingulate cortex represents action-state predictions and causally mediates model-based reinforcement learning in a two-step decision task. biorxiv. https://doi.org/10.1101/126292

Amiez C, Joseph JP, Procyk E (2006) Reward encoding in the monkey anterior cingulate cortex. Cereb Cortex 16:1040–1055

Apergis-Schoute AM, Gillan CM, Fineberg NA, Fernandez-Egea E, Sahakian BJ, Robbins TW (2017) Neural basis of impaired safety signaling in obsessive compulsive disorder. Proc Natl Acad Sci U S A 114:3216–3221

Arnold PD, Sicard T, Burroughs E, Richter MA, Kennedy JL (2006) Glutamate transporter gene SLC1A1 associated with obsessive-compulsive disorder. Arch Gen Psychiatry 63:769–776

Aruga J, Yokota N, Mikoshiba K (2003) Human SLITRK family genes: genomic organization and expression profiling in normal brain and brain tumor tissue. Gene 315:87–94

Atmaca M, Yildirim H, Ozdemir H, Ozler S, Kara B, Ozler Z, Kanmaz E, Mermi O, Tezcan E (2008) Hippocampus and amygdalar volumes in patients with refractory obsessive-compulsive disorder. Prog Neuro-Psychopharmacol Biol Psychiatry 32:1283–1286

Baldan LC, Williams KA, Gallezot JD, Pogorelov V, Rapanelli M, Crowley M, Anderson GM, Loring E, Gorczyca R, Billingslea E, Wasylink S, Panza KE, Ercan-Sencicek AG, Krusong K, Leventhal BL, Ohtsu H, Bloch MH, Hughes ZA, Krystal JH, Mayes L, de Araujo I, Ding YS, State MW, Pittenger C (2014) Histidine decarboxylase deficiency causes tourette syndrome: parallel findings in humans and mice. Neuron 81:77–90

Bari A, Robbins TW (2013) Inhibition and impulsivity: behavioral and neural basis of response control. Prog Neurobiol 108:44–79

Barr LC, Goodman WK, Price LH, McDougle CJ, et al (1992) The serotonin hypothesis of obsessive compulsive disorder: implications of pharmacologic challenge studies. J Clin Psychiatry 53(4, Suppl):17–28

Bellini S, Fleming KE, De M, McCauley JP, Petroccione MA, D'Brant LY, Tkachenko A, Kwon S, Jones LA, Scimemi A (2018) Neuronal glutamate transporters control dopaminergic signaling and compulsive behaviors. J Neurosci 38:937–961

Berlin HA, Schulz KP, Zhang S, Turetzky R, Rosenthal D, Goodman W (2015) Neural correlates of emotional response inhibition in obsessive-compulsive disorder: a preliminary study. Psychiatry Res 234:259–264

Bernier R, Golzio C, Xiong B, Stessman HA, Coe BP, Penn O, Witherspoon K, Gerdts J, Baker C, Vulto-van Silfhout AT, Schuurs-Hoeijmakers JH, Fichera M, Bosco P, Buono S, Alberti A, Failla P, Peeters H, Steyaert J, Lelm V, Francescatto L, Mefford HC, Rosenfeld JA, Bakken T,

O'Roak BJ, Pawlus M, Moon R, Shendure J, Amaral DG, Lein E, Rankin J, Romano C, de Vries BBA, Katsanis N, Eichler EE (2014) Disruptive CHD8 mutations define a subtype of autism early in development. Cell 158:263–276

Bernstein JG, Garrity PA, Boyden ES (2012) Optogenetics and thermogenetics: technologies for controlling the activity of targeted cells within intact neural circuits. Curr Opin Neurobiol 22:61–71

Berridge KC, Aldridge JW, Houchard KR, Zhuang X (2005) Sequential super-stereotypy of an instinctive fixed action pattern in hyper-dopaminergic mutant mice: a model of obsessive compulsive disorder and Tourette's. BMC Biol 3:4

Beucke JC, Sepulcre J, Talukdar T, Linnman C, Zschenderlein K, Endrass T, Kaufmann C, Kathmann N (2013) Abnormally high degree connectivity of the orbitofrontal cortex in obsessive-compulsive disorder. JAMA Psychiat 70:619–629

Bienvenu OJ, Wang Y, Shugart YY, Welch JM, Grados MA, Fyer AJ, Rauch SL, McCracken JT, Rasmussen SA, Murphy DL, Cullen B, Valle D, Hoehn-Saric R, Greenberg BD, Pinto A, Knowles JA, Piacentini J, Pauls DL, Liang KY, Willour VL, Riddle M, Samuels JF, Feng G, Nestadt G (2009) Sapap3 and pathological grooming in humans: results from the OCD collaborative genetics study. Am J Med Genet B Neuropsychiatr Genet 150B:710–720

Bjorn-Yoshimoto WE, Underhill SM (2016) The importance of the excitatory amino acid transporter 3 (EAAT3). Neurochem Int 98:4–18

Bloch MH, Leckman JF (2009) Clinical course of Tourette syndrome. J Psychosom Res 67:497–501

Bloch MH, Green C, Kichuk SA, Dombrowski PA, Wasylink S, Billingslea E, Landeros-Weisenberger A, Kelmendi B, Goodman WK, Leckman JF, Coric V, Pittenger C (2013) Long-term outcome in adults with obsessive-compulsive disorder. Depress Anxiety 30:716–722

Bloch Y, Arad S, Levkovitz Y (2016) Deep TMS add-on treatment for intractable Tourette syndrome: a feasibility study. World J Biol Psychiatry 17:557–561

Bohlhalter S, Goldfine A, Matteson S, Garraux G, Hanakawa T, Kansaku K, Wurzman R, Hallett M (2006) Neural correlates of tic generation in Tourette syndrome: an event-related functional MRI study. Brain 129:2029–2037

Bohn I, Giertler C, Hauber W (2003) Orbital prefrontal cortex and guidance of instrumental behaviour in rats under reversal conditions. Behav Brain Res 143:49–56

Boulougouris V, Robbins TW (2010) Enhancement of spatial reversal learning by 5-HT2C receptor antagonism is neuroanatomically specific. J Neurosci 30:930–938

Boulougouris V, Dalley JW, Robbins TW (2007) Effects of orbitofrontal, infralimbic and prelimbic cortical lesions on serial spatial reversal learning in the rat. Behav Brain Res 179:219–228

Boulougouris V, Glennon JC, Robbins TW (2008) Dissociable effects of selective 5-HT2A and 5-HT2C receptor antagonists on serial spatial reversal learning in rats. Neuropsychopharmacology 33:2007–2019

Braun AR, Stoetter B, Randolph C, Hsiao JK, Vladar K, Gernert J, Carson RE, Herscovitch P, Chase TN (1993) The functional neuroanatomy of Tourette's syndrome: an FDG-PET study. I. Regional changes in cerebral glucose metabolism differentiating patients and controls. Neuropsychopharmacology 9:277–291

Bravo-Rivera C, Roman-Ortiz C, Brignoni-Perez E, Sotres-Bayon F, Quirk GJ (2014) Neural structures mediating expression and extinction of platform-mediated avoidance. J Neurosci 34:9736–9742

Bravo-Rivera C, Roman-Ortiz C, Montesinos-Cartagena M, Quirk GJ (2015) Persistent active avoidance correlates with activity in prelimbic cortex and ventral striatum. Front Behav Neurosci 9:184

Breiter HC, Rauch SL, Kwong KK, Baker JR, Weisskoff RM, Kennedy DN, Kendrick AD, Davis TL, Jiang A, Cohen MS, Stern CE, Belliveau JW, Baer L, O'Sullivan RL, Savage CR, Jenike MA, Rosen BR (1996) Functional magnetic resonance imaging of symptom provocation in obsessive-compulsive disorder. Arch Gen Psychiatry 53:595–606

Brigman JL, Rothblat LA (2008) Stimulus specific deficit on visual reversal learning after lesions of medial prefrontal cortex in the mouse. Behav Brain Res 187:405–410

Broocks A, Pigott TA, Hill JL, Canter S, Grady TA, L'Heureux F, Murphy DL (1998) Acute intravenous administration of ondansetron and m-CPP, alone and in combination, in patients with obsessive-compulsive disorder (OCD): behavioral and biological results. Psychiatry Res 79:11–20

Brown JW, Braver TS (2005) Learned predictions of error likelihood in the anterior cingulate cortex. Science 307:1118–1121

Burguiere E, Monteiro P, Feng G, Graybiel AM (2013) Optogenetic stimulation of lateral orbitofronto-striatal pathway suppresses compulsive behaviors. Science 340:1243–1246

Burguiere E, Monteiro P, Mallet L, Feng G, Graybiel AM (2015) Striatal circuits, habits, and implications for obsessive-compulsive disorder. Curr Opin Neurobiol 30:59–65

Burke KA, Takahashi YK, Correll J, Brown PL, Schoenbaum G (2009) Orbitofrontal inactivation impairs reversal of Pavlovian learning by interfering with 'disinhibition' of responding for previously unrewarded cues. Eur J Neurosci 30:1941–1946

Camarena B, Aguilar A, Loyzaga C, Nicolini H (2004) A family-based association study of the 5-HT-1Dbeta receptor gene in obsessive-compulsive disorder. Int J Neuropsychopharmacol 7:49–53

Campbell KM, de Lecea L, Severynse DM, Caron MG, McGrath MJ, Sparber SB, Sun LY, Burton FH (1999) OCD-like behaviors caused by a neuropotentiating transgene targeted to cortical and limbic D1+ neurons. J Neurosci 19:5044–5053

Capecchi MR (1997) Hox genes and mammalian development. Cold Spring Harb Symp Quant Biol 62:273–281

Cappi C, Oliphant ME, Peter Z, Zai G, do Rosario MC, Sullivan CAW, Gupta AR, Hoffman EJ, Virdee M, Olfson E, Abdallah SB, Willsey AJ, Shavitt RG, Miguel EC, Kennedy JL, Richter MA, Fernandez TV (2020) De novo damaging DNA coding mutations are associated with obsessive-compulsive disorder and overlap with Tourette's disorder and autism. Biol Psychiatry 87:1035–1044

Castellan Baldan L, Williams KA, Gallezot JD, Pogorelov V, Rapanelli M, Crowley M, Anderson GM, Loring E, Gorczyca R, Billingslea E, Wasylink S, Panza KE, Ercan-Sencicek AG, Krusong K, Leventhal BL, Ohtsu H, Bloch MH, Hughes ZA, Krystal JH, Mayes L, de Araujo I, Ding YS, State MW, Pittenger C (2014) Histidine decarboxylase deficiency causes tourette syndrome: parallel findings in humans and mice. Neuron 81:77–90

Chamberlain SR, Menzies L, Hampshire A, Suckling J, Fineberg NA, del Campo N, Aitken M, Craig K, Owen AM, Bullmore ET, Robbins TW, Sahakian BJ (2008) Orbitofrontal dysfunction in patients with obsessive-compulsive disorder and their unaffected relatives. Science 321:421–422

Chaouloff F, Courvoisier H, Moisan MP, Mormede P (1999) GR 127935 reduces basal locomotor activity and prevents RU 24969-, but not D-amphetamine-induced hyperlocomotion, in the Wistar-Kyoto hyperactive (WKHA) rat. Psychopharmacology 141:326–331

Cheetham SC, Heal DJ (1993) Evidence that RU 24969-induced locomotor activity in C57/B1/6 mice is specifically mediated by the 5-HT1B receptor. Br J Pharmacol 110:1621–1629

Chen SK, Tvrdik P, Peden E, Cho S, Wu S, Spangrude G, Capecchi MR (2010) Hematopoietic origin of pathological grooming in Hoxb8 mutant mice. Cell 141:775–785

Chou-Green JM, Holscher TD, Dallman MF, Akana SF (2003) Compulsive behavior in the 5-HT2C receptor knockout mouse. Physiol Behav 78:641–649

Christian CJ, Lencz T, Robinson DG, Burdick KE, Ashtari M, Malhotra AK, Betensky JD, Szeszko PR (2008) Gray matter structural alterations in obsessive-compulsive disorder: relationship to neuropsychological functions. Psychiatry Res 164:123–131

Chudasama Y, Daniels TE, Gorrin DP, Rhodes SE, Rudebeck PH, Murray EA (2013) The role of the anterior cingulate cortex in choices based on reward value and reward contingency. Cereb Cortex 23:2884–2898

Clarke HF, Dalley JW, Crofts HS, Robbins TW, Roberts AC (2004) Cognitive inflexibility after prefrontal serotonin depletion. Science 304:878–880

Clarke HF, Walker SC, Crofts HS, Dalley JW, Robbins TW, Roberts AC (2005) Prefrontal serotonin depletion affects reversal learning but not attentional set shifting. J Neurosci 25:532–538

Clarke HF, Walker SC, Dalley JW, Robbins TW, Roberts AC (2007) Cognitive inflexibility after prefrontal serotonin depletion is behaviorally and neurochemically specific. Cereb Cortex 17:18–27

Corbit VL, Manning EE, Gittis AH, Ahmari SE (2019) Strengthened inputs from secondary motor cortex to striatum in a mouse model of compulsive behavior. J Neurosci 39:2965–2975

Corbit VL, Piantadosi SC, Wood J, Liu G, Choi C, Witten IB, Gittis AH, Ahmari SE (2020) Dissociable roles of central striatum and anterior lateral motor area in initiating and sustaining naturalistic behavior. bioRxiv: 2020.01.08.899070

D'Cruz AM, Ragozzino ME, Mosconi MW, Pavuluri MN, Sweeney JA (2011) Human reversal learning under conditions of certain versus uncertain outcomes. NeuroImage 56:315–322

Dajani DR, Uddin LQ (2015) Demystifying cognitive flexibility: implications for clinical and developmental neuroscience. Trends Neurosci 38:571–578

Dalton GL, Wang NY, Phillips AG, Floresco SB (2016) Multifaceted contributions by different regions of the orbitofrontal and medial prefrontal cortex to probabilistic reversal learning. J Neurosci 36:1996–2006

de Brouwer G, Fick A, Harvey BH, Wolmarans W (2019) A critical inquiry into marble-burying as a preclinical screening paradigm of relevance for anxiety and obsessive-compulsive disorder: mapping the way forward. Cogn Affect Behav Neurosci 19:1–39

de Vries FE, Cath DC, Hoogendoorn AW, van Oppen P, Glas G, Veltman DJ, van den Heuvel OA, van Balkom AJ (2016) Tic-related versus Tic-free obsessive-compulsive disorder: clinical picture and 2-year natural course. J Clin Psychiatry 77:e1240–e1e47

de Wit SJ, de Vries FE, van der Werf YD, Cath DC, Heslenfeld DJ, Veltman EM, van Balkom AJ, Veltman DJ, van den Heuvel OA (2012) Presupplementary motor area hyperactivity during response inhibition: a candidate endophenotype of obsessive-compulsive disorder. Am J Psychiatry 169:1100–1108

de Wolmarans W, Stein DJ, Harvey BH (2016) Of mice and marbles: novel perspectives on burying behavior as a screening test for psychiatric illness. Cogn Affect Behav Neurosci 16:551–560

Deisseroth K (2015) Optogenetics: 10 years of microbial opsins in neuroscience. Nat Neurosci 18:1213–1225

Delgado-Acevedo C, Estay SF, Radke AK, Sengupta A, Escobar AP, Henriquez-Belmar F, Reyes CA, Haro-Acuna V, Utreras E, Sotomayor-Zarate R, Cho A, Wendland JR, Kulkarni AB, Holmes A, Murphy DL, Chavez AE, Moya PR (2019) Behavioral and synaptic alterations relevant to obsessive-compulsive disorder in mice with increased EAAT3 expression. Neuropsychopharmacology 44:1163–1173

Dias R, Robbins TW, Roberts AC (1996) Dissociation in prefrontal cortex of affective and attentional shifts. Nature 380:69–72

Dickel DE, Veenstra-VanderWeele J, Cox NJ, Wu X, Fischer DJ, Van Etten-Lee M, Himle JA, Leventhal BL, Cook EH Jr, Hanna GL (2006) Association testing of the positional and functional candidate gene SLC1A1/EAAC1 in early-onset obsessive-compulsive disorder. Arch Gen Psychiatry 63:778–785

Diehl MM, Bravo-Rivera C, Rodriguez-Romaguera J, Pagan-Rivera PA, Burgos-Robles A, Roman-Ortiz C, Quirk GJ (2018) Active avoidance requires inhibitory signaling in the rodent prelimbic prefrontal cortex. eLife 7

Diehl MM, Bravo-Rivera C, Quirk GJ (2019) The study of active avoidance: a platform for discussion. Neurosci Biobehav Rev 107:229–237

Djodari-Irani A, Klein J, Banzhaf J, Joel D, Heinz A, Harnack D, Lagemann T, Juckel G, Kupsch A, Morgenstern R, Winter C (2011) Activity modulation of the globus pallidus and the nucleus entopeduncularis affects compulsive checking in rats. Behav Brain Res 219:149–158

Drysdale AT, Grosenick L, Downar J, Dunlop K, Mansouri F, Meng Y, Fetcho RN, Zebley B, Oathes DJ, Etkin A, Schatzberg AF, Sudheimer K, Keller J, Mayberg HS, Gunning FM, Alexopoulos GS, Fox MD, Pascual-Leone A, Voss HU, Casey BJ, Dubin MJ, Liston C (2017) Resting-state connectivity biomarkers define neurophysiological subtypes of depression. Nat Med 23:28–38

Dulawa SC, Geyer MA (2000) Effects of strain and serotonergic agents on prepulse inhibition and habituation in mice. Neuropharmacology 39:2170–2179

Dvorkin A, Perreault ML, Szechtman H (2006) Development and temporal organization of compulsive checking induced by repeated injections of the dopamine agonist quinpirole in an animal model of obsessive-compulsive disorder. Behav Brain Res 169:303–311

Dvorkin A, Silva C, McMurran T, Bisnaire L, Foster J, Szechtman H (2010) Features of compulsive checking behavior mediated by nucleus accumbens and orbital frontal cortex. Eur J Neurosci 32:1552–1563

Eagle DM, Noschang C, d'Angelo LS, Noble CA, Day JO, Dongelmans ML, Theobald DE, Mar AC, Urcelay GP, Morein-Zamir S, Robbins TW (2014) The dopamine D2/D3 receptor agonist quinpirole increases checking-like behaviour in an operant observing response task with uncertain reinforcement: a novel possible model of OCD. Behav Brain Res 264:207–229

Ehmer I, Crown L, van Leeuwen W, Feenstra M, Willuhn I, Denys D (2020) Evidence for distinct forms of compulsivity in the SAPAP3 mutant-mouse model for obsessive-compulsive disorder. eNeuro 7

Ercan-Sencicek AG, Stillman AA, Ghosh AK, Bilguvar K, O'Roak BJ, Mason CE, Abbott T, Gupta A, King RA, Pauls DL, Tischfield JA, Heiman GA, Singer HS, Gilbert DL, Hoekstra PJ, Morgan TM, Loring E, Yasuno K, Fernandez T, Sanders S, Louvi A, Cho JH, Mane S, Colangelo CM, Biederer T, Lifton RP, Gunel M, State MW (2010) L-histidine decarboxylase and Tourette's syndrome. N Engl J Med 362:1901–1908

Fellows LK, Farah MJ (2003) Ventromedial frontal cortex mediates affective shifting in humans: evidence from a reversal learning paradigm. Brain 126:1830–1837

Fernandez TV, Sanders SJ, Yurkiewicz IR, Ercan-Sencicek AG, Kim YS, Fishman DO, Raubeson MJ, Song Y, Yasuno K, Ho WS, Bilguvar K, Glessner J, Chu SH, Leckman JF, King RA, Gilbert DL, Heiman GA, Tischfield JA, Hoekstra PJ, Devlin B, Hakonarson H, Mane SM, Gunel M, State MW (2012) Rare copy number variants in tourette syndrome disrupt genes in histaminergic pathways and overlap with autism. Biol Psychiatry 71:392–402

Fullana MA, Zhu X, Alonso P, Cardoner N, Real E, Lopez-Sola C, Segalas C, Subira M, Galfalvy H, Menchon JM, Simpson HB, Marsh R, Soriano-Mas C (2017) Basolateral amygdala-ventromedial prefrontal cortex connectivity predicts cognitive behavioural therapy outcome in adults with obsessive-compulsive disorder. J Psychiatry Neurosci 42:378–385

Geller DA (2006) Obsessive-compulsive and spectrum disorders in children and adolescents. Psychiatr Clin North Am 29:353–370

Geramita MA, Yttri EA, Ahmari SE (2020) The two-step task, avoidance, and OCD. J Neurosci Res 98:1007–1019

Geyer MA, Dulawa SC (2003) Assessment of murine startle reactivity, prepulse inhibition, and habituation. Curr Protoc Neurosci 8:17

Ghahremani DG, Monterosso J, Jentsch JD, Bilder RM, Poldrack RA (2010) Neural components underlying behavioral flexibility in human reversal learning. Cereb Cortex 20:1843–1852

Ghods-Sharifi S, Haluk DM, Floresco SB (2008) Differential effects of inactivation of the orbitofrontal cortex on strategy set-shifting and reversal learning. Neurobiol Learn Mem 89:567–573

Gilbert AR, Mataix-Cols D, Almeida JR, Lawrence N, Nutche J, Diwadkar V, Keshavan MS, Phillips ML (2008) Brain structure and symptom dimension relationships in obsessive-compulsive disorder: a voxel-based morphometry study. J Affect Disord 109:117–126

Gillan CM, Robbins TW (2014) Goal-directed learning and obsessive-compulsive disorder. Philos Trans R Soc Lond B Biol Sci 369

Gillan CM, Papmeyer M, Morein-Zamir S, Sahakian BJ, Fineberg NA, Robbins TW, de Wit S (2011) Disruption in the balance between goal-directed behavior and habit learning in obsessive-compulsive disorder. Am J Psychiatry 168:718–726

Gillan CM, Morein-Zamir S, Urcelay GP, Sule A, Voon V, Apergis-Schoute AM, Fineberg NA, Sahakian BJ, Robbins TW (2014) Enhanced avoidance habits in obsessive-compulsive disorder. Biol Psychiatry 75:631–638

Gillan CM, Apergis-Schoute AM, Morein-Zamir S, Urcelay GP, Sule A, Fineberg NA, Sahakian BJ, Robbins TW (2015) Functional neuroimaging of avoidance habits in obsessive-compulsive disorder. Am J Psychiatry 172:284–293

Gillan CM, Robbins TW, Sahakian BJ, van den Heuvel OA, van Wingen G (2016) The role of habit in compulsivity. Eur Neuropsychopharmacol 26:828–840

Gonzalez LF, Henriquez-Belmar F, Delgado-Acevedo C, Cisternas-Olmedo M, Arriagada G, Sotomayor-Zarate R, Murphy DL, Moya PR (2017) Neurochemical and behavioral characterization of neuronal glutamate transporter EAAT3 heterozygous mice. Biol Res 50:29

Gottlich M, Kramer UM, Kordon A, Hohagen F, Zurowski B (2014) Decreased limbic and increased fronto-parietal connectivity in unmedicated patients with obsessive-compulsive disorder. Hum Brain Mapp 35:5617–5632

Gottlich M, Kramer UM, Kordon A, Hohagen F, Zurowski B (2015) Resting-state connectivity of the amygdala predicts response to cognitive behavioral therapy in obsessive compulsive disorder. Biol Psychol 111:100–109

Gottwald J, de Wit S, Apergis-Schoute AM, Morein-Zamir S, Kaser M, Cormack F, Sule A, Limmer W, Morris AC, Robbins TW, Sahakian BJ (2018) Impaired cognitive plasticity and goal-directed control in adolescent obsessive-compulsive disorder. Psychol Med 48:1900–1908

Grados MA, Mathews CA (2009) Clinical phenomenology and phenotype variability in Tourette syndrome. J Psychosom Res 67:491–496

Graybiel AM, Rauch SL (2000) Toward a neurobiology of obsessive-compulsive disorder. Neuron 28:343–347

Greer JM, Capecchi MR (2002) Hoxb8 is required for normal grooming behavior in mice. Neuron 33:23–34

Gremel CM, Costa RM (2013) Orbitofrontal and striatal circuits dynamically encode the shift between goal-directed and habitual actions. Nat Commun 4:2264

Grimmond S, Larder R, Van Hateren N, Siggers P, Hulsebos TJ, Arkell R, Greenfield A (2000) Cloning, mapping, and expression analysis of a gene encoding a novel mammalian EGF-related protein (SCUBE1). Genomics 70:74–81

Groman SM, Keistler C, Keip AJ, Hammarlund E, DiLeone RJ, Pittenger C, Lee D, Taylor JR (2019) Orbitofrontal circuits control multiple reinforcement-learning processes. Neuron 103:734–746.e3

Grutzmann R, Endrass T, Kaufmann C, Allen E, Eichele T, Kathmann N (2016) Presupplementary motor area contributes to altered error monitoring in obsessive-compulsive disorder. Biol Psychiatry 80:562–571

Gu BM, Park JY, Kang DH, Lee SJ, Yoo SY, Jo HJ, Choi CH, Lee JM, Kwon JS (2008) Neural correlates of cognitive inflexibility during task-switching in obsessive-compulsive disorder. Brain 131:155–164

Gudasheva TA, Tarasiuk AV, Sazonova NM, Povarnina PY, Antipova TA, Seredenin SB (2017) A novel dimeric dipeptide mimetic of the BDNF selectively activates the MAPK-Erk signaling pathway. Dokl Biochem Biophys 476:291–295

Hadjas LC, Luscher C, Simmler LD (2019) Aberrant habit formation in the Sapap3-knockout mouse model of obsessive-compulsive disorder. Sci Rep 9:12061

Hall D, Dhilla A, Charalambous A, Gogos JA, Karayiorgou M (2003) Sequence variants of the brain-derived neurotrophic factor (BDNF) gene are strongly associated with obsessive-compulsive disorder. Am J Hum Genet 73:370–376

Hampson M, Tokoglu F, King RA, Constable RT, Leckman JF (2009) Brain areas coactivating with motor cortex during chronic motor tics and intentional movements. Biol Psychiatry 65:594–599

Harrison BJ, Soriano-Mas C, Pujol J, Ortiz H, Lopez-Sola M, Hernandez-Ribas R, Deus J, Alonso P, Yucel M, Pantelis C, Menchon JM, Cardoner N (2009) Altered corticostriatal functional connectivity in obsessive-compulsive disorder. Arch Gen Psychiatry 66:1189–1200

Hemmings SM, Kinnear CJ, Van der Merwe L, Lochner C, Corfield VA, Moolman-Smook JC, Stein DJ (2008) Investigating the role of the brain-derived neurotrophic factor (BDNF) val66met variant in obsessive-compulsive disorder (OCD). World J Biol Psychiatry 9:126–134

Hervig ME, Fiddian L, Piilgaard L, Bozic T, Blanco-Pozo M, Knudsen C, Olesen SF, Alsio J, Robbins TW (2020) Dissociable and paradoxical roles of rat medial and lateral orbitofrontal cortex in visual serial reversal learning. Cereb Cortex 30:1016–1029

Hesse S, Muller U, Lincke T, Barthel H, Villmann T, Angermeyer MC, Sabri O, Stengler-Wenzke K (2005) Serotonin and dopamine transporter imaging in patients with obsessive-compulsive disorder. Psychiatry Res 140:63–72

Hirschtritt ME, Bloch MH, Mathews CA (2017) Obsessive-compulsive disorder: advances in diagnosis and treatment. JAMA 317:1358–1367

Hoenig K, Hochrein A, Quednow BB, Maier W, Wagner M (2005) Impaired prepulse inhibition of acoustic startle in obsessive-compulsive disorder. Biol Psychiatry 57:1153–1158

Hollander E, DeCaria CM, Nitescu A, Gully R, Suckow RF, Cooper TB, Gorman JM, Klein DF, Liebowitz MR (1992) Serotonergic function in obsessive-compulsive disorder. Behavioral and neuroendocrine responses to oral m-chlorophenylpiperazine and fenfluramine in patients and healthy volunteers. Arch Gen Psychiatry 49:21–28

Holmseth S, Dehnes Y, Huang YH, Follin-Arbelet VV, Grutle NJ, Mylonakou MN, Plachez C, Zhou Y, Furness DN, Bergles DE, Lehre KP, Danbolt NC (2012) The density of EAAC1 (EAAT3) glutamate transporters expressed by neurons in the mammalian CNS. J Neurosci 32:6000–6013

Holstege JC, de Graaff W, Hossaini M, Cardona Cano S, Jaarsma D, van den Akker E, Deschamps J (2008) Loss of Hoxb8 alters spinal dorsal laminae and sensory responses in mice. Proc Natl Acad Sci U S A 105:6338–6343

Hornak J, O'Doherty J, Bramham J, Rolls ET, Morris RG, Bullock PR, Polkey CE (2004) Reward-related reversal learning after surgical excisions in orbito-frontal or dorsolateral prefrontal cortex in humans. J Cogn Neurosci 16:463–478

Hou J, Wu W, Lin Y, Wang J, Zhou D, Guo J, Gu S, He M, Ahmed S, Hu J, Qu W, Li H (2012) Localization of cerebral functional deficits in patients with obsessive-compulsive disorder: a resting-state fMRI study. J Affect Disord 138:313–321

Husted DS, Shapira NA, Goodman WK (2006) The neurocircuitry of obsessive-compulsive disorder and disgust. Prog Neuro-Psychopharmacol Biol Psychiatry 30:389–399

International Obsessive Compulsive Disorder Foundation Genetics Collaborative and O. C. D. Collaborative Genetics Association Studies (2018) Revealing the complex genetic architecture of obsessive-compulsive disorder using meta-analysis. Mol Psychiatry 23:1181–1188

Jimenez-Gomez C, Osentoski A, Woods JH (2011) Pharmacological evaluation of the adequacy of marble burying as an animal model of compulsion and/or anxiety. Behav Pharmacol 22:711–713

Jones B, Mishkin M (1972) Limbic lesions and the problem of stimulus--reinforcement associations. Exp Neurol 36:362–377

Kalanithi PS, Zheng W, Kataoka Y, DiFiglia M, Grantz H, Saper CB, Schwartz ML, Leckman JF, Vaccarino FM (2005) Altered parvalbumin-positive neuron distribution in basal ganglia of individuals with Tourette syndrome. Proc Natl Acad Sci U S A 102:13307–13312

Kalueff AV, Stewart AM, Song C, Berridge KC, Graybiel AM, Fentress JC (2016) Neurobiology of rodent self-grooming and its value for translational neuroscience. Nat Rev Neurosci 17:45–59

Kataoka Y, Kalanithi PS, Grantz H, Schwartz ML, Saper C, Leckman JF, Vaccarino FM (2010) Decreased number of parvalbumin and cholinergic interneurons in the striatum of individuals with Tourette syndrome. J Comp Neurol 518:277–291

Kim JJ, Lee MC, Kim J, Kim IY, Kim SI, Han MH, Chang KH, Kwon JS (2001) Grey matter abnormalities in obsessive-compulsive disorder: statistical parametric mapping of segmented magnetic resonance images. Br J Psychiatry 179:330–334
Kim CK, Adhikari A, Deisseroth K (2017) Integration of optogenetics with complementary methodologies in systems neuroscience. Nat Rev Neurosci 18:222–235
Kimchi EY, Torregrossa MM, Taylor JR, Laubach M (2009) Neuronal correlates of instrumental learning in the dorsal striatum. J Neurophysiol 102:475–489
Koizumi T, Puccetti A, Migliorini P, Barrett KJ, Schwartz RS (1991) Molecular heterogeneity of auto-anti-idiotypic antibodies in MLR-lpr/lpr mice. Eur J Immunol 21:2185–2193
Kringelbach ML, Rolls ET (2003) Neural correlates of rapid reversal learning in a simple model of human social interaction. NeuroImage 20:1371–1383
Kwon JS, Kim JJ, Lee DW, Lee JS, Lee DS, Kim MS, Lyoo IK, Cho MJ, Lee MC (2003) Neural correlates of clinical symptoms and cognitive dysfunctions in obsessive-compulsive disorder. Psychiatry Res 122:37–47
Lagemann T, Rentzsch J, Montag C, Gallinat J, Jockers-Scherubl M, Winter C, Reischies FM (2012) Early orbitofrontal hyperactivation in obsessive-compulsive disorder. Psychiatry Res 202:257–263
Leckman JF, Bloch MH, King RA (2009) Symptom dimensions and subtypes of obsessive-compulsive disorder: a developmental perspective. Dialogues Clin Neurosci 11:21–33
Leckman JF, Denys D, Simpson HB, Mataix-Cols D, Hollander E, Saxena S, Miguel EC, Rauch SL, Goodman WK, Phillips KA, Stein DJ (2010) Obsessive-compulsive disorder: a review of the diagnostic criteria and possible subtypes and dimensional specifiers for DSM-V. Depress Anxiety 27:507–527
LeDoux J, Daw ND (2018) Surviving threats: neural circuit and computational implications of a new taxonomy of defensive behaviour. Nat Rev Neurosci 19:269–282
LeDoux JE, Moscarello J, Sears R, Campese V (2017) The birth, death and resurrection of avoidance: a reconceptualization of a troubled paradigm. Mol Psychiatry 22:24–36
Lee YJ, Koo BH, Seo WS, Kim HG, Kim JY, Cheon EJ (2017) Repetitive transcranial magnetic stimulation of the supplementary motor area in treatment-resistant obsessive-compulsive disorder: an open-label pilot study. J Clin Neurosci 44:264–268
Lennington JB, Coppola G, Kataoka-Sasaki Y, Fernandez TV, Palejev D, Li Y, Huttner A, Pletikos M, Sestan N, Leckman JF, Vaccarino FM (2016) Transcriptome analysis of the human striatum in Tourette syndrome. Biol Psychiatry 79:372–382
Lerner TN, Ye L, Deisseroth K (2016) Communication in neural circuits: tools, opportunities, and challenges. Cell 164:1136–1150
Lynch CJ, Gunning FM, Liston C (2020) Causes and consequences of diagnostic heterogeneity in depression: paths to discovering novel biological depression subtypes. Biol Psychiatry
Maia TV, Cooney RE, Peterson BS (2008) The neural bases of obsessive-compulsive disorder in children and adults. Dev Psychopathol 20:1251–1283
Manning EE, Ahmari SE (2018) How can preclinical mouse models be used to gain insight into prefrontal cortex dysfunction in obsessive-compulsive disorder? Brain Neurosci Adv 2:2398212818783896
Manning EE, Dombrovski AY, Torregrossa MM, Ahmari SE (2019) Impaired instrumental reversal learning is associated with increased medial prefrontal cortex activity in Sapap3 knockout mouse model of compulsive behavior. Neuropsychopharmacology 44:1494–1504
Manning EE, Wang AY, Saikali LM, Winner A, Ahmari SE (2020). Disruption of prepulse inhibition is associated with severity of compulsive behavior and NAc dopamine system changes in SAPAP3-KO mice. bioRxiv: 694935
Mantovani A, Simpson HB, Fallon BA, Rossi S, Lisanby SH (2010) Randomized sham-controlled trial of repetitive transcranial magnetic stimulation in treatment-resistant obsessive-compulsive disorder. Int J Neuropsychopharmacol 13:217–227

Mantovani A, Rossi S, Bassi BD, Simpson HB, Fallon BA, Lisanby SH (2013) Modulation of motor cortex excitability in obsessive-compulsive disorder: an exploratory study on the relations of neurophysiology measures with clinical outcome. Psychiatry Res 210:1026–1032

Marras A, Fineberg N, Pallanti S (2016) Obsessive compulsive and related disorders: comparing DSM-5 and ICD-11. CNS Spectr 21:324–333

Martin JR, Bos M, Jenck F, Moreau J, Mutel V, Sleight AJ, Wichmann J, Andrews JS, Berendsen HH, Broekkamp CL, Ruigt GS, Kohler C, Delft AM (1998) 5-HT2C receptor agonists: pharmacological characteristics and therapeutic potential. J Pharmacol Exp Ther 286:913–924

Martínez-Rivera FJ, Bravo-Rivera C, Velázquez-Díaz CD, Montesinos-Cartagena M, Quirk GJ (2018) Prefrontal circuits signaling active avoidance retrieval and extinction. Psychopharmacology (Berl). https://doi.org/10.1007/s00213-018-5012-7

Mataix-Cols D, Wooderson S, Lawrence N, Brammer MJ, Speckens A, Phillips ML (2004) Distinct neural correlates of washing, checking, and hoarding symptom dimensions in obsessive-compulsive disorder. Arch Gen Psychiatry 61:564–576

Mataix-Cols D, Frost RO, Pertusa A, Clark LA, Saxena S, Leckman JF, Stein DJ, Matsunaga H, Wilhelm S (2010) Hoarding disorder: a new diagnosis for DSM-V? Depress Anxiety 27:556–572

Mattheisen M, Samuels JF, Wang Y, Greenberg BD, Fyer AJ, McCracken JT, Geller DA, Murphy DL, Knowles JA, Grados MA, Riddle MA, Rasmussen SA, McLaughlin NC, Nurmi EL, Askland KD, Qin HD, Cullen BA, Piacentini J, Pauls DL, Bienvenu OJ, Stewart SE, Liang KY, Goes FS, Maher B, Pulver AE, Shugart YY, Valle D, Lange C, Nestadt G (2015) Genome-wide association study in obsessive-compulsive disorder: results from the OCGAS. Mol Psychiatry 20:337–344

McAlonan K, Brown VJ (2003) Orbital prefrontal cortex mediates reversal learning and not attentional set shifting in the rat. Behav Brain Res 146:97–103

McDougle CJ, Goodman WK, Price LH (1994) Dopamine antagonists in tic-related and psychotic spectrum obsessive compulsive disorder. J Clin Psychiatry 55(Suppl):24–31

Meder D, Kolling N, Verhagen L, Wittmann MK, Scholl J, Madsen KH, Hulme OJ, Behrens TEJ, Rushworth MFS (2017) Simultaneous representation of a spectrum of dynamically changing value estimates during decision making. Nat Commun 8:1942

Mei Y, Monteiro P, Zhou Y, Kim JA, Gao X, Fu Z, Feng G (2016) Adult restoration of Shank3 expression rescues selective autistic-like phenotypes. Nature 530:481–484

Menzies L, Chamberlain SR, Laird AR, Thelen SM, Sahakian BJ, Bullmore ET (2008a) Integrating evidence from neuroimaging and neuropsychological studies of obsessive-compulsive disorder: the orbitofronto-striatal model revisited. Neurosci Biobehav Rev 32:525–549

Menzies L, Williams GB, Chamberlain SR, Ooi C, Fineberg N, Suckling J, Sahakian BJ, Robbins TW, Bullmore ET (2008b) White matter abnormalities in patients with obsessive-compulsive disorder and their first-degree relatives. Am J Psychiatry 165:1308–1315

Miller KJ, Botvinick MM, Brody CD (2018) Value representations in the orbitofrontal cortex drive learning, not choice. biorxiv. https://doi.org/10.1101/245720

Mundo E, Richter MA, Sam F, Macciardi F, Kennedy JL (2000) Is the 5-HT(1Dbeta) receptor gene implicated in the pathogenesis of obsessive-compulsive disorder? Am J Psychiatry 157:1160–1161

Nagahama Y, Okada T, Katsumi Y, Hayashi T, Yamauchi H, Oyanagi C, Konishi J, Fukuyama H, Shibasaki H (2001) Dissociable mechanisms of attentional control within the human prefrontal cortex. Cereb Cortex 11:85–92

Nagarajan N, Jones BW, West PJ, Marc RE, Capecchi MR (2018) Corticostriatal circuit defects in Hoxb8 mutant mice. Mol Psychiatry 23:1868–1877

Nahum L, Simon SR, Sander D, Lazeyras F, Schnider A (2011) Neural response to the behaviorally relevant absence of anticipated outcomes and the presentation of potentially harmful stimuli: a human fMRI study. Cortex 47:191–201

Neale BM, Kou Y, Liu L, Ma'ayan A, Samocha KE, Sabo A, Lin CF, Stevens C, Wang LS, Makarov V, Polak P, Yoon S, Maguire J, Crawford EL, Campbell NG, Geller ET, Valladares O,

Schafer C, Liu H, Zhao T, Cai G, Lihm J, Dannenfelser R, Jabado O, Peralta Z, Nagaswamy U, Muzny D, Reid JG, Newsham I, Wu Y, Lewis L, Han Y, Voight BF, Lim E, Rossin E, Kirby A, Flannick J, Fromer M, Shakir K, Fennell T, Garimella K, Banks E, Poplin R, Gabriel S, DePristo M, Wimbish JR, Boone BE, Levy SE, Betancur C, Sunyaev S, Boerwinkle E, Buxbaum JD, Cook EH Jr, Devlin B, Gibbs RA, Roeder K, Schellenberg GD, Sutcliffe JS, Daly MJ (2012) Patterns and rates of exonic de novo mutations in autism spectrum disorders. Nature 485:242–245

Nonogaki K, Strack AM, Dallman MF, Tecott LH (1998) Leptin-independent hyperphagia and type 2 diabetes in mice with a mutated serotonin 5-HT2C receptor gene. Nat Med 4:1152–1156

Noonan MP, Mars RB, Rushworth MF (2011) Distinct roles of three frontal cortical areas in reward-guided behavior. J Neurosci 31:14399–14412

O'Doherty J, Kringelbach ML, Rolls ET, Hornak J, Andrews C (2001) Abstract reward and punishment representations in the human orbitofrontal cortex. Nat Neurosci 4:95–102

O'Roak BJ, Vives L, Fu W, Egertson JD, Stanaway IB, Phelps IG, Carvill G, Kumar A, Lee C, Ankenman K, Munson J, Hiatt JB, Turner EH, Levy R, O'Day DR, Krumm N, Coe BP, Martin BK, Borenstein E, Nickerson DA, Mefford HC, Doherty D, Akey JM, Bernier R, Eichler EE, Shendure J (2012) Multiplex targeted sequencing identifies recurrently mutated genes in autism spectrum disorders. Science 338:1619–1622

Overmier JB, Brackbill RM (1977) On the independence of stimulus evocation of fear and fear evocation of responses. Behav Res Ther 15:51–56

Owen SF, Berke JD, Kreitzer AC (2018) Fast-spiking interneurons supply feedforward control of bursting, calcium, and plasticity for efficient learning. Cell 172:683–695.e15

Pallanti S, Marras A, Salerno L, Makris N, Hollander E (2016) Better than treated as usual: transcranial magnetic stimulation augmentation in selective serotonin reuptake inhibitor-refractory obsessive-compulsive disorder, mini-review and pilot open-label trial. J Psychopharmacol 30:568–578

Papakosta VM, Kalogerakou S, Kontis D, Anyfandi E, Theochari E, Boulougouris V, Papadopoulos S, Panagis G, Tsaltas E (2013) 5-HT2C receptor involvement in the control of persistence in the reinforced spatial alternation animal model of obsessive-compulsive disorder. Behav Brain Res 243:176–183

Paul S, Beucke JC, Kaufmann C, Mersov A, Heinzel S, Kathmann N, Simon D (2019) Amygdala-prefrontal connectivity during appraisal of symptom-related stimuli in obsessive-compulsive disorder. Psychol Med 49:278–286

Pauls DL (2008) The genetics of obsessive compulsive disorder: a review of the evidence. Am J Med Genet C Semin Med Genet 148C:133–139

Pauls DL, Abramovitch A, Rauch SL, Geller DA (2014) Obsessive-compulsive disorder: an integrative genetic and neurobiological perspective. Nat Rev Neurosci 15:410–424

Peca J, Feliciano C, Ting JT, Wang W, Wells MF, Venkatraman TN, Lascola CD, Fu Z, Feng G (2011) Shank3 mutant mice display autistic-like behaviours and striatal dysfunction. Nature 472:437–442

Peghini P, Janzen J, Stoffel W (1997) Glutamate transporter EAAC-1-deficient mice develop dicarboxylic aminoaciduria and behavioral abnormalities but no neurodegeneration. EMBO J 16:3822–3832

Perani D, Garibotto V, Gorini A, Moresco RM, Henin M, Panzacchi A, Matarrese M, Carpinelli A, Bellodi L, Fazio F (2008) In vivo PET study of 5HT(2A) serotonin and D(2) dopamine dysfunction in drug-naive obsessive-compulsive disorder. NeuroImage 42:306–314

Piantadosi SC, Ahmari SE (2015) Using optogenetics to dissect the neural circuits underlying OCD and related disorders. Curr Treat Options Psychiatry 2:297–311

Piantadosi SC, Chamberlain BL, Glausier JR, Lewis DA, Ahmari SE (2019) Lower excitatory synaptic gene expression in orbitofrontal cortex and striatum in an initial study of subjects with obsessive compulsive disorder. Mol Psychiatry

Pico-Perez M, Ipser J, Taylor P, Alonso P, Lopez-Sola C, Real E, Segalas C, Roos A, Menchon JM, Stein DJ, Soriano-Mas C (2019) Intrinsic functional and structural connectivity of emotion regulation networks in obsessive-compulsive disorder. Depress Anxiety 36:110–120

Pinto A, Greenberg BD, Murphy DL, Nestadt G, Rasmussen SA (2009) Using individual items to clarify OCD symptom structure: the case for five factors. Am J Psychiatry 166:728–729; author reply 29-31

Piras F, Piras F, Chiapponi C, Girardi P, Caltagirone C, Spalletta G (2015) Widespread structural brain changes in OCD: a systematic review of voxel-based morphometry studies. Cortex 62:89–108

Pogarell O, Hamann C, Popperl G, Juckel G, Chouker M, Zaudig M, Riedel M, Moller HJ, Hegerl U, Tatsch K (2003) Elevated brain serotonin transporter availability in patients with obsessive-compulsive disorder. Biol Psychiatry 54:1406–1413

Posner J, Marsh R, Maia TV, Peterson BS, Gruber A, Simpson HB (2014) Reduced functional connectivity within the limbic cortico-striato-thalamo-cortical loop in unmedicated adults with obsessive-compulsive disorder. Hum Brain Mapp 35:2852–2860

Pujol J, Soriano-Mas C, Alonso P, Cardoner N, Menchon JM, Deus J, Vallejo J (2004) Mapping structural brain alterations in obsessive-compulsive disorder. Arch Gen Psychiatry 61:720–730

Rajasethupathy P, Ferenczi E, Deisseroth K (2016) Targeting neural circuits. Cell 165:524–534

Rapanelli M, Frick LR, Pogorelov V, Ota KT, Abbasi E, Ohtsu H, Pittenger C (2014) Dysregulated intracellular signaling in the striatum in a pathophysiologically grounded model of Tourette syndrome. Eur Neuropsychopharmacol 24:1896–1906

Rapanelli M, Frick L, Pogorelov V, Ohtsu H, Bito H, Pittenger C (2017) Histamine H3R receptor activation in the dorsal striatum triggers stereotypies in a mouse model of tic disorders. Transl Psychiatry 7:e1013

Rauch SL, Jenike MA, Alpert NM, Baer L, Breiter HC, Savage CR, Fischman AJ (1994) Regional cerebral blood flow measured during symptom provocation in obsessive-compulsive disorder using oxygen 15-labeled carbon dioxide and positron emission tomography. Arch Gen Psychiatry 51:62–70

Rauch SL, Savage CR, Alpert NM, Fischman AJ, Jenike MA (1997) The functional neuroanatomy of anxiety: a study of three disorders using positron emission tomography and symptom provocation. Biol Psychiatry 42:446–452

Reess TJ, Rus OG, Schmidt R, de Reus MA, Zaudig M, Wagner G, Zimmer C, van den Heuvel MP, Koch K (2016) Connectomics-based structural network alterations in obsessive-compulsive disorder. Transl Psychiatry 6:e882

Remijnse PL, Nielen MM, Uylings HB, Veltman DJ (2005) Neural correlates of a reversal learning task with an affectively neutral baseline: an event-related fMRI study. NeuroImage 26:609–618

Remijnse PL, Nielen MM, van Balkom AJ, Cath DC, van Oppen P, Uylings HB, Veltman DJ (2006) Reduced orbitofrontal-striatal activity on a reversal learning task in obsessive-compulsive disorder. Arch Gen Psychiatry 63:1225–1236

Remijnse PL, Nielen MM, van Balkom AJ, Hendriks GJ, Hoogendijk WJ, Uylings HB, Veltman DJ (2009) Differential frontal-striatal and paralimbic activity during reversal learning in major depressive disorder and obsessive-compulsive disorder. Psychol Med 39:1503–1518

Rodriguez MM, Overshiner C, Leander JD, Li X, Morrow D, Conway RG, Nelson DL, Briner K, Witkin JM (2017) Behavioral effects of a novel Benzofuranyl-Piperazine serotonin-2C receptor agonist suggest a potential therapeutic application in the treatment of obsessive-compulsive disorder. Front Psych 8:89

Rodriguez-Romaguera J, Greenberg BD, Rasmussen SA, Quirk GJ (2016) An avoidance-based rodent model of exposure with response prevention therapy for obsessive-compulsive disorder. Biol Psychiatry 80:534–540

Rolls ET, Hornak J, Wade D, McGrath J (1994) Emotion-related learning in patients with social and emotional changes associated with frontal lobe damage. J Neurol Neurosurg Psychiatry 57:1518–1524

Romanelli RJ, Wu FM, Gamba R, Mojtabai R, Segal JB (2014) Behavioral therapy and serotonin reuptake inhibitor pharmacotherapy in the treatment of obsessive-compulsive disorder: a systematic review and meta-analysis of head-to-head randomized controlled trials. Depress Anxiety 31:641–652

Rotge JY, Guehl D, Dilharreguy B, Tignol J, Bioulac B, Allard M, Burbaud P, Aouizerate B (2009) Meta-analysis of brain volume changes in obsessive-compulsive disorder. Biol Psychiatry 65:75–83

Rotge JY, Langbour N, Guehl D, Bioulac B, Jaafari N, Allard M, Aouizerate B, Burbaud P (2010) Gray matter alterations in obsessive-compulsive disorder: an anatomic likelihood estimation meta-analysis. Neuropsychopharmacology 35:686–691

Roth BL (2016) DREADDs for neuroscientists. Neuron 89:683–694

Rudebeck PH, Murray EA (2008) Amygdala and orbitofrontal cortex lesions differentially influence choices during object reversal learning. J Neurosci 28:8338–8343

Rus OG, Reess TJ, Wagner G, Zimmer C, Zaudig M, Koch K (2017) Functional and structural connectivity of the amygdala in obsessive-compulsive disorder. Neuroimage Clin 13:246–255

Rygula R, Walker SC, Clarke HF, Robbins TW, Roberts AC (2010) Differential contributions of the primate ventrolateral prefrontal and orbitofrontal cortex to serial reversal learning. J Neurosci 30:14552–14559

Sanders AR, Duan J, Gejman PV (2002) DNA variation and psychopharmacology of the human serotonin receptor 1B (HTR1B) gene. Pharmacogenomics 3:745–762

Sanders SJ, Murtha MT, Gupta AR, Murdoch JD, Raubeson MJ, Willsey AJ, Ercan-Sencicek AG, DiLullo NM, Parikshak NN, Stein JL, Walker MF, Ober GT, Teran NA, Song Y, El-Fishawy P, Murtha RC, Choi M, Overton JD, Bjornson RD, Carriero NJ, Meyer KA, Bilguvar K, Mane SM, Sestan N, Lifton RP, Gunel M, Roeder K, Geschwind DH, Devlin B, State MW (2012) De novo mutations revealed by whole-exome sequencing are strongly associated with autism. Nature 485:237–241

Saxena S (2008) Recent advances in compulsive hoarding. Curr Psychiatry Rep 10:297–303

Saxena S, Brody AL, Schwartz JM, Baxter LR (1998) Neuroimaging and frontal-subcortical circuitry in obsessive-compulsive disorder. Br J Psychiatry Suppl:26–37

Saxena S, Brody AL, Maidment KM, Dunkin JJ, Colgan M, Alborzian S, Phelps ME, Baxter LR Jr (1999) Localized orbitofrontal and subcortical metabolic changes and predictors of response to paroxetine treatment in obsessive-compulsive disorder. Neuropsychopharmacology 21:683–693

Saxena S, Bota RG, Brody AL (2001) Brain-behavior relationships in obsessive-compulsive disorder. Semin Clin Neuropsychiatry 6:82–101

Schoenbaum G, Nugent SL, Saddoris MP, Setlow B (2002) Orbitofrontal lesions in rats impair reversal but not acquisition of go, no-go odor discriminations. Neuroreport 13:885–890

Scimemi A, Tian H, Diamond JS (2009) Neuronal transporters regulate glutamate clearance, NMDA receptor activation, and synaptic plasticity in the hippocampus. J Neurosci 29:14581–14595

Shanahan NA, Holick Pierz KA, Masten VL, Waeber C, Ansorge M, Gingrich JA, Geyer MA, Hen R, Dulawa SC (2009) Chronic reductions in serotonin transporter function prevent 5-HT1B-induced behavioral effects in mice. Biol Psychiatry 65:401–408

Shanahan NA, Velez LP, Masten VL, Dulawa SC (2011) Essential role for orbitofrontal serotonin 1B receptors in obsessive-compulsive disorder-like behavior and serotonin reuptake inhibitor response in mice. Biol Psychiatry 70:1039–1048

Shapira NA, Liu Y, He AG, Bradley MM, Lessig MC, James GA, Stein DJ, Lang PJ, Goodman WK (2003) Brain activation by disgust-inducing pictures in obsessive-compulsive disorder. Biol Psychiatry 54:751–756

Shmelkov SV, Hormigo A, Jing D, Proenca CC, Bath KG, Milde T, Shmelkov E, Kushner JS, Baljevic M, Dincheva I, Murphy AJ, Valenzuela DM, Gale NW, Yancopoulos GD, Ninan I, Lee FS, Rafii S (2010) Slitrk5 deficiency impairs corticostriatal circuitry and leads to obsessive-compulsive-like behaviors in mice. Nat Med 16:598–602, 1p following 02

Shugart YY, Wang Y, Samuels JF, Grados MA, Greenberg BD, Knowles JA, McCracken JT, Rauch SL, Murphy DL, Rasmussen SA, Cullen B, Hoehn-Saric R, Pinto A, Fyer AJ, Piacentini J, Pauls DL, Bienvenu OJ, Riddle MA, Liang KY, Nestadt G (2009) A family-based association study of the glutamate transporter gene SLC1A1 in obsessive-compulsive disorder in 378 families. Am J Med Genet B Neuropsychiatr Genet 150B:886–892

Silverman JL, Nithianantharajah J, Der-Avakian A, Young JW, Sukoff Rizzo SJ (2020) Lost in translation: at the crossroads of face validity and translational utility of behavioral assays in animal models for the development of therapeutics. Neurosci Biobehav Rev 116:452–453

Simon D, Adler N, Kaufmann C, Kathmann N (2014) Amygdala hyperactivation during symptom provocation in obsessive-compulsive disorder and its modulation by distraction. Neuroimage Clin 4:549–557

Simpson HB (2010) Pharmacological treatment of obsessive-compulsive disorder. Curr Top Behav Neurosci 2:527–543

Sipes TA, Geyer MA (1994) Multiple serotonin receptor subtypes modulate prepulse inhibition of the startle response in rats. Neuropharmacology 33:441–448

Smith KS, Graybiel AM (2013) A dual operator view of habitual behavior reflecting cortical and striatal dynamics. Neuron 79:361–374

Smith KS, Virkud A, Deisseroth K, Graybiel AM (2012) Reversible online control of habitual behavior by optogenetic perturbation of medial prefrontal cortex. Proc Natl Acad Sci U S A 109:18932–18937

Song M, Giza J, Proenca CC, Jing D, Elliott M, Dincheva I, Shmelkov SV, Kim J, Schreiner R, Huang SH, Castren E, Prekeris R, Hempstead BL, Chao MV, Dictenberg JB, Rafii S, Chen ZY, Rodriguez-Boulan E, Lee FS (2015) Slitrk5 mediates BDNF-dependent TrkB receptor trafficking and signaling. Dev Cell 33:690–702

Song M, Mathews CA, Stewart SE, Shmelkov SV, Mezey JG, Rodriguez-Flores JL, Rasmussen SA, Britton JC, Oh YS, Walkup JT, Lee FS, Glatt CE (2017) Rare synaptogenesis-impairing mutations in SLITRK5 are associated with obsessive compulsive disorder. PLoS One 12: e0169994

Stein DJ, Arya M, Pietrini P, Rapoport JL, Swedo SE (2006) Neurocircuitry of disgust and anxiety in obsessive-compulsive disorder: a positron emission tomography study. Metab Brain Dis 21:267–277

Stein DJ, Fineberg NA, Bienvenu OJ, Denys D, Lochner C, Nestadt G, Leckman JF, Rauch SL, Phillips KA (2010) Should OCD be classified as an anxiety disorder in DSM-V? Depress Anxiety 27:495–506

Steinman SA, Ahmari SE, Choo T, Kimeldorf MB, Feit R, Loh S, Risbrough V, Geyer MA, Steinglass JE, Wall M, Schneier FR, Fyer AJ, Simpson HB (2016) Prepulse inhibition deficits only in females with obsessive-compulsive disorder. Depress Anxiety 33:238–246

Stern E, Silbersweig DA, Chee KY, Holmes A, Robertson MM, Trimble M, Frith CD, Frackowiak RS, Dolan RJ (2000) A functional neuroanatomy of tics in Tourette syndrome. Arch Gen Psychiatry 57:741–748

Stern ER, Welsh RC, Gonzalez R, Fitzgerald KD, Abelson JL, Taylor SF (2013) Subjective uncertainty and limbic hyperactivation in obsessive-compulsive disorder. Hum Brain Mapp 34:1956–1970

Sternson SM, Roth BL (2014) Chemogenetic tools to interrogate brain functions. Annu Rev Neurosci 37:387–407

Stewart SE, Geller DA, Jenike M, Pauls D, Shaw D, Mullin B, Faraone SV (2004) Long-term outcome of pediatric obsessive-compulsive disorder: a meta-analysis and qualitative review of the literature. Acta Psychiatr Scand 110:4–13

Stewart SE, Mayerfeld C, Arnold PD, Crane JR, O'Dushlaine C, Fagerness JA, Yu D, Scharf JM, Chan E, Kassam F, Moya PR, Wendland JR, Delorme R, Richter MA, Kennedy JL, Veenstra-VanderWeele J, Samuels J, Greenberg BD, McCracken JT, Knowles JA, Fyer AJ, Rauch SL, Riddle MA, Grados MA, Bienvenu OJ, Cullen B, Wang Y, Shugart YY, Piacentini J, Rasmussen S, Nestadt G, Murphy DL, Jenike MA, Cook EH, Pauls DL, Hanna GL, Mathews

CA (2013) Meta-analysis of association between obsessive-compulsive disorder and the 3′ region of neuronal glutamate transporter gene SLC1A1. Am J Med Genet B Neuropsychiatr Genet 162B:367–379

Subira M, Cano M, de Wit SJ, Alonso P, Cardoner N, Hoexter MQ, Kwon JS, Nakamae T, Lochner C, Sato JR, Jung WH, Narumoto J, Stein DJ, Pujol J, Mataix-Cols D, Veltman DJ, Menchon JM, van den Heuvel OA, Soriano-Mas C, O. C. D. Brain Imaging Consortium (2016) Structural covariance of neostriatal and limbic regions in patients with obsessive-compulsive disorder. J Psychiatry Neurosci 41:115–123

Sun T, Song Z, Tian Y, Tian W, Zhu C, Ji G, Luo Y, Chen S, Wang L, Mao Y, Xie W, Zhong H, Zhao F, Luo MH, Tao W, Wang H, Li J, Li J, Zhou J, Wang K, Zhang Z (2019) Basolateral amygdala input to the medial prefrontal cortex controls obsessive-compulsive disorder-like checking behavior. Proc Natl Acad Sci U S A 116:3799–3804

Swedo SE, Leonard HL, Garvey M, Mittleman B, Allen AJ, Perlmutter S, Lougee L, Dow S, Zamkoff J, Dubbert BK (1998) Pediatric autoimmune neuropsychiatric disorders associated with streptococcal infections: clinical description of the first 50 cases. Am J Psychiatry 155:264–271

Swerdlow NR, Benbow CH, Zisook S, Geyer MA, Braff DL (1993) A preliminary assessment of sensorimotor gating in patients with obsessive compulsive disorder. Biol Psychiatry 33:298–301

Swerdlow NR, Geyer MA, Braff DL (2001) Neural circuit regulation of prepulse inhibition of startle in the rat: current knowledge and future challenges. Psychopharmacology 156:194–215

Szechtman H, Sulis W, Eilam D (1998) Quinpirole induces compulsive checking behavior in rats: a potential animal model of obsessive-compulsive disorder (OCD). Behav Neurosci 112:1475–1485

Tanaka KF, Ahmari SE, Leonardo ED, Richardson-Jones JW, Budreck EC, Scheiffele P, Sugio S, Inamura N, Ikenaka K, Hen R (2010) Flexible accelerated STOP tetracycline operator-knockin (FAST): a versatile and efficient new gene modulating system. Biol Psychiatry 67:770–773

Tang C, Pawlak AP, Prokopenko V, West MO (2007) Changes in activity of the striatum during formation of a motor habit. Eur J Neurosci 25:1212–1227

Taylor S (2011) Etiology of obsessions and compulsions: a meta-analysis and narrative review of twin studies. Clin Psychol Rev 31:1361–1372

Taylor S (2013) Molecular genetics of obsessive-compulsive disorder: a comprehensive meta-analysis of genetic association studies. Mol Psychiatry 18:799–805

Tecott LH, Sun LM, Akana SF, Strack AM, Lowenstein DH, Dallman MF, Julius D (1995) Eating disorder and epilepsy in mice lacking 5-HT2c serotonin receptors. Nature 374:542–546

Tezcan D, Tumkaya S, Bora E (2017) Reversal learning in patients with obsessive-compulsive disorder (OCD) and their unaffected relatives: is orbitofrontal dysfunction an endophenotype of OCD? Psychiatry Res 252:231–233

Thorsen AL, Hagland P, Radua J, Mataix-Cols D, Kvale G, Hansen B, van den Heuvel OA (2018a) Emotional processing in obsessive-compulsive disorder: a systematic review and meta-analysis of 25 functional neuroimaging studies. Biol Psychiatry Cogn Neurosci Neuroimaging 3:563–571

Thorsen AL, Kvale G, Hansen B, van den Heuvel OA (2018b) Symptom dimensions in obsessive-compulsive disorder as predictors of neurobiology and treatment response. Curr Treat Options Psychiatry 5:182–194

Thorsen AL, de Wit SJ, de Vries FE, Cath DC, Veltman DJ, van der Werf YD, Mataix-Cols D, Hansen B, Kvale G, van den Heuvel OA (2019) Emotion regulation in obsessive-compulsive disorder, unaffected siblings, and unrelated healthy control participants. Biol Psychiatry Cogn Neurosci Neuroimaging 4:352–360

Trainor PA, Krumlauf R (2000) Patterning the cranial neural crest: hindbrain segmentation and Hox gene plasticity. Nat Rev Neurosci 1:116–124

Trankner D, Boulet A, Peden E, Focht R, Van Deren D, Capecchi M (2019) A microglia sublineage protects from sex-linked anxiety symptoms and obsessive compulsion. Cell Rep 29:791–799.e3

Tsaltas E, Kontis D, Chrysikakou S, Giannou H, Biba A, Pallidi S, Christodoulou A, Maillis A, Rabavilas A (2005) Reinforced spatial alternation as an animal model of obsessive-compulsive disorder (OCD): investigation of 5-HT2C and 5-HT1D receptor involvement in OCD pathophysiology. Biol Psychiatry 57:1176–1185

Tye KM, Deisseroth K (2012) Optogenetic investigation of neural circuits underlying brain disease in animal models. Nat Rev Neurosci 13:251–266

Ullrich M, Bundschu K, Benz PM, Abesser M, Freudinger R, Fischer T, Ullrich J, Renne T, Walter U, Schuh K (2011) Identification of SPRED2 (sprouty-related protein with EVH1 domain 2) as a negative regulator of the hypothalamic-pituitary-adrenal axis. J Biol Chem 286:9477–9488

Ullrich M, Weber M, Post AM, Popp S, Grein J, Zechner M, Guerrero Gonzalez H, Kreis A, Schmitt AG, Uceyler N, Lesch KP, Schuh K (2018) OCD-like behavior is caused by dysfunction of thalamo-amygdala circuits and upregulated TrkB/ERK-MAPK signaling as a result of SPRED2 deficiency. Mol Psychiatry 23:444–458

Vaghi MM, Hampshire A, Fineberg NA, Kaser M, Bruhl AB, Sahakian BJ, Chamberlain SR, Robbins TW (2017) Hypoactivation and dysconnectivity of a Frontostriatal circuit during goal-directed planning as an endophenotype for obsessive-compulsive disorder. Biol Psychiatry Cogn Neurosci Neuroimaging 2:655–663

Valerius G, Lumpp A, Kuelz AK, Freyer T, Voderholzer U (2008) Reversal learning as a neuropsychological indicator for the neuropathology of obsessive compulsive disorder? A behavioral study. J Neuropsychiatry Clin Neurosci 20:210–218

van den Boom BJG, Mooij AH, Miseviciute I, Denys D, Willuhn I (2019) Behavioral flexibility in a mouse model for obsessive-compulsive disorder: impaired Pavlovian reversal learning in SAPAP3 mutants. Genes Brain Behav 18:e12557

van den Heuvel OA, Veltman DJ, Groenewegen HJ, Dolan RJ, Cath DC, Boellaard R, Mesina CT, van Balkom AJ, van Oppen P, Witter MP, Lammertsma AA, van Dyck R (2004) Amygdala activity in obsessive-compulsive disorder with contamination fear: a study with oxygen-15 water positron emission tomography. Psychiatry Res 132:225–237

van den Heuvel OA, Remijnse PL, Mataix-Cols D, Vrenken H, Groenewegen HJ, Uylings HB, van Balkom AJ, Veltman DJ (2009) The major symptom dimensions of obsessive-compulsive disorder are mediated by partially distinct neural systems. Brain 132:853–868

van der Wee NJ, Stevens H, Hardeman JA, Mandl RC, Denys DA, van Megen HJ, Kahn RS, Westenberg HM (2004) Enhanced dopamine transporter density in psychotropic-naive patients with obsessive-compulsive disorder shown by [123I]{beta}-CIT SPECT. Am J Psychiatry 161:2201–2206

Veenstra-VanderWeele J, Kim SJ, Gonen D, Hanna GL, Leventhal BL, Cook EH Jr (2001) Genomic organization of the SLC1A1/EAAC1 gene and mutation screening in early-onset obsessive-compulsive disorder. Mol Psychiatry 6:160–167

Vickers SP, Clifton PG, Dourish CT, Tecott LH (1999) Reduced satiating effect of d-fenfluramine in serotonin 5-HT(2C) receptor mutant mice. Psychopharmacology 143:309–314

Wan Y, Ade KK, Caffall Z, Ilcim Ozlu M, Eroglu C, Feng G, Calakos N (2014) Circuit-selective striatal synaptic dysfunction in the Sapap3 knockout mouse model of obsessive-compulsive disorder. Biol Psychiatry 75:623–630

Wang W, Li C, Chen Q, van der Goes MS, Hawrot J, Yao AY, Gao X, Lu C, Zang Y, Zhang Q, Lyman K, Wang D, Guo B, Wu S, Gerfen CR, Fu Z, Feng G (2017) Striatopallidal dysfunction underlies repetitive behavior in Shank3-deficient model of autism. J Clin Invest 127:1978–1990

Welch JM, Lu J, Rodriguiz RM, Trotta NC, Peca J, Ding JD, Feliciano C, Chen M, Adams JP, Luo J, Dudek SM, Weinberg RJ, Calakos N, Wetsel WC, Feng G (2007) Cortico-striatal synaptic defects and OCD-like behaviours in Sapap3-mutant mice. Nature 448:894–900

Wendland JR, Moya PR, Timpano KR, Anavitarte AP, Kruse MR, Wheaton MG, Ren-Patterson RF, Murphy DL (2009) A haplotype containing quantitative trait loci for SLC1A1 gene expression and its association with obsessive-compulsive disorder. Arch Gen Psychiatry 66:408–416

Wood J, Ahmari SE (2015) A framework for understanding the emerging role of Corticolimbic-ventral striatal networks in OCD-associated repetitive behaviors. Front Syst Neurosci 9:171
Xu M, Kobets A, Du JC, Lennington J, Li L, Banasr M, Duman RS, Vaccarino FM, DiLeone RJ, Pittenger C (2015) Targeted ablation of cholinergic interneurons in the dorsolateral striatum produces behavioral manifestations of Tourette syndrome. Proc Natl Acad Sci U S A 112:893–898
Xu M, Li L, Pittenger C (2016) Ablation of fast-spiking interneurons in the dorsal striatum, recapitulating abnormalities seen post-mortem in Tourette syndrome, produces anxiety and elevated grooming. Neuroscience 324:321–329
Yadin E, Friedman E, Bridger WH (1991) Spontaneous alternation behavior: an animal model for obsessive-compulsive disorder? Pharmacol Biochem Behav 40:311–315
Yang RB, Ng CK, Wasserman SM, Colman SD, Shenoy S, Mehraban F, Komuves LG, Tomlinson JE, Topper JN (2002) Identification of a novel family of cell-surface proteins expressed in human vascular endothelium. J Biol Chem 277:46364–46373
Yim YS, Kwon Y, Nam J, Yoon HI, Lee K, Kim DG, Kim E, Kim CH, Ko J (2013) Slitrks control excitatory and inhibitory synapse formation with LAR receptor protein tyrosine phosphatases. Proc Natl Acad Sci U S A 110:4057–4062
Yin HH, Knowlton BJ, Balleine BW (2004) Lesions of dorsolateral striatum preserve outcome expectancy but disrupt habit formation in instrumental learning. Eur J Neurosci 19:181–189
Yin HH, Ostlund SB, Knowlton BJ, Balleine BW (2005) The role of the dorsomedial striatum in instrumental conditioning. Eur J Neurosci 22:513–523
Yin HH, Knowlton BJ, Balleine BW (2006) Inactivation of dorsolateral striatum enhances sensitivity to changes in the action-outcome contingency in instrumental conditioning. Behav Brain Res 166:189–196
Yucel M, Harrison BJ, Wood SJ, Fornito A, Wellard RM, Pujol J, Clarke K, Phillips ML, Kyrios M, Velakoulis D, Pantelis C (2007) Functional and biochemical alterations of the medial frontal cortex in obsessive-compulsive disorder. Arch Gen Psychiatry 64:946–955
Yun JY, Boedhoe PSW, Vriend C, Jahanshad N, Abe Y, Ameis SH, Anticevic A, Arnold PD, Batistuzzo MC, Benedetti F, Beucke JC, Bollettini I, Bose A, Brem S, Calvo A, Cheng Y, Cho KIK, Ciullo V, Dallaspezia S, Denys D, Feusner JD, Fouche JP, Gimenez M, Gruner P, Hibar DP, Hoexter MQ, Hu H, Huyser C, Ikari K, Kathmann N, Kaufmann C, Koch K, Lazaro L, Lochner C, Marques P, Marsh R, Martinez-Zalacain I, Mataix-Cols D, Menchon JM, Minuzzi L, Morgado P, Moreira P, Nakamae T, Nakao T, Narayanaswamy JC, Nurmi EL, O'Neill J, Piacentini J, Piras F, Piras F, Reddy YCJ, Sato JR, Simpson HB, Soreni N, Soriano-Mas C, Spalletta G, Stevens MC, Szeszko PR, Tolin DF, Venkatasubramanian G, Walitza S, Wang Z, van Wingen GA, Xu J, Xu X, Zhao Q, Thompson PM, Stein DJ, van den Heuvel OA, Kwon JS, Enigma-Ocd Working Group (2020) Brain structural covariance networks in obsessive-compulsive disorder: a graph analysis from the ENIGMA consortium. Brain 143:684–700
Zike ID, Chohan MO, Kopelman JM, Krasnow EN, Flicker D, Nautiyal KM, Bubser M, Kellendonk C, Jones CK, Stanwood G, Tanaka KF, Moore H, Ahmari SE, Veenstra-VanderWeele J (2017) OCD candidate gene SLC1A1/EAAT3 impacts basal ganglia-mediated activity and stereotypic behavior. Proc Natl Acad Sci U S A 114:5719–5724
Zitterl W, Aigner M, Stompe T, Zitterl-Eglseer K, Gutierrez-Lobos K, Schmidl-Mohl B, Wenzel T, Demal U, Zettinig G, Hornik K, Thau K (2007) [123I]-beta-CIT SPECT imaging shows reduced thalamus-hypothalamus serotonin transporter availability in 24 drug-free obsessive-compulsive checkers. Neuropsychopharmacology 32:1661–1668
Zohar J, Insel TR (1987) Obsessive-compulsive disorder: psychobiological approaches to diagnosis, treatment, and pathophysiology. Biol Psychiatry 22:667–687

Neurocognitive Endophenotypes of OCD

Matilde M. Vaghi

Contents

Abstract In this chapter, I address the concept of endophenotypes for obsessive-compulsive disorder (OCD). Endophenotypes are objective and heritable quantitative traits hypothesized to be more biologically tractable than distal clinical phenotypes. This approach has been adopted to gain a better understanding of psychiatric conditions in general. It is theorized that endophenotypes will particularly assist in clarifying both the diagnostic status and aetiological origins of complex neuropsychiatric conditions such as OCD. At the cognitive level, separable constructs of relevance for OCD have been identified. The prevailing model for OCD assumes the development of abnormalities within fronto-striatal neural circuits leading to impairment of executive functions and their neuropsychological subcomponents. Here, I address whether this model can guide towards the identification of endophenotypes for this condition and discuss possible implications.

Keywords Endophenotype · Executive functions · Obsessive-compulsive disorder

M. M. Vaghi (✉)
Max Planck UCL Centre for Computational Psychiatry and Ageing Research, London, UK

Curr Topics Behav Neurosci (2021) 49: 97–124
https://doi.org/10.1007/7854_2020_197
Published Online: 10 March 2021

1 The Rationale for Endophenotypes

Obsessive-compulsive disorder (OCD) is a complex, heterogeneous neuropsychiatric disorder with multiple genetic, epigenetic, and environmental factors contributing to its development.

Currently, diagnosis for OCD is based on clusters of symptoms. This aligns with the nosological tradition in psychiatry whereby diagnosis and characterization of psychiatric disorders are based on expert consensus. However, unlike the rest of medicine, psychiatry has little or no objective diagnostic tests and it still misses a firm biological footing to support clinical diagnosis. In other words, characterization of psychiatric diseases is almost exclusively informed by clinical symptoms and diagnostic schemes that are not supported or validated by biological markers.

This approach is formalized by the major international diagnostic classifications, such as The Diagnostic and Statistical Manual of Mental Disorders (DSM). Some of these concerns have been addressed in the most recent version of the manual. Specifically, for OCD, changes have been introduced in line with advances made in neurobiology and cognitive neuroscience. Accordingly, OCD has been removed from the anxiety disorders section to its own category of 'obsessive-compulsive and related disorders', reflecting the consensus that anxiety is not the defining feature of OCD (Fineberg et al. 2011b). However, the current approach draws lines between disorder in a way that lacks biological specificity. Indeed, the finding of different degrees of response to pharmacological or psychological therapies among groups of patients sharing the diagnosis of OCD suggests a considerable degree of within-diagnosis heterogeneity (Mataix-Cols et al. 1999), reflecting different aetiologies not captured by the current diagnostic system.

One major obstacle for understanding the biological basis of disorders such as OCD is the issue of comorbidity, i.e. the co-occurrence of more than one psychiatric diagnosis, which is a very common situation. For example, in the case of OCD, approximately half of the patients diagnosed with OCD also meet criteria for comorbid anxiety or mood disorder (Hofmeijer-Sevink et al. 2012; Pinto et al. 2006). In a more recent study, about three quarters of individuals with OCD in an epidemiological sample experienced either an anxiety or an affective disorder (Fineberg et al. 2013). Therefore, multiple diagnoses typically co-exist conferring a significant amount of additional heterogeneity among patients.

A second problem of the current diagnostic system is that the symptomatology manifested by different patients within the same clinical category varies a great deal – resulting in the possibility that two patients with the same diagnosis might often 'look' very different. In some extreme cases, two individuals might have no overlapping symptoms whatsoever, but still receive the same diagnosis. This is particularly true, for example, for depression where several hundred unique combinations of changes in mood, appetite, sleep, energy, cognition, and motor activity are possible (Fried and Nesse 2015). In the case of OCD, the presence of obsessions and compulsions is a pathognomonic diagnostic feature that is almost always shared across patients. However, the content of the obsessions (e.g. contamination, aggressive or taboo themes) and the form of the compulsions (e.g. cleaning, checking, praying) can vary enormously from person to person.

Some authors have addressed the variety of phenomenological manifestations (Bloch et al. 2008) by using factor analysis of large datasets of patients' symptoms in order to attempt to identify more homogeneous subtypes. Accordingly, patients have been clustered based on their presenting symptoms, producing a multidimensional model of OCD that includes separate groups of patients manifesting symmetry/ordering, hoarding, contamination/cleaning, and obsessions/checking symptoms (van den Heuvel et al. 2009; Mataix-Cols et al. 2005). The model has proven helpful in distinguishing hoarding as a relatively specific phenotype, showing a consistently poorer response to treatment compared to other forms of OCD (Cullen et al. 2007; Mataix-Cols et al. 1999). These results have also impacted clinical diagnosis as in the most recent version of the DSM, (DSM – 5), hoarding symptoms no longer exclusively contribute to the diagnosis of OCD and are additionally captured by a separate diagnosis named 'hoarding disorder'. The multidimensional model is also supported by evidence from twin studies, suggesting a genetic influence on the manifestation of the various OCD subtypes (Iervolino et al. 2011). Although the model represents a step forward in categorizing the heterogeneity of OCD, the neurobiological validity of these factors has been difficult to demonstrate in studies so far, mostly due to a lack of overlap in the identified neural substrates (Alvarenga et al. 2012; Harrison et al. 2013; van den Heuvel et al. 2009). This does not invalidate the dimensional model for OCD, but speaks against a clear-cut underlying neurobiological correlate.

Finally, the phenotype is far removed from the pathophysiology and aetiology, severely limiting progress towards the understanding of the underlying diathesis as well as progress towards animal models of the condition.

In the attempt to overcome these shortcomings, the U.S. National Institute of Mental Health launched the Research Domain Criteria initiative in 2009 (Insel et al. 2010). This research framework aims at 'integrating many levels of information (from genomics to circuits and behaviour) in order to explore basic dimensions of functioning that span the full range of human behaviour from normal to abnormal' (https://www.nimh.nih.gov/research-priorities/rdoc/index.shtml).The ambition is to establish new and biologically relevant trans-diagnostic traits that may play a role in multiple disorders, as we currently define them (Robbins et al. 2012), as a basis for more effective intervention.

Endophenotypes might represent a valuable tool to contribute to this enterprise, as they are thought to represent defined and quantifiable measures, which compared to diagnosis are closer to the genetic underpinning of the disorder. In their seminal paper, Gottesman and Shields (1973) introduced the endophenotype concept into psychiatry. They borrowed this term from insect biology to allude to a phenotype that is internal and microscopic and therefore not detectable by the naked eye (Gottesman and Shields 1973). Since then, psychiatry has used the endophenotype concept to refer to the notion of an internal process.

Endophenotypes can include neurophysiological, biochemical, endocrinological, neuropsychological, cognitive, and neuroimaging phenotypes as well as metabolic markers. Overall, they represent intermediate measures of the disease, lying somewhere between the 'distal' phenotype and the 'proximal' genotype. An important feature of endophenotypes, as classically defined (Gottesman and Shields 1973), is that they are state-independent markers and therefore autonomous from the clinical symptoms. Accordingly, it is required that endophenotypes should be heritable, co-segregate with a psychiatric illness, yet be present when the disease is not, and found in non-affected family members at a higher rate than in the healthy population. However, more recently, to take into account the importance of epigenetic and developmental changes (Visscher et al. 2008), the criterion of state independence has been modified so that an endophenotype can be present only at a certain age or after a challenge (Hasler et al. 2006). In summary, an endophenotype is a heritable quantitative trait associated with increased genetic risk for a disorder.

Endophenotypes achieved prominence in research when it became clear that genes and the environment combine to confer susceptibility to the development of diseases and that classification according to standard clinical criteria (phenotypes) alone might impede genetic dissection (Gottesman and Gould 2003). The premise is that endophenotypes more closely correspond to a given gene or genetic profile than loosely defined diagnostic categories. That is, whereas there exists a range of divergent and convergent correspondences between genes and behaviours, identifying the mediating endophenotype will help to clarify these correspondences.

The underlying hypothesis is that endophenotypes might represent specialized and putatively more elementary phenomena compared to the behavioural manifestation of the disease. In fact, endophenotypes are postulated to involve fewer genes and ultimately more defined neuronal circuits, compared to clinical diagnoses defined by a range of manifest symptoms. Therefore, by studying the underlying quantitative traits it is possible to investigate aspects more directly related to biology. Biological traits are expected to be closer to the genetic substrate, with enhanced penetrance and the real possibility of observing these traits in generally 'at risk' but unaffected individuals. Even if it has been suggested that it is unlikely that endophenotypes are related to a simple genetic architecture (see Flint and Munafò (2007) for further discussion on these topics), they still might represent a more reliable marker than the phenotype.

It is conceivable that some of these endophenotypes could have trans-diagnostic validity. In other words, they can be common to different nosographical entities (Robbins et al. 2012). For example, disorders on the compulsivity spectrum such as OCD, eating disorders, or addiction might share similar underlying cognitive endophenotypes in spite of different diagnoses or behavioural manifestations (Robbins et al. 2012). Accordingly, the identification of endophenotypes might represent a viable tool to enable the early detection of individuals at risk of developing an illness – or a spectrum of illnesses – before it becomes clinically apparent. Focusing on measurable traits also represents a viable strategy for the development of animal models, useful for examining the aetiological factors of these associated phenotypes allowing the direct investigation of disease processes. Finally, identifying endophenotypes may advance our understanding of disorder-related aetiological factors, leading to the development of novel and more effective treatments. In summary, the endophenotype approach is used to identify cognitive or biological traits of clinical significance for the manifestation of the disorder that can be expected to further our understanding of the genetics and biology behind the disorder.

Even though the true relationship between genes, endophenotypes, and psychiatric manifestations is likely to be difficult to unfold, several models have been proposed (Kendler and Neale 2010). The simplest ones are the liability-index model and the mediational model. The former assumes that a common set of genes independently codes for the endophenotype and the associated psychiatric disorder. Accordingly, changes in the endophenotype or psychiatric disorder do not necessarily result in changes in the other component. In this framework, for example, we might observe changes in the symptomatology upon treatment yet the endophenotype remains relatively unchanged. In contrast, in the mediational model, any variation in the psychiatric manifestation is associated and potentially determined by variation in the endophenotype. Therefore, a trait that improves after treatment might be accounted for by this model.

2 Genetics of OCD

Family studies are aimed at assessing whether a disease or trait 'runs in families'. They rely on directed structured interview to obtain information from individuals to detect the presence or absence of familial aggregation for the disease or trait. The general conclusion from this kind of study is that rates of OCD are significantly greater in close relatives, who have five-fold higher risk on average of being affected with the same disease (Grabe et al. 2006; Nestadt et al. 2000; Pauls et al. 1995). There is evidence to suggest that familiarity is also linked to specific symptom dimensions, as family members are more likely to have the same types of obsessions and compulsions as the proband. In other words, members of the same family tend to have obsessions or compulsions centred around the same dimension, such as ordering or checking (Alsobrook II et al. 1999; Miguel et al. 2005). While controlled family studies can inform on whether OCD aggregates in families, they are unable to disentangle genetic from environmental factors contributing to the observed familiarity. For example, the presence of symptoms might be due to shared genetics but also to factors involving observational 'vicarious' learning.

In contrast, 'twin studies' comparing concordance rates among monozygotic versus dizygotic twin pairs, or those studying twin pairs reared together versus apart, can theoretically address this limitation, as the relative contribution that the environment makes to the presence of the disorder can be evaluated (Kendler et al. 1993). Evidence from twin studies also supports a genetic aetiology for OCD (Mataix-Cols et al. 2013). A comprehensive review of twin research of OCD has shown that, in children, obsessive-compulsive symptoms are highly heritable with genetic influences in the range of 45% to 65%. Slightly lower genetic influence is estimated from studies of adults with OCD, with heritability ranging between 27% and 47% (van Grootheest et al. 2005). While the number of twin studies is insufficient to allow for accurate heritability estimates, population-based twin studies estimate the heritability of obsessive-compulsive symptoms to be 40–50% with a similar contribution from shared environmental factors (Hudziak et al. 2004; Jonnal et al. 2000; Taylor 2011). These results suggest that both genetic predisposition and environmental factors contribute to the emergence of OCD. In this book (chapter "Genetics of OCD and Related Disorders; Searching for Shared Factors") a more extensive analysis has been conducted investigating heritability of OCD and related disorders.

Due to evidence of heritability, candidate gene studies have been conducted looking for risk genes. There is evidence to suggest a role for genes related, for example, to serotonin and dopamine neurotransmission, as detailed in chapter "Genetics of OCD and Related Disorders; Searching for Shared Factors". Genetic linkage studies have identified several chromosomal regions with suggestive evidence of linkage with OCD (Hanna et al. 2002, 2007; Mathews et al. 2012; Ross et al. 2011; Shugart et al. 2006). However, there is little overlap between the studies, which is probably related to unsuitably low sample sizes for investigating highly polygenic traits. Nevertheless, a number of early linkage studies identified a particular gene (SLCL1A1) on chromosome 9, implicated in regulating the glutamate uptake transporter (Arnold et al. 2006; Dickel et al. 2006). More recent genome-wide association study (GWAS) techniques allow the gathering of additional information on common genes with small effects or copy-number variants. So far, only two GWAS of OCD have been conducted (Mattheisen et al. 2015; Stewart et al. 2013b) by independent consortia. In the first study, no single nucleotide polymorphism (SNP) was found to be associated with OCD at a genome-wide significance level. However, in the second study, the highest association was detected for a SNP in proximity to the protein tyrosine phosphate receptor gene, promoting the differentiation of glutamatergic synapses (see also chapter "Genetics of OCD and Related Disorders; Searching for Shared Factors"). A combined meta-analysis identified several other glutamatergic system genes to be associated with OCD (GRID2, DLAP1) (Arnold et al. 2018). Cumulatively, these results identify genes implicated in the regulation of glutamate neurotransmission in the origins of OCD (Arnold et al. 2004; Stewart et al. 2013a).

3 The Theoretical Model for Defining Endophenotypes for OCD

Neuroimaging and neuropsychological studies have produced a remarkable consensus in relation to the understanding of the neural basis of OCD. Initially, a rapid growth in the number of imaging studies in patients with OCD placed emphasis on dysfunctions within the orbitofronto-striato-thalamic circuits, which rapidly became the dominant model for OCD (Graybiel and Rauch 2000; Saxena et al. 1998). This model has been later revised to integrate several lines of new evidence, which included whole-brain analyses of structural and functional neuroimaging data as well as neuropsychological studies (Menzies et al. 2008). Hence the model has been updated to account for abnormalities within, inter alia, the parietal and the frontal regions as well. Since then, a broader 'cortico-striato-thalamo-cortical (CSTC)' model of OCD has prevailed.

There is evidence that the brain areas showing structural or functional abnormalities in OCD are implicated in the performance of tasks comprising the so-called executive functions (for example (Cole et al. 2013)) – an umbrella term generally used to refer to cognitive functions regulating behaviour (Miyake et al. 2000). Despite occasional inconsistency, possibly due to the use of different test instruments or variation across OCD patient groups, different meta-analyses have confirmed that patients with OCD show deficits in neuropsychological domains related to executive functions (Abramovitch et al. 2013; Shin et al. 2014; Snyder et al. 2015). Individuals with OCD exhibited significantly impaired performance, with a medium effect size, on tasks measuring most aspects of executive functions, such as planning, response inhibition, set-shifting, and working memory. Interestingly, it has also been shown that, in patients with OCD, the extent of some of the deficits in executive function, such as cognitive flexibility and goal-directed planning, relates directly to the degree of dysfunctional connectivity within distinct fronto-striatal circuits at rest (Vaghi et al. 2017b).

A recent meta-analysis of paediatric neuropsychological studies in OCD identified a degree of overlap in cognitive impairment with adulthood OCD. Accordingly, although the results are limited by the scarcity of primary data in paediatric OCD, there is evidence that some cognitive dysfunctions are already manifest in children, while others might appear later in the disorder trajectory. For example, paediatric OCD has been associated with intolerance of uncertainty as well as impaired planning ability and hyperactivity of the fronto-parietal network during working memory tasks. However, other domains such as cognitive flexibility or motor inhibition appear to be relatively spared in the early manifestations of the disease and may only manifest later (Marzuki et al. 2020).

A plausible hypothesis is that obsessions and compulsions stem from an altered capacity in performing various executive functions, which results in insufficient cognitive behavioural control over maladaptive behaviours (Chamberlain et al. 2005). However, it is not agreed whether such executive impairments are trait- or state-dependent. In a longitudinal study, executive functions were tested before and after symptomatic remission. During the symptomatic phase, executive functions such as set-shifting and inhibition were impaired compared to an anxiety control group. At follow-up, when the symptoms had remitted, performance on these measures remained unchanged, suggesting they represented trait phenomena (Bannon et al. 2006). Other studies have also shown that executive deficits remain even in those whose OCD symptoms have remitted (Rao et al. 2008; Sharma et al. 2014), though some have not (Bolton et al. 2000; Kahn et al. 1984). Therefore, there is some evidence to suggest that the cognitive deficits found in OCD are not merely a side effect of the core psychiatric symptoms or medication, and might represent valid endophenotypes.

It is known from twin studies that some of the classic executive functions, such as behavioural inhibition, updating, and shifting, are heritable (Friedman et al. 2008). A quantitative meta-analysis has recently been conducted to investigate whether there is reliable evidence supporting deficits in executive functioning in unaffected first-degree relatives of patients with OCD, with the aim of corroborating the hypothesis that there is a genetic basis to these deficits. The meta-analysis included a variety of studies where first-degree unaffected relatives of patients were tested on a set of executive function tasks. The results are consistent with a small (Hedge's $g = 0.25$), but significant, overall impairment in executive function among the unaffected relatives of people with OCD (Zartaloudi et al. 2019). These results refer to an overall deficit, obtained by pooling standardized difference across different executive function subdomains. The authors also investigated the role of potential moderator factors. Relatively greater deficits were identified in those of older age. In contrast, the impact of gender and severity of symptoms was negligible. The latter analysis however should be taken with caution as information about symptom severity was missing in most of the primary studies, probably because first-degree unaffected relatives are generally symptom-free.

Overall, considering the meta-analyses conducted in patients with OCD or their first-degree unaffected relatives, the emerging picture suggests a mild to moderate deficit in some of the key executive functions of patients with OCD, such as planning, response inhibition, and cognitive flexibility (Table 1). Moreover, impairment in executive functioning extends to unaffected first-degree relatives, albeit to a smaller extent, suggesting a likely genetic contribution to these deficits. Looking more specifically at the different subdomains of executive functions that are affected, at least 5 plausible endophenotypes have been identified; motor inhibition, cognitive flexibility, planning (and goal-directed behaviour), working memory, and error monitoring. They are all relevant to the dominant CTSC theory, which proposes OCD to be characterized by poor executive control. It is, therefore, hypothesized that OCD is associated with dysfunction within neural circuits implicated in these specific cognitive processes and that these deficits are at least in part inherited and thus occur to some extent in first-degree unaffected relatives as well.

4 Motor Inhibition

Impulsivity is a multifaceted construct. Traditionally, impulsive actions have been defined as 'poorly conceived, prematurely expressed, unduly risky, or inappropriate for the situation and that often result in undesirable consequences' (Daruna and Barnes 1993). The inability of patients with OCD to stop repetitive compulsive rituals once they have been initiated has been hypothesized to be linked to deficits in motor inhibition (Chamberlain et al. 2005). In the context of OCD, the construct of motor inhibition also bears relevance to the implementation of goal-directed actions, which is partly dependent on the successful inhibition of competing behaviours.

Table 1 Summary of the candidate cognitive endophenotypes of OCD discussed in this chapter

Construct	Cognitive assay	Proposed neural substrate	Impaired in OCD patients	Presence in unaffected relatives	Specificity for OCD	Relationship to symptoms
Motor inhibition	– Go/No-Go – Stop Signal-Task – Antisaccade Errors	– Cingulo-opercular network – Supplementary motor area – Parietal regions	– **Yes** Chamberlain et al. (2006), Penadés et al. (2007), Lipszyc and Schachar (2010), Abramovitch et al. (2013), Sohn et al. (2014), and Lei et al. (2015) – **No** Kalanthroff et al. (2017), Bohne et al. (2008), Fan et al. (2018)	– **Yes** Chamberlain et al. (2007), Rajender et al. (2011), Lennertz et al. (2012)	– **OCD > Panic disorder** Bannon et al. (2002) – **OCD > Eating disorder** Boisseau et al. (2012) – **OCD < trichotillomania** Chamberlain et al. (2006)	– **No** Bannon et al. (2002), Kalanthroff et al. (2017)
Cognitive flexibility	– Wisconsin card sorting test – Intra-/extra-dimensional set-shifting – Reversal learning	– Ventrolateral prefrontal cortex - Orbitofrontal cortex	– **Yes** Hymas et al. (1991), Okasha et al. (2000), Chamberlain et al. (2006), Abramovitch et al. (2013), Shin et al. (2014), Snyder et al. (2015), Vaghi et al. (2017a, b) – **No** Abbruzzese et al. (1995, 1997), Watkins et al. (2005), Moritz et al. (2009), Gillan et al, (2020)	– **Yes** Remijnse et al. (2006), Chamberlain et al. (2007), Chamberlain et al. (2008), Viswanath et al. (2009), Cavedini et al. (2010), Rajender et al. (2011), Ozcan et al. (2016)	– **OCD > trichotillomania** Chamberlain et al. (2006) – **OCD > depressed** patients Remijnse et al. (2006)	– **No** Bannon et al. (2006), Chamberlain et al. (2006), Rao et al. (2008), Vaghi et al. (2017a, b)
Goal-directed planning	– Tower of London – Tower of Hanoi – Delis–	– Dorsolateral prefrontal cortex – Parietal network	– **Yes** Abramovitch et al. (2013), Shin et al. (2014), Abramovitch and	– **Yes** Delorme et al. (2007), Cavedini et al. (2010), Zhang et al. (2015a, b), Vaghi et al. (2017a, b), Bey		– **No** Bey et al. (2018)

(continued)

Table 1 (continued)

Construct	Cognitive assay	Proposed neural substrate	Impaired in OCD patients	Presence in unaffected relatives	Specificity for OCD	Relationship to symptoms
	Kaplan Executive function System Tower test – Stockings of Cambridge		Cooperman (2015), Vaghi et al. (2017a, b)	et al. (2018) – **No** Rajender et al. (2011), Lennertz et al. (2012)		
Working memory	– N-back working memory task	Fronto-parietal network	– **Yes** Purcell et al. (1998), Savage et al. (1999), Van der Wee at al., (2003), Chamberlain et al. (2007), and Lennertz et al. (2012)	– **Yes** Zartaloudi et al. (2019) – **No** Viswanath et al. (2009)		
Error monitoring	– Forced choice reaction time task	– Anterior cingulate cortex	– **Yes** Gehring et al. (2000), Johannes et al. (2001), Ruchsow et al. (2005), Hajcak et al. (2008), Stern et al. (2010), Xiao et al. (2011), Riesel et al. (2011, 2015), Hanna et al. (2012), Carrasco et al. (2013), Klawohn et al. (2014), Liu et al. (2014), Grützmann et al. (2016), Grützmann et al. 2007 – **No** Nieuwenhuis et al. (2005), Agam et al. (2014), Weinberg et al. (2015), Mathews et al. (2016)	– **Yes** Carrasco et al. (2013), Riesel et al. (2011)		– **Yes** Gehring et al. (2000) – **No** Hajcak et al. (2008), Stern et al. (2010), Endrass and Ullsperger et al. (2014), Riesel et al. (2014, 2015), Huyser et al. (2011)

See text for more detailed discussion

This hypothesis was supported by a series of studies yielding significantly impaired inhibition processes in OCD patients compared to healthy subjects using different experimental procedures to assess inhibitory control. In the laboratory, several experimental tasks have been developed to measure individual variability in the ability to refrain from inappropriate, impulsive motor responses. The two most widely used paradigms are the Go/No-Go and the Stop-Signal Task (Logan et al. 1984). In both paradigms, subjects are required to withhold a motor response to a visual stimulus when a No-Go or a Stop-Signal is presented, respectively. More precisely, on a random selection of trials the signal indicating to refrain from the motor response is delivered with a variable delay from the target stimulus. In the case of the Go/No-Go task, an index of inhibitory control is obtained by measuring the probability of executing a response on a no-go trial. A more sophisticated approach is used in the Stop-Signal Task, where estimation of the time needed to withhold a response is obtained via staircase functions (see recent guidelines from (Verbruggen et al. 2019)). According to a recent meta-analysis (Cieslik et al. 2015), successful motor inhibition requires an intact cingulo-opercular network with additional involvement of the supplementary motor area and the inferior frontal gyrus and parietal regions.

Poorer motor inhibition has been found in patients with OCD compared to various clinical control groups including patients with panic disorder (Bannon et al. 2002) or eating disorder (Boisseau et al. 2012). Interestingly, the extent of the deficit was not related to clinical symptomatology (Bannon et al. 2002), or medication regime (Bannon et al. 2002; Kalanthroff et al. 2017), suggesting that motor inhibition might represent a state-independent trait. Even if some degree of diagnostic specificity has been shown, for example, in the studies mentioned above, these impairments are not exclusive to OCD (Lipszyc and Schachar 2010). For instance, patients with trichotillomania are more severely impaired in terms of motor inhibition compared to patients with OCD (Chamberlain et al. 2006). Nevertheless, patients with OCD do show deficits compared to controls (Chamberlain et al. 2006), regardless of symptom dimensions (Lei et al. 2015), hence suggesting a possible core deficit. Poorer motor inhibition in patients with OCD compared to healthy volunteers has been independently replicated both in a relatively limited sample (Penadés et al. 2007) and a larger one (Sohn et al. 2014). While it should be noted that a few studies also failed to demonstrate behavioural evidence of deficient motor inhibition either using the Go/No-Go (Bohne et al. 2008) or the Stop-Signal Task (Fan et al. 2018; Kalanthroff et al. 2017), two meta-analyses provide convergent evidence for a motor inhibition deficit in OCD (Abramovitch et al. 2013; Lipszyc and Schachar 2010), with a relatively small effect size for commission errors, indexing worse motor inhibition, on the Go/No-Go or the Stop-Signal Task (Abramovitch et al. 2013).

The fact that the extent of the impairment is independent of global symptom severity and clinical dimension (Lei et al. 2015) supports the hypothesis that motor inhibition might represent a candidate endophenotype for OCD. Accordingly, there is additional evidence of behavioural and brain abnormalities linked to motor inhibition in first-degree unaffected relatives of OCD patients. Direct testing of this hypothesis has shown that the behavioural impairment is observed not only in diagnosed patients, but also in their first-degree unaffected relatives (Chamberlain et al. 2007; Rajender et al. 2011). The hypothesis that motor inhibition might represent a candidate endophenotype is also supported by studies investigating this question with different experimental tasks. For example, the anti-saccade task requires inhibitory processes to suppress the reflexive saccade towards a peripheral stimulus (Hutton and Ettinger 2006). Both patients with OCD and their first-degree unaffected relatives show more anti-saccade errors compared to healthy control subjects (Lennertz et al. 2012).

Importantly, the extent of the impairment in motor inhibition in patients and their first-degree unaffected relatives is associated with similar structural changes in cortico-striatal regions. Specifically, there is reduced grey matter volume in orbitofrontal and right inferior frontal regions and increased grey matter volume in cingulate, parietal, and striatal regions (Menzies et al. 2007) that correlates with the magnitude of the motor inhibition deficit. In a separate study, monitoring functional brain activity during the Stop-Signal Task, patients with OCD and their siblings showed increased activity in the left pre-supplementary motor area during successful motor inhibition relative to comparison subjects (de Wit et al. 2012). This finding has been interpreted as a possible compensatory mechanism generated by the associated OCD-related behavioural impairment (de Wit et al. 2012). A follow-up analysis of the same dataset addressed functional connectivity and showed abnormal negative coupling in patients between the inferior frontal gyrus and amygdala during successful inhibition (van Velzen et al. 2015). Similar results were evident, but to a lesser degree, in first-degree relatives of patients compared with control participants (van Velzen et al. 2015). Reduced functional connectivity during the Stop-Signal Task between frontal and posterior brain regions, including the frontal cortex, occipital cortex, and cerebellum, was also found more recently in patients with OCD and their unaffected relatives (Hampshire et al. 2019). The findings, therefore, collectively support neurobiological models for OCD implicating the loss of inhibitory control of motor acts, mediated not only by frontal cortical regions but also by distant posterior brain regions.

A variant of the Stop-Signal Task has been used to test the hypothesis that deficient response inhibition could be extended to non-motor aspects of the disorder (i.e. recurrent thoughts) (Morein-Zamir et al. 2010). However, only evidence of cognitive rigidity, rather than inhibition failure in suppressing cognitive responses, was found (Morein-Zamir et al. 2010) and no investigation in this direction was pursued in first-degree unaffected relatives.

5 Cognitive Flexibility

Cognitive flexibility was originally hypothesized to be relevant for the understanding of OCD by reference to the perseverative and repetitive behaviours that characterize the disorder (Chamberlain et al. 2005). According to this model, ‘getting stuck' in repetitive thoughts and behaviours that are no longer relevant or needed may reflect a more general difficulty in flexibly shifting attention. One recognized form of cognitive flexibility is indexed by 'set-shifting' abilities, measuring the ability to switch attention away from a stimulus dimension (e.g. colour) to another (e.g. shape) when contingencies change. The classic test for this form of cognitive flexibility is the Wisconsin Card Sorting Test (Berg 1948). Over the course of this test, the subject is presented with a series of cards with stimuli which differ in colour, shape, and number. During administration of the task, the rule for a correct response changes. The test assesses the subject's ability to dynamically switch attention to different properties of the stimulus under this contingency, as a measure of cognitive flexibility.

By using this classic test, studies have produced mixed results, with some reporting cognitive flexibility deficits in patients with OCD (Hymas et al. 1991; Okasha et al. 2000), while others have not (Abbruzzese et al. 1995, 1997; Gillan et al. 2020; Moritz et al. 2009). The task has been refined conceptually in the form of the Intra-Dimensional/Extra-Dimensional (IED) Set-Shifting Task. In this revised form, the presence of different stages allows us to disentangle the ability to generalize a learnt rule (intra-dimensional stage) from the ability to shift attention to a new dimension (extra-dimensional stage). The fact that the task, as part of a battery of tests (CANTAB) (Robbins et al. 1998), may be administered via a computerized device, in a controlled fashion, represents another advance. Additionally, the tests are standardized and mostly amenable to cross-species administration and comparison (Dias et al. 1996; Hampshire et al. 2012). The IED task has been used in several studies investigating cognitive flexibility abilities of OCD patients, with evidence of a selective deficits in patients at the extra-dimensional stage, i.e. the stage specifically measuring cognitive flexibility (Watkins et al. 2005). This type of deficit distinguishes patients with OCD from those affected by trichotillomania who, instead, show intact cognitive flexibility (Chamberlain et al. 2006). In addition, cognitive flexibility impairment in OCD is independent of medication status (Chamberlain et al. 2006; Vaghi et al. 2017b).

On brain imaging, the extent of the cognitive flexibility deficit indexed by the IED task appears to be related to reduced functional connectivity between areas of the striatum and the lateral prefrontal cortex (Vaghi et al. 2017b). The fact that this circuit is relevant for cognitive flexibility has been supported by translational work across marmosets and humans (Dias et al. 1996; Hampshire et al. 2012). More generally, several meta-analyses have recently provided a valuable synthesis of the various studies in the literature using different types of neuropsychological task to assess cognitive flexibility. Overall, there is support for a pattern of cognitive inflexibility in OCD, with variable effect sizes generally in the medium range (Abramovitch et al. 2013; Shin et al. 2014; Snyder et al. 2015). A more focused analysis, which only included studies that made use of the computerized IED Set-Shifting Task, is presented in this volume (chapter "Cognitive Inflexibility in OCD and Related Disorders") and also supports the notion of impaired attentional set-shifting in OCD patients.

Evidence that cognitive flexibility represents a candidate endophenotype can be derived from different tiers of evidence. In a study involving only controls and unaffected relatives of patients with OCD, it was shown that the latter have poorer cognitive flexibility (Viswanath et al. 2009). Studies that included not only unaffected relatives of OCD patients and controls but also patients show that both patients and first-degree unaffected relatives have poorer attentional set-shifting abilities than controls (Cavedini et al. 2010; Chamberlain et al. 2007; Ozcan et al. 2016; Rajender et al. 2011). Finally, convergent evidence has been produced showing that impaired cognitive flexibility is independent of OCD symptom severity and detectable even in remitted patients (Bannon et al. 2006; Rao et al. 2008) (Table 1).

Cognitive inflexibility has been found, in patients with OCD, to extend to the ability to adapt behaviour after negative feedback, classically measured with either the reversal learning task or the delayed alternation task. At a behavioural level, patients and their relatives show deficits when compared to controls, even though the deficit is expressed to a milder extent in first-degree unaffected relatives (Tezcan et al. 2017). Behavioural impairment accompanied by decreased activation in the lateral OFC as well as the right caudate nucleus is found in OCD patients (Remijnse et al. 2006). A separate study showed that this neural phenotype was unique to OCD when compared to patients affected by depression (Remijnse et al. 2009). Further work has found that these aberrancies extend to unaffected relatives as well.

Both patients with OCD and their unaffected relatives also displayed hypo-activation of the lateral OFC during a reversal learning task (Chamberlain et al. 2008). In this last study, the authors were not able to show a behavioural deficit in unaffected relatives nor in patients. However, the negative finding might be due to the relative small sample size and to pre-training on the task before the neuroimaging experiment.

6 Goal-Directed Planning

Another domain of executive function that has been extensively studied in OCD is goal-directed planning. A series of neuropsychological tasks has been developed to probe planning abilities in a way that tests subjects' ability to select the correct sub-goal by anticipating the consequence of one course of action on another. The general structure of these tasks involves moving beads or disks from peg to peg, with the aim of achieving the 'goal' of matching the pattern to a predetermined model using the lowest number of moves. Goal-directed planning engages a large-scale brain network most prominently involving the dorsolateral prefrontal cortex as well as the parietal cortex (Baker et al. 1996). The network recruited overlaps substantially with the fronto-striatal circuitry implicated in OCD, according to prevailing neurobiological model (Menzies et al. 2008; Saxena and Rauch 2000), providing the rationale for a possible impairment of these capacities in OCD.

The following tasks have been used to probe planning: Tower of London (Shallice 1982), Tower of Hanoi (Hofstadter 1985), Delis–Kaplan Executive function System Tower test (Delis et al. 2004), and Stockings of Cambridge (Owen et al. 1995). Results from two quantitative meta-analyses (Abramovitch et al. 2013; Shin et al. 2014) as well as a qualitative review (Abramovitch and Cooperman 2015) show that patients with OCD exhibit deficits in these goal-directed planning functions, mostly assessed with the Tower of London task (Shallice 1982), with moderate to large effect sizes. At the neural level, patients with OCD also exhibited hypoactivation of fronto-striatal regions including the dorsolateral prefrontal cortex and the caudate nucleus in a functional magnetic resonance version of the Tower of London task (van den Heuvel et al. 2005).

A number of studies have assessed planning functions in siblings of OCD patients, with relatively mixed results. Delorme and colleagues (Delorme et al. 2007) only included in their study first-degree unaffected relatives of OCD patients and control subjects. They found that unaffected first-degree relatives showed reduced accuracy and longer response times compared to healthy controls. This finding has been independently replicated by a separate study that extended the comparison to patients, unaffected relatives, and healthy controls (Cavedini et al. 2010). A recent replication in a large sample of OCD patients (N = 102) and their first-degree unaffected relatives ($N = 76$) showed that both groups achieved significantly fewer goals within the optimal number of moves than control subjects (Bey et al. 2018). These findings were also independent of medication or symptom severity.

However, other studies have failed to identify impairment in goal-directed abilities in first-degree unaffected relatives of OCD patients. For example, two studies have found deficient performance in patients with OCD but not in unaffected relatives compared to control subjects (Lennertz et al. 2012; Rajender et al. 2011). It is possible that the impairment in first-degree relatives is subtle and difficult to capture. Consistent with this hypothesis, Zhang and colleagues showed that the performance of first-degree relatives was intermediate with, and not significantly different from, either that of patients or controls (Zhang et al. 2015a), in line with an 'endophenotype' pattern. The relatives' intermediate values are in fact consistent with the hypothesis of susceptibility for the disorder but not full-blown expression of it. In another study, increased response times in both OCD patients and relatives were found but only with respect to the most challenging aspects of the task, while no significant group differences were observed for simpler levels of the task (Zhang et al. 2015a, b).

Importantly, most of these studies involved a small number of subjects and consequently might not have been adequately powered to identify relevant differences in the first-degree relatives group. Replication using bigger samples would offer the opportunity to clarify this issue. Additionally, it might be that equivalent performance with healthy controls on the task (i.e. accuracy) is nevertheless underpinned by subtle differences either behaviourally or at the level of neural implementation that the standard task measures do not readily capture. For example, although behavioural accuracy was unimpaired, differences in response times were identified in patients and their relatives compared to controls in another study employing a fMRI version of the Tower of London task, i.e. both OCD subjects and their relatives took longer to solve each problem. Furthermore, both the patients and first-degree unaffected relatives demonstrated a similar pattern of hypo-activation in the dorsolateral prefrontal cortex (Vaghi et al. 2017a, b), specifically during goal-directed planning, providing evidence for goal-directed planning being an endophenotype. That goal-directed planning might represent a core problem is in line with recent computational analysis of OCD suggesting that OCD patients have a difficulty predicting how events unfold as a consequence of their own actions (Fradkin et al. 2020).

7 Working Memory

Working memory represents another cognitive domain that depends on fronto-parietal circuitry (Owen et al. 2005), which has been shown by some studies to be affected in OCD (Purcell et al. 1998; Savage et al. 1999; van der Wee et al. 2003). Working memory is a cognitive function referring to the ability to hold information in mind over a short period of time for immediate use. A classic executive working memory task is the n-back task, measuring a subject's ability to retain spatial information and to manipulate remembered items of information. Participants are presented with a series of items in sequence on the computer screen and have to decide whether each item matches the one that appeared a variable number of items (i.e. n-items) ago. The difficulty of the task can be parametrically varied by asking the subject to compare the current item with an increasing number of trials ago (i.e. 2, 3, or 4 trials ago). However, the n-back task also requires the capacity to inhibit irrelevant responses due to the high degree of interference it imposes.

Behavioural studies provide limited evidence for impairment in spatial working memory in OCD patients (Chamberlain et al. 2007; Purcell et al. 1998). There is also only limited evidence available of altered processing of working memory at the neural level, though abnormalities seem to be present. For example, in a fMRI study (van der Wee et al. 2003) patients with OCD showed poor performance at the highest level of the n-back task, which was associated with elevated activity in the anterior cingulate cortex.

With respect to first-degree unaffected relatives, impairment is generally found across different studies broadly assessing visuospatial working memory (Zartaloudi et al. 2019). However, this domain of working memory has been relatively underexplored as a possible endophenotype. Behaviourally, the performance of siblings of patients with OCD was comparable to that of controls in a battery of neuropsychological tests that assessed auditory working memory, spatial working memory as well as non-verbal working memory (Viswanath et al. 2009). Another study included controls, patients, and first-degree relatives; although the performance of first-degree relatives of patients with OCD was impaired compared to controls, first-degree relatives had intermediate values which were not statistically significant difference from those of controls nor patients, spatial working memory or verbal memory (Lennertz et al. 2012). One study has so far investigated the neural correlates of working memory as a possible endophenotype for OCD. Using a classic n-back task, both patients with OCD and their siblings showed task related hyper-activity in the fronto-parietal network compared with healthy participants (de Vries et al. 2014).

8 Error Monitoring

Central to the pathophysiology of OCD are feelings of incompleteness and doubt generating a need for correction, completion, or behavioural repetition. One hypothesis is that a dysfunctional action-monitoring system in OCD might explain the subjective feeling (egodystonia) that the outcomes of actions are in conflict with the internal representations of those actions and goals (Pitman 1987). This 'cybernetic' model assumes a mismatch between the perceptual input and the internal reference signal, resulting in a deficit in the feedback associated with the successful performance of a goal-directed behaviour. Deficient feedback in turn leads to compulsive responding, as the individual repeats the behaviour in an attempt to gain the missing feedback. The resulting heightened error processing, generating exaggerated feelings that 'something is wrong', has been proposed as a key mechanism triggering hyper corrective OCD behaviours. Based on these theoretical propositions, a 'signal attenuation' animal model of OCD has been developed (Joel 2006). In this model, the goal-directed behaviour consists of instrumental lever pressing for food. The feedback for a successful response is a compound signal of a light and a tone. The model attenuates the signalling properties of the compound stimulus, which is assumed to mirror the feedback deficit underlying compulsive behaviours. Interestingly, in the model, instrumental lever pressing develops a perseverative quality, which is normalized by drugs effective in OCD (Joel et al. 2004) as well as by high frequency stimulation of the sub-thalamic nucleus (Klavir et al. 2009). Even though the model has been validated via multiple avenues, it is nevertheless amenable to alternative theoretical explanations as the signal attenuation is implemented as a special form of extinction (Fineberg et al. 2011a).

In humans, action monitoring can be studied by examining the electroencephalogram (EEG), isolating the event-related potential component known as error-related negativity (ERN). The ERN appears as a negative deflecting peak in the waveform between 50 and 100 ms following the commission of an error during a forced-choice reaction time task (Falkenstein et al. 1991; Gehring et al. 2016). The neural generator of the ERN is identified at the level of the anterior cingulate cortex, which is implicated in the processing of conflict and the generation of error signals (Bush and Posner 2000; Ullsperger et al. 2014). A finding further supported by coupling functional imaging to the EEG, with evidence for the anterior cingulate cortex as the source of the ERN (Debener et al. 2005).

Several studies have found an enhanced ERN in OCD (Carrasco et al. 2013; Endrass et al. 2008; Endrass et al. 2014; Gehring et al. 2000; Grützmann et al. 2016; Hajcak et al. 2008; Hanna et al. 2012; Johannes et al. 2001; Klawohn et al. 2014; Liu et al. 2014; Riesel et al. 2015; Riesel et al. 2011; Ruchsow et al. 2005, 2007; Stern et al. 2010; Xiao et al. 2011) but see (Agam et al. 2014; Mathews et al. 2016; Nieuwenhuis et al. 2005; Weinberg et al. 2015). Similarly the ERN has been investigated in other disorders as further described in the chapter "Electroencepha lographic Correlates of Obsessive-Compulsive Disorder", which also addresses the theories developed to explain the functional significance of ERN.

Importantly, an increased ERN has been repeatedly found in patients with OCD across all symptom dimensions and appears unrelated to symptom severity (Endrass and Ullsperger 2014; Riesel et al. 2014). Evidence of abnormal error processing is also supported by functional magnetic imaging studies of patients with OCD that point to enhanced error and conflict related activity reflected by the anterior cingulate cortex (Fitzgerald et al. 2005; Ursu et al. 2003) (See chapter "Electroencephalo graphic Correlates of Obsessive-Compulsive Disorder).

There is increasing evidence that the ERN might represent a candidate endophenotype for OCD. Studies in both paediatric (Carrasco et al. 2013) and adult (Riesel et al. 2011) populations have shown an increased ERN in affected patients and their first-degree relatives. Further evidence pointing to the ERN as a candidate endophenotype comes from studies investigating the stability of this neural marker over time and its state independence. Studies in paediatric and adult OCD patients show that an enhanced ERN persisted even after symptom reduction following therapy (Hajcak et al. 2008; Huyser et al. 2011; Riesel et al. 2015). Additionally, in the adult study there were no ERN differences between treatment responder and non-responder patients (Riesel et al. 2015) and no correlation between symptom changes and variations in ERN amplitude.

The magnitude of the ERN has shown a heritability of about 50% (Anokhin et al. 2008), pointing to a plausible genetic contribution. However, increased error monitoring is not unique to OCD and characterizes other psychiatric disorders such as depression, social anxiety, and generalized anxiety disorder that are often comorbid with OCD (Endrass et al. 2014; Endrass and Ullsperger 2014; Riesel et al. 2017; Weinberg et al. 2012). This observation suggests that overactive performance monitoring may constitute an endophenotype that crosses the diagnostic border of OCD and represents an intermediate marker of a meaningful dimension of psychological dysfunction that exists regardless of symptom categories (Gillan et al. 2017; Riesel et al. 2019). Alternatively, it may represent a marker of affective or anxiety-related comorbidity.

A recent study attempted to investigate this further by recording EEGs from a large sample of healthy participants. The aim was to assess the magnitude of the ERN, generated by the performance of a 'Flanker' task, – a task that is recognized to produce a large number of erroneous responses and generally used to investigate response inhibition and relate this to self-reported psychiatric symptoms, collected via questionnaires. The results of this study failed to identify an association between the ERN and symptoms of OCD, trait anxiety, depression, social anxiety, impulsivity, eating disorders, alcohol addiction, schizotypy, or apathy within this healthy population group. However, the authors suggested that the negative result might be ascribed to the unreliability of the relatively mild symptoms reported by the healthy sample (Seow et al. 2019). Investigation of one or more clinical samples using similar methodology could help clarify if the ERN is valuable for understanding OCD. Specifically, further exploration of ERN integrity across various relevant disorders meeting the diagnostic threshold and their unaffected relatives may help determine the 'trans-diagnostic' boundaries of the dimension of psychological dysfunction to which this endophenotype relates.

9 Conclusions and Future Directions

In patients with OCD, clinical manifestations of obsessive ruminations and compulsive rituals are accompanied by general cognitive propensities of rigid, disinhibited behaviour, and executive dysfunction. Interestingly, these cognitive difficulties and executive dysfunctions are shared to some extent by first-degree unaffected relatives, supporting the hypothesis that they represent candidate endophenotypes for the disorder. Of the domains investigated here, motor inhibition, visuospatial working memory, and goal-directed planning are also identified as candidate endophenotype via a quantitative meta-analysis (Zartaloudi et al. 2019). These results are in line with theories implicating failure of top-down cortical inhibition.

In future research, the similarities between patients and first-degree unaffected relatives might be leveraged to identify homogeneous underlying neural mechanisms, in a way that is not confounded, for example, by the role of medication. However, it remains to be clarified how and why unaffected relatives who share cognitive traits with OCD probands do not manifest symptoms. Therefore, further insight also needs to be gained in understanding the environmental or genetic mechanisms that might elicit OCD in related vulnerable individuals. Such understanding can be leveraged to identify whether and how treatments could intervene to prevent disease onset.

It is however clear that this field of research is characterized by a relative paucity of studies, of limited sample size. Much work is also needed to examine whether these cognitive endophenotypes can be generalized to other, related disorders on the impulsive-compulsive spectrum. Importantly, the fact that an endophenotype might be 'trans-diagnostic' is a significantly underappreciated feature. In fact, endophenotypes could be leveraged to develop insights on genetic and neural vulnerabilities shared between disorders that have different symptomatic manifestations but might be supported by similar cognitive, genetic, and neural mechanisms.

A further step would be to link endophenotypes to treatment outcomes – a very few studies have already leveraged this knowledge. For example, it has been shown that patients with OCD with better cognitive and executive abilities responded better to either cognitive behavioural therapy or pharmacological interventions (D'Alcante et al. 2012). However, further studies with larger samples of OCD patients are needed to investigate the value of the role of endophenotypes to target therapies to individual subjects.

Future research may benefit from characterization of large samples of patients' and unaffected relatives on the basis of their neurocognitive endophenotypes (as opposed to their phenotypes), achieved using well-defined experimental tasks probing specific cognitive (e.g. dimensions of executive function) and neurophysiological (e.g. ERN) domains. Such classification could be leveraged to define more homogeneous clusters of patients characterized by overall similar genetic and neural substrates. Identification of biologically homogenous groups in this way could be expected to guide the field towards a more precise targeting of existing therapies as well as towards the development of new ones.

References

Abbruzzese M, Ferri S, Scarone S (1995) Wisconsin card sorting test performance in obsessive-compulsive disorder: no evidence for involvement of dorsolateral prefrontal cortex. Psychiatry Res 58(1):37–43

Abbruzzese M, Ferri S, Scarone S (1997) The selective breakdown of frontal functions in patients with obsessive–compulsive disorder and in patients with schizophrenia: a double dissociation experimental finding. Neuropsychologia 35(6):907–912

Abramovitch A, Cooperman A (2015) The cognitive neuropsychology of obsessive-compulsive disorder: A critical review. J Obsess Compuls Related Dis 5:24–36

Abramovitch A, Abramowitz JS, Mittelman A (2013) The neuropsychology of adult obsessive–compulsive disorder: A meta-analysis. Clin Psychol Rev 33(8):1163–1171

Agam Y, Greenberg JL, Isom M, Falkenstein MJ, Jenike E, Wilhelm S et al (2014) Aberrant error processing in relation to symptom severity in obsessive-compulsive disorder: a multimodal neuroimaging study. NeuroImage Clin 5:141–151

Alsobrook JP II, Leckman JF, Goodman WK, Rasmussen SA, Pauls DL (1999) Segregation analysis of obsessive-compulsive disorder using symptom-based factor scores. Am J Med Genet 88(6):669–675

Alvarenga PG, do Rosário MC, Batistuzzo MC, Diniz JB, Shavitt RG, Duran FLS et al (2012) Obsessive-compulsive symptom dimensions correlate to specific gray matter volumes in treatment-naïve patients. J Psychiatr Res 46(12):1635–1642

Anokhin AP, Golosheykin S, Heath AC (2008) Heritability of frontal brain function related to action monitoring. Psychophysiology 45(4):524–534

Arnold PD, Rosenberg DR, Mundo E, Tharmalingam S, Kennedy JL, Richter MA (2004) Association of a glutamate (NMDA) subunit receptor gene (GRIN2B) with obsessive-compulsive disorder: a preliminary study. Psychopharmacology 174(4):530–538

Arnold PD, Sicard T, Burroughs E, Richter MA, Kennedy JL (2006) Glutamate transporter gene SLC1A1 associated with obsessive-compulsive disorder. Arch Gen Psychiatry 63(7):769–776

Arnold PD, Askland KD, Barlassina C, Bellodi L, Bienvenu OJ, Black D et al (2018) Revealing the complex genetic architecture of obsessive–compulsive disorder using meta-analysis. Mol Psychiatry 23(5):1181–1188

Baker SC, Rogers RD, Owen AM, Frith CD, Dolan RJ, Frackowiak RS et al (1996) Neural systems engaged by planning: a pet study of the tower of London task. Neuropsychologia 34(6):515–526

Bannon S, Gonsalvez CJ, Croft RJ, Boyce PM (2002) Response inhibition deficits in obsessive–compulsive disorder. Psychiatry Res 110(2):165–174

Bannon S, Gonsalvez CJ, Croft RJ, Boyce PM (2006) Executive functions in obsessive-compulsive disorder: state or trait deficits? Aust N Z J Psychiatry 40(11–12):1031–1038
Berg EA (1948) A simple objective technique for measuring flexibility in thinking. J Gen Psychol 39(1):15–22
Bey K, Kaufmann C, Lennertz L, Riesel A, Klawohn J, Heinzel S et al (2018) Impaired planning in patients with obsessive-compulsive disorder and unaffected first-degree relatives: evidence for a cognitive endophenotype. J Anxiety Disord 57:24–30
Bloch MH, Landeros-Weisenberger A, Rosario MC, Pittenger C, Leckman JF (2008) Meta-analysis of the symptom structure of obsessive-compulsive disorder. Am J Psychiatr 165(12):1532–1542
Bohne A, Savage CR, Deckersbach T, Keuthen NJ, Wilhelm S (2008) Motor inhibition in trichotillomania and obsessive-compulsive disorder. J Psychiatr Res 42(2):141–150
Boisseau CL, Thompson-Brenner H, Caldwell-Harris C, Pratt E, Farchione T, Harrison Barlow D (2012) Behavioral and cognitive impulsivity in obsessive–compulsive disorder and eating disorders. Psychiatry Res 200(2):1062–1066
Bolton D, Raven P, Madronal-Luque R, Marks IM (2000) Neurological and neuropsychological signs in obsessive compulsive disorder: interaction with behavioural treatment. Behav Res Ther 38(7):695–708
Bush L, Posner (2000) Cognitive and emotional influences in anterior cingulate cortex. Trends Cogn Sci 4(6):215–222
Carrasco M, Harbin SM, Nienhuis JK, Fitzgerald KD, Gehring WJ, Hanna GL (2013) Increased error-related brain activity in youth with obsessive-compulsive disorder and unaffected siblings. Depress Anxiety 30(1):39–46
Cavedini P, Zorzi C, Piccinni M, Cavallini MC, Bellodi L (2010) Executive dysfunctions in obsessive-compulsive patients and unaffected relatives: searching for a new intermediate phenotype. Biol Psychiatry 67(12):1178–1184
Chamberlain SR, Blackwell AD, Fineberg NA, Robbins TW, Sahakian BJ (2005) The neuropsychology of obsessive compulsive disorder: the importance of failures in cognitive and behavioural inhibition as candidate endophenotypic markers. Neurosci Biobehav Rev 29 (3):399–419
Chamberlain SR, Fineberg NA, Blackwell AD, Robbins TW, Sahakian BJ (2006) Motor inhibition and cognitive flexibility in obsessive-compulsive disorder and trichotillomania. Am J Psychiatry 163(7):1282–1284
Chamberlain SR, Fineberg NA, Menzies LA, Blackwell AD, Bullmore ET, Robbins TW et al (2007) Impaired cognitive flexibility and motor inhibition in unaffected first-degree relatives of patients with obsessive-compulsive disorder. Am J Psychiatry 164(2):335–338
Chamberlain SR, Menzies L, Hampshire A, Suckling J, Fineberg NA, del Campo N et al (2008) Orbitofrontal dysfunction in patients with obsessive-compulsive disorder and their unaffected relatives. Science 321(5887):421–422
Cieslik EC, Mueller VI, Eickhoff CR, Langner R, Eickhoff SB (2015) Three key regions for supervisory attentional control: evidence from neuroimaging meta-analyses. Neurosci Biobehav Rev 48:22–34
Cole MW, Reynolds JR, Power JD, Repovs G, Anticevic A, Braver TS (2013) Multi-task connectivity reveals flexible hubs for adaptive task control. Nat Neurosci 16(9):1348–1355
Cullen B, Brown CH, Riddle MA, Grados M, Bienvenu OJ, Hoehn-Saric R et al (2007) Factor analysis of the Yale-Brown obsessive compulsive scale in a family study of obsessive-compulsive disorder. Depress Anxiety 24(2):130–138
D'Alcante CC, Diniz JB, Fossaluza V, Batistuzzo MC, Lopes AC, Shavitt RG et al (2012) Neuropsychological predictors of response to randomized treatment in obsessive–compulsive disorder. Prog Neuropsychopharmacol Biol Psychiatry 39(2):310–317
Daruna JH, Barnes PA (1993) A neurodevelopmental view of impulsivity. The impulsive client: theory, research, and treatment. American Psychological Association, Washington, pp 23–37

de Vries FE, de Wit SJ, Cath DC, van der Werf YD, van der Borden V, van Rossum TB et al (2014) Compensatory frontoparietal activity during working memory: an endophenotype of obsessive-compulsive disorder. Biol Psychiatry 76(11):878–887
de Wit SJ, de Vries FE, van der Werf YD, Cath DC, Heslenfeld DJ, Veltman EM et al (2012) Presupplementary motor area hyperactivity during response inhibition: a candidate endophenotype of obsessive-compulsive disorder. Am J Psychiatry 169(10):1100–1108
Debener S, Ullsperger M, Siegel M, Fiehler K, von Cramon DY, Engel AK (2005) Trial-by-trial coupling of concurrent electroencephalogram and functional magnetic resonance imaging identifies the dynamics of performance monitoring. J Neurosci Off J Soc Neurosci 25 (50):11730–11737
Delis DC, Kramer JH, Kaplan E, Holdnack J (2004) Reliability and validity of the Delis-Kaplan executive function system: an update. J Int Neuropsychol Soc 10(2):301–303
Delorme R, Goussé V, Roy I, Trandafir A, Mathieu F, Mouren-Siméoni M-C et al (2007) Shared executive dysfunctions in unaffected relatives of patients with autism and obsessive-compulsive disorder. Eur Psychiatry 22(1):32–38
Dias R, Robbins TW, Roberts AC (1996) Dissociation in prefrontal cortex of affective and attentional shifts. Nature 380(6569):69–72
Dickel DE, Veenstra-VanderWeele J, Cox NJ, Wu X, Fischer DJ, Etten-Lee MV et al (2006) Association testing of the positional and functional candidate gene SLC1A1/EAAC1 in early-onset obsessive-compulsive disorder. Arch Gen Psychiatry 63(7):778–785
Endrass T, Ullsperger M (2014) Specificity of performance monitoring changes in obsessive-compulsive disorder. Neurosci Biobehav Rev 46(Pt 1):124–138
Endrass T, Klawohn J, Schuster F, Kathmann N (2008) Overactive performance monitoring in obsessive-compulsive disorder: ERP evidence from correct and erroneous reactions. Neuropsychologia 46(7):1877–1887
Endrass T, Riesel A, Kathmann N, Buhlmann U (2014) Performance monitoring in obsessive-compulsive disorder and social anxiety disorder. J Abnorm Psychol 123(4):705–714
Falkenstein M, Hohnsbein J, Hoormann J, Blanke L (1991) Effects of crossmodal divided attention on late ERP components: II. Error processing in choice reaction tasks. Electroencephalogr Clin Neurophysiol 78(6):447–455
Fan S, Cath DC, van der Werf YD, de Wit S, Veltman DJ, van den Heuvel OA (2018) Trans-diagnostic comparison of response inhibition in Tourette's disorder and obsessive-compulsive disorder. World J Biol Psychiatry 19(7):527–537
Fineberg N, Chamberlain S, Hollander E, Boulougouris V, Robbins T (2011a) Translational approaches to obsessive-compulsive disorder: from animal models to clinical treatment. Br J Pharmacol 164(4):1044–1061
Fineberg N, Saxena S, Zohar J, Craig K, Sirovatka PJ (2011b) Obsessive-compulsive disorder: boundary issues. Obsessive-Compulsive Spectrum Disorders: refining the research agenda for dsm-v. pp 1–32
Fineberg NA, Hengartner MP, Bergbaum C, Gale T, Rössler W, Angst J (2013) Lifetime comorbidity of obsessive-compulsive disorder and sub-threshold obsessive-compulsive symptomatology in the community: impact, prevalence, socio-demographic and clinical characteristics. Int J Psychiatry Clin Pract 17(3):188–196
Fitzgerald KD, Welsh RC, Gehring WJ, Abelson JL, Himle JA, Liberzon I et al (2005) Error-related hyperactivity of the anterior cingulate cortex in obsessive-compulsive disorder. Biol Psychiatry 57(3):287–294
Flint J, Munafò MR (2007) The endophenotype concept in psychiatric genetics. Psychol Med 37:163–180
Fradkin I, Adams RA, Parr T, Roiser JP, Huppert JD (2020) Searching for an anchor in an unpredictable world: a computational model of obsessive compulsive disorder. Psychol Rev. http://doi.apa.org/getdoi.cfm?doi=10.1037/rev0000188
Fried EI, Nesse RM (2015) Depression is not a consistent syndrome: an investigation of unique symptom patterns in the STAR*D study. J Affect Disord 172:96–102

Friedman NP, Miyake A, Young SE, DeFries JC, Corley RP, Hewitt JK (2008) Individual differences in executive functions are almost entirely genetic in origin. J Exp Psychol Gen 137(2):201–225
Gehring WJ, Himle J, Nisenson LG (2000) Action-monitoring dysfunction in obsessive-compulsive disorder. Psychol Sci 11(1):1–6
Gehring WJ, Goss B, Coles MGH, Meyer DE, Donchin E (2016) A neural system for error detection and compensation. Psychol Sci. https://doi.org/10.1111/j.1467-9280.1993.tb00586.x
Gillan CM, Fineberg NA, Robbins TW (2017) A trans-diagnostic perspective on obsessive-compulsive disorder. Psychol Med 47(9):1528–1548
Gillan CM, Kalanthroff E, Evans M, Weingarden HM, Jacoby RJ, Gershkovich M et al (2020) Comparison of the association between goal-directed planning and self-reported compulsivity vs obsessive-compulsive disorder diagnosis. JAMA Psychiat 77(1):77–85
Gottesman II, Gould TD (2003) The endophenotype concept in psychiatry: etymology and strategic intentions. Am J Psychiatr 160(4):636–645
Gottesman II, Shields J (1973) Genetic theorizing and schizophrenia. Br J Psychiatry 122 (566):15–30
Grabe HJ, Ruhrmann S, Ettelt S, Buhtz F, Hochrein A, Schulze-Rauschenbach S et al (2006) Familiality of obsessive-compulsive disorder in nonclinical and clinical subjects. Am J Psychiatr 163(11):1986–1992
Graybiel AM, Rauch SL (2000) Toward a neurobiology of obsessive-compulsive disorder. Neuron 28(2):343–347
Grützmann R, Endrass T, Kaufmann C, Allen E, Eichele T, Kathmann N (2016) Presupplementary motor area contributes to altered error monitoring in obsessive-compulsive disorder. Biol Psychiatry 80(7):562–571
Hajcak G, Franklin ME, Foa EB, Simons RF (2008) Increased error-related brain activity in pediatric obsessive-compulsive disorder before and after treatment. Am J Psychiatry 165 (1):116–123
Hampshire A, Chaudhry AM, Owen AM, Roberts AC (2012) Dissociable roles for lateral orbitofrontal cortex and lateral prefrontal cortex during preference driven reversal learning. NeuroImage 59–20(4):4102–4112
Hampshire A, Zadel A, Sandrone S, Soreq E, Fineberg N, Bullmore ET et al (2019) Inhibition-related cortical Hypoconnectivity as a candidate vulnerability marker for obsessive-compulsive disorder. Biol Psychiatry Cogn Neurosci Neuroimaging. https://www.biologicalpsychiatrycnni.org/article/S2451-9022(19)30265-4/abstract
Hanna GL, Veenstra-VanderWeele J, Cox NJ, Boehnke M, Himle JA, Curtis GC et al (2002) Genome-wide linkage analysis of families with obsessive-compulsive disorder ascertained through pediatric probands. Am J Med Genet 114(5):541–552
Hanna GL, Veenstra-VanderWeele J, Cox NJ, Van Etten M, Fischer DJ, Himle JA et al (2007) Evidence for a susceptibility locus on chromosome 10p15 in early-onset obsessive-compulsive disorder. Biol Psychiatry 62(8):856–862
Hanna GL, Carrasco M, Harbin SM, Nienhuis JK, LaRosa CE, Chen P et al (2012) Error-related negativity and tic history in pediatric obsessive-compulsive disorder. J Am Acad Child Adolesc Psychiatry 51(9):902–910
Harrison BJ, Pujol J, Cardoner N, Deus J, Alonso P, López-Solà M et al (2013) Brain corticostriatal systems and the major clinical symptom dimensions of obsessive-compulsive disorder. Biol Psychiatry 73(4):321–328
Hasler G, Drevets WC, Gould TD, Gottesman II, Manji HK (2006) Toward constructing an endophenotype strategy for bipolar disorders. Biol Psychiatry 60(2):93–105
Hofmeijer-Sevink MK, Batelaan NM, van Megen HJGM, Penninx BW, Cath DC, van den Hout MA et al (2012) Clinical relevance of comorbidity in anxiety disorders: a report from the Netherlands Study of Depression and Anxiety (NESDA). J Affect Disord 137(1–3):106–112
Hofstadter DR (1985) Metamagical themas: questing for the essence of mind and pattern. Basic Books, New York. http://archive.org/details/metamagicalthema0000hofs

Hudziak JJ, van Beijsterveldt CEM, Althoff RR, Stanger C, Rettew DC, Nelson EC et al (2004) Genetic and environmental contributions to the child behavior ChecklistObsessive-compulsive scale: a cross-cultural twin study. Arch Gen Psychiatry 61(6):608–616

Hutton SB, Ettinger U (2006) The antisaccade task as a research tool in psychopathology: a critical review. Psychophysiology 43(3):302–313

Huyser C, Veltman DJ, Wolters LH, de Haan E, Boer F (2011) Developmental aspects of error and high-conflict-related brain activity in pediatric obsessive-compulsive disorder: a fMRI study with a Flanker task before and after CBT. J Child Psychol Psychiatry 52:1251–1260

Hymas N, Lees A, Bolton D, Epps K, Head D (1991) The neurology of obsessional slowness. Brain 114(5):2203–2233

Iervolino AC, Rijsdijk FV, Cherkas L, Fullana MA, Mataix-Cols D (2011) A multivariate twin study of obsessive-compulsive symptom dimensions. Arch Gen Psychiatry 68(6):637–644

Insel T, Cuthbert B, Garvey M, Heinssen R, Pine DS, Quinn K et al (2010) Research domain criteria (RDoC): toward a new classification framework for research on mental disorders. Am J Psychiatry 167(7):748–751

Joel D (2006) The signal attenuation rat model of obsessive-compulsive disorder: a review. Psychopharmacology 186(4):487–503

Joel, D., Ben-Amir, E., Doljansky, J., & Flaisher, S. (2004). "Compulsive" lever-pressing in rats is attenuated by the serotonin re-uptake inhibitors paroxetine and fluvoxamine but not by the tricyclic antidepressant desipramine or the anxiolytic diazepam. Behav Pharmacol, 15(3), 241–252

Johannes S, Wieringa BM, Nager W, Rada D, Dengler R, Emrich HM et al (2001) Discrepant target detection and action monitoring in obsessive-compulsive disorder. Psychiatry Res 108 (2):101–110

Jonnal AH, Gardner CO, Prescott CA, Kendler KS (2000) Obsessive and compulsive symptoms in a general population sample of female twins. Am J Med Genet 96(6):791–796

Kahn RS, Westenberg HGM, Jolles J (1984) Zimeldine treatment of obsessive-compulsive disorder. Acta Psychiatr Scand 69(3):259–261

Kalanthroff E, Teichert T, Wheaton MG, Kimeldorf MB, Linkovski O, Ahmari SE et al (2017) The role of response inhibition in medicated and Unmedicated obsessive-compulsive disorder patients: evidence from the stop-signal task. Depress Anxiety 34(3):301–306

Kendler KS, Neale MC (2010) Endophenotype: a conceptual analysis. Mol Psychiatry 15 (8):789–797

Kendler KS, Neale SMC, Kessler RC, Heath AC, Eaves LJ (1993) A test of the equalenvironment assumption in twin studies of psychiatric illness. Behav Genet:21–27

Klavir O, Flash S, Winter C, Joel D (2009) High frequency stimulation and pharmacological inactivation of the subthalamic nucleus reduces 'compulsive' lever-pressing in rats. Exp Neurol 215(1):101–109

Klawohn J, Riesel A, Grützmann R, Kathmann N, Endrass T (2014) Performance monitoring in obsessive-compulsive disorder: a temporo-spatial principal component analysis. Cogn Affect Behav Neurosci 14(3):983–995

Lei H, Zhu X, Fan J, Dong J, Zhou C, Zhang X et al (2015) Is impaired response inhibition independent of symptom dimensions in obsessive-compulsive disorder? Evidence from ERPs. Sci Rep 5(1):1–9

Lennertz L, Rampacher F, Vogeley A, Schulze-Rauschenbach S, Pukrop R, Ruhrmann S et al (2012) Antisaccade performance in patients with obsessive–compulsive disorder and unaffected relatives: further evidence for impaired response inhibition as a candidate endophenotype. Eur Arch Psychiatry Clin Neurosci 262(7):625–634

Lipszyc J, Schachar R (2010) Inhibitory control and psychopathology: a meta-analysis of studies using the stop signal task. J Int Neuropsychol Soc 16(6):1064–1076

Liu Y, Hanna GL, Carrasco M, Gehring WJ, Fitzgerald KD (2014) Altered relationship between electrophysiological response to errors and gray matter volumes in an extended network for

error-processing in pediatric obsessive-compulsive disorder. Hum Brain Mapp 35 (4):1143–1153
Logan GD, Cowan WB, Davis KA (1984) On the ability to inhibit simple and choice reaction time responses: a model and a method. J Exp Psychol Hum Percept Perform 10(2):276–291
Marzuki AA, Pereira de Souza AMFL, Sahakian BJ, Robbins TW (2020) Are candidate neurocognitive endophenotypes of OCD present in paediatric patients? A systematic review. Neurosci Biobehav Rev 108:617–645
Mataix-Cols D, Rauch SL, Manzo PA, Jenike MA, Baer L (1999) Use of factor-analyzed symptom dimensions to predict outcome with serotonin reuptake inhibitors and placebo in the treatment of obsessive-compulsive disorder. Am J Psychiatry 156(9):1409–1416
Mataix-Cols D, do Rosario-Campos MC, Leckman JF (2005) A multidimensional model of obsessive-compulsive disorder. Am J Psychiatr 162(2):228–238
Mataix-Cols D, Boman M, Monzani B, Rück C, Serlachius E, Långström N et al (2013) Population-based, multigenerational family clustering study of obsessive-compulsive disorder. JAMA Psychiat 70(7):709–717
Mathews CA, Badner JA, Andresen JM, Sheppard B, Himle JA, Grant JE et al (2012) Genome-wide linkage analysis of obsessive-compulsive disorder implicates chromosome 1p36. Biol Psychiatry Adv Psychiatric Genom 72(8):629–636
Mathews CA, Perez VB, Roach BJ, Fekri S, Vigil O, Kupferman E et al (2016) Error-related brain activity dissociates hoarding disorder from obsessive-compulsive disorder. Psychol Med 46 (2):367–379
Mattheisen M, Samuels JF, Wang Y, Greenberg BD, Fyer AJ, McCracken JT et al (2015) Genome-wide association study in obsessive-compulsive disorder: results from the OCGAS. Mol Psychiatry 20(3):337–344
Menzies L, Achard S, Chamberlain SR, Fineberg N, Chen C-H, del Campo N et al (2007) Neurocognitive endophenotypes of obsessive-compulsive disorder. Brain 130(12):3223–3236
Menzies L, Chamberlain SR, Laird AR, Thelen SM, Sahakian BJ, Bullmore ET (2008) Integrating evidence from neuroimaging and neuropsychological studies of obsessive-compulsive disorder: The orbitofronto-striatal model revisited. Neurosci Biobehav Rev 32(3):525–549
Miguel EC, Leckman JF, Rauch S, do Rosario-Campos MC, Hounie AG, Mercadante MT et al (2005) Obsessive-compulsive disorder phenotypes: implications for genetic studies. Mol Psychiatry 10(3):258–275
Miyake A, Friedman NP, Emerson MJ, Witzki AH, Howerter A, Wager TD (2000) The unity and diversity of executive functions and their contributions to complex “frontal lobe” tasks: a latent variable analysis. Cogn Psychol 41(1):49–100
Morein-Zamir S, Fineberg NA, Robbins TW, Sahakian BJ (2010) Inhibition of thoughts and actions in obsessive-compulsive disorder: extending the endophenotype? Psychol Med 40(2):263–272
Moritz S, Kloss M, von Eckstaedt FV, Jelinek L (2009) Comparable performance of patients with obsessive-compulsive disorder (OCD) and healthy controls for verbal and nonverbal memory accuracy and confidence: time to forget the forgetfulness hypothesis of OCD? Psychiatry Res 166(2–3):247–253
Nestadt G, Samuels J, Riddle M, Bienvenu OJ 3rd, Liang KY, LaBuda M et al (2000) A family study of obsessive-compulsive disorder. Arch Gen Psychiatry 57(4):358–363
Nieuwenhuis S, Nielen MM, Mol N, Hajcak G, Veltman DJ (2005) Performance monitoring in obsessive-compulsive disorder. Psychiatry Res 134(2):111–122
Okasha A, Rafaat M, Mahallawy N, El Nahas G, El Dawla AS, Sayed M et al (2000) Cognitive dysfunction in obsessive-compulsive disorder. Acta Psychiatr Scand 101(4):281–285
Owen AM, Sahakian BJ, Hodges JR, Summers BA, Polkey CE, Robbins TW (1995) Dopamine-dependent frontostriatal planning deficits in early Parkinson’s disease. Neuropsychology 9 (1):126–140
Owen AM, McMillan KM, Laird AR, Bullmore E (2005) N-back working memory paradigm: a meta-analysis of normative functional neuroimaging studies. Hum Brain Mapp 25(1):46–59

Ozcan H, Ozer S, Yagcioglu S (2016) Neuropsychological, electrophysiological and neurological impairments in patients with obsessive compulsive disorder, their healthy siblings and healthy controls: identifying potential endophenotype(s). Psychiatry Res 240:110–117

Pauls DL, Alsobrook JP, Goodman W, Rasmussen S, Leckman JF (1995) A family study of obsessive-compulsive disorder. Am J Psychiatry 152(1):76–84

Penadés R, Catalán R, Rubia K, Andrés S, Salamero M, Gastó C (2007) Impaired response inhibition in obsessive compulsive disorder. Eur Psychiatry 22(6):404–410

Pinto A, Mancebo MC, Eisen JL, Pagano ME, Rasmussen SA (2006) The Brown longitudinal obsessive compulsive study: clinical features and symptoms of the sample at intake. J Clin Psychiatry 67(5):703–711

Pitman RK (1987) A cybernetic model of obsessive-compulsive psychopathology. Compr Psychiatry 28(4):334–343

Purcell R, Maruff P, Kyrios M, Pantelis C (1998) Cognitive deficits in obsessive-compulsive disorder on tests of frontal-striatal function. Biol Psychiatry 43(5):348–357

Rajender G, Bhatia MS, Kanwal K, Malhotra S, Singh TB, Chaudhary D (2011) Study of neurocognitive endophenotypes in drug-naïve obsessive-compulsive disorder patients, their first-degree relatives and healthy controls. Acta Psychiatr Scand 124(2):152–161

Rao NP, Reddy YCJ, Kumar KJ, Kandavel T, Chandrashekar CR (2008) Are neuropsychological deficits trait markers in OCD? Prog Neuro-Psychopharmacol Biol Psychiatry 32(6):1574–1579

Remijnse PL, Nielen MMA, van Balkom AJLM, Cath DC, van Oppen P, Uylings HBM et al (2006) Reduced orbitofrontal-striatal activity on a reversal learning task in obsessive-compulsive disorder. Arch Gen Psychiatry 63(11):1225–1236

Remijnse PL, Nielen MMA, van Balkom AJLM, Hendriks G-J, Hoogendijk WJ, Uylings HBM et al (2009) Differential frontal-striatal and paralimbic activity during reversal learning in major depressive disorder and obsessive-compulsive disorder. Psychol Med 39(9):1503–1518

Riesel A, Endrass T, Kaufmann C, Kathmann N (2011) Overactive error-related brain activity as a candidate endophenotype for obsessive-compulsive disorder: evidence from unaffected first-degree relatives. Am J Psychiatry 168(3):317–324

Riesel A, Kathmann N, Endrass T (2014) Overactive performance monitoring in obsessive-compulsive disorder is independent of symptom expression. Eur Arch Psychiatry Clin Neurosci 264(8):707–717

Riesel A, Endrass T, Auerbach LA, Kathmann N (2015) Overactive performance monitoring as an Endophenotype for obsessive-compulsive disorder: evidence from a treatment study. Am J Psychiatr 172(7):665–673

Riesel A, Goldhahn S, Kathmann N (2017) Hyperactive performance monitoring as a transdiagnostic marker: results from health anxiety in comparison to obsessive–compulsive disorder. Neuropsychologia 96:1–8

Riesel A, Klawohn J, Grützmann R, Kaufmann C, Heinzel S, Bey K et al (2019) Error-related brain activity as a transdiagnostic endophenotype for obsessive-compulsive disorder, anxiety and substance use disorder. Psychol Med 49(7):1207–1217

Robbins TW, James M, Owen AM, Sahakian BJ, Lawrence AD, McInnes L et al (1998) A study of performance on tests from the CANTAB battery sensitive to frontal lobe dysfunction in a large sample of normal volunteers: implications for theories of executive functioning and cognitive aging. Cambridge Neuropsychological Test Automated Battery. J Int Neuropsychol Soc 4 (5):474–490

Robbins T, Gillan CM, Smith DG, de Wit S, Ersche KD (2012) Neurocognitive endophenotypes of impulsivity and compulsivity: towards dimensional psychiatry. Trends Cogn Sci 16(1):81–91

Ross J, Badner J, Garrido H, Sheppard B, Chavira DA, Grados M et al (2011) Genomewide linkage analysis in Costa Rican families implicates chromosome 15q14 as a candidate region for OCD. Hum Genet 130(6):795–805

Ruchsow M, Grön G, Reuter K, Spitzer M, Hermle L, Kiefer M (2005) Error-related brain activity in patients with obsessive- compulsive disorder and in healthy controls. J Psychophysiol 19 (4):298–304

Ruchsow M, Reuter K, Hermle L, Ebert D, Kiefer M, Falkenstein M (2007) Executive control in obsessive-compulsive disorder: event-related potentials in a Go/Nogo task. J Neural Transm 114(12):1595–1601

Savage CR, Baer L, Keuthen NJ, Brown HD, Rauch SL, Jenike MA (1999) Organizational strategies mediate nonverbal memory impairment in obsessive–compulsive disorder. Biol Psychiatry 45(7):905–916

Saxena S, Rauch SL (2000) Functional neuroimaging and the neuroanatomy of obsessive-compulsive disorder. Psychiatr Clin N Am 23(3):563–586

Saxena S, Brody AL, Schwartz JM, Baxter LR (1998) Neuroimaging and frontal-subcortical circuitry in obsessive-compulsive disorder. Br J Psychiatry 35:26–37

Seow TXF, Benoit E, Dempsey C, Jennings M, Maxwell A, McDonough M et al (2019) Null results from a dimensional study of error-related negativity (ERN) and self-reported psychiatric symptoms. BioRxiv 732594

Shallice T (1982) Specific impairments of planning. Philos Trans R Soc Lond Ser B Biol Sci 298 (1089):199–209

Sharma E, Thennarasu K, Reddy YCJ (2014) Long-term outcome of obsessive-compulsive disorder in adults: a meta-analysis. J Clin Psychiatry 75(9):1019–1027

Shin NY, Lee TY, Kim E, Kwon JS (2014) Cognitive functioning in obsessive-compulsive disorder: a meta-analysis. Psychol Med 44(06):1121–1130

Shugart YY, Samuels J, Willour VL, Grados MA, Greenberg BD, Knowles JA et al (2006) Genomewide linkage scan for obsessive-compulsive disorder: evidence for susceptibility loci on chromosomes 3q, 7p, 1q, 15q, and 6q. Mol Psychiatry 11(8):763–770

Snyder HR, Kaiser RH, Warren SL, Heller W (2015) Obsessive-compulsive disorder is associated with broad impairments in executive function: a meta-analysis. Clin Psychol Sci 3(2):301–330

Sohn SY, Kang JI, Namkoong K, Kim SJ (2014) Multidimensional measures of impulsivity in obsessive-compulsive disorder: cannot wait and stop. PLoS One 9(11):e111739

Stern ER, Liu Y, Gehring WJ, Lister JJ, Yin G, Zhang J et al (2010) Chronic medication does not affect hyperactive error responses in obsessive-compulsive disorder. Psychophysiology 47 (5):913–920

Stewart SE, Mayerfeld C, Arnold PD, Crane JR, O'Dushlaine C, Fagerness JA et al (2013a) Meta-analysis of association between obsessive-compulsive disorder and the 3′ region of neuronal glutamate transporter gene SLC1A1. Am J Med Genet Part B 162B(4):367–379

Stewart SE, Yu D, Scharf JM, Neale BM, Fagerness JA, Mathews CA et al (2013b) Genome-wide association study of obsessive-compulsive disorder. Mol Psychiatry 18(7):788–798

Taylor S (2011) Etiology of obsessions and compulsions: a meta-analysis and narrative review of twin studies. Clin Psychol Rev 31(8):1361–1372

Tezcan D, Tumkaya S, Bora E (2017) Reversal learning in patients with obsessive-compulsive disorder (OCD) and their unaffected relatives: is orbitofrontal dysfunction an endophenotype of OCD? Psychiatry Res 252:231–233

Ullsperger M, Danielmeier C, Jocham G (2014) Neurophysiology of performance monitoring and adaptive behavior. Physiol Rev 94(1):35–79

Ursu S, Stenger VA, Shear MK, Jones MR, Carter CS (2003) Overactive action monitoring in obsessive-compulsive disorder: evidence from functional magnetic resonance imaging. Psychol Sci 14(4):347–353

Vaghi MM, Hampshire A, Fineberg NA, Kaser M, Brühl AB, Sahakian BJ et al (2017a) Hypoactivation and dysconnectivity of a frontostriatal circuit during goal-directed planning as an Endophenotype for obsessive-compulsive disorder. Biol Psychiatry 2(8):655–663

Vaghi MM, Vértes PE, Kitzbichler MG, Apergis-Schoute AM, van der Flier FE, Fineberg NA et al (2017b) Specific frontostriatal circuits for impaired cognitive flexibility and goal-directed planning in obsessive-compulsive disorder: evidence from resting-state functional connectivity. Biol Psychiatry 81(8):708–717 van den Heuvel OA, Veltman DJ, Groenewegen HJ, Cath DC, van Balkom AJLM, van Hartskamp J et al (2005) Frontal-striatal dysfunction during planning in obsessive-compulsive disorder. Arch Gen Psychiatry 62(3):301–309

van den Heuvel OA, Remijnse PL, Mataix-Cols D, Vrenken H, Groenewegen HJ, Uylings HBM et al (2009) The major symptom dimensions of obsessive-compulsive disorder are mediated by partially distinct neural systems. Brain J Neurol 132(Pt 4):853–868

van der Wee NJA, Ramsey NF, Jansma JM, Denys DA, van Megen HJGM, Westenberg HMG et al (2003) Spatial working memory deficits in obsessive compulsive disorder are associated with excessive engagement of the medial frontal cortex. NeuroImage 20(4):2271–2280

van Grootheest DS, Cath DC, Beekman AT, Boomsma DI (2005) Twin studies on obsessive-compulsive disorder: a review. Twin Res Hum Genet 8(5):450–458

van Velzen LS, de Wit SJ, Ćurčić-Blake B, Cath DC, de Vries FE, Veltman DJ et al (2015) Altered inhibition-related frontolimbic connectivity in obsessive-compulsive disorder. Hum Brain Mapp 36(10):4064–4075

Verbruggen F, Aron AR, Band GP, Beste C, Bissett PG, Brockett AT et al (2019) A consensus guide to capturing the ability to inhibit actions and impulsive behaviors in the stop-signal task. elife 8

Visscher PM, Hill WG, Wray NR (2008) Heritability in the genomics era – concepts and misconceptions. Nat Rev Genet 9(4):255–266

Viswanath B, Janardhan Reddy YC, Kumar KJ, Kandavel T, Chandrashekar CR (2009) Cognitive endophenotypes in OCD: a study of unaffected siblings of probands with familial OCD. Prog Neuro-Psychopharmacol Biol Psychiatry 33(4):610–615

Watkins LH, Sahakian BJ, Robertson MM, Veale DM, Rogers RD, Pickard KM et al (2005) Executive function in Tourette's syndrome and obsessive–compulsive disorder. Psychol Med 35(4):571–582

Weinberg A, Riesel A, Hajcak G (2012) Integrating multiple perspectives on error-related brain activity: the ERN as a neural indicator of trait defensive reactivity. Motiv Emot 36(1):84–100

Weinberg A, Kotov R, Proudfit GH (2015) Neural indicators of error processing in generalized anxiety disorder, obsessive-compulsive disorder, and major depressive disorder. J Abnorm Psychol 124(1):172–185

Xiao Z, Wang J, Zhang M, Li H, Tang Y, Wang Y et al (2011) Error-related negativity abnormalities in generalized anxiety disorder and obsessive-compulsive disorder. Prog Neuro-Psychopharmacol Biol Psychiatry 35(1):265–272

Zartaloudi E, Laws K, Bramon E (2019) Endophenotypes of executive functions in obsessive compulsive disorder? A meta-analysis in unaffected relatives. Psychiatr Genet 29(6):211–219

Zhang J, Yang X, Yang Q (2015a) Neuropsychological dysfunction in adults with early-onset obsessive-compulsive disorder: the search for a cognitive endophenotype. Rev Brasileira de Psiquiatria 37(2):126–132

Zhang L, Dong Y, Ji Y, Zhu C, Yu F, Ma H et al (2015b) Dissociation of decision making under ambiguity and decision making under risk: a neurocognitive endophenotype candidate for obsessive–compulsive disorder. Prog Neuro-Psychopharmacol Biol Psychiatry 57:60–68

Cognitive Inflexibility in OCD and Related Disorders

Samuel R. Chamberlain, Jeremy E. Solly, Roxanne W. Hook, Matilde M. Vaghi, and Trevor W. Robbins

Contents

Abstract Cognitive inflexibility is suggested by the hallmark symptoms of obsessive-compulsive disorder (OCD), namely the occurrence of repetitive thoughts and/or behaviours that persist despite being functionally impairing and egodystonic to the individual. As well as being implied by the top-level symptoms, cognitive inflexibility in OCD, and some related conditions, has also been objectively quantified in case–control studies using computerised cognitive tasks. This chapter begins by considering the objective measurement of different aspects of cognitive flexibility using neuropsychological paradigms, with a focus on neural and neurochemical

S. R. Chamberlain (✉), J. E. Solly, and R. W. Hook
Department of Psychiatry, University of Cambridge, Addenbrooke's Hospital, Cambridge, UK

Cambridgeshire and Peterborough NHS Foundation Trust, Cambridge, UK
e-mail: src33@cam.ac.uk

M. M. Vaghi
Max Planck University College London (UCL) Centre for Computational Psychiatry and Ageing Research, London, UK

T. W. Robbins
Department of Psychology, University of Cambridge, Cambridge, UK

Curr Topics Behav Neurosci (2021) 49: 125–145
https://doi.org/10.1007/7854_2020_198
Published Online: 6 February 2021

substrates. It moves on to conduct a systematic review and meta-analysis of findings from a widely deployed flexibility task: the Intra-Dimensional/Extra-Dimensional Set-Shift Task (IDED). By pooling data from 11 studies (335 OCD patients and 311 controls), we show that Extra-Dimensional (ED) shift deficits are a robust and reproducible finding (effect size medium–large) in OCD across the literature, and that this deficit is not attributable to group differences in age or IQ. The OCD ED deficit is then discussed in terms of dysfunction of fronto-striatal pathways (as exemplified, for example, by functional connectivity data), and the putative role of different neurotransmitters. We consider evidence that impaired ED shifting constitutes a candidate vulnerability marker (or 'endophenotype') for OCD. The available literature is then surveyed as to ED findings in other obsessive-compulsive (OC) related disorders (e.g. hoarding, body-dysmorphic disorder, and trichotillomania), as well as in non-OC disorders (schizophrenia and anxiety symptoms in general). Lastly, we consider more recent, emerging developments in the quantification of compulsivity using cognitive tasks and questionnaires, as well as key directions for future research, including the need to refine compulsivity and its composite cognitive processes.

Keywords Cognitive flexibility · Compulsivity · Endophenotype · Obsessive-compulsive disorder · Obsessive-compulsive related disorders · OCD · Intra-dimensional/extra-dimensional set-shift task · Set-shifting

1 Introduction

Cognitive flexibility can be defined simply as 'adjusting to change'. It involves the mental ability to switch or shift from thinking about one conceptual representation to another, especially in response to changes in rules and environmental feedback, though sometimes spontaneously. It is one of a triad of the so-called executive functions that optimise behavioural success; the other two being the updating of working memory and inhibitory response control (Friedman et al. 2006). Whereas working memory is strongly related to IQ, cognitive inflexibility shows a weaker association (Friedman et al. 2006). In its most general sense, cognitive flexibility can be related to learning and neural plasticity. Modern reinforcement learning algorithms explicitly draw a distinction between exploitation and exploration strategies in responding to environmental contingencies (Cohen et al. 2007). It is perhaps not surprising then that cognitive inflexibility is a relevant cause of impairment in many forms of psychopathology, in the case of obsessive-compulsive disorder (OCD), in which both subjective and behavioural responses are prone to repetition, whether of thoughts (obsessions) or actual behaviour (compulsions). However, the failure to modify reinforcement learning contingencies appears distinct from the failure of executive operation that can occur when subjects are required to 'multi-task' and switch optimally from one task to another, as in 'task-set switching' paradigms.

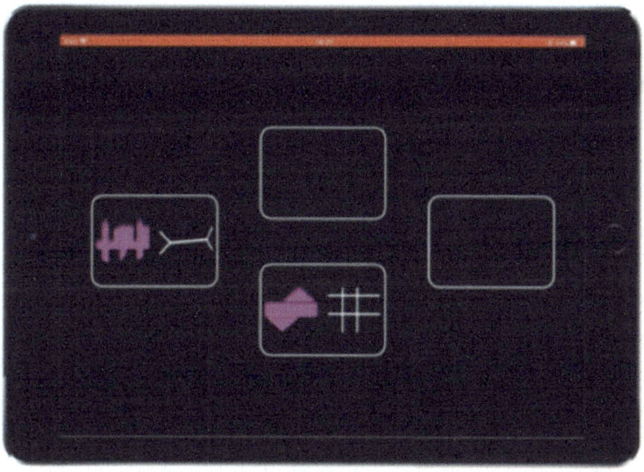

Fig. 1 CANTAB IDED task. © Copyright 2019 Cambridge Cognition Limited. All rights reserved. Accessed online at www.cambridgecognition.com

In clinical neuropsychology, cognitive flexibility has been indexed by tests involving either executive switching such as the Trail-Making Test (TMT) (Lezak et al. 2004), which requires switches of established response rules, or learning to respond to changing feedback, as in the Wisconsin Card Sort Task (WCST), which requires subjects to learn to sort cards according to a rule deduced from trial-and-error feedback, and then crucially to switch that rule when the feedback changes. Classically, patients with lesions of the frontal lobe, whilst learning the initial WCST rule normally, demonstrate impaired cognitive flexibility by being unable to shift away from the rule when the feedback changes and instead tend to perseverate in their original choice (Milner 1963).

The WCST has been adapted into a number of simpler computer-based tests comprising the so-called Intra-Dimensional/Extra-Dimensional Set-Shift Task (IDED) from the Cambridge Neuropsychological Test Automated Battery (CANTAB). The constituent elements of the IDED task are further 'decomposed' to provide a suite of detailed stages suitable for testing experimental animals, as well as human patients, using a sensitive computerised touch-screen format. At each stage of the IDED test, participants are instructed to choose the correct (as determined by the computer program) alternative from either two single (in the early task stages) or two compound stimuli (in the later task stages) presented on the computer screen (Fig. 1). Each stimulus has two relevant dimensions: lines and shapes (e.g. white lines and pink shapes). After the participant chooses one of the two stimuli, the computer provides feedback in the form of 'correct' or 'incorrect' signals on-screen. Through trial and error, the participants attempt to learn the underlying rule about which stimulus is correct. Once a participant successfully learns a given rule and responds by repeatedly selecting the correct stimulus, the computer changes the rule and the participant is then tested on their ability to adapt their responses accordingly.

More specifically, the first stage of the IDED involves testing simple discrimination learning, in which, for example, one visual shape is rewarded and the other punished. Following such learning, these initial contingencies are reversed. This tests 'reversal learning', as the participant has to learn to reverse their choice according to the rule reversal to obtain the reward. This also has the effect of making it clear to the participant that several versions of a particular perceptual dimension are potentially relevant to reward. At the next stage, the discriminative stimuli are compounded into two visual dimensions (lines are superimposed on the shapes), but only one of the dimensions is relevant for acquiring reward. The ability to generalise learning to new stimuli by learning a rule (e.g. to choose one of the shapes) is then tested with entirely novel exemplars. This is the Intra-Dimensional Shift (IDS), also followed by a reversal stage. By this stage it is usually clear to the participant that shapes are the important components of the stimuli to monitor and they exhibit rapid learning.

The final and crucial Extra-Dimensional Shift (EDS) stage occurs when new compound stimuli are again introduced, but at this stage, for the first time, the previously irrelevant visual dimension (e.g. lines) is suddenly made relevant, and the previously relevant visual dimension (e.g. shapes) is made irrelevant. The participant now has to show flexibility by shifting their focus of attention away from the relevant or 'pre-potent' dimension (shape) to the stimulus that has previously been irrelevant (line). This stage is thus equivalent to a rule-shifting stage of the WCST. It is customarily followed by a final reversal learning stage. Thus the EDS stage examines a crucial component of cognitive flexibility: being able to 'unlearn' a previously held rule and shift attention to a new aspect of the environment that was not previously relevant. In contrast, during the IDS stage, which may be used as a control stage for the EDS, the participant must keep their attentional focus on the previously relevant stimulus dimension as new stimuli are introduced.

Use of this test paradigm has led to the discovery that there are different prefrontal cortical circuits for different types of cognitive flexibility. Thus, excitotoxic lesions of the lateral prefrontal cortex in marmoset monkeys selectively impaired EDS, but not IDS or reversal learning, whereas similar lesions of the orbitofrontal cortex (OFC) impaired reversal learning, but not EDS (Dias et al. 1996). These findings suggest that there exists a hierarchical set of processes governing cognitive flexibility; from abstract, rule-based choices as in the EDS, to exemplar-based choices with explicit reward and punishment, as in reversal learning. This general conclusion has also been shown to hold for both the rodent and human prefrontal cortex, using variants of the IDED task, which, for example, may depend on shifting between stimulus modalities (smell and somatosensory texture in rodents, Birrell and Brown 2000), or between abstract qualities such as provided by photographs in humans in a functional neuroimaging setting (e.g. of faces and houses, Hampshire and Owen 2006). Further analyses in experimental animals have revealed further important details about the relevant neural networks. Thus, excitotoxic lesions of the striatum have also been shown to impair reversal learning, indicating the important role of orbitofronto-striatal pathways (Clarke et al. 2008; Jackson et al. 2019).

Additionally, there appear to be dissociable neurochemical neuromodulatory influences upon these two key aspects of cognitive flexibility, EDS and reversal learning (Robbins and Roberts 2007). Most notably, reversal learning is especially sensitive to modulation by serotonin, whether in rodents, monkeys, or humans. For example, Clarke et al. (2004, 2005) found that selective depletion of serotonin in the

OFC using the neurotoxin 5,7 dihydroxytryptamine selectively impaired reversal learning rather than EDS by inducing perseverative responding. Moreover, OFC dopamine (DA) depletion had no significant effects. In addition, rodent studies of reversal learning show that poor reversal performance can be remediated by treatment with relatively high doses of selective serotonin reuptake inhibitors (SSRIs) (Danet et al. 2010; Barlow et al. 2015). By contrast, EDS appears to be more sensitive to catecholamine manipulation (DA or noradrenaline, NA) (Robbins and Roberts 2007).

Patients with lateral frontal lobe damage tend to be impaired at EDS (Owen et al. 1991), whereas those with frontal lobe dementia, which causes hypo-perfusion of the OFC, have difficulties with reversal learning (Rahman et al. 1999). Deficits in EDS have also been found for patients with Parkinson's disease (Downes et al. 1989) and gene-positive but clinically unimpaired Huntington's disease (HD) (Lawrence et al. 1998). In the latter case, advanced HD patients exhibit instead profound deficits in reversal learning, probably because of progressive spread of striatal degeneration (Sprengelmeyer et al. 1995). Patients with schizophrenia display a range of impairments on the IDED, including impairments at the earliest stages of discrimination learning and reversal (Pantelis et al. 1999; Leeson et al. 2009). However, there is little evidence to suggest that performance on the CANTAB task could be used to define a cognitive endophenotype for this disorder (Ceaser et al. 2008).

Cognitive inflexibility might be considered as a likely cognitive endophenotype for OCD and related disorders, as it corresponds with the tendency of these disorders to produce persistent obsessions and compulsions. Indeed, there have been many studies on this aspect of dysexecutive functioning in OCD involving a range of different tests of cognitive flexibility, from traditional instruments including the WCST and TMT, to the CANTAB IDED test. However, there has also been considerable disagreement about the reliability of the cognitive flexibility deficits in OCD and also the degree to which they may represent more general impairments in executive functioning or cognitive control, including, for example, inhibitory response control (Menzies et al. 2007; Chamberlain et al. 2005) and working memory. Consequently, in this survey we address the issue by examining the reliability of deficits in just one of the test procedures used to assess OCD, the CANTAB IDED paradigm. This is because meta-analysis is best performed when the data are collected in a standardised test and this test is delivered in a computerised fashion, thus minimising possible variations in test procedure.

2 Methods

2.1 Search Strategy

The aim of this study was to conduct a systematic review and meta-analysis of studies examining CANTAB IDED performance in adult OCD patients versus controls. The PubMed database was searched using the following string, which

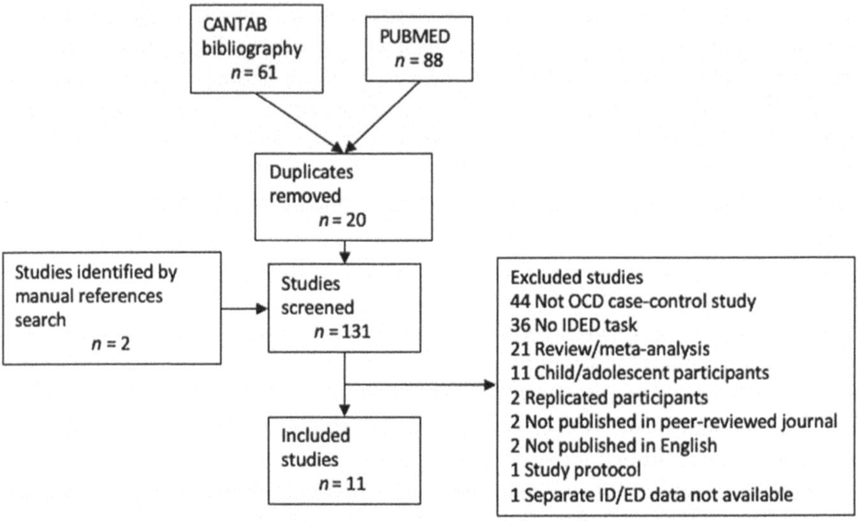

Fig. 2 Search strategy for the meta-analysis. Studies were identified from PubMed, the CANTAB online bibliography, and manual reading of reference lists

was determined by consensus among the authors: [("Obsessive-Compulsive Disorder" OR "Obsessiv*" OR "Compulsiv*" OR "OCD") AND ("CANTAB" OR "IDED" OR "SST" OR "set-shifting")]. Additionally, the online CANTAB bibliography (www.camcog.com) was searched using the categorisation header of 'Obsessive-Compulsive Disorder'. The study was not pre-registered. An overview of the process is presented in Fig. 2. These searches, carried out on 17/6/2019, yielded 129 studies after removal of duplicates. The reference lists of reviews and meta-analyses identified in the search, as well as the references lists of two further papers (Eng et al. 2015; Del Casale et al. 2011), were also screened for relevant studies. This resulted in two further studies being considered. The authors examined full texts to exclude papers that did not meet the inclusion criteria. Overall, 131 studies were considered and 11 were included in the meta-analysis.

2.2 Inclusion and Exclusion Criteria

We included all studies that: (a) were published in peer-reviewed journals between 1996 and May 2019; (b) were written in English or provided a translation; (c) included only adult (age >18) participants; (d) included a direct comparison between patients with a primary diagnosis of OCD and healthy controls; (e) examined attentional set-shifting using the CANTAB IDED task; (f) presented data separately for the Intra-Dimensional (ID) and Extra-Dimensional (ED) stages of the IDED task; and (g) included information necessary for calculating effect sizes. Where a paper had not reported separate results for the ID and ED stages of the task or necessary information to calculate effect sizes, we contacted the authors by email to request this information. We excluded studies that contained participants whose same data had been included in previous studies.

2.3 *Data Collection and Analysis*

Data from each study were extracted from the papers or provided by the given paper's authors. For both OCD and control groups, measures that were extracted comprised: (a) number of participants; (b) mean age; (c) gender distribution (reported as % male and % female); (d) mean IQ and the instrument used; (e) mean score for depressive symptoms and the instrument used; (f) mean ID errors; and (g) mean ED errors. For the OCD group only, additional data extracted comprised: (a) mean Yale–Brown Obsessive-Compulsive Scale (YBOCS) score; (b) comorbidity with other mental disorders, and (c) number of participants taking psychiatric medication at the time of study participation.

Data were analysed using statistical software R version 3.6.0 and RStudio Desktop 1.2.1335. Meta-analysis was performed using the 'metafor' package and a random-effects model was used. The R code used is available from the authors on request. To compare OCD and control groups for ID errors and ED errors, we used the mean and standard deviation to calculate the standardised mean difference (SMD) between groups. For Watkins et al. (2005), the Cohen effect size for ID and ED errors was computed manually. Statistical significance was defined as two-tailed $p < 0.05$. We used meta-regression to address potential effects of group differences in IQ and age on the IDED results. Each meta-regression model included group differences (SMD) for IQ and age, since both may affect aspects of cognition, and age may also affect IQ. Potential publication bias and heterogeneity were assessed using funnel plots. To confirm that no one study was responsible for a significant result, the 'leave-1-out' function was used to iteratively fit the model, leaving out one study at a time.

IQ data were not collected by Martoni et al. (2018), so this study was excluded from the moderator analysis examining effect of IQ on ID and ED findings. ID errors were not reported by Nedeljkovic et al. (2009) or Bersani et al. (2013), so these studies were excluded from the ID errors analysis.

3 Results

A total of 11 studies were included in the meta-analysis (Table 1), comprising a total sample of 335 OCD patients and 311 controls. The mean ages of participants were similar between studies, ranging from 32.1 to 41.5 years for OCD groups and 31.3 to 40.8 years for control groups. The percentage of participants who were male varied between 20.0–69.7% for OCD groups and 29.2–73.1% for control groups. Patient

Table 1 Sample characteristics for studies contributing to the meta-analysis

Study	Age		Gender, male		Sample size		Further patient details		
	PAT Mean (SD)	CON Mean (SD)	PAT N [%]	CON N [%]	PAT, N	CON, N	YBOCS Mean (SD)	Comorbidities reported	Medications reported
Veale et al. (1996)	36.1 (11.13)	32.3 (7.62)	17 [42.5]	16 [44.4]	40	36	NA	None	NA
Purcell et al. (1998)	40.6 (13.6)	40.8 (12.9)	10 [33.3]	12 [40]	30	30	24.1 (8.1)	Depression/anxiety symptoms	23 receiving medication
Nielen and Den Boer (2003)	37.2 (2.4)	31.3 (2.1)	8 [42.1]	7 [29.2]	19	24	24.1 (5.8)	None	Medication-free for 4 weeks
Watkins et al. (2005)	38.2 (13.4)	36.6 (12.8)	8 [40]	9 [45]	20	20	19.3 (14.3)	None	8 taking SSRIs, 1 taking clomipramine, 1 taking moclobemide and trifluoperazine. 10 unmedicated.
Fenger et al. (2005)	39 (15.6)	32.8 (10.4)	7 [46.7]	8 [47.1]	15	17	27.2 (6.2)	None	Medication-free for 4 weeks
Simpson et al. (2006)	41.47 (9.82)	37.86 (12.73)	15 [50]	15 [42.9]	30	35	19.5 (6.91)	None	11 taking psychotropic medication
Chamberlain et al. (2007)	32.1 (11.9)	33.1 (10.5)	4 [20]	7 [35]	20	20	22.7 (6.6)	No Axis 1 disorders	18 taking SSRIs, 2 unmedicated
Nedeljkovic et al. (2009)[a]	37.8 (13.3)	37.5 (12.8)	23 [39]	23 [39]	59	59	23.89 (6.77)	None	41 medicated, mostly with serotonergic medication
Bersani et al. (2013)	36.4 (12.3)	36.85 (13.27)	6 [24]	10 [47.6]	25	21	22.2 (7.05)	None	23 taking SSRIs and/or clomipramine along with benzodiazepines. 2 medication-free
Vaghi et al. (2017)	36.1 (10.7)	37.51 (12.05)	21 [47.7]	22 [51.2]	44	43	22 (5.31)	NA	27 medicated (all but one taking SSRIs), 17 unmedicated
Martoni et al. (2018)	32.42 (15.03)	34.38 (15.18)	23 [69.7]	19 [73.1]	33	26	26.04 (8.81)	One each of: Anorexia nervosa, Tourette syndrome, skin picking disorder, gambling disorder, panic disorder, major depressive disorder, trichotillomania, hoarding disorder	12 medication-free, 12 taking SSRIs, 9 taking two medications

[a]Weighted means for OCD subtypes – data pooled from OCD groups

medication status differed across samples. Mean YBOCS score for OCD participants ranged from 19.3 to 27.2 and percentage of OCD participants who were medicated during the studies ranged from 0 to 92%.

Results from the meta-analysis are provided in Fig. 3. OCD patients made significantly more ED errors than controls, with a moderate–large effect size (SMD $= 0.62$, $p = 0.0004$, 95% CI [0.28, 0.95]). The meta-regression model indicated that the ED deficit was not significantly affected by group differences in age or IQ (both $p > 0.06$). Patients also made significantly more ID errors than controls, but with a small effect size (SMD $= 0.36$, $p = 0.0005$, 95% CI [0.16, 0.56]). Again, the meta-regression model indicated that this was not significantly affected by group differences in age or IQ (both $p > 0.2$). Inspection of funnel plots indicated that there was no evidence of publication bias in any of the analyses, but that two studies appeared to be relative outliers for ED errors specifically (Fig. 4). Repeating the ED meta-analysis excluding these two outliers showed that ED set-shifting was still impaired with medium–large effect size (SMD $= 0.58$, $p < 0.0001$, 95% CI [0.41, 0.74]). Again, the moderator analysis did not indicate a significant influence of group differences in age or IQ (both $p > 0.29$).

Leave-1-out cross-validation indicated that the ID and ED findings retained statistical significance in all cases.

4 Discussion

Problems with cognitive inflexibility are suggested by the core symptoms of established OCD, and are also likely to contribute to other obsessive-compulsive (OC) related disorders characterised by repetitive stereotyped symptoms. Cognitive inflexibility may represent a mechanism by which patients with these disorders are prevented from suppressing their compulsive cycles of repetitive acts and thoughts, and if so may represent a target for therapeutic intervention. By conducting a systematic review and meta-analysis of the available case–control literature, this chapter confirmed that OCD was associated with deficits in EDS with a medium–large effect size. IDS was impaired to a milder extent (small effect size). These meta-analytic findings (total sample size 335 cases, 331 controls) confirm that, contrary to some conclusions in the literature (Simpson et al. 2006; Nielen and Den Boer 2003), EDS deficits are a robust finding in OCD. These meta-analytic results were not significantly affected by group differences in IQ, or age, in meta-regression.

Figure 5 illustrates the percentage of subjects attempting each stage of the IDED task who passed each of the stages of the IDED task, using data from studies for which this information was available (Veale et al. 1996; Watkins et al. 2005; Fenger et al. 2005; Vaghi et al. 2017; Purcell et al. 1998). This graphical overview also highlights the selectivity of the ED impairment in OCD, with a particularly pronounced difference on this stage of the task.

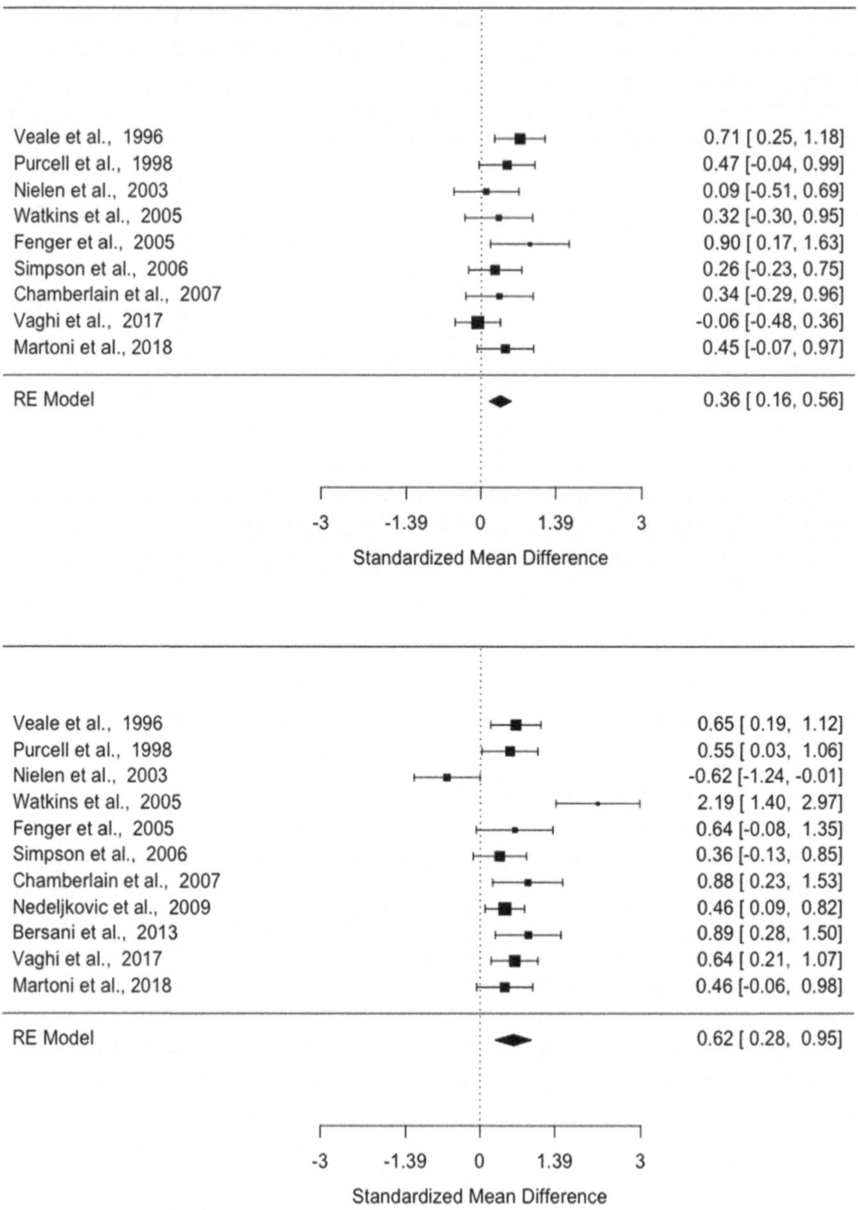

Fig. 3 Results from meta-analysis of performance on the IDED task. Top: Intra-Dimensional (ID) stage error performance; bottom: Extra-Dimensional (ED) stage error performance. Values to the right of null (0) indicate relatively worse performance in OCD patients compared to controls

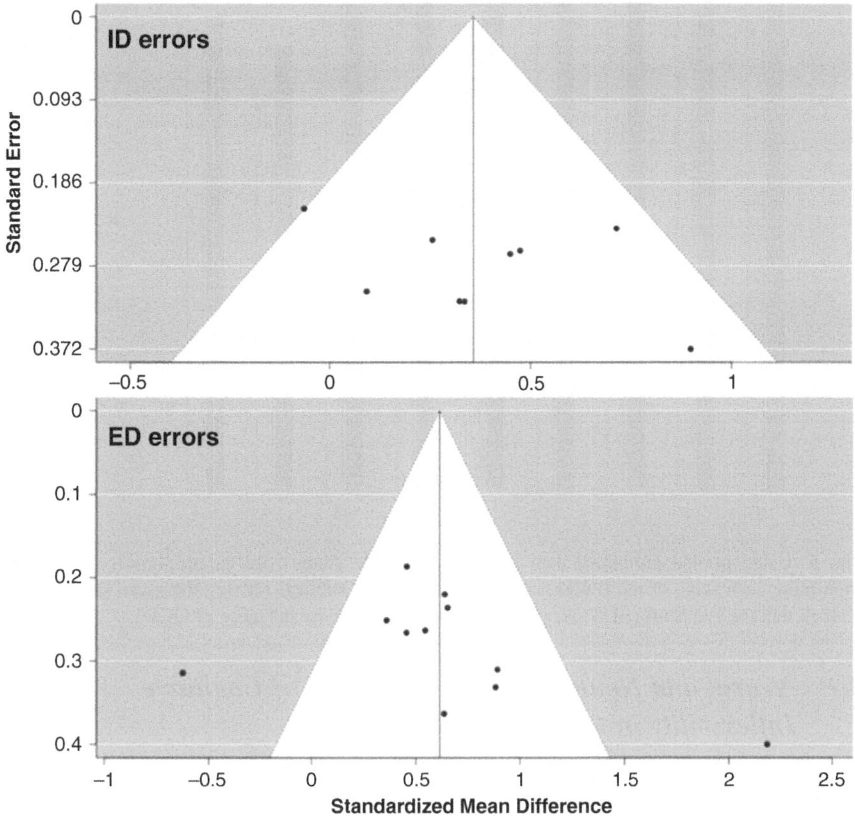

Fig. 4 Funnel plots. ID errors test for plot asymmetry, $p = 0.30$. ED errors test for plot asymmetry, $p = 0.33$. There was no evidence for publication bias. Two studies were potential outliers on the funnel plot for ED errors. The ED meta-analysis findings were extremely similar when these two papers were excluded, i.e. there was still a significant ED deficit (medium–large effect size)

ED deficits have also been associated with a dimensional measure of OC symptoms (the Padua inventory), in a cohort of non-treatment seeking young adults recruited from the general community ($N = 515$ participants) (Chamberlain et al. 2016). Specifically, OC symptoms correlated with ED errors ($r = 0.2, p < 0.001$) but not ID errors ($r < 0.01, p > 0.9$) in the sample; and furthermore, the EDS association retained significance in regression modelling controlling for other variables, including age, education, gender, substance use, and impulsivity (as measured using an inhibitory control cognitive task) (Chamberlain et al. 2016). These results collectively highlight the centrality of EDS deficits in understanding OCD both at the level of disorder, and dimensionally in the general population.

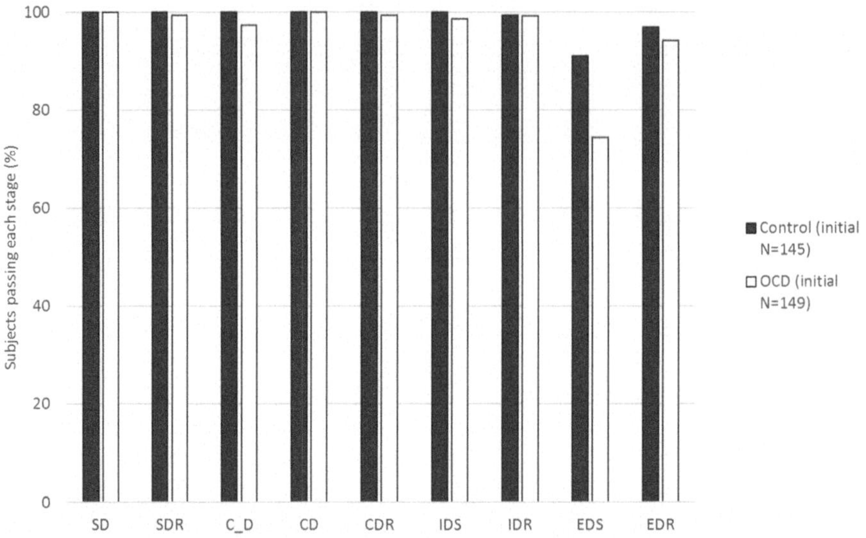

Fig. 5 Using pooled published data from the available studies, the graph shows, of subjects attempting each stage of the IDED task, the percentage of subjects passing the given stage. It can be seen that the ED Shift (EDS) stage is particularly affected by presence of OCD

4.1 Neural and Neurochemical Substrates of Cognitive Inflexibility in OCD

As described in detail in the introduction to this chapter, multiple tiers of translational research highlight the role of specific aspects of fronto-striatal circuitry, and associated monoaminergic transmitter systems, in the regulation of different aspects of cognitive flexibility (Kehagia et al. 2010; Walker et al. 2009; Clarke et al. 2005; Clarke et al. 2004). Anatomically, preclinical research indicates a particular role for the lateral prefrontal cortices in ED shifting, which is dissociable from neural regions involved in other processes such as reversal of responses or inhibition of pre-potent motor responses (Robbins 2007). In a recent study, relationships between cognitive flexibility on the IDED task and functional brain connectivity were evaluated in OCD patients ($N = 44$) and healthy controls ($N = 43$) (Vaghi et al. 2017). It was found that reduced functional connectivity between the ventrolateral prefrontal cortex and dorsal caudate nucleus was linked with poorer cognitive flexibility in the OCD patients (Fig. 6).

In a study of six patients examining effects of Deep Brain Stimulation (DBS) in severe treatment-refractory OCD, stimulation of the sub-thalamic nucleus was found to improve EDS flexibility, whereas stimulation of the ventral capsule/ventral striatum was not (Tyagi et al. 2019). The sub-thalamic nucleus acts as a node within the basal ganglia circuitry, receiving input from areas of frontal cortex via a 'hyperdirect' circuit and providing a potential mechanism for cortical influences to rapidly suppress behaviours programmed within the basal ganglia that are no longer

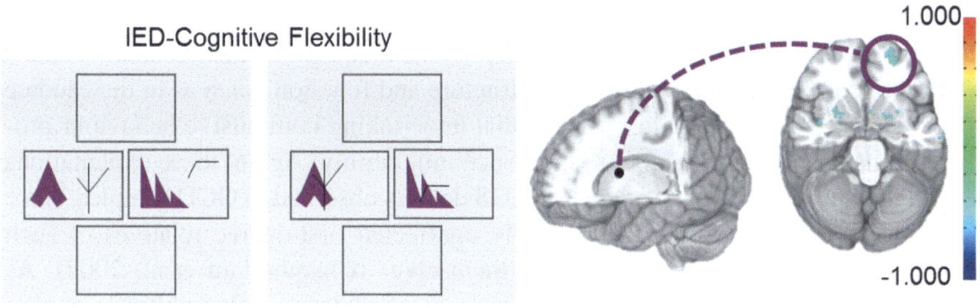

Fig. 6 Reduced connectivity between the left ventrolateral prefrontal cortex and left dorsal caudate was associated with worse ED shifting in OCD patients (Vaghi et al. 2017). Figure adapted under the Creative Commons Attribution License (CC BY), with permission of the authors, from Vaghi et al. (2017)

appropriate (Greenhouse et al. 2013). DBS to the subthalamic nucleus may, therefore, restore cognitive flexibility in OCD by remediating a dysfunctional hyperdirect circuit, and thereby enable patients to suppress their obsessive OC.

In terms of the neurochemical nature of the ED impairment observed in OCD, studies have yet to address this issue directly in humans, such as by using radioligands in conjunction with positron emission tomography. Preclinical literature indicates that serotonin does not play a cardinal role in EDS, but does play a role in other aspects of flexible learning (Robbins and Roberts 2007). In keeping with these data, several cognitive studies indicate that ED deficits in OCD occur irrespective of whether or not patients are receiving treatment with serotonin reuptake inhibitors, which constitute the first-line pharmacological treatment option for OCD (Chamberlain et al. 2007; Vaghi et al. 2017; Fineberg et al. 2015). While the effect of serotonin reuptake inhibitors on cognitive flexibility in the course of OCD treatment has not so far been adequately investigated, this nevertheless raises the prospect that ED problems in OCD may be more related to dysfunction of neurochemical systems other than serotonin and therefore require other treatment approaches.

4.2 *Cognitive Inflexibility as a Candidate Endophenotype for OCD*

Given that EDS deficits are common in OCD and appear unaffected by serotonin reuptake inhibitors, it is important to question the exact relationship between the symptoms and the deficit. Theoretically, some aspects of cognitive dysfunction in OCD could arise secondary to the symptoms – for example, being distracted by intrusive thoughts during neuropsychological assessment could then manifest as various types of cognitive dysfunction. This would seem unlikely in relation to the EDS deficit as it is selective for this stage and distractibility would be expected to manifest as impairment across most stages of the task, which was not observed

(Fig. 5). Another theoretical possibility is that repetitive engagement in compulsions could lead to changes in the basal ganglia, in turn fuelling the manifestation of EDS deficits due to plastic changes in brain structure and function (such as in the caudate nucleus). A parallel possibility may be that undertaking compulsive behaviour produces enhanced attentional focus, and hence inflexibility. Again, these explanations would seem implausible to account for EDS deficits observed in OCD samples, since impaired EDS is also found in clinically unaffected first-degree relatives of such patients, who have not had symptoms themselves (Chamberlain et al. 2007). As such, EDS deficits may constitute a candidate vulnerability marker (or 'endophenotype') existing in people at risk of developing the disorder.

4.3 *Cognitive Inflexibility in Other Related Clinical Settings and Disorders*

While this meta-analysis focused on adult OCD, it may also be informative to consider literature in childhood/adolescent OCD. Research on this topic is extremely sparse and there are insufficient data for meta-analysis using the IDED task. In a sample of adolescent participants, OCD was associated with significant impairment in the stages prior to ED shifting (including IDS), indicative of difficulties forming the attentional set, and EDS was not significantly impaired (Gottwald et al. 2018). This is to be expected since if the attentional set is not formed, then EDS impairment would not be detectable. This raises the interesting possibility that cognitive flexibility, and its dysfunction in OCD, may be contingent on the neurodevelopmental stage of the population being examined. There may be more generalised learning problems in younger OCD patients, masking an EDS deficit, though future work is needed to explore this issue in more detail due to the relative poverty of studies. Future work examining cognition in OCD longitudinally would be valuable, perhaps making use of population-level cohorts and dimensional measures of OCD, though such cohort(s) would need to include suitable neuropsychological tests, an issue often overlooked when designing such studies.

The DSM-5 category of Obsessive-Compulsive Related Disorders has OCD as the archetypal compulsive disorder, but it is salient to also consider IDED findings in the other conditions in this class. In one study, OCD patients with hoarding ($n = 24$) had significantly higher EDS errors than controls ($n = 28$), whereas a group of pure 'Hoarding Disorder' cases ($n = 22$, with no OCD) showed performance intermediate between controls and OCD patients with hoarding (Morein-Zamir et al. 2014). This, combined with the findings from our meta-analysis, indicates that EDS deficits appear to generalise to hoarding within OCD, whereas deficits may be milder or non-existent in pure Hoarding Disorder. For Body Dysmorphic Disorder (BDD), a study in patients with this primary diagnosis ($n = 12$) found significant EDS impairment with large effect size, compared to controls ($n = 16$) (Jefferies-Sewell et al. 2017). Turning to trichotillomania (hair-pulling disorder), the available literature suggests this condition is not characterised by an EDS deficit. For example, a study in comorbidity free patients with trichotillomania ($n = 17$) showed intact EDS versus controls ($n = 20$), whereas OCD patients ($n = 20$) were impaired versus controls; both clinical groups had a significant deficit in another cognitive domain, namely response inhibition (inhibition of pre-potent motor responses) (Chamberlain et al. 2006).

Collectively, these data indicate (1) that EDS deficits generalise across OCD (including patients with predominantly hoarding symptoms); and (2) some (BDD) but not all (trichotillomania) OC Related Conditions have similar EDS deficits. EDS deficits may also be found in conditions with compulsive features, not currently listed in the same diagnostic category. For example, gambling disorder is listed in DSM-5 category of Substance-Related and Addictive Disorders. In a meta-analysis of case–control cognitive studies in gambling disorder, inflexible performance was found on a variety of tests, including the CANTAB IDED, though relatively few studies had used this task (van Timmeren et al. 2018). More research work is needed to delineate overlapping and distinct cognitive problems across these conditions whose symptoms indicate compulsivity, using standardised neuropsychological testing approaches.

Contrasting with the overall profile of IDED task performance in OCD (i.e. selective EDS impairment), schizophreniform disorders are typically characterised by generalised impairment across many stages of the paradigm. For example, patients with first-episode psychosis ($n = 262$) compared to healthy control ($n = 76$) showed deficits on 7 of 9 task stages, including not only EDS deficits, but also relating to reversal learning and other aspects of performance (Leeson et al. 2009). In a study conducted in patients with schizophrenia ($n = 32$), significant deficits were found versus controls ($n = 44$) across multiple stages of the task, including aspects of reversal, compound stimulus discrimination, and EDS (Elliott et al. 1995). Similarly, another study found that patients with schizophrenia ($n = 51$) had deficits on discrimination, reversal, IDS, and EDS stages, versus controls ($n = 18$) (Pantelis et al. 1999). As noted previously, unlike in OCD, it appears that EDS impairment does not constitute a candidate endophenotype for schizophrenia (Ceaser et al. 2008). Additionally, it has been found that patients with Schizo-OCD (i.e. OCD and schizophrenia; $n = 12$) had selective EDS impairment compared to patients with schizophrenia ($n = 16$) (Patel et al. 2010). The selective EDS impairment associated with OCD is unlikely to be attributable to anxiety. EDS deficits have been shown in OCD patients with no comorbid anxiety (or other) disorders, extending also to their first-degree relatives who had no disorders (Chamberlain et al. 2007). In a trans-diagnostic study of young people, those with clinically significant anxiety disorder symptoms were not impaired on IDED task performance; in fact, there was some evidence they had better performance than people with normal levels of anxiety on the task (Murphy et al. 2018).

4.4 *One Thing or Many? Towards Rich, Delineated Measures of Compulsivity*

In contrast to impulsivity, for which many cognitive tasks and trans-diagnostic scales exist and are known to quantify delineated underlying components of impulsivity (Ioannidis et al. 2019; Winstanley et al. 2006; Meda et al. 2009; MacKillop et al. 2016; Stanford et al. 2016), the literature on measures of compulsivity for use in human studies is more circumscribed (Robbins et al. 2019). The current chapter has focused on the IDED task due to its translational heritage and ability to fractionate distinct aspects of learning and flexible responding. We acknowledge nonetheless that aspects of flexible responding have been examined using a variety of paradigms in OCD. For example, a meta-analysis confirmed inflexible responding in OCD using the WCST (Snyder et al. 2015), which is conceptually related to the IDED test. In terms of fractionating different processes, apart from ED shifting, reversal learning has been examined, but is often insensitive to deficits in disorders, likely in part due to ceiling effects – this being a relatively 'simple' cognitive process in humans. Furthermore, inflexibility deficits in OCD do not necessarily extend to other types of paradigm such as task-switching (Moritz et al. 2004). In general, the concept of cognitive flexibility needs to be subjected to more detailed analysis across a range of tasks, to determine if it is unitary or has distinct sub-components.

In addition to developing understanding of cognitive processes germane to compulsivity, it will be important to consider further whether cognitive problems in OCD are specific or general; put differently, are some cognitive domains more impaired in OCD than others, not only within the context of compulsivity, but also other domains (e.g. working memory, planning, inhibitory control)? Because cognition is influenced by many factors such as age, education, IQ and comorbidities (e.g. anxiety and depression), this issue is challenging to address using the existing body of literature, due to studies often failing to measure and control for confounding group differences. In our own studies we have attempted to address this issue by carefully controlling for potential confounds (age, education level, IQ), and comorbidities (e.g. using comorbidity free samples), and results suggest EDS deficits are particularly prominent in OCD, and extend to first-degree relatives who do not have symptoms.

Promising lines of recent work have focused on developing other measures relevant to the study of compulsivity, including habit-learning (alternatively, dysfunction in goal-related control) paradigms (viz. outcome devaluation and contingency degradation) (Gillan et al. 2015; de Wit et al. 2018), a modified variant of the Value-Modulated Attentional Capture (VMAC task) (Albertella et al. 2019), and a trans-diagnostic compulsivity scale (Chamberlain and Grant 2018). Further development, refinement, and deployment of such newer measures, both in case–control and population level studies, would be extremely valuable. In addition to being relevant to OCD, an expert Delphi consensus has explicitly listed compulsivity as an important construct in a comprehensive theoretical framework for addictions (substance and behavioural) research (Yucel et al. 2019).

4.5 Concluding Remarks

In conclusion, this chapter has confirmed using meta-analysis that EDS deficits on a computerised neuropsychological test (CANTAB IDED) constitute a robust finding in OCD. This deficit has been linked, in a neuroimaging study, to aberrant connectivity of fronto-striatal circuitry, between the ventrolateral prefrontal cortex and basal ganglia (caudate nucleus) (Vaghi et al. 2017). Difficulty modulating behaviour in light of changing environmental information is likely to contribute to the functional impairments observed in OCD and other related disorders of compulsivity (Chamberlain and Grant 2019). As such, compulsivity on neuropsychological tests, such as manifested by ED impairment, constitutes key treatment targets for novel interventions, particularly given that existing first-line pharmacological treatments for OCD do not appear to fully ameliorate such deficits (Vaghi et al. 2017; Chamberlain et al. 2007).

Disclosures Dr. Chamberlain consults for Promentis and Ieso Digital Health. He receives a stipend from Elsevier for editorial work at Comprehensive Psychiatry, and at Neuroscience & Biobehavioral Reviews.

Dr. Chamberlain's research is funded by a Wellcome Trust Clinical Fellowship (110,049/Z/15/Z). Professor Robbins' research is funded by a Wellcome Trust Grant (104,631/Z/14/Z). Professor Robbins undertakes consulting work for Cambridge Cognition, Unilever, and Greenfield Bioventures; he has received research grants from Shionogi, GlaxoSmithKline, and Small Pharma; he receives royalties from Cambridge Cognition; and has 484 received editorial honoraria from Springer Nature and Elsevier. The other authors 485 have no disclosures.

References

Albertella L, Watson P, Yucel M, Le Pelley ME (2019) Persistence of value-modulated attentional capture is associated with risky alcohol use. Addict Behav Rep 10:100195

Barlow RL, Alsio J, Jupp B, Rabinovich R, Shrestha S, Roberts AC, Robbins TW, Dalley JW (2015) Markers of serotonergic function in the orbitofrontal cortex and dorsal raphe nucleus predict individual variation in spatial-discrimination serial reversal learning. Neuropsychopharmacology 40:1619–1630

Bersani G, Quartini A, Ratti F, Pagliuca G, Gallo A (2013) Olfactory identification deficits and associated response inhibition in obsessive-compulsive disorder: on the scent of the orbitofronto-striatal model. Psychiatry Res 210:208–214

Birrell JM, Brown VJ (2000) Medial frontal cortex mediates perceptual attentional set shifting in the rat. J Neurosci 20:4320–4324

Ceaser AE, Goldberg TE, Egan MF, Mcmahon RP, Weinberger DR, Gold JM (2008) Set-shifting ability and schizophrenia: a marker of clinical illness or an intermediate phenotype? Biol Psychiatry 64:782–788

Chamberlain SR, Grant JE (2018) Initial validation of a transdiagnostic compulsivity questionnaire: the Cambridge-Chicago compulsivity trait scale. CNS Spectr:1–7
Chamberlain SR, Grant JE (2019) Relationship between quality of life in young adults and impulsivity/compulsivity(). Psychiatry Res 271:253–258
Chamberlain SR, Blackwell AD, Fineberg NA, Robbins TW, Sahakian BJ (2005) The neuropsychology of obsessive compulsive disorder: the importance of failures in cognitive and behavioural inhibition as candidate endophenotypic markers. Neurosci Biobehav Rev 29:399–419
Chamberlain SR, Fineberg NA, Blackwell AD, Robbins TW, Sahakian BJ (2006) Motor inhibition and cognitive flexibility in obsessive-compulsive disorder and trichotillomania. Am J Psychiatry 163:1282–1284
Chamberlain SR, Fineberg NA, Menzies LA, Blackwell AD, Bullmore ET, Robbins TW, Sahakian BJ (2007) Impaired cognitive flexibility and motor inhibition in unaffected first-degree relatives of patients with obsessive-compulsive disorder. Am J Psychiatry 164:335–338
Chamberlain SR, Leppink EW, Redden SA, Grant JE (2016) Are obsessive-compulsive symptoms impulsive, compulsive or both? Compr Psychiatry 68:111–118
Clarke HF, Dalley JW, Crofts HS, Robbins TW, Roberts AC (2004) Cognitive inflexibility after prefrontal serotonin depletion. Science 304:878–880
Clarke HF, Walker SC, Crofts HS, Dalley JW, Robbins TW, Roberts AC (2005) Prefrontal serotonin depletion affects reversal learning but not attentional set shifting. J Neurosci 25:532–538
Clarke HF, Robbins TW, Roberts AC (2008) Lesions of the medial striatum in monkeys produce perseverative impairments during reversal learning similar to those produced by lesions of the orbitofrontal cortex. J Neurosci 28:10972–10982
Cohen JD, McClure SM, Yu AJ (2007) Should I stay or should I go? How the human brain manages the trade-off between exploitation and exploration. Philos Trans R Soc Lond Ser B Biol Sci 362:933–942
Danet M, Lapiz-Bluhm S, Morilak DA (2010) A cognitive deficit induced in rats by chronic intermittent cold stress is reversed by chronic antidepressant treatment. Int J Neuropsychopharmacol 13:997–1009
De Wit S, Kindt M, Knot SL, Verhoeven AAC, Robbins TW, Gasull-Camos J, Evans M, Mirza H, Gillan CM (2018) Shifting the balance between goals and habits: five failures in experimental habit induction. J Exp Psychol Gen 147:1043–1065
Del Casale A, Kotzalidis GD, Rapinesi C, Serata D, Ambrosi E, Simonetti A, Pompili M, Ferracuti S, Tatarelli R, Girardi P (2011) Functional neuroimaging in obsessive-compulsive disorder. Neuropsychobiology 64:61–85
Dias R, Robbins TW, Roberts AC (1996) Dissociation in prefrontal cortex of affective and attentional shifts. Nature 380:69–72
Downes JJ, Roberts AC, Sahakian BJ, Evenden JL, Morris RG, Robbins TW (1989) Impaired extra-dimensional shift performance in medicated and unmedicated Parkinson's disease: evidence for a specific attentional dysfunction. Neuropsychologia 27:1329–1343
Elliott R, McKenna PJ, Robbins TW, Sahakian BJ (1995) Neuropsychological evidence for frontostriatal dysfunction in schizophrenia. Psychol Med 25:619–630
Eng GK, Sim K, Chen SH (2015) Meta-analytic investigations of structural grey matter, executive domain-related functional activations, and white matter diffusivity in obsessive compulsive disorder: an integrative review. Neurosci Biobehav Rev 52:233–257
Fenger MM, Gade A, Adams KH, Hansen ES, Bolwig TG, Knudsen GM (2005) Cognitive deficits in obsessive-compulsive disorder on tests of frontal lobe functions. Nord J Psychiatry 59:39–44
Fineberg NA, Reghunandanan S, Simpson HB, Phillips KA, Richter MA, Matthews K, Stein DJ, Sareen J, Brown A, Sookman D, Accreditation Task Force Of The Canadian Institute For Obsessive Compulsive, D (2015) Obsessive-compulsive disorder (Ocd): practical strategies for pharmacological and somatic treatment in adults. Psychiatry Res 227:114–125

Friedman NP, Miyake A, Corley RP, Young SE, Defries JC, Hewitt JK (2006) Not all executive functions are related to intelligence. Psychol Sci 17:172–179

Gillan CM, Robbins TW, Sahakian BJ, Van Den Heuvel OA, Van Wingen G (2015) The role of habit in compulsivity. Eur Neuropsychopharmacol 26(5):828–840

Gottwald J, De Wit S, Apergis-Schoute AM, Morein-Zamir S, Kaser M, Cormack F, Sule A, Limmer W, Morris AC, Robbins TW, Sahakian BJ (2018) Impaired cognitive plasticity and goal-directed control in adolescent obsessive-compulsive disorder. Psychol Med 48:1900–1908

Greenhouse I, Gould S, Houser M, Aron AR (2013) Stimulation of contacts in ventral but not dorsal subthalamic nucleus normalizes response switching in Parkinson's disease. Neuropsychologia 51:1302–1309

Hampshire A, Owen AM (2006) Fractionating attentional control using event-related FMRI. Cereb Cortex 16:1679–1689

Ioannidis K, Hook R, Wickham K, Grant JE, Chamberlain SR (2019) Impulsivity in gambling disorder and problem gambling: a meta-analysis. Neuropsychopharmacology 44:1354–1361

Jackson SAW, Horst NK, Axelsson SFA, Horiguchi N, Cockcroft GJ, Robbins TW, Roberts AC (2019) Selective role of the putamen in serial reversal learning in the marmoset. Cereb Cortex 29:447–460

Jefferies-Sewell K, Chamberlain SR, Fineberg NA, Laws KR (2017) Cognitive dysfunction in body dysmorphic disorder: new implications for nosological systems and neurobiological models. CNS Spectr 22:51–60

Kehagia AA, Murray GK, Robbins TW (2010) Learning and cognitive flexibility: frontostriatal function and monoaminergic modulation. Curr Opin Neurobiol 20:199–204

Lawrence AD, Hodges JR, Rosser AE, Kershaw A, Ffrench-Constant C, Rubinsztein DC, Robbins TW, Sahakian BJ (1998) Evidence for specific cognitive deficits in preclinical Huntington's disease. Brain 121(Pt 7):1329–1341

Leeson VC, Robbins TW, Matheson E, Hutton SB, Ron MA, Barnes TR, Joyce EM (2009) Discrimination learning, reversal, and set-shifting in first-episode schizophrenia: stability over six years and specific associations with medication type and disorganization syndrome. Biol Psychiatry 66:586–593

Lezak MD, Howieson DB, Loring DW, Hannay HJ, Fischer JS (2004) Neuropsychological assessment. Oxford University Press, New York

Mackillop J, Weafer JJCG, Oshri A, Palmer A, De Wit H (2016) The latent structure of impulsivity: impulsive choice, impulsive action, and impulsive personality traits. Psychopharmacology 233:3361–3370

Martoni RM, Risso G, Giuliani M, De Filippis R, Cammino S, Cavallini C, Bellodi L (2018) Evaluating proactive strategy in patients with OCD during stop signal task. J Int Neuropsychol Soc 24:703–714

Meda SA, Stevens MC, Potenza MN, Pittman B, Gueorguieva R, Andrews MM, Thomas AD, Muska C, Hylton JL, Pearlson GD (2009) Investigating the behavioral and self-report constructs of impulsivity domains using principal component analysis. Behav Pharmacol 20:390–399

Menzies L, Achard S, Chamberlain SR, Fineberg N, Chen CH, Del Campo N, Sahakian BJ, Robbins TW, Bullmore E (2007) Neurocognitive endophenotypes of obsessive-compulsive disorder. Brain 130:3223–3236

Milner B (1963) Effects of different brain lesions on card sorting: the role of the frontal lobes. Arch Neurol 9:90–100

Morein-Zamir S, Papmeyer M, Pertusa A, Chamberlain SR, Fineberg NA, Sahakian BJ, Mataix-Cols D, Robbins TW (2014) The profile of executive function in OCD hoarders and hoarding disorder. Psychiatry Res 215:659–667

Moritz S, Hubner M, Kluwe R (2004) Task switching and backward inhibition in obsessive-compulsive disorder. J Clin Exp Neuropsychol 26:677–683

Murphy YE, Luke A, Brennan E, Francazio S, Christopher I, Flessner CA (2018) An investigation of executive functioning in pediatric anxiety. Behav Modif 42:885–913

Nedeljkovic M, Kyrios M, Moulding R, Doron G, Wainwright K, Pantelis C, Purcell R, Maruff P (2009) Differences in neuropsychological performance between subtypes of obsessive-compulsive disorder. Aust N Z J Psychiatry 43:216–226

Nielen MM, Den Boer JA (2003) Neuropsychological performance of OCD patients before and after treatment with fluoxetine: evidence for persistent cognitive deficits. Psychol Med 33:917–925

Owen AM, Roberts AC, Polkey CE, Sahakian BJ, Robbins TW (1991) Extra-dimensional versus intra-dimensional set shifting performance following frontal lobe excisions, temporal lobe excisions or amygdalo-hippocampectomy in man. Neuropsychologia 29:993–1006

Pantelis C, Barber FZ, Barnes TR, Nelson HE, Owen AM, Robbins TW (1999) Comparison of set-shifting ability in patients with chronic schizophrenia and frontal lobe damage. Schizophr Res 37:251–270

Patel DD, Laws KR, Padhi A, Farrow JM, Mukhopadhaya K, Krishnaiah R, Fineberg NA (2010) The neuropsychology of the schizo-obsessive subtype of schizophrenia: a new analysis. Psychol Med 40:921–933

Purcell R, Maruff P, Kyrios M, Pantelis C (1998) Neuropsychological deficits in obsessive-compulsive disorder: a comparison with unipolar depression, panic disorder, and normal controls. Arch Gen Psychiatry 55:415–423

Rahman S, Sahakian BJ, Hodges JR, Rogers RD, Robbins TW (1999) Specific cognitive deficits in mild frontal variant frontotemporal dementia. Brain 122(Pt 8):1469–1493

Robbins TW (2007) Shifting and stopping: fronto-striatal substrates, neurochemical modulation and clinical implications. Philos Trans R Soc Lond Ser B Biol Sci 362:917–932

Robbins TW, Roberts AC (2007) Differential regulation of fronto-executive function by the monoamines and acetylcholine. Cereb Cortex 17(Suppl 1):i151–i160

Robbins TW, Vaghi MM, Banca P (2019) Obsessive-compulsive disorder: puzzles and prospects. Neuron 102:27–47

Simpson HB, Rosen W, Huppert JD, Lin SH, Foa EB, Liebowitz MR (2006) Are there reliable neuropsychological deficits in obsessive-compulsive disorder? J Psychiatr Res 40:247–257

Snyder HR, Kaiser RH, Warren SL, Heller W (2015) Obsessive-compulsive disorder is associated with broad impairments in executive function: a meta-analysis. Clin Psychol Sci 3:301–330

Sprengelmeyer R, Lange H, Homberg V (1995) The pattern of attentional deficits in Huntington's disease. Brain 118(Pt 1):145–152

Stanford MS, Mathias CW, Dougherty DM, Lake SL, Anderson NE, Patton JH (2016) Fifty years of the Barratt impulsiveness scale: an update and review. Personal Individ Differ 47:385–395

Tyagi H, Apergis-Schoute AM, Akram H, Foltynie T, Limousin P, Drummond LM, Fineberg NA, Matthews K, Jahanshahi M, Robbins TW, Sahakian BJ, Zrinzo L, Hariz M, Joyce EM (2019) A randomized trial directly comparing ventral capsule and anteromedial subthalamic nucleus stimulation in obsessive-compulsive disorder: clinical and imaging evidence for dissociable effects. Biol Psychiatry 85:726–734

Vaghi MM, Vertes PE, Kitzbichler MG, Apergis-Schoute AM, Van Der Flier FE, Fineberg NA, Sule A, Zaman R, Voon V, Kundu P, Bullmore ET, Robbins TW (2017) Specific frontostriatal circuits for impaired cognitive flexibility and goal-directed planning in obsessive-compulsive disorder: evidence from resting-state functional connectivity. Biol Psychiatry 81:708–717

Van Timmeren T, Daams JG, Van Holst RJ, Goudriaan AE (2018) Compulsivity-related neurocognitive performance deficits in gambling disorder: a systematic review and meta-analysis. Neurosci Biobehav Rev 84:204–217

Veale DM, Sahakian BJ, Owen AM, Marks IM (1996) Specific cognitive deficits in tests sensitive to frontal lobe dysfunction in obsessive-compulsive disorder. Psychol Med 26:1261–1269

Walker SC, Robbins TW, Roberts AC (2009) Differential contributions of dopamine and serotonin to orbitofrontal cortex function in the marmoset. Cereb Cortex 19:889–898

Watkins LH, Sahakian BJ, Robertson MM, Veale DM, Rogers RD, Pickard KM, Aitken MR, Robbins TW (2005) Executive function in Tourette's syndrome and obsessive-compulsive disorder. Psychol Med 35:571–582

Winstanley CA, Eagle DM, Robbins TW (2006) Behavioral models of impulsivity in relation to ADHD: translation between clinical and preclinical studies. Clin Psychol Rev 26:379–395

Yucel M, Oldenhof E, Ahmed SH, Belin D, Billieux J, Bowden-Jones H, Carter A, Chamberlain SR, Clark L, Connor J, Daglish M, Dom G, Dannon P, Duka T, Fernandez-Serrano MJ, Field M, Franken I, Goldstein RZ, Gonzalez R, Goudriaan AE, Grant JE, Gullo MJ, Hester R, Hodgins DC, Le Foll B, Lee RSC, Lingford-Hughes A, Lorenzetti V, Moeller SJ, Munafo MR, Odlaug B, Potenza MN, Segrave R, Sjoerds Z, Solowij N, Van Den Brink W, Van Holst RJ, Voon V, Wiers R, Fontenelle LF, Verdejo-Garcia A (2019) A transdiagnostic dimensional approach towards a neuropsychological assessment for addiction: an international Delphi consensus study. Addiction 114:1095–1109

Recent Developments in the Habit Hypothesis of OCD and Compulsive Disorders

Claire M. Gillan

Contents

Abstract This chapter aims to familiarise the reader with a diverse and fast-growing literature concerning the role that habits play in obsessive-compulsive disorder (OCD). Core concepts will be introduced, including how the balance between habits and a more deliberate form of action selection (goal-directed control) has traditionally been measured and how cross-species translation, neuroscience tools, and computational modelling have been used to build on these basic principles and reveal core mechanisms under study today. Next, the application of these methods to the study of OCD and related disorders will be detailed, converging on a theory that enhanced habit expression, and indeed compulsions in OCD, might arise from deficits in goal-directed control systems. These basic findings will be contextualised in terms of major tide changes in the field, including the shift from categorical disease frameworks to dimensional ones. Mechanistically, recent research concerning how goal-directed deficits arise, perhaps through failures in the construction of a mental model, are discussed along with studies critically evaluating our ability to measure habits in humans, in a laboratory setting. The chapter ends with a nod to the future, focusing on the need for clinically oriented, longitudinal, and

C. M. Gillan (✉)
School of Psychology, Trinity College Dublin, Dublin, Ireland
e-mail: gillancl@tcd.ie

Curr Topics Behav Neurosci (2021) 49: 147–167
https://doi.org/10.1007/7854_2020_199
Published Online: 6 February 2021

intervention-based research that aim to translate what is now a wealth of cross-sectional mechanistic insights to the clinic.

Keywords Compulsivity · Habit · Metacognition · OCD; Goal-directed · Reinforcement learning

1 Introduction

Habit-learning is one of our most fundamental capacities, but by virtue of their automatic nature, the importance of habits in our lives is often overlooked. Imagine if getting dressed each day took as much effort as learning the rules of chess, or if each time we changed gears on a car, we first had to generate a mental map of all possible options? Without our capacity to automate routine and well-practiced actions, the human brain would be consumed with the simplest rituals of everyday life, leaving no room for sophisticated behaviour and thought, conversation, creativity, and problem-solving (Dickinson 1985). Studied extensively in rodents (Dickinson 1980; Adams and Dickinson 1981a; Dickinson 1985; Balleine and Dickinson 1998), habits are responses that are automatically triggered by stimuli ('stimulus–response' behaviours), rather than being intentional and 'goal-directed', i.e. driven by a knowledge of (Tolman 1948) and desire for the consequences of action (Dickinson and Balleine 1994).

The key benefit of automatic behaviours is that they can be performed without engaging higher order cognitive processes (Poldrack et al. 2005), allowing us to multi-task, to think while we act, and to chain simple habits to learn more elaborate sequences like driving a car (Graybiel 1998; Dezfouli and Balleine 2013; Cushman and Morris 2015). With the clear benefits of habit come some drawbacks; habits are a cognitive short-cut and as such, they can lead us to make mistakes or 'slips of habit', like starting to type an old password before catching yourself. As this chapter will outline, in the extreme, habits are thought to play a major role in OCD and a range of compulsive disorders where behaviours become somewhat autonomous from our true goals or intentions (Gillan and Robbins 2014).

The notion that dysfunction in the brain's habit system contributes to the compulsive behaviours that characterise OCD had its original basis in three related sources of evidence. The first is clinical; like habits, compulsive behaviours persist despite adverse consequences, are rigid and repetitive in form, and are experienced as out-of-control by the patient (Graybiel and Rauch 2000; Fineberg et al. 2010). Take the example of a person with OCD with checking compulsions who feels compelled to check that their car door is locked multiple times a day: this repetitive act requires one to leave one's place of work at regular intervals, and as a result, is having a negative impact on their ability to maintain employment. One of the most striking features of OCD is that this compulsion persists even though the person is able to report that the act is unnecessary, or even absurd, because they *know* the car is locked (though there are variations in insight in OCD (Jacob et al. 2014)). In other words, these compulsive behaviours are 'ego-dystonic' and represent an internal conflict between what we know (reflection) and what we feel (instinct).

The second line of evidence is neurobiological; as outlined in more detail in other chapters, abnormalities within circuits that run from the frontal lobes to the striatum, and via direct and indirect pathways to the thalamus and back to the frontal cortex, the 'fronto-striatal loops', are characteristic of OCD (Alexander et al. 1986; Saxena and Rauch 2000; Menzies et al. 2008; Milad and Rauch 2012). These same circuits, and in particular some specific regions subsumed by them (i.e. the caudate, putamen, orbitofrontal cortex), play a key role in habit formation, habit expression, and 'goal-directed' control of habit (Graybiel and Rauch 2000; Dolan and Dayan 2013). Although these regions serve many functions, coupled with the clinical parallels, dysfunction in habit-forming processes has thus emerged as a likely cognitive candidate.

The final source of evidence comes from a neighbouring literature focused on how drug use becomes habitual over time, as behaviour transitions from impulsive to compulsive following sustained reinforcement (Robbins and Everitt 1999). Rather than focusing, at least originally, on trait susceptibility to habit formation, this body of work concerns the process through which normal habit-forming mechanisms are co-opted by drugs of abuse and lead addicted individuals to persist in self-injurious behaviour, despite the major negative consequences they suffer. Given the phenomenological parallel with compulsive behaviours in OCD, lessons from this field pointed to a potential role for habit systems in the maintenance, if not the provenance of symptoms.

Over the past two decades, the role of habit in OCD has become a complex and rapidly developing area of research, with several ongoing lines of research that aim to translate basic research into clinical applications. In this chapter, I will start from the beginning, first delineating how experimentalists operationalise habit formation alongside something called 'goal-directed control'; a cognitive capacity that curtails our habits. Next, I will outline the state of the art, in terms of empirical evidence for a disruption in the balance between these systems in OCD. In doing so, the specific mechanisms thought to give rise to this 'imbalance' will be outlined, drawing on brain imaging studies, translational models in rodents, and more recent computational methods of learning and behaviour. The chapter aims to highlight not just the major advances in this area, but also key gaps that research is only just beginning to address; these include recent efforts to move beyond categorical disease frameworks to dimensional and transdiagnostic perspectives, research examining how habits relate to obsessional beliefs or the 'mental model' of patients, and the limitations in our current methods for capturing habits in healthy individuals and causally inducing them. The chapter ends with a critical discussion of the clinical utility of this framework, focusing on the need for well-powered longitudinal, predictive research that can translate mechanistic insights to the clinic.

2 Outcome Devaluation and the Balance Between Habits and Goal-Directed Behaviours

Habits are behaviours that are elicited by stimuli, automatic and fast, rather than carefully selected to best satisfy our current wants and needs. Habits are thought to develop through practice, and as such flourish within our *routines* – the often-repeated acts of everyday life, such as dressing, bathing, commuting, or cooking are performed fluidly by engaging multiple micro-habits without the need for concentration (e.g. buttoning; scrubbing, driving, and whisking). Goal-directed actions, on the other hand, are slow and considered, designed to meet our ever-changing needs, using not just instinct, but planning, simulation, and reflection to achieve the best possible outcome (Dickinson 1985). Guided initially by translational research and more recently human functional neuroimaging, distinct and mechanistically precise definitions of these processes have been developed (Dolan and Dayan 2013). The putamen and its connection to the premotor cortex are thought to be critical in the development and maintenance of stimulus–response habits (Yin et al. 2004; Tricomi et al. 2009; de Wit et al. 2012), while the caudate and orbitofrontal cortex are critical for linking actions to their consequences and using motivation to drive flexible goal-directed actions (Yin et al. 2005; Valentin et al. 2007; Tanaka et al. 2008; Wunderlich et al. 2012). Despite the wide agreement with respect to the neuroanatomical and mechanistic distinction between these processes, their unique contribution to choice has been difficult to distinguish behaviourally. Studies have largely relied on a reciprocal definition, where habits and goal-directed control are operationalised as two sides of a coin. That is, behaviour is deemed habitual if it lacks the characteristics of goal-directed control, and vice versa.

The most commonly used paradigm for measuring the relative contribution of goal-directed versus habitual associations to choice is the 'outcome devaluation test'. Goal-directed actions are intended to achieve a current goal, and it thus follows that when a goal is no longer available, or indeed valuable, behaviour should be terminated. This premise is operationalised in the outcome devaluation test, whereby rewards that were formerly associated with an action are reduced in value (i.e. devalued) and the experimenter tests if the probability of action reduces in-kind. In the first demonstration of this effect, Adams (1982) trained rats to perform an action to earn food pellets, before reducing the value of the pellets through taste-aversion (pairing ingestion of the pellets with nausea-inducing lithium chloride injections). He found that, to the surprise of some, rats updated their behaviour in the operant chamber in response to this change in motivation for the pellets. That is, they reduced response rates, showing goal-directed control over their behaviour. However, in a separate group of rats who had received more training with the rewarding response, prior to devaluation, this sensitivity to devaluation was absent; through repetition, the rats, Adams deduced, had automated the response and now performed it without thought for the consequences. Adams' intuition is that these results reflect changes in the habit itself, i.e. the increased strength of stimulus–response associative links that arises through 'over-training' of those associations.

However, based upon the results from an outcome devaluation test alone, we cannot eliminate the possibility that it is not increased strength of the habit, but a reduction in goal-directed processing that drives a tendency towards continued responding in the over-trained rats. Couched recently as a 'zero-sum game' (Robbins and Costa 2017; Schreiner et al. 2019), either alternative may produce exactly the same behavioural result: more responding towards devalued outcomes. It is for this reason that many researchers conservatively refer to the 'balance' between these systems when studying devaluation performance (Gillan et al. 2011; de Wit et al. 2012, 2018; Gremel and Costa 2013; Linnebank et al. 2018). It is, therefore, from this starting point that I will begin the discussion of the evidence in OCD.

3 Evidence for Imbalance in OCD

The first direct evidence for a disruption in the balance between goal-directed control and habit in OCD came in 2011 (Gillan et al. 2011). Using a human translation of the outcome devaluation procedure (de Wit et al. 2009; de Wit and Dickinson 2009), OCD patients were asked to learn through 'trial and error' which keyboard responses should be performed when presented with certain stimuli in order to earn rewards (Gillan et al. 2011). After training, some of these rewards were devalued (stripped of points) and subjects were subsequently tested for the presence of an imbalance between goal-directed control and habit, defined as redundant responding towards now-devalued rewards. Compared to healthy controls, patients with OCD were more likely to continue responding to stimuli with a devalued outcome on this test, despite the fact that those responses resulted in no gain. These results were striking in a number of ways, but perhaps most notable was the observation that these habits arose in an abstract setting, one that bore no relation to patients' ongoing obsessive themes and symptoms. This finding did not accord with most cognitive models of OCD that considered obsessions primary and compulsive behaviours secondary or 'epiphenomenal' to those obsessions (Salkovskis 1985)

The presence of an imbalance between goal-directed control and habit in OCD was subsequently probed in the context of avoidance (Gillan et al. 2014). This was an important test, because compulsions in OCD have the goal of avoiding a feared event, not obtained a desired one, and at this time, habits in the case of avoidance had not yet been demonstrated in any species. In this experiment, patients and controls performed responses to stimuli in order to avoid the delivery of a mild electric shock to their wrist. Following training, shocks were devalued by rendering their delivery impossible by disconnecting the administering cables from one of the wrists in full view of the subject. In this aversive setting, patients with OCD again showed an increased tendency to respond to avoid devalued outcomes, i.e. shocks that could no longer be delivered. Importantly, they also showed intact explicit knowledge of the fact they could not be delivered, and patients did not show any excess of physiological fear responses to stimuli. Together, these findings point to a purely behavioural habit-forming proclivity.

To understand the brain-based mechanism that leaves individuals vulnerable to performing outcome-insensitive behaviours, researchers have since sought to combine behavioural assays of habit with neuroscience techniques to reveal if devaluation failures reflect impairments in the goal-directed system or enhancements in the habitual system. To focus first on the former, a distinct set of brain regions are involved in maintaining our ability to make considered, prospective choices. For example, activity in the orbitofrontal cortex is sensitive to whether an action will be followed by a valued or devalued outcome (Valentin et al. 2007). The caudate and orbitofrontal cortex both show sensitivity to instrumental contingency – the degree of association that exists between actions and rewards (Tricomi et al. 2004; Liljeholm et al. 2011) and the prospective value of goal-directed choice (Wunderlich et al. 2012). In rodents, lesions to the dorsomedial part of the striatum (functionally overlapping with the caudate) render even modestly trained actions habitual (Yin et al. 2005; Gremel and Costa 2013), as do lesion to the medial OFC in mice (Gremel and Costa 2013).

Building on these basic insights, functional activity in patients with OCD and healthy controls was examined during the performance of avoidance habits, with a focus on establishing the role, if any, of goal-directed brain regions in devaluation insensitivity in OCD (Gillan et al. 2015a). This study revealed that hyperactivity in the caudate nucleus in OCD at the time of choice was correlated with individual differences in devaluation sensitivity and self-reported urges to perform habitual responses (Gillan et al. 2015a). These data suggest that devaluation failures in OCD might be caused by alterations in the brain systems that support purposeful goal-directed control over action, rather than an excess of habitual associations. Notably, we did not find evidence for altered functioning in brain regions linked to habit formation like the putamen – discussed in more detail later.

In parallel to this imaging work, new paradigms and analytic approaches have been developed that aim to distinguish the contribution of the goal-directed system from that of habit in behavioural terms. The most well-established methodology is the 'model-based, model-free' dichotomy and associated two-step decision-making task (Daw et al. 2005, 2011), which formalises goal-directed control as 'model-based planning'. Model-based planning is operationalised as the extent to which decision-makers use a mental model of action-state relationships to guide choice. Model-based planning is a prospective and goal-directed comparison of available options, and indeed neuroimaging data shows that model-based planners mentally simulate the intended goal at the time of choice (Doll et al. 2015). Converging with the basic research on habit translated from rodents, poor model-based planning characterises individuals that are likely to behave habitually in a devaluation test (Gillan et al. 2015b) and individuals high in model-based planning have larger caudate and medial orbitofrontal volumes (Voon et al. 2014), converging on core commonalities across these frameworks of goal-directed/model-based processing.

The advantage of studying model-based planning is that it can be assayed independently of the contribution of potentially concurrent S-R habit formation, which confounds behaviour in devaluation studies. In other words, it provides a neat method for testing if goal-directed control is specifically compromised in OCD, absent of any habit. Voon and colleagues were the first to test this; they found that compared to healthy controls, patients with OCD had deficits in model-based planning (Voon et al. 2014). Considered alongside the neuroimaging data and behavioural links to devaluation performance, together these data suggest it is possible that goal-directed deficits drive habitual tendencies in OCD.

Studies broadly agree that a distinction between these processes is apt, but the nature of their interaction is an active area of study and debate. For example, some researchers have focused on the notion of 'third party' arbitration between habit and goal-directed control systems at the time of choice (Lee et al. 2014; Kool et al. 2017), while others have begun to outline a hierarchical organisation of goals and habits (Dezfouli and Balleine 2013; Cushman and Morris 2015). In both formulations, the tendency to prioritise either system is amenable to circumstance, where, for example, in hierarchical frameworks, the agent can 'automatically' initiate a sequence of goal-directed actions (Cushman and Morris 2015), while in others, goal-directed control can be used to initiate an habitual sequence of action (Dezfouli and Balleine 2013). This line of research is fast developing but has yet to see clinical investigation. Future work will likely focus on testing some of these ideas in patients with OCD, with the goal of finding neural targets that are the most amenable to intervention.

4 From Categorical to Dimensional Perspectives on Goals and Habits

Habits are automatic behaviours, executed without top-down 'control' and often in inappropriate settings or in a manner disconnected from goals. Compulsive behaviours observed clinically parallel the phenomenology of habit in several key ways. Compulsions involve the experience of an urge to act, the sometimes inexplicable 'force-from-within' experience that typifies OCD. More formally, compulsions in OCD have been defined as stereotyped behaviours, performed according to rigid rules, and designed (at least at the outset) to reduce or avoid unpleasant or unwanted consequences (Chamberlain and Menzies 2009). Like habits, compulsions in OCD follow a set routine and notably do not actually help the individual to achieve the intended goal. OCD patients usually understand that their routines are excessive or even senseless, but the drive to act persists. As outlined in the Introduction, these clinical parallels do not stop with OCD. Compulsions have been described in a range of disorders and although manifest in different ways, they appear to share many of these core characteristics (Fineberg et al. 2010). For example, in those with substance addiction, the individual experiences an overwhelming urge to take drugs (as opposed to washing their hands), despite it being contrary to their long-term goals. Compulsive drug-taking, like habits, can be triggered by exposure to familiar stimuli (cues) and shows a similar disconnection to goals, for example, persisting despite diminishing returns and major adverse consequences (Robbins and Everitt 1999). Although research in rodents has primarily focused on the consequences of prolonged drug use and how that 'hijacks' the brain's habit system, it is possible that, like OCD, addicted individuals have a more generalised trait-like tendency to form habits and that this conspires with the acute effects of the drug to make certain individuals more vulnerable. Research has recently shown that these parallels are not just superficial; failures in goal-directed control are evident in addicted individuals, with evidence coming from both devaluation and model-based planning experiments (Sjoerds et al. 2013; Voon et al. 2014; Ersche et al. 2016).

Ersche and colleagues studied both aversive and appetitive habit formation using devaluation, utilising the same paradigms as were used to study OCD, and found evidence for appetitive devaluation failures in individuals addicted to cocaine relative to controls, although they did not find similar effects in avoidance (Ersche et al. 2016). These results must be considered in the context of another finding from this study – that addicted individuals also exhibited much more general learning failures across both appetitive and aversive paradigms repetitive, which may confound interpretations of the devaluation effects. Notably, generalised deficits of this kind were not present in the OCD studies described earlier. In alcohol dependent individuals, Sjoerds and colleagues found evidence for impairments in action–outcome associative knowledge (they did not probe devaluation sensitivity per se) (Sjoerds et al. 2013) and in a separate study, diminished model-based planning (Sebold et al. 2014). Unlike those addicted to cocaine, these learning impairments did not extend to basic discriminative learning and so provide in some respects stronger (or at least, potentially less confounded) evidence for deficits in goal-directed systems. Finally, Voon et al. studied the ratio between model-based and model-free learning (referred to as ω) in abstinent methamphetamine abusers and found evidence for a shift from model-based towards model-free behaviour relative to healthy controls.

In the case of addiction research in particular, it is challenging to study cause and effect, i.e. whether these biases were present prior to the instigation of drug use or are a partial consequence of drug-abuse. Conducting a direct comparison across compulsive patient groups cannot fully resolve this but offers context. Voon and colleagues found that methamphetamine-addicted individuals did not differ from patients with OCD and individuals with binge-eating disorder, who both also showed shifts in ω relative to controls (Voon et al. 2014). Extending further into the compulsive spectrum, these deficits also extend to pathological gamblers (Wyckmans et al. 2019).

These studies begin to paint a compelling picture of common disturbances in goal-directed control in disorders that involve *compulsivity*. This accounts for uncontrollable, goal-insensitive and sometimes self-injurious behaviours characteristic of compulsivity and, moreover, goes some way towards explaining comorbidity across compulsive disorders. However, when we broaden our frame of view, the picture becomes cloudier. Deficits in goal-directed behaviour have been studied in a range of disorders that vary markedly in their putatively 'compulsive' characteristics and the results challenge the validity of this model as one that uniquely characterises compulsive behaviour. For example, similar deficits in goal-directed control have been observed in schizophrenia (Morris et al. 2014; Culbreth et al. 2016), social anxiety disorder (Alvares et al. 2014), and Tourette's syndrome (Delorme et al. 2016).

This lack of specificity is dishearteningly common in case–control investigations in psychiatric populations, where an initially promising neurocognitive model is thought to explain the key symptoms of a DSM disorder and is later found to be common to many or all (e.g. the same has happened for response inhibition, delay discounting, and error-related negativity (Lipszyc and Schachar 2010; Bickel et al. 2012; Gillan et al. 2017). Upon observing this, many have argued (myself included) that this outcome is an inevitability, given the way we define mental illness (Hyman 2007, 2010; Kapur et al. 2012; Robbins et al. 2012; Gillan et al. 2017). DSM-5 definitions of disorder classes (APA 2013) are both overlapping in definition and internally heterogeneous in symptom presentation (Hyman 2010), comorbidity is substantial (Kessler et al. 2005) and the same treatments (e.g. antidepressants) have efficacy across multiple disorder classes (Schatzberg 2000), but do not work for all patients within the same diagnostic category (Chekroud et al. 2017a). Therefore, continuing to use these categories for research will likely preclude the discovery of specific cognitive-clinical relationships needed to develop new clinical tools (Kapur et al. 2012) and ignores the individual in favour of the average (Chekroud et al. 2017b).

A key challenge we face is in the absence of a clear psychiatric 'ground truth' is how can we achieve clarity on mechanisms that underpin clinical phenomena. One relatively simplistic approach is to test for associations with specific symptoms, rather than syndromes. Because patients present with multiple comorbid and potentially confounding symptoms, large samples are required to disentangle them and identify specific relationships, if any exist. Internet-based testing has become a powerful new tool for addressing questions of this kind (Gillan and Daw 2016) and the results have thus far been illuminating for compulsivity, in particular. In a general population sample of N = 1,413 individuals recruited and tested entirely via the Internet, the extent to which the association between model-based planning and compulsive symptomatology is specific, rather than generalised, was addressed (Gillan et al. 2016). Following the data from categorical research outlined above, when a broad range of psychiatric symptoms were assessed in the same sample, several associations with model-based planning were observed. Eating disorders, impulsivity, OCD, and alcohol addiction were all associated with diminished model-planning, but no (significant) relationships were observed with schizotypy, apathy, depression, trait anxiety, or social anxiety (listed in order of effect size). Given that questionnaires assessing these symptoms are moderately-highly correlated with one-another and that some are internally heterogeneous (e.g. schizotypy measures positive and negative symptoms), a factor analysis was conducted to test if a single dimension of mental health exists that could explain these data more parsimoniously. Three factors were identified, 'anxious-depression', 'compulsive behaviour and intrusive thought', and 'social withdrawal' and based on the loadings of individual questionnaire items on these factors, subjects could then be given a score for where they lie on each factor ('dimension'). When testing the association between these transdiagnostic dimensions and model-based planning, only the compulsive factor showed a significant association, and it was larger and more specific than any other formulation. This finding has since been independently replicated (Patzelt et al. 2019) and suggests that this transdiagnostic compulsive dimension of mental health might be driving the seemingly general pattern that emerges from case–control comparison studies.

This premise was recently tested directly in patients with clinically confirmed diagnoses of OCD, generalised anxiety disorder (GAD), or a combination. Two hundred and eighty-five patients were recruited online and completed a clinical diagnostic interview by telephone, followed by an assessment of their self-report symptoms and model-based planning ability via a web-browser (Gillan et al. 2020). Consistent with the general trend of in-person research in diagnosed patients, model-based planning did not differ based on diagnostic information – a diagnosis of OCD was not linked to any decrement in performance compared to a diagnosis of GAD. Examining self-report symptoms, however, an entirely different picture emerged. Of the three transdiagnostic dimensions studied here across all patients, regardless of diagnosis, 'obsessionality', 'compulsivity', and 'general distress', only compulsivity was linked to deficits in model-based planning. This lack of association with general distress accords with the results of recent causal interventions where acute anxiety induced through the inhalation of CO_2 was found to be ineffective at inducing model-based deficits in healthy subjects across multiple experiments (Gillan et al. 2020). Speaking to the importance of individual-focused research, compulsivity, although higher in OCD relative to GAD patients, was by no means exclusive to those with an OCD diagnosis. Moreover, not all patients with OCD had high levels of compulsivity, some are primarily obsessional, anxious, or depressed. These individual differences highlight the importance of moving beyond diagnosis-based frameworks towards studying the individual if we are to effectively link mechanisms to clinical phenomena and ultimately translate our basic understanding of goal-directed processes into novel interventions or early detection tools.

5 Goal-Directed Control and the Mental Model

Returning now to the mechanistic question, the studies detailed hitherto measure a construct called goal-directed control. While we know a great deal about this process and its links to habit expression, the reader must be cautious in interpreting it as a unique, or 'special case', of executive function. Goal-directed control is without doubt a multi-faceted process, requiring by definition (1) a 'mental model' of the relationship between the response and the reinforcer and (2) the ability to use this model, and current motivations, to make optimal decisions. These functions rely themselves on working memory (Otto et al. 2013a, b, Culbreth et al. 2016) and cognitive control (Otto et al. 2014) and more generally, model-based planning has been shown to correlate with intelligence quotient (IQ) (Culbreth et al. 2016; Gillan et al. 2016).

In order to be more targeted in our understanding of the mechanism, we must ask if patients with OCD have difficulty in forming and maintaining a mental model (e.g. learning what actions lead to what consequences, holding that in memory, etc.), or are their deficits are confined to the implementation/execution of that model (if-then scenarios, mental simulation, etc.)? Are these processes functionally distinct, or reliant on the same neural architecture? This area of research is underdeveloped, but recent studies have begun to shed some light. Vaghi and colleagues conducted a study that aimed in part to address the former. Patients with OCD were required to learn whether a contingency existed between making a button press and receiving a reward, and express that in their behaviour (decision to respond or not) and using causality ratings (Vaghi et al. 2018). Causality judgements were not distinguishable between groups, suggesting that their mental model might be intact. Behaviour was similarly unaffected in OCD, at least in 4/5 conditions studied (there was some evidence that patients with OCD tended to over-respond in one condition though, which might indicate patients thought their actions were more causal than they were).

Following on from this work, Vaghi and colleagues examined 'predictive inference' in OCD, a paradigm that can better dissociate behaviour from confidence judgements. Specifically, they examined how well behaviour and confidence ratings responded to environmental feedback (Vaghi et al. 2017). In this task, subjects learned to catch a flying particle using a bucket by observing its typical path over many trials. Subjects can make adjustments to the positioning of the bucket on each trial and typically do so, depending on the feedback they received on the prior trial. Overall confidence ratings and the size of the action taken (how far the bucket is moved) are correlated, indicating that when subjects move the bucket more, they tend to be less confident in their ability to catch the particle. This normal 'coupling' between action and confidence was found to be diminished in OCD patients compared to health controls. Drilling into the mechanism, the authors found that reduced confidence-action coupling in OCD likely arises due to differences in how actions, and not confidence, adjust to feedback trial-by-trial. Specifically, compared to controls, patients with OCD updated their bucket position more in response to small prediction errors and at the same time, updated their actions less in instances where major alterations to behaviour were required. In contrast, their confidence ratings were not distinguishable from controls. Together these two studies suggest that the mental model may be intact in OCD, but that patients fail to *implement what they know* in behaviour.

This broad conclusion must be somewhat tempered by the fact that these studies examined behavioural updates with respect to 1-step reinforcement learning (i.e. feedback-driven behavioural updates), which are not the kind of multi-stage decisions that are characteristic of goal-directed control ('model-based' planning). Prior studies investigating simple feedback-driven learning in OCD have for the most part failed to find differences between patients and controls. For example, 'learning rate', a computational parameter which measures how quickly someone updates their behaviour using recent evidence (prediction errors) appears to be intact in OCD (Voon et al. 2014; Hauser et al. 2017) and is uncorrelated with OCD symptoms in the general population (Gillan et al. 2016). Similarly, 'model-free learning', the extent to which responses are repeated following reward, does not differ between OCD patients and controls (Voon et al. 2014) and is also not correlated with OCD symptoms in the general population (Gillan et al. 2016).

In terms of evidence that the meta-model is intact, similar caution should be taken at this stage. While Vaghi and colleagues found no deficits in confidence/contingency judgements in OCD, studies examining prerequisites for goal-directed behaviour, i.e. explicit action–outcome contingency knowledge, have identified impairments in OCD that correlate with patients' devaluation failures, when the task is challenging enough to remove floor effects (Gillan et al. 2011). Metacognitive judgements have been studied in OCD for quite some time in relation to confidence in *memory*, and the majority of studies have demonstrated *decreased* confidence in patients with OCD, though many studies show no differences between patients or controls (Hoven et al. 2019). This lack of consistency in the literature may be explained by task differences and the fact that metacognition is domain specific – e.g. metacognitive accuracy for perceptual versus memory tasks appears to be neuroanatomically distinct (Fleming et al. 2014). A second factor is that confidence ratings in reinforcement learning tasks are influenced by behaviour itself. That means that confidence is confounded by differences in the experience of participants that arises from the choices they make on the task (e.g. the level of contingency they observe, reinforcement received, etc.), and indeed their own overall performance. Recently, methods have been developed that account for this analytically by adjusting confidence based on differences in performance (Fleming and Lau 2014), mostly in the context of tasks where the information subjects receive is not affected by their choices (i.e. perceptual decision-making tasks).

This approach was recently applied to a study comparing those with high vs low OCD symptoms, and consistent with Vaghi et al., those with high OCD scores showed no 'bias' in overall confidence ratings on a perceptual decision-making task. That is, once performance was accounted for, those high in OCD symptoms were just as confident as those with low symptom scores. However, in this study, the metacognitive model was far from intact; individuals high in OCD symptoms had lower 'metacognitive efficiency' (Hauser et al. 2017), a signal detection measure that captures how closely confidence aligns with actual variations in performance. This means subjects were less accurate at differentiating trials where they got the answer right versus wrong, signalling a failure to reflect accurately on one's own performance.

Similar findings have recently emerged from studies examining metacognitive performance in compulsivity using the Internet-based transdiagnostic methodology outlined above (Rouault et al. 2018; Seow and Gillan 2020). In the same perceptual-decision-making task used by Hauser et al. (2017), OCD symptoms were not found to be related to confidence biases or metacognitive deficits in a large general population sample (Rouault et al. 2018). Notably though, depression and anxiety, two common OCD comorbidities, were linked to decreases in confidence. When subjects' responses were re-factored into the three transdiagnostic dimensions previously studied in the context of goal-directed control (Gillan et al. 2016), anxious-depression was indeed associated with decreased confidence, and controlling for this revealed that compulsivity was in fact linked to increased confidence judgements. Moreover, a trend for an impairment in metacognitive efficiency was linked to compulsivity, while individuals high in anxious-depression tended to show enhancements. This suggested that in case–control studies where OCD patients typically have co-occurring anxiety and depression, these comorbidities might mask impairments in the meta-model that are specifically linked to compulsivity.

This premise was recently tested (Seow and Gillan 2020) using the same predictive inference task used by Vaghi and colleagues (Vaghi et al. 2017). Using the transdiagnostic method in a large general population sample collected online, compulsivity was again found to be associated with increased confidence and anxious-depression decreased confidence. Additionally, compulsivity was associated with a host of metacognitive impairments that impinge on the quality of the mental model acquired – these include failures to update confidence based on prediction errors and in response to large errors that signal change-points in the task (Seow and Gillan 2020).

Understanding how OCD patients' mental model of the world becomes altered may be crucial for delineating the mechanism through which compulsions dynamically interact with obsessive beliefs to form vicious cycles. But to fully understand this, studies of metacognition must find a bridge with studies that directly assess goal-directed control and habit expression. This might help us to resolve some of the apparently discrepant findings outlined above and also begin to shed light on dynamic symptom interactions seen in the clinic. Do habit forming tendencies confer risk for developing obsessional beliefs, or are obsessions and compulsions two manifestations of the same underlying dysfunction in goal-directed control (Gillan and Sahakian 2015)?

While there is little doubt that strong links exist between these symptoms, study designs that can temporally dissociate them are prohibitively challenging to undertake. The nearest we have got are through studies examining habit formation in the lab while also assessing changes in explicit beliefs about that same task. In research described earlier concerning the development of avoidance habits, i.e. habits that form when one performs an action to prevent an unpleasant event from taking place (Gillan et al. 2014), subjects were tested for habits twice, once at a very early stage of training (no group difference observed) and again following over-training (group difference observed). They were asked to reflect on why they continued to respond in the final test and curiously several patients provided what we classed as post hoc explanations for their behaviour. For example, they thought that they would still be shocked if they failed to perform the response, or that they might now be shocked on the alternative wrist. Although plausible explanations taken at face value, it is notable that they were absent at the earlier probe test, before habits had been established.

In a follow-up study, subjects were asked at regular intervals how likely it was that stimulus X would deliver a shock if no response is performed (Gillan et al. 2015a). Intriguingly, both patients and controls alike reported a small amount of uncertainty in this regard following the disconnection of electrodes (if anything nominally greater uncertainty was observed in controls, but not statistically different). It was only *following the performance of the habit* that OCD patients showed an elevation in this measure relative to controls, with controls reducing that uncertainty to zero. In OCD, this suggests that the performance of habits may serve to keep them alive, solidifying and strengthening them over time as habits become more rigid, elaborate, and compulsive (Gillan and Sahakian 2015).

6 What Has Habit Got to Do With It?

Convergent data from classic translational tasks like outcome devaluation, and, more recently, from formal computational and mathematical models and neuroimaging studies suggest there are deficits in goal-directed processing in OCD. So what then is the role of the brain's habit system? In terms of basic research, over-training of simple stimulus–response behaviours in humans has been linked to increased blood oxygen level dependent (BOLD) signal in the posterior putamen (Tricomi et al. 2009) and activity in this same region is sensitive to values associated with extensively-trained choices (Wunderlich et al. 2012), making it a good candidate for the source of S-R habitual associative learning in the brain. Perhaps most compellingly, linking the putamen causally to devaluation *performance*, lesions to the functionally homologous dorsolateral striatum in rodents have been shown to restore devaluation sensitivity in rodents that would otherwise behave habitually (Yin et al. 2004). Together these cross-species data suggest it might be possible to assay the functioning of one's habit system based on activity in the putamen at the time of choice.

However, direct links between human neuroimaging findings with respect to the putamen and the extent of habitual responses performed by the same subjects have been notably rare. For example, functional activity in OCD patients was examined during the performance of avoidance habits with the goal of testing not just for abnormalities characteristic of goal-directed processing (as described earlier in the chapter), but also for any alterations in the putamen that might explain the observed devaluation insensitive behaviour in OCD (Gillan et al. 2015a). No evidence emerged for the involvement of the putamen in task performance and no association was observed with devaluation failures. However, we must of course acknowledge that an absence of evidence for accelerated S-R automaticity in OCD is certainly not evidence for its absence. While there is little evidence in functional brain activity, one study showed that devaluation performance correlates with the structural integrity of white matter tracts connecting the putamen and premotor cortex (de Wit et al. 2012).

This issue is particularly important now, as recent papers have eloquently described the limitations that currently exist in our understanding of basic habit-learning processes. For example, one of the cornerstone findings from the rodent literature on habit is that rats that are trained for longer are more likely to fail a devaluation test (Adams and Dickinson 1981b; Adams 1982). One study has shown this effect in humans (Tricomi et al. 2009), but since then, no study has replicated it. In the only published failure-to-replicate, 5 experiments are reported that incorporate both direct and conceptual replication protocols, with each one failing to show that outcome devaluation sensitivity is reduced by over-training of S-R behaviours in humans (de Wit et al. 2018). In a similar vein, computational modelling approaches designed with the intention of capturing individual differences in S-R learning have been largely unsuccessful. For example, 'model-free' forms of decision-making have been defined, which rely exclusively on past experience (Daw et al. 2011), but they fail to predict habitual responding in humans (Gillan et al. 2015b). Some have suggested model-free learning might simply be an imperfect formalisation of habit-learning, which captures noisy behaviour and/or may even reflect misunderstandings of the task rules (Shahar et al. 2019; da Silva and Hare 2020). Whatever the cause, there is now broad consensus that revision to this framework is warranted if we are to capture individual differences in how humans acquire stimulus–response behaviours. Indeed it is clear that we know a lot less about habit formation in humans than we once thought (Robbins and Costa 2017; De Houwer et al. 2018).

One explanation for the failure of existing methods to capture the unique signature of habit is that the human capacity for goal-directed control might eclipse the contribution of S-R automaticity to choice on standard behavioural paradigms (rendering S-R associative learning effectively *invisible*)(Luque et al. 2019). To address this, researchers have built on the observation that the goal-directed system needs more time to become effective (Keramati et al. 2016) and shifted focus toward the importance of *timing* in parsing distinct contributions of well-trained S-R links to choice behaviour (Hardwick et al. 2019; Luque et al. 2019). For example, reaction time (RT) switch costs have been suggested to be a more sensitive measure of the strength of S-R links than actual choice (Luque et al. 2019). Moreover, if the time pressure is such that individuals *cannot* recruit the slower, goal-directed control system, over-training effects can be revealed in the choices of humans (i.e. habit expression) (Hardwick et al. 2019). RT-based paradigms such as this offer a compelling route forward for measuring and manipulating S-R automatic learning in humans, but currently are without any neurophysiological exposition or clinical investigation. It is anticipated that the next wave of research in this area will attempt to finally resolve this issue and test if the basic acquisition of habit is directly affected in OCD, or if the entirety of the evidence points to deficits in goal-directed processing. If this is the case, habits are still part of the picture, but may develop downstream of these failures. Compulsions (like any other behaviour we perform regularly) may begin to engage habit systems as they are performed repetitively over time, explaining observations such as hyperactivity in the putamen during 'compulsion provocation' in OCD (Banca et al. 2015).

7 Where Do We Go from Here?

It is anticipated that the next wave of research will focus on the issues outlined above. First, whether we can isolate a specific brain process that gives rise to the broad class of goal-directed deficits observed clinically. The ultimate goal of which is to facilitate the development of interventions that target specific and precisely defined mental functions (Huys et al. 2016). Examples may include brain stimulation that might target caudate function indirectly, e.g. through prefrontal stimulation, and can be implemented on a more individualised basis, i.e. provided to only those patients with goal-directed deficits. Or the use of behavioural/neural assessments to guide pharmacological interventions, for example, SSRI augmentation with D2 antagonists in OCD might be indicated for only a subset of patients, discernible from a cognitive test. Second, the field will surely shift focus to begin to address the question of how habits causally relate to beliefs and indeed obsessions, requiring us to solve the extant issue with achieving robust experimental habit inductions in the lab. As this work necessitates the study of sequential causal interactions, it will, therefore, likely begin to bridge a growing literature on network theories of mental illness that are growing in influence and popularity (Borsboom 2017). Understanding how obsessions and compulsions interact, at a mechanistic level, will bring clinical insights of particular importance for behavioural therapies; how to effectively interrupt the reinforcing cycle of OCD and prevent escalation of illness. Such an approach may prove key for developing early interventions that seek to limit duration of illness and ideally, prevent transition to illness altogether (Fineberg et al. 2019).

To achieve these goals, it is clear that major structural change to how we conduct research in psychiatry more broadly is required. Large-scale, longitudinal, and interventional studies must replace a large proportion of the small, cross-sectional, and observational ones that dominate the field in order to bridge the bench to bedside gap (Browning et al. 2020). These study designs are logistically challenging, time-consuming, and expensive, but essential if we are to establish clinical utility of what is now a well-developed field. As cross-sectional research will no doubt continue to elaborate on the basic mechanisms outlined in this chapter in the coming years, we are missing a fundamental understanding of whether failures in goal-directed control carry predictive, or just descriptive, utility. This gap constitutes, in my view, the most exciting frontier for the next wave of research in this area.

References

Adams CD (1982) Variations in the sensitivity of instrumental responding to Reinforcer devaluation. Q J Exp Psychol Sect B Comp Physiol Psychol 34(May):77–98

Adams C, Dickinson A (1981a) In: Spear N, Miller R (eds) Actions and habits. Information processing in animal: memory mechanisms. Lawrence Erlbaum Associates, New Jersey

Adams CD, Dickinson A (1981b) Instrumental responding following reinforcer devaluation. Q J Exp Psychol Sect B Comp Physiol Psychol 33(MAY):109–121

Alexander GE, DeLong MR, Strick PL (1986) Parallel organization of functionally segregated circuits linking basal ganglia and cortex. Annu Rev Neurosci 9:357–381

Alvares GA, Balleine BW, Guastella AJ (2014) Impairments in goal-directed actions predict treatment response to cognitive-behavioral therapy in social anxiety disorder. PLoS One 9(4): e94778

APA (2013) Diagnostic and statistical manual of mental disorders. American Psychiatric Publishing, Arlington

Balleine BW, Dickinson A (1998) Goal-directed instrumental action: contingency and incentive learning and their cortical substrates. Neuropharmacology 37(4–5):407–419

Banca P, Voon V, Vestergaard MD, Philipiak G, Almeida I, Pocinho F, Relvas J, Castelo-Branco M (2015) Imbalance in habitual versus goal directed neural systems during symptom provocation in obsessive-compulsive disorder. Brain 138(Pt 3):798–811

Bickel WK, Jarmolowicz DP, Mueller ET, Koffarnus MN, Gatchalian KM (2012) Excessive discounting of delayed reinforcers as a trans-disease process contributing to addiction and other disease-related vulnerabilities: emerging evidence. Pharmacol Ther 134(3):287–297

Borsboom D (2017) A network theory of mental disorders. World Psychiatry 16(1):5–13

Browning M, Carter C, Chatham C, Den Ouden H, Gillan C, Baker J, Chekroud A, Cools R, Dayan P, Gold J, Hartley C, Kepecs A, Lawson R, Mourao-Miranda J, Phillips M, Pizzagalli D, Powers A, Rindskopf D, Roiser J, Schmack K, Schiller D, Sebold M, Stephan K, Frank M, Huys Q, Paulus M (2020) Realising the clinical potential of computational psychiatry: report from the Banbury Centre meeting, February 2019. Biol Psychiatry 88:e5–e10

Chamberlain SR, Menzies L (2009) Endophenotypes of obsessive-compulsive disorder: rationale, evidence and future potential. Expert Rev Neurother 9(8):1133–1146

Chekroud AM, Gueorguieva R, Krumholz HM, Trivedi MH, Krystal JH, McCarthy G (2017a) Reevaluating the efficacy and predictability of antidepressant treatments: a symptom clustering approach. JAMA Psychiat

Chekroud AM, Lane CE, Ross DA (2017b) Computational psychiatry: embracing uncertainty and focusing on individuals, not averages. Biol Psychiatry 82(6):e45–e47

Culbreth AJ, Westbrook A, Daw ND, Botvinick M, Barch D (2016) Reduced model-based decision-making in schizophrenia. J Abnorm Psychol 125:777

Cushman F, Morris A (2015) Habitual control of goal selection in humans. Proc Natl Acad Sci U S A 112:13817–13822

da Silva CF, Hare TA (2020) Humans primarily use model-based inference in the two-stage task. Nat Hum Behav 4(10):1053–1066

Daw ND, Niv Y, Dayan P (2005) Uncertainty-based competition between prefrontal and dorsolateral striatal systems for behavioral control. Nat Neurosci 8(12)

Daw ND, Gershman SJ, Seymour B, Dayan P, Dolan RJ (2011) Model-based influences on Humans' choices and striatal prediction errors. Neuron 69(6)

De Houwer J, Tanaka A, Moors A, Tibboel H (2018) Kicking the habit: why evidence for habits in humans might be overestimated. Motiv Sci 4:50–59

de Wit S, Dickinson A (2009) Associative theories of goal-directed behaviour: a case for animal-human translational models. Psychol Res 73(4):463–476

de Wit S, Corlett PR, Aitken MR, Dickinson A, Fletcher PC (2009) Differential engagement of the ventromedial prefrontal cortex by goal-directed and habitual behavior toward food pictures in humans. J Neurosci 29(36):11330–11338

de Wit S, Watson P, Harsay HA, Cohen MX, van de Vijver I, Ridderinkhof KR (2012) Corticostriatal connectivity underlies individual differences in the balance between habitual and goal-directed action control. J Neurosci 32(35):12066–12075

de Wit S, Kindt M, Knot SL, Verhoeven AAC, Robbins TW, Gasull-Camos J, Evans M, Mirza H, Gillan CM (2018) Shifting the balance between goals and habits: five failures in experimental habit induction. J Exp Psychol Gen 147(7):1043–1065

Delorme C, Salvador A, Valabrègue R, Roze E, Palminteri S, Vidailhet M, de Wit S, Robbins T, Hartmann A, Worbe Y (2016) Enhanced habit formation in Gilles de la Tourette syndrome. Brain 139(Pt 2):605–615

Dezfouli A, Balleine BW (2013) Actions, action sequences and habits: evidence that goal-directed and habitual action control are hierarchically organized. PLoS Comput Biol 9(12):e1003364

Dickinson A (1980) Contemporary animal learning theory. Cambridge University Press, Cambridge

Dickinson A (1985) Actions and habits: the development of Behavioural autonomy. Philos Trans R Soc Lond Ser B Biol Sci 308(1135):67–78

Dickinson A, Balleine B (1994) Motivational control of goal-directed action. Anim Learn Behav 22 (1):1–18

Dolan RJ, Dayan P (2013) Goals and habits in the brain. Neuron 80(2):312–325

Doll BB, Duncan KD, Simon DA, Shohamy D, Daw ND (2015) Model-based choices involve prospective neural activity. Nat Neurosci 18(5):767–772

Ersche KD, Gillan CM, Jones PS, Williams GB, Luitjen M, de Wit S, Ward LHE, Sahakian BJ, Bullmore ET, Robbins TW (2016) Carrots and sticks fail to change behavior in cocaine addiction. Science 352(6292):1468–1471

Fineberg NA, Potenza MN, Chamberlain SR, Berlin HA, Menzies L, Bechara A, Sahakian BJ, Robbins TW, Bullmore ET, Hollander E (2010) Probing compulsive and impulsive behaviors, from animal models to Endophenotypes: a narrative review. Neuropsychopharmacology 35:591–604

Fineberg NA, Dell'Osso B, Albert U, Maina G, Geller D, Carmi L, Sireau N, Walitza S, Grassi G, Pallanti S, Hollander E, Brakoulias V, Menchon JM, Marazziti D, Ioannidis K, Apergis-Schoute A, Stein DJ, Cath DC, Veltman DJ, Van Ameringen M, Fontenelle LF, Shavitt RG, Costa D, Diniz JB, Zohar J (2019) Early intervention for obsessive compulsive disorder: an expert consensus statement. Eur Neuropsychopharmacol 29(4):549–565

Fleming SM, Lau HC (2014) How to measure metacognition. Front Hum Neurosci 8:443

Fleming SM, Ryu J, Golfinos JG, Blackmon KE (2014) Domain-specific impairment in metacognitive accuracy following anterior prefrontal lesions. Brain 137(Pt 10):2811–2822

Gillan CM, Daw ND (2016) Taking psychiatry research online. Neuron 91(1):19–23

Gillan CM, Robbins TW (2014) Goal-directed learning and obsessive-compulsive disorder. Philos Trans R Soc Lond Ser B Biol Sci 369(1655)

Gillan CM, Sahakian BJ (2015) Which is the driver, the obsessions or the compulsions, in OCD? Neuropsychopharmacology 40(1):247–248

Gillan CM, Papmeyer M, Morein-Zamir S, Sahakian BJ, Fineberg NA, Robbins TW, de Wit S (2011) Disruption in the balance between goal-directed behavior and habit learning in obsessive-compulsive disorder. Am J Psychiatry 168(7):718–726

Gillan CM, Morein-Zamir S, Urcelay GP, Sule A, Voon V, Apergis-Schoute AM, Fineberg NA, Sahakian BJ, Robbins TW (2014) Enhanced avoidance habits in obsessive-compulsive disorder. Biol Psychiatry 75(8):631–638

Gillan CM, Apergis-Schoute AM, Morein-Zamir S, Urcelay GP, Sule A, Fineberg NA, Sahakian BJ, Robbins TW (2015a) Functional neuroimaging of avoidance habits in obsessive-compulsive disorder. Am J Psychiatry 172(3):284–293

Gillan CM, Otto AR, Phelps EA, Daw ND (2015b) Model-based learning protects against forming habits. Cogn Affect Behav Neurosci 15:523–536

Gillan CM, Kosinski M, Whelan R, Phelps EA, Daw ND (2016) Characterizing a psychiatric symptom dimension related to deficits in goal-directed control. elife 5:e11305

Gillan CM, Fineberg NA, Robbins TW (2017) A trans-diagnostic perspective on obsessive-compulsive disorder. Psychol Med 47(9):1528–1548

Gillan C, Kalanthroff E, Evans M, Weingarden H, Jacoby R, Gershkovich M, Snorrason I, Campeas R, Cervoni C, Crimarco N, Sokol Y, Garnaat S, Mclaughlin N, Phelps E, Pinto A, Boisseau C, Wilhelm S, Daw N, Simpson H (2019) Comparison of the association between

goal-directed planning and self-reported compulsivity vs obsessive-compulsive disorder diagnosis. JAMA Psychiatry 77:77–85
Gillan C, Vaghi M, Hezemans F, van Ghesel Grothe S, Dafflon J, Brühl A, Savulich G, Robbins T (2020) Experimentally-induced and real-world acute anxiety have no effect on goal-directed behaviour
Graybiel AM (1998) The basal ganglia and chunking of action repertoires. Neurobiol Learn Mem 70(1–2):119–136
Graybiel AM, Rauch SL (2000) Toward a neurobiology of obsessive-compulsive disorder. Neuron 28(2):343–347
Gremel CM, Costa RM (2013) Orbitofrontal and striatal circuits dynamically encode the shift between goal-directed and habitual actions. Nat Commun 4:2264–2264
Hardwick R, Forrence A, Krakauer J, Haith A (2019) Time-dependent competition between habitual and goal-directed response preparation. Nat Hum Behav
Hauser TU, Iannaccone R, Dolan RJ, Ball J, Hättenschwiler J, Drechsler R, Rufer M, Brandeis D, Walitza S, Brem S (2017) Increased fronto-striatal reward prediction errors moderate decision making in obsessive-compulsive disorder. Psychol Med 47(7):1246–1258
Hoven M, Lebreton M, Engelmann JB, Denys D, Luigjes J, van Holst RJ (2019) Abnormalities of confidence in psychiatry: an overview and future perspectives. Transl Psychiatry 9(1):268
Huys QJ, Maia TV, Frank MJ (2016) Computational psychiatry as a bridge from neuroscience to clinical applications. Nat Neurosci 19(3):404–413
Hyman SE (2007) Can neuroscience be integrated into the DSM-V? Nat Rev Neurosci 8 (9):725–732
Hyman SE (2010) The diagnosis of mental disorders: the problem of reification. Annu Rev Clin Psychol 6:155–179
Jacob ML, Larson MJ, Storch EA (2014) Insight in adults with obsessive-compulsive disorder. Compr Psychiatry 55(4):896–903
Kapur S, Phillips AG, Insel TR (2012) Why has it taken so long for biological psychiatry to develop clinical tests and what to do about it? Mol Psychiatry 17(12):1174–1179
Keramati M, Smittenaar P, Dolan RJ, Dayan P (2016) Adaptive integration of habits into depth-limited planning defines a habitual-goal-directed spectrum. Proc Natl Acad Sci U S A 113 (45):12868–12873
Kessler RC, Chiu WT, Demler O, Walters EE (2005) Prevalence, severity, and comorbidity of 12-month DSM-IV disorders in the National Comorbidity Survey Replication. Arch Gen Psychiatry 62(6)
Kool W, Gershman SJ, Cushman FA (2017) Cost-benefit arbitration between multiple reinforcement-learning systems. Psychol Sci:956797617708288
Lee SW, Shimojo S, O'Doherty JP (2014) Neural computations underlying arbitration between model-based and model-free learning. Neuron 81(3):687–699
Liljeholm M, Tricomi E, O'Doherty JP, Balleine BW (2011) Neural correlates of instrumental contingency learning: differential effects of action-reward conjunction and disjunction. J Neurosci 31(7):2474–2480
Linnebank FE, Kindt M, de Wit S (2018) Investigating the balance between goal-directed and habitual control in experimental and real-life settings. Learn Behav 46(3):306–319
Lipszyc J, Schachar R (2010) Inhibitory control and psychopathology: a meta-analysis of studies using the stop signal task. J Int Neuropsychol Soc 16(6):1064–1076
Luque D, Molinero S, Watson P, López FJ, Le Pelley ME (2019) Measuring habit formation through goal-directed response switching. J Exp Psychol Gen
Menzies L, Chamberlain SR, Laird AR, Thelen SM, Sahakian BJ, Bullmore ET (2008) Integrating evidence from neuroimaging and neuropsychological studies of obsessive-compulsive disorder: the orbitofronto-striatal model revisited. Neurosci Biobehav Rev 32(3):525–549
Milad MR, Rauch SL (2012) Obsessive-compulsive disorder: beyond segregated cortico-striatal pathways. Trends Cogn Sci 16(1):43–51

Morris RW, Quail S, Griffiths KR, Green MJ, Balleine BW (2014) Corticostriatal control of goal-directed action is impaired in schizophrenia. Biol Psychiatry
Otto AR, Gershman SJ, Markman AB, Daw ND (2013a) The curse of planning: dissecting multiple reinforcement-learning systems by taxing the central executive. Psychol Sci 24(5):751–761
Otto AR, Raio CM, Chiang A, Phelps EA, Daw ND (2013b) Working-memory capacity protects model-based learning from stress. Proc Natl Acad Sci U S A 110(52):20941–20946
Otto AR, Skatova A, Madlon-Kay S, Daw ND (2014) Cognitive control predicts use of model-based reinforcement learning. J Cogn Neurosci:1–15
Patzelt EH, Kool W, Millner AJ, Gershman SJ (2019) Incentives boost model-based control across a range of severity on several psychiatric constructs. Biol Psychiatry 85(5):425–433
Poldrack RA, Sabb FW, Foerde K, Tom SM, Asarnow RF, Bookheimer SY, Knowlton BJ (2005) The neural correlates of motor skill automaticity. J Neurosci 25(22):5356–5364
Robbins TW, Costa RM (2017) Habits. Curr Biol 27(22):R1200–R1206
Robbins TW, Everitt BJ (1999) Drug addiction: bad habits add up. Nature 398(6728):567–570
Robbins TW, Gillan CM, Smith DG, de Wit S, Ersche KD (2012) Neurocognitive endophenotypes of impulsivity and compulsivity: towards dimensional psychiatry. Trends Cogn Sci 16(1):81–91
Rouault M, Seow T, Gillan CM, Fleming SM (2018) Psychiatric symptom dimensions are associated with dissociable shifts in metacognition but not task performance. Biol Psychiatry 84 (6):443–451
Salkovskis PM (1985) Obsessional-compulsive problems: a cognitive-behavioural analysis. Behav Res Ther 23(5):571–583
Saxena S, Rauch SL (2000) Functional neuroimaging and the neuroanatomy of obsessive-compulsive disorder. Psychiatr Clin North Am 23(3):563–586
Schatzberg AF (2000) New indications for antidepressants. J Clin Psychiatry 61(Suppl 11):9–17
Schreiner DC, Renteria R, Gremel CM (2019) Fractionating the all-or-nothing definition of goal-directed and habitual decision-making. J Neurosci Res
Sebold M, Deserno L, Nebe S, Schad DJ, Garbusow M, Hägele C, Keller J, Jünger E, Kathmann N, Smolka MN, Smolka M, Rapp MA, Schlagenhauf F, Heinz A, Huys QJ (2014) Model-based and model-free decisions in alcohol dependence. Neuropsychobiology 70(2):122–131
Seow X, Gillan C (2020) Transdiagnostic phenotyping reveals a range of metacognitive deficits associated with compulsivity. Sci Rep
Shahar N, Moran R, Hauser TU, Kievit RA, McNamee D, Moutoussis M, Dolan RJ, N. Consortium (2019) Credit assignment to state-independent task representations and its relationship with model-based decision making. Proc Natl Acad Sci U S A 116(32):15871–15876
Sjoerds Z, de Wit S, van den Brink W, Robbins TW, Beekman AT, Penninx BW, Veltman DJ (2013) Behavioral and neuroimaging evidence for overreliance on habit learning in alcohol-dependent patients. Transl Psychiatry 3:e337
Tanaka SC, Balleine BW, O'Doherty JP (2008) Calculating consequences: brain systems that encode the causal effects of actions. J Neurosci 28(26):6750–6755
Tolman EC (1948) Cognitive maps in rats and men. Psychol Rev 55(4)
Tricomi EM, Delgado MR, Fiez JA (2004) Modulation of caudate activity by action contingency. Neuron 41(2):281–292
Tricomi E, Balleine BW, O'Doherty JP (2009) A specific role for posterior dorsolateral striatum in human habit learning. Eur J Neurosci 29(11):2225–2232
Vaghi MM, Luyckx F, Sule A, Fineberg NA, Robbins TW, De Martino B (2017) Compulsivity reveals a novel dissociation between action and confidence. Neuron 96(2):348–354.e344
Vaghi MM, Cardinal RN, Apergis-Schoute AM, Fineberg NA, Sule A, Robbins TW (2018) Action-outcome knowledge dissociates from behavior in obsessive-compulsive disorder following contingency degradation. Biol Psychiatry Cogn Neurosci Neuroimaging
Valentin VV, Dickinson A, O'Doherty JP (2007) Determining the neural substrates of goal-directed learning in the human brain. J Neurosci 27(15):4019–4026

Voon V, Derbyshire K, Rück C, Irvine MA, Worbe Y, Enander J, Schreiber LR, Gillan C, Fineberg NA, Sahakian BJ, Robbins TW, Harrison NA, Wood J, Daw ND, Dayan P, Grant JE, Bullmore ET (2014) Disorders of compulsivity: a common bias towards learning habits. Mol Psychiatry

Wunderlich K, Dayan P, Dolan RJ (2012) Mapping value based planning and extensively trained choice in the human brain. Nat Neurosci 15(5)

Wyckmans F, Otto AR, Sebold M, Daw N, Bechara A, Saeremans M, Kornreich C, Chatard A, Jaafari N, Noël X (2019) Reduced model-based decision-making in gambling disorder. Sci Rep 9(1):19625

Yin HH, Knowlton BJ, Balleine BW (2004) Lesions of dorsolateral striatum preserve outcome expectancy but disrupt habit formation in instrumental learning. Eur J Neurosci 19(1):181–189

Yin HH, Ostlund SB, Knowlton BJ, Balleine BW (2005) The role of the dorsomedial striatum in instrumental conditioning. Eur J Neurosci 22(2):513–523

Electroencephalographic Correlates of Obsessive-Compulsive Disorder

Ana Maria Frota Lisbôa Pereira de Souza

Contents

Abstract This chapter reviews EEG research in Obsessive-Compulsive Disorder (OCD), focusing on Event-Related Potentials (ERPs) such as the Contingent Negative Variation, N2, Error-Related Negativity, the feedback Error-Related Negativity and the Readiness Potential and their neural bases. The functional significance, utility and correlation of these ERPs with OCD symptoms will be discussed, alongside novel theories for integrating the research findings. I will consider hypotheses including goal-directed behaviour, overreliance on habits, dissociations between action and knowledge, and excessive intolerance of uncertainty in the context of EEG studies, thus providing a comprehensive framework of the electroencephalographic literature concerning OCD.

Keywords Anterior cingulate cortex · Contingent negative variation · Error-related negativity · Evoked potential · Readiness potential

A. M. Frota Lisbôa Pereira de Souza (✉)
Department of Psychology, Behavioural and Clinical Neuroscience Institute (BCNI), University of Cambridge, Cambridge, UK

Curr Topics Behav Neurosci (2021) 49: 169–199
https://doi.org/10.1007/7854_2020_200
Published Online: 16 February 2021

> The study of error is not only in the highest degree prophylactic, but it serves as a stimulating introduction to the study of truth.
>
> Walter Lippmann

1 Background

Since German Psychiatrist Hans Berger's (1873–1941) first recordings of brain electrical activity in 1924 (Luck and Kappenman 2012), electroencephalographic measures have been increasingly applied in research in the fields of psychiatry, neurology and neurosciences. Described as "a technique for recording electrical activity of the human brain from the surface of the head" (Millet 2002), electroencephalography (EEG) has achieved status and followers due to its relative low-cost, high tolerability and non-invasive characteristics (Kappenman and Luck 2016). It is imperative, though, to understand this technique and the processes underlying the generation of the EEG signal, in order to fully comprehend its results, contributions and limitations.

2 What Can an Electroencephalographic Approach Provide to the Understanding of OCD?

The technique of measuring neuronal signalling is called Electroencephalography (EEG) and has been widely used in research for diagnostic purposes and uncovering cognitive processes (Kappenman and Luck 2016). It provides a rapid method for measuring brain activity, which has considerable temporal resolution that can help dissect cognitive processes prior to behavioural manifestation. This velocity is particularly relevant since cognitive processes may happen in fractions of seconds and offers a remarkable advantage in comparison with other gold standards of neuroimaging such as functional Magnetic Resonance Imaging (fMRI). Whilst fMRI data provide excellent spatial resolution, the Blood Oxygen Level Dependant (BOLD) response on which they rely typically requires a few seconds to be completed (Glover 2011), proving its unsuitability for measuring rapid cognitive processes (Luck and Kappenman 2012).

Although electroencephalographic recordings provide excellent temporal resolution, the method is not without its pitfalls. The signal recorded by the electrodes placed on the scalp is a result of the summation of the activity of populations of neurons that have fired in response to a stimulus or spontaneously and have crossed several layers of brain tissue and especially a very resistant skull (Burle et al. 2015). Due to the fact that it has been originated several centimetres below the scalp, where it is recorded, the signal originated from the EEG carries activity from underlying brain sources and does not necessarily correspond to the area underneath the channel where it is maximal (Cacioppo et al. 2016). This diffusion results in a signal that is potentially generated in a different location from the one in which it is captured, perhaps similar to sensing an electric shock when touching a power plug, which could lead to the conclusion that the current had originated there (instead of from the power outlet). EEG recordings, therefore, are not ideal for deriving conclusions concerning their spatial source (Luck and Kappenman 2012), although powerful methods of mathematical transformations can be applied to the EEG signal, minimising this limitation (Michel and Brunet 2019).

Another disadvantage associated with this technique is the low signal-to-noise ratio. To validate robust conclusions, a high number of trials and a long duration of the recordings are necessary. EEG experiments, therefore, require careful design and laborious data analysis (Keil et al. 2014). Nevertheless, EEG recordings can provide important information about the brain and specific cognitive processes, for instance, how humans adapt to errors, detect conflict and receive feedback.

2.1 Event-Related Potentials

The electrical signal captured in the scalp resulting from a specific sensory, motor or cognitive event or stimulus is called an Event-Related Potential (ERP). The first studies of ERP date from 1935, when Pauline and Hallowell Davis recorded data from awake humans (Luck 2005). It was only later, in 1964, that the first ERP waveform was discovered, namely the Contingent Negative Variation (CNV), by Walter and colleagues (Walter et al. 1964).

ERP research is now one of the most widely used methods for mapping cognitive processes. By creating time windows (epochs) that are "locked" to specific events (e.g. a button press) it is possible to average electrical activity in the brain across multiple trials, uncovering a resulting waveform (Luck 2005). It was through this process that many cognitive phenomena have been elucidated, including attention, memory, conflict detection and error monitoring (Luck and Kappenman 2012). Given the particular advantages of temporal resolution and high tolerability, ERP research has long been used as a tool for diagnostic purposes of both neurological (e.g. sleep disorders and epilepsy) and psychiatric disorders (e.g. schizophrenia, attention deficit hyperactivity disorder, obsessive-compulsive disorder, among others) (Kappenman and Luck 2016).

3 Obsessive-Compulsive Disorder and Electrophysiology

Obsessive-Compulsive Disorder (OCD) is a highly debilitating mental disorder, characterised by persistent, intrusive and distressing obsessions and/or compulsions, affecting individual's social, occupational and other areas of functioning lives and causing considerable distress (American Psychiatric Association 2013). It is estimated to affect between 1 and 3% of the population around the world (Ruscio et al.

2010) and can frequently present a chronic course when not treated (Riesel et al. 2015).

The main characteristics of OCD are the presence of obsessions or compulsions, the first being defined as "recurrent and persistent thoughts, impulses, or images that are experienced, at some time during the disturbance, as intrusive and inappropriate and that cause marked anxiety or distress" and the latter consisting of "repetitive behaviours (e.g. hand washing, ordering, checking) or mental acts (e.g. praying, counting, repeating words silently) that the person feels driven to perform in response to an obsession or according to rules that must be applied rigidly" (American Psychiatric Association 2013). The level of insight the individuals present regarding their compulsions can vary from high to absent (Gillan et al. 2017), and, although in many cases the subjects are aware of the poor connection between the compulsion and the obsession (i.e. checking if the oven is turned off several times to avoid harm to a beloved relative), they yet feel an urge to perform the compulsion, characterising OCD as an ego-dystonic disorder (Vaghi et al. 2017a).

The neurobiology of OCD is not yet consistently established, however, extensive research indicates a dysfunction in the cortico-striato-thalamo-cortical (CSTC) circuits of the brain in the aetiology of this disease (Rauch et al. 1994). More specifically, an overactivation of the orbitofrontal (OFC) and anterior cingulate (ACC) cortices, as well as the caudate, insula and amygdala, and an hypoactivation of the dorsolateral prefrontal cortex (DLPFC) appear to be present in individuals with OCD (Apergis-Schoute et al. 2018; Baxter et al. 1988; Robbins et al. 2019; Milad and Rauch 2012). However, some evidence also indicates the influence of the Supplementary Motor Area (SMA) and of the Pre-Supplementary Motor Area (pre-SMA) in the inhibitory control dysfunction perceived in OCD patients, which could potentially be related to their lack of control over the performance of compulsions (Bonini et al. 2014; de Wit et al. 2012; Grützmann et al. 2016; Lee et al. 2017). OCD has been, therefore, conceptualised as a disorder of self-control and behavioural inhibition, and patients tend to perform their compulsions despite negative consequences (Banca et al. 2015). A possible candidate neural centre for the deficient inhibitory control of actions (compulsions) and thoughts (obsessions) in OCD is located in the basal ganglia. A recent meta-analysis has suggested that this brain region, extensively studied for its role in motor inhibitory control, might also be involved in higher-order cognitive processes, for instance, memory retrieval and thought inhibition (Guo et al. 2018).

Another of the recent theories advanced to explain the acquisition and maintenance of compulsions in OCD is the imbalance between goal-directed and habit learning systems (Banca et al. 2015; Dayan et al. 2017; Gillan and Robbins 2014; Gillan et al. 2011, 2016, 2017; Vaghi et al. 2017b; Voon et al. 2015a). This approach postulates that patients with OCD tend to rely on habits rather than on goal-directed behaviours, despite appearing aware of more appropriate actions (Vaghi et al. 2017a, b), and that the SMA, pre-SMA and ACC are involved in this process by affecting inhibitory control and thereby facilitating the acquisition of compulsive habits (Grützmann et al. 2016; Lee et al. 2017; Riesel et al. 2015). It is possible, thus, to infer that these mechanisms are connected to what are called "slips of action", a commonly seen feature of OCD in laboratory studies. Those behaviours occur in situations when subjects are performing tasks and suddenly commit errors due to the lack of ability to suppress a response that was previously reinforced, hypothetically indicating a reliability on habits (Gillan and Robbins 2014). Whilst the ACC is responsible for inhibitory control, the SMA and pre-SMA are important for "readiness to act", which could explain increased performance of "slips of action" if these mechanisms are impaired in OCD. It is important to emphasise, though, that the extent to which compulsive responding in OCD is instrumental remains the subject of current investigation (see the chapter "Recent Developments in the Habit Hypothesis of OCD and Compulsive Disorders").

In an attempt to clarify the underlying mechanisms depicted above (i.e. inhibitory control, "slips of action", habitual behaviour) electroencephalography has been widely used in patients with OCD, with techniques ranging from: (1) Quantitative EEG (QEEG) (also referred to as "brain mapping"). This is an analytical technique that describes EEG parameters related to band power, synchronisation and activation patterns in the brain during electroencephalographic recordings, resulting in a map of the brain (Tong and Thakor 2009); (2) intracerebral electroencephalography (iEEG). This technique involves placing electrodes directly onto the exposed surface of the brain in order to monitor activity from the cortex, providing excellent anatomical precision (Parvizi and Kastner 2018); and (3) ERP (Bonini et al. 2014; Perera et al. 2019; Riesel et al. 2015).

A recent review of the electrophysiological literature of OCD has suggested frontal asymmetries in alpha and theta band power in these patients (Perera et al. 2019). Asymmetries are defined as the difference between right and left activity over frontal regions of the brain (Davidson et al. 1990). The band power suggests which of the areas (left or right) is more strongly active. Band types have been extensively studied in psychiatric disorders and present some clues regarding cognitive functions (Perera et al. 2019). Whilst alpha band asymmetries have been linked to avoidance motivation in OCD (Ischebeck et al. 2014), theta band asymmetries are thought to suggest impairments in active inhibition during cognitive tasks, a previously reported robust cognitive marker of OCD deficits (Menzies et al. 2007; Min et al. 2011; Riesel et al. 2015). Perhaps not surprisingly, research evidence proposes that this theta activity is mainly generated in the ACC, a widely recognised area for mediating inhibitory impairments in OCD (Wang et al. 2005). On the other hand, task related alpha asymmetries have also been associated with difficulties in suppressing distractors or task-irrelevant details. Taken together, these results might elucidate the strains experienced by patients with OCD in suppressing obsessive thoughts (Crawford et al. 1995; Perera et al. 2019).

In addition, several ERP components have been studied in order to investigate the cognitive processes associated with the disorder (Fig. 1). Examples include the Contingent Negative Variation (CNV) (Shen et al. 2017), the Readiness Potential (RP) (Dayan et al. 2017), the N2 (Dayan-Riva et al. 2020; Dieterich et al. 2017; Riesel et al. 2017a), the Correct-Related Negativity (CRN) (Olvet and Hajcak 2008; Riesel et al. 2015), the error Positivity (Pe) (Klawohn et al. 2014), the feedback Error-Related Negativity (fERN) (Hajcak et al. 2006; Holroyd et al. 2006) and perhaps the most widely recognised, the Error-Related Negativity (ERN)

ERP	Main generators	Cognitive process	Paradigms	Timeframes	OCD findings
Contingent negative variation (CNV)	–Pre-motor cortex –Basal ganglia	–Expectancy –Goal-directed behaviour –Motor preparation	Reaction-time tasks with warning and imperative stimuli	~260–470 ms between warning signal and imperative stimuli	Enhanced amplitudes
Readiness potential (RP)	–Pre-SMA –SMA –ACC –Basal Ganglia	Motor preparation	Motor tasks (reaction - time tasks)	Up to 1.5 s prior to movement	Enhanced amplitudes
N2	ACC	Conflict detection	–Mismatch detection -Cognitive control (inhibitory tasks)	~ 200–350 ms post stimuli	Mixed results with studies describing enhanced and diminished amplitudes
Error-related negativity (ERN)	-ACC -SMA	Error monitoring	Inhibitory control	~50–100 ms post erroneous response	Enhanced amplitudes
Feedback error - related negativity (fERN)	ACC	Error detection	Inhibitory control tasks with feedback	~250–300 ms after feedback	Diminished amplitudes

Fig. 1 OCD-related ERP components and the brain regions generating the signals, associated cognitive processes and experimental paradigms

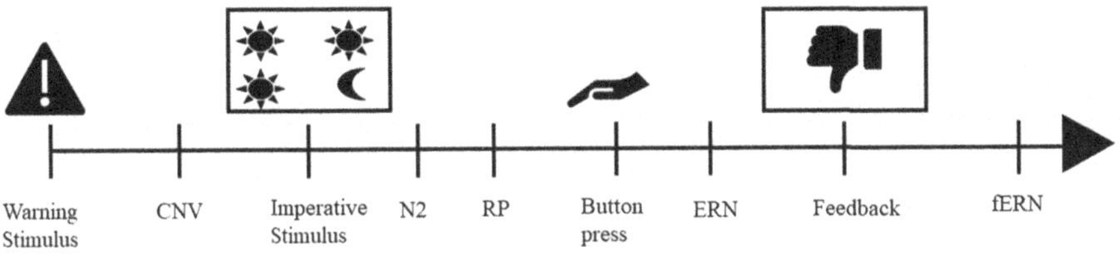

Fig. 2 Time course of CNV, N2, RP, ERN and fERN: a simple guide

(Grützmann et al. 2016; Klawohn et al. 2014; Moser et al. 2013; Olvet and Hajcak 2008; Riesel 2019; Weinberg et al. 2015a).

Figure 2 illustrates an example of an experimental paradigm that generates some of the above-mentioned components and their time course. In the example, a warning signal is delivered that alerts the individual to the need for fast responding towards an upcoming stimulus. The participant is then presented with stimuli that can either be homogeneous (e.g. Four identical images) or heterogeneous (e.g. one image differs from the rest) and needs to press a button indicating if they are the same or not. Feedback is provided after the response is executed. In the scenario depicted below, the warning stimulus will elicit the CNV. The participant is then presented with heterogeneous stimuli, which evoke the N2. The instruction to respond leads to motor preparation and consequently the RP. The participant then choses the incorrect response, giving rise to the ERN. Finally, the negative feedback causes the generation of the fERN. It is important to consider, though, that not all these ERPs are elicited in the same tasks, since different cognitive processes are implied. This illustration has been created for didactic reasons purely and does not represent accurate timeframes.

3.1 Contingent Negative Variation and OCD

The ability to make decisions and predict outcomes is essential for a successful life where environmental feedback informs cognitive processes that will act towards established goals. Indeed, when tested on laboratory-based tasks, both goal-directed learning and cognitive flexibility seem to be impaired in patients with OCD (Gillan and Robbins 2014; Gillan et al. 2011; Menzies et al. 2007; Murray et al. 2019; Robbins et al. 2019), resulting in non-optimal choices and higher prevalence of errors (Banca et al. 2015; Vaghi et al. 2017b).

The CNV is mainly generated in frontal regions of the brain and can be measured by the EEG between a warning signal and an imperative stimulus during reaction-time tasks (Linssen et al. 2011). It has been investigated as an index of goal-directed behaviour (Shen et al. 2017), providing possible markers of expectation, anticipation, orientation or intention, as well as readiness to act or motivation (Tecce 1972). Therefore, as OCD involves a loss of control over thoughts and behaviours, the CNV is an ERP component that would be expected to be affected in OCD. Figure 3 depicts

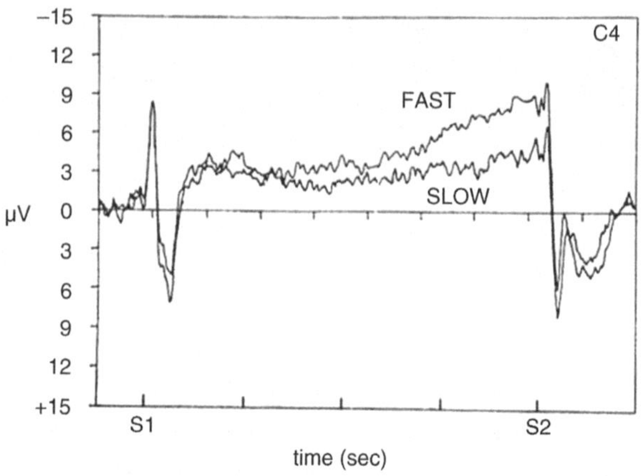

Fig. 3 Example of CNV component (Luck and Kappenman 2012). Figure reproduced with permission of the Licensor through PLSclear

the CNV prior to a foot movement. S1 represents the warning signal and S2 the imperative stimulus.

One of the generators of the CNV appears to be in the basal ganglia (Fan et al. 2007), which has dopamine as one of its main neuromodulators. Dopaminergic brain activity has been proposed as a possible mechanism underlying this ERP (Amabile et al. 1986; Linssen et al. 2011). Indeed, various dopaminergic disorders such as Gilles de la Tourette Syndrome and Parkinson's Disease present abnormal CNV amplitudes (Weate et al. 1993; van Woerkom et al. 1988), and drugs such as methylphenidate (a dopamine re-uptake inhibitor) seem effective in modulating its amplitude (Linssen et al. 2011). The role of dopamine in OCD will be discussed later in the chapter.

Research into CNV in OCD, however, is not vast. A few studies have reported enhanced CNV amplitudes in OCD patients (Khanna 1988; McCallum and Walter 1968; Sartory and Master 1984; Timsit-Berthier et al. 1973), consistent with enhanced negative prediction errors (those will be discussed in the context of reinforcement learning in this chapter) (Murray et al. 2019). Although the CNV has been implicated in motor preparation (i.e. preparation to act), this has not so far been investigated to any large extent in OCD by previous studies. Nevertheless, motor preparation in OCD will be considered in the context of other event-related potentials below.

3.2 *Mismatch Detection and the N2*

The discrepancy between an expected and an actual outcome generates an ERP termed N2. As the name suggests, this negative deflection peaks around 200 ms after a conflict is detected by the brain and is usually measured experimentally through

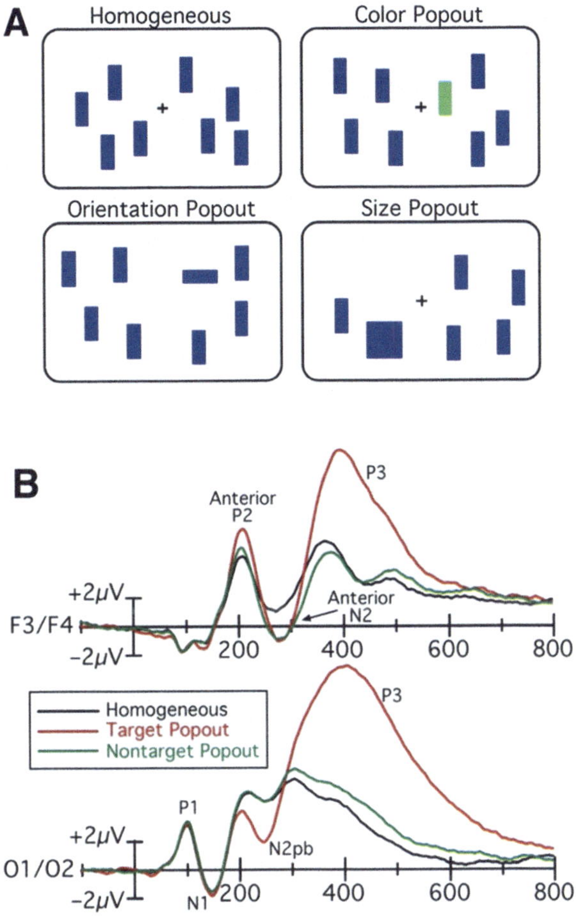

Fig. 4 Example of the N2 component.
Figure reproduced with permission of the Licensor through PLSclear

inhibitory control, mismatch detection and probabilistic learning tasks (Azizian et al. 2006; Dieterich et al. 2017; Folstein and Van Petten 2008; Riesel et al. 2017a).

The discovery of the N2 dates from 1965, when Sutton and colleagues designed an experiment to investigate uncertainty on sensory potentials. In their study, a certain condition predicted that a cue was followed by a specific target, that could be a click or a light flash, whereas in the uncertain condition it was not possible to predict which target would follow. The researchers identified a negativity peak around 200 ms only in the uncertain condition, in opposition to the robust positivity at 300 ms (P300) that supplants correctly predicted cues (Sutton et al. 1965). The figure below (Fig. 4) depicts a classic paradigm for the measurement of the N2. The stimuli were either eight homogeneous blue vertical bars or seven blue bars and a discriminant "pop-out" one, that could differ in colour, size or orientation. Homogeneous arrays were present on 50% of the trials, with the three "pop-out" targets appearing 16.7% of the time. One of the "pop-out" stimuli was selected as the target prior to the beginning of the trial, and participants were instructed to press one of the two keys indicating whether the target was present or not. An enlarged N2 wave was observed on frontal sites to either target and non-target "pop-outs" relative to homogeneous arrays (Luck and Hillyard 1994).

In OCD, studies seem to report both enhanced (Ciesielski et al. 2011; Riesel et al. 2017a; Ruchsow et al. 2007) and diminished (Dayan-Riva et al. 2020; Dieterich et al. 2017) amplitudes of N2. However, it is important to note that different paradigms may actually be capturing diverse cognitive processes. Research that suggests diminished N2 is often based on paradigms that require response inhibition, a mechanism well-known to be impaired in OCD (Chamberlain et al. 2005), whereas enhanced N2 amplitudes are seen in response to conflict processing paradigms, such as the Flanker Task (Riesel et al. 2017a). Despite the contradictory results, research seems to be unanimous regarding the role of the N2 as a marker of inhibitory control and mismatch detection. It appears that the N2 is part of a monitoring network, which detects conflict and signals the need for cognitive control, even in situations where monitoring would not be as necessary or beneficial (Riesel et al. 2017a). A perceived increased need for control might also explain the reason why conflict is highly monitored in OCD (Riesel et al. 2019) and reflects another key feature of the disorder: intolerance of uncertainty.

3.2.1 Uncertainty as a Generator of the N2 and OCD Symptoms

Uncertainty appears to play an important role in the generation of the N2 and is also a marker of OCD (Dieterich et al. 2017; Scholl and Rushworth 2017) as well as other anxiety-related disorders. Patients with OCD have been shown to present difficulties when making decisions, especially in uncertain contexts (Holaway et al. 2006; Marzuki et al. 2020; Morein-Zamir et al. 2020). In fact, unpredictability seems to be evaluated negatively, is not well tolerated by patients with OCD, and increases demand for attentional control, anxiety levels and neural markers of conflict detection (i.e. N2) (Dieterich et al. 2017). In addition, stress is thought to reduce Reward Positivity, an ERP that measures sensitivity to reward (Burani et al. 2020), which could explain non-optimal decision-making in unpredictable – and consequently stressful – tasks by OCD patients. Confidence levels, on the other hand, are not necessarily predictive of most advantageous choices and patients with OCD seem to discard acquired knowledge and instead explore the tasks further. In a predictive inference learning task in which uncertainty was a major element determining performance, Vaghi et al. (2017a) reported a dissociation between action and confidence, with OCD patients demonstrating similar confidence rates to healthy volunteers but not acting in accordance to the optimal choices. The same pattern has been shown in children with OCD undertaking the Information Sampling Task (Clark et al. 2006). Patients seem to discount subjective costs of time spent performing the task, such as fatigue and impatience in order to turn more cards before making a decision (Hauser et al. 2017).

Intolerance of uncertainty might also be one of the underlying mechanisms driving checking behaviour in OCD. Since patients with this disorder present enhanced amplitudes of N2 in response to conflict processing paradigms implicating overactive conflict/mismatch detection mechanisms (Riesel et al. 2017a), it is possible to infer that checking compulsions may be generated to serve a protective role in avoiding a heightened sense of error commission. The features of behavioural rigidity and perfectionism, also key markers of OC symptoms (American Psychiatric Association 2013), might be added to this model as a means of controlling the environment and alleviating anxiety facing uncertainty (Moser et al. 2013). Indeed, a study has attempted to explain OCD symptomatology through the Bayesian Brain Framework (Fradkin et al. 2020; Knill and Pouget 2004). This framework proposes that the brain makes probabilistic inferences regarding states through sensory feedback, predictions and beliefs weighted according to their perceived uncertainty. For example, someone with checking symptoms of OCD might put more weight on sensory feedback (the action of checking and seeing that the stove is turned off) than on the "uncertain" belief that they have already turned off the stove. According to Fradkin et al. (2020), the core symptom of OCD is excessive uncertainty regarding state transitions, which could be described as the ability to understand and predict changes (or the lack of them) from one state to the next. The authors propose that this "transition uncertainty" could explain the impairment in planning, excessive checking and overreliance on sensory feedback seen in this condition.

A second possible theory to explain this process derives from the ego-dystonic nature of the disorder (Banca et al. 2015; Vaghi et al. 2017a). Although patients are aware of the poor connection between their obsessions and compulsions, it is plausible to hypothesise that compulsions arise from the need to control an uncertain environment. For instance, one could aim to protect a beloved relative by turning the light switches on and off, behaviour characterised as magical thinking (Einstein and Menzies 2004; West and Willner 2011). One of the most commonly known strategies for dealing with uncertainty and the consequential anxiety derived from it is the creation of, and engagement in rituals (Graybiel 2008; Hobson et al. 2018). From children's pre-bed routines to the habits of athletes prior to competitions, rituals have served the purpose of controlling an uncertain environment by providing order to an unpredictable world (Tonna et al. 2019). An enhanced need for control might therefore explain many symptoms in OCD, for instance, checking (to avoid making a mistake), hoarding (to avoid the loss of something that maybe be useful in the future), mental rituals (to avoid harm), and many others, behaviours that describe what is termed "illusion of control" (Reuven-Magril et al. 2008). However, the role of perceived control in OCD remains controversial, as Gillan et al. (2014) showed that patients reported reduced estimates on a traditional illusion of control task.

Rituals are described as predefined sequences of symbolic actions that are performed in a strict and repetitive way and lack instrumental purpose (Brooks et al. 2016; Hobson et al. 2017). This definition, similar to the meaning of compulsion, differs in one key aspect of the latter: rituals are symbolically meaningful. If compulsions are derived from an imbalance between goal-directed and habitual behaviours, with the preponderance of the latter, one could argue that they have served a purpose formerly. Indeed, evolutionary studies propose that the acquisition of habits serves an important role in automatizing and expediting instrumental actions which could potentially lead to higher survival rates (Eilam et al. 2006). For instance, routinization would allow for less attention required to perform habitual tasks, fewer errors due to training (Banca et al. 2020) and more resources available for monitoring potential threats (Fentress 1976) and detecting conflict.

It is possible that in OCD the anxiety reported when compulsions are not performed is a misrepresentation of an overactive striatum and enhanced action tendencies rather than an actual fear response. The relationship between anxiety and motor behaviour does seem to present some clues regarding the aetiology of OCD symptomatology. Two recent studies have attempted to manipulate anxiety and electrophysiological amplitudes of error monitoring through the introduction of rituals in healthy volunteers (Brooks et al. 2016; Hobson et al. 2017). Participants were asked to perform novel random ritualistic motor behaviours prior to an anxiety-inducing task. Electroencephalographic recordings were measured at baseline and following the execution of the ritual, showing decreased amplitudes. The authors propose that the execution of rituals acts as a buffer against anxiety and by increasing confidence in performance, which in turn diminishes the sensitivity to self-generated errors (Hobson et al. 2017). This study is particularly important for differentiating compulsions from rituals, introducing a completely novel set of ritualised behaviours. Despite the lack of research in OCD patients, this could shed light on the reason why rituals are perceived by patients as anxiety-reducing.

3.3 Error-Related Negativity as an Endophenotype of OCD

After the 1990s and the "decade of the brain" (Bush 1990), much biological research entered the "age of the endophenotypes". An endophenotype constitutes a heritable trait that is associated with an augmented genetic risk for a disorder (Chamberlain and Menzies 2009). Traits classified as endophenotypes mediate the relationship between genes and a behavioural phenotype (Olvet and Hajcak 2008) are state independent and occur in unaffected relatives at a higher rate than in the general population (Gottesman and Gould 2003). Additionally, endophenotypes are thought to be impervious to treatment, despite the improvement of symptoms (Hajcak et al. 2008; Riesel et al. 2011, 2015). For a more comprehensive view of endophenotypes, see Box 1 and the chapter "Neurocognitive Endophenotypes of OCD".

Box 1 ERN and the RDoC

Albeit not new, the interest for endophenotypes is likely to have risen as a consequence of the proposition of the Research Domain Criteria (RDoC) framework (Cuthbert and Insel 2013; Insel et al. 2010; Insel 2014; Sanislow

(continued)

Box 1 (continued)
et al. 2010). Launched in 2008 by the National Institute of Mental Health (NIMH), the RDoC aims at combatting the categorical classification that has dominated psychiatric diagnosis for the past 150 years and incorporating neuroscience findings to the formulation of a clinical hypothesis (de Souza et al. 2018). Within the RDoC framework, researchers are encouraged to investigate disorders such as OCD dimensionally and trans-diagnostically (Gillan et al. 2017), providing insights into mechanisms underlying mental disorders (Cuthbert and Insel 2013; Insel et al. 2010; Insel 2014; Krueger et al. 2018; Sanislow et al. 2010).

The RDoC matrix currently comprises five domains (negative valence systems, positive valence systems, cognitive systems, social processes and arousal and regulatory systems) and eight units of analysis (i.e. genes, molecules, cells, circuits, physiology, behaviour, self-report and paradigms), thought to encompass all processes responsible for impacting behaviour and cognition, though more domains could be added (Cuthbert and Insel 2013; Etkin and Cuthbert 2014; Insel et al. 2010; Sanislow et al. 2010). Due to its involvement in different processes, the ERN is a physiological unit of analysis of three RDoC domains: (1) *Performance Monitoring* in the Cognitive domain; (2) *Sustained Threat* in the Negative Valence Systems domain and (3) *Reward Learning* in the Positive Valence Systems domain (de Souza et al. 2018). Thus, the ERN is thought to be implicated in the recognition of threat, learning from feedback and performance monitoring, which are thought to be impaired in OCD (Apergis-Schoute et al. 2017; Riesel 2019; Vaghi et al. 2017b).

The Error-Related Negativity (ERN) is a component of the ERP that reaches a maximum amplitude between 50 and 100 ms after the commission of an error in forced-choice inhibitory control tasks (e.g. Stroop Colour and Word Task (SCWT, Stroop 1935); Flanker Task (Eriksen and Eriksen 1974); and the Go/No-Go Task) (Luck and Kappenman 2012; Olvet and Hajcak 2008; Weinberg et al. 2015a). It was independently discovered by two research groups in the early 1990s. In Germany, Falkenstein et al. (1991) reported what they called the Error Negativity (Ne), whereas in the United States, Gehring et al. (1993) described the Error-Related Negativity (ERN) (Falkenstein et al. 1991; Gehring et al. 1993; Weinberg et al. 2015a; Wessel 2012). Within less than three decades of its discovery, the ERN has become the most widely investigated index of error processing (Wessel 2012) and is implicated in multiple forms of psychopathology (Riesel et al. 2017b). Figure 5 depicts an example of the ERN component, created for illustrative purposes.

The validity of the ERN as an endophenotype for psychiatric disorders has been extensively investigated and remains unrefuted, having been proven in different conditions (Gillan et al. 2017; Moser et al. 2013; Olvet and Hajcak 2008; Perera et al. 2019; Riesel et al. 2011, 2017a, b; Weinberg et al. 2015a). Several studies have

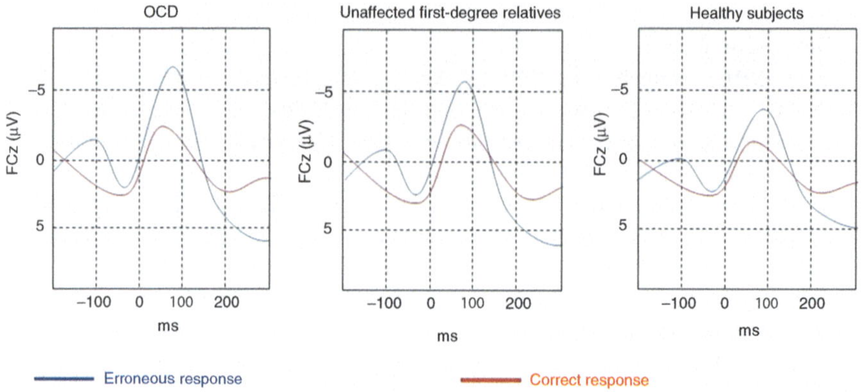

Fig. 5 Example of the ERN component in OCD, unaffected first-degree relatives of OCD patients and healthy subjects

attempted to contrast ERN amplitudes in disorders such as attention deficit hyperactivity disorder (ADHD) (Marquardt et al. 2018), psychosis (Foti et al. 2012), substance abuse and addictions (Franken et al. 2007; Riesel et al. 2019), depression (Holmes and Pizzagalli 2008), generalised anxiety disorder (GAD) (Hajcak et al. 2003a; Moser et al. 2013), among many others (Olvet and Hajcak 2008). Conditions such as ADHD, psychosis, substance abuse, addictions and other disorders characterised by impulsivity seem to be marked by reduced amplitudes of ERN when compared to healthy volunteers, whereas disorders in the spectrum of anxiety and OCD show larger amplitudes (Olvet and Hajcak 2008). Depression appears as a contradictory case, with studies showing both enhanced (Holmes and Pizzagalli 2008) and diminished (Weinberg et al. 2015b) ERN amplitudes. It seems, then, that ERN amplitudes are less related to diagnostic conditions than to individual differences on the evaluation of mistakes (Hajcak et al. 2005), fitting well within the trans-diagnostic RDoC framework (Weinberg et al. 2015a) (see Box 1).

Amongst different psychiatric disorders, OCD is the disorder in which the ERN has been most extensively studied. The findings of enhanced amplitudes of the ERN in unaffected family members of OCD patients and its imperviousness to the effects of treatment indicate that the ERN represents an endophenotype of the disorder (Olvet and Hajcak 2008; Riesel 2019). Indeed, a recent systematic review of the literature has investigated whether the six proposed endophenotypes for OCD (impairments in *Decision-Making, Action Monitoring, Inhibition, Memory* (working, verbal and nonverbal), *Reversal Learning* (either behavioural or associated with Orbitofrontal Cortex (OFC) dysfunction) and *Cognitive Flexibility* (Menzies et al. 2007)) would also apply in adolescents with the disorder (Marzuki et al. 2020). The most robust findings suggested increased action monitoring as a shared trait between paediatric and adult OCD, potentially indicating this as a deficit with an onset that precedes that of the others. Indeed, research has evidenced that an abnormally enhanced action monitoring tendency, as measured through EEG in unaffected children as young as 6 years old re-tested at the age of nine, could predict the later development of anxiety disorders (Meyer et al. 2015).

In order to fully comprehend the ERN's role as an endophenotype, though, it is important to clarify its functional significance and biological basis. Many theories have been developed to elucidate this question, with the three major accepted ones described below.

3.3.1 Error Detection/Comparator Theory

Detecting mistakes and adapting behaviours efficiently is essential in a changing environment (Riesel 2019; Weinberg et al. 2015a). The ability to detect and respond to the mismatch between a motor response and the expected outcome gave rise to the "Comparator Theory" as developed in the first studies of ERN (Falkenstein et al. 1991; Gehring et al. 1993). Essentially, the ERN was thought to reflect the comparison between the correct response and that of the individual, and to be generated when the two were not the same (i.e. when a perceived error had occurred) (Coles et al. 2001). This theory indicates that the error processing system represented by the ERN might even be triggered by correct responses, if they are interpreted as incorrect by the subject, i.e. not necessarily indicating mistakes objectively, but the individual's subjective evaluation of them (Coles et al. 2001; Hajcak et al. 2005; Vidal et al. 2003). This could also explain why some disorders are marked by enhanced ERN amplitudes (e.g. anxiety, OCD) and others by blunted (e.g. addiction, ADHD), despite individuals with these disorders being unimpaired in detecting errors. The ERN may thus reflect the "motivational salience" of errors rather than a literal representation of them (Olvet and Hajcak 2008).

Researchers have further proposed that the ERN represents an effective response to errors, processed by the limbic system for action regulation (Luu et al. 2003). Indeed, a plethora of studies has been conducted in an attempt to manipulate the subjective value of committing errors, ranging from monetary rewards (Hajcak et al. 2005), to punishment (Endrass et al. 2010) and social evaluation (Moser et al. 2013), among others. The detection of errors, conflict and the loss of reward could provoke an emotional response (Luu and Pederson 2004), which is particularly aversive in the case of perfectionism (a trait commonly seen in obsessive-compulsive and related disorders), for instance (Perrone-McGovern et al. 2017). Motivational state is especially relevant for interpreting the ERN, as disengagement from the task is predictive of a smaller ERN, as seen in the case of depression (Weinberg et al. 2015b). Therefore, the ERN has also been conceptualised as a putative marker of negative affect (Hajcak et al. 2004).

It is important to emphasise, though, that the ERN is a task dependent measure, being elicited in speeded reaction-time experiments (Riesel 2019; Riesel et al. 2013). Thus, it is possible that most of the errors committed in the classic ERN-generating tasks (e.g. SCWT, Flanker Task and the Go/No-Go Task) are caused by a premature response of the subject, before stimulus evaluation is complete (Coles et al. 2001).

3.3.2 Conflict Monitoring Theory

The Conflict Monitoring Theory is derived from the view that the Error Detection Comparator Theory was implausible, since the brain would have to possess the information of what the intended (correct) response was and be able to compare this with the actual response. If this was the case and the brain had access to this information, then why would the error be committed? (Carter 1998; Luck and Kappenman 2012). It suggests that conflicts are generated from the activation of multiple competing responses, which alerts the brain to the ongoing dispute. Therefore, it does not postulate that the brain possesses all the information to respond, but rather signals the need for increased control in high-conflict trials. Following those, performance will be improved through feedback and learning, and less control will be required (Luck and Kappenman 2012). The locus of this conflict monitoring is the main generator of the ERN, the Anterior Cingulate Cortex (ACC), specifically its dorsal portion (dACC) (Falkenstein et al. 1991; Gehring et al. 1993; Riesel 2019; Weinberg et al. 2015a).

The functional role of the ACC has sparked an effervescent debate, with some authors proposing that it monitors conflicts in information processing (Botvinick 2007; Botvinick et al. 2004; Shenhav et al. 2013) and others that it informs optimal decision-making and guides voluntary actions (Behrens et al. 2007; Kennerley et al. 2006). Given the fact that an enhanced ERN is a convincing endophenotype of OCD (Olvet and Hajcak 2008; Riesel 2019; Riesel et al. 2011; Weinberg et al. 2015a) and that several neuroimaging studies have suggested an overactivation of the ACC in OCD patients (Apergis-Schoute et al. 2017; Gillan et al. 2017; Robbins et al. 2019), it is imperative that this region is studied in more detail.

OCD deficits of goal-directed behaviour have been extensively reported in the literature (Banca et al. 2015; Gillan et al. 2011, 2017; Gillan and Robbins 2014). In fact, a predisposition for the formation of habits and preference for model-free over goal-directed model-based learning seems a potential candidate endophenotype of the disorder (Gillan and Robbins 2014; Gillan et al. 2011, 2016; Vaghi et al. 2017b; Voon et al. 2015a, b) (see the chapter "Recent Developments in the Habit Hypothesis of OCD and Compulsive Disorders"). If that is the case, it is possible to infer that the overactive dACC is not informing goal-directed behaviours, but rather monitoring them through the ERN.

One possible explanation for the dissociation between knowledge and action in OCD can be derived from the neurobiological evidence relating to habitual behaviour. Neuroimaging studies show that the basal ganglia are overactive in OCD and regions responsible for the formation of habits such as the striatum (comprising caudate and putamen) are particularly affected (Banca et al. 2015; Fineberg et al. 2018; Gillan et al. 2015). The striatum seems exceptionally important in this context, given that its neurons are thought to fire at the beginning and end of a behavioural routine. Thus striatal overactivity could explain the exaggerated urge to perform compulsions and why such sequences must be performed until completion (Graybiel and Grafton 2015; Guo et al. 2018). It is thus also possible to hypothesise that an over-functioning striatum contributes to increased habit formation in OCD by failing to interrupt motor sequences regardless of the dACC signalling that they are erroneous. It is conceivable, therefore, that the ACC is trying to engage cognitive control and inform optimal decision-making (Behrens et al. 2007; Kennerley et al. 2006) as is the case with the generation of the N2, however, in patients with OCD, this activity is being overridden by striatally driven habitual behaviour (Gillan et al. 2011, 2015).

A further possible explanation concerns the function of the ERN. The fact that it signals the need for employment of cognitive control does not necessarily result in this being achieved. In fact, the ERN may be an early warning signal of conflict generated by the ACC, which is then followed by a cascade of events, including activation of defensive systems and of the dorsolateral prefrontal cortex (DLPFC), a region responsible for the employment of planning and cognitive flexibility that is also found to be impaired in OCD (Robbins et al. 2019; Weinberg et al. 2016). In the case of the ERN, conflict monitoring responses are observed in patients following the commission of errors (Hajcak et al. 2003b), representing their evaluation, but the compensatory behaviours that follow depend on intermediary processes (Weinberg et al. 2016).

3.3.3 Reinforcement Learning Theory

Proposed by Holroyd and Coles (2002), the Reinforcement Learning Theory (RL-ERN) suggests that a monitoring system in the basal ganglia produces an error signal when adverse events occur that are worse than expected (Hajcak et al. 2006; Holroyd et al. 2006; Luck and Kappenman 2012). The system is modulated by the midbrain dopamine system that sends the signal to the ACC, where it is hypothetically used to improve performance by adapting motor behaviour (Holroyd and Coles 2002; Luck and Kappenman 2012). Perhaps not surprisingly, this system relies heavily on feedback and predictions. Dopamine neurons are thought to demonstrate increased rates of firing resulting from unexpected or better than expected results, events termed positive prediction errors. On the contrary, firing rates diminish following the omission of expected rewards, generating negative prediction errors as measured by the BOLD response during fMRI. These cingulate negative prediction errors appear to be enhanced in OCD (Murray et al. 2019) and can be modulated by dopaminergic drugs, including the dopamine receptor antagonist amisulpride, which was found to ameliorate the excessive negative prediction error in OCD (Murray et al. 2019). Albeit not considered first-line treatments for OCD, dopamine receptor blocking agents can be used for refractory cases with positive results, as shown by meta-analysis (Veale et al. 2014). It is therefore possible that one of the therapeutic effects of dopamine antagonists in OCD involves the suppression of exaggerated cingulate negative prediction errors, which could be tested electrophysiologically in future studies using the ERN.

Feedback Error-Related Negativity (fERN) and Evaluation of Errors

Another ERP component generated by the ACC and closely related to prediction errors, with a later time course than the ERN, is the feedback Error-Related Negativity (fERN). This potential reaches maximum amplitude following negative feedback from tasks when the subject does not know the correct answer and has his predictions unexpectedly violated, therefore differentiating it from the ERN (Potts et al. 2011). Many researchers propose that the fERN is in fact the same component as the ERN, simply occurring at a later stage (Nieuwenhuis 2004; Nieuwenhuis et al. 2004, 2005). The ERN would reflect a mistake committed when the correct response is known (i.e. simple reaction-time paradigms), whereas the fERN indicates a prediction error that has solely come to awareness through later feedback. As the subject progresses with a task and learns the rules, the ERN starts being elicited and feedback is no longer necessary (Potts et al. 2011). The fERN is particularly relevant in tasks with high difficulty levels or probabilistic tasks, when feedback is essential and participants are unable to be confident in the accuracy of their judgement (Luck and Kappenman 2012). The RL-ERN suggests that the ERN/fERN reflect the arrival of a negative reward prediction error signal to the ACC and the subsequent employment of the dopamine systems and motor areas to adapt behaviour (Holroyd and Coles 2002).

Concerning OCD patients, however, this relationship presents some complexities. Although it has been established that individuals with OCD are oversensitive to errors and to punishment, with subjects expressing a tendency to switch more after negative feedback on probabilistic reversal learning tasks (Kanen et al. 2019), it is still unclear why patients perform the tasks sub-optimally. A recent study has shown that fERN amplitudes are actually blunted in OCD patients, regardless of the enhanced ERN (Endrass et al. 2013). As proposed by Hajcak et al. (2005), the ERN seems to be sensitive to the subjective significance of errors, rather than to their external evaluation, and has been thought to reflect the degree to which errors are considered threatening (Weinberg et al. 2016). Thus the ERN is signalling more endogenous (self-perceived) than exogenous (externally signalled) errors (Weinberg et al. 2016). This could potentially contribute to perseveration deficits as well as the impairment of overt goal-directed behaviour in OCD.

Several studies have aimed to manipulate the magnitude of gains and losses in electroencephalographic tasks with OCD participants, as well as to correlate the ERN with symptom severity (Hajcak et al. 2005; Nieuwenhuis et al. 2005; Riesel et al. 2017a, b, 2019; Weinberg et al. 2015a). Perhaps not surprisingly, the ERN does not seem to be affected by external variables, such as reward and punishment, but rather expresses the internal evaluation of errors (Weinberg et al. 2016). Not even highly efficacious drug treatments like the Selective Serotonin Re-uptake Inhibitors (SSRIs), the first-line treatment for OCD, are able to attenuate ERN amplitudes (Endrass et al. 2010; Stern et al. 2010).

Interestingly, a recent study has attempted to manipulate the magnitude of the ERN through a computerised cognitive-behavioural treatment targeting error-sensitivity. Two groups of participants were tested at baseline and after a computerised intervention consisting of information and quizzes about either general health (control group) or error-sensitivity (experimental group). Results suggested that the experimental group showed diminished amplitudes of ERN (Meyer et al. 2020). For this reason, amongst the units of analyses of the RDoC, many researchers have suggested its suitability as a marker of sustained threat (Ladouceur 2016).

Moreover, it is disputed whereas the functional role of ERN is as a signal of cognitive control (Azizian et al. 2006; Scholl and Rushworth 2017) or reward learning (Holroyd and Coles 2002; Nieuwenhuis et al. 2004). Nevertheless, these processes are probably both relevant to OCD. A clear deficit concerning the execution of optimal actions can be seen in this disorder. One possible explanation for this is the motor inhibition deficit seen in this population resulting in rigidity that hinders patients' ability to flexibly adapt behaviour and marks out OCD and other obsessive-compulsive related disorders (American Psychiatric Association 2013) (see the chapter "Cognitive Inflexibility in OCD and Related Disorders"). In any case, motor systems, as represented by the employment of compensatory behaviour following errors and by perseveration, seem to play a role in the aetiology of OCD symptoms (Grützmann et al. 2016).

3.4 The Readiness Potential

Another ERP poorly studied in OCD, although undoubtedly important in this disorder, is the *Bereitschaftspotential* or the Readiness Potential (RP). First reported in 1965 as a marker of initiated voluntary action (Kornhuber and Deecke 2016), this brain component is initiated around 1.5 ms before the onset of motor responses and represents motor preparation or the decision to initiate movement. The figure (Fig. 6) below depicts the RP (Luck and Kappenman 2012).

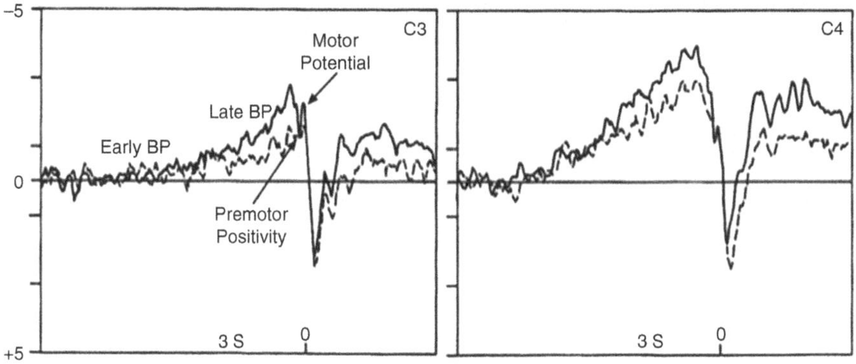

Fig. 6 Example of the RP. Left: *Bereitschaftspotential* (BP) recorded prior to the right finger movement (solid line) is larger over the left hemisphere (C3) than that recorded prior to the left finger movement (dotted line). Right: A BP recorded prior to the right foot movement (solid line) is larger over the right hemisphere (C4) than that recorded prior to the left foot movement (dotted line). Time zero is movement onset. Figure reproduced with permission of the Licensor through PLSclear

The Supplementary Motor Area (SMA) has been indicated as the main generator of the RP (Cunnington et al. 2003; Deecke and Kornhuber 1978) and many studies have been conducted to address the individual's degree of control over movements (Libet 1985). In a famous experiment, Libet and colleagues examined conscious intention of movement whilst brain activity was being recorded. Their findings suggested that motor preparation preceded the subjective intention to move by an average of 800 ms, casting doubt and debate about humans' sense of agency and free will. If that is the case, it may be possible to argue that compulsions generated in OCD might be involuntary, just as tics in Tourette's Syndrome or tremor in Parkinson's Disorder. Indeed, recent studies have investigated the RP in OCD patients, finding enhanced amplitudes of the RP in this population and proposing that OCD patients have increased action tendencies (Dayan et al. 2014, 2017; Dayan-Riva et al. 2020; Kalanthroff et al. 2017). It is possible that the RP and the ERN interact in patients with OCD, producing their deficits in goal-directed behaviour and the dissociation between knowledge and action.

A few studies have also suggested that the ERN might be generated in the SMA and pre-SMA (Bonini et al. 2014; Grützmann et al. 2016; Luck and Kappenman 2012). Both the ACC and the motor areas of the brain are considered part of a motor preparation network (Nguyen et al. 2014), which could explain the increased action tendencies in OCD (Dayan et al. 2017). Given the evidence for impaired goal-directed behaviour and habitual learning in OCD (see the chapter "Recent Developments in the Habit Hypothesis of OCD and Compulsive Disorders") and the fast and somewhat involuntary character of the RP (Libet et al. 1983; Libet 1985), it is possible to infer that enhanced action tendencies might be the cause of excessive reliance on habitual behaviour in OCD (Dayan et al. 2014, 2017). The automatic character of habits (Hardwick et al. 2019; Robbins and Costa 2017) might explain perseveration in patients and the consequent commission of errors once task parameters change, leading to enhanced ERN and N2 amplitudes (Riesel et al. 2017a, b).

Abnormal SMA function has been proposed as another endophenotype of OCD (van den Heuvel et al. 2016). The authors propose that the pre-SMA is a key region for the inhibition network, from which the stop signal is conveyed to the motor cortex through CSTC projections, finally arriving at the DLPFC (van Velzen et al. 2014). A dysfunction in this region leads to altered recruitment of the dorsal cognitive control system, which then contributes to maladaptive behaviours, including habits (van den Heuvel et al. 2016). However, the relationship between the ERN, the RP and OCD remains unclear.

4 Summary and Future Perspectives

Electrophysiological research in OCD is extensive and rich, but unfortunately far from being at consensus. An enhanced ERN, possibly reflecting an effective response to error, or the need for cognitive control, or behavioural monitoring under conditions of conflict or threat, has been established as a robust endophenotype of the disorder (Gillan et al. 2017; Hajcak et al. 2008; Perera et al. 2019; Riesel 2019; Riesel et al. 2011; Weinberg et al. 2015a), however, other key ERP components with neuropsychological credentials of relevance for OCD, such as the N2, the Readiness Potential, the Contingent Negative Variation and the feedback Error-Related Negativity remain relatively neglected. Research has failed to relate the clinical presentation of symptoms to these neurophysiological markers, proposed by the Research Domain Criteria framework (Cuthbert and Insel 2013; Insel et al. 2010; Sanislow et al. 2010) as being necessary to understand the causal relationships and origins of OCD symptomatology.

The study of electrophysiological correlates of OCD endophenotypes nevertheless has potential to improve the understanding and treatment of the disorder. For instance, the impact of drugs on impaired mechanisms, as demonstrated by Murray et al. (2019) on the attenuation of cingulate prediction errors in OCD following amisulpride administration, could be endorsed electrophysiologically. Recent research has proposed high-density electroencephalography as a new and powerful tool for uncovering neural processes and a possible candidate for gold standard measure in neurosciences, both regarding temporal and spatial resolution, as new analytic techniques are developed (Sohrabpour et al. 2020).

Although much has been learnt about OCD, our understanding of the links between different dimensions of the disorder involving the integration of results from different study modalities remains incomplete. For instance, why do OCD patients perform sub-optimally on some behavioural tasks despite being highly sensitive to errors? If these individuals are as sensitive to the commission of errors as research suggests (Fullana et al. 2004; Kaufmann et al. 2013) and given the fact that both subjective knowledge of optimal actions and confidence in performance do not seem to be impaired in OCD (Vaghi et al. 2017a), why do they go on to commit errors? Perhaps the dACC and the ERN are signalling the error and the need for adaptation as an alarm that will only be turned off once the error is remediated (Luck and Kappenman 2012), which does not seem to occur in OCD due to the cognitive inflexibility, habitual behaviour and rigidity that interferes with executive functioning (Gillan et al. 2011, 2016; Voon et al. 2015a, b). Do learning deficits drive this dysfunction or do automatic and impulsive motor components, perhaps override goal-directed behaviour? Moreover, what explains the finding that random motor movements, as reported in Dayan et al.'s (2017) study, attenuate ERN amplitudes, although many treatments do not? Is it the belief in rituals that diminishes the ERN or an impaired motor mechanism that perceives urges and the need for compensatory behaviour? Is it uncertainty that drives the disorder, suggesting that compulsions reflect an attempt to control the environment or are they caused by excessive doubt regarding state transitions? What is the relationship between the basal ganglia, the SMA and the ACC in OCD? How do we best integrate EEG measures with methods borrowed from cognitive neuroscience, such as imaging and neuropsychopharmacology, to better understand OCD? Whilst these questions remain unanswered, the "technique for recording electrical activity of the human brain from the surface of the head" (Millet 2002) might eventually cast some further light on these issues.

Acknowledgements I thank the Coordination for the Improvement of Higher Education Personnel (CAPES) for financial support and Maité Crespo-Garcia, Aleya Marzuki. Naomi Fineberg and Trevor Robbins for their comments on earlier drafts.

References

Amabile G, Fattapposta F, Pozzessere G, Albani G, Sanarelli L, Rizzo PA, Morocutti C (1986) Parkinson disease: electrophysiological (CNV) analysis related to pharmacological treatment. Electroencephalogr Clin Neurophysiol 64(6):521–524. https://doi.org/10.1016/0013-4694(86)90189-6

American Psychiatric Association (2013) Diagnostic and statistical manual of mental disorders, 5th edn. American Psychiatric Association, Arlington

Apergis-Schoute AM, Gillan CM, Fineberg NA, Fernandez-Egea E, Sahakian BJ, Robbins TW (2017) Neural basis of impaired safety signaling in obsessive compulsive disorder. Proc Natl Acad Sci 114(12):3216–3221. https://doi.org/10.1073/pnas.1609194114

Apergis-Schoute AM, Bijleveld B, Gillan CM, Fineberg NA, Sahakian BJ, Robbins TW (2018) Hyperconnectivity of the ventromedial prefrontal cortex in obsessive-compulsive disorder. Brain Neurosci Adv 2:2398212818808710. https://doi.org/10.1177/2398212818808710

Azizian A, Freitas AL, Parvaz MA, Squires NK (2006) Beware misleading cues: perceptual similarity modulates the N2/P3 complex. Psychophysiology 43(3):253–260. https://doi.org/10.1111/j.1469-8986.2006.00409.x

Banca P, Voon V, Vestergaard MD, Philipiak G, Almeida I, Pocinho F, Relvas J, Castelo-Branco M (2015) Imbalance in habitual versus goal directed neural systems during symptom provocation in obsessive-compulsive disorder. Brain 138(3):798–811. https://doi.org/10.1093/brain/awu379

Banca P, McNamee D, Piercy T, Luo Q, Robbins TW (2020) A mobile phone app for the generation and characterization of motor habits. Front Psychol 10

Baxter LR, Schwartz JM, Mazziotta JC, Phelps ME, Pahl JJ, Guze BH, Fairbanks L (1988) Cerebral glucose metabolic rates in nondepressed patients with obsessive-compulsive disorder. Am J Psychiatry 145(12):1560–1563. https://doi.org/10.1176/ajp.145.12.1560

Behrens TEJ, Woolrich MW, Walton ME, Rushworth MFS (2007) Learning the value of information in an uncertain world. Nat Neurosci 10(9):1214–1221. https://doi.org/10.1038/nn1954

Bonini F, Burle B, Liegeois-Chauvel C, Regis J, Chauvel P, Vidal F (2014) Action monitoring and medial frontal cortex: leading role of supplementary motor area. Science 343(6173):888–891. https://doi.org/10.1126/science.1247412

Botvinick MM (2007) Conflict monitoring and decision making: reconciling two perspectives on anterior cingulate function. Cogn Affect Behav Neurosci 7(4):356–366. https://doi.org/10.3758/CABN.7.4.356

Botvinick MM, Cohen JD, Carter CS (2004) Conflict monitoring and anterior cingulate cortex: an update. Trends Cogn Sci 8(12):539–546. https://doi.org/10.1016/j.tics.2004.10.003

Brooks AW, Schroeder J, Risen J, Gino F, Galinsky AD, Norton MI, Schweitzer M (2016) Don't stop believing: rituals improve performance by decreasing anxiety. http://www.hbs.edu/faculty/Pages/item.aspx?num=51401

Burani K, Gallyer A, Ryan J, Jordan C, Joiner T, Hajcak G (2020) Acute stress reduces reward-related neural activity: evidence from the reward positivity. BioRxiv, 2020.01.08.898593. https://doi.org/10.1101/2020.01.08.898593

Burle B, Spieser L, Roger C, Casini L, Hasbroucq T, Vidal F (2015) Spatial and temporal resolutions of EEG: is it really black and white? A scalp current density view. Int J Psychophysiol 97(3):210–220. https://doi.org/10.1016/j.ijpsycho.2015.05.004

Bush G (1990) Presidential proclamation 6158, library of congress website. http://www.loc.gov/loc/brain/proclaim.html

Cacioppo JT, Tassinary LG, Berntson GG (2016) Handbook of psychophysiology, 4th edn. Cambridge University Press, Cambridge. https://doi.org/10.1017/9781107415782

Carter CS (1998) Anterior cingulate cortex, error detection, and the online monitoring of performance. Science 280(5364):747–749. https://doi.org/10.1126/science.280.5364.747

Chamberlain SR, Menzies L (2009) Endophenotypes of obsessive-compulsive disorder: rationale, evidence and future potential. Expert Rev Neurother 9(8):1133–1146. https://doi.org/10.1586/ern.09.36

Chamberlain SR, Blackwell AD, Fineberg NA, Robbins TW, Sahakian BJ (2005) The neuropsychology of obsessive compulsive disorder: the importance of failures in cognitive and behavioural inhibition as candidate endophenotypic markers. Neurosci Biobehav Rev 29 (3):399–419. https://doi.org/10.1016/j.neubiorev.2004.11.006

Ciesielski KT, Rowland LM, Harris RJ, Kerwin AA, Reeve A, Knight JE (2011) Increased anterior brain activation to correct responses on high-conflict Stroop task in obsessive–compulsive disorder. Clin Neurophysiol 122(1):107–113. https://doi.org/10.1016/j.clinph.2010.05.027

Clark L, Robbins TW, Ersche KD, Sahakian BJ (2006) Reflection impulsivity in current and former substance users. Biol Psychiatry 60(5):515–522. https://doi.org/10.1016/j.biopsych.2005.11.007

Coles MGH, Scheffers MK, Holroyd CB (2001) Why is there an ERN/Ne on correct trials? Response representations, stimulus-related components, and the theory of error-processing. Biol Psychol 56(3):173–189. https://doi.org/10.1016/S0301-0511(01)00076-X

Crawford HJ, Knebel TL, Vendemia JM, Kaplan L, Ratcliff B (1995) EEG activation patterns during tracking and decision-making tasks. Differences between low and high sustained attention adults. In: International symposium on aviation psychology, 8 th, Columbus, OH. pp 886–890

Cunnington R, Windischberger C, Deecke L, Moser E (2003) The preparation and readiness for voluntary movement: a high-field event-related fMRI study of the Bereitschafts-BOLD response. NeuroImage 20(1):404–412. https://doi.org/10.1016/S1053-8119(03)00291-X

Cuthbert BN, Insel TR (2013) Toward the future of psychiatric diagnosis: the seven pillars of RDoC. BMC Med 11(1). https://doi.org/10.1186/1741-7015-11-126

Davidson RJ, Ekman P, Saron CD, Senulis JA, Friesen WV (1990) Approach-withdrawal and cerebral asymmetry: emotional expression and brain physiology: I. J Pers Soc Psychol 58 (2):330–341. https://doi.org/10.1037/0022-3514.58.2.330

Dayan A, Berger A, Anholt GE (2014) Enhanced action tendencies in high versus low obsessive-compulsive symptoms: an event-related potential study. Psychiatry Res 224(2):133–138. https://doi.org/10.1016/j.pscychresns.2014.07.007

Dayan A, Berger A, Anholt GE (2017) Enhanced action tendencies in obsessive-compulsive disorder: an ERP study. Behav Res Ther 93:13–21. https://doi.org/10.1016/j.brat.2017.03.005

Dayan-Riva A, Berger A, Anholt GE (2020) Affordances, response conflict, and enhanced-action tendencies in obsessive-compulsive disorder: an ERP study. Psychol Med:1–16. https://doi.org/10.1017/S0033291719003866

de Souza AMFLP, de Nonohay RG, Gauer G (2018) Frustrating feedback effects on error-related negativity (ERN) as a dimensional marker of negative affect. Clin Neuropsychiatry J Treat Eval 15(5):275–283

de Wit SJ, de Vries FE, van der Werf YD, Cath DC, Heslenfeld DJ, Veltman EM, van Balkom AJLM, Veltman DJ, van den Heuvel OA (2012) Presupplementary motor area hyperactivity during response inhibition: a candidate endophenotype of obsessive-compulsive disorder. Am J Psychiatry 169(10):1100–1108. https://doi.org/10.1176/appi.ajp.2012.12010073

Deecke L, Kornhuber HH (1978) An electrical sign of participation of the mesial 'supplementary' motor cortex in human voluntary finger movement. Brain Res 159(2):473–476. https://doi.org/10.1016/0006-8993(78)90561-9

Dieterich R, Endrass T, Kathmann N (2017) Uncertainty increases neural indices of attention in obsessive-compulsive disorder. Depress Anxiety 34(11):1018–1028. https://doi.org/10.1002/da.22655

Eilam D, Zor R, Szechtman H, Hermesh H (2006) Rituals, stereotypy and compulsive behavior in animals and humans. Neurosci Biobehav Rev 30(4):456–471. https://doi.org/10.1016/j.neubiorev.2005.08.003

Einstein DA, Menzies RG (2004) The presence of magical thinking in obsessive compulsive disorder. Behav Res Ther 42(5):539–549. https://doi.org/10.1016/S0005-7967(03)00160-8

Endrass T, Schuermann B, Kaufmann C, Spielberg R, Kniesche R, Kathmann N (2010) Performance monitoring and error significance in patients with obsessive-compulsive disorder. Biol Psychol 84(2):257–263. https://doi.org/10.1016/j.biopsycho.2010.02.002

Endrass T, Koehne S, Riesel A, Kathmann N (2013) Neural correlates of feedback processing in obsessive–compulsive disorder. J Abnorm Psychol 122(2):387–396. https://doi.org/10.1037/a0031496

Eriksen BA, Eriksen CW (1974) Effects of noise letters upon the identification of a target letter in a nonsearch task. Percept Psychophys 16(1):143–149. https://doi.org/10.3758/BF03203267

Etkin A, Cuthbert B (2014) Beyond the DSM: development of a transdiagnostic psychiatric neuroscience course. Acad Psychiatry 38(2):145–150. https://doi.org/10.1007/s40596-013-0032-4

Falkenstein M, Hohnsbein J, Hoormann J, Blanke L (1991) Effects of crossmodal divided attention on late ERP components. II. Error processing in choice reaction tasks. Electroencephal Clin Neurophysiol 78(6):447–455. https://doi.org/10.1016/0013-4694(91)90062-9

Fan J, Kolster R, Ghajar J, Suh M, Knight RT, Sarkar R, McCandliss BD (2007) Response anticipation and response conflict: an event-related potential and functional magnetic resonance imaging study. J Neurosci Off J Soc Neurosci 27(9):2272–2282. https://doi.org/10.1523/JNEUROSCI.3470-06.2007

Fentress JC (1976) Dynamic boundaries of patterned behaviour: interaction and self-organization. In: Growing points in ethology. Cambridge University Press, Cambridge

Fineberg NA, Apergis-Schoute AM, Vaghi MM, Banca P, Gillan CM, Voon V, Chamberlain SR, Cinosi E, Reid J, Shahper S, Bullmore ET, Sahakian BJ, Robbins TW (2018) Mapping compulsivity in the DSM-5 obsessive compulsive and related disorders: cognitive domains, neural circuitry, and treatment. Int J Neuropsychopharmacol 21(1):42–58. https://doi.org/10.1093/ijnp/pyx088

Folstein JR, Van Petten C (2008) Influence of cognitive control and mismatch on the N2 component of the ERP: a review. Psychophysiology 45(1):152–170. https://doi.org/10.1111/j.1469-8986.2007.00602.x

Foti D, Kotov R, Bromet E, Hajcak G (2012) Beyond the broken error-related negativity: functional and diagnostic correlates of error processing in psychosis. Biol Psychiatry 71(10):864–872. https://doi.org/10.1016/j.biopsych.2012.01.007

Fradkin I, Adams RA, Parr T, Roiser JP, Huppert JD (2020) Searching for an anchor in an unpredictable world: a computational model of obsessive compulsive disorder. Psychol Rev. https://doi.org/10.1037/rev0000188

Franken IHA, van Strien JW, Franzek EJ, van de Wetering BJ (2007) Error-processing deficits in patients with cocaine dependence. Biol Psychol 75(1):45–51. https://doi.org/10.1016/j.biopsycho.2006.11.003

Fullana MA, Mataix-Cols D, Caseras X, Alonso P, Menchón JM, Vallejo J, Torrubia R (2004) High sensitivity to punishment and low impulsivity in obsessive-compulsive patients with hoarding symptoms. Psychiatry Res 129(1):21–27. https://doi.org/10.1016/j.psychres.2004.02.017

Gehring WJ, Goss B, Coles MGH, Meyer DE, Donchin E (1993) A neural system for error detection and compensation. Psychol Sci 4(6):385–390. https://doi.org/10.1111/j.1467-9280.1993.tb00586.x

Gillan CM, Robbins TW (2014) Goal-directed learning and obsessive-compulsive disorder. Philos Trans R Soc B Biol Sci 369(1655):20130475–20130475. https://doi.org/10.1098/rstb.2013.0475

Gillan CM, Papmeyer M, Morein-Zamir S, Sahakian BJ, Fineberg NA, Robbins TW, de Wit S (2011) Disruption in the balance between goal-directed behavior and habit learning in obsessive-compulsive disorder. Am J Psychiatr 168(7):718–726

Gillan CM, Morein-Zamir S, Durieux AMS, Fineberg NA, Sahakian BJ, Robbins TW (2014) Obsessive–compulsive disorder patients have a reduced sense of control on the illusion of control task. Front Psychol 5. https://doi.org/10.3389/fpsyg.2014.00204

Gillan CM, Apergis-Schoute AM, Morein-Zamir S, Urcelay GP, Sule A, Fineberg NA, Sahakian BJ, Robbins TW (2015) Functional neuroimaging of avoidance habits in obsessive-compulsive disorder. Am J Psychiatr 172(3):284–293. https://doi.org/10.1176/appi.ajp.2014.14040525

Gillan CM, Robbins TW, Sahakian BJ, van den Heuvel OA, van Wingen G (2016) The role of habit in compulsivity. Eur Neuropsychopharmacol 26(5):828–840. https://doi.org/10.1016/j.euroneuro.2015.12.033

Gillan CM, Fineberg NA, Robbins TW (2017) A trans-diagnostic perspective on obsessive-compulsive disorder. Psychol Med 47(09):1528–1548. https://doi.org/10.1017/S0033291716002786

Glover GH (2011) Overview of functional magnetic resonance imaging. Neurosurg Clin N Am 22 (2):133–139. https://doi.org/10.1016/j.nec.2010.11.001

Gottesman II, Gould TD (2003) The endophenotype concept in psychiatry: etymology and strategic intentions. Am J Psychiatry 160(4):636–645. https://doi.org/10.1176/appi.ajp.160.4.636

Graybiel AM (2008) Habits, rituals, and the evaluative brain. Annu Rev Neurosci 31:359–387. https://doi.org/10.1146/annurev.neuro.29.051605.112851

Graybiel AM, Grafton ST (2015) The striatum: where skills and habits meet. Cold Spring Harb Perspect Biol 7(8):a021691. https://doi.org/10.1101/cshperspect.a021691

Grützmann R, Endrass T, Kaufmann C, Allen E, Eichele T, Kathmann N (2016) Presupplementary motor area contributes to altered error monitoring in obsessive-compulsive disorder. Biol Psychiatry 80(7):562–571. https://doi.org/10.1016/j.biopsych.2014.12.010

Guo Y, Schmitz TW, Mur M, Ferreira CS, Anderson MC (2018) A supramodal role of the basal ganglia in memory and motor inhibition: meta-analytic evidence. Neuropsychologia 108:117–134. https://doi.org/10.1016/j.neuropsychologia.2017.11.033

Hajcak G, McDonald N, Simons RF (2003a) Anxiety and error-related brain activity. Biol Psychol 64(1–2):77–90. https://doi.org/10.1016/s0301-0511(03)00103-0

Hajcak G, McDonald N, Simons RF (2003b) To err is autonomic: error-related brain potentials, ANS activity, and post-error compensatory behavior. Psychophysiology 40(6):895–903. https://doi.org/10.1111/1469-8986.00107

Hajcak G, McDonald N, Simons RF (2004) Error-related psychophysiology and negative affect. Brain Cogn 56(2):189–197. https://doi.org/10.1016/j.bandc.2003.11.001

Hajcak G, Moser JS, Yeung N, Simons RF (2005) On the ERN and the significance of errors. Psychophysiology 42(2):151–160. https://doi.org/10.1111/j.1469-8986.2005.00270.x

Hajcak G, Moser JS, Holroyd CB, Simons RF (2006) The feedback-related negativity reflects the binary evaluation of good versus bad outcomes. Biol Psychol 71(2):148–154. https://doi.org/10.1016/j.biopsycho.2005.04.001

Hajcak G, Franklin ME, Foa EB, Simons RF (2008) Increased error-related brain activity in pediatric obsessive-compulsive disorder before and after treatment. Am J Psychiatry 165 (1):116–123. https://doi.org/10.1176/appi.ajp.2007.07010143

Hardwick RM, Forrence AD, Krakauer JW, Haith AM (2019) Time-dependent competition between goal-directed and habitual response preparation. Nat Hum Behav 3(12):1252–1262. https://doi.org/10.1038/s41562-019-0725-0

Hauser TU, Iannaccone R, Dolan RJ, Ball J, Hättenschwiler J, Drechsler R, Rufer M, Brandeis D, Walitza S, Brem S (2017) Increased fronto-striatal reward prediction errors moderate decision making in obsessive-compulsive disorder. Psychol Med 47(7):1246–1258. https://doi.org/10.1017/S0033291716003305

Hobson NM, Bonk D, Inzlicht M (2017) Rituals decrease the neural response to performance failure. PeerJ 5:e3363. https://doi.org/10.7717/peerj.3363

Hobson NM, Schroeder J, Risen JL, Xygalatas D, Inzlicht M (2018) The psychology of rituals: an integrative review and process-based framework. Personal Soc Psychol Rev 22(3):260–284. https://doi.org/10.1177/1088868317734944

Holaway RM, Heimberg RG, Coles ME (2006) A comparison of intolerance of uncertainty in analogue obsessive-compulsive disorder and generalized anxiety disorder. J Anxiety Disord 20 (2):158–174. https://doi.org/10.1016/j.janxdis.2005.01.002

Holmes AJ, Pizzagalli DA (2008) Spatiotemporal dynamics of error processing dysfunctions in major depressive disorder. Arch Gen Psychiatry 65(2):179–188. https://doi.org/10.1001/archgenpsychiatry.2007.19

Holroyd CB, Coles MGH (2002) The neural basis of human error processing: reinforcement learning, dopamine, and the error-related negativity. Psychol Rev 109(4):679–709. https://doi.org/10.1037/0033-295X.109.4.679

Holroyd CB, Hajcak G, Larsen JT (2006) The good, the bad and the neutral: electrophysiological responses to feedback stimuli. Brain Res 1105(1):93–101. https://doi.org/10.1016/j.brainres.2005.12.015

Insel TR (2014) The NIMH research domain criteria (RDoC) project: precision medicine for psychiatry. Am J Psychiatr 171(4):395–397. https://doi.org/10.1176/appi.ajp.2014.14020138

Insel T, Cuthbert B, Garvey M, Heinssen R, Pine DS, Quinn K, Sanislow C, Wang P (2010) Research domain criteria (RDoC): toward a new classification framework for research on mental disorders. Am J Psychiatry 167(7):748–751. https://doi.org/10.1176/appi.ajp.2010.09091379

Ischebeck M, Endrass T, Simón D, Kathmann N (2014) Altered frontal EEG asymmetry in obsessive-compulsive disorder. Psychophysiology 51(7):596–601. https://doi.org/10.1111/psyp.12214

Kalanthroff E, Henik A, Simpson HB, Todder D, Anholt GE (2017) To do or not to do? Task control deficit in obsessive-compulsive disorder. Behav Ther 48(5):603–613. https://doi.org/10.1016/j.beth.2017.01.004

Kanen JW, Ersche KD, Fineberg NA, Robbins TW, Cardinal RN (2019) Computational modelling reveals contrasting effects on reinforcement learning and cognitive flexibility in stimulant use disorder and obsessive-compulsive disorder: remediating effects of dopaminergic D2/3 receptor agents. Psychopharmacology 236(8):2337–2358. https://doi.org/10.1007/s00213-019-05325-w

Kappenman ES, Luck SJ (2016) Best practices for event-related potential research in clinical populations. Biol Psychiatry Cogn Neurosci Neuroimag 1(2):110–115. https://doi.org/10.1016/j.bpsc.2015.11.007

Kaufmann C, Beucke JC, Preuße F, Endrass T, Schlagenhauf F, Heinz A, Juckel G, Kathmann N (2013) Medial prefrontal brain activation to anticipated reward and loss in obsessive-compulsive disorder. NeuroImage Clin 2:212–220. https://doi.org/10.1016/j.nicl.2013.01.005

Keil A, Debener S, Gratton G, Junghöfer M, Kappenman ES, Luck SJ, Luu P, Miller GA, Yee CM (2014) Committee report: publication guidelines and recommendations for studies using electroencephalography and magnetoencephalography. Psychophysiology 51(1):1–21. https://doi.org/10.1111/psyp.12147

Kennerley SW, Walton ME, Behrens TEJ, Buckley MJ, Rushworth MFS (2006) Optimal decision making and the anterior cingulate cortex. Nat Neurosci 9(7):940–947. https://doi.org/10.1038/nn1724

Khanna S (1988) Obsessive-compulsive disorder: is there a frontal lobe dysfunction? Biol Psychiatry 24(5):602–613. https://doi.org/10.1016/0006-3223(88)90171-0

Klawohn J, Riesel A, Grützmann R, Kathmann N, Endrass T (2014) Performance monitoring in obsessive–compulsive disorder: a temporo-spatial principal component analysis. Cogn Affect Behav Neurosci 14(3):983–995. https://doi.org/10.3758/s13415-014-0248-0

Knill DC, Pouget A (2004) The Bayesian brain: the role of uncertainty in neural coding and computation. Trends Neurosci 27(12):712–719. https://doi.org/10.1016/j.tins.2004.10.007

Kornhuber HH, Deecke L (2016) Brain potential changes in voluntary and passive movements in humans: readiness potential and reafferent potentials. Pflugers Arch Eur J Physiol 468 (7):1115–1124. https://doi.org/10.1007/s00424-016-1852-3

Krueger RF, Kotov R, Watson D, Forbes MK, Eaton NR, Ruggero CJ, Simms LJ, Widiger TA, Achenbach TM, Bach B, Bagby RM, Bornovalova MA, Carpenter WT, Chmielewski M, Cicero DC, Clark LA, Conway C, DeClercq B, DeYoung CG et al (2018) Progress in achieving quantitative classification of psychopathology. World Psychiatry 17(3):282–293. https://doi.org/10.1002/wps.20566

Ladouceur CD (2016) The error-related negativity: a transdiagnostic marker of sustained threat? Psychophysiology 53(3):389–392. https://doi.org/10.1111/psyp.12585

Lee Y-J, Koo B-H, Seo W-S, Kim H-G, Kim J-Y, Cheon E-J (2017) Repetitive transcranial magnetic stimulation of the supplementary motor area in treatment-resistant obsessive-compulsive disorder: an open–label pilot study. J Clin Neurosci 44:264–268. https://doi.org/10.1016/j.jocn.2017.06.057

Libet B (1985) Unconscious cerebral initiative and the role of conscious will in voluntary action. Behav Brain Sci 8(4):529–539. https://doi.org/10.1017/S0140525X00044903

Libet B, Gleason CA, Wright EW, Pearl DK (1983) Time of conscious intention to act in relation to onset of cerebral activity (readiness-potential). The unconscious initiation of a freely voluntary act. Brain J Neurol 106(Pt 3):623–642. https://doi.org/10.1093/brain/106.3.623

Linssen AMW, Vuurman EFPM, Sambeth A, Nave S, Spooren W, Vargas G, Santarelli L, Riedel WJ (2011) Contingent negative variation as a dopaminergic biomarker: evidence from dose-related effects of methylphenidate. Psychopharmacology 218(3):533–542. https://doi.org/10.1007/s00213-011-2345-x

Luck SJ (2005) An introduction to the event-related potential technique. MIT Press, Cambridge

Luck SJ, Hillyard SA (1994) Electrophysiological correlates of feature analysis during visual search. Psychophysiology 31(3):291–308. https://doi.org/10.1111/j.1469-8986.1994.tb02218.x

Luck SJ, Kappenman ES (2012) Oxford library of psychology. The Oxford handbook of event-related potential components. Oxford University Press

Luu P, Pederson SM (2004) The anterior cingulate cortex: regulating actions in context. In: Posner MI (ed) Cognitive neuroscience of attention. Guilford Press, New York

Luu P, Tucker DM, Derryberry D, Reed M, Poulsen C (2003) Electrophysiological responses to errors and feedback in the process of action regulation. Psychol Sci 14(1):47–53. https://doi.org/10.1111/1467-9280.01417

Marquardt L, Eichele H, Lundervold AJ, Haavik J, Eichele T (2018) Event-related-potential (ERP) correlates of performance monitoring in adults with attention-deficit hyperactivity disorder (ADHD). Front Psychol 9:485. https://doi.org/10.3389/fpsyg.2018.00485

Marzuki AA, Pereira de Souza AMFL, Sahakian BJ, Robbins TW (2020) Are candidate neurocognitive endophenotypes of OCD present in paediatric patients? A systematic review. Neurosci Biobehav Rev 108:617–645. https://doi.org/10.1016/j.neubiorev.2019.12.010

McCallum WC, Walter WG (1968) The effects of attention and distraction on the contingent negative variation in normal and neurotic subjects. Electroencephalogr Clin Neurophysiol 25(4):319–329. https://doi.org/10.1016/0013-4694(68)90172-7

Menzies L, Achard S, Chamberlain SR, Fineberg N, Chen C-H, del Campo N, Sahakian BJ, Robbins TW, Bullmore E (2007) Neurocognitive endophenotypes of obsessive-compulsive disorder. Brain 130(12):3223–3236. https://doi.org/10.1093/brain/awm205

Meyer A, Proudfit GH, Torpey-Newman DC, Kujawa A, Klein DN (2015) Enhanced error-related brain activity in children predicts the onset of anxiety disorders between the ages of 6 and 9. J Abnorm Psychol 124(2):266–274. https://doi.org/10.1037/abn0000044

Meyer A, Gibby B, Wissemann K, Klawohn J, Hajcak G, Schmidt NB (2020) A brief, computerized intervention targeting error sensitivity reduces the error-related negativity. Cogn Affect Behav Neurosci 20(1):172–180. https://doi.org/10.3758/s13415-019-00760-w

Michel CM, Brunet D (2019) EEG source imaging: a practical review of the analysis steps. Front Neurol 10. https://doi.org/10.3389/fneur.2019.00325

Milad MR, Rauch SL (2012) Obsessive-compulsive disorder: beyond segregated cortico-striatal pathways. Trends Cogn Sci 16(1):43–51. https://doi.org/10.1016/j.tics.2011.11.003

Millet D (2002) ISHN 2002 annual meeting – abstract 24. http://www.bri.ucla.edu/nha/ishn/ab24-2002.htm. Accessed 4 Nov 2019

Min B-K, Kim SJ, Park JY, Park H-J (2011) Prestimulus top-down reflection of obsessive-compulsive disorder in EEG frontal theta and occipital alpha oscillations. Neurosci Lett 496 (3):181–185. https://doi.org/10.1016/j.neulet.2011.04.018

Morein-Zamir S, Shapher S, Gasull-Camos J, Fineberg NA, Robbins TW (2020) Avoid jumping to conclusions under uncertainty in obsessive compulsive disorder. PLoS One 15(1):e0225970. https://doi.org/10.1371/journal.pone.0225970

Moser JS, Moran TP, Schroder HS, Donnellan MB, Yeung N (2013) On the relationship between anxiety and error monitoring: a meta-analysis and conceptual framework. Front Hum Neurosci 7. https://doi.org/10.3389/fnhum.2013.00466

Murray GK, Knolle F, Ersche KD, Craig KJ, Abbott S, Shabbir SS, Fineberg NA, Suckling J, Sahakian BJ, Bullmore ET, Robbins TW (2019) Dopaminergic drug treatment remediates exaggerated cingulate prediction error responses in obsessive-compulsive disorder. Psychopharmacology 236(8):2325–2336. https://doi.org/10.1007/s00213-019-05292-2

Nguyen VT, Breakspear M, Cunnington R (2014) Reciprocal interactions of the SMA and cingulate cortex sustain premovement activity for voluntary actions. J Neurosci 34(49):16397–16407. https://doi.org/10.1523/JNEUROSCI.2571-14.2014

Nieuwenhuis S (2004) Sensitivity of electrophysiological activity from medial frontal cortex to utilitarian and performance feedback. Cereb Cortex 14(7):741–747. https://doi.org/10.1093/cercor/bhh034

Nieuwenhuis S, Holroyd CB, Mol N, Coles MGH (2004) Reinforcement-related brain potentials from medial frontal cortex: origins and functional significance. Neurosci Biobehav Rev 28 (4):441–448. https://doi.org/10.1016/j.neubiorev.2004.05.003

Nieuwenhuis S, Nielen MM, Mol N, Hajcak G, Veltman DJ (2005) Performance monitoring in obsessive-compulsive disorder. Psychiatry Res 134(2):111–122. https://doi.org/10.1016/j.psychres.2005.02.005

Olvet DM, Hajcak G (2008) The error-related negativity (ERN) and psychopathology: toward an endophenotype. Clin Psychol Rev 28(8):1343–1354. https://doi.org/10.1016/j.cpr.2008.07.003

Parvizi J, Kastner S (2018) Human intracranial EEG: promises and limitations. Nat Neurosci 21 (4):474–483. https://doi.org/10.1038/s41593-018-0108-2

Perera MPN, Bailey NW, Herring SE, Fitzgerald PB (2019) Electrophysiology of obsessive compulsive disorder: a systematic review of the electroencephalographic literature. J Anxiety Disord 62:1–14. https://doi.org/10.1016/j.janxdis.2018.11.001

Perrone-McGovern K, Simon-Dack S, Esche A, Thomas C, Beduna K, Rider K, Spurling A, Matsen J (2017) The influence of emotional intelligence and perfectionism on error-related negativity: an event related potential study. Personal Individ Differ 111:65–70. https://doi.org/10.1016/j.paid.2017.02.009

Potts GF, Martin LE, Kamp S-M, Donchin E (2011) Neural response to action and reward prediction errors: comparing the error-related negativity to behavioral errors and the feedback-related negativity to reward prediction violations: response to action and reward prediction errors. Psychophysiology 48(2):218–228. https://doi.org/10.1111/j.1469-8986.2010.01049.x

Rauch SL, Jenike MA, Alpert NM, Baer L, Breiter HCR, Savage CR, Fischman AJ (1994) Regional cerebral blood flow measured during symptom provocation in obsessive-compulsive disorder using oxygen 15 – labeled carbon dioxide and positron emission tomography. Arch Gen Psychiatry 51(1):62–70. https://doi.org/10.1001/archpsyc.1994.03950010062008

Reuven-Magril O, Dar R, Liberman N (2008) Illusion of control and behavioral control attempts in obsessive-compulsive disorder. J Abnorm Psychol 117(2):334–341. https://doi.org/10.1037/0021-843X.117.2.334

Riesel A (2019) The erring brain: error-related negativity as an endophenotype for OCD – a review and meta-analysis. Psychophysiology 56(4):e13348. https://doi.org/10.1111/psyp.13348

Riesel A, Endrass T, Kaufmann C, Kathmann N (2011) Overactive error-related brain activity as a candidate endophenotype for obsessive-compulsive disorder: evidence from unaffected first-degree relatives. Am J Psychiatry 168(3):317–324. https://doi.org/10.1176/appi.ajp.2010.10030416

Riesel A, Weinberg A, Endrass T, Meyer A, Hajcak G (2013) The ERN is the ERN is the ERN? Convergent validity of error-related brain activity across different tasks. Biol Psychol 93 (3):377–385. https://doi.org/10.1016/j.biopsycho.2013.04.007

Riesel A, Endrass T, Auerbach LA, Kathmann N (2015) Overactive performance monitoring as an endophenotype for obsessive-compulsive disorder: evidence from a treatment study. Am J Psychiatr 172(7):665–673. https://doi.org/10.1176/appi.ajp.2014.14070886

Riesel A, Klawohn J, Kathmann N, Endrass T (2017a) Conflict monitoring and adaptation as reflected by N2 amplitude in obsessive-compulsive disorder. Psychol Med 47(8):1379–1388. https://doi.org/10.1017/S0033291716003597

Riesel A, Goldhahn S, Kathmann N (2017b) Hyperactive performance monitoring as a transdiagnostic marker: results from health anxiety in comparison to obsessive-compulsive disorder. Neuropsychologia 96:1–8. https://doi.org/10.1016/j.neuropsychologia.2016.12.029

Riesel A, Klawohn J, Grützmann R, Kaufmann C, Heinzel S, Bey K, Lennertz L, Wagner M, Kathmann N (2019) Error-related brain activity as a transdiagnostic endophenotype for obsessive-compulsive disorder, anxiety and substance use disorder. Psychol Med 49 (07):1207–1217. https://doi.org/10.1017/S0033291719000199

Robbins TW, Costa RM (2017) Habits. Curr Biol 27(22):R1200–R1206. https://doi.org/10.1016/j.cub.2017.09.060

Robbins TW, Vaghi MM, Banca P (2019) Obsessive-compulsive disorder: puzzles and prospects. Neuron 102(1):27–47. https://doi.org/10.1016/j.neuron.2019.01.046

Ruchsow M, Reuter K, Hermle L, Ebert D, Kiefer M, Falkenstein M (2007) Executive control in obsessive-compulsive disorder: event-related potentials in a Go/Nogo task. J Neural Trans 114 (12):1595–1601. https://doi.org/10.1007/s00702-007-0779-4

Ruscio AM, Stein DJ, Chiu WT, Kessler RC (2010) The epidemiology of obsessive-compulsive disorder in the national comorbidity survey replication. Mol Psychiatry 15(1):53–63. https://doi.org/10.1038/mp.2008.94

Sanislow CA, Pine DS, Quinn KJ, Kozak MJ, Garvey MA, Heinssen RK, Wang PS-E, Cuthbert BN (2010) Developing constructs for psychopathology research: research domain criteria. J Abnorm Psychol 119(4):631–639. https://doi.org/10.1037/a0020909

Sartory G, Master D (1984) Contingent negative variation in obsessional-compulsive patients. Biol Psychol 18(4):253–267. https://doi.org/10.1016/0301-0511(84)90057-7

Scholl J, Rushworth MFS (2017) Obsessing about uncertainty? Neuron 96(2):250–252. https://doi.org/10.1016/j.neuron.2017.09.045

Shen A, Zhao W, Han B, Zhang Q, Zhang Z, Chen X, Zhai J, Chen M, Du B, Deng X, Ji F, Wang C, Xiang Y-T, Wu H, Dong Q, Chen C, Li J (2017) The contribution of the contingent negative variation (CNV) to goal maintenance. Schizophr Res. https://doi.org/10.1016/j.schres.2017.09.038

Shenhav A, Botvinick MM, Cohen JD (2013) The expected value of control: an integrative theory of anterior cingulate cortex function. Neuron 79(2):217–240. https://doi.org/10.1016/j.neuron.2013.07.007

Sohrabpour A, Cai Z, Ye S, Brinkmann B, Worrell G, He B (2020) Noninvasive electromagnetic source imaging of spatiotemporally distributed epileptogenic brain sources. Nat Commun 11 (1):1–15. https://doi.org/10.1038/s41467-020-15781-0

Stern ER, Liu Y, Gehring WJ, Lister JJ, Yin G, Zhang J, Fitzgerald KD, Himle JA, Abelson JL, Taylor SF (2010) Chronic medication does not affect hyperactive error responses in obsessive-compulsive disorder. Psychophysiology. https://doi.org/10.1111/j.1469-8986.2010.00988.x

Stroop JR (1935) Studies of interference in serial verbal reactions. J Exp Psychol 18:643–662. https://doi.org/10.1037/h0054651

Sutton S, Braren M, Zubin J, John ER (1965) Evoked-potential correlates of stimulus uncertainty. Science 150(3700):1187–1188. https://doi.org/10.1126/science.150.3700.1187

Tecce JJ (1972) Contingent negative variation (CNV) and psychological processes in man. Psychol Bull 77(2):73–108. https://doi.org/10.1037/h0032177

Timsit-Berthier M, Delaunoy J, Koninckx N, Rousseau JC (1973) Slow potential changes in psychiatry. I. Contingent negative variation. Electroencephalogr Clin Neurophysiol 35 (4):355–361. https://doi.org/10.1016/0013-4694(73)90191-0

Tong S, Thakor NV (2009) Quantitative EEG analysis methods and clinical applications. Artech House

Tonna M, Marchesi C, Parmigiani S (2019) The biological origins of rituals: an interdisciplinary perspective. Neurosci Biobehav Rev 98:95–106. https://doi.org/10.1016/j.neubiorev.2018.12.031

Vaghi MM, Luyckx F, Sule A, Fineberg NA, Robbins TW, De Martino B (2017a) Compulsivity reveals a novel dissociation between action and confidence. Neuron 96(2):348–354.e4. https://doi.org/10.1016/j.neuron.2017.09.006

Vaghi MM, Vértes PE, Kitzbichler MG, Apergis-Schoute AM, van der Flier FE, Fineberg NA, Sule A, Zaman R, Voon V, Kundu P, Bullmore ET, Robbins TW (2017b) Specific frontostriatal circuits for impaired cognitive flexibility and goal-directed planning in obsessive-compulsive disorder: evidence from resting-state functional connectivity. Biol Psychiatry 81(8):708–717. https://doi.org/10.1016/j.biopsych.2016.08.009

van den Heuvel OA, van Wingen G, Soriano-Mas C, Alonso P, Chamberlain SR, Nakamae T, Denys D, Goudriaan AE, Veltman DJ (2016) Brain circuitry of compulsivity. Eur Neuropsychopharmacol 26(5):810–827. https://doi.org/10.1016/j.euroneuro.2015.12.005

van Velzen LS, Vriend C, de Wit SJ, van den Heuvel OA (2014) Response inhibition and interference control in obsessive–compulsive spectrum disorders. Front Hum Neurosci 8. https://doi.org/10.3389/fnhum.2014.00419

van Woerkom TC, Fortgens C, van de Wetering BJ, Martens CM (1988) Contingent negative variation in adults with Gilles de la Tourette syndrome. J Neurol Neurosurg Psychiatry 51 (5):630–634. https://doi.org/10.1136/jnnp.51.5.630

Veale D, Miles S, Smallcombe N, Ghezai H, Goldacre B, Hodsoll J (2014) Atypical antipsychotic augmentation in SSRI treatment refractory obsessive-compulsive disorder: a systematic review and meta-analysis. BMC Psychiatry 14:317. https://doi.org/10.1186/s12888-014-0317-5

Vidal F, Burle B, Bonnet M, Grapperon J, Hasbroucq T (2003) Error negativity on correct trials: a reexamination of available data. Biol Psychol 64(3):265–282. https://doi.org/10.1016/S0301-0511(03)00097-8

Voon V, Baek K, Enander J, Worbe Y, Morris LS, Harrison NA, Robbins TW, Rück C, Daw N (2015a) Motivation and value influences in the relative balance of goal-directed and habitual behaviours in obsessive-compulsive disorder. Transl Psychiatry 5(11):e670–e670. https://doi.org/10.1038/tp.2015.165

Voon V, Derbyshire K, Rück C, Irvine MA, Worbe Y, Enander J, Schreiber LR, Gillan C, Fineberg NA, Sahakian BJ (2015b) Disorders of compulsivity: a common bias towards learning habits. Mol Psychiatry 20(3):345–352

Walter WG, Cooper R, Aldridge VJ, McCallum WC, Winter AL (1964) Contingent negative variation: an electric sign of sensori-motor association and expectancy in the human brain. Nature 203(4943):380–384. https://doi.org/10.1038/203380a0

Wang C, Ulbert I, Schomer DL, Marinkovic K, Halgren E (2005) Responses of human anterior cingulate cortex microdomains to error detection, conflict monitoring, stimulus-response mapping, familiarity, and orienting. J Neurosci Off J Soc Neurosci 25(3):604–613. https://doi.org/10.1523/JNEUROSCI.4151-04.2005

Weate SJ, Newell SA, Bogner JE, Andrews JM, Drake ME (1993) Contingent negative variation in Gilles de la Tourette syndrome. Clin EEG 24(4):188–191. https://doi.org/10.1177/155005949302400410

Weinberg A, Dieterich R, Riesel A (2015a) Error-related brain activity in the age of RDoC: a review of the literature. Int J Psychophysiol 98(2, part 2):276–299. https://doi.org/10.1016/j.ijpsycho.2015.02.029

Weinberg A, Kotov R, Proudfit GH (2015b) Neural indicators of error processing in generalised anxiety disorder, obsessive-compulsive-disorder, and major depressive disorder. J Abnorm Psychol 124(1):172–185

Weinberg A, Meyer A, Hale-Rude E, Perlman G, Kotov R, Klein DN, Hajcak G (2016) Error-related negativity (ERN) and sustained threat: conceptual framework and empirical evaluation in an adolescent sample. Psychophysiology 53(3):372–385. https://doi.org/10.1111/psyp.12538

Wessel JR (2012) Error awareness and the error-related negativity: evaluating the first decade of evidence. Front Hum Neurosci 6. https://doi.org/10.3389/fnhum.2012.00088

West B, Willner P (2011) Magical thinking in obsessive-compulsive disorder and generalized anxiety disorder. Behav Cogn Psychother 39(4):399–411. https://doi.org/10.1017/S1352465810000883

Structural Imaging in OCD

D. J. Veltman

Contents

Abstract Anatomical imaging in OCD using magnetic resonance imaging (MRI) has been performed since the late 1980s. MRI research was further stimulated with the advent of automated image processing techniques such as voxel-based morphometry (VBM) and surface-based methods (e.g., FreeSurfer) which allow for detailed whole-brain data analyses. Early studies suggesting involvement of corticostriatal circuitry (particularly orbitofrontal cortex and ventral striatum) have been complemented by meta-analyses and pooled analyses indicating additional involvement of posterior brain regions, in particular parietal cortex. Recent large-scale meta-analyses from the ENIGMA consortium have revealed greater pallidum and smaller hippocampus volume in adult OCD, coupled with parietal cortical thinning. Frontal cortical thinning was only observed in medicated patients. Previous reports of symptom dimension-specific alterations were not confirmed. In paediatric OCD, thalamus enlargement has been a consistent finding. Studies investigating white matter volume (VBM) or integrity (using diffusion tensor imaging (DTI)) have shown mixed results, with recent DTI meta-analyses mainly showing involvement of posterior cortical-subcortical tracts in addition to subcortical-prefrontal connections. To which extent these abnormalities are unique to OCD or common to other

D. J. Veltman (✉)
Department of Psychiatry, Amsterdam UMC, Amsterdam, The Netherlands
e-mail: dj.veltman@amsterdamumc.nl

Curr Topics Behav Neurosci (2021) 49: 201–229
https://doi.org/10.1007/7854_2020_209
Published Online: 19 February 2021

psychiatric disorders is unclear, as few comparative studies have been performed. Overall, neuroanatomical alterations in OCD appear to be subtle and may vary with time, stressing the need for adequately powered longitudinal studies. Although multivariate approaches using machine learning methodologies have so far been disappointing in distinguishing individual OCD patients from healthy controls, including multimodal data in such analyses may aid in further establishing a neurobiological profile of OCD.

Keywords Cortical thickness · DTI · MRI · VBM

1 Introduction

Within the newly formed DSM-5 category of obsessive-compulsive and related disorders (APA 2013), OCD has received by far the most attention from neuroimaging researchers. In contrast, imaging studies in body dysmorphic disorder (Madsen et al. 2015), trichotillomania (Isobe et al. 2018), skin picking disorder (Harries et al. 2017), and hoarding disorder (Suñol et al. 2020) are scarce. Moreover, comparative studies between obsessive-compulsive and related disorders as defined in DSM-5 have been all but absent, as OCD was grouped together with anxiety disorders in previous versions of the DSM (APA 1980, 1994).

Neuroimaging in OCD has been performed since the late 1980s with the advent of magnetic resonance imaging (MRI) and nuclear imaging: positron emission tomography (PET) and single photon emission computed tomography (SPECT). Of these, PET and SPECT are currently mainly used for molecular imaging, in functional imaging (i.e., measuring blood flow and/or glucose utilization), PET and SPECT have been all but superseded by functional MRI (fMRI) since the late 1990s. For anatomical imaging, MRI is still the method of choice, as it is non-invasive and highly versatile, allowing for high resolution imaging of e.g. regional brain volume (T1-weighted MRI), macroscopic lesions (T2-weighted MRI), and white matter organization (Diffusion Tensor Imaging, DTI). Typically, spatial resolution of anatomical T1 MR images is 1–1.5 mm when using currently widely available 3T scanners. DTI involves assessment of water diffusion directionality, its parameters of fractional anisotropy (FA), radial and axial diffusivity (RD, AD) are thought to reflect white matter health. Most DTI studies have reported FA, which is reportedly sensitive to fibre coherence and organization, myelination levels, and axonal integrity (Kochunov et al. 2020). DTI accuracy depends not only on spatial resolution (usually 2–2.5 mm) but also on the strength (b-factor) and number of MR diffusion gradients applied (typically 30–60, six being the minimum).

In this chapter, we will focus on structural (anatomical) imaging studies in OCD and related disorders. Functional and molecular imaging in OCD are covered elsewhere in this issue.

2 Early Studies in OCD

Early anatomical MRI studies in OCD mostly used region of interest (ROI) approaches, necessitated by time-consuming manual drawing of a limited number of relevant brain structures, which moreover requires trained observers to enhance interrater reliability. Also, the modest signal-to-noise ratio provided by standard 1.5 Tesla MR scanners employed during the 1990s tended to provide suboptimal resolution and image contrast compared to currently available hardware. These studies have reported smaller caudate nucleus volume (Robinson et al. 1995; Rosenberg et al. 1997), furthermore smaller orbitofrontal and amygdala volume (Szeszko et al. 1999) although findings have not been wholly consistent (Aylward et al. 1996; Stein et al. 1997). Overall, these studies complemented clinical investigations (often case reports) highlighting the link between lesions in the basal ganglia (e.g., Thobois et al. 2004) or orbitofrontal cortex and (secondary) OCD (Ames et al. 1994). In paediatric OCD, thalamus enlargement in medication-naïve patients has repeatedly been reported (Gilbert et al. 2000; Rosenberg et al. 2000).

Whole-brain approaches using statistical parametric mapping were introduced in OCD research (Kim et al. 2001) following the development of volume-based morphology (VBM; Ashburner and Friston 2000; Good et al. 2001). VBM involves warping individual scans (segmented into white matter (WM), grey matter (GM), and cerebrospinal fluid (CSF)) to anatomical standard space, thereby transforming regional volume alterations into density differences, permitting group comparisons on a voxel-by-voxel basis. The first large-scale OCD study employing VBM (Pujol et al. 2004) compared 72 patients and 72 control subjects, reporting smaller orbitofrontal, dorsomedial prefrontal and anterior insula volume, whereas ventral striatum (ventral putamen and nucleus accumbens) and cerebellum were found to be larger in patients. Striatal enlargement was associated with age-related GM conservation not present in controls. These results were mostly confirmed in a study by van den Heuvel et al. (2009), who likewise reported smaller orbitofrontal and dorsomedial, as well as lateral prefrontal, GM volume in OCD. In contrast, in medication-naïve paediatric OCD, not only increased putamen but also increased orbitofrontal GM volume has been observed using VBM (Szeszko et al. 2008). Thus, these findings were generally in agreement with existing frontostriatal models of OCD neurobiology, postulating that hyperactivity in orbitofrontal-striatal circuitry underlies OCD symptoms, due to an imbalance between direct and indirect cortico-striato-thalamo-cortical (CSTC) pathways (Saxena et al. 1998).

3 Meta-Analyses and Pooled Analyses

Early structural imaging studies were summarized in several meta-analyses, including those by Radua and Mataix-Cols (2009), Radua et al. (2010), and Rotge et al. (2010). Radua and Mataix-Cols (2009) included 12 samples with overall 401 OCD patients and 376 healthy controls (HC). OCD patients were found to have larger basal ganglia GM volume, particularly in samples with severe OCD patients, whereas dorsomedial and anterior cingulate GM volume was smaller in OCD patients. Antidepressant use did not alter these findings. In the Rotge et al. (2010) meta-analysis, 10 studies with 343 OCD patients and 318 HC were included, results showing greater basal ganglia (putamen) and anterior PFC grey matter density (GMD) in patients, whereas GMD was smaller in dorsolateral prefrontal, orbitofrontal, and parietal cortices. No significant differences were observed between children and adults with OCD. Increased putamen volume was also reported in another meta-analysis in overall 419 OCD patients relative to 415 HC (Eng et al. 2015). These authors furthermore reported increased GMD of various prefrontal and parietal regions as well as in the thalamus and cerebellum. Decreased GMD in OCD was found in right inferior frontal and temporal cortex.

Literature-based meta-analyses rely on summary statistics extracted from published studies, which are likely to differ with regard to image pre-processing and data analysis workflow. To circumvent this problem, a different approach was adopted by de Wit et al. (2014), who performed a pooled VBM analysis using imaging data from six sites (the OCD Brain Imaging Consortium (OBIC), which included the samples described in the Pujol et al. 2004 and van den Heuvel et al. 2009 studies), hence a mega-analysis rather than a meta-analysis. In 412 adult OCD patients compared to 368 HC the authors observed smaller volumes of both prefrontal GM and WM. Also, OCD patients showed a relative age-related preservation of GM volume in putamen, insula, and orbitofrontal cortex, coupled with a more rapid age-related volume loss in bilateral temporal cortex, explained as resulting from altered neuroplasticity associated with chronic compulsive behaviours, or compensatory processes related to cognitive dysfunction. These age-dependent alterations may thus account for previous inconsistent findings in these regions.

A re-analysis of the data from the de Wit et al. study was published by Fouche et al. (2017), who used FreeSurfer software rather than VBM. FreeSurfer is a surface-based method which, unlike VBM, allows separate assessment of cortical thickness and surface area in native space in addition to subcortical volumes (https://surfer.nmr.mgh.harvard.edu). Since cortical thickness and surface area are genetically independent (Winkler et al. 2010), using FreeSurfer is preferable when investigating possible genetic factors underlying morphological alterations. Moreover, FreeSurfer cortical thickness measurements are not sensitive to T1 signal variations, unlike VBM (Chung et al. 2017) and are less dependent on tissue contrast for measuring subcortical volumes (Wonderlick et al. 2009). However, it has been shown that segmentation methods in VBM and FreeSurfer are likely to produce similar results to manual drawing of ROIs, provided definitions of anatomic boundaries are the same (Grimm et al. 2015). The results of the Fouche et al. paper were partly consistent with the earlier VBM results by de Wit et al. (2014), showing lower cortical thickness (but not surface area) in ventrolateral and dorsomedial prefrontal cortex, as well as greater age-related decline in parietal-temporal cortical thickness. Overall hippocampal volume was found to be smaller in OCD, whereas the earlier finding of striatal enlargement could not be replicated, possibly because small regional differences can be picked up by voxel-based methods but are not reflected in overall volume.

4 Meta-and Mega-Analyses: ENIGMA

The OBIC consortium subsequently became the basis for the OCD Working Group as part of ENIGMA (Enhancing NeuroImaging Genetics through Meta-Analysis). The ENIGMA consortium was founded by Paul Thompson and coworkers in 2009 with the goal of addressing fundamental questions in neuroscience by harnessing the ever-increasing availability of existing imaging data as well as widely used automated analysis tools (Thompson et al. 2014). Within ENIGMA, each participating centre employs identical, standardized pre-processing, and data analysis tools, the resulting summary statistics being uploaded for central meta-analysis. As the acronym implies, ENIGMA's first projects aimed at bringing together researchers with genome-wide data and brain scans, to achieve the necessary power for genome-wide association studies (GWAS). Most genetic polymorphisms associated with brain metrics account for less than 1% of the overall variance in any given brain measure; moreover, as GWAS analyses typically involve around one million independent tests per subject, the required significance threshold is ca. 5×10^{-8} to adequately control the risk of Type I error. This collaboration has resulted in a series of papers detailing genetic variants implicated in subcortical and cortical architecture (Stein et al. 2012; Hibar et al. 2015; Adams et al. 2016; Satizabal et al. 2019; Grasby et al. 2020) as well as studies investigating general features such as brain asymmetry (Guadalupe et al. 2017; Kong et al. 2018).

In ENIGMA, around a third of data were obtained from patients with a range of psychiatric conditions, although results from the primary GWAS studies were not affected by excluding patients (Thompson et al. 2017). Consequently, researchers were quick to recognize the opportunities presented by these data, and started working groups (WG's) for performing large-scale patients vs. controls comparisons and investigating effects of possible moderating factors (disease severity and duration, medication use, gender, and comorbidities). Following the establishment (in 2012) of WG's for schizophrenia (SZ), bipolar disorder (BD), major depression (MDD), and attention deficit hyperactivity disorder (ADHD), a WG for OCD was begun in 2013, led by Odile van den Heuvel and Dan Stein (van den Heuvel et al. 2020). Similar to the main other disease working groups, the first two projects involved comparisons between patients and controls regarding subcortical volumes (Boedhoe et al. 2017) and cortical thickness surface area and thickness (Boedhoe et al. 2018). However, unlike these other WG's, both meta- and mega-analyses were performed. In the first paper, 1,830 OCD patients and 1,759 HC were included. Meta/mega-analytical results revealed lower hippocampal and larger pallidum volumes in adult OCD patients relative to HC, these effects were driven by medicated patients. Lateral ventricle volume was also larger in medicated patients, suggesting additional cortical GM/WM volume decreases. Between unmedicated and medicated OCD patients or between unmedicated patients and HC, no significant differences were found. None of the subcortical volumes was significantly different between paediatric OCD patients and controls, although unmedicated patients showed larger thalamic and smaller nucleus accumbens volumes. Mega-analyses and meta-analyses revealed similar results. Further moderator analyses showed that in adults, early-onset OCD was characterized by a larger pallidum volume, whereas late-onset OCD patients (and, at least partly, those with comorbid depression) showed a smaller hippocampal volume. The overall paediatric sample lacked power and/or additional data for detailed moderator analyses (Boedhoe et al. 2017). The authors noted that the association between lower hippocampal volume and comorbid depression/medication use in adult OCD suggests that this finding is nonspecific. Indeed, lower hippocampal volume has also been reported by other ENIGMA disease WG's, notably schizophrenia (van Erp et al. 2016), bipolar disorder (Hibar et al. 2016), MDD (Schmaal et al. 2016), and posttraumatic stress disorder (Logue et al. 2018), consistent with extensive research linking hippocampal atrophy to traumatic stress (e.g., Pitman et al. 2012). In contrast, pallidum enlargement in early-onset adult OCD patients was thought to reflect repetitive symptoms unique to OCD. In line with this hypothesis, putamen enlargement associated with subclinical OCD traits has similarly been found (Kubota et al. 2016). However, as noted by the authors, basal ganglia (i.e., pallidum) enlargement was also reported by the ENIGMA schizophrenia WG, explained as resulting from antipsychotic medication (van Erp et al. 2016), in line with findings of caudate enlargement in MDD patients using antipsychotic drugs (Schmaal et al. 2016).

A second study reported cortical findings in 1,905 OCD patients and 1,760 healthy controls (Boedhoe et al. 2018). Overall, adult OCD patients revealed lower temporal cortical surface area and inferior parietal cortical thinning. These effects were driven by medicated patients (roughly half of the overall sample), moreover, medicated patients showed widespread cortical thinning in frontal, temporal, and parietal regions compared to unmedicated patients. Parietal thinning was also observed in paediatric OCD patients, again predominantly in medicated patients, who also revealed additional frontal surface area reductions. In contrast, parietal thinning was not found in early-onset adult OCD, probably due to insufficient power.

Thus, these two largest meta/mega-analyses to date corroborated earlier studies in demonstrating subcortical (pallidum) and parietal involvement in (adult) OCD. Cortical thinning was more prominent than surface area reductions, which were limited to temporal regions. Prefrontal cortical thinning was found only in medicated patients but not in the overall sample, which is surprising given that the entire OBIC sample (6 sites out of 37) was included in these studies. In paediatric OCD, the finding of thalamic enlargement was likewise in keeping with earlier studies, suggesting thalamic enlargement is an early marker for OCD. This effect was mainly driven by unmedicated patients, in agreement with an early study showing SSRI treatment may normalize thalamic enlargement (Gilbert et al. 2000). Overall, medication use was found to be a major moderator of brain metrics, not only in OCD but in other ENIGMA WG's as well. In MDD, robust and widespread cortical thinning was observed mainly in medicated patients (Schmaal et al. 2017), similar to OCD. In bipolar disorder, lower thalamus volume was not observed in lithium users (Hibar et al. 2016); also, lithium appeared to protect against cortical thinning in BD, unlike mood stabilizers (Hibar et al. 2018). As noted earlier, pallidum enlargement in schizophrenia was found to be associated with medication use; in addition, lower caudate and putamen volume was not detected in those on antipsychotics (van Erp et al. 2016), but widespread cortical thinning was correlated with antipsychotics dose (chlorpromazine equivalents) (van Erp et al. 2018). The mechanisms underlying medication-induced atrophy and/or neuroplasticity are as yet insufficiently known. Furthermore, in OCD (and also MDD) medication use is likely to be associated with disease severity, although medication effects were still apparent after adjusting for disease severity (Boedhoe et al. 2018). A methodological drawback in ENIGMA studies published to date is that disease severity and medication use were generally recorded only at the time of scanning, and that few centres have detailed longitudinal records with regard to illness severity and duration/number of episodes, and overall medication use (cumulative lifetime dose).

5 OCD Subdimensions

OCD is a clinically heterogeneous disorder, and imaging researchers have sought to identify structural (and functional) brain correlates of OCD subdimensions (i.e., symmetry and ordering – hoarding – contamination and cleaning – aggressive and checking – sexual and religious obsessions). In the Pujol et al. (2004) study, the authors failed to observe correlations between subdimension symptom scores and GM alterations within their ROI's. However, they observed a decrease in GM volume in the right amygdala region in a subgroup of patients with prominent aggressive obsessions and checking compulsions. This early result was partially confirmed and extended by van den Heuvel et al. (2009), who found a significant negative correlation between scores on the 'harm/checking' subdimension and bilateral temporal GM/WM volume. Furthermore, scores on the 'contamination/washing' and 'symmetry/ordering' subdimensions were negatively correlated with bilateral caudate and left insula/parietal GM volume, respectively. These results therefore indicated that OCD is characterized by both common (mainly prefrontal) and symptom-specific neuroanatomical alterations, as was also reported for task-related (Mataix-Cols et al. 2004) and resting-state functional MRI (Harrison et al. 2009). Findings have been inconsistent, however, with Okada et al. (2015) reporting negative correlations between aggressive and checking symptom scores and cerebellum GM volume as well as contamination/washing and right insula GM volume in their sample of 37 OCD patients and 37 matched controls. Hirose et al. (2017), on the other hand, reported negative correlations between 'washing' scores and thalamus GM volume, and hoarding scores and left angular gyrus WM volume. In a recent review of structural imaging studies on subdimensions of OCD, it was concluded that taken together, there is some, albeit mixed, evidence for negative correlations between sexual/religious/aggressive and/or checking obsessions and GM volume in various brain regions (Thorsen et al. 2018). The authors furthermore concluded that evidence for altered brain morphology and ordering/symmetry ratings is scant and conflicting, and that the use of different instruments (Y-BOCS-SC and Padua Inventory-revised) may have contributed to the overall inconsistent findings. Moreover, none of the above-discussed results was replicated in the meta- and mega-analyses performed within the context of ENIGMA. Regression analyses in adult patients for whom subdimension scores were available ($N = 1{,}151$ for subcortical volumes, $N = 1{,}214$ for cortical thickness/surface area) did not reveal any significant associations between subcortical volumes or cortical thickness/surface area. Therefore, it is questionable if OCD subtypes and/or subdimensions can be differentiated based on brain morphology.

6 The Role of Comorbidity

In addition to clinical subdimensions, the presence or absence of various comorbid disorders represents another major source of heterogeneity in OCD. However, even relatively large single-centre studies such as the Rus et al. (2017) study (with 75 OCD patients and 75 healthy controls) or the Shim et al. (2009) study (110 OCD patients, 169 healthy controls) either did not control for comorbidity or failed to observe significant associations. In the ENIGMA OCD subcortical meta/mega-analysis, adult OCD patients with a comorbid lifetime diagnosis ($N = 325$) had smaller hippocampal volumes than controls, whereas OCD patients without comorbid depression ($N = 1{,}041$) had both larger pallidum and smaller hippocampal volumes compared to controls. Although the effect size for lower hippocampal volume was greater in OCD patients with comorbid depression ($d = -0.27$) compared to those without ($d = -0.16$), a direct comparison between these groups did not reveal significant subcortical volume differences (Boedhoe et al. 2017). Similarly, no associations between cortical thickness/surface area and the presence of either comorbid MDD or anxiety disorders were observed. The authors noted that their subgroups with established comorbid MDD or anxiety disorders were likely too small (around 200–300 patients) and may represent an underestimation of true comorbidity (Boedhoe et al. 2018). Studies investigating common and unique neuroanatomical alterations in 'pure' OCD and anxiety disorders or MDD have likewise been rare, at least for anatomical imaging. Several of these studies included only modest samples of MDD (Fallucca et al. 2011), posttraumatic stress disorder and social anxiety disorder (Cheng et al. 2015), or schizophrenia (Kwon et al. 2003) in addition to OCD. A recent study (Gong et al. 2019) conducted as part of a larger Chinese cohort study sought to establish a transdiagnostic neuroanatomical signature of psychiatric illness in 404 patients (OCD 92, MDD 92, PTSD 91, psychosis 129) compared with 210 controls and 20 unaffected first-degree relatives, using VBM. The four psychiatric groups were all characterized by greater bilateral putamen GM volume, and putamen volume was positively correlated with symptom severity.

Crucially, all patients were medication-naïve, moreover, putamen enlargement was also observed in first-degree relatives. Consequently, this finding could not be explained as resulting from medication use but may represent a transdiagnostic familial marker of psychopathology. Validating such transdiagnostic markers (structural or functional) as 'psychiatric phenotypes' may greatly aid in identifying underlying disease mechanisms and eventually develop targeted treatments (Gillan et al. 2017). Whether putamen enlargement is a general marker of psychopathology is unclear, however. In their meta-analysis comparing VBM studies in OCD vs. other anxiety disorders, Radua et al. (2010) included 26 datasets with 639 patients and 737 healthy controls. Lower dorsomedial prefrontal and anterior cingulate GM volume was a common feature of OCD and (other) anxiety disorders, mostly PTSD and panic disorder (PD). In contrast, OCD patients were characterized by larger lentiform/caudate nuclei, whereas PTSD/PD patients had smaller left lentiform nuclei compared with controls. Another exceptionally large meta-analysis (193 studies, 15,892 individuals) revealed that GM loss converged across diagnostic groups (schizophrenia, bipolar disorder, MDD, OCD, anxiety disorders, substance use disorder) in three regions: dorsal anterior cingulate and bilateral anterior insula, but not basal ganglia (Goodkind et al. 2015), interpreted as the neural substrate of executive dysfunction.

In ENIGMA, cross-disorder comparisons have been slow to materialize in view of the challenges to obtain individual data from dozens of centres (necessary for pooled analyses) and the need to harmonize not only image pre-processing pipelines but also clinical variables for various disorders (e.g., diagnostic instruments, medication use). However, in a recent ENIGMA study (Boedhoe et al. 2020) data from 151 centres were combined to compare OCD ($N = 2{,}323$) with two other common neurodevelopmental disorders, attention deficit hyperactivity disorder (ADHD; $N = 2{,}271$), and autism spectrum disorder (ASD; 1,777) as well as controls ($N = 5{,}827$), across three age groups (children, <12 years, adolescents, 12–18 years, adults, >18 years). Overall results did not reveal clear evidence for shared differences in subcortical volumes or cortical thickness/surface area in children and adolescents with either OCD, ADHD, or ASD compared with controls. In adults, smaller hippocampal volume was found not only in OCD but also, albeit to a lesser extent, in ASD. No shared cortical abnormalities were observed in adults. Analyses of disease-specific abnormalities revealed smaller intracranial volume (ICV) in children and adolescents with ADHD compared with OCD and ASD. Adolescents with OCD had lower surface area of the medial orbitofrontal cortex compared with ADHD. In adults, frontal cortical thickness was greater in ASD compared with OCD and ADHD, but there were no disease-specific differences in subcortical volumes or regional cortical thickness. Comparisons between OCD and mood/anxiety disorders are scarce (Fallucca et al. 2011; Cheng et al. 2015) and have not yet been performed in ENIGMA. Summarizing, increased lentiform/pallidum volume has been repeatedly observed in OCD, but whether this finding is specific to OCD or generalizes to other psychiatric disorders (in particular anxiety/mood disorders) awaits further investigation.

7 White Matter: VBM and DTI

To date, most anatomical neuroimaging studies in OCD have focused on grey matter volume alterations. White matter abnormalities in OCD have been assessed either using DTI, its main parameter fractional anisotropy (FA) usually being interpreted as a measure of local WM integrity, or VBM, which provides voxel-by-voxel measurements of WM density in anatomical standard space as a proxy for regional WM volume, in addition to GM. Whereas in the Pujol et al. (2004) study discussed earlier no significant differences between patients and controls were found, van den Heuvel et al. (2009) reported smaller prefrontal WM volume adjacent to cortical regions showing GM decreases relative to controls. In line with these findings, two early DTI studies reported lower FA mainly in the anterior cingulate region (Szeszko et al. 2005; Cannistraro et al. 2007), but Menzies et al. (2008) found increased medial frontal FA coupled with decreased parietal FA in 30 OCD patients and first-degree relatives compared to controls. Widespread increased WM FA coupled with greater caudate and putamen volume was also reported by Zarei et al. (2011) in their sample of 26 adolescents with OCD compared with 26 controls. For WM volume findings have likewise been conflicting, as lower prefrontal and parietal WM volume associated with Y-BOCS scores in adult OCD (Park and Jeong 2015) and lower prefrontal and anterior cingulate WM in paediatric OCD (Lázaro et al. 2014) have also been reported, similar to the van den Heuvel et al. 2009 study but in contrast to Menzies et al. (2008) and Zarei et al. (2011).

A combined meta-analysis of WM volume and FA in OCD was performed by Radua et al. (2014), who included 22 samples with overall 537 patients and 575 controls. OCD patients showed widespread WM abnormalities, but mainly in anterior midline tracts (i.e., interhemispheric connections in corpus callosum and anterior cingulate region). Both increased WM volume and lower FA was observed, indicating increased fibre crossings in these regions. These effects were driven by adult medicated patients. A second meta-analysis (Eng et al. 2015) included many of the same samples as the Radua et al. study, combining measurements of GM volume, executive performance (task-fMRI), and FA. Results were mostly supportive of corticostriatal-thalamocortical (CSTC) involvement but parietal and cerebellar regions were also implicated. FA data were mixed, showing both increased and decreased diffusivity within single WM tracts such as the superior longitudinal fasciculus, a major association pathway interconnecting frontal, parietal, and temporal regions, and the corpus callosum. Reduced diffusivity was observed in the posterior corona radiata, containing both ascending and descending fibres connecting basal ganglia (including the thalamus) and parietal/frontal cortical regions, thought to underlie executive deficits in OCD. Thus, while indicating possibly widespread WM involvement in the pathophysiology of OCD, results have been inconsistent so far.

Similar to earlier VBM reports, these meta-analyses relied on published results, limiting accounting for various clinical and demographic factors as well as methodological heterogeneity. To address these issues, in a recent ENIGMA meta-analysis

DTI metrics from 700 adult OCD patients were compared to 645 controls, and separately in 174 paediatric patients and 144 controls (Piras et al. 2019). All participating sites used a standardized workflow based on tract-based spatial statistics (TBSS) providing tract-average FA-values for major WM tracts (Kochunov et al. 2020). Of 25 investigated regions, five showed lower FA in adult patients: the genu of the corpus callosum, the posterior corona radiata, the uncinate fasciculus, and the sagittal stratum (SS), also including the posterior thalamic radiation (PTR). However, only the sagittal stratum/posterior thalamic radiation findings survived additional multiple comparisons correction. Meta-regressions showed that lower FA in the SS was associated with younger age of onset, illness duration, and medication use, but not illness severity. In contrast, lower FA in the PTR was not associated with clinical variables. These corticosubcortical tracts (PTR and SS) contain efferent fibres from posterior cingulate, parietal, temporal, and occipital regions to various brainstem structures as well as afferents from the thalamus to the cortex (https://neurosurgery.directory/2019/08/01/sagittal-stratum/), again implicating posterior cortical regions in the pathophysiology of OCD. Lower FA was associated with increased radial diffusivity (RD); although only at trend level, this finding indicates that lower FA reflects disruption of myelin sheaths rather than increased crossings (Piras et al. 2019). In paediatric OCD patients, no FA abnormalities were detected in any of the regions studied, which may have been due to low power. Nevertheless, the absence of FA abnormalities in paediatric OCD and their association with disease duration, age of onset, and medication use in adult OCD suggest that these alterations are a consequence of having the disorder rather than a cause.

8 Discussion

In conclusion, 30 years of anatomical imaging in OCD have produced a wealth of data regarding subcortical and cortical involvement in the pathophysiology of OCD. Early studies focussing on prefrontal-striatal regions have been complemented with reports showing involvement of widespread posterior cortical areas including parietal and (to a lesser extent) temporal cortex, as well as limbic regions and cerebellum, representative of the affective and visuospatial domain (Hazari et al. 2019). These findings have led to extended models of OCD, beyond CSTC circuitry, as parietal involvement may account, at least in part, for cognitive deficits involving attention, set shifting, planning, and response inhibition, as seen in OCD patients (Boedhoe et al. 2018). White matter pathology has been less extensively investigated, but studies to date show a pattern somewhat similar to the GM literature in OCD, i.e. stressing the involvement of posterior cortical-subcortical pathways in addition to subcortical-prefrontal connections. Both traditional meta-analyses and the pooled mega-analyses performed as part of OBIC and ENIGMA have demonstrated greater volume of parts of the basal ganglia as well as lower GM volume/cortical thinning in prefrontal areas, although this latter effect was observed only in medicated patients, and in parietal cortex. Also, smaller hippocampal volume was found only in OCD patients with comorbid depression, and is likely not specific for OCD. Pallidum enlargement may differentiate OCD from (other) anxiety disorders, but since similar findings have been observed in other psychiatric disorders, this result awaits confirmation in large-scale comparative studies. In the same vein, although analyses of imaging data in paediatric OCD have indicated both common and unique alterations compared to adult OCD (see also Hu et al. 2017), direct comparisons have been rare.

Overall, neuroanatomical alterations in OCD appear to be subtle and results of individual studies have been inconsistent. Meta-analyses may address methodological problems in single-centre studies, in particular small sample sizes and patient heterogeneity regarding demographic and clinical variables associated with brain morphology (e.g., age, choice of diagnostic instrument, illness duration, medication status). Small sample sizes will obviously limit statistical power to model such factors or to detect subtle neurobiological effects. Moreover, small studies may lead to reporting false positives or overestimate effect sizes, which may also result from homogeneous samples due to stringent inclusion criteria. That is, observed effects are frequently exaggerated in an underpowered study, since realistic estimates (closer to the true effect size) are less likely to reach statistical significance and hence go undetected. Thus, the 'lucky' scientist who makes such a discovery is cursed by finding inflated estimates: the so-called winner's curse, contributing to the replication crisis in neuroimaging (Button et al. 2013).

As noted earlier, consortium studies such as ENIGMA represent an advance compared to standard meta-analyses relying on published summary statistics with its risk of publication bias, as ENIGMA has developed standardized analysis pipelines for analysing multimodal imaging data and has included previously unpublished results. ENIGMA is not without limitations, however, as its data have been collected in the past and there is considerable methodological heterogeneity with regard to imaging hardware and scanning sequences. Also, in large cohorts clinical heterogeneity represents both a strength and a weakness. On one hand, it helps to create an 'ecologically valid' sample while allowing for stratification based on clinical variables. On the other hand, analyses of subgroups or additional meta-regression analyses are frequently hampered by heterogeneous and limited phenotyping of participants (demographics and clinical variables) across sites. Furthermore, ENIGMA studies to date have relied on cross-sectional data, as few sites have performed repeated measurements, for example, pre-post therapy (Quidé et al. 2012). Consequently, there is still a need for carefully designed, prospective longitudinal studies.

Although results of meta- and mega-analyses in ENIGMA have generally been in line with previous reports, these effects have been modest, with the possible exception of schizophrenia (van Erp et al. 2016, 2018). For subcortical volumes and cortical thickness/surface area, these modest results may have been due to the atlas-based regional approach commonly used in ENIGMA (the Desikan-Killiany atlas, Desikan et al. 2006), which may obscure small abnormalities which can be identified with vertex-based analyses (in FreeSurfer) or with VBM. Put differently, ENIGMA studies to date may have underestimated true effects. Subcortical volume measurements can be refined using shape analyses, as has been done for the hippocampus in OCD (Zhang et al. 2019). However, given that ENIGMA studies to date have generally been data-driven and hypothesis-free, more fine-grained whole-brain methods will require additional safeguards against Type I error, which may be problematic even in these large consortium samples. In ENIGMA, effect sizes in mood and anxiety disorders including OCD have been typically around 0.1–0.2 for atlas-based volume measurements (0.2–0.3 for DTI; Kochunov et al. 2020), implying that to obtain an alpha of 0.05 and a power of 0.8 for Cohen's $d = 0.2$ and single comparisons, sample sizes of at least 400 subjects are necessary, whereas for $d = 0.1$ thousands of subjects are required.

The above-reviewed studies all relied on mass-univariate analyses, involving multiple parallel tests at a regional or vertex/voxel level. Given that OCD is likely a network disorder, researchers have additionally sought to characterize distributed brain alterations by investigating structural covariance networks in OCD (e.g., Pujol et al. 2011), which can be expressed as single metrics (e.g., clustering, modularity, small-worldness). A recent ENIGMA meta-analysis in OCD (Yun et al. 2020) indicated lower overall network segregation together with reorganization of brain hubs such as caudate nucleus and thalamus, and suggested altered brain development and maturation trajectories in cingulate and orbitofrontal regions. Spatially distributed effects can also be assessed using multivariate pattern analysis (MVPA), with its promise of clinical applications based on classifying individuals rather than establishing group differences. Multivariate approaches using machine learning methodologies have been used to distinguish OCD patients from healthy controls (e.g., Hu et al. 2016; Zhou et al. 2018a, b), predict outcome (Hoexter et al. 2015; Askland et al. 2015), and symptom severity (Hoexter et al. 2013) based on brain morphology. Classification accuracy (commonly expressed as area under the curve, AUC) in individual studies has varied between 0.72 and 0.93 for single-centre samples with low methodological and clinical heterogeneity (Bruin et al. 2019b). However, a recent ENIGMA study in 2,304 OCD patients and 2,068 controls showed that multi-site MVPA is as yet unable to reliably differentiate between cases and controls, as site-stratified accuracy ranged between 0.58 and 0.62 AUC, which, although statistically significant, falls far short of clinical applicability (Bruin et al. 2019a). Out of sample validation on new data performed even worse, hardly different from chance level. Of note, accuracy for distinguishing medicated from unmedicated OCD patients was high (AUC = 0.86), again stressing the importance of medication use as a moderator of brain morphology. Future studies therefore need to control carefully for medication use. Machine learning should furthermore aim to include multimodal data, not only WM/GM volume and DTI (Zhou et al. 2018a, b) but also resting-state fMRI which may be more sensitive (Yang et al. 2019) and task-related fMRI, results of which may also corroborate and extend morphological findings (Norman et al. 2016; Picó-Pérez et al. 2020).

Finally, given the limitations of retrospective data analysis noted earlier, there is still a clear need for adequately powered, hence multi-centre, prospective studies using uniform imaging methods (scanner type, sequences) and harmonized assessments (demographics, clinical variables, neuropsychological tests). Such large-scale studies, especially when involving follow-up measurements, are inherently costly. However, a recently funded study involving five specialized OCD centres across the globe is currently ongoing (Simpson et al. 2020) with the aim of identifying a general 'brain signature' of OCD. Ultimately, establishing a multimodal profile of OCD will not only enhance our understanding of the disorder but also inform targeted therapies.

References

Adams HH, Hibar DP, Chouraki V, Stein JL, Nyquist PA, Rentería ME, Trompet S, Arias-Vasquez A, Seshadri S, Desrivières S, Beecham AH, Jahanshad N, Wittfeld K, Van der Lee SJ, Abramovic L, Alhusaini S, Amin N, Andersson M, Arfanakis K, Aribisala BS, Armstrong NJ, Athanasiu L, Axelsson T, Beiser A, Bernard M, Bis JC, Blanken LM, Blanton SH, Bohlken MM, Boks MP, Bralten J, Brickman AM, Carmichael O, Chakravarty MM, Chauhan G, Chen Q, Ching CR, Cuellar-Partida G, Braber AD, Doan NT, Ehrlich S, Filippi I, Ge T, Giddaluru S, Goldman AL, Gottesman RF, Greven CU, Grimm O, Griswold ME, Guadalupe T, Hass J, Haukvik UK, Hilal S, Hofer E, Hoehn D, Holmes AJ, Hoogman M, Janowitz D, Jia T, Kasperaviciute D, Kim S, Klein M, Kraemer B, Lee PH, Liao J, Liewald DC, Lopez LM, Luciano M, Macare C, Marquand A, Matarin M, Mather KA, Mattheisen M, Mazoyer B, McKay DR, McWhirter R, Milaneschi Y, Mirza-Schreiber N, Muetzel RL, Maniega SM, Nho K, Nugent AC, Loohuis LM, Oosterlaan J, Papmeyer M, Pappa I, Pirpamer L, Pudas S, Pütz B, Rajan KB, Ramasamy A, Richards JS, Risacher SL, Roiz-Santiañez R, Rommelse N, Rose EJ, Royle NA, Rundek T, Sämann PG, Satizabal CL, Schmaal L, Schork AJ, Shen L, Shin J, Shumskaya E, Smith AV, Sprooten E, Strike LT, Teumer A, Thomson R, Tordesillas-Gutierrez D, Toro R, Trabzuni D, Vaidya D, Van der Grond J, Van der Meer D, Van Donkelaar MM, Van Eijk KR, Van Erp TG, Van Rooij D, Walton E, Westlye LT, Whelan CD, Windham BG, Winkler AM, Woldehawariat G, Wolf C, Wolfers T, Xu B, Yanek LR, Yang J, Zijdenbos A, Zwiers MP, Agartz I, Aggarwal NT, Almasy L, Ames D, Amouyel P, Andreassen OA, Arepalli S, Assareh AA, Barral S, Bastin ME, Becker DM, Becker JT, Bennett DA, Blangero J, van Bokhoven H, Boomsma DI, Brodaty H, Brouwer RM, Brunner HG, Buckner RL, Buitelaar JK, Bulayeva KB, Cahn W, Calhoun VD, Cannon DM, Cavalleri GL, Chen C, Cheng CY, Cichon S, Cookson MR, Corvin A, Crespo-Facorro B, Curran JE, Czisch M, Dale AM, Davies GE, De Geus EJ, De Jager PL, de Zubicaray GI, Delanty N, Depondt C, DeStefano AL, Dillman A, Djurovic S, Donohoe G, Drevets WC, Duggirala R, Dyer TD, Erk S, Espeseth T, Evans DA, Fedko IO, Fernández G, Ferrucci L, Fisher SE, Fleischman DA, Ford I, Foroud TM, Fox PT, Francks C, Fukunaga M, Gibbs JR, Glahn DC, Gollub RL, Göring HH, Grabe HJ, Green RC, Gruber O, Gudnason V, Guelfi S, Hansell NK, Hardy J, Hartman CA, Hashimoto R, Hegenscheid K, Heinz A, Le Hellard S, Hernandez DG, Heslenfeld DJ, Ho BC, Hoekstra PJ, Hoffmann W, Hofman A, Holsboer F, Homuth G, Hosten N, Hottenga JJ, Hulshoff Pol HE, Ikeda M, Ikram MK, Jack CR Jr, Jenkinson M, Johnson R, Jönsson EG, Jukema JW, Kahn RS, Kanai R, Kloszewska I, Knopman DS, Kochunov P, Kwok JB, Lawrie SM, Lemaître H, Liu X, Longo DL, Longstreth WT Jr, Lopez OL, Lovestone S, Martinez O, Martinot JL, Mattay VS, McDonald C, McIntosh AM, McMahon KL, McMahon FJ, Mecocci P, Melle I, Meyer-Lindenberg A, Mohnke S, Montgomery GW, Morris DW, Mosley TH, Mühleisen TW, Müller-Myhsok B, Nalls MA, Nauck M, Nichols TE, Niessen WJ, Nöthen MM, Nyberg L, Ohi K, Olvera RL, Ophoff RA, Pandolfo M, Paus T, Pausova Z, Penninx BW, Pike GB, Potkin SG, Psaty BM, Reppermund S, Rietschel M, Roffman JL, Romanczuk-Seiferth N, Rotter JI, Ryten M, Sacco RL, Sachdev PS, Saykin AJ, Schmidt R, Schofield PR, Sigurdsson S,

Simmons A, Singleton A, Sisodiya SM, Smith C, Smoller JW, Soininen H, Srikanth V, Steen VM, Stott DJ, Sussmann JE, Thalamuthu A, Tiemeier H, Toga AW, Traynor BJ, Troncoso J, Turner JA, Tzourio C, Uitterlinden AG, Hernández MC, Van der Brug M, Van der Lugt A, Van der Wee NJ, Van Duijn CM, Van Haren NE, Van T Ent D, Van Tol MJ, Vardarajan BN, Veltman DJ, Vernooij MW, Völzke H, Walter H, Wardlaw JM, Wassink TH, Weale ME, Weinberger DR, Weiner MW, Wen W, Westman E, White T, Wong TY, Wright CB, Zielke HR, Zonderman AB, Deary IJ, DeCarli C, Schmidt H, Martin NG, De Craen AJ, Wright MJ, Launer LJ, Schumann G, Fornage M, Franke B, Debette S, Medland SE, Ikram MA, Thompson PM (2016) Novel genetic loci underlying human intracranial volume identified through genome-wide association. Nat Neurosci 19(12):1569–1582. https://doi.org/10.1038/nn.4398. Epub 2016 Oct 3. PubMed PMID: 27694991; PubMed Central PMCID: PMC5227112

American Psychiatric Association (1980) Diagnostic and statistical manual of mental disorders, 3rd edn. American Psychiatric Publications, Washington, DC

American Psychiatric Association (1994) Diagnostic and statistical manual of mental disorders, 4th edn. American Psychiatric Publications, Washington, DC

American Psychiatric Association (2013) Diagnostic and statistical manual of mental disorders, 5th edn. American Psychiatric Publications, Washington, DC

Ames D, Cummings JL, Wirshing WC, Quinn B, Mahler M (1994) Repetitive and compulsive behavior in frontal lobe degenerations. J Neuropsychiatry Clin Neurosci 6(2):100–113. Review. PubMed PMID: 8044031

Ashburner J, Friston KJ (2000) Voxel-based morphometry--the methods. NeuroImage 11(6 Pt 1):805–821. Review. PubMed PMID: 10860804

Askland KD, Garnaat S, Sibrava NJ, Boisseau CL, Strong D, Mancebo M, Greenberg B, Rasmussen S, Eisen J (2015) Prediction of remission in obsessive compulsive disorder using a novel machine learning strategy. Int J Methods Psychiatr Res 24(2):156–169. https://doi.org/10.1002/mpr.1463. Epub 2015 May 21. PubMed PMID: 25994109; PubMed Central PMCID: PMC5466447

Aylward EH, Harris GJ, Hoehn-Saric R, Barta PE, Machlin SR, Pearlson GD (1996) Normal caudate nucleus in obsessive-compulsive disorder assessed by quantitative neuroimaging. Arch Gen Psychiatry 53(7):577–584. Review. PubMed PMID: 8660124.

Boedhoe PS, Schmaal L, Abe Y, Ameis SH, Arnold PD, Batistuzzo MC, Benedetti F, Beucke JC, Bollettini I, Bose A, Brem S, Calvo A, Cheng Y, Cho KI, Dallaspezia S, Denys D, Fitzgerald KD, Fouche JP, Giménez M, Gruner P, Hanna GL, Hibar DP, Hoexter MQ, Hu H, Huyser C, Ikari K, Jahanshad N, Kathmann N, Kaufmann C, Koch K, Kwon JS, Lazaro L, Liu Y, Lochner C, Marsh R, Martínez-Zalacaín I, Mataix-Cols D, Menchón JM, Minuzzi L, Nakamae T, Nakao T, Narayanaswamy JC, Piras F, Piras F, Pittenger C, Reddy YC, Sato JR, Simpson HB, Soreni N, Soriano-Mas C, Spalletta G, Stevens MC, Szeszko PR, Tolin DF, Venkatasubramanian G, Walitza S, Wang Z, van Wingen GA, Xu J, Xu X, Yun JY, Zhao Q, ENIGMA OCD Working Group, Thompson PM, Stein DJ, van den Heuvel OA (2017) Distinct subcortical volume alterations in pediatric and adult OCD: a worldwide meta- and mega-analysis. Am J Psychiatry 174(1):60–69. https://doi.org/10.1176/appi.ajp.2016.16020201. Epub 2016 Sep 9. Erratum in: Am J Psychiatry. 2016 Oct 1;173(10):1049. Am J Psychiatry. 2017 Feb 1;174(2):190. PubMed PMID: 27609241; PubMed Central PMCID: PMC5344782

Boedhoe PSW, Schmaal L, Abe Y, Alonso P, Ameis SH, Anticevic A, Arnold PD, Batistuzzo MC, Benedetti F, Beucke JC, Bollettini I, Bose A, Brem S, Calvo A, Calvo R, Cheng Y, Cho KIK, Ciullo V, Dallaspezia S, Denys D, Feusner JD, Fitzgerald KD, Fouche JP, Fridgeirsson EA, Gruner P, Hanna GL, Hibar DP, Hoexter MQ, Hu H, Huyser C, Jahanshad N, James A, Kathmann N, Kaufmann C, Koch K, Kwon JS, Lazaro L, Lochner C, Marsh R, Martínez-Zalacaín I, Mataix-Cols D, Menchón JM, Minuzzi L, Morer A, Nakamae T, Nakao T, Narayanaswamy JC, Nishida S, Nurmi E, O'Neill J, Piacentini J, Piras F, Piras F, Reddy YCJ, Reess TJ, Sakai Y, Sato JR, Simpson HB, Soreni N, Soriano-Mas C, Spalletta G, Stevens MC, Szeszko PR, Tolin DF, van Wingen GA, Venkatasubramanian G, Walitza S, Wang Z, Yun JY, ENIGMA-OCD Working Group, Thompson PM, Stein DJ, van den Heuvel OA, ENIGMA

OCD Working Group (2018) Cortical abnormalities associated with pediatric and adult obsessive-compulsive disorder: findings from the ENIGMA Obsessive-Compulsive Disorder Working Group. Am J Psychiatry 175(5):453–462. https://doi.org/10.1176/appi.ajp.2017.17050485. Epub 2017 Dec 15. PubMed PMID: 29377733; PubMed Central PMCID: PMC7106947

Boedhoe PS, van Rooij D, Hoogman M, Twisk JW, Schmaal L, Abe Y, Alonso P, Ameis SH, Anikin A, Anticevic A, Aherson P, Arango C, Arnold PD, Assogna F, Auzias G, Banaschewski T, Baranov A, Batistuzzo MC, Baumeister S, Baur-Streubel R, Behrmann M, Bellgrove MA, Benedetti F, Beucke JC, Biederman J, Bollettini I, Bose A, Bralten J, Bramati IE, Brandeis D, Brem S, Brennan BP, Busatto GF, Calderoni S, Calvo A, Calvo R, Castellanos FX, Cercignani M, Chaim-Avancini TM, Chantiluke KC, Cheng Y, Cho KI, Christakou A, Coghill D, Conzelmann A, Cubillo AI, Dale AM, Dallaspezia S, Daly E, Denys D, Deruelle C, Martino AD, Dinstein I, Doyle AE, Durston S, Earl EA, Ecker C, Ehrlich S, Ely BA, Epstein JN, Ethofer T, Fair DA, Fallgatter AJ, Faraone SV, Fedor J, Feng X, Feusner JD, Fitzgerald J, Fitzgerald KD, Fouche J, Freitag CM, Fridgeirsson EA, Frodl T, Gabel MC, Gallagher L, Gogberashvili T, Gori I, Gruner P, Gürsel DA, Haar S, Haavik J, Hall GB, Harrison NA, Hartman CA, Heslenfeld DJ, Hirano Y, Hoekstra PJ, Hoexter MQ, Hohmann S, Høvik MF, Hu H, Huyser C, Jahanshad N, Jalbrzikowski M, James A, Janssen J, Jaspers-Fayer F, Jernigan TL, Kapilushniy D, Kardatzki B, Karkashadze G, Kathmann N, Kaufmann C, Kelly C, Khadka S, King JA, Koch K, Kohls G, Kohls K, Kuno M, Kuntsi J, Kvale G, Kwon JS, Lázaro L, Lera-Miguel S, Lesch K, Hoekstra L, Liu Y, Lochner C, Louza MR, Luna B, Lundervold AJ, Malpas CB, Marques P, Marsh R, Martínez-Zalacaín I, Mataix-Cols D, Mattos P, McCarthy H, McGrath J, Mehta MA, Menchón JM, Mennes M, Martinho MM, Moreira PS, Morer A, Morgado P, Muratori F, Murphy CM, Murphy DG, Nakagawa A, Nakamae T, Nakao T, Namazova-Baranova L, Narayanaswamy JC, Nicolau R, Nigg JT, Novotny SE, Nurmi EL, Weiss EO, O'Gorman Tuura RL, O'Hearn K, O'Neill J, Oosterlaan J, Oranje B, Paloyelis Y, Parellada M, Pauli P, Perriello C, Piacentini J, Piras F, Piras F, Plessen KJ, Puig O, Ramos-Quiroga JA, Janardhan Reddy Y, Reif A, Reneman L, Retico A, Rosa PG, Rubia K, Rus OG, Sakai Y, Schrantee A, Schwarz L, Schweren LJ, Seitz J, Shaw P, Shook D, Silk TJ, Simpson HB, Skokauskas N, Soliva Vila JC, Solovieva A, Soreni N, Soriano-Mas C, Spalletta G, Stern ER, Stevens MC, Stewart SE, Sudre G, Szeszko PR, Tamm L, Taylor MJ, Tolin DF, Tosetti M, Tovar-Moll F, Tsuchiyagaito A, van Erp TG, van Wingen GA, Vance A, Venkatasubramanian G, Vilarroya O, Vives-Gilabert Y, von Polier GG, Walitza S, Wallace GL, Wang Z, Wolfers T, Yoncheva YN, Yun J, Zanetti MV, Zhou F, Ziegler GC, Zierhut KC, Zwiers MP, Thompson PM, Stein DJ, Buitelaar J, Franke B, van den Heuvel OA (2020) Subcortical brain volume, regional cortical thickness and cortical surface area across attention-deficit/hyperactivity disorder (ADHD), autism spectrum disorder (ASD), and obsessive-compulsive disorder (OCD) – findings from the ENIGMA-ADHD, -ASD, and -OCD working groups. Am J Psychiatry 177:834–843

Bruin W, Denys D, van Wingen G (2019a) Diagnostic neuroimaging markers of obsessive-compulsive disorder: initial evidence from structural and functional MRI studies. Prog Neuro-Psychopharmacol Biol Psychiatry 91:49–59. https://doi.org/10.1016/j.pnpbp.2018.08.005. Epub 2018 Aug 11. PubMed PMID: 30107192

Bruin WB, Taylor L, Thomas RM, Shock JP, Zhutovsky P, Abe Y, Alonso P, Ameis SH, Anticevic A, Arnold PD, Assogna F, Benedetti F, Beucke JC, Boedhoe PSW, Bollettini I, Bose A, Brem S, Brennan BP, Buitelaar JK, Calvo R, Cheng Y, Cho KIK, Dallaspezia S, Denys D, Ely BA, Feusner JD, Fitzgerald KD, Fouche J-P, Fridgeirsson EA, Gruner P, Gürsel DA, Hauser TU, Hirano Y, Hoexter MQ, Hu H, Huyser C, Ivanov I, James A, Jaspers-Fayer F, Kathmann N, Kaufmann C, Koch K, Kuno M, Kvale G, Kwon JS, Liu Y, Lochner C, Lázaro L, Marques P, Marsh R, Martínez-Zalacaín I, Mataix-Cols D, Menchón JM, Minuzzi L, Moreira PS, Morer A, Morgado P, Nakagawa A, Nakamae T, Nakao T, Narayanaswamy JC, Nurmi EL, O'Neill J, Pariente JC, Perriello C, Piacentini J, Piras F, Piras F, Reddy YCJ, Rus-Oswald OG, Sakai Y, Sato JR, Schmaal L, Shimizu E, Simpson HB, Soreni N, Soriano-Mas C, Spalletta G,

Stern ER, Stevens MC, Stewart SE, Szeszko PR, Tolin DF, Venkatasubramanian G, Wang Z, Yun J-Y, van Rooij D, ENIGMA-OCD Consortium, Thompson PM, van den Heuvel OA, Stein DJ, van Wingen GA (2019b) Structural neuroimaging biomarkers for obsessive-compulsive disorder in the ENIGMA-OCD consortium: medication matters. medRxiv. https://doi.org/10.1101/19012567

Button KS, Ioannidis JP, Mokrysz C, Nosek BA, Flint J, Robinson ES, Munafò MR (2013) Empirical evidence for low reproducibility indicates low pre-study odds. Nat Rev Neurosci 14(12):877. https://doi.org/10.1038/nrn3475-c6. Epub 2013 Oct 23. PubMed PMID: 24149186

Cannistraro PA, Makris N, Howard JD, Wedig MM, Hodge SM, Wilhelm S, Kennedy DN, Rauch SL (2007) A diffusion tensor imaging study of white matter in obsessive-compulsive disorder. Depress Anxiety 24(6):440–446. PubMed PMID: 17096398

Cheng B, Huang X, Li S, Hu X, Luo Y, Wang X, Yang X, Qiu C, Yang Y, Zhang W, Bi F, Roberts N, Gong Q (2015) Gray matter alterations in post-traumatic stress disorder, obsessive-compulsive disorder, and social anxiety disorder. Front Behav Neurosci 9:219. https://doi.org/10.3389/fnbeh.2015.00219. eCollection 2015. PubMed PMID: 26347628; PubMed Central PMCID: PMC4542504

Chung S, Wang X, Lui YW (2017) Influence of T1-weighted signal intensity on FSL voxel-based morphometry and FreeSurfer cortical thickness. AJNR Am J Neuroradiol 38(4):726–728. https://doi.org/10.3174/ajnr.A5053. Epub 2016 Dec 29. PubMed PMID: 28034997; PubMed Central PMCID: PMC5389905

de Wit SJ, Alonso P, Schweren L, Mataix-Cols D, Lochner C, Menchón JM, Stein DJ, Fouche JP, Soriano-Mas C, Sato JR, Hoexter MQ, Denys D, Nakamae T, Nishida S, Kwon JS, Jang JH, Busatto GF, Cardoner N, Cath DC, Fukui K, Jung WH, Kim SN, Miguel EC, Narumoto J, Phillips ML, Pujol J, Remijnse PL, Sakai Y, Shin NY, Yamada K, Veltman DJ, van den Heuvel OA (2014) Multicenter voxel-based morphometry mega-analysis of structural brain scans in obsessive-compulsive disorder. Am J Psychiatry 171(3):340–349. https://doi.org/10.1176/appi.ajp.2013.13040574. PMID: 24220667

Desikan RS, Ségonne F, Fischl B, Quinn BT, Dickerson BC, Blacker D, Buckner RL, Dale AM, Maguire RP, Hyman BT, Albert MS, Killiany RJ (2006) An automated labeling system for subdividing the human cerebral cortex on MRI scans into gyral based regions of interest. NeuroImage 31(3):968–980. Epub 2006 Mar 10. PubMed PMID: 16530430

Eng GK, Sim K, Chen SH (2015) Meta-analytic investigations of structural grey matter, executive domain-related functional activations, and white matter diffusivity in obsessive compulsive disorder: an integrative review. Neurosci Biobehav Rev 52:233–257. https://doi.org/10.1016/j.neubiorev.2015.03.002. Epub 2015 Mar 10. PMID: 25766413

Fallucca E, MacMaster FP, Haddad J, Easter P, Dick R, May G, Stanley JA, Rix C, Rosenberg DR (2011) Distinguishing between major depressive disorder and obsessive-compulsive disorder in children by measuring regional cortical thickness. Arch Gen Psychiatry 68(5):527–533. https://doi.org/10.1001/archgenpsychiatry.2011.36. PubMed PMID: 21536980; PubMed Central PMCID: PMC4777351

Fouche JP, du Plessis S, Hattingh C, Roos A, Lochner C, Soriano-Mas C, Sato JR, Nakamae T, Nishida S, Kwon JS, Jung WH, Mataix-Cols D, Hoexter MQ, Alonso P, OCD Brain Imaging Consortium, de Wit SJ, Veltman DJ, Stein DJ, van den Heuvel OA (2017) Cortical thickness in obsessive-compulsive disorder: multisite mega-analysis of 780 brain scans from six centres. Br J Psychiatry 210(1):67–74. https://doi.org/10.1192/bjp.bp.115.164020. Epub 2016 May 19. PubMed PMID: 27198485

Gilbert AR, Moore GJ, Keshavan MS, Paulson LA, Narula V, Mac Master FP, Stewart CM, Rosenberg DR (2000) Decrease in thalamic volumes of pediatric patients with obsessive-compulsive disorder who are taking paroxetine. Arch Gen Psychiatry 57(5):449–456. PubMed PMID: 10807485

Gillan CM, Fineberg NA, Robbins TW (2017) A trans-diagnostic perspective on obsessive-compulsive disorder. Psychol Med 47(9):1528–1548. https://doi.org/10.1017/S0033291716002786. Epub 2017 Mar 27. Review. PubMed PMID: 28343453; PubMed Central PMCID: PMC5964477

Gong Q, Scarpazza C, Dai J, He M, Xu X, Shi Y, Zhou B, Vieira S, McCrory E, Ai Y, Yang C, Zhang F, Lui S, Mechelli A (2019) A transdiagnostic neuroanatomical signature of psychiatric illness. Neuropsychopharmacology 44(5):869–875. https://doi.org/10.1038/s41386-018-0175-9. Epub 2018 Aug 8. PubMed PMID: 30127342; PubMed Central PMCID: PMC6461829

Good CD, Johnsrude IS, Ashburner J, Henson RN, Friston KJ, Frackowiak RS (2001) A voxel-based morphometric study of ageing in 465 normal adult human brains. NeuroImage 14(1 Pt 1):21–36. PubMed PMID: 11525331

Goodkind M, Eickhoff SB, Oathes DJ, Jiang Y, Chang A, Jones-Hagata LB, Ortega BN, Zaiko YV, Roach EL, Korgaonkar MS, Grieve SM, Galatzer-Levy I, Fox PT, Etkin A (2015) Identification of a common neurobiological substrate for mental illness. JAMA Psychiat 72(4):305–315. https://doi.org/10.1001/jamapsychiatry.2014.2206. PubMed PMID: 25651064; PubMed Central PMCID: PMC4791058

Grasby KL, Jahanshad N, Painter JN, Colodro-Conde L, Bralten J, Hibar DP, Lind PA, Pizzagalli F, Ching CRK, McMahon MAB, Shatokhina N, Zsembik LCP, Thomopoulos SI, Zhu AH, Strike LT, Agartz I, Alhusaini S, Almeida MAA, Alnæs D, Amlien IK, Andersson M, Ard T, Armstrong NJ, Ashley-Koch A, Atkins JR, Bernard M, Brouwer RM, Buimer EEL, Bülow R, Bürger C, Cannon DM, Chakravarty M, Chen Q, Cheung JW, Couvy-Duchesne B, Dale AM, Dalvie S, de Araujo TK, de Zubicaray GI, de Zwarte SMC, den Braber A, Doan NT, Dohm K, Ehrlich S, Engelbrecht HR, Erk S, Fan CC, Fedko IO, Foley SF, Ford JM, Fukunaga M, Garrett ME, Ge T, Giddaluru S, Goldman AL, Green MJ, Groenewold NA, Grotegerd D, Gurholt TP, Gutman BA, Hansell NK, Harris MA, Harrison MB, Haswell CC, Hauser M, Herms S, Heslenfeld DJ, Ho NF, Hoehn D, Hoffmann P, Holleran L, Hoogman M, Hottenga JJ, Ikeda M, Janowitz D, Jansen IE, Jia T, Jockwitz C, Kanai R, Karama S, Kasperaviciute D, Kaufmann T, Kelly S, Kikuchi M, Klein M, Knapp M, Knodt AR, Krämer B, Lam M, Lancaster TM, Lee PH, Lett TA, Lewis LB, Lopes-Cendes I, Luciano M, Macciardi F, Marquand AF, Mathias SR, Melzer TR, Milaneschi Y, Mirza-Schreiber N, Moreira JCV, Mühleisen TW, Müller-Myhsok B, Najt P, Nakahara S, Nho K, Olde Loohuis LM, Orfanos DP, Pearson JF, Pitcher TL, Pütz B, Quidé Y, Ragothaman A, Rashid FM, Reay WR, Redlich R, Reinbold CS, Repple J, Richard G, Riedel BC, Risacher SL, Rocha CS, Mota NR, Salminen L, Saremi A, Saykin AJ, Schlag F, Schmaal L, Schofield PR, Secolin R, Shapland CY, Shen L, Shin J, Shumskaya E, Sønderby IE, Sprooten E, Tansey KE, Teumer A, Thalamuthu A, Tordesillas-Gutiérrez D, Turner JA, Uhlmann A, Vallerga CL, van der Meer D, van Donkelaar MMJ, van Eijk L, van Erp TGM, van Haren NEM, van Rooij D, van Tol MJ, Veldink JH, Verhoef E, Walton E, Wang M, Wang Y, Wardlaw JM, Wen W, Westlye LT, Whelan CD, Witt SH, Wittfeld K, Wolf C, Wolfers T, Wu JQ, Yasuda CL, Zaremba D, Zhang Z, Zwiers MP, Artiges E, Assareh AA, Ayesa-Arriola R, Belger A, Brandt CL, Brown GG, Cichon S, Curran JE, Davies GE, Degenhardt F, Dennis MF, Dietsche B, Djurovic S, Doherty CP, Espiritu R, Garijo D, Gil Y, Gowland PA, Green RC, Häusler AN, Heindel W, Ho BC, Hoffmann WU, Holsboer F, Homuth G, Hosten N, Jack CR Jr, Jang M, Jansen A, Kimbrel NA, Kolskår K, Koops S, Krug A, Lim KO, Luykx JJ, Mathalon DH, Mather KA, Mattay VS, Matthews S, Mayoral Van Son J, McEwen SC, Melle I, Morris DW, Mueller BA, Nauck M, Nordvik JE, Nöthen MM, O'Leary DS, Opel N, Martinot MP, Pike GB, Preda A, Quinlan EB, Rasser PE, Ratnakar V, Reppermund S, Steen VM, Tooney PA, Torres FR, Veltman DJ, Voyvodic JT, Whelan R, White T, Yamamori H, Adams HHH, Bis JC, Debette S, Decarli C, Fornage M, Gudnason V, Hofer E, Ikram MA, Launer L, Longstreth WT, Lopez OL, Mazoyer B, Mosley TH, Roshchupkin GV, Satizabal CL, Schmidt R, Seshadri S, Yang Q, Alzheimer's Disease Neuroimaging Initiative, CHARGE Consortium, EPIGEN Consortium, IMAGEN Consortium, SYS Consortium, Parkinson's Progression Markers Initiative, Alvim MKM, Ames D, Anderson TJ, Andreassen OA, Arias-Vasquez A, Bastin ME, Baune BT, Beckham JC, Blangero J, Boomsma DI, Brodaty H, Brunner HG, Buckner RL, Buitelaar JK, Bustillo JR, Cahn W, Cairns MJ, Calhoun V, Carr VJ, Caseras X, Caspers S, Cavalleri GL, Cendes F, Corvin A, Crespo-Facorro B, Dalrymple-Alford JC, Dannlowski U, de Geus EJC, Deary IJ, Delanty N, Depondt C, Desrivières S, Donohoe G, Espeseth T, Fernández G, Fisher SE, Flor H, Forstner AJ, Francks C, Franke B, Glahn DC, Gollub RL, Grabe HJ, Gruber O, Håberg AK, Hariri AR, Hartman CA, Hashimoto R, Heinz A, Henskens FA, Hillegers MHJ, Hoekstra PJ, Holmes AJ,

Hong LE, Hopkins WD, Hulshoff Pol HE, Jernigan TL, Jönsson EG, Kahn RS, Kennedy MA, Kircher TTJ, Kochunov P, Kwok JBJ, Le Hellard S, Loughland CM, Martin NG, Martinot JL, McDonald C, McMahon KL, Meyer-Lindenberg A, Michie PT, Morey RA, Mowry B, Nyberg L, Oosterlaan J, Ophoff RA, Pantelis C, Paus T, Pausova Z, Penninx BWJH, Polderman TJC, Posthuma D, Rietschel M, Roffman JL, Rowland LM, Sachdev PS, Sämann PG, Schall U, Schumann G, Scott RJ, Sim K, Sisodiya SM, Smoller JW, Sommer IE, St Pourcain B, Stein DJ, Toga AW, Trollor JN, Van der Wee NJA, van 't Ent D, Völzke H, Walter H, Weber B, Weinberger DR, Wright MJ, Zhou J, Stein JL, Thompson PM, Medland SE, Enhancing NeuroImaging Genetics through Meta-Analysis Consortium (ENIGMA)—Genetics working group (2020) The genetic architecture of the human cerebral cortex. Science 367(6484). pii: eaay6690. https://doi.org/10.1126/science.aay6690. PubMed PMID: 32193296

Grimm O, Pohlack S, Cacciaglia R, Winkelmann T, Plichta MM, Demirakca T, Flor H (2015) Amygdalar and hippocampal volume: a comparison between manual segmentation, Freesurfer and VBM. J Neurosci Methods 253:254–261. https://doi.org/10.1016/j.jneumeth.2015.05.024. Epub 2015 Jun 6. PubMed PMID: 26057114

Guadalupe T, Mathias SR, van Erp TGM, Whelan CD, Zwiers MP, Abe Y, Abramovic L, Agartz I, Andreassen OA, Arias-Vásquez A, Aribisala BS, Armstrong NJ, Arolt V, Artiges E, Ayesa-Arriola R, Baboyan VG, Banaschewski T, Barker G, Bastin ME, Baune BT, Blangero J, Bokde ALW, Boedhoe PSW, Bose A, Brem S, Brodaty H, Bromberg U, Brooks S, Büchel C, Buitelaar J, Calhoun VD, Cannon DM, Cattrell A, Cheng Y, Conrod PJ, Conzelmann A, Corvin A, Crespo-Facorro B, Crivello F, Dannlowski U, de Zubicaray GI, de Zwarte SMC, Deary IJ, Desrivières S, Doan NT, Donohoe G, Dørum ES, Ehrlich S, Espeseth T, Fernández G, Flor H, Fouche JP, Frouin V, Fukunaga M, Gallinat J, Garavan H, Gill M, Suarez AG, Gowland P, Grabe HJ, Grotegerd D, Gruber O, Hagenaars S, Hashimoto R, Hauser TU, Heinz A, Hibar DP, Hoekstra PJ, Hoogman M, Howells FM, Hu H, Hulshoff Pol HE, Huyser C, Ittermann B, Jahanshad N, Jönsson EG, Jurk S, Kahn RS, Kelly S, Kraemer B, Kugel H, Kwon JS, Lemaitre H, Lesch KP, Lochner C, Luciano M, Marquand AF, Martin NG, Martínez-Zalacaín I, Martinot JL, Mataix-Cols D, Mather K, McDonald C, McMahon KL, Medland SE, Menchón JM, Morris DW, Mothersill O, Maniega SM, Mwangi B, Nakamae T, Nakao T, Narayanaswaamy JC, Nees F, Nordvik JE, Onnink AMH, Opel N, Ophoff R, Paillère Martinot ML, Papadopoulos Orfanos D, Pauli P, Paus T, Poustka L, Reddy JY, Renteria ME, Roiz-Santiáñez R, Roos A, Royle NA, Sachdev P, Sánchez-Juan P, Schmaal L, Schumann G, Shumskaya E, Smolka MN, Soares JC, Soriano-Mas C, Stein DJ, Strike LT, Toro R, Turner JA, Tzourio-Mazoyer N, Uhlmann A, Hernández MV, van den Heuvel OA, van der Meer D, van Haren NEM, Veltman DJ, Venkatasubramanian G, Vetter NC, Vuletic D, Walitza S, Walter H, Walton E, Wang Z, Wardlaw J, Wen W, Westlye LT, Whelan R, Wittfeld K, Wolfers T, Wright MJ, Xu J, Xu X, Yun JY, Zhao J, Franke B, Thompson PM, Glahn DC, Mazoyer B, Fisher SE, Francks C (2017) Human subcortical brain asymmetries in 15,847 people worldwide reveal effects of age and sex. Brain Imaging Behav 11(5):1497–1514. https://doi.org/10.1007/s11682-016-9629-z. PubMed PMID: 27738994; PubMed Central PMCID: PMC5540813

Harries MD, Chamberlain SR, Redden SA, Odlaug BL, Blum AW, Grant JE (2017) A structural MRI study of excoriation (skin-picking) disorder and its relationship to clinical severity. Psychiatry Res Neuroimaging 269:26–30. https://doi.org/10.1016/j.pscychresns.2017.09.006. Epub 2017 Sep 8. PubMed PMID: 28918268; PubMed Central PMCID: PMC5604737

Harrison BJ, Soriano-Mas C, Pujol J, Ortiz H, López-Solà M, Hernández-Ribas R, Deus J, Alonso P, Yücel M, Pantelis C, Menchon JM, Cardoner N (2009) Altered corticostriatal functional connectivity in obsessive-compulsive disorder. Arch Gen Psychiatry 66 (11):1189–1200. https://doi.org/10.1001/archgenpsychiatry.2009.152. PubMed PMID: 19884607

Hazari N, Narayanaswamy JC, Venkatasubramanian G (2019) Neuroimaging findings in obsessive-compulsive disorder: a narrative review to elucidate neurobiological underpinnings. Indian J Psychiatry 61(Suppl 1):S9–S29. https://doi.org/10.4103/psychiatry.IndianJPsychiatry_525_18. Review. PubMed PMID: 30745673; PubMed Central PMCID: PMC6343409

Hibar DP, Stein JL, Renteria ME, Arias-Vasquez A, Desrivières S, Jahanshad N, Toro R, Wittfeld K, Abramovic L, Andersson M, Aribisala BS, Armstrong NJ, Bernard M, Bohlken MM, Boks MP, Bralten J, Brown AA, Chakravarty MM, Chen Q, Ching CR, Cuellar-Partida G, den Braber A, Giddaluru S, Goldman AL, Grimm O, Guadalupe T, Hass J, Woldehawariat G, Holmes AJ, Hoogman M, Janowitz D, Jia T, Kim S, Klein M, Kraemer B, Lee PH, Olde Loohuis LM, Luciano M, Macare C, Mather KA, Mattheisen M, Milaneschi Y, Nho K, Papmeyer M, Ramasamy A, Risacher SL, Roiz-Santiañez R, Rose EJ, Salami A, Sämann PG, Schmaal L, Schork AJ, Shin J, Strike LT, Teumer A, van Donkelaar MM, van Eijk KR, Walters RK, Westlye LT, Whelan CD, Winkler AM, Zwiers MP, Alhusaini S, Athanasiu L, Ehrlich S, Hakobjan MM, Hartberg CB, Haukvik UK, Heister AJ, Hoehn D, Kasperaviciute D, Liewald DC, Lopez LM, Makkinje RR, Matarin M, Naber MA, McKay DR, Needham M, Nugent AC, Pütz B, Royle NA, Shen L, Sprooten E, Trabzuni D, van der Marel SS, van Hulzen KJ, Walton E, Wolf C, Almasy L, Ames D, Arepalli S, Assareh AA, Bastin ME, Brodaty H, Bulayeva KB, Carless MA, Cichon S, Corvin A, Curran JE, Czisch M, de Zubicaray GI, Dillman A, Duggirala R, Dyer TD, Erk S, Fedko IO, Ferrucci L, Foroud TM, Fox PT, Fukunaga M, Gibbs JR, Göring HH, Green RC, Guelfi S, Hansell NK, Hartman CA, Hegenscheid K, Heinz A, Hernandez DG, Heslenfeld DJ, Hoekstra PJ, Holsboer F, Homuth G, Hottenga JJ, Ikeda M, Jack CR Jr, Jenkinson M, Johnson R, Kanai R, Keil M, Kent JW Jr, Kochunov P, Kwok JB, Lawrie SM, Liu X, Longo DL, McMahon KL, Meisenzahl E, Melle I, Mohnke S, Montgomery GW, Mostert JC, Mühleisen TW, Nalls MA, Nichols TE, Nilsson LG, Nöthen MM, Ohi K, Olvera RL, Perez-Iglesias R, Pike GB, Potkin SG, Reinvang I, Reppermund S, Rietschel M, Romanczuk-Seiferth N, Rosen GD, Rujescu D, Schnell K, Schofield PR, Smith C, Steen VM, Sussmann JE, Thalamuthu A, Toga AW, Traynor BJ, Troncoso J, Turner JA, Valdés Hernández MC, van 't Ent D, van der Brug M, van der Wee NJ, van Tol MJ, Veltman DJ, Wassink TH, Westman E, Zielke RH, Zonderman AB, Ashbrook DG, Hager R, Lu L, McMahon FJ, Morris DW, Williams RW, Brunner HG, Buckner RL, Buitelaar JK, Cahn W, Calhoun VD, Cavalleri GL, Crespo-Facorro B, Dale AM, Davies GE, Delanty N, Depondt C, Djurovic S, Drevets WC, Espeseth T, Gollub RL, Ho BC, Hoffmann W, Hosten N, Kahn RS, Le Hellard S, Meyer-Lindenberg A, Müller-Myhsok B, Nauck M, Nyberg L, Pandolfo M, Penninx BW, Roffman JL, Sisodiya SM, Smoller JW, van Bokhoven H, van Haren NE, Völzke H, Walter H, Weiner MW, Wen W, White T, Agartz I, Andreassen OA, Blangero J, Boomsma DI, Brouwer RM, Cannon DM, Cookson MR, de Geus EJ, Deary IJ, Donohoe G, Fernández G, Fisher SE, Francks C, Glahn DC, Grabe HJ, Gruber O, Hardy J, Hashimoto R, Hulshoff Pol HE, Jönsson EG, Kloszewska I, Lovestone S, Mattay VS, Mecocci P, McDonald C, McIntosh AM, Ophoff RA, Paus T, Pausova Z, Ryten M, Sachdev PS, Saykin AJ, Simmons A, Singleton A, Soininen H, Wardlaw JM, Weale ME, Weinberger DR, Adams HH, Launer LJ, Seiler S, Schmidt R, Chauhan G, Satizabal CL, Becker JT, Yanek L, van der Lee SJ, Ebling M, Fischl B, Longstreth WT Jr, Greve D, Schmidt H, Nyquist P, Vinke LN, van Duijn CM, Xue L, Mazoyer B, Bis JC, Gudnason V, Seshadri S, Ikram MA, Alzheimer's Disease Neuroimaging Initiative, CHARGE Consortium, EPIGEN, IMAGEN, SYS, Martin NG, Wright MJ, Schumann G, Franke B, Thompson PM, Medland SE (2015) Common genetic variants influence human subcortical brain structures. Nature 520(7546):224–229. https://doi.org/10.1038/nature14101. Epub 2015 Jan 21. PubMed PMID: 25607358; PubMed Central PMCID: PMC4393366

Hibar DP, Westlye LT, van Erp TG, Rasmussen J, Leonardo CD, Faskowitz J, Haukvik UK, Hartberg CB, Doan NT, Agartz I, Dale AM, Gruber O, Krämer B, Trost S, Liberg B, Abé C, Ekman CJ, Ingvar M, Landén M, Fears SC, Freimer NB, Bearden CE, Costa Rica/Colombia Consortium for Genetic Investigation of Bipolar Endophenotypes, Sprooten E, Glahn DC, Pearlson GD, Emsell L, Kenney J, Scanlon C, McDonald C, Cannon DM, Almeida J, Versace A, Caseras X, Lawrence NS, Phillips ML, Dima D, Delvecchio G, Frangou S, Satterthwaite TD, Wolf D, Houenou J, Henry C, Malt UF, Bøen E, Elvsåshagen T, Young AH, Lloyd AJ, Goodwin GM, Mackay CE, Bourne C, Bilderbeck A, Abramovic L, Boks MP, van Haren NE, Ophoff RA, Kahn RS, Bauer M, Pfennig A, Alda M, Hajek T, Mwangi B,

Soares JC, Nickson T, Dimitrova R, Sussmann JE, Hagenaars S, Whalley HC, McIntosh AM, Thompson PM, Andreassen OA (2016) Subcortical volumetric abnormalities in bipolar disorder. Mol Psychiatry 21(12):1710–1716. https://doi.org/10.1038/mp.2015.227. Epub 2016 Feb 9. PubMed PMID: 26857596; PubMed Central PMCID: PMC5116479

Hibar DP, Westlye LT, Doan NT, Jahanshad N, Cheung JW, Ching CRK, Versace A, Bilderbeck AC, Uhlmann A, Mwangi B, Krämer B, Overs B, Hartberg CB, Abé C, Dima D, Grotegerd D, Sprooten E, Bøen E, Jimenez E, Howells FM, Delvecchio G, Temmingh H, Starke J, Almeida JRC, Goikolea JM, Houenou J, Beard LM, Rauer L, Abramovic L, Bonnin M, Ponteduro MF, Keil M, Rive MM, Yao N, Yalin N, Najt P, Rosa PG, Redlich R, Trost S, Hagenaars S, Fears SC, Alonso-Lana S, van Erp TGM, Nickson T, Chaim-Avancini TM, Meier TB, Elvsåshagen T, Haukvik UK, Lee WH, Schene AH, Lloyd AJ, Young AH, Nugent A, Dale AM, Pfennig A, McIntosh AM, Lafer B, Baune BT, Ekman CJ, Zarate CA, Bearden CE, Henry C, Simhandl C, McDonald C, Bourne C, Stein DJ, Wolf DH, Cannon DM, Glahn DC, Veltman DJ, Pomarol-Clotet E, Vieta E, Canales-Rodriguez EJ, Nery FG, Duran FLS, Busatto GF, Roberts G, Pearlson GD, Goodwin GM, Kugel H, Whalley HC, Ruhe HG, Soares JC, Fullerton JM, Rybakowski JK, Savitz J, Chaim KT, Fatjó-Vilas M, Soeiro-de-Souza MG, Boks MP, Zanetti MV, Otaduy MCG, Schaufelberger MS, Alda M, Ingvar M, Phillips ML, Kempton MJ, Bauer M, Landén M, Lawrence NS, van Haren NEM, Horn NR, Freimer NB, Gruber O, Schofield PR, Mitchell PB, Kahn RS, Lenroot R, Machado-Vieira R, Ophoff RA, Sarró S, Frangou S, Satterthwaite TD, Hajek T, Dannlowski U, Malt UF, Arolt V, Gattaz WF, Drevets WC, Caseras X, Agartz I, Thompson PM, Andreassen OA (2018) Cortical abnormalities in bipolar disorder: an MRI analysis of 6503 individuals from the ENIGMA Bipolar Disorder Working Group. Mol Psychiatry 23(4):932–942. https://doi.org/10.1038/mp.2017.73. Epub 2017 May 2. PubMed PMID: 28461699; PubMed Central PMCID: PMC5668195

Hirose M, Hirano Y, Nemoto K, Sutoh C, Asano K, Miyata H, Matsumoto J, Nakazato M, Matsumoto K, Masuda Y, Iyo M, Shimizu E, Nakagawa A (2017) Relationship between symptom dimensions and brain morphology in obsessive-compulsive disorder. Brain Imaging Behav 11(5):1326–1333. https://doi.org/10.1007/s11682-016-9611-9. PubMed PMID: 27730476

Hoexter MQ, Miguel EC, Diniz JB, Shavitt RG, Busatto GF, Sato JR (2013) Predicting obsessive-compulsive disorder severity combining neuroimaging and machine learning methods. J Affect Disord 150(3):1213–1216. https://doi.org/10.1016/j.jad.2013.05.041. Epub 2013 Jun 14. PubMed PMID: 23769292

Hoexter MQ, Diniz JB, Lopes AC, Batistuzzo MC, Shavitt RG, Dougherty DD, Duran FL, Bressan RA, Busatto GF, Miguel EC, Sato JR (2015) Orbitofrontal thickness as a measure for treatment response prediction in obsessive-compulsive disorder. Depress Anxiety 32(12):900–908. https://doi.org/10.1002/da.22380. Epub 2015 May 29. PubMed PMID: 26032588

Hu X, Liu Q, Li B, Tang W, Sun H, Li F, Yang Y, Gong Q, Huang X (2016) Multivariate pattern analysis of obsessive-compulsive disorder using structural neuroanatomy. Eur Neuropsychopharmacol 26(2):246–254. https://doi.org/10.1016/j.euroneuro.2015.12.014. Epub 2015 Dec 11. PubMed PMID: 26708318

Hu X, Du M, Chen L, Li L, Zhou M, Zhang L, Liu Q, Lu L, Mreedha K, Huang X, Gong Q (2017) Meta-analytic investigations of common and distinct grey matter alterations in youths and adults with obsessive-compulsive disorder. Neurosci Biobehav Rev 78:91–103. https://doi.org/10.1016/j.neubiorev.2017.04.012. Epub 2017 Apr 23. Review. PubMed PMID: 28442404

Isobe M, Redden SA, Keuthen NJ, Stein DJ, Lochner C, Grant JE, Chamberlain SR (2018) Striatal abnormalities in trichotillomania: a multi-site MRI analysis. Neuroimage Clin 17:893–898. https://doi.org/10.1016/j.nicl.2017.12.031. Epub 2017 Dec 22. PubMed PMID: 29515968; PubMed Central PMCID: PMC5836997

Kim JJ, Lee MC, Kim J, Kim IY, Kim SI, Han MH, Chang KH, Kwon JS (2001) Grey matter abnormalities in obsessive-compulsive disorder: statistical parametric mapping of segmented magnetic resonance images. Br J Psychiatry 179:330–334. PubMed PMID: 11581113

Kochunov P et al (2020) ENIGMA-DTI: translating reproducible white matter deficits into personalized vulnerability metrics in cross-diagnostic psychiatric research. Hum Brain Mapp. https://doi.org/10.1002/hbm.24998

Kong XZ, Mathias SR, Guadalupe T, ENIGMA Laterality Working Group, Glahn DC, Franke B, Crivello F, Tzourio-Mazoyer N, Fisher SE, Thompson PM, Francks C (2018) Mapping cortical brain asymmetry in 17,141 healthy individuals worldwide via the ENIGMA Consortium. Proc Natl Acad Sci U S A 115(22):E5154–E5163. https://doi.org/10.1073/pnas.1718418115. Epub 2018 May 15. PubMed PMID: 29764998; PubMed Central PMCID: PMC5984496

Kubota Y, Sato W, Kochiyama T, Uono S, Yoshimura S, Sawada R, Sakihama M, Toichi M (2016) Putamen volume correlates with obsessive compulsive characteristics in healthy population. Psychiatry Res Neuroimaging 249:97–104. https://doi.org/10.1016/j.pscychresns.2016.01.014. Epub 2016 Jan 25. PubMed PMID: 26849956

Kwon JS, Shin YW, Kim CW, Kim YI, Youn T, Han MH, Chang KH, Kim JJ (2003) Similarity and disparity of obsessive-compulsive disorder and schizophrenia in MR volumetric abnormalities of the hippocampus-amygdala complex. J Neurol Neurosurg Psychiatry 74(7):962–964. PubMed PMID: 12810792; PubMed Central PMCID: PMC1738533

Lázaro L, Ortiz AG, Calvo A, Ortiz AE, Moreno E, Morer A, Calvo R, Bargallo N (2014) White matter structural alterations in pediatric obsessive-compulsive disorder: relation to symptom dimensions. Prog Neuro-Psychopharmacol Biol Psychiatry 54:249–258. https://doi.org/10.1016/j.pnpbp.2014.06.009. Epub 2014 Jun 28. PubMed PMID: 24977330

Logue MW, van Rooij SJH, Dennis EL, Davis SL, Hayes JP, Stevens JS, Densmore M, Haswell CC, Ipser J, Koch SBJ, Korgaonkar M, Lebois LAM, Peverill M, Baker JT, Boedhoe PSW, Frijling JL, Gruber SA, Harpaz-Rotem I, Jahanshad N, Koopowitz S, Levy I, Nawijn L, O'Connor L, Olff M, Salat DH, Sheridan MA, Spielberg JM, van Zuiden M, Winternitz SR, Wolff JD, Wolf EJ, Wang X, Wrocklage K, Abdallah CG, Bryant RA, Geuze E, Jovanovic T, Kaufman ML, King AP, Krystal JH, Lagopoulos J, Bennett M, Lanius R, Liberzon I, McGlinchey RE, McLaughlin KA, Milberg WP, Miller MW, Ressler KJ, Veltman DJ, Stein DJ, Thomaes K, Thompson PM, Morey RA (2018) Smaller hippocampal volume in posttraumatic stress disorder: a multisite ENIGMA-PGC study: subcortical volumetry results from posttraumatic stress disorder consortia. Biol Psychiatry 83(3):244–253. https://doi.org/10.1016/j.biopsych.2017.09.006. Epub 2017 Sep 20. PubMed PMID: 29217296; PubMed Central PMCID: PMC5951719

Madsen SK, Zai A, Pirnia T, Arienzo D, Zhan L, Moody TD, Thompson PM, Feusner JD (2015) Cortical thickness and brain volumetric analysis in body dysmorphic disorder. Psychiatry Res 232(1):115–122. https://doi.org/10.1016/j.pscychresns.2015.02.003. Epub 2015 Feb 26. PubMed PMID: 25797401; PubMed Central PMCID: PMC4404218

Mataix-Cols D, Wooderson S, Lawrence N, Brammer MJ, Speckens A, Phillips ML (2004) Distinct neural correlates of washing, checking, and hoarding symptom dimensions in obsessive-compulsive disorder. Arch Gen Psychiatry 61(6):564–576. PubMed PMID: 15184236

Menzies L, Williams GB, Chamberlain SR, Ooi C, Fineberg N, Suckling J, Sahakian BJ, Robbins TW, Bullmore ET (2008) White matter abnormalities in patients with obsessive-compulsive disorder and their first-degree relatives. Am J Psychiatry 165(10):1308–1315. https://doi.org/10.1176/appi.ajp.2008.07101677. Epub 2008 Jun 2. PubMed PMID: 18519525

Norman LJ, Carlisi C, Lukito S, Hart H, Mataix-Cols D, Radua J, Rubia K (2016) Structural and functional brain abnormalities in attention-deficit/hyperactivity disorder and obsessive-compulsive disorder: a comparative meta-analysis. JAMA Psychiatry 73(8):815–825. https://doi.org/10.1001/jamapsychiatry.2016.0700. Erratum in: JAMA Psychiatry. 2017 Oct 1;74(10):1079. JAMA Psychiatry. 2018 Mar 1;75(3):303. PubMed PMID: 27276220

Okada K, Nakao T, Sanematsu H, Murayama K, Honda S, Tomita M, Togao O, Yoshiura T, Kanba S (2015) Biological heterogeneity of obsessive-compulsive disorder: a voxel-based morphometric study based on dimensional assessment. Psychiatry Clin Neurosci 69(7):411–421. https://doi.org/10.1111/pcn.12269. Epub 2015 Feb 9. PubMed PMID: 25556718

Park SE, Jeong GW (2015) Cerebral white matter volume changes in patients with obsessive-compulsive disorder: voxel-based morphometry. Psychiatry Clin Neurosci 69(11):717–723. https://doi.org/10.1111/pcn.12317. Epub 2015 Jun 26. PubMed PMID: 25966931

Picó-Pérez M, Moreira PS, de Melo Ferreira V, Radua J, Mataix-Cols D, Sousa N, Soriano-Mas C, Morgado P (2020) Modality-specific overlaps in brain structure and function in obsessive-compulsive disorder: multimodal meta-analysis of case-control MRI studies. Neurosci Biobehav Rev 112:83–94. https://doi.org/10.1016/j.neubiorev.2020.01.033. [Epub ahead of print] Review. PubMed PMID: 32006553

Piras F, Piras F, Abe Y, Agarwal SM, Anticevic A, Ameis S, Arnold P, Bargalló N, Batistuzzo MC, Benedetti F, Beucke J, Boedhoe PS, Bollettini I, Brem S, Calvo A, Kevin Cho KI, Dallaspezia S, Dickie E, Ely BA, Fan S, Fouche J, Gruner P, Gürsel DA, Hauser T, Hirano Y, Hoexter MQ, Iorio M, James A, Reddy J, Kaufmann C, Koch K, Kochunov P, Kwon JS, Lazaro L, Lochner C, Marsh R, Nakagawa A, Nakamae T, Narayanaswamy JC, Sakai Y, Shimizu E, Simon D, Simpson HB, Soreni N, Stämpfli P, Stern ER, Szeszko P, Takahashi J, Venkatasubramanian G, Wang Z, Yun J, Stein DJ, Jahanshad N, Thompson PM, van den Heuvel OA, Spalletta G (2019) White matter microstructure and its relation to clinical features of obsessive-compulsive disorder: findings from the ENIGMA OCD Working Group. bioRxiv:855916. https://doi.org/10.1101/855916

Pitman RK, Rasmusson AM, Koenen KC, Shin LM, Orr SP, Gilbertson MW, Milad MR, Liberzon I (2012) Biological studies of post-traumatic stress disorder. Nat Rev Neurosci 13(11):769–787. https://doi.org/10.1038/nrn3339. Epub 2012 Oct 10. Review. PubMed PMID: 23047775; PubMed Central PMCID: PMC4951157

Pujol J, Soriano-Mas C, Alonso P, Cardoner N, Menchón JM, Deus J, Vallejo J (2004) Mapping structural brain alterations in obsessive-compulsive disorder. Arch Gen Psychiatry 61 (7):720–730. PubMed PMID: 15237084

Pujol J, Soriano-Mas C, Gispert JD, Bossa M, Reig S, Ortiz H, Alonso P, Cardoner N, López-Solà M, Harrison BJ, Deus J, Menchón JM, Desco M, Olmos S (2011) Variations in the shape of the frontobasal brain region in obsessive-compulsive disorder. Hum Brain Mapp 32 (7):1100–1108. https://doi.org/10.1002/hbm.21094. Epub 2010 Jul 6. PubMed PMID: 20607751

Quidé Y, Witteveen AB, El-Hage W, Veltman DJ, Olff M (2012) Differences between effects of psychological versus pharmacological treatments on functional and morphological brain alterations in anxiety disorders and major depressive disorder: a systematic review. Neurosci Biobehav Rev 36(1):626–644. https://doi.org/10.1016/j.neubiorev.2011.09.004. Epub 2011 Sep 24. Review. PubMed PMID: 21963442

Radua J, Mataix-Cols D (2009) Voxel-wise meta-analysis of grey matter changes in obsessive-compulsive disorder. Br J Psychiatry 195(5):393–402. https://doi.org/10.1192/bjp.bp.108.055046. Review. PubMed PMID: 19880927

Radua J, van den Heuvel OA, Surguladze S, Mataix-Cols D (2010) Meta-analytical comparison of voxel-based morphometry studies in obsessive-compulsive disorder vs other anxiety disorders. Arch Gen Psychiatry 67(7):701–711. https://doi.org/10.1001/archgenpsychiatry.2010.70. PubMed PMID: 20603451

Radua J, Grau M, van den Heuvel OA, Thiebaut de Schotten M, Stein DJ, Canales-Rodríguez EJ, Catani M, Mataix-Cols D (2014) Multimodal voxel-based meta-analysis of white matter abnormalities in obsessive-compulsive disorder. Neuropsychopharmacology 39 (7):1547–1557. https://doi.org/10.1038/npp.2014.5. Epub 2014 Jan 10. PubMed PMID: 24407265; PubMed Central PMCID: PMC4023155

Robinson D, Wu H, Munne RA, Ashtari M, Alvir JM, Lerner G, Koreen A, Cole K, Bogerts B (1995) Reduced caudate nucleus volume in obsessive-compulsive disorder. Arch Gen Psychiatry 52(5):393–398. PubMed PMID: 7726720

Rosenberg DR, Keshavan MS, O'Hearn KM, Dick EL, Bagwell WW, Seymour AB, Montrose DM, Pierri JN, Birmaher B (1997) Frontostriatal measurement in treatment-naive children with obsessive-compulsive disorder. Arch Gen Psychiatry 54(9):824–830

Rosenberg DR, Benazon NR, Gilbert A, Sullivan A, Moore GJ (2000) Thalamic volume in pediatric obsessive-compulsive disorder patients before and after cognitive behavioral therapy. Biol Psychiatry 48(4):294–300. Erratum in: Biol Psychiatry 2001 Aug 15;50(4):312. PubMed PMID: 10960160

Rotge JY, Langbour N, Guehl D, Bioulac B, Jaafari N, Allard M, Aouizerate B, Burbaud P (2010) Gray matter alterations in obsessive-compulsive disorder: an anatomic likelihood estimation meta-analysis. Neuropsychopharmacology 35(3):686–691. https://doi.org/10.1038/npp.2009.175. Epub 2009 Nov 4. PubMed PMID: 19890260; PubMed Central PMCID: PMC3055616

Rus OG, Reess TJ, Wagner G, Zaudig M, Zimmer C, Koch K (2017) Structural alterations in patients with obsessive-compulsive disorder: a surface-based analysis of cortical volume, surface area and thickness. J Psychiatry Neurosci 42(6):395–403. PubMed PMID: 28832321; PubMed Central PMCID: PMC5662461

Satizabal CL, Adams HHH, Hibar DP, White CC, Knol MJ, Stein JL, Scholz M, Sargurupremraj M, Jahanshad N, Roshchupkin GV, Smith AV, Bis JC, Jian X, Luciano M, Hofer E, Teumer A, van der Lee SJ, Yang J, Yanek LR, Lee TV, Li S, Hu Y, Koh JY, Eicher JD, Desrivières S, Arias-Vasquez A, Chauhan G, Athanasiu L, Rentería ME, Kim S, Hoehn D, Armstrong NJ, Chen Q, Holmes AJ, den Braber A, Kloszewska I, Andersson M, Espeseth T, Grimm O, Abramovic L, Alhusaini S, Milaneschi Y, Papmeyer M, Axelsson T, Ehrlich S, Roiz-Santiañez R, Kraemer B, Håberg AK, Jones HJ, Pike GB, Stein DJ, Stevens A, Bralten J, Vernooij MW, Harris TB, Filippi I, Witte AV, Guadalupe T, Wittfeld K, Mosley TH, Becker JT, Doan NT, Hagenaars SP, Saba Y, Cuellar-Partida G, Amin N, Hilal S, Nho K, Mirza-Schreiber N, Arfanakis K, Becker DM, Ames D, Goldman AL, Lee PH, Boomsma DI, Lovestone S, Giddaluru S, Le Hellard S, Mattheisen M, Bohlken MM, Kasperaviciute D, Schmaal L, Lawrie SM, Agartz I, Walton E, Tordesillas-Gutierrez D, Davies GE, Shin J, Ipser JC, Vinke LN, Hoogman M, Jia T, Burkhardt R, Klein M, Crivello F, Janowitz D, Carmichael O, Haukvik UK, Aribisala BS, Schmidt H, Strike LT, Cheng CY, Risacher SL, Pütz B, Fleischman DA, Assareh AA, Mattay VS, Buckner RL, Mecocci P, Dale AM, Cichon S, Boks MP, Matarin M, Penninx BWJH, Calhoun VD, Chakravarty MM, Marquand AF, Macare C, Kharabian Masouleh S, Oosterlaan J, Amouyel P, Hegenscheid K, Rotter JI, Schork AJ, Liewald DCM, de Zubicaray GI, Wong TY, Shen L, Sämann PG, Brodaty H, Roffman JL, de Geus EJC, Tsolaki M, Erk S, van Eijk KR, Cavalleri GL, van der Wee NJA, McIntosh AM, Gollub RL, Bulayeva KB, Bernard M, Richards JS, Himali JJ, Loeffler M, Rommelse N, Hoffmann W, Westlye LT, Valdés Hernández MC, Hansell NK, van Erp TGM, Wolf C, Kwok JBJ, Vellas B, Heinz A, Olde Loohuis LM, Delanty N, Ho BC, Ching CRK, Shumskaya E, Singh B, Hofman A, van der Meer D, Homuth G, Psaty BM, Bastin ME, Montgomery GW, Foroud TM, Reppermund S, Hottenga JJ, Simmons A, Meyer-Lindenberg A, Cahn W, Whelan CD, van Donkelaar MMJ, Yang Q, Hosten N, Green RC, Thalamuthu A, Mohnke S, Hulshoff Pol HE, Lin H, Jack CR Jr, Schofield PR, Mühleisen TW, Maillard P, Potkin SG, Wen W, Fletcher E, Toga AW, Gruber O, Huentelman M, Davey Smith G, Launer LJ, Nyberg L, Jönsson EG, Crespo-Facorro B, Koen N, Greve DN, Uitterlinden AG, Weinberger DR, Steen VM, Fedko IO, Groenewold NA, Niessen WJ, Toro R, Tzourio C, Longstreth WT Jr, Ikram MK, Smoller JW, van Tol MJ, Sussmann JE, Paus T, Lemaître H, Schroeter ML, Mazoyer B, Andreassen OA, Holsboer F, Depondt C, Veltman DJ, Turner JA, Pausova Z, Schumann G, van Rooij D, Djurovic S, Deary IJ, McMahon KL, Müller-Myhsok B, Brouwer RM, Soininen H, Pandolfo M, Wassink TH, Cheung JW, Wolfers T, Martinot JL, Zwiers MP, Nauck M, Melle I, Martin NG, Kanai R, Westman E, Kahn RS, Sisodiya SM, White T, Saremi A, van Bokhoven H, Brunner HG, Völzke H, Wright MJ, van 't Ent D, Nöthen MM, Ophoff RA, Buitelaar JK, Fernández G, Sachdev PS, Rietschel M, van Haren NEM, Fisher SE, Beiser AS, Francks C, Saykin AJ, Mather KA, Romanczuk-Seiferth N, Hartman CA, DeStefano AL, Heslenfeld DJ, Weiner MW, Walter H, Hoekstra PJ, Nyquist PA, Franke B, Bennett DA, Grabe HJ, Johnson AD, Chen C, van Duijn CM, Lopez OL, Fornage M, Wardlaw JM, Schmidt R, DeCarli C, De Jager PL, Villringer A, Debette S, Gudnason V, Medland SE, Shulman JM, Thompson PM, Seshadri S, Ikram MA (2019) Genetic architecture of subcortical brain structures in 38,851 individuals. Nat

Genet 51(11):1624–1636. https://doi.org/10.1038/s41588-019-0511-y. Epub 2019 Oct 21. PubMed PMID: 31636452; PubMed Central PMCID: PMC7055269

Saxena S, Brody AL, Schwartz JM, Baxter LR (1998) Neuroimaging and frontal-subcortical circuitry in obsessive-compulsive disorder. Br J Psychiatry Suppl 35):26–37. Review. PubMed PMID: 9829024

Schmaal L, Veltman DJ, van Erp TG, Sämann PG, Frodl T, Jahanshad N, Loehrer E, Tiemeier H, Hofman A, Niessen WJ, Vernooij MW, Ikram MA, Wittfeld K, Grabe HJ, Block A, Hegenscheid K, Völzke H, Hoehn D, Czisch M, Lagopoulos J, Hatton SN, Hickie IB, Goya-Maldonado R, Krämer B, Gruber O, Couvy-Duchesne B, Rentería ME, Strike LT, Mills NT, de Zubicaray GI, McMahon KL, Medland SE, Martin NG, Gillespie NA, Wright MJ, Hall GB, MacQueen GM, Frey EM, Carballedo A, van Velzen LS, van Tol MJ, van der Wee NJ, Veer IM, Walter H, Schnell K, Schramm E, Normann C, Schoepf D, Konrad C, Zurowski B, Nickson T, McIntosh AM, Papmeyer M, Whalley HC, Sussmann JE, Godlewska BR, Cowen PJ, Fischer FH, Rose M, Penninx BW, Thompson PM, Hibar DP (2016) Subcortical brain alterations in major depressive disorder: findings from the ENIGMA Major Depressive Disorder working group. Mol Psychiatry 21(6):806–812. https://doi.org/10.1038/mp.2015.69. Epub 2015 Jun 30. PubMed PMID: 26122586; PubMed Central PMCID: PMC4879183

Schmaal L, Hibar DP, Sämann PG, Hall GB, Baune BT, Jahanshad N, Cheung JW, van Erp TGM, Bos D, Ikram MA, Vernooij MW, Niessen WJ, Tiemeier H, Hofman A, Wittfeld K, Grabe HJ, Janowitz D, Bülow R, Selonke M, Völzke H, Grotegerd D, Dannlowski U, Arolt V, Opel N, Heindel W, Kugel H, Hoehn D, Czisch M, Couvy-Duchesne B, Rentería ME, Strike LT, Wright MJ, Mills NT, de Zubicaray GI, McMahon KL, Medland SE, Martin NG, Gillespie NA, Goya-Maldonado R, Gruber O, Krämer B, Hatton SN, Lagopoulos J, Hickie IB, Frodl T, Carballedo A, Frey EM, van Velzen LS, Penninx BWJH, van Tol MJ, van der Wee NJ, Davey CG, Harrison BJ, Mwangi B, Cao B, Soares JC, Veer IM, Walter H, Schoepf D, Zurowski B, Konrad C, Schramm E, Normann C, Schnell K, Sacchet MD, Gotlib IH, MacQueen GM, Godlewska BR, Nickson T, McIntosh AM, Papmeyer M, Whalley HC, Hall J, Sussmann JE, Li M, Walter M, Aftanas L, Brack I, Bokhan NA, Thompson PM, Veltman DJ (2017) Cortical abnormalities in adults and adolescents with major depression based on brain scans from 20 cohorts worldwide in the ENIGMA Major Depressive Disorder Working Group. Mol Psychiatry 22(6):900–909. https://doi.org/10.1038/mp.2016.60. Epub 2016 May 3. PubMed PMID: 27137745; PubMed Central PMCID: PMC5444023

Shim G, Jung WH, Choi JS, Jung MH, Jang JH, Park JY, Choi CH, Kang DH, Kwon JS (2009) Reduced cortical folding of the anterior cingulate cortex in obsessive-compulsive disorder. J Psychiatry Neurosci 34(6):443–449. PubMed PMID: 19949720; PubMed Central PMCID: PMC2783435

Simpson HB, van den Heuvel OA, Miguel EC, Reddy YCJ, Stein DJ, Lewis-Fernández R, Shavitt RG, Lochner C, Pouwels PJW, Narayanawamy JC, Venkatasubramanian G, Hezel DM, Vriend C, Batistuzzo MC, Hoexter MQ, de Joode NT, Costa DL, de Mathis MA, Sheshachala K, Narayan M, van Balkom AJLM, Batelaan NM, Venkataram S, Cherian A, Marincowitz C, Pannekoek N, Stovezky YR, Mare K, Liu F, Otaduy MCG, Pastorello B, Rao R, Katechis M, Van Meter P, Wall M (2020) Toward identifying reproducible brain signatures of obsessive-compulsive profiles: rationale and methods for a new global initiative. BMC Psychiatry 20(1):68. https://doi.org/10.1186/s12888-020-2439-2. PubMed PMID: 32059696; PubMed Central PMCID: PMC7023814

Stein DJ, Coetzer R, Lee M, Davids B, Bouwer C (1997) Magnetic resonance brain imaging in women with obsessive-compulsive disorder and trichotillomania. Psychiatry Res 74 (3):177–182. PubMed PMID: 9255863

Stein JL, Medland SE, Vasquez AA, Hibar DP, Senstad RE, Winkler AM, Toro R, Appel K, Bartecek R, Bergmann Ø, Bernard M, Brown AA, Cannon DM, Chakravarty MM, Christoforou A, Domin M, Grimm O, Hollinshead M, Holmes AJ, Homuth G, Hottenga JJ, Langan C, Lopez LM, Hansell NK, Hwang KS, Kim S, Laje G, Lee PH, Liu X, Loth E, Lourdusamy A, Mattingsdal M, Mohnke S, Maniega SM, Nho K, Nugent AC, O'Brien C,

Papmeyer M, Pütz B, Ramasamy A, Rasmussen J, Rijpkema M, Risacher SL, Roddey JC, Rose EJ, Ryten M, Shen L, Sprooten E, Strengman E, Teumer A, Trabzuni D, Turner J, van Eijk K, van Erp TG, van Tol MJ, Wittfeld K, Wolf C, Woudstra S, Aleman A, Alhusaini S, Almasy L, Binder EB, Brohawn DG, Cantor RM, Carless MA, Corvin A, Czisch M, Curran JE, Davies G, de Almeida MA, Delanty N, Depondt C, Duggirala R, Dyer TD, Erk S, Fagerness J, Fox PT, Freimer NB, Gill M, Göring HH, Hagler DJ, Hoehn D, Holsboer F, Hoogman M, Hosten N, Jahanshad N, Johnson MP, Kasperaviciute D, Kent JW Jr, Kochunov P, Lancaster JL, Lawrie SM, Liewald DC, Mandl R, Matarin M, Mattheisen M, Meisenzahl E, Melle I, Moses EK, Mühleisen TW, Nauck M, Nöthen MM, Olvera RL, Pandolfo M, Pike GB, Puls R, Reinvang I, Rentería ME, Rietschel M, Roffman JL, Royle NA, Rujescu D, Savitz J, Schnack HG, Schnell K, Seiferth N, Smith C, Steen VM, Valdés Hernández MC, Van den Heuvel M, van der Wee NJ, Van Haren NE, Veltman JA, Völzke H, Walker R, Westlye LT, Whelan CD, Agartz I, Boomsma DI, Cavalleri GL, Dale AM, Djurovic S, Drevets WC, Hagoort P, Hall J, Heinz A, Jack CR Jr, Foroud TM, Le Hellard S, Macciardi F, Montgomery GW, Poline JB, Porteous DJ, Sisodiya SM, Starr JM, Sussmann J, Toga AW, Veltman DJ, Walter H, Weiner MW, Alzheimer's Disease Neuroimaging Initiative, EPIGEN Consortium, IMAGEN Consortium, Saguenay Youth Study Group, Bis JC, Ikram MA, Smith AV, Gudnason V, Tzourio C, Vernooij MW, Launer LJ, DeCarli C, Seshadri S, Cohorts for Heart and Aging Research in Genomic Epidemiology Consortium, Andreassen OA, Apostolova LG, Bastin ME, Blangero J, Brunner HG, Buckner RL, Cichon S, Coppola G, de Zubicaray GI, Deary IJ, Donohoe G, de Geus EJ, Espeseth T, Fernández G, Glahn DC, Grabe HJ, Hardy J, Hulshoff Pol HE, Jenkinson M, Kahn RS, McDonald C, McIntosh AM, McMahon FJ, McMahon KL, Meyer-Lindenberg A, Morris DW, Müller-Myhsok B, Nichols TE, Ophoff RA, Paus T, Pausova Z, Penninx BW, Potkin SG, Sämann PG, Saykin AJ, Schumann G, Smoller JW, Wardlaw JM, Weale ME, Martin NG, Franke B, Wright MJ, Thompson PM, Enhancing Neuro Imaging Genetics through Meta-Analysis Consortium (2012) Identification of common variants associated with human hippocampal and intracranial volumes. Nat Genet 44(5):552–561. https://doi.org/10.1038/ng.2250. PubMed PMID: 22504417; PubMed Central PMCID: PMC3635491

Suñol M, Martínez-Zalacaín I, Picó-Pérez M, López-Solà C, Real E, Fullana MÀ, Pujol J, Cardoner N, Menchón JM, Alonso P, Soriano-Mas C (2020) Differential patterns of brain activation between hoarding disorder and obsessive-compulsive disorder during executive performance. Psychol Med 50(4):666–673. https://doi.org/10.1017/S0033291719000515. Epub 2019 Mar 25. PubMed PMID: 30907337

Szeszko PR, Robinson D, Alvir JM, Bilder RM, Lencz T, Ashtari M, Wu H, Bogerts B (1999) Orbital frontal and amygdala volume reductions in obsessive-compulsive disorder. Arch Gen Psychiatry 56(10):913–919. PubMed PMID: 10530633

Szeszko PR, Ardekani BA, Ashtari M, Malhotra AK, Robinson DG, Bilder RM, Lim KO (2005) White matter abnormalities in obsessive-compulsive disorder: a diffusion tensor imaging study. Arch Gen Psychiatry 62(7):782–790. PubMed PMID: 15997020

Szeszko PR, Christian C, Macmaster F, Lencz T, Mirza Y, Taormina SP, Easter P, Rose M, Michalopoulou GA, Rosenberg DR (2008) Gray matter structural alterations in psychotropic drug-naive pediatric obsessive-compulsive disorder: an optimized voxel-based morphometry study. Am J Psychiatry 165(10):1299–1307. https://doi.org/10.1176/appi.ajp.2008.08010033. Epub 2008 Apr 15. PubMed PMID: 18413702

Thobois S, Jouanneau E, Bouvard M, Sindou M (2004) Obsessive-compulsive disorder after unilateral caudate nucleus bleeding. Acta Neurochir 146(9):1027–1031; discussion 1031. Epub 2004 Jul 9. PubMed PMID: 15340815

Thompson PM, Stein JL, Medland SE, Hibar DP, Vasquez AA, Renteria ME, Toro R, Jahanshad N, Schumann G, Franke B, Wright MJ, Martin NG, Agartz I, Alda M, Alhusaini S, Almasy L, Almeida J, Alpert K, Andreasen NC, Andreassen OA, Apostolova LG, Appel K, Armstrong NJ, Aribisala B, Bastin ME, Bauer M, Bearden CE, Bergmann O, Binder EB, Blangero J, Bockholt HJ, Bøen E, Bois C, Boomsma DI, Booth T, Bowman IJ, Bralten J, Brouwer RM, Brunner HG, Brohawn DG, Buckner RL, Buitelaar J, Bulayeva K, Bustillo JR, Calhoun VD, Cannon DM,

Cantor RM, Carless MA, Caseras X, Cavalleri GL, Chakravarty MM, Chang KD, Ching CR, Christoforou A, Cichon S, Clark VP, Conrod P, Coppola G, Crespo-Facorro B, Curran JE, Czisch M, Deary IJ, de Geus EJ, den Braber A, Delvecchio G, Depondt C, de Haan L, de Zubicaray GI, Dima D, Dimitrova R, Djurovic S, Dong H, Donohoe G, Duggirala R, Dyer TD, Ehrlich S, Ekman CJ, Elvsåshagen T, Emsell L, Erk S, Espeseth T, Fagerness J, Fears S, Fedko I, Fernández G, Fisher SE, Foroud T, Fox PT, Francks C, Frangou S, Frey EM, Frodl T, Frouin V, Garavan H, Giddaluru S, Glahn DC, Godlewska B, Goldstein RZ, Gollub RL, Grabe HJ, Grimm O, Gruber O, Guadalupe T, Gur RE, Gur RC, Göring HH, Hagenaars S, Hajek T, Hall GB, Hall J, Hardy J, Hartman CA, Hass J, Hatton SN, Haukvik UK, Hegenscheid K, Heinz A, Hickie IB, Ho BC, Hoehn D, Hoekstra PJ, Hollinshead M, Holmes AJ, Homuth G, Hoogman M, Hong LE, Hosten N, Hottenga JJ, Hulshoff Pol HE, Hwang KS, Jack CR Jr, Jenkinson M, Johnston C, Jönsson EG, Kahn RS, Kasperaviciute D, Kelly S, Kim S, Kochunov P, Koenders L, Krämer B, Kwok JB, Lagopoulos J, Laje G, Landen M, Landman BA, Lauriello J, Lawrie SM, Lee PH, Le Hellard S, Lemaître H, Leonardo CD, Li CS, Liberg B, Liewald DC, Liu X, Lopez LM, Loth E, Lourdusamy A, Luciano M, Macciardi F, Machielsen MW, Macqueen GM, Malt UF, Mandl R, Manoach DS, Martinot JL, Matarin M, Mather KA, Mattheisen M, Mattingsdal M, Meyer-Lindenberg A, McDonald C, McIntosh AM, McMahon FJ, McMahon KL, Meisenzahl E, Melle I, Milaneschi Y, Mohnke S, Montgomery GW, Morris DW, Moses EK, Mueller BA, Muñoz Maniega S, Mühleisen TW, Müller-Myhsok B, Mwangi B, Nauck M, Nho K, Nichols TE, Nilsson LG, Nugent AC, Nyberg L, Olvera RL, Oosterlaan J, Ophoff RA, Pandolfo M, Papalampropoulou-Tsiridou M, Papmeyer M, Paus T, Pausova Z, Pearlson GD, Penninx BW, Peterson CP, Pfennig A, Phillips M, Pike GB, Poline JB, Potkin SG, Pütz B, Ramasamy A, Rasmussen J, Rietschel M, Rijpkema M, Risacher SL, Roffman JL, Roiz-Santiañez R, Romanczuk-Seiferth N, Rose EJ, Royle NA, Rujescu D, Ryten M, Sachdev PS, Salami A, Satterthwaite TD, Savitz J, Saykin AJ, Scanlon C, Schmaal L, Schnack HG, Schork AJ, Schulz SC, Schür R, Seidman L, Shen L, Shoemaker JM, Simmons A, Sisodiya SM, Smith C, Smoller JW, Soares JC, Sponheim SR, Sprooten E, Starr JM, Steen VM, Strakowski S, Strike L, Sussmann J, Sämann PG, Teumer A, Toga AW, Tordesillas-Gutierrez D, Trabzuni D, Trost S, Turner J, Van den Heuvel M, van der Wee NJ, van Eijk K, van Erp TG, van Haren NE, van 't Ent D, van Tol MJ, Valdés Hernández MC, Veltman DJ, Versace A, Völzke H, Walker R, Walter H, Wang L, Wardlaw JM, Weale ME, Weiner MW, Wen W, Westlye LT, Whalley HC, Whelan CD, White T, Winkler AM, Wittfeld K, Woldehawariat G, Wolf C, Zilles D, Zwiers MP, Thalamuthu A, Schofield PR, Freimer NB, Lawrence NS, Drevets W, Alzheimer's Disease Neuroimaging Initiative, EPIGEN Consortium, IMAGEN Consortium, Saguenay Youth Study (SYS) Group (2014) The ENIGMA Consortium: large-scale collaborative analyses of neuroimaging and genetic data. Brain Imaging Behav 8 (2):153–182. https://doi.org/10.1007/s11682-013-9269-5. Review. PubMed PMID: 24399358; PubMed Central PMCID: PMC4008818

Thompson PM, Andreassen OA, Arias-Vasquez A, Bearden CE, Boedhoe PS, Brouwer RM, Buckner RL, Buitelaar JK, Bulayeva KB, Cannon DM, Cohen RA, Conrod PJ, Dale AM, Deary IJ, Dennis EL, de Reus MA, Desrivieres S, Dima D, Donohoe G, Fisher SE, Fouche JP, Francks C, Frangou S, Franke B, Ganjgahi H, Garavan H, Glahn DC, Grabe HJ, Guadalupe T, Gutman BA, Hashimoto R, Hibar DP, Holland D, Hoogman M, Pol HEH, Hosten N, Jahanshad N, Kelly S, Kochunov P, Kremen WS, Lee PH, Mackey S, Martin NG, Mazoyer B, McDonald C, Medland SE, Morey RA, Nichols TE, Paus T, Pausova Z, Schmaal L, Schumann G, Shen L, Sisodiya SM, Smit DJA, Smoller JW, Stein DJ, Stein JL, Toro R, Turner JA, van den Heuvel MP, van den Heuvel OL, van Erp TGM, van Rooij D, Veltman DJ, Walter H, Wang Y, Wardlaw JM, Whelan CD, Wright MJ, Ye J, ENIGMA Consortium (2017) ENIGMA and the individual: predicting factors that affect the brain in 35 countries worldwide. Neuroimage 145(Pt B):389–408. https://doi.org/10.1016/j.neuroimage.2015.11.057. Epub 2015 Dec 4. Review. PubMed PMID: 26658930; PubMed Central PMCID: PMC4893347

Thorsen AL, Kvale G, Hansen B, van den Heuvel OA (2018) Symptom dimensions in obsessive-compulsive disordèr as predictors of neurobiology and treatment response. Curr Treat Options Psychiatry 5(1):182–194. https://doi.org/10.1007/s40501-018-0142-4. Epub 2018 Feb 23. PubMed PMID: 30237966; PubMed Central PMCID: PMC6141210

van den Heuvel OA, Remijnse PL, Mataix-Cols D, Vrenken H, Groenewegen HJ, Uylings HB, van Balkom AJ, Veltman DJ (2009) The major symptom dimensions of obsessive-compulsive disorder are mediated by partially distinct neural systems. Brain 132(Pt 4):853–868. https://doi.org/10.1093/brain/awn267. Epub 2008 Oct 24. PubMed PMID: 18952675

van den Heuvel OA, Boedhoe PSW, Bertolin S, Bruin WB, Francks C, Ivanov I, Jahanshad N, Kong XZ, Kwon JS, O'Neill J, Paus T, Patel Y, Piras F, Schmaal L, Soriano-Mas C, Spalletta G, van Wingen GA, Yun JY, Vriend C, Simpson HB, van Rooij D, Hoexter MQ, Hoogman M, Buitelaar JK, Arnold P, Beucke JC, Benedetti F, Bollettini I, Bose A, Brennan BP, De Nadai AS, Fitzgerald K, Gruner P, Grünblatt E, Hirano Y, Huyser C, James A, Koch K, Kvale G, Lazaro L, Lochner C, Marsh R, Mataix-Cols D, Morgado P, Nakamae T, Nakao T, Narayanaswamy JC, Nurmi E, Pittenger C, Reddy YCJ, Sato JR, Soreni N, Stewart SE, Taylor SF, Tolin D, Thomopoulos SI, Veltman DJ, Venkatasubramanian G, Walitza S, Wang Z, Thompson PM, Stein DJ, ENIGMA-OCD Working Group (2020) An overview of the first 5 years of the ENIGMA obsessive-compulsive disorder working group: the power of worldwide collaboration. Hum Brain Mapp. https://doi.org/10.1002/hbm.24972. [Epub ahead of print]. Review. PubMed PMID: 32154629

van Erp TG, Hibar DP, Rasmussen JM, Glahn DC, Pearlson GD, Andreassen OA, Agartz I, Westlye LT, Haukvik UK, Dale AM, Melle I, Hartberg CB, Gruber O, Kraemer B, Zilles D, Donohoe G, Kelly S, McDonald C, Morris DW, Cannon DM, Corvin A, Machielsen MW, Koenders L, de Haan L, Veltman DJ, Satterthwaite TD, Wolf DH, Gur RC, Gur RE, Potkin SG, Mathalon DH, Mueller BA, Preda A, Macciardi F, Ehrlich S, Walton E, Hass J, Calhoun VD, Bockholt HJ, Sponheim SR, Shoemaker JM, van Haren NE, Pol HE, Ophoff RA, Kahn RS, Roiz-Santiañez R, Crespo-Facorro B, Wang L, Alpert KI, Jönsson EG, Dimitrova R, Bois C, Whalley HC, McIntosh AM, Lawrie SM, Hashimoto R, Thompson PM, Turner JA (2016) Subcortical brain volume abnormalities in 2028 individuals with schizophrenia and 2540 healthy controls via the ENIGMA consortium. Mol Psychiatry 21(4):585. https://doi.org/10.1038/mp.2015.118. Epub 2015 Aug 18. PubMed PMID: 26283641; PubMed Central PMCID: PMC5751698

van Erp TGM, Walton E, Hibar DP, Schmaal L, Jiang W, Glahn DC, Pearlson GD, Yao N, Fukunaga M, Hashimoto R, Okada N, Yamamori H, Bustillo JR, Clark VP, Agartz I, Mueller BA, Cahn W, de Zwarte SMC, Hulshoff Pol HE, Kahn RS, Ophoff RA, van Haren NEM, Andreassen OA, Dale AM, Doan NT, Gurholt TP, Hartberg CB, Haukvik UK, Jørgensen KN, Lagerberg TV, Melle I, Westlye LT, Gruber O, Kraemer B, Richter A, Zilles D, Calhoun VD, Crespo-Facorro B, Roiz-Santiañez R, Tordesillas-Gutiérrez D, Loughland C, Carr VJ, Catts S, Cropley VL, Fullerton JM, Green MJ, Henskens FA, Jablensky A, Lenroot RK, Mowry BJ, Michie PT, Pantelis C, Quidé Y, Schall U, Scott RJ, Cairns MJ, Seal M, Tooney PA, Rasser PE, Cooper G, Shannon Weickert C, Weickert TW, Morris DW, Hong E, Kochunov P, Beard LM, Gur RE, Gur RC, Satterthwaite TD, Wolf DH, Belger A, Brown GG, Ford JM, Macciardi F, Mathalon DH, O'Leary DS, Potkin SG, Preda A, Voyvodic J, Lim KO, McEwen S, Yang F, Tan Y, Tan S, Wang Z, Fan F, Chen J, Xiang H, Tang S, Guo H, Wan P, Wei D, Bockholt HJ, Ehrlich S, Wolthusen RPF, King MD, Shoemaker JM, Sponheim SR, De Haan L, Koenders L, Machielsen MW, van Amelsvoort T, Veltman DJ, Assogna F, Banaj N, de Rossi P, Iorio M, Piras F, Spalletta G, McKenna PJ, Pomarol-Clotet E, Salvador R, Corvin A, Donohoe G, Kelly S, Whelan CD, Dickie EW, Rotenberg D, Voineskos AN, Ciufolini S, Radua J, Dazzan P, Murray R, Reis Marques T, Simmons A, Borgwardt S, Egloff L, Harrisberger F, Riecher-Rössler A, Smieskova R, Alpert KI, Wang L, Jönsson EG, Koops S, Sommer IEC, Bertolino A, Bonvino A, Di Giorgio A, Neilson E, Mayer AR, Stephen JM, Kwon JS, Yun JY, Cannon DM, McDonald C, Lebedeva I, Tomyshev AS, Akhadov T, Kaleda V, Fatouros-Bergman H, Flyckt L, Karolinska Schizophrenia Project, Busatto GF, Rosa PGP, Serpa MH,

Zanetti MV, Hoschl C, Skoch A, Spaniel F, Tomecek D, Hagenaars SP, McIntosh AM, Whalley HC, Lawrie SM, Knöchel C, Oertel-Knöchel V, Stäblein M, Howells FM, Stein DJ, Temmingh HS, Uhlmann A, Lopez-Jaramillo C, Dima D, McMahon A, Faskowitz JI, Gutman BA, Jahanshad N, Thompson PM, Turner JA (2018) Cortical brain abnormalities in 4474 individuals with Schizophrenia and 5098 control subjects via the Enhancing Neuro Imaging Genetics Through Meta Analysis (ENIGMA) Consortium. Biol Psychiatry 84(9):644–654. https://doi.org/10.1016/j.biopsych.2018.04.023. Epub 2018 May 14. PubMed PMID: 29960671; PubMed Central PMCID: PMC6177304

Winkler AM, Kochunov P, Blangero J, Almasy L, Zilles K, Fox PT, Duggirala R, Glahn DC (2010) Cortical thickness or grey matter volume? The importance of selecting the phenotype for imaging genetics studies. NeuroImage 53(3):1135–1146. https://doi.org/10.1016/j.neuroimage.2009.12.028. Epub 2009 Dec 16. PubMed PMID: 20006715; PubMed Central PMCID: PMC2891595

Wonderlick JS, Ziegler DA, Hosseini-Varnamkhasti P, Locascio JJ, Bakkour A, van der Kouwe A, Triantafyllou C, Corkin S, Dickerson BC (2009) Reliability of MRI-derived cortical and subcortical morphometric measures: effects of pulse sequence, voxel geometry, and parallel imaging. NeuroImage 44(4):1324–33. https://doi.org/10.1016/j.neuroimage.2008.10.037. Epub 2008 Nov 7. PubMed PMID: 19038349; PubMed Central PMCID: PMC2739882

Yang X, Hu X, Tang W, Li B, Yang Y, Gong Q, Huang X (2019) Multivariate classification of drug-naive obsessive-compulsive disorder patients and healthy controls by applying an SVM to resting-state functional MRI data. BMC Psychiatry 19(1):210. https://doi.org/10.1186/s12888-019-2184-6. PubMed PMID: 31277632; PubMed Central PMCID: PMC6612132

Yun JY, Boedhoe PSW, Vriend C, Jahanshad N, Abe Y, Ameis SH, Anticevic A, Arnold PD, Batistuzzo MC, Benedetti F, Beucke JC, Bollettini I, Bose A, Brem S, Calvo A, Cheng Y, Cho KIK, Ciullo V, Dallaspezia S, Denys D, Feusner JD, Fouche JP, Giménez M, Gruner P, Hibar DP, Hoexter MQ, Hu H, Huyser C, Ikari K, Kathmann N, Kaufmann C, Koch K, Lazaro L, Lochner C, Marques P, Marsh R, Martínez-Zalacaín I, Mataix-Cols D, Menchón JM, Minuzzi L, Morgado P, Moreira P, Nakamae T, Nakao T, Narayanaswamy JC, Nurmi EL, O'Neill J, Piacentini J, Piras F, Piras F, Reddy YCJ, Sato JR, Simpson HB, Soreni N, Soriano-Mas C, Spalletta G, Stevens MC, Szeszko PR, Tolin DF, Venkatasubramanian G, Walitza S, Wang Z, van Wingen GA, Xu J, Xu X, Zhao Q, ENIGMA-OCD Working Group, Thompson PM, Stein DJ, van den Heuvel OA, Kwon JS (2020) Brain structural covariance networks in obsessive-compulsive disorder: a graph analysis from the ENIGMA Consortium. Brain 143(2):684–700. https://doi.org/10.1093/brain/awaa001. PubMed PMID: 32040561; PubMed Central PMCID: PMC7009583

Zarei M, Mataix-Cols D, Heyman I, Hough M, Doherty J, Burge L, Winmill L, Nijhawan S, Matthews PM, James A (2011) Changes in gray matter volume and white matter microstructure in adolescents with obsessive-compulsive disorder. Biol Psychiatry 70(11):1083–1090. https://doi.org/10.1016/j.biopsych.2011.06.032. Epub 2011 Sep 8. PubMed PMID: 21903200

Zhang L, Hu X, Lu L, Li B, Hu X, Bu X, Li H, Tang S, Yang Y, Roberts N, Sweeney JA, Gong Q, Huang X (2019) Abnormalities of hippocampal shape and subfield volumes in medication-free patients with obsessive-compulsive disorder. Hum Brain Mapp 40(14):4105–4113. https://doi.org/10.1002/hbm.24688. Epub 2019 Jun 12. PubMed PMID: 31188536

Zhou C, Cheng Y, Ping L, Xu J, Shen Z, Jiang L, Shi L, Yang S, Lu Y, Xu X (2018a) Support vector machine classification of obsessive-compulsive disorder based on whole-brain volumetry and diffusion tensor imaging. Front Psych 9:524. https://doi.org/10.3389/fpsyt.2018.00524. eCollection 2018. PubMed PMID: 30405461; PubMed Central PMCID: PMC6206075

Zhou C, Xu J, Ping L, Zhang F, Chen W, Shen Z, Jiang L, Xu X, Cheng Y (2018b) Cortical thickness and white matter integrity abnormalities in obsessive-compulsive disorder: a combined multimodal surface-based morphometry and tract-based spatial statistics study. Depress Anxiety 35(8):742–751. https://doi.org/10.1002/da.22758. Epub 2018 May 7. PubMed PMID: 29734487

Magnetic Resonance Spectroscopy (MRS) and Positron Emission Tomography (PET) Imaging in Obsessive-Compulsive Disorder

Marjan Biria, Lucia-Manuela Cantonas, and Paula Banca

Contents

Abstract Obsessive-compulsive disorder (OCD) is characterised by structural and functional deficits in the cortico-striato-thalamic-cortical (CSTC) circuitry and abnormal neurochemical changes are thought to modulate these deficits. The hypothesis that an imbalanced concentration of the brain neurotransmitters, in particular glutamate (Glu) and gamma-amino-butyric acid (GABA), could impair the normal functioning of the CSTC, thus leading to OCD symptoms, has been tested in humans using magnetic resonance spectroscopy (MRS) and positron emission tomography (PET). This chapter summarises these neurochemical findings and represents an attempt to condense such scattered literature. We also discuss potential challenges in the field that may explain the inconsistent findings and suggest ways to overcome them. There is some convergent research from MRS pointing towards abnormalities in the brain concentration of neurometabolite markers of neuronal integrity, such as

M. Biria (✉) and P. Banca
Department of Psychology, University of Cambridge, Cambridge, UK

L.-M. Cantonas
Department of Basic Neuroscience, University of Geneva, Geneva, Switzerland

Curr Topics Behav Neurosci (2021) 49: 231–268
https://doi.org/10.1007/7854_2020_201
Published Online: 10 March 2021

N-acetylaspartate (NAA) and choline (Cho). Lower NAA levels have been found in dorsal and rostral ACC of OCD patients (as compared to healthy volunteers), which increase after CBT and SSRI treatment, and higher Cho concentration has been reported in the thalamus of the OCD brain. However, findings for other neurometabolites are very inconsistent. Studies have reported abnormalities in the concentrations of creatine (Cr), GABA, glutamate (Glu), glutamine (Gln), Ins (myo-inositol), and serotonin (5-HT), but most of the results were not replicated. The question remains whether the NAA and Cho findings are genuinely the only neurochemical abnormalities in OCD or whether the lack of consistent findings for the other neurometabolites is caused by the lower magnetic field (1–3 Tesla (T)) used by the studies conducted so far, their small sample sizes or a lack of proper control for medication effects. To answer these questions and to further inform the biological underpinning of the symptoms and the cognitive problems at the basis of OCD we need better controlled studies using clear medicated vs unmedicated groups, larger sample sizes, stronger magnetic fields (e.g. at 7 T), and more consistency in the definition of the regions of interest.

Keywords Brain neurochemistry · GABA · Glutamate · Magnetic resonance spectroscopy · Neurotransmitters · Obsessive-compulsive disorder · Positron emission tomography · Serotonin

1 Introduction

Cortico-striato-thalamic-cortical (CSTC) circuitry is consistently found to be abnormal in structure, neurochemistry, and function in patients with OCD and provides the prevailing neural model of the disorder (Robbins et al. 2019; Sanjaya Saxena and Rauch 2000). The main circuits involve excitatory (glutamatergic) projections from medial prefrontal regions, such as orbitofrontal and cingulate cortex, to ventral and dorsal striatal nuclei, which, in turn send gamma-amino-butyric acid (GABA)-ergic projections to the globus pallidus. Further GABAergic projections are sent to the thalamus, which then conveys glutamatergic projections back to the prefrontal cortex, thus closing the 'loops'. These circuits have been functionally divided into three broad networks engaging different domains of processing, the dorsal cognitive circuit (dorsolateral prefrontal cortex (dlPFC) – dorsal caudate – thalamus), the limbic circuit (ventral striatal – thalamus – prefrontal cortex), and the sensorimotor circuit (orbitofrontal cortex (OFC)/supplementary motor area (SMA)/dorsal anterior cingulate cortex (dACC) – putamen – thalamus) loops (Milad and Rauch 2012).

Patterns of neural activity within these circuits are dynamic and change constantly to adapt to new cognitive and motor demands. These different behavioural states are mainly regulated by glutamatergic projection neurons, which trigger neural communication, and by the GABAergic projection neurons, which in turn inhibit communication by decreasing the likelihood that the neurons will fire, and therefore perturbations in glutamate or GABA neurotransmission may translate into neuronal hyper- or hypoactivity.

The prefrontal–subcortical circuits mediate different behavioural functions, and imbalanced glutamate and GABA levels within this widespread network might translate into symptomatic behavioural expression observed in OCD. Striatal hypoactivation during inhibition and interference tasks and prefrontal hypoactivation in the anterior cingulate, medial frontal and dorsolateral prefrontal cortices during switching tasks (Eng et al. 2015), as well as hypoactivation of the caudate and prefrontal areas alongside with hyperactivation of the subthalamic nuclei and putamen during symptoms provocation in patients with OCD (Banca et al. 2015) might suggest abnormal glutamate and GABA expression. However, such complex behavioural deficits cannot be explained simply by glutamate\GABA perturbation. Modulatory systems, and particularly the serotonergic and dopaminergic systems, are also tuning the information flow through these circuits and may thus contribute to dysregulation of behaviours like response inhibition, cognitive flexibility, or working memory in OCD (Robbins et al. 2019).

Animal models reveal modulatory effects of dopaminergic neurons on postsynaptic cortical and thalamic glutamatergic release and synaptic plasticity on GABAergic medium spiny inhibitory neurons of the striatum; glutamate signalling is enhanced through dopamine D1 receptors in striato-nigral neurons while reduced through dopamine D2 receptors in striato-pallidal neurons (Surmeier et al. 2007). In humans, functional neuroimaging combined with pharmacological manipulation reveals increased functional connectivity between caudate nuclei and thalamic nuclei and ventral midbrain following treatment with the dopaminergic drug sulpiride (Honey et al. 2003).

In addition, modulation of glutamate transmission by prefrontal serotonergic system (5-HT) has been described. In animal models, activation of frontal glutamate NMDA receptors inhibits local glutamate release and stimulates subcortical striatal glutamate release through both pre- and postsynaptic mechanisms, namely 5-HT_{1A} and 5-HT_{2A} receptors (Ciranna 2006). Serotonin can also stimulate receptors in GABA interneurons to modulate inhibitory inputs onto pyramidal excitatory neurons (Celada et al. 2013). In humans, 5-HT has been shown to modulate cognitive behaviours, such as enhancing cognitive flexibility (Kehagia et al. 2010) and response inhibition (Cools et al. 2008). These findings are important and suggest that fine regulation of excitation and inhibition is modulated by monoamine systems that might be crucially involved in the abnormal functioning of CSTC circuits in patients with OCD (Sanjaya Saxena and Rauch 2000).

More than two decades ago Saxena et al. (1998) hypothesised that hyperactivity in orbitofrontal-subcortical circuits underlies the symptoms of OCD (S. Saxena et al. 1998). Since then, a large number of structural and functional neuroimaging data (reviewed and discussed in more detail in chapters "Structural Imaging in OCD" and "Functional Brain Imaging and OCD") have substantiated and extended this hypothesis. Dysfunction in the orbitofrontal-striatal circuits and connected structures, such as the ACC, nucleus accumbens and thalamus, has been found to contribute to the pathology of OCD (Robbins et al. 2019) and abnormal neurochemical changes have been hypothesised to modulate the structural and functional abnormalities within these circuits (Graat et al. 2017).

Structural and functional studies have illuminated the neurobiological underpinnings of OCD but the literature on neurochemical changes is more sparse. Glutamate (Glu) and gamma-amino-butyric acid (GABA) are evidently two important neurotransmitters for the normal functioning of the fronto-striatal-thalamic circuits. Hypothetically, a loss of brain volume and abnormal function could be explained by hypo-functioning of GABA interneurons that results in disinhibition of pyramidal cells and hence increased glutamatergic neurotransmission within the main cortico-striatal-thalamic circuits. This glutamatergic over-activity may lead to excitotoxic brain tissue damage, reduced neuropil (i.e. the dense mesh of dendrites, axon terminals, and glia), and synaptic loss (Hardingham and Do 2016), which ultimately results in cognitive and behavioural impairments underpinning OCD symptoms.

Although such hypotheses have been explicitly investigated in disorders such as schizophrenia, similar studies have been lacking in OCD. OCD is also treated with drugs that modulate the functioning of the CTSC circuitry especially selective serotonin reuptake inhibitors (SSRIs) (Soomro et al. 2008) and dopamine (DA) D2 receptor antagonists (Dold et al. 2015) leading to classical hypotheses of disturbed serotonin (5-hydroxytryptamine, 5-HT) or dopamine function. In general, ligand-based positron emission tomography (PET) has been used to evaluate such hypotheses, whereas magnetic resonance spectroscopy (MRS) has been employed to quantify changes in glutamate and GABA. We will initially survey and update the use of PET methods, which have been extensively reviewed elsewhere, before focusing on our main topic for this chapter, magnetic resonance spectroscopy (MRS).

Magnetic resonance spectroscopy is a non-invasive technique that measures the brain physiology and chemistry revealing unique information about chemical composition across various brain tissue in volumes less than 1 ml (Brownell et al. 1998). The MRS technique allows in vivo quantification of neurometabolites, such as choline, creatine, N-acetylaspartate (a marker of neuronal health), glutamate (the main excitatory neurotransmitter), or GABA (the main inhibitory neurotransmitter) and thus provides direct evidence of biochemical levels and tissue integrity across cortical and subcortical in both healthy and clinical population. Nevertheless, MRS has a low signal-to-noise ratio translated into relatively low spatial resolution and chemical discrimination due to low concentration of metabolites as compared to water when magnetic fields lower than 3 T are applied (Alger 2010). On the contrary, PET imaging does not depend on intrinsic tissue features, but rather relies on external factors injected into the bloodstream named radiotracers that can be metabolised in the brain. A great advantage of PET imaging as compared to MRS is its versatility. PET can measure neurotransmitter receptor density, drug occupancy, endogenous neurotransmitter release, and glucose metabolism depending on the injected radiotracer and thus provides information about many different neural processes and abnormal changes across cortical and subcortical brain areas at a spatial resolution of 5–10 mm with a very low temporal resolution of tens of seconds to minutes (Sossi 2018). However, the development of radiotracers is a complex and expensive process. Further, as the radiotracer must be intravenously administrated it exposes the patient to some levels of radiation.

Consequently, MRS, a measure of biochemical brain environment and a direct measure of excitatory\inhibitory levels, complements PET imaging, a measure of monoamine receptor availability and drug modulation, when studying the functioning of the CTSC circuitry.

2 Positron Emission Tomography (PET)

PET has enlightened the field of biological psychiatry and provided considerable evidence for the hyperactive cortico-striatal model of OCD. It uses radioactive substances (named radiotracers) that enable visualisation and measurement of metabolic processes in the brain. The radiotracers are injected into the bloodstream, reaching the brain through arterial flow after crossing the blood–brain barrier (Sossi 2018). A great advantage of PET is the study of brain neurophysiology by measuring the glucose metabolism and the activity and availability of neurotransmitter receptors, while an important disadvantage is the low temporal resolution (tens of seconds to minutes). Therefore, for dynamic measures of brain responses to various behavioural tasks, functional MRI is preferred by researchers due to slightly higher temporal resolutions and flexibility regarding task implementation. However, the fMRI signal is related to hemodynamic responses and is still much slower than the underlying neural processes (Glover 2011).

Commonly used radiotracers for investigating the neurochemical dysfunctions of OCD include fluorodeoxyglucose, which measures cerebral glucose metabolism, [^{11}C] MDL, used to measure the availability of the 5-HT$_{2A}$ group of serotonin receptors, [^{11}C] Raclopride and [^{11}C]-SCH23390, used to measure the availability of dopamine D$_{2/3}$ and D1 receptors, respectively, and [11C]ABP688, used to measure metabotropic glutamate receptors in patients with OCD and healthy controls (Baxter 1990). For more than two decades, PET has provided evidence for abnormal brain activity in patients with OCD, specifically considerably increased cerebral glucose metabolism and neurotransmitter dysfunction in the frontal cortices and basal ganglia (Whiteside et al. 2004). Narrative reviews of studies employing PET scans implicate the orbitofrontal cortex (OFC), caudate nuclei, and thalamus as major contributors to OCD neuropathology (Newberg et al. 2011; Saxena et al. 1998; Sanjaya Saxena and Rauch 2000). The results of a meta-analysis also highlight both the orbital gyrus and the caudate nucleus as the main areas of robust difference in patients with OCD as compared to controls (Whiteside et al. 2004). The observed pattern of results suggests that imaging focal regions of interest such as prefrontal areas and striatal regions is crucial to detect subtle abnormalities present in the brain of patients with OCD. However, as Pittenger counsels in chapter "Pharmacotherapeutic Strategies and New Targets in OCD", caution is needed in drawing inferences about specific pathophysiology based on increased or decreased metabolic activity or receptor binding potential, as abnormalities may reflect increased or decreased functionality, they may be causal or consequential or even an epiphenomenon of little relevance to the disease process itself.

Perani et al. (2008) measured in vivo postsynaptic 5-HT_{2A} and D_2 receptors binding potential or availability using [^{11}C] MDL and [^{11}C]Raclopride, respectively. They reported a significant reduction in availability of the 5-HT_{2A} receptor in the dorsolateral prefrontal cortex (dlPFC), frontopolar, middle anterior cingulate cortex, and temporal and parietal association cortices and a significant reduction of D_2 receptor availability in the striatum in drug-naïve OCD patients as compared to controls. Further, they also found a negative association between the availability of 5-HT_{2A} receptors in orbitofrontal cortex (OFC) and dlPFC and the severity of OCD symptoms (Perani et al. 2008). These findings were not replicated in a study conducted by Simpson et al. (2011), who found no significant group differences in 5-HT_{2A} receptor availability in a sample of antipsychotic drug free patients, but found a significant negative correlation between [^{11}C]MDL binding in the OFC and the age of OCD onset. However, 37% patients with OCD had a history of SSRI treatment (Simpson et al. 2011). Additionally, Adams et al. (2005) conducted a study using the radioligand [^{18}F]Altanserin (a 5-HT2A antagonist), yet with lower selectivity for $5HT_{2A}$ receptors as compared with [^{11}C]MDL) and found higher 5-HT2A receptor availability in the caudate nuclei (Adams et al. 2005).

The 5-HT_{1B} receptor, a presynaptic autoreceptor located on serotonergic neurons, has also been a focus of interest in patients with OCD. Pittenger et al. (2016) investigated the binding potential of 5-HT1B receptor in relation to diagnosis, symptomatology, and prepulse inhibition in unmedicated patients with OCD using (11)C-P943 as the radioligand. They reported no diagnosis effect on receptor availability or on the prepulse inhibition in patients with OCD as compared to healthy controls (Pittenger et al. 2016).

Therefore, taken together, these results do not reveal clear alterations of the availability of serotonergic pre- or postsynaptic receptors in the feedback loop between the OFC and subcortical regions such as the striatum.

Patients with OCD seem to show abnormal density of the serotonin transporter 5-HTT (or SERT). Three studies using a highly SERT-selective radiotracer [^{11}C] DASB, reported reduced availability of the SERT in the thalamus, midbrain (Reimold et al. 2007), paralimbic ACC (Hesse et al. 2011), and insular cortex (Matsumoto et al. 2010) in drug naïve and drug free patients with OCD. Furthermore, Hesse et al. (2011) observed lower SERT availability in ventral striatum, the amygdala and the hypothalamus in drug-naïve patients with late, but not early onset OCD, suggesting differential abnormalities between early and late onset groups regarding the limbic serotonergic system (Hesse et al. 2011). One study found additional negative correlations between SERT availability in the thalamus and symptom severity (Reimold et al. 2007), while two studies found no significant association between symptomatology and SERT availability (Hesse et al. 2011; Matsumoto et al. 2010). Additionally, Wong et al. (2008) demonstrated SERT reductions, measured using the [^{11}C]McN5652 radiotracer, co-occurring with increased DA release and 5-HT_{2A} receptor availability (measured using [^{11}C] MDL100907 radiotracer) in the ventral striatum in patients with Tourette Syndrome and OCD as compared with healthy controls. These results were proposed as the basis for a neurochemical model of dopamine-serotonin interaction in OCD (Wong et al. 2008). On the contrary, Simpson et al. (2003) using the [^{11}C]McN5652 radiotracer observed no significant differences in SERT availability (Simpson et al. 2003) between patients with OCD and comorbid depression and healthy controls.

Consequently, the evidence from multiple studies show decreased SERT binding in various brain areas of interest in patients with OCD. Since the serotonin transporter (SERT) plays a critical role in regulating serotonin neurotransmission, abnormal function of these transporter proteins might be implicated alongside with DA and glutamate dysregulations in wide cognitive and emotional processes reported to be impaired in OCD (Robbins et al. 2019) controls.

With regard to investigations of DA receptor function in OCD, Olver et al. (2009) reported decreased availability of postsynaptic D1 receptors in the caudate nucleus and putamen (using D_1 receptor antagonist [^{11}C]-SCH23390) (Olver et al. 2009) and Denys et al. (2013) reported decreased availability of postsynaptic D2 and D3 receptors restricted to the putamen (using as a radioligand [^{11}C]Raclopride) (Denys et al. 2013). Decreased postsynaptic DA receptor availability might reflect increased endogenous dopamine levels within striatal regions, changes that have been also linked to reduced firing pattern of presynaptic midbrain dopaminergic neurons and altered cue-dependent learning and incentive motivation in animal models (Krabbe et al. 2015). Interestingly, dopamine receptor abnormalities were not associated with the symptom severity nor the age of onset in the aforementioned studies.

In addition, a complementary study by Figee et al. (2014) in severely ill and highly treatment resistant patients with OCD measured the effects of high frequency stimulation (with deep brain stimulation electrodes) of nucleus accumbens and anterior limb of the internal capsule on the dopamine $D_{2/3}$ receptor availability in the striatum (caudate and putamen) with single photon emission computed tomography (SPECT) using [^{123}I]iodobenzamide as radioligand (Figee et al. 2014). The authors observed an average of 10.2% decrease of $D_{2/3}$ dopaminergic receptor availability in the striatum and 20.1% decrease in the stimulated area, as well as reduced $D_{2/3}$ receptor availability bilaterally in the putamen during acute stimulation, suggesting an acute increase in dopamine release in these brain regions competing with the ligand for binding at the receptor.

Given the increased number of magnetic resonance spectroscopy studies suggesting a potential role of glutamate signalling dysfunction in OCD within the CSTC circuits (Wu et al. 2012), Akkus et al. (2014) using [11C]ABP688 (a highly selective radioligand for mGluR5 receptors) measured the distribution volume ratio of metabotropic glutamate receptor 5 and found no significant global difference between patients with OCD and healthy controls. The authors observed, however, a significant positive correlation between the Y-BOCS obsession score and mGluR5 ratio in the amygdala, anterior cingulate cortex (ACC), and medial orbitofrontal cortex (mOFC) (Akkus et al. 2014).

Consequently, PET studies have demonstrated abnormal receptor binding using candidate neurotransmitters like 5HT, DA, and Glu, suggesting complex neurotransmitter changes in patients with OCD. However, the results are not consistent and need further investigation using bigger samples.

PET imaging has also been used to investigate the impact of cognitive behavioural therapy (CBT) and medication on OCD symptomatology. Findings from small studies suggest that specific patterns of cerebral metabolism in the OFC and caudate nucleus may predispose patients with OCD to preferentially respond either to CBT or to medication with an SSRI. For example, lower glucose metabolism in the OFC and caudate nucleus was found to be associated with better response to medication (paroxetine), while higher metabolism within these areas was associated with improved symptoms in response to CBT (Schwartz et al. 1996) and worsening after medication administration (fluoxetine) (Brody et al. 1998). Furthermore, lower pre-treatment glucose metabolism in the OFC predicted better outcome after the paroxetine treatment (S. Saxena et al. 1999).

In summary, functional neuroimaging using PET enables precise measurement of different brain neurotransmitters and their receptors, which would not be accessible in vivo otherwise, and represents a complement to the MRS methodology described below. However, strong conclusions are premature and more research, e.g. using multimodal imaging methods, is needed to better understand the relationship between the concentrations of DA and 5-HT with glutamate and GABA and thereby improve our knowledge of the functional and structural changes of the CSTC pathways in OCD.

3 Magnetic Resonance Spectroscopy (MRS)

Magnetic resonance spectroscopy (MRS) is a non-invasive neuroimaging technique that allows the in vivo quantification of different brain metabolites by measuring magnetic resonance signal amplitudes generated by atomic nuclei in living tissues. MRS is particularly well suited for studying OCD, as it is capable of quantifying concentrations of glutamate and GABA in localised regions of the human brain. Proton magnetic resonance spectroscopy (1H-MRS) is the most widely used MRS technique. Using the magnetic field and a brief radiofrequency pulse, 1H-MRS generates resonance signals with peaks unique to each brain neurometabolite (see Fig. 1). The scanner generates a signal from the protons of the hydrogen atoms of the molecules and the strength of this signal indicates the molecular concentration of the specific neurochemical in the brain (Brennan et al. 2013).

The signal sensed by the scanner usually contains considerable field variability. To control for this, researchers usually obtain a reference signal which can be internal or external. An internal reference is generated in the brain such as a neurometabolite or water, whereas an external reference signal is generated outside the brain, using a phantom. The metabolites are then expressed in terms of their

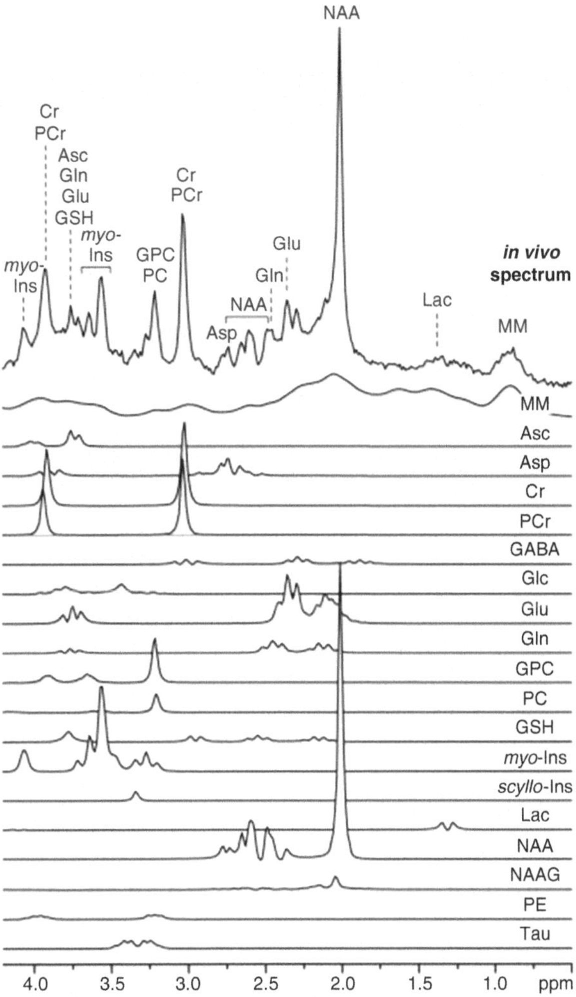

Fig. 1 Shows the LCModel analysis of an in vivo ^{1}H MR spectrum acquired from a human brain at 3T with voxel placed at the gray-matter-rich occipital cortex (Reprinted from Xin and Tkáč 2017)

ratios to these references. See Box 1 for further technical details regarding MRS methodology.

Box 1 Technical Aspects of MRS
Water Suppression

In the healthy human brain, grey and white matter are, respectively, made up of 83% and 70% water (Turner 1986). In contrast with water, concentrations of brain metabolites are 103–104 times smaller and much harder to detect. The commonly used solution to this problem is to directly suppress

(continued)

Box 1 (continued)

the water signal from the MR spectra by exciting the water resonance using a frequency selective radio frequency pulse. For this purpose, differences in chemical shifts of water and other neurometabolites are used to suppress water signals, while the signals of other brain neurochemicals remain unaffected. This is called the CHEmical Shift Selective (CHESS) method for water suppression (Haase et al. 1986) and is the most commonly used solution to detect neurometabolites in the brain that would otherwise be camouflaged by the high concentrations of water.

Magnetic Field Homogeneity and Shimming

When the magnetic field does not vary a lot within a target voxel, a single frequency or peak will be observed. For example, see the simulated NAA peak in Fig. 1 and how narrow it is. However, when the magnetic field does vary a lot, instead of a single narrow peak, a distribution of frequencies will be measured. The broadening of the spectral peaks is a sign of inhomogeneity of the magnetic field and the loss of spectral sensitivity and resolution. Various radiofrequency shimming methods have been developed to correct for the inhomogeneity of the magnetic field while acquiring data (Gruetter 1993; Juchem et al. 2010, 2011; Shen et al. 1999). Unfortunately, this cannot be corrected for once the data is acquired. Therefore it is of utmost importance to make sure the magnetic field is homogeneous while acquiring the data. Poor shimming is one common cause of bad data quality.

Poor shimming or inhomogeneity of magnetic field can also be numerically detected by looking at the full width at half maximum (FWHM) parameter, which is the spectral width at the half amplitude of the signal. If the shimming is poor, the peaks will be broad and the FWHM will be larger. Smaller FWHM indicates better shimming. For example, in a pilot single voxel MRS study at 7 T conducted in our lab, better shimming and data quality was observed when FWHM value was <17 Hz. If the shimming is found to be poor, indicating inhomogeneity of magnetic field, this can sometimes be improved by changing the shim currents or by repositioning of the voxel to be farther from factors leading to this inhomogeneity. The latter will of course be at the cost of less spatial consistency across subjects. Overall, this technique requires careful assessment of the settings at the right time, considering the specific measurements being investigated, in order to optimise the quality of the collected data.

Single Voxel Versus Multivoxel MRS

Based on the number of voxels or the volume of interest (VOI) studied, MRS can be divided into single voxel spectroscopy and magnetic resonance spectroscopic imaging (MRSI). Single voxel spectroscopy is widely used for assessing in vivo brain metabolite changes, as it is usually easier to perform in a shorter time period and tends to have a higher signal-to-noise ratio. However, this technique only evaluates the metabolite profiles within a priori selected

(continued)

Box 1 (continued)
areas/voxels of the brain, with a high risk of missing the target region where disorder-related differences in neurochemicals might be observed. As such, strong a priori hypotheses about which regions are most relevant for investigation is essential.

In contrast, MRSI simultaneously measures metabolite concentrations in many voxels covering more areas of the brain, thus having a higher spatial coverage (a top-down approach). However, longer acquisition times, more complex processing, and a lower signal-to-noise ratio are important considerations (Boer and Klomp 2014). In MRSI, neighbouring voxels can contribute to the spectrum and the larger volumes make shimming and water suppression more difficult (Kraguljac et al. 2012). The spectroscopy signal typically has poor quality due to inhomogeneity of the magnetic field. Another issue is that the water signal is higher than all other metabolites, obscuring metabolites of interest. To solve this problem, a very good water suppression technique is required (Nixon et al. 2008).

Referencing to Standard Values

In MRS there are two ways to report metabolite ratios, i.e. the ratio of the neurochemical species of interest to a 'standard' metabolite. The first method, more commonly found in OCD studies, is internal referencing using a neurometabolite with a stable concentration. Creatine (Cr) is often used as an internal reference, as it is assumed to have a constant concentration in both the healthy and diseased brain and also under different physiological conditions (Bagory et al. 2007). It also has the advantage of being routinely collected in MRS studies. However, one must be aware that this implicit assumption of constant Cr concentrations is often not substantiated, and using creatine as reference has its own disadvantages: (1) The fact that the Cr measurements can be contaminated by other neurochemicals (e.g. choline) or by noise and (2) The Cr concentration is less uniform throughout the brain than once thought (Alger 2010).

The second approach to reporting metabolite values, less commonly found in the OCD studies reviewed for this chapter, is to report absolute concentrations. This terminology can be confusing when incorrectly interpreted as true values rather than ratios. Absolute concentration simply refers to comparison with either brain water concentration (internal reference) or other metabolites in a phantom (external reference) for which the concentration is already known (Bagory et al. 2007).

Using water as an internal reference has the advantage that it can be obtained very quickly, with a very strong signal. Nevertheless, this determination usually requires an additional step as the water concentration must be recorded separately from other neurochemicals and without water suppression.

(continued)

Box 1 (continued)
It is also suggested that the water signal levels may change due to disease, brain location, or hydration levels (Alger 2010).

Each signal referencing technique has its own pros and cons, and careful thought must go into choosing a sufficiently optimal reference according to the specific needs of the study while taking into account different factors such as the sample size, study design, and other possibilities such as recording the water signal separately.

3.1 MRS in OCD

The most commonly studied neurometabolites using the 1H-MRS in OCD are choline (Cho), creatine (Cr), GABA, Glu, glutamine (Gln), myo-inositol (Ins), and N-acetylaspartate (NAA). These neurometabolites have important implications for normal brain functioning and therefore have all been used in exploratory MRS studies in OCD. Nevertheless, GABA and Glu (see below) are probably more relevant to study, given their implications for neuropathology within the fronto-striatal-thalamic circuits, that are implicated in OCD.

Choline is involved in membrane synthesis and degradation and is thought to be a marker of cell membrane turnover (Miller 1991). *Creatine* and *phosphocreatine* are important for cerebral intracellular energy production or cerebral bioenergetics (Govindaraju et al. 2000). *GABA* and *Glu* are the main inhibitory and excitatory neurotransmitters. GABA mediates inhibitory modulation, generally via local inter-neurons (Hardingham and Do 2016). Whereas Glu, in addition to playing a central metabolic role in the brain (Krebs 1935), is involved in processes such as fast synaptic transmission and plasticity (Hardingham and Do 2016). In addition to an intermediate in neurotransmitter metabolism, *Gln* is the main precursor of Glu and GABA (Hertz and Zielke 2004), although the de novo synthesis of Glu through tricarboxylic acid (TCA) cycle is also possible (de Graaf et al. 2011). In the glutamate–glutamine cycle, the synthesis of Glu is dependent on Gln and vice versa (Ramadan et al. 2013). Additionally, the structure of Gln is very similar to Glu and at a magnetic field below 3 T, their resonance overlaps, thus preventing independent measurement. This is also true to some extent for GABA. Traditionally Glu and Gln have been measured together as a combined peak termed 'Glx'. However, Glx may not be a precise measure of neurochemicals of interest as Glu and Gln changes might be in opposite direction and the Glx may still remain unchanged. For example, a meta-analysis in schizophrenia showed no Glx changes but reduced glutamate ratios (Marsman et al. 2013). Thus a separate quantification of these neurochemicals is very important and the more recently developed 7 T ultra-high magnetic field does enable this (Choi et al. 2010; Yang et al. 2008). To the best of our knowledge, no ultra-high field studies have been published so far in OCD populations. *Myo-Inositol (Ins)* is suggested to be a marker of glial cell proliferation, a precursor of intracellular second messengers, and a major osmolyte in central nervous system, important for the integrity of cells (Fisher et al. 2002). Lastly, *NAA* appears to be a marker of neuronal density (Ebisu et al. 1994).

Since the CSCT circuitry is strongly implicated in the pathophysiology of OCD, MRS studies to date have focused on regions of interest within the cognitive, affective, or motor loops mentioned above. We will therefore review first the specific anatomical regions within the CSTC circuits that have been investigated in OCD using MRS. These include the rostral and dorsal anterior cingulate cortex (rACC and dACC), dlPFC, OFC and medial prefrontal cortex (mPFC), striatum (putamen and caudate), lenticular nuclei (putamen and globus pallidus), thalamus, posterior cingulate cortex (PCC), frontal and parietal white matter (WM), and hippocampus. Next, we will report the neurometabolite associations with symptom severity and treatments effect including the outcomes of CBT and medication with SSRI.

It is important to note that the majority of these studies were based on small sample sizes (fewer than 15 participants per group), many failed to control for the effects of medication and those that did were lacking a large sample size, reducing the reliability of their findings. In addition, they acquired the MRS data at lower magnetic field strengths (1.5–3 T), which prevented accurate separation of Glu from Gln, accurate measurement of GABA or even regional separation of the caudate nucleus from the putamen. These limitations may well contribute to the inconsistent findings we review below. Additionally, given the lack of reliability of the MRSI technique, especially its lower signal-to-noise ratio, we decided to exclude some of the MRSI studies performed using 1.5 T scanners from this review. However, we made an exception for MRSI studies of NAA, as of all the neurochemicals it has the highest peak and so we believe it may still be detected with high reliability using the 1.5 T MRSI.

3.2 *MRS and OCD: Regional Findings*

3.2.1 Neurometabolite Concentrations in the Rostral Anterior Cingulate Cortex

One of the functions of the rACC is to integrate action values in control networks (Rushworth et al. 2004). It has extensive connections with prefrontal cortical areas, such as OFC and the vmPFC, as well as the amygdala (Haber and Behrens 2014) and has been identified as an important region for the development of OCD symptomatology. The rostral ACC, as defined in the articles reviewed for this chapter, refers to volumes of interest (VOI) in the perigenual and subgenual ACC and also its border with mOFC and mPFC (Fig. 2a). This region, together with the OFC, has an important role in mediating contingency learning, relevant for goal directed behaviour (Jackson et al. 2016) which has been shown to be compromised in patients with OCD (Gillan et al. 2014) and may therefore account for their rigid behaviour. The rACC is a good example of a region that lacks anatomical consistency within the

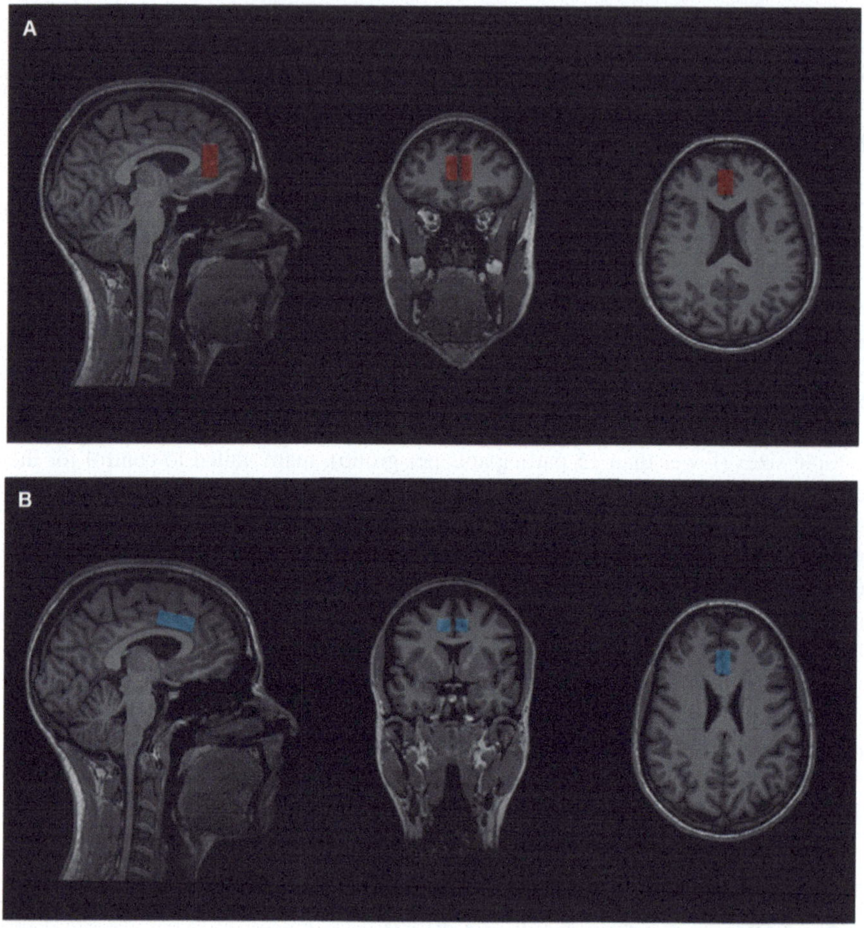

Fig. 2 The areas of the anterior cingulate cortex (ACC) coverage in magnetic resonance spectroscopy (MRS) studies reviewed in this chapter. (**a**) depicts the location for the rostracl ACC (rACC) voxels, and (**b**) shows the location for dorsal ACC voxels (dACC) in sagittal, coronal, and horizontal planes

MRS literature, with most of its investigations also including some parts of the mPFC or OFC when defining their target rACC voxel. Therefore, findings concerning the rACC must be considered with caution.

With the exception of one study by Zhu et al. (2015), which describes lower Glu levels in the rostral ACC of OCD patients, as compared to that of healthy volunteers (Zhu et al. 2015), most studies report no difference in Glu or Gln concentrations within this region (Brennan et al. 2016; Li et al. 2019; O'Neill et al. 2016; Wang et al. 2018; Zheng et al. 2020; Zurowski et al. 2012). Of note, Zhu et al. (2015) was one of the few studies testing only unmedicated patients, which of course could explain these different findings (Zhu et al. 2015).

Findings on Glx are still more inconsistent. Despite most studies reporting no between-group differences in the combined measures of Glu and Gln (Glx) (Brennan et al. 2016; Parmar et al. 2019; H. B. Simpson et al. 2012; Zhang et al. 2016; Zheng et al. 2020; Zhu et al. 2015), two studies found higher Glx levels in OCD patients compared with healthy controls (de Salles Andrade et al. 2019; O'Neill et al. 2016), whereas one study (Yücel et al. 2008) found lower Glx concentrations in OCD patients. All these studies were all carried out using a 3 T scanner, and the differences do not seem to be related to medication type or sample size.

Only three studies have measured GABA levels in the rACC of OCD patients. Results are also erratic, with one study showing higher GABA and GABA/Glu ratios in the rACC of patients with OCD (Li et al. 2019) and two others reporting lower GABA levels in OCD, as compared to the healthy control group (H. B. Simpson et al. 2012; Zhang et al. 2016).

Inconsistent results have also been found with regard to concentrations of the total NAA (tNAA) in the rACC. Three studies reported no between-group differences (de Salles Andrade et al. 2019; Parmar et al. 2019; Zheng et al. 2020), while two other studies found lower tNAA concentrations in OCD patients (O'Neill et al. 2013; Zhang et al. 2016). Despite the inconsistency of findings for tNAA, there might be a trend for lower NAA levels in unmedicated patients as the study by Zhang et al. (2016) had the largest sample size of patients with OCD ($N = 78$) of all the studies reviewed in this chapter (Zhang et al. 2016). In the Discussion section it will be seen that a sample size of 39 or larger is required to detect reliable differences in NAA.

Consistent results have been found in the rACC only for metabolites such as Cho, Ins, and Cr, with no abnormalities reported by nine studies in adults with moderate to severe OCD symptoms (Andrade et al. 2019; O'Neill et al. 2013; Parmar et al. 2019; Yücel et al. 2007; Yücel et al. 2008; Zheng et al. 2020; Zhu et al. 2015; Zurowski et al. 2012). Only one study reported higher Cho levels in rACC, which may be a less reliable finding since the investigators used a multivoxel MRSI technique (Hatchondo et al. 2017).

Some studies have also been conducted in paediatric OCD, where the pattern of inconsistency remains. Two studies reported higher Glu and Glx levels in the rACC of OCD patients (Naaijen et al. 2017; Rosenberg et al. 2004) whereas three others found no between-group differences in Glu or Glx concentration levels (O'Neill et al. 2017; Ortiz et al. 2015; Whiteside et al. 2012a). Nevertheless, despite no Glx differences between groups, Ortiz et al. (2015) found a positive correlation between Glx ratios and duration of illness longer than 2 years (Ortiz et al. 2015). Young patients with an OCD onset longer than 2 years showed significantly higher Glx levels compared to those with a more recent onset of less than 2 years (Ortiz et al. 2015).

With regard to NAA measures within the rACC of children with OCD, Yalçın et al. (2011) observed lower NAA levels in the left rACC (Yalçın et al. 2011), whereas others reported increased NAA in left rACC (O'Neill et al. 2012; Whiteside et al. 2012a) or no NAA abnormalities (Rosenberg et al. 2004). No differences were found in Cho levels in OCD patients compared to healthy children. Similar to adulthood OCD, there have been no abnormalities reported for Ins (Rosenberg et al. 2004; Whiteside et al. 2012a) and Cr (Rosenberg et al. 2004).

To summarise, MRS in OCD does not demonstrate abnormalities in Cho, Ins, or Cr within the rACC. Despite mixed findings for Glx, GABA, and tNAA, the majority of studies do not observe significant differences in the concentration levels of the main neurometabolites within the rACC between adult OCD and healthy controls, particularly for Glu and Gln. However, in children with OCD, there might be a trend towards higher Glx levels in the rACC and the concentrations seem to be higher in individuals with a longer illness duration.

3.2.2 Neurometabolite Concentrations in the Dorsal Anterior Cingulate Cortex

The dACC (Fig. 2b) is important for error monitoring as shown by electrophysiological studies and tends to be hyperactive in OCD patients during resting state functional neuroimaging (McGovern and Sheth 2017). This hyperactivity is also associated with increased error related negativity (ERN) following making an incorrect response, measured in electrophysiological studies using tasks provoking conflict (Mathews et al. 2012). Chapter "Electroencephalographic Correlates of Obsessive-Compulsive Disorder" of this book discusses the ERN findings in OCD in more detail.

With the exception of a single study in female patients with OCD reporting lower Glx concentrations in the dACC (Yücel et al. 2008), no abnormalities in Glu (Fan et al. 2017; O'Neill et al. 2016; Starck et al. 2008), Gln (Ebert et al. 1997; Fan et al. 2017; Starck et al. 2008), or Glx (Bédard and Chantal 2011; O'Neill et al. 2016; Starck et al. 2008; Yücel et al. 2007; Zhang et al. 2016) concentrations have been described in OCD.

Less consistent are findings in GABA measurements of the dACC of OCD patients. While Zhang et al. (2016) reported lower GABA levels in the dACC of OCD patients (Zhang et al. 2016), Starck et al. (2008) did not find any GABA abnormalities (Starck et al. 2008).

Less contradictory findings are reported for NAA concentrations in the dACC, with only two studies reporting no differences in NAA concentrations between OCD and healthy volunteers (Bédard and Chantal 2011; Starck et al. 2008), while two other studies in medicated patients found lower NAA or tNAA concentrations in OCD patients (Ebert et al. 1997; Yücel et al. 2007). The lower NAA concentrations were also reported in treatment naïve (Jang et al. 2006) and unmedicated OCD patients (Tükel et al. 2015; Tükel et al. 2014).

Similarly to the rACC, no abnormalities were found for Cho, Ins (Bédard and Chantal 2011; Ebert et al. 1997; Fan et al. 2017; Kitamura et al. 2006; O'Neill et al. 2016; Starck et al. 2008; Tükel et al. 2014; Yücel et al. 2007, 2008; Zhang et al. 2016), and Cr (Fan et al. 2017; Kitamura et al. 2006; O'Neill et al. 2016; Starck et al. 2008; Yücel et al. 2007, 2008; Zhang et al. 2016) in the dACC.

The only MRS study in paediatric OCD patients using dACC, performed as a VOI study, did not report concentration differences for any of the aforementioned metabolites (Glx, Glu, Gln, GABA, NAA or tNAA, Cr, Ins, and Cho) between OCD and healthy volunteers (Ljungberg et al. 2017).

To summarise, no altered concentrations of Glu, Gln, or Glx have been found in the dorsal ACC of OCD patients. Nevertheless, this region is characterised by lower NAA concentrations, found in both medicated and unmedicated patients with OCD. As for the rostral ACC, no clear abnormalities have been found in Cho, Ins, and Cr within the dACC. Possible GABA changes in this region have not yet been investigated.

3.2.3 Neurometabolite Concentrations in the Basal Ganglia

The basal ganglia (BG) are a crucial component of the cortico-striatal circuitry that underlies the pathophysiology of OCD. For this reason, many studies have focused on investigating its neurochemical basis in these structures. The striatum, comprising the caudate nucleus and the putamen, represents the main input region of the BG, whereas the globus pallidus represents the main output region of this constellation of nuclei. Several meta-analyses found increased volumes in different components of these deep structures, most consistently in putamen and globus pallidus (de Wit et al. 2014; Kong et al. 2019). Further, abnormal functional connectivity has been reported during resting state fMRI, with increased connectivity between the ventral striatum and the orbitofrontal cortex that predicted overall symptom severity (Harrison et al. 2013) and reduced ventrolateral PFC-caudate and dl-PFC-putamen functional connectivity, respectively, during attentional set-shifting and planning (Vaghi et al. 2017, 2017). Neurochemical investigations have therefore been conducted on the dorsal and ventral components of the striatum, given their different functions and distinct circuitries, as well as lenticular nuclei (putamen and globus pallidus).

Neurochemical concentrations in the ventral striatum have been investigated in two studies using both multi- (H. B. Simpson et al. 2015) and single voxel approaches (Zurowski et al. 2012). The results were similar in showing no neurochemical concentration abnormality. Simpson et al. (2015) investigated specifically Glu levels in unmedicated adults with OCD and reported no Glu differences and no volumetric differences as compared to healthy controls (Simpson et al. 2015). Additionally, Zurowski et al. (2012) investigated the NAA and myo-inositol levels in the right ventral striatum, in unmedicated OCD patients compared to healthy controls and reported no diagnosis effects on neurometabolite concentrations (Zurowski et al. 2012).

Less consistent are MRS data for the dorsal striatum. Ebert et al. (1997) showed decreased concentrations of NAA in the left striatum of unmedicated patients with OCD, which correlated with illness severity (Ebert et al. 1997). Conversely, Hatchondo et al. (2017) reported increased Cho and NAA ratios in a sample of adult patients (Hatchondo et al. 2017). Paediatric OCD studies report no abnormalities in the concentration of most of the dorsal striatal metabolites, with the exception of Cho, for which the total concentration was found to be diminished in the left striatum of medication naïve OCD (Lazaro et al. 2012).

Within the dorsal striatum, several authors have argued they could technically isolate the caudate and putamen and assess their respective neurochemistry. However, following our aforementioned explanations in the Introduction, we recommend that the findings reported below are interpreted with caution.

Three studies have found reduced NAA concentrations and NAA/Cr ratios in the caudate nuclei of unmedicated patients with moderate to severe OCD symptoms (Bartha et al. 1998; Chen et al. 2017; Tükel et al. 2015). Tükel et al. (2015) found reduced caudate NAA/Cr ratio in the OCD group (Tükel et al. 2015). However, other studies have not observed such changes in NAA (Simpson et al. 2015; Tükel et al. 2014; Whiteside et al. 2012b). In fact, they reported no differences, not only in the concentration of NAA, but also in the levels of Glu, Glx, and Cr concentrations between healthy controls and medicated and unmedicated adult patients with moderate obsessive and compulsive symptoms.

Paediatric studies with OCD patients also report inconclusive results for the caudate nucleus. One study reported no differences in tNAA, tCho, Glx, and myo-inositol between young patients with OCD and healthy participants (Ljungberg et al. 2017). In addition, Rosenberg et al. (2000) observed increased Glx in patients with OCD compared to typically developing children (Rosenberg et al. 2000). Whiteside et al. (2012a, 2012b) reported lower levels of NAA and Cr at baseline in non-medicated patients compared to medicated patients and controls (Whiteside et al. 2012a). Nevertheless, these studies included small samples and acquired the 1H-MR spectra on a 1.5-T MRI scanner.

Only 3 studies have investigated specifically the neurochemistry of the putamen in patients with OCD, two using MRSI and one using MRS. All reported no significant differences in either absolute values of Glu levels (H. B. Simpson et al. 2015) or the NAA, Cho, and myo-inositol ratios between patients and controls (H. B. Simpson et al. 2015; Tükel et al. 2015; Tükel et al. 2014). One study has, however, focused on the lenticular nuclei (putamen and globus pallidus) but revealed no abnormalities in the NAA and Cho concentrations in medicated adult patients with severe symptoms (Ohara et al. 1999).

Two studies reported no differences in the concentration of the metabolites NAA, Cho, and Cr between medicated adult patients with moderate OCD symptoms and healthy individuals in the basal ganglia (Kitamura et al. 2006; Sumitani et al. 2007). Nevertheless, the results should be considered carefully as they might have been influenced by selection of the ROI. In the study of Kitamura et al. (2006) the basal ganglia were defined as putamen and globus pallidus accompanied by the adjacent white matter (Kitamura et al. 2006), while in the study of Sumitani et al. (2007) the basal ganglia were measured across both the caudate and putamen (Sumitani et al. 2007).

Altogether, no robust and consistent neurochemical abnormalities in the basal ganglia of patients with OCD have emerged.

3.2.4 Neurometabolite Concentrations in the Thalamus

The thalamus is another important element of the CSTC circuitry, serving as a hub between the striatum and the prefrontal cortex. The thalamic nuclei are heavily interconnected with the orbitofrontal cortex and are known to be involved in cognitive flexibility and goal-directed actions, as measured behaviourally by reversal learning and action-outcome contingency learning tasks (Alcaraz et al. 2018; Parnaudeau et al. 2015; Pergola et al. 2018). Furthermore, a meta-analysis of structural alterations published in 2009 (Rotge et al. 2009) has proven increased bilateral thalamic volumes in OCD patients that correlated with the severity of obsessive-compulsive symptoms.

Proton MRS studies investigating the neurochemical underpinnings of thalamus in OCD have also revealed inconsistent findings. Zhu et al. (2015) observed lower levels of Glu concentrations in the right thalamus of unmedicated adult patients with moderate OCD symptoms. These authors also found increased thalamus Cho bilaterally in patients (Zhu et al. 2015). More recently, Parmar et al. (2019) also showed increased total choline (tCho), Ins, and Glx concentrations in the medial thalamus of unmedicated adults with an illness duration of less than 5 years (Parmar et al. 2019). However, increased levels of thalamic Cho/tCho were confirmed by one study in unmedicated patients with moderate OCD symptoms (Chen et al. 2017) and two studies on adult medicated patients with severe and moderate OCD symptoms (Fan et al. 2017; Hatchondo et al. 2017). Nevertheless, four other studies have not found any reliable between-group differences in the concentrations of Cho, Glu, Glx, NAA, tNAA, myo-inositol, and Cr between medicated patients with OCD and healthy volunteers (Bédard and Chantal 2011; Gnanavel et al. 2014; Kitamura et al. 2006; O'Neill et al. 2016).

Increased Cho concentrations in the medial thalamus of OCD patients were also observed in paediatric drug-naïve patients with severe OCD symptoms (Rosenberg et al. 2001; Smith et al. 2003). In addition to higher Cho levels, paediatric OCD patients also expressed increased Cr concentrations (Mirza et al. 2006) and reduced NAA/Cho ratios (Fitzgerald et al. 2000) in medial thalamic nuclei.

In conclusion, MRS investigations of the neurochemistry of the thalamus in OCD converge to suggest that enhanced thalamic cholinergic concentrations might be a marker of abnormality in adult and paediatric populations diagnosed with OCD. Nevertheless, it would perhaps be wise to consider these promising results as preliminary hypotheses that require future investigation.

3.2.5 Neurometabolite Concentrations in Other Brain Regions

Fewer studies have investigated neurochemical imbalances related to OCD in other brain regions. Below we briefly describe a few findings in the following areas: DLPFC, mPFC, OFC, PCC, and frontal and parietal WM.

The DLPFC is part of the CSTC circuit and plays a role in executive functioning, attention, and memory, all being impaired in patients with OCD (Vahabzadeh and McDougle 2014). Additionally, DLPFC is reported to be hypoactive in OCD which may underlie the deficits in executive functioning observed in this patient group (McGovern and Sheth 2017; Vaghi et al. 2017, 2017). A positive correlation has been found between the grey matter volume of the right DLPFC (and right OFC) and the severity of sexual and religious obsessions (Alvarenga et al. 2012). Furthermore, this region has been successfully used as a target region for treating OCD using rTMS (Sachdev et al. 2007; Sachdev et al. 2001), which additionally supports its role in the pathophysiology of this disorder. Sachdev et al. (2001) highlighted the DLPFC in the right hemisphere as being a more effective target for treating OCD symptoms, as compared to the same region in the left hemisphere (Sachdev et al. 2001). This is probably the reason why the three MRS studies examining metabolite imbalance in this area in OCD chose their VOIs in the right hemisphere (Moon and Jeong 2018; Park et al. 2017; Simpson et al. 2012). The first study found no differences in GABA and Glx measures between unmedicated OCD patients and healthy volunteers (H. B. Simpson et al. 2012). The second study, however, found reduced GABA and increased Ins in medicated patients with OCD compared with healthy controls (Park et al. 2017). A third study, also in medicated patients, found increased GABA and reduced NAA and Cho levels in DLPFC of patients with OCD (Moon and Jeong 2018). However, half of the patients in the latter study were taking psychedelic drugs in addition to their antidepressant medication, which may well have contributed to the differences observed.

The prefrontal cortex is crucial for human cognitive performance and has been proved to be dysfunctional in OCD. Its lateral and medial connections to the striatum were found to be functionally autonomous and context-dependent in OCD (Robbins et al. 2019). Hypoactivation in the lateral portions of the OFC and hyperactivation of the vmPFC including medial OFC in OCD patients were found at rest, during reversal learning paradigms (Apergis-Schoute et al. 2018; Remijnse et al. 2006) and in first-degree unaffected relatives (Chamberlain et al. 2008). However, this pattern of activation is intriguingly apparently reversed when patients are exposed to symptom provocation paradigms (Banca et al. 2015; Morgiève et al. 2014).

Studies investigating the potential neurochemical abnormalities in OCD within the OFC are very scarce, due to the close proximity of this brain region to the sinuses and hence to bone and air. This can in turn lead to increased inhomogeneity of the magnetic field which makes this region a very difficult target for spectroscopy. One study that tested medicated patients found no between-group differences in NAA, Glx, Ins, and Cho concentrations in the OFC (Bédard and Chantal 2011). However, another that tested unmedicated OCD patients reported higher NAA concentrations in the OCD group, as compared to healthy controls, and no differences in Cho levels (Chen et al. 2017). Finally, Zurowski et al. (2012) found no abnormalities in the OFC Ins levels in adult OCD patients (Zurowski et al. 2012). None of these 3 studies investigated GABA within the OFC.

Within the PFC (including the mPFC), one study that tested unmedicated patients found lower Glu concentrations in patients with OCD but no differences in the levels of Glx, NAA, tNAA, Cho, and Ins (Zhu et al. 2015). However, another study that tested medicated adult OCD patients found lower NAA levels in prefrontal cortex (Jang et al. 2006).

A few studies also explored metabolite ratios in posterior cingulate cortex and hippocampus, brain areas less commonly explored in OCD. Two studies investigated the neurochemical concentrations in the PCC and found no abnormalities in Glu, Gln, GABA, Cho, Ins, and NAA concentration levels in adult patients with OCD and no correlations with the OCD symptoms (Brennan et al. 2016; Jang et al. 2006). O'Neill et al. (2017) confirmed the lack of abnormalities for Glu levels within the ventral PCC in paediatric patients before and after cognitive-behavioural interventions. However, the author observed that lower baseline Glu within this area was associated with greater clinical response (O'Neill et al. 2017).

Measurements in the hippocampus have shown that reduced levels of choline normalised NAA (NAA/Cho) co-occurred with smaller volumes of the hippocampus (Atmaca et al. 2009). However, this finding should be interpreted with caution as this was an MRSI study at 1.5 T with a higher noise in the signal, thus making this result less reliable.

Finally, potential abnormalities in the white matter were also investigated in the frontal and parietal lobes of the OCD brain. While two studies found no abnormalities in tNAA and Cho concentration levels in the prefrontal WM of medicated adolescent and adult OCD patients (Kitamura et al. 2006; Sumitani et al. 2007), another study investigating the same region in paediatric OCD patients reported higher tNAA and Cho levels in patients (Weber et al. 2014). Age and medication status of the patients are likely driving these inconsistent findings. No abnormalities for Glu, Ins, and Cr were found in the same region in paediatric OCD patients (Weber et al. 2014). Finally, lower Cr and NAA levels in the right orbital white matter (Whiteside et al. 2012b) and high concentrations of Cho in the parietal white matter were also reported in adult OCD patients (Kitamura et al. 2006).

To summarise, no neurochemical abnormalities were found in the PCC of OCD patients. Similarly, no significant differences between OCD and healthy volunteers were found for most of the neurometabolites within the frontal regions of the brain such as PFC and OFC, with the exception of possibly higher NAA concentrations in the OFC and lower glutamate in the PFC. Contradictory results were, however, found in the DLPFC and the frontal and parietal white matter.

4 Associations Between Neurometabolite Concentrations and Clinical Symptoms

Studies using MRS could additionally inform us about how differences in neurometabolite concentrations could relate with different clinical symptoms. The MRS studies conducted so far using 1.5-3 T scanners report mainly correlations between neurometabolite changes in some frontal areas and depressive and anxiety symptoms. We will discuss these findings in more detail below.

4.1 OCD Symptom Severity

Symptom severity as measured with the Yale-Brown Obsessive-Compulsive Scale (Y-BOCS; Goodman et al. 1989) was found to be correlated positively with concentrations of the following neurochemicals: (1) Cho in the parietal WM (Kitamura et al. 2006); (2) Glx in the rostral and dorsal ACC but only in female OCD patients (Yücel et al. 2008); (3) NAA ratios in dACC with the obsessions subscale (Yücel et al. 2008). Several studies, however, reported negative associations between the Y-BOCS scores and (1) the NAA in dACC (Ebert et al. 1997; Tükel et al. 2014) and caudate nucleus (Tükel et al. 2015); (2) Cr and Cho ratios of the rostral and dorsal ACC (O'Neill et al. 2016); (3) GABA and tNAA in the OFC including the rACC (Zhang et al. 2016); (4) Ins concentrations in the left OFC (Bédard and Chantal 2011); and (5) tCho levels in medial thalamus (Parmar et al. 2019) and Glx ratios in thalamus, specifically the compulsion subscore of the Y-BOCS (Zhu et al. 2015). There are also studies reporting no significant correlations between Y-BOCS scores and the different neurometabolites in dACC (Starck et al. 2008), posterior cingulate cortex (Brennan et al. 2016), rACC (Andrade et al., 2019; Li et al. 2019), striatum and thalamus (Hatchondo et al. 2017), dorsal caudate, dorsal putamen, and ventral striatum (Simpson et al. 2012), MPFC (Zhu et al. 2015), or DLPFC (Simpson et al. 2015).

In paediatric OCD, only one study has found a positive association between the children's Y-BOCS (CY-BOCS) scores and the concentration of Ins, Cr, and NAA in the right prefrontal white matter (Weber et al. 2014). The remaining studies did not observe any associations between the OCD symptoms as measured by the CY-BOCS and the specific neurochemical within the ACC, dorsal (Ljungberg et al. 2017) and rostral (Naaijen et al. 2017).

There is no clear trend towards specific associations between the neurometabolite concentrations and symptom severity as measured with Y-BOCS, especially in paediatric samples. The findings do not seem to be robust and thus more studies are needed to provide more reliable evidence.

4.2 Anxiety and Depression Scales (in OCD)

In adult patients with OCD, thalamic Glu levels were positively correlated with anxiety scores (Fan et al. 2017). In another study, Whiteside et al. (2006) found decreased caudate myo-inositol ratio, which was linked to anxiety traits (measured with STAI-T) but after correcting for anxiety, the differences observed became insignificant (Whiteside et al. 2006). In another group of adults patient, anxiety symptoms were negatively correlated with (1) NAA and tNAA levels in the left OFC, dACC (Bédard and Chantal 2011), and rACC (Zhang et al. 2016) and (2) Glu concentration in rACC (Zheng et al. 2020). No correlations were found between concentrations of NAA, Cho, Ins in the dACC and anxiety scores (Tükel et al. 2014).

In addition, severity of depression was negatively correlated with Cho and Cr levels in rACC and dACC (O'Neill et al. 2016). The OFC concentrations of NAA were also negatively associated with depressive symptoms (Zhang et al. 2016). There were no correlations reported between concentrations of GABA and Glu, Glx, tNAA, or Cho in the rACC and clinical variables such as anxiety, depression (de Salles Andrade et al. 2019; Li et al. 2019). Furthermore, no associations were observed between concentrations of NAA, Cho, Ins in the dACC and depression (Tükel et al. 2014).

In summary, the most robust finding seems to be the negative association between the rACC, dACC, and OFC concentrations of NAA and tNAA and comorbid anxiety symptoms. In other words, the higher the NAA level in the OFC, which included in most studies a large part of rACC, the lower the anxiety symptoms in OCD patients. Additionally, the higher the Cho and Cr levels in the anterior cingulate and the higher the NAA levels in the OFC, the lower the severity of depression symptoms seem to be in OCD patients. However, more studies are needed to confirm these findings.

5 Treatment Effects on Neurometabolite Concentrations in OCD

5.1 Cognitive Behavioural Therapy

Cognitive behavioural therapy (CBT), a treatment that ameliorates OCD symptoms in about 60% patients (Rufer et al. 2006), has also been shown to alter brain function (Olatunji et al. 2014). Several studies have shown decreased resting state glucose metabolism or blood flow in the right caudate of OCD patients with a successful treatment response (Linden 2006). Therefore, it is plausible to hypothesise that CBT may also induce changes in the concentration of neurometabolites in areas of brain of relevance for OCD.

Lower NAA levels in the rACC were found at baseline in a group of OCD patients, which were then elevated after 20 sessions of successful CBT treatment (90 min per session) as measured by reductions in their Y-BOCS scores (O'Neill et al. 2013). Additionally, baseline levels of NAA in the rACC could predict treatment outcome, with higher NAA levels at baseline correlating with better response to CBT (O'Neill et al. 2013). Importantly, these findings should be interpreted with caution, as the latter study had a sample of 10 OCD patients only and used a 1.5 T scanner and an MRSI method.

However, Zurowski et al. (2012) found no differences in NAA and Ins concentrations in rACC, OFC, and right ventral striatum, neither at baseline between OCD patients and healthy controls nor after 24 sessions of CBT (40 min per session) within patients, while the Ins levels in the OFC were negatively correlated with treatment outcome as measured with the Y-BOCS (Zurowski et al. 2012). Although these findings are in contradiction to the findings by O'Neill et al. (2013), Zurowski et al. (2012) had a slightly larger sample of unmedicated OCD patients ($N = 16$) and used a single voxel imaging technique on a 3 T scanner. One other study looking at caudate NAA observed a significant increase in patients with OCD after 16 successful CBT sessions leading to a decrease in OCD and anxiety symptoms (Whiteside et al. 2012b). Although it is important to note that the latter study included mixed medicated and unmedicated patients and a small sample size and employed a 1.5 T MRS scanner, we must also keep in mind the larger peak of the NAA may still be detected even with weaker techniques.

In paediatric OCD, opposite results were found for NAA levels in the rACC. Higher levels of NAA in the left rACC were found at baseline in a group of 5 children with OCD that were then reduced after successful CBT treatment with a mean 32.8% reduction in their CY-BOCS scores (O'Neill et al. 2012). Such opposite findings between adult and paediatric OCD could, of course, simply be due to age differences between the two samples, but such conclusions must not be drawn too hastily given that the sample size in this paediatric study was only five patients and the technique employed (MRSI) is generally less reliable (see Discussion below). Another paediatric study from the same group found no baseline differences for Glu levels in the rACC and PCC. However, the investigators reported a decrease in Glu concentration after 12–14 weekly CBT sessions, though only in the rACC (O'Neill et al. 2017). The CBT group also showed a 42.8% reduction in their CY-BOCS scores compared to the 8.6% drop in the wait-list control group. The latter study had the largest sample of paediatric OCD ($N = 44$; mostly unmedicated) and employed a 3 T MRS scanner. This Glu finding must also be interpreted with caution as it is difficult to separate Glu from Gln at 3 T.

The impact of the cognitive-behavioural intervention in the caudate neurochemistry was also assessed in 2 other paediatric studies (Benazon et al. 2003; Whiteside et al. 2012a). Whiteside et al. (2012a, 2012b) reported decreased Glx in the right caudate after the 16 sessions of successful behavioural treatment and lower levels of NAA and Cr at baseline in non-medicated patients compared to medicated patients and controls (Whiteside et al. 2012a), while Benazon et al. (2003) reported no change in metabolite concentrations after 12 sessions of behavioural therapy that led to improvements of OCD, depressive, and anxiety symptoms (Benazon et al. 2003). In the PCC, Glu levels at baseline were predictive of better treatment outcome, given its associations with a reduction in the Y-BOCS score (O'Neill et al. 2017).

In summary, no clear conclusion can be drawn based on the studies reviewed above. There might be a trend of reduced Glu and Glx after CBT treatment in, respectively, rACC and caudate and elevated NAA levels in rACC. However, better MRS studies are required to confirm these results.

5.2 SSRI Medication

It would be expected that SSRI would also influence the neurometabolite ratios within the different areas of the brain. There are two ways to study these medication effects: 1) with a within-group longitudinal design and measuring the neurometabolite concentrations before and after the SSRI treatment or 2) by having a between-group cross-sectional design and comparing medicated and unmedicated patients with one another.

Three longitudinal studies have focused on the effect of SSRI treatment on the neurometabolite concentrations of the ACC (dorsal and rostral). Two of them showed lower NAA levels in dACC at baseline, with an increase after 12 weeks of successful sertraline (Tükel et al. 2015) and citalopram (Jang et al. 2006) treatment. It is noteworthy that the latter studies had samples, respectively, of 19 and 13 unmedicated (mostly drug naïve) OCD patients using a 1.5 T scanner. A third 3 T study using a slightly larger sample (N = 21) found no neurochemical differences between OCD and healthy volunteers at baseline and no changes after 12 weeks of escitalopram treatment in OCD patients (Parmar et al. 2019). Although it is important to note that only half of the patients in the latter study responded to their SSRI treatment, whereas the other half showed no improvement in their OCD symptoms after 12 weeks of escitalopram treatment. However, they did find reduced tCho and Glx in medial thalamus after the escitalopram treatment, which were higher at baseline. In the study by Tükel et al. (2015), the lower caudate NAA levels remained unchanged after 12 weeks of successful treatment with sertraline (Tükel et al. 2015).

In a drug-naïve paediatric OCD study, the lower concentration of dorsal striatal tCho in OCD patients was not changed following 6 months of fluoxetine treatment, despite the reduction of OCD and depressive symptoms (Lazaro et al. 2012). Lastly, higher caudate Glx concentrations in a paediatric sample with OCD significantly decreased after 12 weeks of successful treatment with paroxetine (Rosenberg et al. 2000). The reduced caudate Glx levels after SSRI treatment seem to be similar to the CBT findings and would be interesting to replicate by performing more controlled studies with a more powerful scanner.

Cross-sectional studies comparing medicated and unmedicated patients have largely failed to observe significant differences in Glu and Gln in the rACC (Brennan et al. 2016); Cr, Cho, NAA, Glx, and Ins in the dorsal and rostral ACC (Yücel et al. 2008); and Glx, mI, NAA Cho, and Cr in the dACC (Yücel et al. 2007). Although the latter study found a trend for lower Glx levels in unmedicated patients, it did not reach statistical significance (Yücel et al. 2007). Only one study has found lower Glu, Glx, Cr, Cho, and Ins concentrations in the dACC of medicated patients when compared to an unmedicated group (O'Neill et al. 2016). There were also no concentration differences reported in a paediatric sample of medicated versus unmedicated patients for Glx, tNAA, mI, and total Cho in the rACC (Ortiz et al. 2015). In general, the cross-sectional studies did not identify clear differences in neurochemical levels between the medicated and unmedicated OCD groups.

In summary, the most consistent findings are increased NAA levels in the rACC and decreased caudate Glx concentrations following SSRI treatment. These findings seem to partly mirror CBT-related changes. It is important to note that the treatment studies had an average sample of 10 patients only. The absence of changes in neurochemical concentrations could be either a true absence or a false negative finding due to various limitations in methodology and design to be discussed below.

6 Discussion

To summarise, following an exhaustive review of the literature, the majority of available MRS studies appear to have failed to find convincing evidence of group differences between patients with OCD and healthy controls for Glu, Gln, and GABA in several of the key brain regions of interest. This null conclusion could mean that the differences were simply not detected due to limitations of the methods used rather than a true absence of differences. There is also no evidence of abnormality in Cr and Ins levels in OCD patients. There are however, a few findings that show some degree of consistency, such as lower NAA levels in dorsal (and to a lesser extent rostral) ACC of OCD patients. Interestingly, the lower NAA level seems to increase after successful treatment with both CBT and SSRI's and is correlated with improvement of OCD symptoms. Since NAA is a marker of neuronal density, it would probably be relevant to further investigate possible correlations with structural data in OCD and their implications to the pathophysiology of this disease. The NAA concentration seems to be sensitive to pathological brain processes and thus appears to be potentially reversible following successful treatment by either CBT or SSRI's.

Another, rather more consistent finding is a higher Cho concentration in the thalamus. Since Cho is a marker of cell membrane turnover, involved in membrane synthesis and degradation, one could hypothesise that a misadjusted concentration of this metabolite could lead to dysfunctional communication between neurons and ultimately unbalanced neurotransmission, which could contribute to OCD deficits. Further studies are warranted to enable further conclusions. Nevertheless, the question remains whether the Cho and NAA are the only neurochemicals impacted by OCD, or whether they can just be better detected due to their higher peaks compared to other neurochemicals? Studies using a higher magnetic field can potentially answer this intriguing question.

The remaining findings for both children and adults are also extremely inconsistent, with changes in both directions or null effects. It is important to highlight that all paediatric studies were universally acquired using the 1H-MR spectra on 1.5-T MRI scanners and so should be interpreted with caution due to the low signal-to-noise ratios. Figure 3 shows the summary of all findings for NAA, tNAA, Cho, tCho, and Glx in different brain regions.

The absence of differences, or mixed findings, could be due to the differences in the scanner field strengths but also the limited overlap in voxel placements. With this

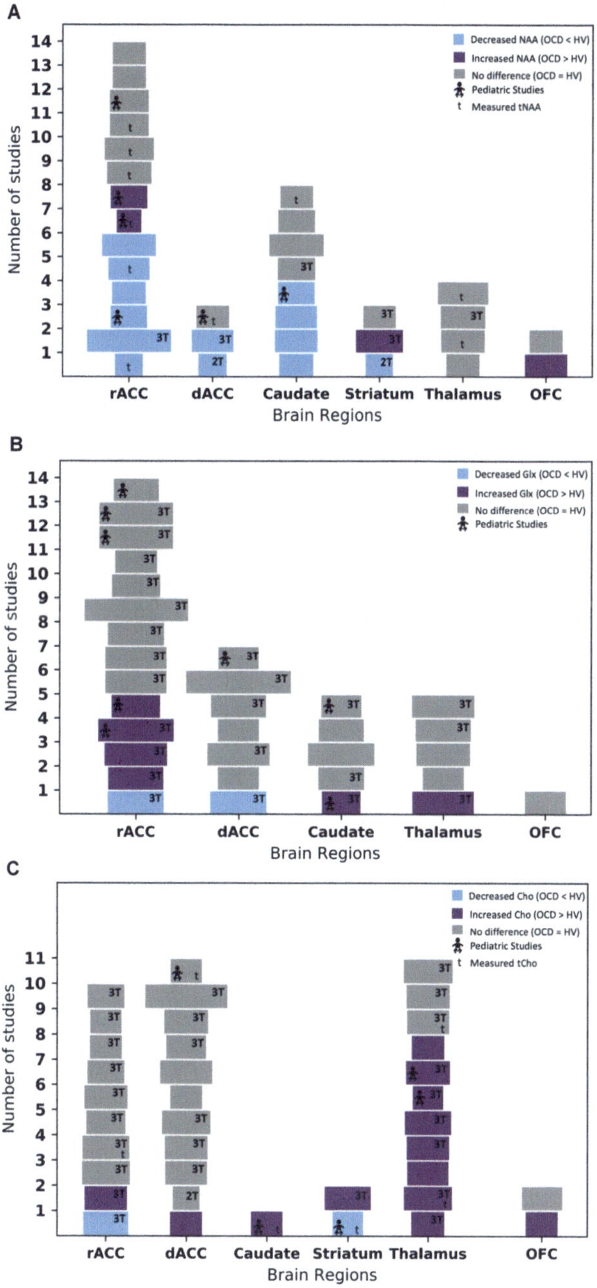

Fig. 3 Depicts a summary of the studies which used 1H-MRS to measure the brain neurochemicals of patients with obsessive-compulsive disorder (OCD) and healthy volunteers (HV). Findings are summarised by brain region: rostral anterior cingulate cortex (rACC), dorsal anterior cingulate cortex (dACC), caudate, thalamus, striatum (which includes both nucleus and putamen), and orbitofrontal cortex (OFC). (**a**) shows the findings for N-acetylaspartate (NAA) and total NAA (tNAA); (**b**) shows the findings for Glx (glutamate + glutamine); and (**c**) shows the findings for

technique, voxels are placed manually and this can cause variability between voxel placement in different scan sessions but also different studies. The lack of overlap between exact location of voxels among studies makes it difficult to arrive at definitive conclusions.

There are also problems arising from high inter-individual variability of brain metabolites or, in the case of patients, contaminating effects of medication, with variations in dose, duration, and dosing regimen, which may have led to long lasting, but variable changes. Indeed, many studies included both medicated and unmedicated patients in the same OCD group. As pharmacotherapy substantially influences neurotransmitter concentrations (Whiteside et al. 2012a) this would be expected to produce a level of confound that is unlikely to be overcome even if sample sizes were larger. Although ideally one should have more clearly defined medicated versus unmedicated groups, at least a sufficiently large sample size would allow the use of medication status as a covariate to inform us more about the role that medication can play on neurotransmitter concentration changes.

A major problem with almost all the studies reviewed in this chapter is the small sample size used, with the majority of the studies recruiting fewer than 20 patients. For example, to detect a 10% change in NAA, which is one of the easiest neurometabolites to measure, with 80% power, a sample size larger than 39 has been suggested per group (Steen et al. 2005). Only 4 studies reviewed here had a sample size meeting this criterion. Additionally, studies have tended to use scanners with a lower field strength and lower powers of spatial resolution that would not allow satisfactory separation of some relevant regions such as the caudate from putamen from one another.

Another potentially confounding variable is the difference in the sequences employed by different studies. Each sequence is optimised for a different purpose to either optimise the detection of some neurochemicals over others or to be sensitive to the size of voxel selected. All these factors could lead to variable data quality and accuracy of detecting specific neurometabolites in different regions. Another confounding variable could be the difference in reference signals used to normalise the data, although the majority of studies has used creatine to calculate the metabolite ratios. Moreover, some studies have performed brain tissue content corrections, whereas others did not.

The lack of consistency of the set-up parameters and the type of MRS sequences used by the different studies provide yet further sources of confound. There are some parameters that can be set before the data collection starts, such as calculating the best transmitter voltage and flip angle for water saturation pulse. But there are also other noise producing parameters such as poor water suppression and shimming (see Box 1) that must be carefully scrutinised immediately after acquisition of

---

Fig. 3 (continued) choline (Cho) and total Cho (tCho). The letter 't' indicates studies measuring the total NAA in (**a**) or total choline in (**b**). The baby icon indicates studies in paediatric population. The power of the magnetic field used in the studies is also indicated: 3 T, 2 T, and if none, it is 1.5 T. The width of each block corresponds to square root of the total sample size (OCD + HV) of each study

spectroscopic findings for each voxel so that, if necessary, the scan for that voxel could be repeated after determining the possible cause of noise in the data.

7 Conclusions

Despite the aforementioned difficulties, the MRS studies so far conducted provide preliminary findings that are quite promising. They have been helpful in establishing current issues causing reduced data reliability. Nevertheless, they indicate that the methodology is promising and could be much more valuable if better controlled study designs could be implemented, preferably using stronger magnetic fields, e.g. 7 T. For example, by deciding to include either medicated or unmedicated patients in one group and not both or including sufficiently large sample sizes that would allow controlling for medication type, dosage, and treatment duration. Once these design issues are overcome, MRS studies are well placed to inform the molecular underpinning of cognitive problems at the basis of OCD. Future developments in the PET arm of the functional neuroimaging methods will further enable achieving such a goal. Researchers are working on developing novel radiotracers that will enable better in vivo assessment of the serotoninergic and glutamatergic systems of the human brain. A future way forward could involve combining the PET and MRS methods to 1) detect changes in the concentrations of Glu and GABA as well as in their receptors and 2) understand the interaction between the different neurotransmitters and their link to the functional and structural changes of the CSTC pathways in OCD. A multimodal neuroimaging approach such as this will help to gain a better picture of the pathophysiology of OCD and how it responds to treatments. In turn, the findings may lead to treatment options that could include not only new drugs but also better targeted brain stimulation, e.g. through transcranial alternating current stimulation (tACS) or repetitive transcranial magnetic stimulation (rTMS).

Acknowledgements MB was supported by her studentship from the Mental Health Research UK. LCM has received support from the Swiss National Science Foundation: Doc-mobility grant No. P1GEP3_191308. PB was supported by the Sir Henry Wellcome Trust Postdoctoral Research Fellowship (204727/Z/16/Z).

References

Adams KH, Hansen ES, Pinborg LH, Hasselbalch SG, Svarer C, Holm S et al (2005) Patients with obsessive-compulsive disorder have increased 5-HT2A receptor binding in the caudate nuclei. Int J Neuropsychopharmacol 8(3):391–401. https://doi.org/10.1017/s1461145705005055

Akkus F, Terbeck S, Ametamey SM, Rufer M, Treyer V, Burger C et al (2014) Metabotropic glutamate receptor 5 binding in patients with obsessive-compulsive disorder. Int J Neuropsychopharmacol 17(12):1915–1922. https://doi.org/10.1017/s1461145714000716

Alcaraz F, Fresno V, Marchand AR, Kremer EJ, Coutureau E, Wolff M (2018) Thalamocortical and corticothalamic pathways differentially contribute to goal-directed behaviors in the rat. eLife 7: e32517. https://doi.org/10.7554/eLife.32517

Alger JR (2010) Quantitative proton magnetic resonance spectroscopy and spectroscopic imaging of the brain: a didactic review. Top Magn Reson Imaging 21(2):115–128. https://doi.org/10.1097/RMR.0b013e31821e568f

Alvarenga PG, do Rosário MC, Batistuzzo MC, Diniz JB, Shavitt RG, Duran FL et al (2012) Obsessive-compulsive symptom dimensions correlate to specific gray matter volumes in treatment-naïve patients. J Psychiatr Res 46(12):1635–1642. https://doi.org/10.1016/j.jpsychires.2012.09.002

Apergis-Schoute AM, Bijleveld B, Gillan CM, Fineberg NA, Sahakian BJ, Robbins TW (2018) Hyperconnectivity of the ventromedial prefrontal cortex in obsessive-compulsive disorder. Brain Neurosci Adv 2:2398212818808710. https://doi.org/10.1177/2398212818808710

Atmaca M, Yildirim H, Ozdemir H, Koc M, Ozler S, Tezcan E (2009) Neurochemistry of the hippocampus in patients with obsessive-compulsive disorder. Psychiatry Clin Neurosci 63 (4):486–490. https://doi.org/10.1111/j.1440-1819.2009.01993.x

Bagory M, Durand-Dubief F, Ibarrola D, Confavreux C, Sappey-Marinier D (2007) "Absolute" quantification in magnetic resonance spectroscopy: validation of a clinical protocol in multiple sclerosis. Conf Proc IEEE Eng Med Biol Soc 2007:3458–3461. https://doi.org/10.1109/iembs.2007.4353075

Banca P, Voon V, Vestergaard MD, Philipiak G, Almeida I, Pocinho F et al (2015) Imbalance in habitual versus goal directed neural systems during symptom provocation in obsessive-compulsive disorder. Brain 138(Pt 3):798–811. https://doi.org/10.1093/brain/awu379

Bartha R, Stein MB, Williamson PC, Drost DJ, Neufeld RW, Carr TJ et al (1998) A short echo 1H spectroscopy and volumetric MRI study of the corpus striatum in patients with obsessive-compulsive disorder and comparison subjects. Am J Psychiatry 155(11):1584–1591. https://doi.org/10.1176/ajp.155.11.1584

Baxter LR (1990) Brain imaging as a tool in establishing a theory of brain pathology in obsessive compulsive disorder. J Clin Psychiatry 51(Suppl):22–25; discussion 26

Bédard MJ, Chantal S (2011) Brain magnetic resonance spectroscopy in obsessive-compulsive disorder: the importance of considering subclinical symptoms of anxiety and depression. Psychiatry Res 192(1):45–54. https://doi.org/10.1016/j.pscychresns.2010.10.008

Benazon NR, Moore GJ, Rosenberg DR (2003) Neurochemical analyses in pediatric obsessive-compulsive disorder in patients treated with cognitive-behavioral therapy. J Am Acad Child Adolesc Psychiatry 42(11):1279–1285. https://doi.org/10.1097/01.chi.0000087562.01900.de

Boer VO, Klomp DWJ (2014) Chapter 1.3 – technical considerations for multivoxel approaches and magnetic resonance spectroscopic imaging. In: Stagg C, Rothman D (eds) Magnetic resonance spectroscopy. Academic Press, San Diego, pp 31–39

Brennan BP, Rauch SL, Jensen JE, Pope HG Jr (2013) A critical review of magnetic resonance spectroscopy studies of obsessive-compulsive disorder. Biol Psychiatry 73(1):24–31. https://doi.org/10.1016/j.biopsych.2012.06.023

Brennan BP, Jensen JE, Perriello C, Pope HG Jr, Jenike MA, Hudson JI et al (2016) Lower posterior cingulate cortex glutathione levels in obsessive-compulsive disorder. Biol Psychiatry Cogn Neurosci Neuroimaging 1(2):116–124. https://doi.org/10.1016/j.bpsc.2015.12.003

Brody AL, Saxena S, Schwartz JM, Stoessel PW, Maidment K, Phelps ME, Baxter LR Jr (1998) FDG-PET predictors of response to behavioral therapy and pharmacotherapy in obsessive compulsive disorder. Psychiatry Res 84(1):1–6. https://doi.org/10.1016/s0925-4927(98)00041-9

Brownell A-L, Jenkins BG, Elmaleh DR, Deacon TW, Spealman RD, Isacson O (1998) Combined PET/MRS brain studies show dynamic and long-term physiological changes in a primate model of Parkinson disease. Nat Med 4(11):1308–1312. https://doi.org/10.1038/3300

Celada P, Puig MV, Artigas F (2013) Serotonin modulation of cortical neurons and networks. Front Integr Neurosci 7:25–25. https://doi.org/10.3389/fnint.2013.00025

Chamberlain SR, Menzies L, Hampshire A, Suckling J, Fineberg NA, del Campo N et al (2008) Orbitofrontal dysfunction in patients with obsessive-compulsive disorder and their unaffected relatives. Science 321(5887):421–422. https://doi.org/10.1126/science.1154433

Chen F, Guo M, Liu T, Jiang XL, Li JJ, Guo JC (2017) Magnetic resonance spectroscopy of brain circuits in patients with first-episode obsessive-compulsive disorder. Iran J Radiol 14(1): e13452. https://doi.org/10.5812/iranjradiol.15566

Choi C, Dimitrov IE, Douglas D, Patel A, Kaiser LG, Amezcua CA, Maher EA (2010) Improvement of resolution for brain coupled metabolites by optimized (1)H MRS at 7T. NMR Biomed 23(9):1044–1052. https://doi.org/10.1002/nbm.1529

Ciranna L (2006) Serotonin as a modulator of glutamate- and GABA-mediated neurotransmission: implications in physiological functions and in pathology. Curr Neuropharmacol 4(2):101–114. https://doi.org/10.2174/157015906776359540

Cools R, Roberts AC, Robbins TW (2008) Serotoninergic regulation of emotional and behavioural control processes. Trends Cogn Sci 12(1):31–40. https://doi.org/10.1016/j.tics.2007.10.011

de Graaf RA, Rothman DL, Behar KL (2011) State of the art direct 13C and indirect 1H-[13C] NMR spectroscopy in vivo. A practical guide. NMR Biomed 24(8):958–972. https://doi.org/10.1002/nbm.1761

de Salles Andrade JB, Ferreira FM, Suo C, Yücel M, Frydman I, Monteiro M et al (2019) An MRI study of the metabolic and structural abnormalities in obsessive-compulsive disorder. Front Hum Neurosci 13:186–186. https://doi.org/10.3389/fnhum.2019.00186

de Wit SJ, Alonso P, Schweren L, Mataix-Cols D, Lochner C, Menchón JM et al (2014) Multicenter voxel-based morphometry mega-analysis of structural brain scans in obsessive-compulsive disorder. Am J Psychiatry 171(3):340–349. https://doi.org/10.1176/appi.ajp.2013.13040574

Denys D, de Vries F, Cath D, Figee M, Vulink N, Veltman DJ et al (2013) Dopaminergic activity in Tourette syndrome and obsessive-compulsive disorder. Eur Neuropsychopharmacol 23 (11):1423–1431. https://doi.org/10.1016/j.euroneuro.2013.05.012

Dold M, Aigner M, Lanzenberger R, Kasper S (2015) Antipsychotic augmentation of serotonin reuptake inhibitors in treatment-resistant obsessive-compulsive disorder: an update meta-analysis of double-blind, randomized, placebo-controlled trials. Int J Neuropsychopharmacol 18(9):pyv047. https://doi.org/10.1093/ijnp/pyv047

Ebert D, Speck O, Konig A, Berger M, Hennig J, Hohagen F (1997) 1H-magnetic resonance spectroscopy in obsessive-compulsive disorder: evidence for neuronal loss in the cingulate gyrus and the right striatum. Psychiatry Res 74(3):173–176. https://doi.org/10.1016/s0925-4927(97)00016-4

Ebisu T, Rooney WD, Graham SH, Weiner MW, Maudsley AA (1994) N-Acetylaspartate as an in vivo marker of neuronal viability in Kainate-induced status Epilepticus: 1H magnetic resonance spectroscopic imaging. J Cereb Blood Flow Metab 14(3):373–382. https://doi.org/10.1038/jcbfm.1994.48

Eng GK, Sim K, Chen SH (2015) Meta-analytic investigations of structural grey matter, executive domain-related functional activations, and white matter diffusivity in obsessive compulsive disorder: an integrative review. Neurosci Biobehav Rev 52:233–257. https://doi.org/10.1016/j.neubiorev.2015.03.002

Fan S, Cath DC, van den Heuvel OA, van der Werf YD, Schöls C, Veltman DJ, Pouwels PJW (2017) Abnormalities in metabolite concentrations in tourette's disorder and obsessive-compulsive disorder-A proton magnetic resonance spectroscopy study. Psychoneuroendocrinology 77:211–217. https://doi.org/10.1016/j.psyneuen.2016.12.007

Figee M, de Koning P, Klaassen S, Vulink N, Mantione M, van den Munckhof P et al (2014) Deep brain stimulation induces striatal dopamine release in obsessive-compulsive disorder. Biol Psychiatry 75(8):647–652. https://doi.org/10.1016/j.biopsych.2013.06.021

Fisher SK, Novak JE, Agranoff BW (2002) Inositol and higher inositol phosphates in neural tissues: homeostasis, metabolism and functional significance. J Neurochem 82(4):736–754. https://doi.org/10.1046/j.1471-4159.2002.01041.x

Fitzgerald KD, Moore GJ, Paulson LA, Stewart CM, Rosenberg DR (2000) Proton spectroscopic imaging of the thalamus in treatment-naive pediatric obsessive-compulsive disorder. Biol Psychiatry 47(3):174–182. https://doi.org/10.1016/s0006-3223(99)00286-3

Gillan CM, Morein-Zamir S, Kaser M, Fineberg NA, Sule A, Sahakian BJ et al (2014) Counterfactual processing of economic action-outcome alternatives in obsessive-compulsive disorder: further evidence of impaired goal-directed behavior. Biol Psychiatry 75(8):639–646. https://doi.org/10.1016/j.biopsych.2013.01.018

Glover GH (2011) Overview of functional magnetic resonance imaging. Neurosurg Clin N Am 22 (2):133–vii. https://doi.org/10.1016/j.nec.2010.11.001

Gnanavel S, Sharan P, Khandelwal S, Sharma U, Jagannathan NR (2014) Neurochemicals measured by (1)H-MR spectroscopy: putative vulnerability biomarkers for obsessive compulsive disorder. MAGMA 27(5):407–417. https://doi.org/10.1007/s10334-013-0427-y

Goodman WK, Price LH, Rasmussen SA, Mazure C, Delgado P, Heninger GR, Charney DS (1989) The Yale-Brown obsessive compulsive scale. II. Validity. Arch Gen Psychiatry 46 (11):1012–1016. https://doi.org/10.1001/archpsyc.1989.01810110054008

Govindaraju V, Young K, Maudsley AA (2000) Proton NMR chemical shifts and coupling constants for brain metabolites. NMR Biomed 13(3):129–153. https://doi.org/10.1002/1099-1492(200005)13:3<129::Aid-nbm619>3.0.Co;2-v

Graat I, Figee M, Denys D (2017) Neurotransmitter dysregulation in OCD. In: Pittenger C (ed) Obsessive-compulsive disorder: phenomenology, pathophysiology, and treatment. Oxford University Press, Oxford, pp 271–288

Gruetter R (1993) Automatic, localized in vivo adjustment of all first- and second-order shim coils. Magn Reson Med 29(6):804–811. https://doi.org/10.1002/mrm.1910290613

Haase A, Frahm J, Merboldt KD, Matthaei D, Hanicke W, Deimling M, Weber H (1986) Chemical-shift-selective (CHESS) FLASH MR imaging. Radiological Society of North America Inc

Haber SN, Behrens TE (2014) The neural network underlying incentive-based learning: implications for interpreting circuit disruptions in psychiatric disorders. Neuron 83(5):1019–1039. https://doi.org/10.1016/j.neuron.2014.08.031

Hardingham GE, Do KQ (2016) Linking early-life NMDAR hypofunction and oxidative stress in schizophrenia pathogenesis. Nat Rev Neurosci 17(2):125–134. https://doi.org/10.1038/nrn.2015.19

Harrison BJ, Pujol J, Cardoner N, Deus J, Alonso P, López-Solà M et al (2013) Brain corticostriatal systems and the major clinical symptom dimensions of obsessive-compulsive disorder. Biol Psychiatry 73(4):321–328. https://doi.org/10.1016/j.biopsych.2012.10.006

Hatchondo L, Jaafari N, Langbour N, Maillochaud S, Herpe G, Guillevin R, Guillevin C (2017) (1) H magnetic resonance spectroscopy suggests neural membrane alteration in specific regions involved in obsessive-compulsive disorder. Psychiatry Res Neuroimaging 269:48–53. https://doi.org/10.1016/j.pscychresns.2017.08.010

Hertz L, Zielke HR (2004) Astrocytic control of glutamatergic activity: astrocytes as stars of the show. Trends Neurosci 27(12):735–743. https://doi.org/10.1016/j.tins.2004.10.008

Hesse S, Stengler K, Regenthal R, Patt M, Becker G-A, Franke A et al (2011) The serotonin transporter availability in untreated early-onset and late-onset patients with obsessive–compulsive disorder. Int J Neuropsychopharmacol 14(5):606–617. https://doi.org/10.1017/S1461145710001604

Honey GD, Suckling J, Zelaya F, Long C, Routledge C, Jackson S et al (2003) Dopaminergic drug effects on physiological connectivity in a human cortico-striato-thalamic system. Brain J Neurol 126(8):1767–1781. https://doi.org/10.1093/brain/awg184

Jackson SAW, Horst NK, Pears A, Robbins TW, Roberts AC (2016) Role of the perigenual anterior cingulate and orbitofrontal cortex in contingency learning in the marmoset. Cereb Cortex 26 (7):3273–3284. https://doi.org/10.1093/cercor/bhw067

Jang JH, Kwon JS, Jang DP, Moon WJ, Lee JM, Ha TH et al (2006) A proton MRSI study of brain N-acetylaspartate level after 12 weeks of citalopram treatment in drug-naive patients with obsessive-compulsive disorder. Am J Psychiatry 163(7):1202–1207. https://doi.org/10.1176/appi.ajp.163.7.1202

Juchem C, Nixon TW, Diduch P, Rothman DL, Starewicz P, de Graaf RA (2010) Dynamic shimming of the human brain at 7 tesla. Concepts Magn Reson Part B Magn Reson Eng 37B (3):116–128. https://doi.org/10.1002/cmr.b.20169

Juchem C, Nixon TW, McIntyre S, Boer VO, Rothman DL, de Graaf RA (2011) Dynamic multi-coil shimming of the human brain at 7 T. J Magnet Resonance 212(2):280–288. https://doi.org/10.1016/j.jmr.2011.07.005

Kehagia AA, Murray GK, Robbins TW (2010) Learning and cognitive flexibility: frontostriatal function and monoaminergic modulation. Curr Opin Neurobiol 20(2):199–204. https://doi.org/10.1016/j.conb.2010.01.007

Kitamura H, Shioiri T, Kimura T, Ohkubo M, Nakada T, Someya T (2006) Parietal white matter abnormalities in obsessive-compulsive disorder: a magnetic resonance spectroscopy study at 3-tesla. Acta Psychiatr Scand 114(2):101–108. https://doi.org/10.1111/j.1600-0447.2006.00858.x

Kong XZ, Boedhoe PSW, Abe Y, Alonso P, Ameis SH, Arnold PD et al (2019) Mapping cortical and subcortical asymmetry in obsessive-compulsive disorder: findings from the ENIGMA consortium. Biol Psychiatry. https://doi.org/10.1016/j.biopsych.2019.04.022

Krabbe S, Duda J, Schiemann J, Poetschke C, Schneider G, Kandel ER et al (2015) Increased dopamine D2 receptor activity in the striatum alters the firing pattern of dopamine neurons in the ventral tegmental area. Proc Natl Acad Sci U S A 112(12):E1498–E1506. https://doi.org/10.1073/pnas.1500450112

Kraguljac NV, Reid M, White D, Jones R, den Hollander J, Lowman D, Lahti AC (2012) Neurometabolites in schizophrenia and bipolar disorder – a systematic review and meta-analysis. Psychiatry Res 203(2–3):111–125. https://doi.org/10.1016/j.pscychresns.2012.02.003

Krebs HA (1935) Metabolism of amino-acids: the synthesis of glutamine from glutamic acid and ammonia, and the enzymic hydrolysis of glutamine in animal tissues. Biochem J 29 (8):1951–1969. https://doi.org/10.1042/bj0291951

Lazaro L, Bargallo N, Andres S, Falcon C, Morer A, Junque C, Castro-Fornieles J (2012) Proton magnetic resonance spectroscopy in pediatric obsessive-compulsive disorder: longitudinal study before and after treatment. Psychiatry Res 201(1):17–24. https://doi.org/10.1016/j.pscychresns.2011.01.017

Li Y, Zhang CC, Kathrin W, Zhang Y, He N, Jin H et al (2019) Investigation of anterior cingulate cortex gamma-aminobutyric acid and glutamate-glutamine levels in obsessive-compulsive disorder using magnetic resonance spectroscopy. BMC Psychiatry 19(1):164. https://doi.org/10.1186/s12888-019-2160-1

Linden DE (2006) How psychotherapy changes the brain--the contribution of functional neuroimaging. Mol Psychiatry 11(6):528–538. https://doi.org/10.1038/sj.mp.4001816

Ljungberg M, Nilsson MKL, Melin K, Jönsson L, Carlsson A, Carlsson Å et al (2017) 1H magnetic resonance spectroscopy evidence for occipital involvement in treatment-naive paediatric obsessive-compulsive disorder. Acta Neuropsychiatr 29(3):179–190. https://doi.org/10.1017/neu.2016.52

Marsman A, van den Heuvel MP, Klomp DWJ, Kahn RS, Luijten PR, Hulshoff Pol HE (2013) Glutamate in schizophrenia: a focused review and meta-analysis of ^{1}H-MRS studies. Schizophr Bull 39(1):120–129. https://doi.org/10.1093/schbul/sbr069

Mathews CA, Perez VB, Delucchi KL, Mathalon DH (2012) Error-related negativity in individuals with obsessive–compulsive symptoms: toward an understanding of hoarding behaviors. Biol Psychol 89(2):487–494. https://doi.org/10.1016/j.biopsycho.2011.12.018

Matsumoto R, Ichise M, Ito H, Ando T, Takahashi H, Ikoma Y et al (2010) Reduced serotonin transporter binding in the insular cortex in patients with obsessive-compulsive disorder: a [11C] DASB PET study. NeuroImage 49(1):121–126. https://doi.org/10.1016/j.neuroimage.2009.07.069

McGovern RA, Sheth SA (2017) Role of the dorsal anterior cingulate cortex in obsessive-compulsive disorder: converging evidence from cognitive neuroscience and psychiatric neurosurgery. J Neurosurg 126(1):132–147. https://doi.org/10.3171/2016.1.Jns15601

Milad MR, Rauch SL (2012) Obsessive-compulsive disorder: beyond segregated cortico-striatal pathways. Trends Cogn Sci 16(1):43–51. https://doi.org/10.1016/j.tics.2011.11.003

Miller BL (1991) A review of chemical issues in 1H NMR spectroscopy: N-acetyl-L-aspartate, creatine and choline. NMR Biomed 4(2):47–52. https://doi.org/10.1002/nbm.1940040203

Mirza Y, O'Neill J, Smith EA, Russell A, Smith JM, Banerjee SP et al (2006) Increased medial thalamic creatine-phosphocreatine found by proton magnetic resonance spectroscopy in children with obsessive-compulsive disorder versus major depression and healthy controls. J Child Neurol 21(2):106–111. https://doi.org/10.1177/08830738060210020201

Moon C-M, Jeong G-W (2018) Associations of neurofunctional, morphometric and metabolic abnormalities with clinical symptom severity and recognition deficit in obsessive–compulsive disorder. J Affect Disord 227:603–612. https://doi.org/10.1016/j.jad.2017.11.059

Morgiève M, N'Diaye K, Haynes WI, Granger B, Clair AH, Pelissolo A, Mallet L (2014) Dynamics of psychotherapy-related cerebral haemodynamic changes in obsessive compulsive disorder using a personalized exposure task in functional magnetic resonance imaging. Psychol Med 44 (7):1461–1473. https://doi.org/10.1017/s0033291713002237

Naaijen J, Zwiers MP, Amiri H, Williams SCR, Durston S, Oranje B et al (2017) Fronto-striatal glutamate in autism Spectrum disorder and obsessive compulsive disorder. Neuropsychopharmacology 42(12):2456–2465. https://doi.org/10.1038/npp.2016.260

Newberg AB, Moss AS, Monti DA, Alavi A (2011) Positron emission tomography in psychiatric disorders. Ann N Y Acad Sci 1228:E13–E25. https://doi.org/10.1111/j.1749-6632.2011.06162.x

Nixon TW, McIntyre S, Rothman DL, de Graaf RA (2008) Compensation of gradient-induced magnetic field perturbations. J Magn Reson 192(2):209–217. https://doi.org/10.1016/j.jmr.2008.02.016

Ohara K, Isoda H, Suzuki Y, Takehara Y, Ochiai M, Takeda H et al (1999) Proton magnetic resonance spectroscopy of lenticular nuclei in obsessive-compulsive disorder. Psychiatry Res 92 (2–3):83–91. https://doi.org/10.1016/s0925-4927(99)00040-2

Olatunji BO, Ferreira-Garcia R, Caseras X, Fullana MA, Wooderson S, Speckens A et al (2014) Predicting response to cognitive behavioral therapy in contamination-based obsessive-compulsive disorder from functional magnetic resonance imaging. Psychol Med 44(10):2125–2137. https://doi.org/10.1017/s0033291713002766

Olver JS, O'Keefe G, Jones GR, Burrows GD, Tochon-Danguy HJ, Ackermann U et al (2009) Dopamine D1 receptor binding in the striatum of patients with obsessive-compulsive disorder. J Affect Disord 114(1–3):321–326. https://doi.org/10.1016/j.jad.2008.06.020

O'Neill J, Piacentini JC, Chang S, Levitt JG, Rozenman M, Bergman L et al (2012) MRSI correlates of cognitive-behavioral therapy in pediatric obsessive-compulsive disorder. Prog Neuro-Psychopharmacol Biol Psychiatry 36(1):161–168. https://doi.org/10.1016/j.pnpbp.2011.09.007

O'Neill J, Gorbis E, Feusner JD, Yip JC, Chang S, Maidment KM et al (2013) Effects of intensive cognitive-behavioral therapy on cingulate neurochemistry in obsessive-compulsive disorder. J Psychiatr Res 47(4):494–504. https://doi.org/10.1016/j.jpsychires.2012.11.010

O'Neill J, Lai TM, Sheen C, Salgari GC, Ly R, Armstrong C et al (2016) Cingulate and thalamic metabolites in obsessive-compulsive disorder. Psychiatry Res Neuroimaging 254:34–40. https://doi.org/10.1016/j.pscychresns.2016.05.005

O'Neill J, Piacentini J, Chang S, Ly R, Lai TM, Armstrong CC et al (2017) Glutamate in pediatric obsessive-compulsive disorder and response to cognitive-behavioral therapy: randomized clinical trial. Neuropsychopharmacology 42(12):2414–2422. https://doi.org/10.1038/npp.2017.77

Ortiz AE, Ortiz AG, Falcon C, Morer A, Plana MT, Bargalló N, Lázaro L (2015) 1H-MRS of the anterior cingulate cortex in childhood and adolescent obsessive-compulsive disorder: a case-control study. Eur Neuropsychopharmacol 25(1):60–68. https://doi.org/10.1016/j.euroneuro.2014.11.007

Park SE, Choi NG, Jeong GW (2017) Metabolic abnormality in the right dorsolateral prefrontal cortex in patients with obsessive-compulsive disorder: proton magnetic resonance spectroscopy. Acta Neuropsychiatr 29(3):164–169. https://doi.org/10.1017/neu.2016.48

Parmar A, Sharan P, Khandelwal SK, Agarwal K, Sharma U, Jagannathan NR (2019) Brain neurochemistry in unmedicated obsessive–compulsive disorder patients and effects of 12-week escitalopram treatment: 1H-magnetic resonance spectroscopy study. Psychiatry Clin Neurosci 73(7):386–393. https://doi.org/10.1111/pcn.12850

Parnaudeau S, Taylor K, Bolkan SS, Ward RD, Balsam PD, Kellendonk C (2015) Mediodorsal thalamus hypofunction impairs flexible goal-directed behavior. Biol Psychiatry 77(5):445–453. https://doi.org/10.1016/j.biopsych.2014.03.020

Perani D, Garibotto V, Gorini A, Moresco RM, Henin M, Panzacchi A et al (2008) In vivo PET study of 5HT2A serotonin and D2 dopamine dysfunction in drug-naive obsessive-compulsive disorder. NeuroImage 42(1):306–314. https://doi.org/10.1016/j.neuroimage.2008.04.233

Pergola G, Danet L, Pitel AL, Carlesimo GA, Segobin S, Pariente J et al (2018) The regulatory role of the human Mediodorsal thalamus. Trends Cogn Sci 22(11):1011–1025. https://doi.org/10.1016/j.tics.2018.08.006

Pittenger C, Adams TG Jr, Gallezot J-D, Crowley MJ, Nabulsi N, James R et al (2016) OCD is associated with an altered association between sensorimotor gating and cortical and subcortical 5-HT1b receptor binding. J Affect Disord 196:87–96. https://doi.org/10.1016/j.jad.2016.02.021

Ramadan S, Lin A, Stanwell P (2013) Glutamate and glutamine: a review of in vivo MRS in the human brain. NMR Biomed 26(12):1630–1646. https://doi.org/10.1002/nbm.3045

Reimold M, Smolka MN, Zimmer A, Batra A, Knobel A, Solbach C et al (2007) Reduced availability of serotonin transporters in obsessive-compulsive disorder correlates with symptom severity – a [11C]DASB PET study. J Neural Transm 114(12):1603–1609. https://doi.org/10.1007/s00702-007-0785-6

Remijnse PL, Nielen MM, van Balkom AJ, Cath DC, van Oppen P, Uylings HB, Veltman DJ (2006) Reduced orbitofrontal-striatal activity on a reversal learning task in obsessive-compulsive disorder. Arch Gen Psychiatry 63(11):1225–1236. https://doi.org/10.1001/archpsyc.63.11.1225

Robbins TW, Vaghi MM, Banca P (2019) Obsessive-compulsive disorder: puzzles and prospects. Neuron 102(1):27–47. https://doi.org/10.1016/j.neuron.2019.01.046

Rosenberg DR, MacMaster FP, Keshavan MS, Fitzgerald KD, Stewart CM, Moore GJ (2000) Decrease in caudate glutamatergic concentrations in pediatric obsessive-compulsive disorder patients taking paroxetine. J Am Acad Child Adolesc Psychiatry 39(9):1096–1103. https://doi.org/10.1097/00004583-200009000-00008

Rosenberg DR, Amponsah A, Sullivan A, MacMillan S, Moore GJ (2001) Increased medial thalamic choline in pediatric obsessive-compulsive disorder as detected by quantitative in vivo spectroscopic imaging. J Child Neurol 16(9):636–641. https://doi.org/10.1177/088307380101600902

Rosenberg DR, Mirza Y, Russell A, Tang J, Smith JM, Banerjee SP et al (2004) Reduced anterior cingulate glutamatergic concentrations in childhood OCD and major depression versus healthy controls. J Am Acad Child Adolesc Psychiatry 43(9):1146–1153. https://doi.org/10.1097/01.chi.0000132812.44664.2d

Rotge JY, Guehl D, Dilharreguy B, Tignol J, Bioulac B, Allard M et al (2009) Meta-analysis of brain volume changes in obsessive-compulsive disorder. Biol Psychiatry 65(1):75–83. https://doi.org/10.1016/j.biopsych.2008.06.019

Rufer M, Fricke S, Moritz S, Kloss M, Hand I (2006) Symptom dimensions in obsessive-compulsive disorder: prediction of cognitive-behavior therapy outcome. Acta Psychiatr Scand 113(5):440–446. https://doi.org/10.1111/j.1600-0447.2005.00682.x

Rushworth MFS, Walton ME, Kennerley SW, Bannerman DM (2004) Action sets and decisions in the medial frontal cortex. Trends Cogn Sci 8(9):410–417. https://doi.org/10.1016/j.tics.2004.07.009

Sachdev PS, McBride R, Loo CK, Mitchell PB, Malhi GS, Croker VM (2001) Right versus left prefrontal transcranial magnetic stimulation for obsessive-compulsive disorder: a preliminary investigation. J Clin Psychiatry 62(12):981–984. https://doi.org/10.4088/jcp.v62n1211

Sachdev PS, Loo CK, Mitchell PB, McFarquhar TF, Malhi GS (2007) Repetitive transcranial magnetic stimulation for the treatment of obsessive compulsive disorder: a double-blind controlled investigation. Psychol Med 37(11):1645–1649. https://doi.org/10.1017/s0033291707001092

Saxena S, Rauch SL (2000) Functional neuroimaging and the NEUROANATOMY of obsessive-compulsive disorder. Psychiatr Clin 23(3):563–586. https://doi.org/10.1016/S0193-953X(05)70181-7

Saxena S, Brody AL, Schwartz JM, Baxter LR (1998) Neuroimaging and frontal-subcortical circuitry in obsessive-compulsive disorder. Br J Psychiatry 35:26–37

Saxena S, Brody AL, Maidment KM, Dunkin JJ, Colgan M, Alborzian S et al (1999) Localized orbitofrontal and subcortical metabolic changes and predictors of response to paroxetine treatment in obsessive-compulsive disorder. Neuropsychopharmacology 21(6):683–693. https://doi.org/10.1016/s0893-133x(99)00082-2

Schwartz JM, Stoessel PW, Baxter LR Jr, Martin KM, Phelps ME (1996) Systematic changes in cerebral glucose metabolic rate after successful behavior modification treatment of obsessive-compulsive disorder. Arch Gen Psychiatry 53(2):109–113. https://doi.org/10.1001/archpsyc.1996.01830020023004

Shen J, Rothman DL, Hetherington HP, Pan JW (1999) Linear projection method for automatic slice shimming. Magn Reson Med 42(6):1082–1088. https://doi.org/10.1002/(sici)1522-2594(199912)42:6<1082::aid-mrm12>3.0.co;2-g

Simpson HB, Lombardo I, Slifstein M, Huang HY, Hwang D-R, Abi-Dargham A et al (2003) Serotonin transporters in obsessive-compulsive disorder: a positron emission tomography study with [11C]McN 5652. Biological Psychiatry 54(12):1414–1421. https://doi.org/10.1016/S0006-3223(03)00544-4

Simpson HB, Slifstein M, Bender J, Xu X, Hackett E, Maher MJ, Abi-Dargham A (2011) Serotonin 2A receptors in obsessive-compulsive disorder: a positron emission tomography study with [11C]MDL 100907. Biological Psychiatry 70(9):897–904. https://doi.org/10.1016/j.biopsych.2011.06.023

Simpson HB, Shungu DC, Bender J Jr, Mao X, Xu X, Slifstein M, Kegeles LS (2012) Investigation of cortical glutamate-glutamine and γ-aminobutyric acid in obsessive-compulsive disorder by proton magnetic resonance spectroscopy. Neuropsychopharmacology 37(12):2684–2692. https://doi.org/10.1038/npp.2012.132

Simpson HB, Kegeles LS, Hunter L, Mao X, Van Meter P, Xu X et al (2015) Assessment of glutamate in striatal subregions in obsessive-compulsive disorder with proton magnetic resonance spectroscopy. Psychiatry Res 232(1):65–70. https://doi.org/10.1016/j.pscychresns.2015.01.009

Smith EA, Russell A, Lorch E, Banerjee SP, Rose M, Ivey J et al (2003) Increased medial thalamic choline found in pediatric patients with obsessive-compulsive disorder versus major depression or healthy control subjects: a magnetic resonance spectroscopy study. Biol Psychiatry 54(12):1399–1405. https://doi.org/10.1016/s0006-3223(03)00474-8

Soomro GM, Altman D, Rajagopal S, Oakley-Browne M (2008) Selective serotonin re-uptake inhibitors (SSRIs) versus placebo for obsessive compulsive disorder (OCD). Cochrane Database Syst Rev 2008(1):CD001765–CD001765. https://doi.org/10.1002/14651858.CD001765.pub3

Sossi V (2018) Chapter one – advances in PET methodology. In: Politis M (ed) International review of neurobiology, vol 141. Academic Press, San Diego, pp 3–30

Starck G, Ljungberg M, Nilsson M, Jönsson L, Lundberg S, Ivarsson T et al (2008) A 1H magnetic resonance spectroscopy study in adults with obsessive compulsive disorder: relationship between metabolite concentrations and symptom severity. J Neural Transm (Vienna) 115(7):1051–1062. https://doi.org/10.1007/s00702-008-0045-4

Steen RG, Hamer RM, Lieberman JA (2005) Measurement of brain metabolites by 1H magnetic resonance spectroscopy in patients with schizophrenia: a systematic review and meta-analysis. Neuropsychopharmacology 30(11):1949–1962. https://doi.org/10.1038/sj.npp.1300850

Sumitani S, Harada M, Kubo H, Ohmori T (2007) Proton magnetic resonance spectroscopy reveals an abnormality in the anterior cingulate of a subgroup of obsessive-compulsive disorder patients. Psychiatry Res 154(1):85–92. https://doi.org/10.1016/j.pscychresns.2006.02.003

Surmeier DJ, Ding J, Day M, Wang Z, Shen W (2007) D1 and D2 dopamine-receptor modulation of striatal glutamatergic signaling in striatal medium spiny neurons. Trends Neurosci 30 (5):228–235. https://doi.org/10.1016/j.tins.2007.03.008

Tükel R, Aydın K, Ertekin E, Özyıldırım S, Taravari V (2014) Proton magnetic resonance spectroscopy in obsessive-compulsive disorder: evidence for reduced neuronal integrity in the anterior cingulate. Psychiatry Res 224(3):275–280. https://doi.org/10.1016/j.pscychresns.2014.08.012

Tükel R, Aydın K, Ertekin E, Özyıldırım S, Barburoğlu M (2015) 1H-magnetic resonance spectroscopy in obsessive-compulsive disorder: effects of 12 weeks of sertraline treatment on brain metabolites. Eur Arch Psychiatry Clin Neurosci 265(3):219–226. https://doi.org/10.1007/s00406-014-0545-1

Turner AJ (1986) Biochemistry and the central nervous system (fifth edition): by H. McIlwain and H S Bachelard. pp 660. Churchill Livingstone, Edinburgh. 1985. £40 ISBN 0-443-01961-4. Biochem Educ 14(1):46–46. https://doi.org/10.1016/0307-4412(86)90054-3

Vaghi MM, Hampshire A, Fineberg NA, Kaser M, Brühl AB, Sahakian BJ et al (2017) Hypoactivation and Dysconnectivity of a Frontostriatal circuit during goal-directed planning as an Endophenotype for obsessive-compulsive disorder. Biol Psychiatry Cogn Neurosci Neuroimaging 2(8):655–663. https://doi.org/10.1016/j.bpsc.2017.05.005

Vahabzadeh A, McDougle CJ (2014) Obsessive–compulsive disorder. In: McManus LM, Mitchell RN (eds) Pathobiology of human disease. Academic Press, San Diego, pp 2512–2521

Wang R, Fan Q, Zhang Z, Chen Y, Zhu Y, Li Y (2018) Anterior thalamic radiation structural and metabolic changes in obsessive-compulsive disorder: a combined DTI-MRS study. Psychiatry Res Neuroimaging 277:39–44. https://doi.org/10.1016/j.pscychresns.2018.05.004

Weber AM, Soreni N, Stanley JA, Greco A, Mendlowitz S, Szatmari P et al (2014) Proton magnetic resonance spectroscopy of prefrontal white matter in psychotropic naïve children and adolescents with obsessive–compulsive disorder. Psychiatry Res Neuroimaging 222(1):67–74. https://doi.org/10.1016/j.pscychresns.2014.02.004

Whiteside SP, Port JD, Abramowitz JS (2004) A meta-analysis of functional neuroimaging in obsessive-compulsive disorder. Psychiatry Res 132(1):69–79. https://doi.org/10.1016/j.pscychresns.2004.07.001

Whiteside SP, Port JD, Deacon BJ, Abramowitz JS (2006) A magnetic resonance spectroscopy investigation of obsessive-compulsive disorder and anxiety. Psychiatry Res 146(2):137–147. https://doi.org/10.1016/j.pscychresns.2005.12.006

Whiteside SP, Abramowitz JS, Port JD (2012a) Decreased caudate N-acetyl-l-aspartic acid in pediatric obsessive-compulsive disorder and the effects of behavior therapy. Psychiatry Res 202(1):53–59. https://doi.org/10.1016/j.pscychresns.2011.11.010

Whiteside SP, Abramowitz JS, Port JD (2012b) The effect of behavior therapy on caudate N-acetyl-l-aspartic acid in adults with obsessive-compulsive disorder. Psychiatry Res 201(1):10–16. https://doi.org/10.1016/j.pscychresns.2011.04.004

Wong DF, Brasić JR, Singer HS, Schretlen DJ, Kuwabara H, Zhou Y et al (2008) Mechanisms of dopaminergic and serotonergic neurotransmission in Tourette syndrome: clues from an in vivo neurochemistry study with PET. Neuropsychopharmacology 33(6):1239–1251. https://doi.org/10.1038/sj.npp.1301528

Wu K, Hanna GL, Rosenberg DR, Arnold PD (2012) The role of glutamate signaling in the pathogenesis and treatment of obsessive-compulsive disorder. Pharmacol Biochem Behav 100 (4):726–735. https://doi.org/10.1016/j.pbb.2011.10.007

Xin L, Tkáč I (2017) A practical guide to in vivo proton magnetic resonance spectroscopy at high magnetic fields. Anal Biochem 529:30–39. https://doi.org/10.1016/j.ab.2016.10.019

Yalçın O, Sener S, Konuş Boyunağa OL, Sarıpınar EG, Oğur T, Güney E et al (2011) Comparing brain magnetic resonance spectroscopy findings of pediatric treatment-naive obsessive-compulsive disorder patients with healthy controls. Turk Psikiyatri Derg 22(4):222–229

Yang S, Hu J, Kou Z, Yang Y (2008) Spectral simplification for resolved glutamate and glutamine measurement using a standard STEAM sequence with optimized timing parameters at 3, 4, 4.7, 7, and 9.4T. Magn Reson Med 59(2):236–244. https://doi.org/10.1002/mrm.21463

Yücel M, Harrison BJ, Wood SJ, Fornito A, Wellard RM, Pujol J et al (2007) Functional and biochemical alterations of the medial frontal cortex in obsessive-compulsive disorder. Arch Gen Psychiatry 64(8):946–955. https://doi.org/10.1001/archpsyc.64.8.946

Yücel M, Wood SJ, Wellard RM, Harrison BJ, Fornito A, Pujol J et al (2008) Anterior cingulate glutamate–glutamine levels predict symptom severity in women with obsessive–compulsive disorder. Aust N Z J Psychiatry 42(6):467–477. https://doi.org/10.1080/00048670802050546

Zhang Z, Fan Q, Bai Y, Wang Z, Zhang H, Xiao Z (2016) Brain gamma-Aminobutyric acid (GABA) concentration of the prefrontal lobe in Unmedicated patients with obsessive-compulsive disorder: a research of magnetic resonance spectroscopy. Shanghai Arch Psychiatry 28(5):263–270. https://doi.org/10.11919/j.issn.1002-0829.216043

Zheng H, Yang W, Zhang B, Hua G, Wang S, Jia F et al (2020) Reduced anterior cingulate glutamate of comorbid skin-picking disorder in adults with obsessive-compulsive disorder. J Affect Disord 265:193–199. https://doi.org/10.1016/j.jad.2020.01.059

Zhu Y, Fan Q, Han X, Zhang H, Chen J, Wang Z et al (2015) Decreased thalamic glutamate level in unmedicated adult obsessive-compulsive disorder patients detected by proton magnetic resonance spectroscopy. J Affect Disord 178:193–200. https://doi.org/10.1016/j.jad.2015.03.008

Zurowski B, Kordon A, Weber-Fahr W, Voderholzer U, Kuelz AK, Freyer T et al (2012) Relevance of orbitofrontal neurochemistry for the outcome of cognitive-behavioural therapy in patients with obsessive-compulsive disorder. Eur Arch Psychiatry Clin Neurosci 262(7):617–624. https://doi.org/10.1007/s00406-012-0304-0

Functional Brain Imaging and OCD

Carles Soriano-Mas

Contents

Abstract In the last 20 years, functional magnetic resonance imaging (fMRI) has been extensively used to investigate system-level abnormalities in the brain of patients with obsessive-compulsive disorder (OCD). In this chapter, we start by reviewing the studies assessing regional brain differences between patients with OCD and healthy controls in task-based fMRI. Specifically, we review studies on executive functioning and emotional processing, protocols in which these patients have been described to show alterations at the behavioral level, as well as research using symptom provocation protocols. Next, we review studies on brain connectivity alterations, focusing on resting-state studies evaluating disruptions in fronto-subcortical functional connectivity and in cortical networks. Likewise, we also

C. Soriano-Mas (✉)
Department of Psychiatry, Bellvitge University Hospital, Bellvitge Biomedical Research Institute-IDIBELL, Barcelona, Spain

Centro de Investigación Biomédica en Red de Salud Mental (CIBERSAM), Barcelona, Spain

Department of Psychobiology and Methodology of Health Sciences, Universitat Autònoma de Barcelona, Barcelona, Spain
e-mail: csoriano@idibell.cat

Curr Topics Behav Neurosci (2021) 49: 269–300
https://doi.org/10.1007/7854_2020_203
Published Online: 19 February 2021

review research on effective connectivity, which, different from functional connectivity, allows for ascertaining the directionality of inter-regional connectivity alterations. We conclude by reviewing the most significant findings on a topic of translational impact, such as the use of different fMRI measurements to predict response across a variety of treatment approaches. Overall, results suggest that there exists a pattern of regions, involving, but not limited to, different nodes of the cortico-striatal-thalamo-cortical circuits, showing robust evidence of functional alteration across studies, although the nature of the alterations critically depends on the specific tasks and their particular demands. Moreover, such findings have been, to date, poorly translated into clinical practice. It is suggested that this may be partially accounted for by the difficulty to integrate into a common framework results obtained under a wide variety of analysis approaches.

Keywords Obsessive-compulsive disorder · Functional magnetic resonance imaging · Task-based fMRI · Functional connectivity · Effective connectivity · Resting state · Brain networks

1 Introduction

In the decade of the 1980s researchers started using Positron Emission Tomography (PET), a nuclear imaging approach, to identify and characterize the abnormal patterns of brain activity associated with mental health disorders, such as obsessive-compulsive disorder (OCD) (Baxter et al. 1987). However, since the development, in the early 1990s, of functional Magnetic Resonance Imaging (fMRI) (Ogawa et al. 1990), most of these researchers have preferred using this technique because of its almost null invasiveness and superior spatial and temporal resolution. fMRI is based on blood-oxygen-level-dependent imaging, or BOLD-contrast imaging, an endogenous hemodynamic signal that reflects changes in blood oxygenation linked to neuronal activity (Raichle and Mintun 2006).

BOLD fMRI is sensitive to differences between brain states, and, therefore, it has been primarily used in the study of regional changes in brain activity (termed "activation") when individuals are required to perform different experimental tasks (sensorimotor, cognitive or emotional paradigms) (Huettel et al. 2004). In a typical fMRI experiment, brain activity during one of these tasks is contrasted against activity during a control condition differing from the task in the experimental question of interest, but identical to it in all other aspects, therefore effectively controlling for activations due to non-specific confounding factors. For example, in an emotional face processing paradigm, the presentation of non-emotional faces, effectively controlling for the non-specific activation related with face processing, would be preferred as a control condition rather than a low-level reference task, such as cross-fixation. Moreover, depending on the duration of the stimulation, fMRI paradigms can be divided into "block" or "event-related" fMRI. On the one hand, in block fMRI, experimental and control stimuli are presented for around 15–20 s, and each condition is repeated a variable number of times, typically between 3 and 6, normally with cross-fixation periods interspersed throughout the experiment to obtain an estimation of baseline activity (i.e., free from any form of stimulation). On the other hand, event-related fMRI is used to assess brain activations evoked by short duration stimuli (such as the stop-signal in motor inhibition protocols), and activations during these events, which should characteristically constitute around a 20% of the trials, are contrasted against activations in a background control condition (following with the same example, the "go" trials of a motor inhibition task).

In the early 2000s, BOLD fMRI also began to be used to detect spontaneous fluctuations of neuronal activity, that is, not evoked by any kind of block- or event-like stimulation. This was possible through the assessment of "resting-state" activity (i.e., participants lying awake and still in the scanner), and, different from task-based fMRI, these experiments did not provide information on regional brain activation, but rather about changes in brain functional connectivity. This is defined as the spatiotemporal dependency or correlation among distributed brain regions, reflecting the coordinated activity of large-scale brain systems or brain networks (Fox and Raichle 2007). At a neuronal level, functional connectivity is interpreted as representing co-fluctuations of neuronal excitability among regions with strong direct or indirect anatomical connections (Raichle 2015). Although resting-state fMRI is particularly well suited for measuring functional connectivity, the existence of inter-regional patterns of connectivity can also be reliably studied in the context of experimental tasks (block or event-related fMRI) (Friston et al. 1997). In this case, it is generally considered that inter-regional connectivity may be modulated by the experimental protocol, typically describing connectivity increases within a particular brain network during task performance (vs. control or baseline trials).

In this chapter, we will review the existing literature evaluating brain function in OCD with task-based fMRI and resting-state or task-related connectivity assessments. While task-based fMRI has been mainly used to ascertain the neurobiological correlates of alleged neuropsychological or emotional processing dysfunctions (including symptom provocation), connectivity studies have predominantly investigated functional abnormalities from a more exploratory perspective, in some cases without a clear link with clinical variables or behavioral-level dysfunctions. Nevertheless, connectivity studies have focused their attention on the assessment of particular brain circuits, such as the cortico-striato-thalamo-cortical (CSTC) circuits or specific resting-state networks, such as the default-mode network (DMN), the dysfunction of which has been supposed to account for the symptom profile of patients with OCD.

2 Task-Based fMRI

In this section, we summarize the results from different studies assessing a variety of neuropsychological and emotional domains of interest to account for the OCD symptom profile. First, we will summarize the results from studies on executive functioning, a broad domain that involves different experimental paradigms that have been shown to capture different aspects of the neurocognitive profile of patients with OCD. Next, we will summarize the results from studies on emotional processing in OCD to close with a section specifically devoted to symptom provocation studies.

2.1 Executive Functioning

The assessment of executive functioning involves a large series of paradigms evaluating a partially overlapping set of cognitive processes purportedly critical for the cognitive control of behavior. Disruption in some of these processes has been linked to specific OCD symptoms. Thus, alterations in cognitive control, including response inhibition, cognitive flexibility, and set-shifting abilities, may partially underlie the poor control over obsessions and compulsions of patients with OCD, although, in most cases, these patients show good insight about their symptoms (Chamberlain et al. 2005). Likewise, such processes may also relate to the perseverative and repetitive nature of OCD symptoms (Chamberlain et al. 2006, 2007).

Response inhibition has been assessed with a number of different protocols, such as response conflict tasks (e.g., Flanker, Stroop or Multisource Interferences tasks), Go/No-go, and stop-signal protocols. Results from fMRI studies evaluating brain activations evoked by these tasks, however, have been somewhat heterogenous. Although prefrontal cortex hypoactivations have been frequently reported, these have been located either in the right lateral (Roth et al. 2007) or the ventromedial prefrontal cortices (Page et al. 2009). Moreover, at the subcortical level, there have been reports of both decreased (Kang et al. 2013; Page et al. 2009) and increased dorsal striatal and thalamic activations (Morein-Zamir et al. 2016), and other studies reported no significant differences with healthy controls at the cortical or subcortical level (Suñol et al. 2020). In a recent meta-analysis, Norman et al. (2019) attempted to identify the most consistent alterations across a series of inhibition paradigms, observing significantly decreased activations in ventromedial and dorsal striatal-thalamic regions. Nevertheless, these results should be interpreted with caution because this analysis combined not only clinically diverse samples, but also a variety of inhibition protocols. Overall, interpretation of findings from inhibition studies in OCD samples is hampered not only by cross-sample variability but also by task-related sources of variance.

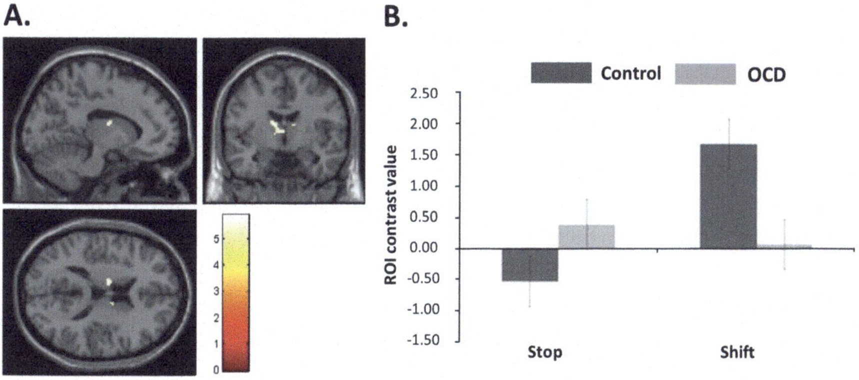

Fig. 1 (**a**) Brain voxels in the thalamus and caudate nucleus showing increased activation for stopping but decreased for shifting in patients with OCD. Color bar represents *t*-value. (**b**) Post-hoc analysis of caudate activity for stop and shift trials in control and patient groups. Error bars represent S.E.M. Adapted with permission from Morein-Zamir et al. Psychol. Med. 2016;46:829–840

In the study of Morein-Zamir et al. (2016), the authors aimed at controlling for such sources of variance exploring, in a sample of comorbidity-free OCD individuals and within the same protocol, both response inhibition (i.e., Go/No-go) and set-shifting abilities (changing the focus of attention between simultaneously presented faces and houses). They found widespread under-activation across frontal-parietal areas in OCD patients for shifting, but not stopping, which suggests that response inhibition hypoactivations in OCD may indeed depend on specific task demands and difficulty. Likewise, the authors also observed that the stopping-related over-activations in caudate and thalamus contrasted with shift-related under-activations in the same structures (see Fig. 1). These results concur with findings from previous reports (Gu et al. 2008; Page et al. 2009), and were interpreted as indicating that regional functional abnormalities in OCD are related to the precise cognitive requirements of tasks, with no tendency for general across-task hypoactivations or hyperactivations of CSTC structures. Cognitive flexibility has also been studied with reversal learning tasks, where decreased fronto-parietal hypoactivations have been more precisely mapped onto the lateral orbitofrontal cortex and the inferior parietal cortex (Chamberlain et al. 2008; Remijnse et al. 2006).

In combination with alterations in response inhibition and task-switching, a significant number of studies have also reported disrupted error processing and monitoring in patients with OCD. Heightened error monitoring, as evidenced by an increased error related negativity (ERN) after following incorrect responses, has been considered to be a biomarker or candidate endophenotype of OCD (Endrass and Ullsperger 2014). Such increased ERN is especially observed in response conflict tasks (Mathews et al. 2012), and, in fMRI experiments, corresponds to an increased activation in dorsal regions of the medial wall, spanning from the dorsal anterior cingulate cortex (ACC) to the pre-supplementary motor area (pre-SMA) (Grützmann et al. 2016). Congruently, in response conflict tasks, different studies have reported regional hyperactivations in this area in patients with OCD (Schlösser et al. 2010; Yücel et al. 2007). Other studies have also reported increased ACC activation during error monitoring in switch-tasks (Suñol et al. 2020), as well as pre-SMA hyperactivations during motor inhibition protocols (de Wit et al. 2012). Likewise, in reward processing paradigms, the processing of negative reward prediction errors has been also shown to hyperactivate the ACC (Hauser et al. 2017). Interestingly, this hyperactivation seems of dopaminergic origin (Murray et al. 2019), possibly related to the abnormal processing of reward anticipation by ventral striatal neurons described in patients with OCD (Figee et al. 2011).

Results from studies using other protocols of executive functioning have been found partially to concur with the above findings. Ursu et al. (2003), for instance, used a modified continuous performance test to describe increased dorsal ACC activation during high-conflict trials in association with error monitoring. In contrast, in working memory tasks (N-back), activation of medial wall structures (i.e., ACC and SMA) has been found to be decreased rather than increased with high memory load trials (Heinzel et al. 2017; Koch et al. 2012). Importantly, in these two studies, while activity in medial wall areas was reported to linearly increase with memory load in healthy controls, patients with OCD failed to activate these regions in more demanding trials. Such results may be tentatively interpreted according to the neuropsychiatric executive overload model of OCD (Abramovitch et al. 2012), in which the obsessive overflow can cause a processing overload in the executive system. Nevertheless, there are also other reports of increased ACC activation with increasing memory load in working memory tasks (van der Wee et al. 2003), which suggests that brain activation abnormalities during working memory tasks may depend on the particularities of each specific protocol. Likewise, other authors have also highlighted the importance of the putative compensatory mechanisms put into practice by patients with OCD in working memory protocols. de Vries et al. (2014) observed that patients displayed left dorsal frontal–including pre-SMA and dorsolateral prefrontal cortex–hyperactivations that allowed them to preserve task performance in low task loads, but that were not sufficient to maintain performance at high task loads. Moreover, working memory alterations may be also secondary to other aspects of executive dysfunction, such as disrupted planning abilities and associated hypoactivations in dorsolateral prefrontal and dorsal caudate areas (van den Heuvel et al. 2005a).

2.2 Emotional Processing

Abnormally increased amygdala activation has been considered the hallmark of abnormal emotional processing in anxiety-related disorders (Etkin and Wager 2007). In this sense, OCD seemed to differ from anxiety disorders since amygdala hyperactivations were not observed in early studies (Britton et al. 2010; Cannistraro et al. 2004; Lawrence et al. 2007), although research with better powered samples did show amygdala hyperactivation in patients with OCD when processing emotional faces (Cardoner et al. 2011; Via et al. 2014). Similar amygdala hyperactivations were also observed in an emotional Stroop protocol with OCD-related words (van den Heuvel et al. 2005b) or during the presentation of negative OCD related (de Wit et al. 2015) or unrelated images (Weidt et al. 2016). Interestingly, in one of these studies (Via et al. 2014), amygdala hyperactivity was correlated with the severity of two specific symptom dimensions (i.e., harm-checking and sexual-religious symptoms). Nevertheless, other studies using a variety of emotional paradigms (Apergis-Schoute et al. 2017; Berlin et al. 2015; Brennan et al. 2015; Gonçalves et al. 2015; Harrison et al. 2012; Park et al. 2016) did not report increases in amygdala activity, or, in agreement with Via et al. (2014), these were restricted to a particular subgroup of participants (i.e., those with sexual-religious symptoms (Harrison et al. 2012)). In other studies, hyperactivations were observed in other limbic areas (i.e., parahippocampal gyrus), but not in the amygdala itself (Hennig-Fast et al. 2015). Therefore, although there is sufficient evidence to suggest that there is occasionally amygdala hyperreactivity in response to emotional stimuli in patients with OCD, this may depend on a number of factors, such as the particularities of the emotional protocol or the clinical features of participants.

Increased activations in other brain areas, such as the ventromedial prefrontal cortex (vmPFC), have been also robustly reported in OCD samples across different emotional paradigms. Thus, moral dilemma (Harrison et al. 2012), aversive image processing (Gonçalves et al. 2015), and retrieval of unpleasant words (Park et al. 2016) have shown to hyperactivate the vmPFC. Similar results were observed throughout the acquisition and reversal phases of an aversive conditioning task (Apergis-Schoute et al. 2017) and during the acquisition of avoidance responses (Gillan et al. 2015). Such sustained excessive vmPFC response may prevent this region from tackling changing contingencies between stimuli and, therefore, from flexibly updating the value of emotional stimuli, probably underpinning the lack of safety signals and the inflexible threat beliefs that characterize patients with OCD (Robbins et al. 2019). Apparently at odds with these notions, Milad et al. (2013) described vmPFC hypoactivation during fear extinction learning, although, paradoxically, they also observed that vmPFC activation was positively correlated with OCD symptom severity.

Emotion regulation is another topic that has been investigated in OCD samples. De Wit et al. (2015) were the first to assess functional brain changes during regulation of OCD-related stimuli, finding increased amygdala activity at baseline, together with a decreased activation of the dorsolateral prefrontal cortex (dlPFC), partially compensated by an increased activation of the dorsomedial prefrontal cortex (dmPFC), during voluntary emotion reappraisal. Interestingly, substantiating the idea of a compensatory dmPFC activation, severity of obsessive-compulsive symptoms was inversely correlated with activation of this region during reappraisal (de Wit et al. 2015). These results have been replicated in a more recent study from the same group (Thorsen et al. 2018), where, additionally, the authors showed that brain activity in the dmPFC of unaffected relatives partially overlapped with that of patients, suggesting that this increased dmPFC activation during emotion regulation might be an endophenotype for OCD.

Importantly, beyond the balance between dlPFC and dmPFC activations during emotional regulation in OCD, recent research suggests that anterior insular cortex and dorsal ACC participate in a domain-general regulatory network, underlying both cognitive and more implicit emotion regulation strategies (Picó-Pérez et al. 2019). Activity in these two areas has been shown to be typically increased in patients with OCD across different emotional processing paradigms, such as guilt induction (Hennig-Fast et al. 2015), negative or ambiguous feedback (Becker et al. 2014), emotional response inhibition (Berlin et al. 2015), or emotional Stroop using OCD-related words (Brennan et al. 2015). Such hyperactivations could therefore be interpreted as the result of an augmented emotion regulatory effort in patients with OCD, probably aimed at regulating the enhanced amygdala reactivity described above. Other authors, however, have described decreased, rather than increased, ACC and anterior insula activations during guilt induction (Basile et al. 2014), although, in agreement with our notions, these results were interpreted in terms of the "neural efficiency hypothesis," therefore likely reflecting increased cerebral efficiency in patients with OCD deriving from their frequent exposure to guilty feelings.

Two meta-analyses have been recently published summarizing the results of emotional processing studies in OCD (Picó-Pérez et al. 2020; Thorsen et al. 2018). These two studies have reported concurring findings involving hyperactivations in the bilateral amygdalae and the right putamen. Moreover, Picó-Pérez et al. (2020) also described increased activations in the right anterior insula, the left angular gyrus, the left ventral putamen, the left precentral gyrus, the medial prefrontal cortex (mPFC), and the left thalamus, while Thorsen et al. (2018) reported hyperactivations in the orbitofrontal/vmPFC, the ACC, and the middle temporal and left inferior occipital cortices. In any case, none of these studies identified significant brain activation decreases in OCD samples during emotional processing. Importantly, moreover, between one third and one half of the original studies scrutinized in these meta-analyses were symptom provocation studies. Given the uniqueness of this kind of studies for reproducing clinical OCD symptoms in experimental contexts, they are next reviewed here.

2.3 *Symptom Provocation*

Overall, the results of the above meta-analyses of emotional processing in OCD are similar to those reported by Breiter et al. (1996), the first symptom provocation study using fMRI, where widespread hyperactivations in medial orbitofrontal, lateral frontal, anterior temporal, ACC, and insular cortex, as well as significant subcortical activations in the caudate, lenticular nucleus (putamen and pallidum), and amygdala were reported. Brain responses in symptom provocation studies, however, seem to depend on a number of confounding factors. Thus, in Adler et al. (2000), using the same protocol as Breiter et al. (1996) in medication-free patients, prefrontal activations were not limited to the orbitofrontal part, but extended to superior frontal and dorsolateral prefrontal cortices. Significant activations in the ACC were also observed, but not in the insular cortex or the striatum, and, within the medial temporal lobe, activations were preferentially observed in the hippocampus rather than in the amygdala. Other studies with medication-free patients also found increased activations after provocation involving both left orbitofrontal and dorsolateral prefrontal cortices (Nakao et al. 2005). Interestingly, in this study, hyperactivations in the left orbitofrontal cortex normalized after symptom improvement, similar to what was found in Morgiève et al. (2014), who observed normalization of left (but not right) orbitofrontal and ACC hyperactivation after completion of a CBT trial. Importantly, Morgiève et al. (2014) also reported that personalized obsession-inducing images evoked stronger hyperactivations in the ACC and bilateral orbitofrontal cortices, something also reported by Baioui et al. (2013), who found larger activations in different components of the CSTC during symptom induction with personalized items in relation to standard OCD-related images.

With the aim of controlling for the clinical heterogeneity of the OCD profile, other research has attempted to evoke dimension-specific symptoms in OCD samples. Early studies of this kind mainly assessed patients or stimuli related with washing obsessions, finding increased insula activations associated to disgust induction (Phillips et al. 2000; Shapira et al. 2003; Schienle et al. 2005). Such notions have been more recently confirmed in a symptom provocation study with odor stimulation. Patients with OCD displayed increased insula activations in front of both unpleasant and pleasant stimuli, and such activations were correlated with clinical severity and disgust sensitivity (Berlin et al. 2017). Mataix-Cols et al. (2004), on the other hand, expanded the focus of interest to other symptom dimensions and performed a symptom provocation study using blocks of images representative of washing, checking, and hoarding dimensions, as well as a general aversive condition. Their results, however, did not concur with those reported above, since they observed vmPFC and caudate, but not insula, activations associated to washing provocation, while lenticular nucleus, thalamus, and dorsal cortical areas were hyperactivated by checking stimuli. Hoarding images evoked activations in the left precentral gyrus and the right orbitofrontal cortex, while the general aversive condition only evoked activity in posterior visual processing regions. These results were partially replicated by Murayama et al. (2013), whom, by using individually tailored stimuli, observed similar vmPFC hyperactivations in patients with washing symptoms, but not in patients checking symptoms. In contrast, An et al. (2009) found analogous increased vmPFC activations in OCD patients with hoarding symptoms exposed to hoarding stimuli. Remarkably, Gilbert et al. (2009) reported decreased, rather than increased, neural activations across the insula and different components of the CSTC circuits during contamination and symmetry provocation in pediatric OCD, suggesting developmental effects on neural systems underlying symptom dimensions.

vmPFC hyperactivations have therefore been observed after the induction of, at least, washing and hoarding symptoms (for a review, see Rotge et al. 2008). Nevertheless, Banca et al. (2015) showed that this might be reversed with some experimental manipulations. Giving the participants the opportunity to terminate provocative stimulus presentation (i.e., avoid aversive stimulation) led to significant hypoactivations in the vmPFC and strong putamen hyperactivations. These avoidance responses were supposed to model the compulsive behavior of patients with OCD, and, therefore, these results were interpreted as reflecting an imbalance between goal-oriented and habit behavior, in agreement with recent models of OCD emphasizing the deficiency of goal-oriented strategies to account for the clinical profile of OCD (Gillan and Robbins 2014; Gillan et al. 2015). The study of Banca et al. (2015) also enabled dissociation between the neural correlates underlying exposure to stimuli from those related to the decision to avoid the stimuli. Importantly, during exposure, and in agreement with some of the above studies using disgust provocation, the authors observed significant insula activations. Amygdala activations, however, were mainly observed after stimulus avoidance, probably reflecting post-rejection appraisal. Other authors, however, have shown significant amygdala hyperactivations during symptom provocation, particularly when using paradigms accounting for the nature of amygdala response (i.e., rapid onsets and fast attenuation of responses) (Simon et al. 2010). This amygdala hyperactivation might constitute a correlate of fear expression in OCD, linking it to other anxiety disorders (Simon et al. 2014).

In summary, what we have learnt from task-based fMRI is that patients with OCD show a rich pattern of abnormal hyper- and hypo-activation across different contexts. Both the lateral orbitofrontal cortex and the vmPFC have been shown to be significantly less activated during cognitive protocols (i.e., cognitive flexibility and response inhibition, respectively), although the particular demands of each protocol, such as task difficulty, may play a relevant role in these findings. On the other hand, abnormal activations in regions of the medial wall may be related with poor performance in other cognitive protocols (i.e., working memory). Conversely, during symptom provocation, the lateral orbitofrontal cortex and the vmPFC have been shown to be hyperactivated in OCD samples. vmPFC alterations have been also observed in other emotional processing protocols, and, interestingly, they may show the opposite sign (i.e., decreases rather increases) when individuals are given the opportunity to avoid aversive stimulation. In this context, individuals with OCD also show increased putamen activity, and the combination of decreased vmPFC activity and putamen hyperactivations has been suggested to underpin compulsive responding. During emotional processing, significantly increased amygdala activity has also been consistently reported, especially in patients with harm-checking or sexual-religious obsessions, resembling what is observed in patients with anxiety disorders. Patients with OCD also show disrupted emotion regulation capacities, and these are probably underpinned by decreased dlPFC cortex activations accompanied by compensatory hyperactivations in the dmPFC, ACC, and anterior insula. Nevertheless, such hyperactivation is not specific for emotion regulation tasks, since abnormally increased insula activity has been also observed after disgust induction, whereas increased ACC response has been characteristically associated with increased error monitoring.

3 Brain Connectivity

Resting-state fMRI research has well recapitulated the functional anatomy of large-scale human brain systems, and it has become the method of choice in the study of brain network dysfunction in neuropsychiatric disorders (Fornito and Bullmore 2010; Greicius 2008). Most of the resting-state fMRI research employs one of the three approaches described below.

Seed-based functional connectivity studies select a particular region of interest (or multiple regions), termed the seed(s), and its BOLD fMRI time-series is extracted. Such time-series can then be correlated with the time-series of other brain regions (or, more typically, of all other voxels in the brain), with the aim of quantifying the degree of temporal correlation between them. With this approach, a critical factor is the selection of the seed region(s) of interest, which ideally needs to be strongly hypothesis-driven. As distinct from seed-based approaches, *independent component analysis* (ICA) is a data-driven technique that allows the separation of complex multivariate signals into different statistically independent components (i.e., networks). This has allowed, for instance, the identification and anatomical delineation of the default mode, salience or central-executive networks. In comparison with seed-based approaches, ICA approaches are more exploratory and do not require strong *a priori* hypotheses about specific regions or networks of interest, although a general challenge of these studies relates to the separation of components of interest from common large-scale noise components. This typically involves using anatomical templates or results from prior studies to inform the selection of the components, which introduces a degree of arbitrariness into the application of this method. Finally, *graph theoretical approaches* have been more recently applied to the assessment of brain networks as derived from fMRI studies. A graph is made up of vertices and edges, and paths through a graph consist of succeeding vertices connected by edges, of different lengths. When applied to fMRI data, vertices correspond to voxels or groups of voxels, and edges to the estimates of functional connectivity between them. Graphs may be characterized by their local and global properties. Local properties refer to the characteristics of individual vertices, whereas global properties refer to the overall characteristics of the graphs. The "degree" of a vertex, for instance, is a local property that allows estimating the functional connectivity strength of each vertex (i.e., number of significant connections with other vertices), and the identification of "hubs," which are regions of high degree (or centrality) within graphs. Measures of modularity and average path length, on the other hand, are global characteristics of a graph, which refer, respectively, to the existence of modules that are highly internal connected but only sparsely connected to one another, and to the mean distance between any pair of vertices, which can be used as proxy for integration within a network. The combination of these measurements can be used to characterize "small-world" properties, which denotes a combination of modularity and efficient long-range connections that maximizes parallel processing, minimizes wiring costs, and increases resilience to local disruptions.

3.1 Resting-State Connectivity in CSTC Circuits

Cortico-striatal-thalamo-cortical (CSTC) circuits connect different portions of the frontal cortex with the basal ganglia and the thalamus (Alexander et al. 1986). At a broad functional level, these circuits appear to support different domains of processing, which has led to the description of distinct "cognitive" (dorsal striatal-prefrontal cortex), "limbic" (ventral striatal-prefrontal cortex), and "sensorimotor" (putamen-motor cortices) loops (Haber 2003). Its dysregulation has been implicated in different neuropsychiatric disorders (Shepherd 2013), and alteration of CSTC circuits remains central to contemporary neurobiological models of OCD (Menzies et al. 2008; Milad and Rauch 2012).

There have been numerous studies characterizing resting-state functional connectivity alterations within these circuits in OCD. Using a seed-based approach, Harrison et al. (2009) were the first to describe alterations in frontal-striatal circuits, involving both increases and decreases in functional connectivity. Patients with OCD had increased functional connectivity between ventral striatal seeds and the orbitofrontal cortex and surrounding ventromedial areas, but reduced functional connectivity between dorsal caudate seeds and the lateral prefrontal cortex and between dorsal putamen seeds and the supplementary motor area. Notably, connectivity between the ventral caudate/nucleus accumbens region and the anterior orbitofrontal cortex correlated with overall symptom severity. This overall pattern of findings was replicated in another study from the same group assessing a larger series of 74 OCD patients and 74 healthy controls (Harrison et al. 2013, see Fig. 2).

The results from Harrison et al. were replicated with unmedicated patients (Sakai et al. 2011). In contrast, Jung et al. (2013) and Posner et al. (2014) reported

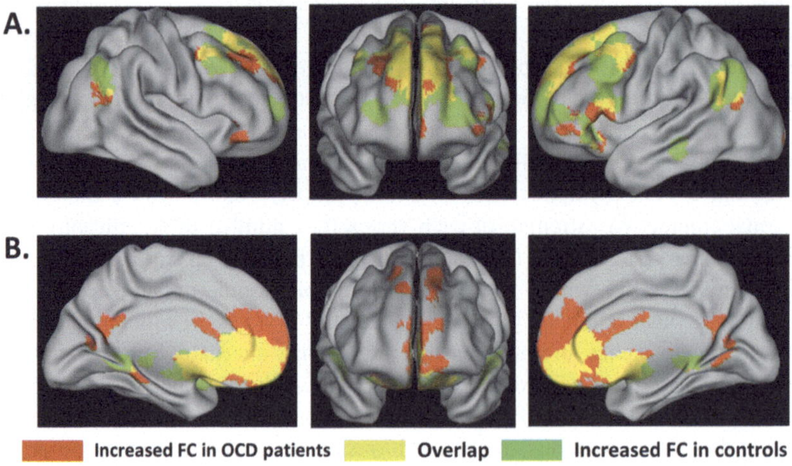

Fig. 2 Regions showing increased and decreased functional connectivity with the dorsal (**a**) and ventral (**b**) caudate in patients with OCD. Adapted with permission from Harrison et al. Biol Psychiatry. 2013;73(4):321–328

decreased functional connectivity between the ventral striatum and the ventromedial prefrontal cortex in unmedicated patients with OCD, and, also contrary to Harrison et al., Posner et al. observed a negative correlation between functional connectivity of the ventral striatum with the vmPFC and symptom severity, which was interpreted as evidence that previous reports were probably affected by the lasting effects of antidepressant treatment. However, studies specifically assessing the effect of Selective Serotonin Reuptake Inhibitors (SSRIs) on ventral striatum-ventromedial prefrontal cortex connectivity have not shown significant effects (McCabe and Mishor 2011).

Using a different methodological approach, Beucke et al. (2013) reported that the putamen of patients with OCD displayed increased functional connectivity exclusively with neighboring regions, although such short-range connectivity between striatal structures was reduced in medicated patients. The results of Anticevic et al. (2014) were in agreement with these findings, since they reported enhanced functional connectivity in the right putamen and within the basal ganglia circuitry, which was reduced in patients on medication. The findings from these two studies could therefore be interpreted as overall being supportive of the data obtained with seed-based approaches showing increased functional connectivity of basal ganglia regions in OCD, although such connectivity, at least with neighboring structures, might be modulated by antidepressant medication. Interestingly, however, Anticevic et al. (2014) described that connectivity was increased throughout the dorsal caudate, putamen, and thalamus, but decreased in the ventral striatum/accumbens. Such dissociation supports the idea that cortico-striatal circuits are differently affected in OCD (Milad and Rauch 2012: Robbins et al. 2019). Notably, however, although Anticevic et al. (2014) showed that the ventral striatum exhibited a global decrease in functional connectivity, these authors also observed a specific increase in ventral striatal functional connectivity with the vmPFC, which is in agreement with most studies using a seed-based approach.

Regarding functional connectivity from other nodes of the basal ganglia, Beucke et al. (2013) also observed that the subthalamic nucleus of OCD patients showed increased functional connectivity with distant structures. Concurring with this last result, Cano et al. (2018) reported increased functional connectivity between the subthalamus and the pre-motor cortex, although these authors also reported decreased connectivity in OCD between the subthalamus and the lenticular nucleus. This pattern of findings was interpreted in the context of an imbalance between direct excitatory and indirect inhibitory CSTC pathways, and suggests that further research with more sensitive methods in terms of spatial resolution is warranted to wholly elucidate the functional connectivity alterations within the different CSTC loops.

In other studies, the seed region of interest has been placed in the orbitofrontal cortex rather than in striatal structures. Results of these studies, however, have been heterogenous. Thus, Meunier et al. (2012), assessed putative resting-state functional connectivity alterations from the orbitofrontal cortex in OCD in comparison with a group of stimulant-dependent individuals and healthy controls, finding decreased functional connectivity levels from medial and right lateral orbitofrontal cortices in OCD, which were correlated with compulsivity (Y-BOCS) scores. In agreement with this study, Anticevic et al. (2014) reported decreased functional connectivity in the left lateral prefrontal cortex of patients with OCD. By contrast, Beucke et al. (2013) reported increased functional connectivity from the medial orbitofrontal cortex associated with symptom severity. Importantly, however, results from Beucke et al. (2013) were driven to a large extent by increased functional connectivity with neighboring orbitofrontal areas, while the results from Anticevic et al. (2014) and Meunier et al. (2012) stem from decreased connectivity with other frontal or sensorimotor regions. Indeed, in another study, Ping et al. (2013) also described short-range functional connectivity increases from the orbitofrontal cortex in OCD, while Hou et al. (2014) described increased functional connectivity from the orbitofrontal cortex with near prefrontal and subcortical areas in OCD and their first-degree relatives, suggesting that such alterations may be used to identify heritable neuroimaging endophenotypes for OCD. Nevertheless, there are still some discrepancies across these studies warranting further research. Beucke et al. (2013), for instance, also identified increased functional connectivity from the medial orbitofrontal cortex with sensorimotor regions, in contrast with the results of Meunier et al. (2012). Such differences may depend on a number of confounding factors, including differences in the analysis methods or the clinical characteristics of the OCD samples. The results of Beucke et al. (2013) were observed in a sample of unmedicated patients, and medication, as mentioned above, has been proposed by some authors to modulate functional connectivity in ventral CSTC circuits. In any case, Meunier et al. (2012), Beucke et al. (2013), and Anticevic et al. (2014) performed specific analyses to ascertain the role of medication on orbitofrontal connectivity, reporting, in all cases, null findings.

3.2 Resting-State Networks

Resting-state networks represent the coordinated activity of distant brain regions purportedly underpinning specific cognitive and behavioral processes. Although neural activity within the components of these networks increases during particular tasks or situations, their coordinated activity may be also observed at rest, when correlations in spontaneous neural activity are detected across the regions of the network. Thus, coordinated activity across the regions of the central-executive or salience networks may be observed both at rest and during executive performance or the identification of salient stimuli, respectively. Likewise, the arguably most investigated resting-state brain network, the default-mode network (DMN), characteristically increases activity during passive imaging conditions, such as wakeful resting states, as compared to when subjects perform specific cognitive tasks or pay attention to the external world. Psychologically, this default activity has been linked to self-referential cognition (i.e., thinking about oneself). In anatomical terms, the DMN comprises the medial prefrontal cortex, the anterior and posterior cingulate cortices, the inferior parietal cortex, and the lateral temporal cortex.

Jang et al. (2010) placed a seed region of interest in the posterior cingulate cortex (PCC) and observed attenuated functional connectivity within the DMN in OCD. Likewise, using independent Component Analysis (ICA), Cheng et al. (2013) analyzed resting-state data in a series of 23 medication-naïve patients and 23 matched healthy controls, observing, in agreement with the above study, decreased functional connectivity strength within the DMN, mainly affecting the precuneus and the PCC. Moreover, Stern et al. (2012) reported reduced negative correlations (i.e., increased connectivity) between different regions of the DMN (i.e., PCC, medial frontal cortex, and inferior parietal lobe) and regions of the central-executive network. Similarly, Peng et al. (2014) found that short-distance functional connectivity from a PCC seed was decreased in OCD patients, while connectivity with regions from outside the DMN was increased. Taken as a whole, the results from these studies suggest that, in OCD, functional connectivity is decreased within the DMN, whereas connectivity between DMN regions and regions from other functional networks may be increased. This pattern of findings led to suggestions that patients with OCD might experience difficulties in disengaging from internally generated thoughts when performing tasks requiring attention to external stimuli, and, therefore, DMN activity could be interfering on information processing by other networks (Stern and Taylor 2014).

More recent research has nevertheless nuanced the above notions. Fan et al. (2017) reported disrupted coupling between the DMN and the salience and central-executive networks in OCD, reflecting the difficulties in switching between task-negative and task-positive processing modes and substantiating the relevance of the "triple-network model" in the pathophysiology of psychiatric disorders (Menon 2011). Gürsel et al. (2018) corroborated these findings in a meta-analysis of resting-state networks in OCD, where they observed disrupted connectivity between the default mode, the salience, and the central-executive networks. In this same study, the authors suggest that functional connectivity alterations involving the central-executive network may link impairments in large-scale brain networks with CSTC alterations, therefore integrating into a common framework two influential pathophysiological models of OCD. More recently, these same authors have shown that functional connectivity alterations in the central-executive network may characterize large-scale brain networks alterations in OCD better than alterations in the salience or default-mode networks (Gürsel et al. 2020), in agreement with results from earlier reports (Göttlich et al. 2014).

3.3 Graph Theory Approaches

Zhang et al. (2011) were the first to apply graph theory approaches to the analysis of fMRI data in OCD. They reported that OCD patients showed significantly higher local clustering within a network involved in top-down control of behavior (including the cingulate gyrus, the thalamus, the precuneus, and the cerebellum), although connectivity with other networks was reduced. Such imbalance between modularized and distributed information processing led the authors to suggest that the top-down control network of OCD patients did not show an optimal small-world architecture. Other studies more precisely mapped such neural architecture alterations onto regions of the cortico-striatal circuits, reporting increased functional connectivity strength in hub regions located throughout these cortico-subcortical loops (Hou et al. 2014; Tian et al. 2015). Indeed, Jung et al. (2017) showed that patients with OCD exhibited disruptions in small-world properties of cortico-striatal circuits. More specifically, patients displayed an imbalance between functional integration and segregation of cortical and subcortical regions, with decreased connectivity strength between the orbitofrontal cortex and dorsomedial striatum, and increased connectivity strength between the medial thalamus and striatal areas. In line with these reports of altered modular organization in cortico-striatal circuits, Vaghi et al. (2017) showed that the striatum constituted a cohesive module in the community structure of the functional connectome in OCD, and found that distant brain regions, such as the cerebellum, may be strongly linked at the functional level with the striatum.

3.4 *Effective Connectivity*

Different from the correlational estimations of functional connectivity, effective connectivity allows describing direct causal influences across distant brain regions. These causal relationships may be assessed during resting state or in relation to task performance. A particular case of the latter is the analysis of the psychophysiological interactions (PPI), which estimates context-dependent changes in effective connectivity, therefore identifying brain regions whose activity depends on the interaction between a psychological context (the task) and the physiological input from another brain area. Effective connectivity studies have been elegantly used by different research groups to shed light on the meaning of some of the above results from task-based activation and functional connectivity studies.

Cardoner et al. (2011), for instance, showed that increased amygdala activation during emotional face processing in OCD was related to increased context-dependent (i.e., specifically during emotional face processing) connectivity between the amygdala and visual face processing areas such as the fusiform gyrus. In another study, Rus et al. (2016) also showed increased amygdala coupling with posterior parieto-occipital regions during a negative affective task. Disruptions of amygdala effective connectivity, however, have not been exclusively observed during emotional tasks and have been not limited to posterior brain regions. Thus, De Vries et al. (2014) revealed increased task-related functional connectivity between frontal regions and bilateral amygdala during performance of a working memory task, which was related to impaired performance and therefore interpreted as a limbic interference onto prefrontal functioning during cognitive protocols. Concurring with this interpretation, negative emotional distraction, which implied amygdala hyperactivation, disrupted the modulatory input from the dlPFC to lateral orbitofrontal cortex during a working memory task in patients with OCD (Han et al. 2016). Conversely, during a reward learning task, modulation of lateral orbitofrontal activity stemmed from the vmPFC (Alves-Pinto et al. 2019), suggesting that cortico-cortical interactions within the prefrontal cortex may depend on the emotional valence of the protocol.

The ACC has also been reported to display abnormal patterns of connectivity during task performance in OCD. In one of these studies, Beucke et al. (2012) demonstrated an abnormal inverse coupling of the anterior mid-cingulate cortex with the ventral striatum and the lateral orbitofrontal cortex during punishment anticipation in OCD. This suggests that the abnormal connectivity described among these regions in seed-based functional connectivity studies may be indeed hierarchically organized with the ACC modulating activity in lateral orbitofrontal and striatal areas. In other contexts, however, such as in the early phases of an aversive conditioning, it has been shown that the vmPFC may be regulating brain activity in the ACC and other regions of the salience network (Apergis-Schoute et al. 2017). It would be of interest to interpret this combination of findings against the background of the hypothesized lateral vs. medial fronto-striatal imbalance in OCD (Robbins et al. 2019).

Modulatory effects of the ACC have also been described in cognitive protocols. For example, during a Stroop task, Schlösser et al. (2010) described significantly stronger task-related modulation of effective connectivity from the dorsal ACC to the left DLPFC in patients with OCD, which was interpreted in terms of an overactive error control system. Likewise, in a working memory experiment, it was observed that patients with OCD showed increased dorsal ACC modulation of activity in cortical, striatal, and thalamic regions (Diwadkar et al. 2015). In a further experiment, this same group showed that such increased ACC modulation of neural activity across different brain regions could indeed be more related to alterations in basic sensorimotor processing than to cognitive processes (Friedman et al. 2017). Such interpretation resonates with other reports suggesting the existence of significant alterations in sensorimotor processing in patients with OCD (Armstrong et al. 2016; Moreira et al. 2019; Pujol et al. 2019). In agreement with such notions, Hampshire et al. (2020) have recently shown that the pattern of task-related connectivity disruptions during stop-signal task performance expands from the prefrontal cortex to primary cortices in temporal, sensorimotor, and occipital brain regions, and seems to preferentially involve cerebellar networks.

It is not straightforward to provide a cohesive summary of the results from functional and effective connectivity studies in OCD. In addition to confounding factors such as across-sample heterogeneity (e.g., medication status), studies have used a wide variety of analysis approaches leading to outcomes that have not been always easy to reconcile. Indeed, some studies seem to principally aim at providing evidence for the potential application or superiority of a specific method in clinical contexts. In consequence, neurobiological models of OCD have been sometimes used as a post-hoc framework for the interpretation of findings rather than to put forward heuristic aims and hypotheses enabling to advance in the understanding of the disorder. Nevertheless, there are some meaningful conclusions to draw from the studies reviewed above. Firstly, functional connectivity between the ventral striatum and the vmPFC is altered in OCD. In most studies, such alterations have been described as an increase in functional connectivity, although some reports suggest this may depend on medication status. Likewise, functional connectivity among the striatal nodes of CSTC circuits seems to be also increased, at least for the major striatal nodes (i.e., putamen), although this may also depend on medication intake. Within the prefrontal cortex, functional connectivity of the lateral prefrontal cortex has been shown to be decreased in OCD, while the orbitofrontal cortex has been reported to display increased functional connectivity with neighboring regions. Secondly, patients with OCD show a disrupted interaction across the three main brain networks (default-mode, salience, and central-executive networks), and the striatal nodes of the CSTC circuits seem to behave as a powerful attractor of neural activity from distant brain circuits. Finally, studies on effective connectivity have shown that increased emotion-related amygdala activity may interfere with neural processing in both perceptual and cognitive networks, while the ACC can play a pivotal role by relaying vmPFC input to the lateral prefrontal cortex and other regions within and outside the prefrontal cortex, probably modulating not only cognitive, but also basic sensorimotor processing.

4 Prediction of Treatment Response

Predicting who is going to respond to a specific treatment is one of the most important challenges for mental health clinicians. Therefore, different authors have used functional neuroimaging to identify potential biomarkers of treatment response that might be eventually of utility in the clinical practice. We will close this chapter by briefly reviewing the results from some of these studies.

In the study of Sanematsu et al. (2010), OCD patients undertook an fMRI symptom provocation protocol before initiating a 12-week fluvoxamine trial. The activations from this baseline fMRI assessment were correlated with the after-treatment percentage of Y-BOCS reduction, and the authors found that larger activations during symptom provocation in the right cerebellum and the left superior temporal gyrus predicted subsequent reduction in OCD symptom severity. More recently, using a different approach, Kim et al. (2019) found that functional connectivity between the raphe nuclei (the main serotoninergic source in the central nervous system) and the left middle temporal gyrus was positively correlated with obsessive-compulsive symptoms in a baseline assessment. Remarkably, the authors also observed that the greater this connectivity, the poorer was symptom reduction after a 16-week treatment trial with selective serotonin reuptake inhibitors.

Other studies have aimed at predicting treatment response to non-pharmacological treatments, such as transcranial magnetic stimulation (TMS) or cognitive behavioral therapy (CBT). Dunlop et al. (2016), for instance, showed that OCD patients responding to a repetitive TMS aimed at the dorsomedial prefrontal cortex exhibited higher connectivity between this region and the ventral striatum at baseline. Moreover, the reduction of such connectivity between pre- and post-treatment assessments predicted symptomatic improvement. In another study, Douw et al. (2020) have shown that greater local connectivity in the dlPFC at baseline predicted better treatment response after a repetitive TMS treatment targeting this area. On the other hand, Yin et al. (2018) investigated the fMRI predictors of response to capsulotomy in patients with OCD. They observed that decreased functional connectivity between the dorsal caudate and the dorsal ACC at baseline discriminated responders from non-responders, while the extent of post-capsulotomy functional connectivity decrease between the ventral striatum and the dorsal ACC was predictive of better response.

Although brain activity in the striatum has also been associated with the response to CBT (Freyer et al. 2011), some studies have suggested that activity and functional connectivity in the amygdala and regions related to emotion regulation may predict response to CBT. Thus, Olatunji et al. (2014) showed that activation in emotional processing regions (i.e., anterior temporal pole and amygdala) during symptom provocation prior to 12 weeks of CBT predicted better treatment response, while activity in areas involved in emotion regulation (i.e., dlPFC) predicted poorer response. Similarly, and in partial agreement with these findings, Fullana et al. (2017) described that decreased functional connectivity between another cortical regulatory region, the vmPFC, and the basolateral amygdala predicted better CBT response. Speculatively, CBT may be more beneficial for those patients less able to regulate their limbic activity before treatment initiation. Contrastingly, however, Göttlich et al. (2015) observed that lower pre-treatment functional connectivity of the basolateral amygdala was associated with a diminished response to CBT, although in that report the connectivity with prefrontal regulatory regions was not specifically assessed.

Other studies, however, found other functional measurements to be more predictive of CBT response than activity or functional connectivity of limbic regions. Reggente et al. (2018), for instance, described that pre-treatment multivariate connectivity within the DMN and the visual network significantly predicted patients' symptoms after 4 weeks of intensive CBT. Moreover, including amygdala activity in such predictive models did not increase their accuracy. Pre-treatment activity within the DMN, as well as within the salience network, during a conflict task also predicted CBT response in another recent study (Pagliaccio et al. 2019). In the same vein, increased baseline functional connectivity between regions of the DMN (i.e., angular gyrus) and task-positive regions (i.e., activated by cognitive or emotion induction protocols, such as the frontal pole) predicted better response to CBT (Cyr et al. 2020).

Although the above section does not comprehensively review all the studies aimed at predicting functional neuroimaging biomarkers of treatment response, they may well serve to exemplify the variety of results obtained across treatment approaches and the lack of converging findings in key regions of prevailing neurobiological models of the disorder, such as the CSTC circuits. This may be in part consequence of the non-specific nature of most treatments for OCD, which may be also effective, for instance, for mood or anxiety disorders. In this sense, perhaps only deep brain stimulation, aimed at ventral nodes of the CSTC circuitry or particular subdivisions of the subthalamic nucleus, may be considered specific for OCD, even for distinct symptom dimensions (Tyagi et al. 2019). Moreover, as discussed above in relation to results from connectivity studies, the variety of methodological approaches undertaken by researchers across the different studies does not allow for a fair comparison of the results from the different publications. As an example, Table 1 summarizes the findings from studies using fMRI estimates to predict response to CBT.

5 Conclusion

Functional magnetic resonance imaging research capitalized on early clinical, neurocognitive, and positron emission tomography research to develop a very influential body of research and a theoretical framework that has dominated neurobiological accounts of OCD in recent years and has had a significant impact on more clinical conceptualizations of the disorder. Nevertheless, although, overall, there have been consistent findings of alterations in CSTC circuits and different cortical networks, to date it has been not possible to integrate the results from the different studies within a common and heuristic framework, which has hampered the translation of research findings into the clinical practice. The lack of consistent neurobiological correlates of clinical heterogeneity in OCD and reliable neuroimaging biomarkers of treatment response are probably the best examples of this. In any case, functional neuroimaging provides an unapparelled opportunity to unveil and understand alterations of the OCD brain and develop biologically informed therapies accounting for the diversity of symptoms. Achieving this goal will require homogenization of analysis methods and utilizing studies using both exploratory data-driven and confirmatory model-based approaches. This should be central to our research agendas in the upcoming years.

Disclosures The author reports no interests to disclose.

Table 1 Summary of the studies assessing the relationship between different brain activity and functional connectivity estimates and response to CBT

Study	Population	Treatment	Protocol	Analysis approach	Main findings
Cyr et al. (2020)	25 medication-free children and adolescents with OCD + 23 healthy controls	12-16 ERP sessions	Pre- and post-treatment **resting-state** assessment	**Whole-brain connectome level analyses** with using the Network-Based Statistics Toolbox	Greater (less altered) connectivity between left angular gyrus and left frontal pole predicted better response to CBT
Thorsen et al. (2020)	34 adult patients + 28 healthy controls	Intensive four-day treatment	Pre- and post-treatment **resting-state** assessment	**Static and dynamic graph measures**	Post-CBT decrease in frontoparietal-limbic connectivity and subgenual ACC flexibility
Pagliaccio et al. (2019)	36 medication-free adults with OCD + 33 healthy controls	17 ERP sessions	Pre-CBT assessment with the **Simon Conflict task**	**Linear mixed-effects models**	Increased activity in left pallidum and 35 cortical regions (including anterior insula and PCC) predicted treatment response
Li et al. (2018)	20 medication-free adult patients + 20 healthy controls	14 ERP sessions	Pre- and post-treatment **resting-state** assessment	**Degree centrality** (a graph theory measure defining the number of links of a node to other nodes of a network) and seed-based functional connectivity	Degree centrality in left dlPFC was reduced after CBT. Increased FC between left dlPFC and right OBF cortex normalized after CBT
van der Straten et al. (2018)	24 medication-free children and adolescents with OCD +	16 ERP and cognitive therapy sessions	Pre-, post-treatment and long-term follow-up (2 years) activations during a planning task	**Task-based fMRI** (GLM analysis with SPM12)	Increased left IFG, MFG, and insula activations before CBT, but not at post-CBT and long-

(continued)

Table 1 (continued)

Study	Population	Treatment	Protocol	Analysis approach	Main findings
	24 healthy controls		(**Tower of London**)		term assessments
Reggente et al. (2018)	42 adult patients	Intensive CBT (20 ERP sessions in 4 weeks)	Pre- and post-treatment **resting-state** assessment	**Machine Learning Multivariate Regression** Analyses	Pre-treatment connectivity patterns within the DMN and visual network predicted post-treatment severity
Moody et al. (2017)	43 adult patients + 24 healthy controls	Intensive CBT (20 ERP sessions in 4 weeks)	Pre- and post-treatment **resting-state** assessment	**Network-based-statistic analyses**	Pre- to post-CBT connectivity increases between the cerebellum and caudate, putamen, dlPFC and vlPFC
Fullana et al. (2017)	73 adult patients + 84 healthy controls	20 ERP sessions	Pre- treatment **resting-state** assessment	**ROI-to-ROI functional connectivity** assessments	Decreased BLA–vmPFC connectivity predicted CBT response
Göttlich et al. (2015)	17 medication-free adult patients + 19 healthy controls	14 ERP sessions including s education and cognitive restructuring techniques	Pre-treatment **resting-state** assessment	**Degree centrality**	Degree centrality of the right BLA correlated with treatment response
Yang et al. (2015)	22 medication-free adult patients + 22 healthy controls	14 ERP sessions	Pre- and post-treatment **resting-state** assessment	**ReHo**, a method to measure local coherence in neighboring voxels, calculation with DPARSF	Response associated with decreased ReHo in right OBF, bilateral middle frontal cortex, left cerebellum and vermis, and increased ReHo in left caudate.
Feusner et al. (2015)	17 adult patients	Intensive CBT (20 ERP	Pre- and post-treatment	**Graph theory metrics** with the Brain	Small-worldness and clustering

(continued)

Table 1 (continued)

Study	Population	Treatment	Protocol	Analysis approach	Main findings
		sessions in 4 weeks)	**resting-state** assessment	Connectivity toolbox	coefficient increases from pre- to post-treatment. Modularity decreases correlated with symptom decrease.
Olatunji et al. (2014)	12 adult patients	24 ERP sessions	Pre-treatment activations to a **symptom provocation** protocol	**Task-based fMRI** (permutation-based nonparametric analysis with XBAM)	Temporal pole and amygdala (positive) and dlPFC (negative) activations correlated with treatment response.
Morgiève et al. 2014	31 adult patients scanned in 4 time points	15 ERP sessions	Pre-, during-, and post-treatment activations to a **symptom provocation**	**Task-based fMRI** (GLM analysis with SPM8)	Progressive decrease in dorsal ACC and left (but not right) OBF cortex activations
Huyser et al. (2011)	25 medication-free children and adolescents with OCD + 25 healthy controls	16 ERP and cognitive therapy sessions	Pre- and post-treatment activations during an **interference Flanker task**	**Task-based fMRI** (GLM analysis with SPM5)	Increased bilateral insula activation before, but not after, CBT in older subjects
Huyser et al. (2010)	25 medication-free children and adolescents with OCD + 25 healthy controls	16 ERP and cognitive therapy sessions	Pre-, post-treatment activations during a planning task (**Tower of London**)	**Task-based fMRI** (GLM analysis with SPM5)	Increased vlPFC, mPFC, insula, and ACC activations with increasing task load before, but not after, CBT

(continued)

Table 1 (continued)

Study	Population	Treatment	Protocol	Analysis approach	Main findings
Freyer et al. (2011)	10 medication-free adult patients + 10 healthy controls	16 ERP sessions	Pre- and post-treatment activations during a **Reversal Learning task**	**Task-based fMRI** (GLM analysis with SPM5)	Decreased caudate activation before, but not after, CBT

ACC Anterior Cingulate Cortex, *BLA* Basolateral Amygdala, *CBT* Cognitive Behavioral Therapy, *dlPFC* dorsolateral Prefrontal Cortex, *DMN* Default Mode Network, *DPARSF* Data Processing Assistant for Resting-State fMRI, *ERP* Exposure and Response Prevention, *fMRI* functional Magnetic Resonance Imaging, *GLM* General Linear Model, *IFG* Inferior Frontal Gyrus, *MFG* Middle Frontal Gyrus, *mPFC* medial Prefrontal Cortex, *OBF* Orbitofrontal, *OCD* Obsessive-Compulsive Disorder, *PCC* Posterior Cingulate Cortex, *ReHo* Regional Homogeneity, *ROI* Region of Interest, *SPM* Statistical Parametric Mapping, *vlPFC* ventrolateral Prefrontal Cortex, *vmPFC* ventromedial Prefrontal Cortex, *XBAM* Brain Activation Map

References

Abramovitch A, Dar R, Hermesh H, Schweiger A (2012) Comparative neuropsychology of adult obsessive-compulsive disorder and attention deficit/hyperactivity disorder: implications for a novel executive overload model of OCD. J Neuropsychol 6:161–191

Adler CM, McDonough-Ryan P, Sax KW, Holland SK, Arndt S, Strakowski SM (2000) fMRI of neuronal activation with symptom provocation in unmedicated patients with obsessive compulsive disorder. J Psychiatr Res 34(4-5):317–324

Alexander GE, DeLong MR, Strick PL (1986) Parallel organization of functionally segregated circuits linking basal ganglia and cortex. Annu Rev Neurosci 9:357–381

Alves-Pinto A, Rus OG, Reess TJ, Wohlschläger A, Wagner G, Berberich G, Koch K (2019) Altered reward-related effective connectivity in obsessive-compulsive disorder: an fMRI study. J Psychiatry Neurosci 44(6):395–406

An SK, Mataix-Cols D, Lawrence NS, Wooderson S, Giampietro V, Speckens A, Brammer MJ, Phillips ML (2009) To discard or not to discard: the neural basis of hoarding symptoms in obsessive-compulsive disorder. Mol Psychiatry 14(3):318–331

Anticevic A, Hu S, Zhang S, Savic A, Billingslea E, Wasylink S, Repovs G, Cole MW, Bednarski S, Krystal JH, Bloch MH, Li CS, Pittenger C, Bernstein et al (2014) Global resting-state functional magnetic resonance imaging analysis identifies frontal cortex, striatal, and cerebellar dysconnectivity in obsessive-compulsive disorder. Biol Psychiatry 75 (8):595–605

Apergis-Schoute AM, Gillan CM, Fineberg NA, Fernandez-Egea E, Sahakian BJ, Robbins TW (2017) Neural basis of impaired safety signaling in obsessive compulsive disorder. Proc Natl Acad Sci U S A 114(12):3216–3221

Armstrong CC, Moody TD, Feusner JD, McCracken JT, Chang S, Levitt JG, Piacentini JC, O'Neill J (2016) Graph-theoretical analysis of resting-state fMRI in pediatric obsessive-compulsive disorder. J Affect Disord 193:175–184

Baioui A, Pilgramm J, Merz CJ, Walter B, Vaitl D, Stark R (2013) Neural response in obsessive-compulsive washers depends on individual fit of triggers. Front Hum Neurosci 7:143

Banca P, Voon V, Vestergaard MD, Philipiak G, Almeida I, Pocinho F, Relvas J, Castelo-Branco M (2015) Imbalance in habitual versus goal directed neural systems during symptom provocation in obsessive-compulsive disorder. Brain 138(Pt 3):798–811

Basile B, Mancini F, Macaluso E, Caltagirone C, Bozzali M (2014) Abnormal processing of deontological guilt in obsessive-compulsive disorder. Brain Struct Funct 219(4):1321–1331
Baxter LR Jr, Phelps ME, Mazziotta JC, Guze BH, Schwartz JM, Selin CE (1987) Local cerebral glucose metabolic rates in obsessive-compulsive disorder. A comparison with rates in unipolar depression and in normal controls. Arch Gen Psychiatry 44(3):211–218
Becker MP, Nitsch AM, Schlösser R, Koch K, Schachtzabel C, Wagner G, Miltner WH, Straube T (2014) Altered emotional and BOLD responses to negative, positive and ambiguous performance feedback in OCD. Soc Cogn Affect Neurosci 9(8):1127–1133
Berlin HA, Schulz KP, Zhang S, Turetzky R, Rosenthal D, Goodman W (2015) Neural correlates of emotional response inhibition in obsessive-compulsive disorder: a preliminary study. Psychiatry Res 234(2):259–264
Berlin HA, Stern ER, Ng J, Zhang S, Rosenthal D, Turetzky R, Tang C, Goodman W (2017) Altered olfactory processing and increased insula activity in patients with obsessive-compulsive disorder: an fMRI study. Psychiatry Res Neuroimaging 262:15–24
Beucke JC, Kaufmann C, Linnman C, Gruetzmann R, Endrass T, Deckersbach T, Dougherty DD, Kathmann N (2012) Altered cingulostriatal coupling in obsessive-compulsive disorder. Brain Connect 2(4):191–202
Beucke JC, Sepulcre J, Talukdar T, Linnman C, Zschenderlein K, Endrass T, Kaufmann C, Kathmann N (2013) Abnormally high degree connectivity of the orbitofrontal cortex in obsessive-compulsive disorder. JAMA Psychiatry 70(6):619–629
Breiter HC, Rauch SL, Kwong KK, Baker JR, Weisskoff RM, Kennedy DN, Kendrick AD, Davis TL, Jiang A, Cohen MS, Stern CE, Belliveau JW, Baer L, O'Sullivan RL, Savage CR, Jenike MA, Rosen BR (1996) Functional magnetic resonance imaging of symptom provocation in obsessive-compulsive disorder. Arch Gen Psychiatry 53(7):595–606
Brennan BP, Tkachenko O, Schwab ZJ, Juelich RJ, Ryan EM, Athey AJ, Pope HG, Jenike MA, Baker JT, Killgore WD, Hudson JI, Jensen JE, Rauch SL (2015) An examination of rostral anterior cingulate cortex function and neurochemistry in obsessive-compulsive disorder. Neuropsychopharmacology 40(8):1866–1876
Britton JC, Stewart SE, Killgore WD, Rosso IM, Price LM, Gold AL, Pine DS, Wilhelm S, Jenike MA, Rauch SL (2010) Amygdala activation in response to facial expressions in pediatric obsessive-compulsive disorder. Depress Anxiety 27(7):643–651
Cannistraro PA, Wright CI, Wedig MM, Martis B, Shin LM, Wilhelm S, Rauch SL (2004) Amygdala responses to human faces in obsessive-compulsive disorder. Biol Psychiatry 56 (12):916–920
Cano M, Alonso P, Martínez-Zalacaín I, Subirà M, Real E, Segalàs C, Pujol J, Cardoner N, Menchón JM, Soriano-Mas C (2018) Altered functional connectivity of the subthalamus and the bed nucleus of the stria terminalis in obsessive-compulsive disorder. Psychol Med 48 (6):919–928
Cardoner N, Harrison BJ, Pujol J, Soriano-Mas C, Hernández-Ribas R, López-Solá M, Real E, Deus J, Ortiz H, Alonso P, Menchón JM (2011) Enhanced brain responsiveness during active emotional face processing in obsessive compulsive disorder. World J Biol Psychiatry 12 (5):349–363
Chamberlain SR, Blackwell AD, Fineberg NA, Robbins TW, Sahakian BJ (2005) The neuropsychology of obsessive compulsive disorder: the importance of failures in cognitive and behavioural inhibition as candidate endophenotypic markers. Neurosci Biobehav Rev 29:399–419
Chamberlain SR, Fineberg NA, Blackwell AD, Robbins TW, Sahakian BJ (2006) Motor inhibition and cognitive flexibility in obsessive-compulsive disorder and trichotillomania. Am J Psychiatry 163:1282–1284
Chamberlain SR, Fineberg NA, Menzies LA, Blackwell AD, Bullmore ET, Robbins TW, Sahakian BJ (2007) Impaired cognitive flexibility and motor inhibition in unaffected first-degree relatives of patients with obsessive-compulsive disorder. Am J Psychiatry 164:335–338

Chamberlain SR, Menzies L, Hampshire A, Suckling J, Fineberg NA, del Campo N, Aitken M, Craig K, Owen AM, Bullmore ET, Robbins TW, Sahakian BJ (2008) Orbitofrontal dysfunction in patients with obsessive-compulsive disorder and their unaffected relatives. Science 321 (5887):421–422

Cheng Y, Xu J, Nie B, Luo C, Yang T, Li H, Lu J, Xu L, Shan B, Xu X (2013) Abnormal resting-state activities and functional connectivities of the anterior and the posterior cortexes in medication-naïve patients with obsessive-compulsive disorder. PLoS One 8(6):e67478

Cyr M, Pagliaccio D, Yanes-Lukin P, Fontaine M, Rynn MA, Marsh R (2020) Altered network connectivity predicts response to cognitive-behavioral therapy in pediatric obsessive-compulsive disorder. Neuropsychopharmacology 45(7):1232-1240

de Vries FE, de Wit SJ, Cath DC, van der Werf YD, van der Borden V, van Rossum TB, van Balkom AJ, van der Wee NJ, Veltman DJ, van den Heuvel OA (2014) Compensatory frontoparietal activity during working memory: an endophenotype of obsessive-compulsive disorder. Biol Psychiatry 76(11):878–887

de Wit SJ, de Vries FE, van der Werf YD, Cath DC, Heslenfeld DJ, Veltman EM, van Balkom AJ, Veltman DJ, van den Heuvel OA (2012) Presupplementary motor area hyperactivity during response inhibition: a candidate endophenotype of obsessive-compulsive disorder. Am J Psychiatry 169(10):1100–1108

de Wit SJ, van der Werf YD, Mataix-Cols D, Trujillo JP, van Oppen P, Veltman DJ, van den Heuvel OA (2015) Emotion regulation before and after transcranial magnetic stimulation in obsessive compulsive disorder. Psychol Med 45(14):3059–3073

Diwadkar VA, Burgess A, Hong E, Rix C, Arnold PD, Hanna GL, Rosenberg DR (2015) Dysfunctional activation and brain network profiles in youth with obsessive-compulsive disorder: a focus on the dorsal anterior cingulate during working memory. Front Hum Neurosci 9:149

Douw L, Quaak M, Fitzsimmons SMDD, de Wit SJ, van der Werf YD, van den Heuvel OA, Vriend C (2020) Static and dynamic network properties of the repetitive transcranial magnetic stimulation target predict changes in emotion regulation in obsessive-compulsive disorder. Brain Stimul 13(2):318–326

Dunlop K, Woodside B, Olmsted M, Colton P, Giacobbe P, Downar J (2016) Reductions in cortico-striatal hyperconnectivity accompany successful treatment of obsessive-compulsive disorder with dorsomedial prefrontal rTMS. Neuropsychopharmacology 41(5):1395–1403

Endrass T, Ullsperger M (2014) Specificity of performance monitoring changes in obsessive-compulsive disorder. Neurosci Biobehav Rev 46(Pt 1):124–138

Etkin A, Wager TD (2007) Functional neuroimaging of anxiety: a meta-analysis of emotional processing in PTSD, social anxiety disorder, and specific phobia. Am J Psychiatry 164 (10):1476–1488

Fan J, Zhong M, Gan J, Liu W, Niu C, Liao H, Zhang H, Yi J, Chan RCK, Tan C, Zhu X (2017) Altered connectivity within and between the default mode, central executive, and salience networks in obsessive-compulsive disorder. J Affect Disord 223:106–114

Feusner JD, Moody T, Lai TM, Sheen C, Khalsa S, Brown J, Levitt J, Alger J, O'Neill J (2015) Brain connectivity and prediction of relapse after cognitive-behavioral therapy in obsessive-compulsive disorder. Front Psychiatry 6:74

Figee M, Vink M, de Geus F, Vulink N, Veltman DJ, Westenberg H, Denys D (2011) Dysfunctional reward circuitry in obsessive-compulsive disorder. Biol Psychiatry 69(9):867–874

Fornito A, Bullmore ET (2010) What can spontaneous fluctuations of the blood oxygenation-level-dependent signal tell us about psychiatric disorders? Curr Opin Psychiatry 23(3):239–249

Fox MD, Raichle ME (2007) Spontaneous fluctuations in brain activity observed with functional magnetic resonance imaging. Nat Rev Neurosci 8(9):700–711

Freyer T, Klöppel S, Tüscher O, Kordon A, Zurowski B, Kuelz AK, Speck O, Glauche V, Voderholzer U (2011) Frontostriatal activation in patients with obsessive-compulsive disorder before and after cognitive behavioral therapy. Psychol Med 41(1):207–216

Friedman AL, Burgess A, Ramaseshan K, Easter P, Khatib D, Chowdury A, Arnold PD, Hanna GL, Rosenberg DR, Diwadkar VA (2017) Brain network dysfunction in youth with obsessive-

compulsive disorder induced by simple uni-manual behavior: the role of the dorsal anterior cingulate cortex. Psychiatry Res Neuroimaging 260:6–15
Friston KJ, Buechel C, Fink GR, Morris J, Rolls E, Dolan RJ (1997) Psychophysiological and modulatory interactions in neuroimaging. Neuroimage 6(3):218–229
Fullana MA, Zhu X, Alonso P, Cardoner N, Real E, López-Solà C, Segalàs C, Subirà M, Galfalvy H, Menchón JM, Simpson HB, Marsh R, Soriano-Mas C (2017) Basolateral amygdala-ventromedial prefrontal cortex connectivity predicts cognitive behavioural therapy outcome in adults with obsessive-compulsive disorder. J Psychiatry Neurosci 42(6):378–385
Gilbert AR, Akkal D, Almeida JRC, Mataix-Cols D, Kalas C, Devlin B, Birmaher B, Phillips ML (2009) Neural correlates of symptom dimensions in pediatric obsessive-compulsive disorder: a functional magnetic resonance imaging study. J Am Acad Child Adolesc Psychiatry 48 (9):936–944
Gillan CM, Robbins TW (2014) Goal-directed learning and obsessive-compulsive disorder. Philos Trans R Soc Lond B Biol Sci 369(1655):20130475
Gillan CM, Apergis-Schoute AM, Morein-Zamir S, Urcelay GP, Sule A, Fineberg NA, Sahakian BJ, Robbins TW (2015) Functional neuroimaging of avoidance habits in obsessive-compulsive disorder. Am J Psychiatry 172(3):284–293
Gonçalves ÓF, Soares JM, Carvalho S, Leite J, Ganho A, Fernandes-Gonçalves A, Frank B, Pocinho F, Relvas J, Carracedo A, Sampaio A (2015) Brain activation of the defensive and appetitive survival systems in obsessive compulsive disorder. Brain Imaging Behav 9 (2):255–263
Göttlich M, Krämer UM, Kordon A, Hohagen F, Zurowski B (2014) Decreased limbic and increased fronto-parietal connectivity in unmedicated patients with obsessive-compulsive disorder. Hum Brain Mapp 35(11):5617–5632
Göttlich M, Krämer UM, Kordon A, Hohagen F, Zurowski B (2015) Resting-state connectivity of the amygdala predicts response to cognitive behavioral therapy in obsessive compulsive disorder. Biol Psychol 111:100–109
Greicius M (2008) Resting-state functional connectivity in neuropsychiatric disorders. Curr Opin Neurol 21(4):424–430
Grützmann R, Endrass T, Kaufmann C, Allen E, Eichele T, Kathmann N (2016) Presupplementary motor area contributes to altered error monitoring in obsessive-compulsive disorder. Biol Psychiatry 80:562–571
Gu BM, Park JY, Kang DH, Lee SJ, Yoo SY, Jo HJ, Choi CH, Lee JM, Kwon JS (2008) Neural correlates of cognitive inflexibility during task-switching in obsessive-compulsive disorder. Brain 131(Pt1):155–164
Gürsel DA, Avram M, Sorg C, Brandl F, Koch K (2018) Frontoparietal areas link impairments of large-scale intrinsic brain networks with aberrant fronto-striatal interactions in OCD: a meta-analysis of resting-state functional connectivity. Neurosci Biobehav Rev 87:151–160
Gürsel DA, Reinholz L, Bremer B, Schmitz-Koep B, Franzmeier N, Avram M, Koch K (2020) Frontoparietal and salience network alterations in obsessive–compulsive disorder: insights from independent component and sliding time window analyses. J Psychiatry Neurosci 45(3):190038
Haber SN (2003) The primate basal ganglia: parallel and integrative networks. J Chem Neuroanat 26(4):317–330
Hampshire A, Zadel A, Sandrone S, Soreq E, Fineberg N, Bullmore ET, Robbins TW, Sahakian BJ, Chamberlain SR (2020) Inhibition-related cortical hypoconnectivity as a candidate vulnerability marker for obsessive-compulsive disorder. Biol Psychiatry Cogn Neurosci Neuroimaging 5 (2):222–230
Han HJ, Jung WH, Yun JY, Park JW, Cho KK, Hur JW, Shin NY, Lee TY, Kwon JS (2016) Disruption of effective connectivity from the dorsolateral prefrontal cortex to the orbitofrontal cortex by negative emotional distraction in obsessive-compulsive disorder. Psychol Med 46 (5):921–932
Harrison BJ, Soriano-Mas C, Pujol J, Ortiz H, López-Solà M, Hernández-Ribas R, Deus J, Alonso P, Yücel M, Pantelis C, Menchon JM, Cardoner N (2009) Altered corticostriatal

functional connectivity in obsessive-compulsive disorder. Arch Gen Psychiatry 66 (11):1189–1200

Harrison BJ, Pujol J, Soriano-Mas C, Hernández-Ribas R, López-Solà M, Ortiz H, Alonso P, Deus J, Menchon JM, Real E, Segalàs C, Contreras-Rodríguez O, Blanco-Hinojo L, Cardoner N (2012) Neural correlates of moral sensitivity in obsessive-compulsive disorder. Arch Gen Psychiatry 69(7):741–749

Harrison BJ, Pujol J, Cardoner N, Deus J, Alonso P, López-Solà M, Contreras-Rodríguez O, Real E, Segalàs C, Blanco-Hinojo L, Menchon JM, Soriano-Mas C (2013) Brain corticostriatal systems and the major clinical symptom dimensions of obsessive-compulsive disorder. Biol Psychiatry 73(4):321–328

Hauser TU, Iannaccone R, Dolan RJ, Ball J, Hättenschwiler J, Drechsler R, Rufer M, Brandeis D, Walitza S, Brem S (2017) Increased fronto-striatal reward prediction errors moderate decision making in obsessive-compulsive disorder. Psychol Med 47(7):1246–1258

Heinzel S, Kaufmann C, Grützmann R, Hummel R, Klawohn J, Riesel A, Bey K, Lennertz L, Wagner M, Kathmann N (2017) Neural correlates of working memory deficits and associations to response inhibition in obsessive compulsive disorder. Neuroimage Clin 17:426–434

Hennig-Fast K, Michl P, Müller J, Niedermeier N, Coates U, Müller N, Engel RR, Möller HJ, Reiser M, Meindl T (2015) Obsessive-compulsive disorder--A question of conscience? An fMRI study of behavioural and neurofunctional correlates of shame and guilt. J Psychiatr Res 68:354–362

Hou JM, Zhao M, Zhang W, Song LH, Wu WJ, Wang J, Zhou DQ, Xie B, He M, Guo JW, Qu W, Li HT (2014) Resting-state functional connectivity abnormalities in patients with obsessive-compulsive disorder and their healthy first-degree relatives. J Psychiatry Neurosci 39 (5):304–311

Huettel SA, Song AW, McCarthy G (2004) Functional magnetic resonance imaging. Sinauer Associates Press, Sunderland

Huyser C, Veltman DJ, Wolters LH, de Haan E, Boer F (2010) Functional magnetic resonance imaging during planning before and after cognitive-behavioral therapy in pediatric obsessive-compulsive disorder. J Am Acad Child Adolesc Psychiatry 49(12):1238–1248

Huyser C, Veltman DJ, Wolters LH, de Haan E, Boer F (2011) Developmental aspects of error and high-conflict-related brain activity in pediatric obsessive-compulsive disorder: a fMRI study with a Flanker task before and after CBT. J Child Psychol Psychiatry 52(12):1251–1260

Jang JH, Kim JH, Jung WH, Choi JS, Jung MH, Lee JM, Choi CH, Kang DH, Kwon JS (2010) Functional connectivity in fronto-subcortical circuitry during the resting state in obsessive-compulsive disorder. Neurosci Lett 474(3):158–162

Jung WH, Kang DH, Kim E, Shin KS, Jang JH, Kwon JS (2013) Abnormal corticostriatal-limbic functional connectivity in obsessive-compulsive disorder during reward processing and resting-state. Neuroimage Clin 3:27–38

Jung WH, Yücel M, Yun JY, Yoon YB, Cho KI, Parkes L, Kim SN, Kwon JS (2017) Altered functional network architecture in orbitofronto-striato-thalamic circuit of unmedicated patients with obsessive-compulsive disorder. Hum Brain Mapp 38(1):109–119

Kang DH, Jang JH, Han JY, Kim JH, Jung WH, Choi JS, Choi CH, Kwon JS (2013) Neural correlates of altered response inhibition and dysfunctional connectivity at rest in obsessive–compulsive disorder. Prog Neuro-Psychopharmacol Biol Psychiatry 40:340–346

Kim M, Kwak S, Yoon YB, Kwak YB, Kim T, Cho KIK, Lee TY, Kwon JS (2019) Functional connectivity of the raphe nucleus as a predictor of the response to selective serotonin reuptake inhibitors in obsessive-compulsive disorder. Neuropsychopharmacology 44(12):2073–2081

Koch K, Wagner G, Schachtzabel C, Peikert G, Schultz CC, Sauer H, Schlösser RG (2012) Aberrant anterior cingulate activation in obsessive-compulsive disorder is related to task complexity. Neuropsychologia 50(5):958–964

Lawrence NS, An SK, Mataix-Cols D, Ruths F, Speckens A, Phillips ML (2007) Neural responses to facial expressions of disgust but not fear are modulated by washing symptoms in OCD. Biol Psychiatry 61(9):1072–1080

Li P, Yang X, Greenshaw AJ, Li S, Luo J, Han H, Liu J, Zhong Z, Guo Z, Xiong H, Yao S, Chen Y, Sun J, Li Z (2018) The effects of cognitive behavioral therapy on resting-state functional brain network in drug-naive patients with obsessive-compulsive disorder. Brain Behav 8(5):e00963

Mataix-Cols D, Wooderson S, Lawrence N, Brammer MJ, Speckens A, Phillips ML (2004) Distinct neural correlates of washing, checking, and hoarding symptom dimensions in obsessive-compulsive disorder. Arch Gen Psychiatry 61(6):564–576

Mathews CA, Perez VB, Delucchi KL, Mathalon DH (2012) Error-related negativity in individuals with obsessive-compulsive symptoms: toward an understanding of hoarding behaviors. Biol Psychol 89(2):487–494

McCabe C, Mishor Z (2011) Antidepressant medications reduce subcortical-cortical resting-state functional connectivity in healthy volunteers. Neuroimage 57(4):1317–1323

Menon V (2011) Large-scale brain networks and psychopathology: a unifying triple network model. Trends Cogn Sci 15(10):483–506

Menzies L, Chamberlain SR, Laird AR, Thelen SM, Sahakian BJ, Bullmore ET (2008) Integrating evidence from neuroimaging and neuropsychological studies of obsessive-compulsive disorder: the orbitofronto-striatal model revisited. Neurosci Biobehav Rev 32(3):525–549

Meunier D, Ersche KD, Craig KJ, Fornito A, Merlo-Pich E, Fineberg NA, Shabbir SS, Robbins TW, Bullmore ET (2012) Brain functional connectivity in stimulant drug dependence and obsessive-compulsive disorder. Neuroimage 59(2):1461–1468

Milad MR, Rauch SL (2012) Obsessive-compulsive disorder: beyond segregated cortico-striatal pathways. Trends Cogn Sci 16(1):43–51

Milad MR, Furtak SC, Greenberg JL, Keshaviah A, Im JJ, Falkenstein MJ, Jenike M, Rauch SL, Wilhelm S (2013) Deficits in conditioned fear extinction in obsessive-compulsive disorder and neurobiological changes in the fear circuit. JAMA Psychiatry 70(6):608–618

Moody TD, Morfini F, Cheng G, Sheen C, Tadayonnejad R, Reggente N, O'Neill J, Feusner JD (2017) Mechanisms of cognitive-behavioral therapy for obsessive-compulsive disorder involve robust and extensive increases in brain network connectivity. Transl Psychiatry 7(9):e1230

Morein-Zamir S, Voon V, Dodds CM, Sule A, van Niekerk J, Sahakian BJ, Robbins TW (2016) Divergent subcortical activity for distinct executive functions: stopping and shifting in obsessive compulsive disorder. Psychol Med 46:829–840

Moreira PS, Marques P, Magalhães R, Esteves M, Sousa N, Soares JM, Morgado P (2019) The resting-brain of obsessive-compulsive disorder. Psychiatry Res Neuroimaging 290:38–41

Morgiève M, N'Diaye K, Haynes WI, Granger B, Clair AH, Pelissolo A, Mallet L (2014) Dynamics of psychotherapy-related cerebral haemodynamic changes in obsessive compulsive disorder using a personalized exposure task in functional magnetic resonance imaging. Psychol Med 44 (7):1461–1473

Murayama K, Nakao T, Sanematsu H, Okada K, Yoshiura T, Tomita M, Masuda Y, Isomura K, Nakagawa A, Kanba S (2013) Differential neural network of checking versus washing symptoms in obsessive-compulsive disorder. Prog Neuropsychopharmacol Biol Psychiatry 40:160–166

Murray GK, Knolle F, Ersche KD, Craig KJ, Abbott S, Shabbir SS, Fineberg NA, Suckling J, Sahakian BJ, Bullmore ET, Robbins TW (2019) Dopaminergic drug treatment remediates exaggerated cingulate prediction error responses in obsessive-compulsive disorder. Psychopharmacology (Berl) 236(8):2325–2336

Nakao T, Nakagawa A, Yoshiura T, Nakatani E, Nabeyama M, Yoshizato C, Kudoh A, Tada K, Yoshioka K, Kawamoto M, Togao O, Kanba S (2005) Brain activation of patients with obsessive-compulsive disorder during neuropsychological and symptom provocation tasks before and after symptom improvement: a functional magnetic resonance imaging study. Biol Psychiatry 57(8):901–910

Norman LJ, Taylor SF, Liu Y, Radua J, Chye Y, De Wit SJ, Huyser C, Karahanoglu FI, Luks T, Manoach D, Mathews C, Rubia K, Suo C, van den Heuvel OA, Yücel M, Fitzgerald K (2019) Error processing and inhibitory control in obsessive-compulsive disorder: a meta-analysis using statistical parametric maps. Biol Psychiatry 85(9):713–725

Ogawa S, Lee TM, Kay AR, Tank DW (1990) Brain magnetic resonance imaging with contrast dependent on blood oxygenation. Proc Natl Acad Sci USA 87(24):9868–9872

Olatunji BO, Ferreira-Garcia R, Caseras X, Fullana MA, Wooderson S, Speckens A, Lawrence N, Giampietro V, Brammer MJ, Phillips ML, Fontenelle LF, Mataix-Cols D (2014) Predicting response to cognitive behavioral therapy in contamination-based obsessive-compulsive disorder from functional magnetic resonance imaging. Psychol Med 44(10):2125–2137

Page LA, Rubia K, Deeley Q, Daly E, Toal F, Mataix-Cols D, Giampietro V, Schmitz N, Murphy DGM (2009) A functional magnetic resonance imaging study of inhibitory control in obsessive-compulsive disorder. Psychiatry Res Neuroimaging 174:202–209

Pagliaccio D, Middleton R, Hezel D, Steinman S, Snorrason I, Gershkovich M, Campeas R, Pinto A, Van Meter P, Simpson HB, Marsh R (2019) Task-based fMRI predicts response and remission to exposure therapy in obsessive-compulsive disorder. Proc Natl Acad Sci USA 116 (41):20346–20353

Park SE, Yang JC, Jeong GW (2016) Neuroanatomical assessment of the impact of negative emotion on implicit memory in patients with obsessive compulsive disorder. Acta Neuropsychiatr 28(4):206–213

Peng ZW, Xu T, He QH, Shi CZ, Wei Z, Miao GD, Jing J, Lim KO, Zuo XN, Chan RC (2014) Default network connectivity as a vulnerability marker for obsessive compulsive disorder. Psychol Med 44(7):1475–1484

Phillips ML, Marks IM, Senior C, Lythgoe D, O'Dwyer AM, Meehan O, Williams SC, Brammer MJ, Bullmore ET, McGuire PK (2000) A differential neural response in obsessive-compulsive disorder patients with washing compared with checking symptoms to disgust. Psychol Med 30 (5):1037–1050

Picó-Pérez M, Alemany-Navarro M, Dunsmoor JE, Radua J, Albajes-Eizagirre A, Vervliet B, Cardoner N, Benet O, Harrison BJ, Soriano-Mas C, Fullana MA (2019) Common and distinct neural correlates of fear extinction and cognitive reappraisal: a meta-analysis of fMRI studies. Neurosci Biobehav Rev 104:102–115

Picó-Pérez M, Moreira PS, de Melo Ferreira V, Radua J, Mataix-Cols D, Sousa N, Soriano-Mas C, Morgado P (2020) Modality-specific overlaps in brain structure and function in obsessive-compulsive disorder: multimodal meta-analysis of case-control MRI studies. Neurosci Biobehav Rev 112:83–94

Ping L, Su-Fang L, Hai-Ying H, Zhang-Ye D, Jia L, Zhi-Hua G, Hong-Fang X, Yu-Feng Z, Zhan-Jiang L (2013) Abnormal spontaneous neural activity in obsessive-compulsive disorder: a resting-state functional magnetic resonance imaging study. PLoS One 8(6):e67262

Posner J, Marsh R, Maia TV, Peterson BS, Gruber A, Simpson HB (2014) Reduced functional connectivity within the limbic cortico-striato-thalamo-cortical loop in unmedicated adults with obsessive-compulsive disorder. Hum Brain Mapp 35(6):2852–2860

Pujol J, Blanco-Hinojo L, Maciá D, Alonso P, Harrison BJ, Martínez-Vilavella G, Deus J, Menchón JM, Cardoner N, Soriano-Mas C (2019) Mapping alterations of the functional structure of the cerebral cortex in obsessive-compulsive disorder. Cereb Cortex 29(11):4753–4762

Raichle ME (2015) The restless brain: how intrinsic activity organizes brain function. Philos Trans R Soc Lond B Biol Sci 370:1668

Raichle ME, Mintun MA (2006) Brain work and brain imaging. Annu Rev Neurosci 29:449–476

Reggente N, Moody TD, Morfini F, Sheen C, Rissman J, O'Neill J, Feusner JD (2018) Multivariate resting-state functional connectivity predicts response to cognitive behavioral therapy in obsessive-compulsive disorder. Proc Natl Acad Sci USA 115(9):2222–2227

Remijnse PL, Nielen MM, van Balkom AJ, Cath DC, van Oppen P, Uylings HB, Veltman DJ (2006) Reduced orbitofrontal-striatal activity on a reversal learning task in obsessive-compulsive disorder. Arch Gen Psychiatry 63(11):1225–1236

Robbins TW, Vaghi MM, Banca P (2019) Obsessive-compulsive disorder: puzzles and prospects. Neuron 102(1):27–47

Rotge JY, Guehl D, Dilharreguy B, Cuny E, Tignol J, Bioulac B, Allard M, Burbaud P, Aouizerate B (2008) Provocation of obsessive-compulsive symptoms: a quantitative voxel-based meta-analysis of functional neuroimaging studies. J Psychiatry Neurosci 33(5):405–412

Roth RM, Saykin AJ, Flashman LA, Pixley HS, West JD, Mamourian AC (2007) Event-related functional magnetic resonance imaging of response inhibition in obsessive-compulsive disorder. Biol Psychiatry 62:901–909

Rus OG, Reess TJ, Wagner G, Zimmer C, Zaudig M, Koch K (2016) Functional and structural connectivity of the amygdala in obsessive-compulsive disorder. Neuroimage Clin 13:246–255

Sakai Y, Narumoto J, Nishida S, Nakamae T, Yamada K, Nishimura T, Fukui K (2011) Corticostriatal functional connectivity in non-medicated patients with obsessive-compulsive disorder. Eur Psychiatry 26(7):463–469

Sanematsu H, Nakao T, Yoshiura T, Nabeyama M, Togao O, Tomita M, Masuda Y, Nakatani E, Nakagawa A, Kanba S (2010) Predictors of treatment response to fluvoxamine in obsessive-compulsive disorder: an fMRI study. J Psychiatr Res 44(4):193–200

Schienle A, Schäfer A, Stark R, Walter B, Vaitl D (2005) Neural responses of OCD patients towards disorder-relevant, generally disgust-inducing and fear-inducing pictures. Int J Psychophysiol 57(1):69–77

Schlösser RGM, Wagner G, Schachtzabel C, Peikert G, Koch K, Reichenbach JR, Sauer H (2010) Fronto-cingulate effective connectivity in obsessive compulsive disorder: a study with fMRI and dynamic causal modeling. Hum Brain Mapp 31:1834–1850

Shapira NA, Liu Y, He AG, Bradley MM, Lessig MC, James GA, Stein DJ, Lang PJ, Goodman WK (2003) Brain activation by disgust-inducing pictures in obsessive-compulsive disorder. Biol Psychiatry 54(7):751–756

Shepherd GM (2013) Corticostriatal connectivity and its role in disease. Nat Rev Neurosci 14 (4):278–291

Simon D, Kaufmann C, Müsch K, Kischkel E, Kathmann N (2010) Fronto-striato-limbic hyperactivation in obsessive-compulsive disorder during individually tailored symptom provocation. Psychophysiology 47(4):728–738

Simon D, Adler N, Kaufmann C, Kathmann N (2014) Amygdala hyperactivation during symptom provocation in obsessive-compulsive disorder and its modulation by distraction. Neuroimage Clin 4:549–557

Stern ER, Taylor SF (2014) Cognitive neuroscience of obsessive-compulsive disorder. Psychiatr Clin North Am 37(3):337–352

Stern ER, Fitzgerald KD, Welsh RC, Abelson JL, Taylor SF (2012) Resting-state functional connectivity between fronto-parietal and default mode networks in obsessive-compulsive disorder. PLoS One 7(5):e36356

Suñol M, Martínez-Zalacaín I, Picó-Pérez M, López-Solà C, Real E, Fullana MA, Pujol J, Cardoner N, Menchón JM, Alonso P, Soriano-Mas C (2020) Differential patterns of brain activation between hoarding disorder and obsessive-compulsive disorder during executive performance. Psychol Med 50(4):666–673

Thorsen AL, Hagland P, Radua J, Mataix-Cols D, Kvale G, Hansen B, van den Heuvel OA (2018) Emotional processing in obsessive-compulsive disorder: a systematic review and meta-analysis of 25 functional neuroimaging studies. Biol Psychiatry Cogn Neurosci Neuroimaging 3 (6):563–571

Thorsen AL, Vriend C, de Wit SJ, Ousdal OT, Hagen K, Hansen B, Kvale G, van den Heuvel OA (2020) Effects of Bergen 4-day treatment on resting-state graph features in obsessive-compulsive disorder. Biol Psychiatry Cogn Neurosci Neuroimaging S2451-9022(20)30026-4

Tian L, Meng C, Jiang Y, Tang Q, Wang S, Xie X, Fu X, Jin C, Zhang F, Wang J (2015) Abnormal functional connectivity of brain network hubs associated with symptom severity in treatment-naive patients with obsessive-compulsive disorder: a resting-state functional MRI study. Prog Neuropsychopharmacol Biol Psychiatry 66:104–111

Tyagi H, Apergis-Schoute AM, Akram H, Foltynie T, Limousin P, Drummond LM, Fineberg NA, Matthews K, Jahanshahi M, Robbins TW, Sahakian BJ, Zrinzo L, Hariz M, Joyce EM (2019) A

randomized trial directly comparing ventral capsule and anteromedial subthalamic nucleus stimulation in obsessive-compulsive disorder: clinical and imaging evidence for dissociable effects. Biol Psychiatry 85(9):726–734
Ursu S, Stenger VA, Shear MK, Jones MR, Carter CS (2003) Overactive action monitoring in obsessive-compulsive disorder: evidence from functional magnetic resonance imaging. Psychol Sci 14(4):347–353
Vaghi MM, Vértes PE, Kitzbichler MG, Apergis-Schoute AM, van der Flier FE, Fineberg NA, Sule A, Zaman R, Voon V, Kundu P, Bullmore ET, Robbins TW (2017) Specific frontostriatal circuits for impaired cognitive flexibility and goal-directed planning in obsessive-compulsive disorder: evidence from resting-state functional connectivity. Biol Psychiatry 81(8):708–717
van den Heuvel OA, Veltman DJ, Groenewegen HJ, Cath DC, van Balkom AJ, van Hartskamp J, Barkhof F, van Dyck R (2005a) Frontal-striatal dysfunction during planning in obsessive-compulsive disorder. Arch Gen Psychiatry 62(3):301–309
van den Heuvel OA, Veltman DJ, Groenewegen HJ, Witter MP, Merkelbach J, Cath DC, van Balkom AJ, van Oppen P, van Dyck R (2005b) Disorder-specific neuroanatomical correlates of attentional bias in obsessive-compulsive disorder, panic disorder, and hypochondriasis. Arch Gen Psychiatry 62(8):922–933
van der Straten A, Huyser C, Wolters L, Denys D, van Wingen G (2018) Long-term effects of cognitive behavioral therapy on planning and prefrontal cortex function in pediatric obsessive-compulsive disorder. Biol Psychiatry Cogn Neurosci Neuroimaging 3(4):320–328
van der Wee NJ, Ramsey NF, Jansma JM, Denys DA, van Megen HJ, Westenberg HM, Kahn RS (2003) Spatial working memory deficits in obsessive compulsive disorder are associated with excessive engagement of the medial frontal cortex. Neuroimage 20(4):2271–2280
Via E, Cardoner N, Pujol J, Alonso P, López-Solà M, Real E, Contreras-Rodríguez O, Deus J, Segalàs C, Menchón JM, Soriano-Mas C, Harrison BJ (2014) Amygdala activation and symptom dimensions in obsessive-compulsive disorder. Br J Psychiatry 204(1):61–68
Weidt S, Lutz J, Rufer M, Delsignore A, Jakob NJ, Herwig U, Bruehl AB (2016) Common and differential alterations of general emotion processing in obsessive-compulsive and social anxiety disorder. Psychol Med 46(7):1427–1436
Yang XY, Sun J, Luo J, Zhong ZX, Li P, Yao SM, Xiong HF, Huang FF, Li ZJ (2015) Regional homogeneity of spontaneous brain activity in adult patients with obsessive-compulsive disorder before and after cognitive behavioural therapy. J Affect Disord 188:243–251
Yin D, Zhang C, Lv Q, Chen X, Zeljic K, Gong H, Zhan S, Jin H, Wang Z, Sun B (2018) Dissociable frontostriatal connectivity: mechanism and predictor of the clinical efficacy of capsulotomy in obsessive-compulsive disorder. Biol Psychiatry 84(12):926–936
Yücel M, Harrison BJ, Wood SJ, Fornito A, Wellard RM, Pujol J, Clarke K, Phillips ML, Kyrios M, Velakoulis D, Pantelis C (2007) Functional and biochemical alterations of the medial frontal cortex in obsessive-compulsive disorder. Arch Gen Psychiatry 64:946–955
Zhang T, Wang J, Yang Y, Wu Q, Li B, Chen L, Yue Q, Tang H, Yan C, Lui S, Huang X, Chan RC, Zang Y, He Y, Gong Q (2011) Abnormal small-world architecture of top-down control networks in obsessive-compulsive disorder. J Psychiatry Neurosci 36(1):23–31

Innovations in the Delivery of Exposure and Response Prevention for Obsessive-Compulsive Disorder

Sapana R. Patel, Jonathan Comer, and Helen Blair Simpson

Contents

Abstract Obsessive-Compulsive Disorder is an important cause of global health-related disability. In the last several decades, exposure and response prevention (EX/RP) has emerged as one of the most evidence-based treatments for adult and pediatric OCD. Recommended as a first-line treatment in practice guidelines for OCD, EX/RP, when expertly delivered, can be superior to serotonin reuptake

S. R. Patel and H. B. Simpson (✉)
Columbia University Vagelos College of Physicians and Surgeons, The New York State Psychiatric Institute, New York, NY, USA
e-mail: Sapana.Patel@nyspi.columbia.edu; Blair.Simpson@nyspi.columbia.edu

J. Comer
Florida International University, Miami, FL, USA
e-mail: jocomer@fiu.edu

Curr Topics Behav Neurosci (2021) 49: 301–329
https://doi.org/10.1007/7854_2020_202
Published Online: 16 February 2021

inhibitor (SRI) medications alone and superior to adding antipsychotic medication to augment SRI treatment response. Despite a robust evidence base, EX/RP is not widely available. Moreover, although effective, only about half of patients who receive a standard course of EX/RP will achieve remission.

This chapter will review innovations in delivering EX/RP, focusing on technology-based methods designed to increase access to EX/RP and translational neuroscience approaches to personalizing and optimizing EX/RP. Technology-based innovations to deliver EX/RP include video conferencing, internet-based treatment, and smartphone apps. Of these, internet-based, clinician-supported treatment has the most evidence base to date. Relevant to all technology-based innovations are the need for advances in the ethical, regulatory and financial aspects of understanding how access to EX/RP may be delivered to individuals of diverse socioeconomic backgrounds in accordance with professional standards and regulations and covered by healthcare.

Advances in our understanding of the neural processes underlying learning and memory have led to new ways to combine EX/RP with medications, behavioral interventions, or neuromodulatory methods, with the goal of enhancing the functioning of brain circuits that subserve fear processing and cognitive control. Among the pharmacological approaches to enhancing EX/RP outcome, both ketamine and cannabinoids show promise in small open trials but are in need of further study. Studies to train cognitive control are at an early stage of development yet provide preliminary evidence that training neural processes may be a new path to personalize treatment. How best to combine EX/RP with different types of neuromodulation is being actively studied.

Together these innovations in the delivery of EX/RP for OCD hold great promise for improving outcomes of care for individuals with OCD by increasing the availability and the individual treatment effects of this already effective treatment.

Keywords Exposure and response prevention · Technology-based interventions · Translational neuroscience

1 Introduction

With lifetime prevalence of 2%, Obsessive-Compulsive Disorder (OCD) is an important cause of global health-related disability (Ruscio et al. 2010). Epidemiological studies find that half of cases of OCD start by age 19 (and a quarter by age 14); after onset, OCD is typically chronic, with waxing and waning symptoms (Kessler et al. 1994; Skoog and Skoog 1999). Given its early onset and typically chronic course. OCD is associated with substantially reduced health-related quality of life and impairments in education, social relations, and family functioning (Asnaani et al. 2017). The societal cost of OCD in adults in the United States is estimated at $10.6 billion per year (DuPont et al. 1995).

Practice guidelines (APA 2007; NICE 2005) recommend two evidence-based first-line treatments for OCD: serotonin reuptake inhibitor (SRI) medications and cognitive behavioral therapy (CBT) involving Exposure and Response (or Ritual) Prevention (termed EX/RP in this chapter). In EX/RP treatment, individuals learn to confront feared situations (i.e., exposures) and refrain from performing compulsions (i.e., response or ritual prevention). EX/RP alone or in combination with medications can be more effective for OCD than treatment with medications alone (Foa et al. 2005). In addition, many patients prefer EX/RP either alone or in combination with medications compared to medications alone (Patel et al. 2017) or as a way to augment SRI response compared to antipsychotic medication (Patel et al. 2017). Yet, EX/RP is grossly underutilized in real world clinical practice (Patel et al. 2014; Blanco et al. 2006; Mancebo et al. 2011; Goodwin et al. 2002).

In this review, we will briefly describe EX/RP and its theoretical basis, review the evidence for using it to treat both pediatric and adult OCD, and outline factors associated with EX/RP outcome. Then, we will review reasons for the underutilization of EX/RP and highlight two innovations in its delivery: (1) using technology to deliver EX/RP with the aim of expanding access to this evidence-based treatment; (2) and using translational neuroscience approaches to enhance learning from EX/RP with the aim of improving its effects in individual patients.

2 Exposure and Response Prevention

2.1 Treatment

In EX/RP treatment, individuals learn to confront situations that trigger their obsessions and compulsions (i.e., exposures) and refrain from performing compulsions (i.e., response or ritual prevention). For example, a patient with a fear of harming others would be coached to confront feared situations (e.g., standing in front of a knife on the kitchen counter), to imagine feared consequences (e.g., stabbing someone and killing them) and to refrain from checking for harm. EX/RP also includes psychoeducation, self/symptom monitoring, homework practice between sessions (i.e., monitoring for rituals and conducting exposures on one's own) and home visits, where the clinician supports exposure practice in the person's home environment (Foa et al. 2012). Different EX/RP protocols vary in length and their emphasis on cognitive and behavioral techniques, but all are similar in recommending these procedures (Foa et al. 2012; Abramowitz et al. 2018; Steketee 1996).

EX/RP manuals provide specific guidelines for how to conduct EX/RP. In clinical practice, these guidelines are tailored to fit patient needs (Abramowitz et al. 2003). Even in practice though, the key is the development of well-constructed exposures and ritual prevention as prior dismantling studies suggest that both are necessary for good outcome (Foa et al. 1980). A review of EX/RP studies by Storch et al. (2009) concluded that it is effective in a variety of settings (e.g., home vs. office), when delivered in different intensities (e.g., weekly vs. intensive treatment), and with pediatric and adolescent populations.

2.2 Theory

There are different ideas for how EX/RP might work (Tryon 2005). Two leading ideas invoke emotional processing theory (EPT) and inhibitory learning theories. These are briefly reviewed below and described in detail elsewhere (McGuire and Storch 2019; Jacoby and Abramowitz 2016; Fradkin et al. 2020).

Emotional processing theory (EPT) (Foa et al. 2006; Foa and Kozak 1986; Foa and McNally 1996) posits that exposure activates a "fear structure." A "fear structure," as first put forth by Lang (1971), is a set of propositions about a stimulus (e.g., a knife), response (e.g., racing heart) and their meaning (e.g., "I could lose control and stab someone with that knife") that are stored in memory. Once this fear structure is activated, information that is incompatible with the original fear structure (e.g., I prepared dinner using a knife in front of my family and did not lose control or harm anyone) must be integrated for corrective learning to occur. This then is the goal of EX/RP. Fear reduction during repeated exposure (i.e., habituation) (Groves and Thompson 1970; Lader and Mathews 1968; Watts 1979) has been thought to be an important index of change in EX/RP and evidence that corrective learning is taking place, such that one's "original reaction toward the stimulus weakens or even disappears" (Eelen et al. 2001).

In contrast to this idea that EX/RP works by breaking fear-based associations via habituation, Craske et al. (2014) have posited that these associations do not break or disappear; rather EX/RP leads to learning of new safe associations. Influenced by translational neuroscience work in animals and healthy volunteers on the learning and extinction of fear, she and others have called for greater acknowledgment of the role of strengthening inhibitory learning in successful implementation of exposure therapy (Craske et al. 2014). Specifically, this means promoting the encoding and long-term recall of newly learned safe associations so that they will inhibit fear-based learning, a process termed "inhibitory learning". In this inhibitory learning theory of EX/RP, exposures work principally by violating expectancies of feared events and strengthening inhibitory extinction learning of safe associations (Apergis-Schoute et al. 2017).

These two theories have been discussed in the literature (Storch et al. 2009; Fradkin et al. 2020; Jacoby and Abramowitz 2016). Most conclude that while an inhibitory learning perspective for EX/RP shifts the focus from anxiety reduction to anxiety tolerance, inhibitory learning theory and EPT recommend similar procedures. These include generating a hierarchy of fears and using it to guide treatment; evoking fear/anxiety/distress during exposures; and teaching patients to eliminate safety signals and behaviors (McGuire and Storch 2019). All are procedures commonly used by skilled EX/RP therapists (McGuire and Storch 2019; Jacoby and Abramowitz 2016; Fradkin et al. 2020).

More recent literature has advanced that a fundamental deficit in OCD is an imbalance of goal-redirected versus habit-like behavior (Gillan et al. 2015; Fineberg et al. 2018b). From this perspective, the key of EX/RP therapy may not be activating a fear structure or strengthening inhibitory learning but breaking habits and restoring goal-directed control (Kalanthroff et al. 2018).

3 Effects of EX/RP

As one of the first-line treatments for OCD, the clinical effects of EX/RP have been studied in clinical trials of adults and children with OCD. We review the robust evidence for this treatment in these populations below, focusing on randomized controlled trials (RCTs).

3.1 Adult OCD

Many RCTs support the efficacy of EX/RP delivered in individual treatment. Early studies demonstrated EX/RP's superiority in reducing patients' OCD symptoms relative to relaxation therapy, anxiety management, or a wait-list condition (Abramowitz 1997, 2006; Fals-Stewart et al. 1993; Freeston et al. 1997; Lindsay et al. 1997). A meta-analysis examining intent-to-treat group data in efficacy trials showed that EX/RP is effective for the majority of patients, with large pre-post effect sizes (e.g., Hedges's $g = 1.39$), (Olatunji et al. 2013) and with a substantial percentage of patients attaining clinically significant improvement (e.g., 69%) (Eddy et al. 2004). Further, EX/RP has also been shown to lead to a decrease in sleep disturbances and improved quality of life (Storch et al. 2008, 2009; Diefenbach et al. 2007).

As RCTs demonstrated that serotonin reuptake inhibitors (SRIs) were also effective for OCD in RCTs, an important question became the comparative efficacy of EX/RP and SRIs and whether combination treatment was superior for OCD symptoms in particular. Only a handful of RCTs have directly addressed this question, and they used different designs and variants of EX/RP (reviewed in Simpson et al. 2004; Fineberg et al. 2018a). In the largest RCT to date, 122 adults with a principal diagnosis of OCD (e.g., without significant comorbid depression) and moderate to severe OCD found that 17 sessions of EX/RP with or without the SRI clomipramine led to greater reduction in OCD symptoms than clomipramine alone and was significantly better than pill placebo (Foa et al. 2005). EX/RP plus clomipramine did not differ in posttreatment symptom severity from those treated with EX/RP alone; these data indicate that medication did not bolster the efficacy of EX/RP at least when started concurrently as it was in this study. One limitation of this study is that the patients receiving either EX/RP alone or EX/RP plus clomipramine knew their treatment randomization. On the other hand, the independent raters who assessed symptom change were blind to randomization. In addition, there was not a psychological control; however, as noted above, prior RCTs have shown that time and attention alone does not treat OCD.

Subsequent RCTs have shown that adults with OCD who have clinically significant symptoms despite an adequate trial of an SRI can also benefit from the addition of EX/RP. For example, Tenneij et al. (2005) found that the addition of EX/RP was superior to continuing the SRI in 96 adults with OCD. Moreover, adding 17 sessions of EX/RP was more effective in augmenting SRI response than adding stress-management therapy, controlling for non-specific time and attention (Simpson et al. 2008); in a later trial, adding 17 sessions of EX/RP was also more effective in augmenting SRI response than adding either risperidone (Simpson et al. 2013), an antipsychotic medication used to treat SRI-resistant OCD, or pill placebo.

A network meta-analysis of 54 RCTs for OCD found that cognitive-behavioral therapy variants (some trials studying EX/RP but others studying cognitive therapy or a combination of these) alone or combined with medications demonstrated greater effects than medications alone. However the authors highlighted the methodological limitations of this literature for such an analysis. Moreover, although initial data suggest that EX/RP (with or without medication) may lead to lower relapse than medication alone when stopped (Simpson et al. 2004), and that the SRI augmenting effects of EX/RP can be maintained for at least 6 months follow-up (Foa et al. 2013, 2015), more research is needed to understand the comparative sustained benefits and health economic analysis of EX/RP, SRIs and their combination (Fineberg et al. 2018a).

In the end though, practice guideline recommends SRIs, EX/RP or their combination as first-line treatments for OCD (APA 2007; NICE 2005). This takes into account the need to tailor treatment options to the individual patient and to take into account their health needs at that moment, their preferences, and the availability of these different treatments (Wheaton et al. 2016).

3.2 Pediatric OCD

Although about half of all OCD cases start by age 18 (Kessler et al. 2005). When treating pediatric OCD, simple adoption of successful methods in the treatment of adult OCD is misguided. The past two decades, however, have witnessed advances in the development and evaluation of developmentally appropriate psychosocial interventions for pediatric OCD (Freeman et al. 2018; Melin et al. 2019). The most successful among these interventions have been family-based (Freeman et al. 2018; Melin et al. 2019; Thompson-Hollands et al. 2014a), placing increased focus on the role of parents in treatment engagement and working to reduce patterns of providing reassurance or helping with or performing compulsions, or family accommodation (Peris et al. 2017; Piacentini et al. 2011). These programs center around EX/RP, but carefully attend to the more limited cognitive and socioemotional capacities characteristic of earlier developmental stages, often emphasize behavioral over cognitive components of treatment, and incorporate parental "scaffolding" and modeling in fostering child exposure engagement. Effective treatment of early child OCD (i.e., <8 years) also helps parents learn to address disruptive child behaviors associated with pediatric OCD (Comer et al. 2017; Freeman et al. 2014).

A series of landmark studies has now demonstrated that multimodal treatment strategies combining EX/RP with SRI medication may be optimal in the treatment of pediatric OCD. For example, the large-scale, multisite RCT, Pediatric OCD Treatment Study (Pediatric OCD Treatment Study [POTS] Team 2004), found that treatment entailing a combination of CBT (consisting of EX/RP) and the selective serotonin reuptake inhibitor (SSRI) sertraline significantly outperformed monotherapy CBT and monotherapy SSRI in youth ages 7–17 with OCD. Both monotherapies, in turn, significantly outperformed pill placebo. Follow-up analyses examining rates of clinical remission (defined as Children's Yale Brown Obsessive-Compulsive Scale [CY-BOCS < 10], the gold standard assessment of OCD in children and adolescents [Scahill et al. 1997]) found across all sites that combination treatment (53.6%) significantly outperformed monotherapy SSRI medication (21.4%) and pill placebo (3.6%), but not monotherapy CBT (39.3%). Safety and acceptability data indicated that all three active treatments were well tolerated across children and adolescents. Further analyses indicated that there were important site differences in these effects for monotherapy. Specifically, one site (known for its expertise in CBT) had better outcomes with CBT than another and that other site (known for its expertise in medication management) had better outcomes with medication alone (POTS) (Pediatric OCD Treatment Study [POTS] Team 2004). There were no site differences in combined treatment, suggesting it is less susceptible to setting-specific variations in treatment outcome.

A subsequent RCT, the POTS II study (Franklin et al. 2011) investigated two different EX/RP augmentation protocols in children and adolescents with a diagnosis of OCD who were partial responders to an adequate trial of SSRI pharmacotherapy. Pediatric outpatients were randomized to receive one of three conditions: augmentation with medication management only, medication management plus instructions in CBT provided by the psychiatrist or medication management plus 14 concurrent CBT sessions. Findings indicated that among patients' ages 7–17 years with OCD, the medication management plus CBT strategy was superior to the other 2 strategies on all outcome measures. In the primary intention-to-treat analysis, 68.6% (95% CI, 53.9–83.3%) in the medication management plus CBT group were considered responders (defined as CY-BOCS reduction of 30% or more from baseline to week 12), which was significantly better than the 34.0% (95% CI, 18.0–50.0%) in the medication management plus instructions in CBT group, and 30.0% (95% CI, 14.9–45.1%) in the medication management only group. The authors concluded that the addition of CBT to medication management compared with medication management alone resulted in a significantly greater response rate, whereas augmentation of medication management with the addition of instructions in CBT did not. It is important to note that the studies comparing CBT with SSRIs (POTS I and POTS II) did not include an adjunctive psychological placebo arm to control for the potentially confounding effect of more intensive face-to-face clinician contact in the CBT arm.

Indeed, a recent systematic review and meta-analysis of CBT for OCD in children and adolescents reviewed nine clinical trials comparing CBT with no intervention and three trials comparing CBT with SSRIs found that CBT may be more effective than no intervention and comparable to SRIs for pediatric OCD (Uhre et al. 2020). However, authors cited uncertainty about estimate of effects due to lack of sufficient data on adverse events and quality of life. In the treatment of OCD in young children below age 7, there is no controlled evidence to support the safety and efficacy of SRI medication (alone or in combination) (Comer et al. 2019). For children with OCD in this age group, studies on family-based EX/RP emphasizing parenting factors and drawing on parental modeling and scaffolding has shown the most promising evidence to date (Comer et al. 2017; Freeman et al. 2014).

In summary, based on the existing evidence, EX/RP appears efficacious in adolescents and young children. Future studies that examining EX/RP effects including quality of life, long-term data, and relapse prevention are warranted.

4 Factors Associated with EX/RP Outcome

Although treatment guidelines recommend EX/RP as a first-line treatment for adults and children with OCD (APA 2007; NICE 2005), only about half of patients who receive a standard course of therapy (e.g., up to 17 sessions) will achieve remission (defined as a Yale-Brown Obsessive Compulsive Scale [YBOCS] the gold standard assessment of OCD in adults (Goodman et al. 1989) consistent with Simpson et al. 2006; Mataix-Cols 2016).

Studies have investigated patient characteristics associated with better EX/RP outcome (Olatunji et al. 2013; Knopp et al. 2013; Maher et al. 2010; Wheaton et al. 2015; Steketee et al. 2019; Simpson et al. 2011, 2012; Öst et al. 2015; Pinto et al. 2011; Solem et al. 2009; Wells et al. 2017; Flygare et al. 2020; D'Alcante et al. 2012). Many variables have been examined: demographic characteristics, OCD severity, dysfunctional beliefs, severity of metacognitions, cognitive inflexibility, degree of insight, comorbidity (severe depression, obsessive-compulsive personality disorder and autism spectrum disorder), treatment history, and functioning/quality of life. Lack of clinician skill in delivering EX/RP is another important factor but has been little studied (Gillihan et al. 2012). Most variables show small effects and limited replication across studies. We have reported one notable exception: patient adherence to EX/RP homework.

To measure patient adherence, the Patient EX/RP Adherence Scale (PEAS), a psychometrically valid scale, was developed to assess between session adherence (Simpson et al. 2010). Using this 3-item scale, ranging from 1 (no adherence) to 7 (excellent adherence), the clinician rates the quantity of exposures attempted, the quality of exposures attempted, and the degree of ritual prevention achieved since the prior session. Higher PEAS scores indicating greater homework adherence was strongly associated with EX/RP outcome in three separate trials (Simpson et al. 2011; Wheaton et al. 2016; Simpson, Foa et al., under review) as well as outcome 6 months later (Simpson et al. 2012). Moreover, early adherence (in sessions 5–9) significantly predicted posttreatment OCD severity (Simpson et al. 2011). These findings have since been replicated and extended to show that patient adherence specifically with response prevention tended to increase as treatment progressed, and differentiated patients (i.e., mean PEAS_C subscale score > 5.3 or at least good adherence) were more likely to achieve posttreatment wellness (i.e., Y-BOCS <12) from those who did not (Wheaton et al. 2016).

In the pediatric OCD literature, data from the POTS study has shown that high levels of family accommodation predict worse EX/RP treatment outcome (Garcia et al. 2010). Adherence to EX/RP has not yet been formally studied in children with OCD, an area in need for further study.

Given the evidence base for the effect of patient adherence in adults and family accommodation in pediatric OCD on EX/RP outcome, clinicians should provide education on the importance of these factors and their impact on EX/RP outcome during treatment planning.

5 Access to EX/RP

EX/RP is an effective treatment for OCD, but many patients do not have access to it. Barriers exist at multiple levels (Patel et al 2014; Blanco et al. 2006; Mancebo et al. 2011; Goodwin et al. 2002; Marques et al. 2010; Glazier et al. 2013; Schwartz et al. 2013). These barriers include both clinician and patient factors.

With regards to clinician factors, few clinicians have extensive experience with diagnosing and treating OCD due to its low prevalence rates in the community as compared with anxiety or depressive disorders (Glazier et al. 2013, 2015; Schwartz et al. 2013); even fewer know how to deliver EX/RP (Mancebo et al. 2006, 2011; Schwartz et al. 2013; Glazier et al. 2015). As a result, EX/RP typically is available only in specialty clinics or major medical centers. Studies of nationwide data document underutilization of EX/RP (Patel et al 2014; Blanco et al. 2006; Goodwin et al. 2002; Mancebo et al. 2006). This lack of experience may be due to: (1) insufficient exposure during clinicians' training to CBT in general and more specifically EX/RP for OCD (Glazier et al. 2013, 2015; Mancebo et al. 2006); (2) insufficient exposure to OCD patients in practice for community clinicians to develop expertise; and (3) clinician's negative beliefs about exposure-based treatments (Harned et al. 2014).

With regards to patient factors, some individuals may never seek help because they do not know that their symptoms are OCD or out of reluctance to disclose one's symptoms due to the nature of their symptoms (Marques et al. 2010; Coles et al. 2014). Additionally, surveys of individuals with OCD have shown that some patients and families reject the treatment as "too difficult or time consuming" (Mancebo et al. 2011; Freeman et al. 2009).

To address the dearth of trained clinicians, group treatment is being used to extend the reach of EX/RP-trained clinicians. For example, Anderson and Rees (2007) conducted an RCT in which 51 individuals with OCD received either individual or group CBT (each using both cognitive techniques like restructuring and behavioral techniques like EX/RP), or a waitlist control condition. No significant difference was found between the group and individual treatments on outcome measures (i.e., Y-BOCS), and both groups were significantly improved compared to the control condition. Another strategy to extend the reach of EX/RP-trained clinicians is through the use of paraprofessionals. For example, Rogers Memorial has trained bachelor's level "Behavioral Specialists" to provide CBT successfully for OCD. The behavioral specialists complete didactic training including job shadowing and ongoing supervision with licensed psychologist. Outcomes in both adults and adolescents have indicated that this model of using paraprofessionals can be effective at reducing OCD severity (Farrell et al. 2019).

To reduce the length of EX/RP, a group in Norway has developed a highly concentrated format called the Bergen 4-Day treatment (B4DT), where treatment is delivered over four consecutive days (Hansen et al. 2018, 2019; Kvale et al. 2018). Described as "individual treatment in a group setting," the treatment is delivered in a group of 3–6 patients with the same number of clinicians. The 1:1 ratio between patients and clinicians allows for individually tailored and clinician-assisted exposures in this group setting; at the same time, the format also enables patients to observe others going through parallel processes. In the first RCT evaluating the effects of the B4DT, 48 patients diagnosed with OCD were randomized to B4DT, self-help (SH), or waiting list (WL) with 16 patients in each condition (Launes et al. 2019). All participants randomized to the B4DT underwent the treatment without any attrition. The B4DT yielded significantly better effects than control conditions on measures of OCD, depression, and generalized anxiety. Using a modification of the international consensus criteria (Mataix-Cols et al. 2016), response was a $\geq 35\%$ reduction of the individual patient's pretreatment Y-BOCS score and remission was response criterion is fulfilled and the posttreatment Y-BOCS score is ≤ 12 points. The response rate was 93.8% in B4DT, 12.5% in SH, and 0% in WL, while remission rate (response criterion is fulfilled and the posttreatment Y-BOCS score is ≤ 12 points) was 62.5%, 6.3%, and 0%, respectively.

6 Technology-Based Innovations in Delivering EX/RP

In the last decade, mental health technologies have been developed aimed at increasing access to evidence-based treatment like EX/RP. These technologies include video conferencing; internet-based treatments; and smartphone apps (Abramowitz 2018; Van Ameringen et al. 2017). Digital methods of delivery may have particular value in situations as the current global COVID-19 pandemic, when face-to face contact is seriously curtailed. Some of the more promising are reviewed briefly below by technology.

6.1 Video Conferencing for Family-Based CBT

Given regional workforce shortages in pediatric EX/RP expertise, along with the documented importance of high-quality CBT for achieving favorable response (Franklin et al. 2011), recent years have witnessed promising technology-based innovations (Comer et al. 2014), in which secure video conferencing platforms are leveraged to remotely provide real-time treatment with a remote clinician. Such formats can overcome geographic disparities in high-quality EX/RP, as well as transportation and stigma-related barriers that can interfere with treatment-seeking at brick and mortar mental health facilities. Such "face-to-face, different place" treatments also offer opportunities to treat individuals with OCD in their natural settings (Comer et al. 2014).

Research studies have now documented the acceptability and feasibility of video conferencing-based formats for remotely delivering EX/RP in real time and have demonstrated successful outcomes (Thompson-Hollands et al. 2014b; Comer et al. 2014; Himle et al. 2006; Goetter et al. 2014; Storch et al. 2011; Vogel et al. 2014). In a pilot RCT, children with OCD received family-based EX/RP in the clinic or via video conferencing. Treatment outcomes were comparable across conditions, with over 70% of children treated via video conferencing achieving clinical response (defined as CGI-I [score of 1 or 2], the CY-BOCS ["subclinical"/"mild" range; i.e., score $\leq$ 15], and no OCD diagnosis on the Anxiety Disorders Interview Schedule for Children and Parents for DSM–IV: ADIS-IV-C/P; Silverman and Albano 1996] at posttreatment and at follow-up) at posttreatment and 80% achieving clinical response at 6-month follow-up (Comer et al. 2017).

6.2 Internet-Based CBT for OCD

Internet-based CBT for OCD has been developed to improve access to EX/RP in different countries with varying formats and degree of clinician support (Bachofen et al. 1999; Greist et al. 1998, 2002; Kenwright et al. 2005; Tolin et al. 2007; Moritz et al. 2011; Wootton et al. 2013; Andersson et al. 2012; Matsumoto et al. 2018). Of these, delivery of EX/RP via internet-based clinician-supported treatment, OCD-NET, a program introduced by Andersson and colleagues at the Karolinska Institutet, has the most evidence base to date. OCD-NET offers the patient the same content and homework tasks as in person EX/RP; the main difference being that treatment is delivered entirely via an integrated clinician-guided, internet-based treatment platform. The patient logs into the platform and in a self-guided way progresses through the 10 treatment modules, typically over 10 weeks. The modules consist of text material, quizzes to test comprehension of material, worksheets and homework assignments. Modules 1–4 comprise psychoeducation, cognitive restructuring of metacognitions, and establishment of an individual EX/RP hierarchy. Modules 5–10 focus on daily in vivo EX/RP exercises. Asynchronously, the clinician makes these 10 modules sequentially available to patients in the platform based on their self-assessments, module progress and clinical judgment (Andersson et al. 2012).

An RCT in Sweden found OCD-NET superior to online supportive therapy (treatment response [2 standard deviations below pretreatment Y-BOCS score: OCD-NET 60% vs. control 6%; between-group Cohen's $d = 1.12$); OCD-NET had low discontinuation rates (12%) (Andersson et al. 2012). A 24-month long-term follow-up analysis of this trial where participants ($n = 93$) were randomized to receive an internet-based booster program versus no booster found sustained long-term effects of OCD-NET and also showed that an internet-based booster program could help further improve long-term outcomes and prevent relapse (Andersson et al. 2014). An ongoing non-inferiority trial is now comparing unguided OCD-NET to clinician-guided-OCD-NET to in person EX/RP (Rück et al. 2018: Clinical Trials.gov: NCT02541968).

The same group at the Karolinska Institutet developed a clinician-guided internet-delivered CBT program for pediatric OCD called BIP OCD (Aspvall et al. 2018). With assistance from parents or legal guardians, a designated clinician communicates with both the child and the parent through the entire treatment. An RCT including 67 patients demonstrated superiority and cost-effectiveness of BIP OCD against a wait-list control on the CY-BOCS (CY-BOCS: time-by-group interaction, $B = -4.53$, $z = -3.74$, $p < 0.001$; Cohen's $d = 0.69$; 95% CI 0.19–1.18) (Lenhard et al. 2017). Patients randomized to BIP OCD also showed further improvement from posttreatment to 3-month follow-up, with a within-group pretreatment to follow-up effect size (Cohen's $d = 1.68$; 95% CI 1.00–2.36). Patient satisfaction with BIP OCD was high. There were no treatment-related adverse events. Average clinician support time was 17.5 min per patient per week. An ongoing non-inferiority trial of stepped care internet-delivered cognitive behavior therapy for 16 weeks in pediatric primary care is underway by this same group (Aspvall et al. 2019: Clinical Trials.gov: NCT03263546).

This Swedish OCD-NET program was recently adapted for testing in the United States (Patel et al. 2018). This included translating the program into English and making adjustments for cultural context. Of the 40 adults with OCD enrolled in a 10-week trial, 28 participants (70%) completed the 10-week treatment; discontinuation was due to loss of interest ($n = 6$), loss of computer access ($n = 2$), loss to follow-up ($n = 3$), and one withdrawn due to worsening of OCD and depressive symptoms. Mixed effects models using the intent-to-treat sample showed that OCD severity (as measured by the Y-BOCS) and depression scores (as measured by the Hamilton Depression Scale (HAMD: Hamilton 1960) significantly decreased from baseline to post treatment at 12 weeks (Y-BOCS: −7.48; $p < 0.001$; HAMD: −2.45; $p = 0.025$) and quality of life (as measure by the Quality of Life Enjoyment Satisfaction Scale (Q-LES-Q: Endicott et al. 1993) scores significantly increased (+6.50; $p = 0.005$). Clinicians reported delivering OCD-NET was feasible; they spent on average 500 min of support per patient (including email time logged onto the platform and telephone calls) compared to >2,500 min per patient (sessions and telephone calls) in 17 sessions of clinic-based face-to-face EX/RP. Finally, of those that completed treatment, 79% percent of patients reported feeling mostly-to-very satisfied; 50% continued to use the platform 4 months after clinician support ended.

To understand the role of low-intensity interventions prior to high-intensity CBT, the Obsessive-Compulsive Treatment Efficacy Trial (OCTET), commissioned by NICE, compared two low-intensity interventions – (computerized CBT (using the commercial program *OC Fighter*) or guided self-help (using the workbook: *Overcoming OCD*) supported by a mental health practitioner – to waiting list for high-intensity CBT. Neither low-intensity intervention led to clinically significant improvements in OCD symptoms compared to waiting list. The authors concluded that more effective low-intensity interventions should be developed and evaluated against high-intensity CBT to see if they can achieve equivalent (or at least non-inferior) outcomes (Lovell et al. 2017; Rück et al. 2018: Clinical Trials.gov: NCT02541968).

6.3 Smartphone Apps

Although less researched, mobile applications ("apps") of CBT hosted on smartphones are another promising strategy to increase access. To date, the strongest support for these apps has been in their supporting the delivery of in person interventions or Internet programs (Firth et al. 2017). Only one published study (Boisseau et al. 2017) explored the feasibility and preliminary efficacy of a standalone self-help app for EX/RP (*LiveOCDFree*) in a 12-week open trial. Twenty-one individuals with mild to moderate OCD [defined by a Y-BOCS score < 23 were enrolled. Participants found using the app feasible and acceptable as measured by retention and the Client Satisfaction Survey (CSQ: Larsen et al. 1979). Fifteen of twenty-one (71%) completed the trial. Of these, 40% had a clinically significant decrease in OCD symptoms (i.e., Y-BOCS $> 35\%$) but only 20% met the authors response criteria (defined as posttreatment Y-BOCS-Self-Report < 14). This observed magnitude of change is consistent with other self-help interventions for OCD, but substantially lower than face-to-face clinician-guided EX/RP as described above.

Gershkovich et al. (2020) recently completed an 8-week pilot clinic trial of 33 adults treated with low-intensity clinician support with a mobile app in an integrated program (called Brief Exposure and Response Prevention Assisted by Mobile app, or BEAM) over 8 weeks. The *nOCD* app (www.treatmyocd.com), was the mobile app used because of its alignment with the standard EX/RP treatment model (Foa et al. 2012) and its focus on accountability and tracking of exposures (Hong et al. 2020). The data suggest that this integrated treatment may be both acceptable and efficacious. Twenty-seven participants (81.8%) completed the program, most participants were very [68.2%] or mostly [31.8%] satisfied. Mixed effects models of the intent-to-treat sample showed that OCD severity (as measured by the Y-BOCS) and depression scores (as measured by the HAMD) significantly decreased from baseline to posttreatment at 8 weeks (Y-BOCS: −8.93;

$p < 0.001$; HAMD: −2.00; $p = 0.004$) and quality of life (as measure by the Q-LES-Q scores significantly increased (+3.01; $p = 0.005$). Of the 33 study entrants, 14 (42%) responded to treatment (defined as Y-BOCS decrease ≥35% with CGI-of 1 or 2), and eight (24%) achieved minimal symptoms (defined as Y-BOCS ≤ 12). The BEAM model requires further investigation, including which elements of the program were most helpful, but it provides another promising way to increase access to empirically supported treatment for OCD.

7 Translational Neuroscience Approaches to Enhancing EX/RP

As technology is transforming the delivery of EX/RP, advances in the neuroscience of learning, memory, and emotion, including the learning and extinction of fear, have led to novel ideas for how to enhance outcomes from exposure-based CBT treatments like EX/RP (Jacoby and Abramowitz 2016; Dunsmoor et al. 2015a; Fullana et al. 2020; Craske et al. 2014). The basic idea is that exposure-based CBT treatments (like EX/RP) teach new ways to regulate fear and anxiety and thus rely upon the brain network that supports the expression and regulation of fear and anxiety. Key brain regions in this network include the ventromedial prefrontal cortex, hippocampus, and amygdala (Apergis-Schoute et al. 2017; Dunsmoor et al. 2015a, b; Milad and Quirk 2012). Therefore, methods that enhance their function and the relevant processes they subserve (e.g., extinction learning and retention, memory consolidation and reconsolidation, safety learning) may enhance outcomes from exposure-based CBT treatments.

EX/RP, however, involves not only exposures but also teaching patients to resist the urge to conduct compulsive rituals. Dismantling studies have shown that it is the combination of exposure and ritual prevention that leads to the best outcome (Foa et al. 1980); moreover, patient adherence to response prevention has been shown to be one of the most powerful predictors of EX/RP outcome (Simpson et al. 2011; Wheaton et al. 2016). Thus, some are testing whether EX/RP outcomes can be improved by training cognitive control and enhancing the functioning of the brain network that subserves it. Key brain regions in this network include the dorsolateral prefrontal cortex (DLPFC), dorsal anterior cingulate cortex (dACC), and presupplementary motor area (preSMA) (McTeague et al. 2017; Niendam et al. 2012). Finally, EX/RP is also being combined with neuromodulatory techniques. This work is just beginning but the idea has ranged from using exposures to "activate" circuits to then modulate them with transcranial magnetic stimulation (TMS) or using TMS to induce "plasticity" such that patients can better learn from EX/RP.

With the intent of enhancing the functioning of brain regions involved in fear processing or cognitive control, EX/RP is being combined with medications, specific behavioral interventions, and neuromodulatory techniques. Below, we review some of these strategies, focusing on studies with empirical data in OCD.

7.1 *Pharmacological Enhancement of EX/RP*

Pharmacological agents have been combined with EX/RP and tested in animal and human studies. We will review three agents: D-cycloserine, intravenous ketamine, and cannabinoids.

D-cycloserine (DCS), a partial agonist at the N-methyl-D-aspartate (NMDA) receptor, has been shown in animal studies to enhance associative learning processes involved in fear extinction (Ledgerwood et al. 2003; Walker et al. 2002; Davis et al. 2006). After a proof of concept study conducted by Ressler and colleagues showed that DCS when combined with exposure therapy could enhance outcomes in patients with height phobia, DCS has been tested in many other psychiatric conditions, including OCD (Ressler et al. 2004).

Four RCTs compared EX/RP to exposure treatment augmented by DCS in adults with OCD. In two of these studies, D-cycloserine reduced the time to response, but did not result in an increased response by the end of the trial (Kushner et al. 2007; Wilhelm et al. 2008); the other two RCTs found no group differences in symptom improvements (Storch et al. 2007; Andersson et al. 2015). Three RCTs compared EX/RP to EX/RP augmented with DCS in pediatric OCD (Farrell et al. 2013; Mataix-Cols et al. 2014; Storch et al. 2016). Only one study (Farrell et al. 2013) found a benefit to DCS augmentation relative to placebo augmentation at one time point: from posttreatment to the 1-month follow-up. A recent systematic review and meta-analysis of DCS augmentation of exposure-based CBT for anxiety, OCD, and PTSD concluded that DCS is associated with only a small augmentation effect only (Mataix-Cols et al. 2017), but the authors note that all of these studies are limited by their including participants without knowing whether they have extinction learning deficits at baseline and administering DCS independent of within-session learning experiences. As a result of the largely negative literature to date, DCS is not currently recommended for use with EX/RP in clinical practice.

Other agents found to enhance extinction learning in rodents have also been combined with EX/RP. For example, intravenous (IV) ketamine, an agent found to enhance plasticity and extinction learning in rodents (Duman 2014; Liu et al. 2012; Gideons et al. 2014), was combined with 10 exposure sessions delivered over 2 weeks (the putative time period when ketamine facilitates extinction learning). Preliminary results from this open trial confirmed prior findings that IV ketamine can rapidly reduce obsessions in unmedicated OCD patients and suggested that the combination of 10 exposure sessions and ketamine might be more effective than either alone (Rodriguez et al. 2013). These preliminary findings need to be formally tested in an RCT. Cannabinoids are also being combined with EX/RP. The rationale is twofold. First, as reviewed by Kayser and colleagues (Kayser et al. 2019, 2020; Kayser 2020), cannabinoids have been shown in both animal and human studies to modulate threat responding and to enhance fear extinction learning. Second, rodent studies suggest that the cannabinoid 1 receptor (the primary cannabinoid receptor in the brain) facilitates appropriate shifting between goal-directed and habitual action strategies (Gremel et al. 2016). Thus, Kayser et al. (2020) conducted a small randomized pilot trial in adults with OCD to assess the acceptability, tolerability, and preliminary efficacy of nabilone, a synthetic cannabinoid, alone or combined with EX/RP. Nabilone had little effect on OCD symptoms on its own. However, those randomized to nabilone and 9 sessions of EX/RP over 4 weeks had a significant decrease in OCD severity. Their degree of change was nearly twice as large as that seen in a comparison sample who received EX/RP alone. These preliminary clinical effects require confirmation in a large RCT. Moreover, whether nabilone's effects in patients with OCD are due to engagement of the brain fear network merits investigation.

7.2 Behavioral Methods for Enhancing EX/RP Outcomes

Neuroscientists have also discovered different behavioral interventions that enhance fear processing in animals and healthy humans (Jacoby and Abramowitz 2016; Fradkin et al. 2020; Craske et al. 2014; Dunsmoor et al. 2015a). Some of these strategies are beginning to be tested in OCD patients. Given the substantial costs associated with clinical trials, a first step has been to use laboratory paradigms to test whether behavioral methods that enhance extinction learning and recall in rodents and healthy volunteers have the same effects in OCD patients. Success in the laboratory would then provide support for combining this intervention with EX/RP.

One such intervention is novelty-facilitated extinction (NFE), where a novel outcome is provided instead of an expected threat. NFE has been shown in both rodents and healthy volunteers to enhance extinction learning and recall in a fear conditioning and extinction laboratory paradigm (Dunsmoor et al. 2015b). Using a similar laboratory paradigm, Steinman and colleagues tested the effects of novelty-facilitated extinction in a transdiagnostic sample of patients with an anxiety disorder or OCD, half of whom were randomized to standard extinction ($n = 27$) and half to novelty-facilitated extinction ($n = 24$) (Steinman et al. 2020). Participants returned 24 h later to test for extinction recall and fear reinstatement. Both groups learned the fear association but variably extinguished. Novelty did not facilitate extinction. Post hoc exploratory analyses suggested that an individual's cognitive style affected the ability to extinguish. This deserves further study.

EX/RP includes not only exposures but also response prevention where patients are coached to inhibit the urge to complete compulsions. Some OCD models suggest that deficits in the ability to inhibit action, or overreliance on stimulus-driven habits, leads to compulsions (Robbins et al. 2012; Gillan et al. 2015). Thus, some have focused on whether training executive control (and in particular the ability to inhibit unwanted or irrelevant stimulus-driven behaviors) can lead to improved EX/RP outcomes. For example, Kalanthroff et al. (2018) developed a program called personalized-computerized inhibitory training (P-CIT), with the goal of improving a patient's ability to inhibit automatic responses while exposed to OCD triggers. This program trained individuals with OCD to inhibit responses on the stop signal task when shown personalized OCD-related pictures. Ten individuals, all who had failed at least 1 adequate EX/RP trial in the past 2 years, were trained with P-CIT for 7 consecutive days and then completed 8 hourly EX/RP sessions over 2.5 weeks while continuing P-CIT training. All but one responded (defined as Y-BOCS decrease of $\geq$35%). These preliminary results require confirmation, and the exact mechanisms of P-CIT's effects require elucidation. However, the work illustrates that training the neural processes that enable response prevention may be an additional way to improve EX/RP outcomes.

7.3 Neuromodulatory Techniques for Enhancing EX/RP Outcomes

Repetitive transcranial magnetic stimulation (rTMS) is a noninvasive brain stimulation technique that has been used safely in various neuropsychiatric disorders, including OCD. High frequency (HF, $\geq$10 Hz) and low frequency (LF, $\leq$1 Hz) rTMS seem to exert excitatory and inhibitory effects on cortical excitability, respectively; rTMS is hypothesized to lead to long-lasting effects via homeostatic plasticity mechanisms (i.e., long-term depression and long-term potentiation) (Karabanov et al. 2015). Given its potential to target specific brain regions and its safety in clinical practice, rTMS is now being combined with exposure-based CBT treatments like EX/RP (Chalah and Ayache 2019).

On its own, rTMS has been shown effective in reducing OCD symptoms when delivered to the DLPFC or preSMA (Zhou et al. 2017). Hypothesizing that this effect might be due to the enhancement of cognitive control, van den Heuvel and colleagues are now conducting an RCT in adults with OCD to test whether HF rTMS to the DLPFC or to the preSMA directly prior to an EX/RP session can not only enhance cognitive control but also improve EX/RP outcomes (ClinicalTrials.gov: NCT03667807).

Recently, a modification to TMS, so called "deep" TMS (dTMS) has been made that allows stimulation of deeper and larger brain structure as compared to conventional rTMS. Using this technique, high frequency dTMS (20 Hz) targeting the medial prefrontal cortex and the anterior cingulate cortex delivered daily over 6 weeks has recently been shown in an RCT to significantly improve OCD symptoms in adults who did not respond adequately to pharmacological and psychological interventions (Carmi et al. 2019). As a result, dTMS is now approved by the Food and Drug Administration in the United States for the treatment of OCD. This study used a 3–5-min individualized symptom provocation before each of the 29 dTMS treatments, and the patient was asked to keep thinking about the specific obsession during the treatment; the rationale provided was to "activate the relevant neuronal circuit." OCD severity decreased significantly in both groups but the group receiving active dTMS ($n = 47$) had a small but significantly greater reduction (a greater Y-BOCS decrease of 2.8 points) and a higher rate of response (defined as Y-BOCS decrease of $\geq$30%: 38% versus 11%). The trial design could not differentiate whether the difference was due to the stimulation per se or to the combination of the stimulation with the brief exposure protocol. Moreover, whether the symptom provocation indeed activated the relevant circuit was not studied. While much has yet to be learned, combining neuromodulatory treatments (like rTMS and dTMS) with EXRP reflects a new and exciting direction in the field.

8 Conclusions

OCD was once considered an untreatable disorder. With the evolution of exposure-based treatments for anxiety and other related disorders, EX/RP for OCD emerged and through multiple RCTs, has shown evidence of efficacy. Now EX/RP is a first-line treatment for both adult and pediatric OCD. In some trials, EX/RP has been shown to be superior to SRI medications alone. In other trials, adding EX/RP to SRIs has been shown to be superior to adding antipsychotic medication to augment SRI treatment response. However, more data are needed on the clinical and cost-effectiveness of EX/RP over the longer term, its ability to prevent relapse or to sustain improved quality of life, and its potential to prevent the full syndrome of OCD in those with emerging obsessive-compulsive symptoms.

Despite its efficacy and acceptability as a patient-preferred treatment, EX/RP is not widely available in practice. Barriers include lack of access by patients and lack of training by clinicians. Exciting new technological methods to deliver EX/RP have been developed, including video conferencing, internet-based CBT, and mobile apps. Of these, internet-based, clinician-supported treatment for adults and children with OCD has the most evidence base to date. Future studies should confirm its effectiveness in larger samples of individuals with OCD across diverse healthcare systems and determine if OCD-NET can produce equivalent and/or clinically meaningful change for OCD patients as compared to face-to-face treatment. Relevant to all technology-based innovations are the need for advances in the ethical, regulatory, and financial aspects of understanding how access to EX/RP may be delivered to individuals of diverse socioeconomic backgrounds in accordance with professional standards and regulations and covered by healthcare (Torous and Roberts 2017).

Even when available, not all achieve remission with EX/RP and even those that do can relapse. Advances in our understanding of the neural processes underlying learning and memory have led to exciting new ways to enhance learning that occurs during EX/RP have been developed. These include combining EX/RP with medications, behavioral interventions, and neuromodulatory methods, with the goal of enhancing the functioning of brain circuits that subserve fear processing and cognitive control. Among the pharmacological approaches to enhancing EX/RP outcome, both ketamine and cannabinoids show promise in small open trials but are in need of further study. Studies to train cognitive control are at an early stage of development yet provide preliminary evidence that training neural processes may be a new path to personalize treatment. How best to combine EX/RP with different types of neuromodulation is being actively studied.

Together these innovations in the delivery of EX/RP for OCD offer great hope to alleviate individual suffering and the public health burden of this disabling illness. The technological innovations have the potential to make EX/RP far more accessible; the translational neuroscience innovations could enhance individual treatment effects, increasing the power of this already effective treatment.

References

Abramowitz JS (1997) Effectiveness of psychological and pharmacological treatments for obsessive-compulsive disorder: a quantitative review. J Consult Clin Psychol 65:44–52

Abramowitz JS (2006) The psychological treatment of obsessive-compulsive disorder. Can J Psychiatr 51:407–416

Abramowitz J (2018) Getting over OCD, second edition: a 10-step workbook for taking back your life, The Guilford self-help workbook series. Guilford, New York

Abramowitz JS, Franklin ME, Cahill SP (2003) Approaches to common obstacles in the exposure-based treatment of obsessive-compulsive disorder. Cogn Behav Pract 10(1):14–22. https://doi.org/10.1016/S1077-7229(03)80004-4

Abramowitz JS, Blakey SM, Reuman L, Buchholz JL (2018) New directions in the cognitive-behavioral treatment of OCD: theory, research, and practice. Behav Ther 49(3):311–322. https://doi.org/10.1016/j.beth.2017.09.002

American Psychiatric Association (2007) Practice guideline for the treatment of patients with obsessive-compulsive disorder. American Psychiatric Association, Arlington, VA. Available online at http://www.psych.org/psych_pract/treatg/pg/prac_guide.cfm

Anderson RA, Rees CS (2007) Group versus individual cognitive-behavioural treatment for obsessive-compulsive disorder: a controlled trial. Behav Res Ther 45(1):123–137. https://doi.org/10.1016/j.brat.2006.01.016

Andersson E, Enander J, Andren P, Hedman E, Ljotsson B, Hursti T, Bergstrom J, Kaldo V, Lindefors N, Andersson G, Ruck C (2012) Internet-based cognitive behaviour therapy for obsessive-compulsive disorder: a randomized controlled trial. Psychol Med 42(10):2193–2203. https://doi.org/10.1017/S0033291712000244

Andersson E, Steneby S, Karlsson K, Ljotsson B, Hedman E, Enander J, Kaldo V, Andersson G, Lindefors N, Ruck C (2014) Long-term efficacy of Internet-based cognitive behavior therapy for obsessive-compulsive disorder with or without booster: a randomized controlled trial. Psychol Med 44(13):2877–2887. https://doi.org/10.1017/S0033291714000543

Andersson E, Hedman E, Enander J, Radu Djurfeldt D, Ljotsson B, Cervenka S, Isung J, Svanborg C, Mataix-Cols D, Kaldo V, Andersson G, Lindefors N, Ruck C (2015) D-cycloserine vs placebo as adjunct to cognitive behavioral therapy for obsessive-compulsive disorder and interaction with antidepressants: a randomized clinical trial. JAMA Psychiat 72(7):659–667. https://doi.org/10.1001/jamapsychiatry.2015.0546

Apergis-Schoute AM, Gillan CM, Fineberg NA, Fernandez-Egea E, Sahakian BJ, Robbins TW (2017) Neural basis of impaired safety signaling in Obsessive Compulsive Disorder. Proc Natl Acad Sci U S A 114(12):3216–3221. https://doi.org/10.1073/pnas.1609194114

Asnaani A, Kaczkurkin AN, Alpert E, McLean CP, Simpson HB, Foa EB (2017) The effect of treatment on quality of life and functioning in OCD. Compr Psychiatry 73:7–14. https://doi.org/10.1016/j.comppsych.2016.10.004

Aspvall K, Andrén P, Lenhard F, Andersson E, Mataix-Cols D, Serlachius E (2018) Internet-delivered cognitive behavioural therapy for young children with obsessive-compulsive disorder: development and initial evaluation of the BIP OCD junior programme. BJPsych open 4 (3):106–112. https://doi.org/10.1192/bjo.2018.10

Aspvall K, Andersson E, Lenhard F, Melin K, Norlin L, Wallin L, Silverberg-Mörse M, Feldman I, Bottai M, Mataix-Cols D, Serlachius E (2019) Stepped care internet-delivered vs face-to-face cognitive-behavior therapy for pediatric obsessive-compulsive disorder: a trial protocol for a randomized noninferiority trial. JAMA Netw Open 2(10):e1913810–e1913810. https://doi.org/10.1001/jamanetworkopen.2019.13810

Bachofen M, Nakagawa A, Marks IM, Park JM, Greist JH, Baer L, Wenzel KW, Parkin JR, Dottl SL (1999) Home self-assessment and self-treatment of obsessive-compulsive disorder using a manual and a computer-conducted telephone interview: replication of a UK-US study. J Clin Psychiatry 60(8):545–549

Blanco C, Olfson M, Stein DJ, Simpson HB, Gameroff MJ, Narrow WH (2006) Treatment of obsessive-compulsive disorder by U.S. psychiatrists. J Clin Psychiatry 67(6):946–951

Boisseau CL, Schwartzman CM, Lawton J, Mancebo MC (2017) App-guided exposure and response prevention for obsessive compulsive disorder: an open pilot trial. Cogn Behav Ther 46(6):447–458. https://doi.org/10.1080/16506073.2017.1321683

Carmi L, Tendler A, Bystritsky A, Hollander E, Blumberger DM, Daskalakis J, Ward H, Lapidus K, Goodman W, Casuto L, Feifel D, Barnea-Ygael N, Roth Y, Zangen A, Zohar J (2019) Efficacy and safety of deep transcranial magnetic stimulation for obsessive-compulsive disorder: a prospective multicenter randomized double-blind placebo-controlled trial. Am J Psychiatr 176 (11):931–938. https://doi.org/10.1176/appi.ajp.2019.18101180

Chalah MA, Ayache SS (2019) Noninvasive brain stimulation and psychotherapy in anxiety and depressive disorders: a viewpoint. Brain Sci 9(4):82. https://doi.org/10.3390/brainsci9040082

Coles ME, Schubert JR, Heimberg RG, Weiss BD (2014) Disseminating treatment for anxiety disorders: step 1: recognizing the problem as a precursor to seeking help. J Anxiety Disord 28 (8):737–740. https://doi.org/10.1016/j.janxdis.2014.07.011

Comer JS, Furr JM, Cooper-Vince CE, Kerns CE, Chan PT, Edson AL, Khanna M, Franklin ME, Garcia AM, Freeman JB (2014) Internet-delivered, family-based treatment for early-onset OCD: a preliminary case series. J Clin Child Adolesc Psychol 43(1):74–87. https://doi.org/10.1080/15374416.2013.855127

Comer JS, Furr JM, Kerns CE, Miguel E, Coxe S, Elkins RM, Carpenter AL, Cornacchio D, Cooper-Vince CE, DeSerisy M, Chou T, Sanchez AL, Khanna M, Franklin ME, Garcia AM, Freeman JB (2017) Internet-delivered, family-based treatment for early-onset OCD: a pilot randomized trial. J Consult Clin Psychol 85(2):178–186. https://doi.org/10.1037/ccp0000155

Comer JS, Hong N, Poznanski B, Silva K, Wilson M (2019) Evidence base update on the treatment of early childhood anxiety and related problems. J Clin Child Adolesc Psychol 48(1):1–15. https://doi.org/10.1080/15374416.2018.1534208

Craske MG, Treanor M, Conway CC, Zbozinek T, Vervliet B (2014) Maximizing exposure therapy: an inhibitory learning approach. Behav Res Ther 58:10–23. https://doi.org/10.1016/j.brat.2014.04.006

D'Alcante C, Diniz BJ, Fossaluza V, Batistuzzo CM, Lopes CA, Shavitt GR, Deckersbach T, Malloy-Diniz L, Miguel CE, Hoexter QM (2012) Neuropsychological predictors of response to randomized treatment in obsessive-compulsive disorder. Prog Neuro-Psychopharmacol Biol Psychiatry 39(2):310–317. https://doi.org/10.1016/j.pnpbp.2012.07.002

Davis M, Ressler K, Rothbaum BO, Richardson R (2006) Effects of D-cycloserine on extinction: translation from preclinical to clinical work. Biol Psychiatry 60(4):369–375. https://doi.org/10.1016/j.biopsych.2006.03.084

Diefenbach GJ, Abramowitz JS, Norberg MM, Tolin DF (2007) Changes in quality of life following cognitive-behavioral therapy for obsessive-compulsive disorder. Behav Res Ther 45(12):3060–3068. https://doi.org/10.1016/j.brat.2007.04.014

Duman RS (2014) Neurobiology of stress, depression, and rapid acting antidepressants: Remodeling synaptic connections. Depress Anxiety 31(4):291–296. https://doi.org/10.1002/da.22227

Dunsmoor JE, Campese VD, Ceceli AO, LeDoux JE, Phelps EA (2015a) Novelty-facilitated extinction: providing a novel outcome in place of an expected threat diminishes recovery of defensive responses. Biol Psychiatry 78(3):203–209. https://doi.org/10.1016/j.biopsych.2014.12.008

Dunsmoor JE, Niv Y, Daw N, Phelps EA (2015b) Rethinking extinction. Neuron 88(1):47–63. https://doi.org/10.1016/j.neuron.2015.09.028

DuPont RL, Rice DP, Shiraki S, Rowland CR (1995) Economic costs of obsessive-compulsive disorder. Med Interface 8(4):102–109

Eddy KT, Dutra L, Bradley R, Westen D (2004) A multidimensional meta-analysis of psychotherapy and pharmacotherapy for obsessive-compulsive disorder. Clin Psychol Rev 24 (8):1011–1030. https://doi.org/10.1016/j.cpr.2004.08.004

Eelen P, Hermans D, Baeyens F (2001) Learning perspectives on anxiety disorders. In: Griez EJ, Faravelli C, Nutt D, Zohar J (eds) Anxiety disorders: an introduction to clinical management and research. Wiley, New York, pp 249–264

Endicott J, Nee J, Harrison W, Blumenthal R (1993) Quality of life enjoyment and satisfaction questionnaire: a new measure. Psychopharmacol Bull 29:321–326

Fals-Stewart W, Marks AP, Schafer J (1993) A comparison of behavioral group therapy and individual behavior therapy in treating obsessive-compulsive disorder. J Nerv Ment Dis 181:189–193

Farrell LJ, Waters AM, Boschen MJ, Hattingh L, McConnell H, Milliner EL, Collings N, Zimmer-Gembeck M, Shelton D, Ollendick TH, Testa C, Storch EA (2013) Difficult-to-treat pediatric obsessive-compulsive disorder: feasibility and preliminary results of a randomized pilot trial of D-cycloserine-augmented behavior therapy. Depress Anxiety 30(8):723–731. https://doi.org/10.1002/da.22132

Farrell NR, Leonard RC, Riemann BC (2019) The "behavioral specialist" model of training novice paraprofessional clinicians: an innovative, cost-effective approach for increasing the scalability of CBT. Behav Ther 42(4):111–117. Retrieved from http://www.abct.org/docs/PastIssue/42n4.pdf

Fineberg NA, Baldwin DS, Drummond LM, Wyatt S, Hanson J, Gopi S, Kaur S, Reid J, Marwah V, Sachdev RA, Pampaloni I, Shahper S, Varlakova Y, Mpavaenda D, Manson C, O'Leary C, Irvine K, Monji-Patel D, Shodunke A, Dyer T, Dymond A, Barton G, Wellsted D (2018a) Optimal treatment for obsessive compulsive disorder: a randomized controlled feasibility study of the clinical-effectiveness and cost-effectiveness of cognitive-behavioural therapy, selective serotonin reuptake inhibitors and their combination in the management of obsessive-compulsive disorder. Int Clin Psychopharmacol 33(6):334–348. https://doi.org/10.1097/YIC.0000000000000237

Fineberg NA, Apergis-Schoute AM, Vaghi MM, Banca P, Gillan CM, Voon V, Chamberlain SR, Cinosi E, Reid J, Shahper S, Bullmore ET, Sahakian BJ, Robbins TW (2018b) Mapping compulsivity in the DSM-5 obsessive compulsive and related disorders: cognitive domains, neural circuitry, and treatment. Int J Neuropsychopharmacol 21(1):42–58. https://doi.org/10.1093/ijnp/pyx088

Firth J, Torous J, Nicholas J, Carney R, Rosenbaum S, Sarris J (2017) Can smartphone mental health interventions reduce symptoms of anxiety? A meta-analysis of randomized controlled trials. J Affect Disord 218:15–22

Flygare O, Andersson E, Ringberg H, Hellstadius A-C, Edbacken J, Enander J, Dahl M, Aspvall K, Windh I, Russell A, Mataix-Cols D, Rück C (2020) Adapted cognitive behavior therapy for obsessive-compulsive disorder with co-occurring autism spectrum disorder: a clinical effectiveness study. Autism 24(1):190–199. https://doi.org/10.1177/1362361319856974

Foa EB, Kozak MJ (1986) Emotional processing of fear: exposure to corrective information. Psychol Bull 99(1):20–35. https://doi.org/10.1037/0033-2909.99.1.20

Foa EB, McNally RJ (1996) Mechanisms of change in exposure therapy. In: Rapee RM (ed) Current controversies in the anxiety disorders. Guilford, New York, pp 329–343

Foa EB, Steketee G, Milby JB (1980) Differential effects of exposure and response prevention in obsessive-compulsive washers. J Consult Clin Psychol 48(1):71–79. https://doi.org/10.1037// 0022-006x.48.1.71

Foa EB, Liebowitz MR, Kozak MJ, Davies S, Campeas R, Franklin ME, Huppert JD, Kjernisted K, Rowan V, Schmidt AB, Simpson HB, Tu X (2005) Randomized, placebo-controlled trial of exposure and ritual prevention, clomipramine, and their combination in the treatment of obsessive-compulsive disorder. Am J Psychiatry 162(1):151–161. https://doi.org/10.1176/ appi.ajp.162.1.151

Foa EB, Huppert JD, Cahill SP (2006) Emotional processing theory: an update. In: Pathological anxiety: emotional processing in etiology and treatment, pp 3–24

Foa EB, Yadin E, Lichner TK (2012) Exposure and response (ritual) prevention for obsessive-compulsive disorder: clinician guide, 2nd edn. Oxford University Press, Oxford, p 182

Foa EB, Simpson HB, Liebowitz MR, Powers MB, Rosenfield D, Cahill S, Campeas R, Franklin M, Hahn C, Hembree EA, Huppert JD, Schmidt AB, Vermes D, Williams MT (2013) Six-month follow-up of a randomized controlled trial augmenting serotonin reuptake inhibitor treatment with exposure and ritual prevention for obsessive-compulsive disorder. J Clin Psychiatry 74 (5):464–469

Foa EB, Simpson HB, Rosenfield D, Liebowitz MR, Cahill S, Huppert JD, Bender J, McLean CP, Maher MJ, Campeas R, Hahn C, Imms P, Pinto A, Powers MB, Rodriguez CI, Van Meter PE, Vermes D, Williams MT (2015) Six-month outcomes from a randomized trial augmenting serotonin reuptake inhibitors with exposure and ritual prevention or risperidone in adults with obsessive-compulsive disorder. J Clin Psychiatry 76(4):440–446

Fradkin I, Cahill SP, Huppert JD (2020) CBT for anxiety disorders: memory reconsolidation theory and its relationship to cognitive, emotional processing, and inhibitory models. In: Neuroscience of enduring change: implications for psychotherapy. pp 241–269

Franklin ME, Sapyta J, Freeman JB, Khanna M, Compton S, Almirall D, Moore P, Choate-Summers M, Garcia A, Edson AL, Foa EB, March JS (2011) Cognitive behavior therapy augmentation of pharmacotherapy in pediatric obsessive-compulsive disorder: the Pediatric OCD treatment study II (POTS II) randomized controlled trial. JAMA 306(11):1224–1232. https://doi.org/10.1001/jama.2011.1344

Freeman J, Sapyta J, Garcia A, Compton S, Khanna M, Flessner C, FitzGerald D, Mauro C, Dingfelder R, Benito K, Harrison J, Curry J, Foa E, March J, Moore P, Franklin M (2014) Family-based treatment of early childhood obsessive-compulsive disorder: the pediatric obsessive-compulsive disorder treatment study for young children (POTS Jr)—a randomized clinical trial. JAMA Psychiat 71(6):689–698. https://doi.org/10.1001/jamapsychiatry.2014.170

Freeman J, Benito K, Herren J, Kemp J, Sung J, Georgiadis C, Arora A, Walther M, Garcia A (2018) Evidence base update of psychosocial treatments for pediatric obsessive-compulsive disorder: evaluating, improving, and transporting what works. J Clin Child Adolesc Psychol 47 (5):669–698. https://doi.org/10.1080/15374416.2018.1496443

Freeman JB, Choate-Summers ML, Garcia AM et al (2009) The pediatric obsessive-compulsive disorder treatment study II: rationale, design and methods. Child Adolesc Psychiatry Ment Health 3:4. https://doi.org/10.1186/1753-2000-3-4

Freeston MH, Ladouceur R, Gagnon F, Thibodeau N, Rhéaume J, Letarte H et al (1997) Cognitive-behavioral treatment of obsessive thoughts: a controlled study. J Consult Clin Psychol 65:405–413

Fullana MA, Dunsmoor JE, Schruers KRJ, Savage HS, Bach DR, Harrison BJ (2020) Human fear conditioning: from neuroscience to the clinic. Behav Res Ther 124:103528. https://doi.org/10. 1016/j.brat.2019.103528

Garcia AM, Sapyta JJ, Moore PS, Freeman JB, Franklin ME, March JS, Foa EB (2010) Predictors and moderators of treatment outcome in the Pediatric Obsessive-Compulsive Treatment Study (POTS I). J Am Acad Child Adolesc Psychiatry 49(10):1024–1033.; quiz 1086. https://doi.org/ 10.1016/j.jaac.2010.06.013

Gershkovich M, Middleton R, Hezel D, Grimaldi S, Renna M, Basaraba C, Patel S, Simpson HB (2020) Integrating exposure and response prevention with a mobile app to treat obsessive-compulsive disorder: feasibility, acceptability, and preliminary effects. Behav Ther. https://doi.org/10.1016/j.beth.2020.05.001

Gideons ES, Kavalali ET, Monteggia LM (2014) Mechanisms underlying differential effectiveness of memantine and ketamine in rapid antidepressant responses. Proc Natl Acad Sci U S A 111 (23):8649–8654. https://doi.org/10.1073/pnas.1323920111

Gillan CM, Apergis-Schoute AM, Morein-Zamir S, Urcelay GP, Sule A, Fineberg NA, Sahakian BJ, Robbins TW (2015) Functional neuroimaging of avoidance habits in obsessive-compulsive disorder. Am J Psychiatry 172(3):284–293. https://doi.org/10.1176/appi.ajp.2014.14040525

Gillihan SJ, Williams MT, Malcoun E, Yadin E, Foa EB (2012) Common pitfalls in exposure and response prevention (EX/RP) for OCD. J Obsess-Compuls Rel Disord 1(4):251–257. https://doi.org/10.1016/j.jocrd.2012.05.002

Glazier K, Calixte RM, Rothschild R, Pinto A (2013) High rates of OCD symptom misidentification by mental health professionals. Ann Clin Psychiatry 25(3):201–209

Glazier K, Swing M, McGinn LK (2015) Half of obsessive-compulsive disorder cases misdiagnosed: vignette-based survey of primary care physicians. J Clin Psychiatry 76(6): e761–e767. https://doi.org/10.4088/JCP.14m09110

Goetter EM, Herbert JD, Forman EM, Yuen EK, Thomas JG (2014) An open trial of videoconference-mediated exposure and ritual prevention for obsessive-compulsive disorder. J Anxiety Disord 28(5):460–462. https://doi.org/10.1016/j.janxdis.2014.05.004

Goodman WK, Price LH, Rasmussen SA, Mazure C, Fleischmann RL, Hill CL, Heninger GR, Charney DS (1989) The Yale-brown obsessive-compulsive scale. I. Development, use, and reliability. Arch Gen Psychiatry 46(11):1006–1011. https://doi.org/10.1001/archpsyc.1989.01810110048007

Goodwin R, Koenen KC, Hellman F, Guardino M, Struening E (2002) Help seeking and access to mental health treatment for obsessive-compulsive disorder. Acta Psychiatr Scand 106 (2):143–149. https://doi.org/10.1034/j.1600-0447.2002.01221.x

Greist JH, Marks IM, Baer L, Parkin JR, Manzo PA, Mantle JM, Wenzel KW, Spierings CJ, Kobak KA, Dottl SL, Bailey TM, Forman L (1998) Self-treatment for obsessive compulsive disorder using a manual and a computerized telephone interview: a U.S.-U.K. study. MD Comput 15 (3):149–157

Greist JH, Marks IM, Baer L, Kobak KA, Wenzel KW, Hirsch MJ, Mantle JM, Clary CM (2002) Behavior therapy for obsessive-compulsive disorder guided by a computer or by a clinician compared with relaxation as a control. J Clin Psychiatry 63(2):138–145

Gremel CM, Chancey JH, Atwood BK, Luo G, Neve R, Ramakrishnan C, Deisseroth K, Lovinger DM, Costa RM (2016) Endocannabinoid modulation of orbitostriatal circuits gates habit formation. Neuron 90(6):1312–1324. https://doi.org/10.1016/j.neuron.2016.04.043

Groves PM, Thompson RF (1970) Habituation: a dual-process theory. Psychol Rev 77(5):419–450. https://doi.org/10.1037/h0029810

Hamilton M (1960) A rating scale for depression. J Neurol Neurosurg Psychiatry 23:56–62. https://doi.org/10.1136/jnnp.23.1.56

Hansen B, Hagen K, Öst L-G, Solem S, Kvale G (2018) The Bergen 4-day OCD treatment delivered in a group setting: 12-month follow-up. Front Psychol 9:639–639. https://doi.org/10.3389/fpsyg.2018.00639

Hansen B, Kvale G, Hagen K, Havnen A, Öst L-G (2019) The Bergen 4-day treatment for OCD: four years follow-up of concentrated ERP in a clinical mental health setting. Cogn Behav Ther 48(2):89–105. https://doi.org/10.1080/16506073.2018.1478447

Harned MS, Korslund KE, Linehan MM (2014) A pilot randomized controlled trial of Dialectical Behavior Therapy with and without the Dialectical Behavior Therapy Prolonged Exposure protocol for suicidal and self-injuring women with borderline personality disorder and PTSD. Behav Res Ther 55:7–17. https://doi.org/10.1016/j.brat.2014.01.008

Himle JA, Fischer DJ, Muroff JR, Van Etten ML, Lokers LM, Abelson JL, Hanna GL (2006) Videoconferencing-based cognitive-behavioral therapy for obsessive-compulsive disorder. Behav Res Ther 44:1821–1829

Hong N, Sanchez AL, Comer JS (2020) Multimedia field test: can users strike out OCD with the NOCD App?. Cogn Behav Pract 27(1):93–99. https://doi.org/10.1016/j.cbpra.2019.02.005

Jacoby RJ, Abramowitz JS (2016) Inhibitory learning approaches to exposure therapy: a critical review and translation to obsessive-compulsive disorder. Clin Psychol Rev 49:28–40. https://doi.org/10.1016/j.cpr.2016.07.001

Kalanthroff E, Steinman SA, Schmidt AB, Campeas R, Simpson HB (2018) Piloting a personalized computerized inhibitory training program for individuals with obsessive-compulsive disorder. Psychother Psychosom 87(1):52–54. https://doi.org/10.1159/000481199

Karabanov A, Ziemann U, Hamada M, George MS, Quartarone A, Classen J, Massimini M, Rothwell J, Siebner HR (2015) Consensus paper: probing homeostatic plasticity of human cortex with non-invasive transcranial brain stimulation. Brain 8(3):442–454. https://doi.org/10.1016/j.brs.2015.01.404

Kayser RR (2020) Pharmacotherapy for treatment-resistant obsessive-compulsive disorder. J Clin Psychiatry 81(5):19ac13182

Kayser RR, Snorrason I, Haney M, Lee FS, Simpson HB (2019) The endocannabinoid system: a new treatment target for obsessive compulsive disorder? Cannabis Cannabinoid Res 4 (2):77–87. https://doi.org/10.1089/can.2018.0049

Kayser RR, Raskin M, Snorrason I, Hezel DM, Haney M, Simpson HB (2020) Cannabinoid augmentation of exposure-based psychotherapy for obsessive-compulsive disorder. J Clin Psychopharmacol 40(2):207–210. https://doi.org/10.1097/jcp.0000000000001179

Kenwright M, Marks I, Graham C, Franses A, Mataix-Cols D (2005) Brief scheduled phone support from a clinician to enhance computer-aided self-help for obsessive-compulsive disorder: randomized controlled trial. J Clin Psychol 61(12):1499–1508. https://doi.org/10.1002/jclp.20204

Kessler RC, McGonagle KA, Zhao S, Nelson CB, Hughes M, Eshleman S, Wittchen HU, Kendler KS (1994) Lifetime and 12-month prevalence of DSM-III-R psychiatric disorders in the United States. Results from the National Comorbidity Survey. Arch Gen Psychiatry 51(1):8–19. https://doi.org/10.1001/archpsyc.1994.03950010008002

Kessler RC, Chiu WT, Demler O, Merikangas KR, Walters EE (2005) Prevalence, severity, and comorbidity of 12-month DSM-IV disorders in the National Comorbidity Survey Replication. Arch Gen Psychiatry 62(6):617–627. https://doi.org/10.1001/archpsyc.62.6.617

Knopp J, Knowles S, Bee P, Lovell K, Bower P (2013) A systematic review of predictors and moderators of response to psychological therapies in OCD: do we have enough empirical evidence to target treatment? Clin Psychol Rev 33(8):1067–1081. https://doi.org/10.1016/j.cpr.2013.08.008

Kushner MG, Kim SW, Donahue C, Thuras P, Adson D, Kotlyar M, McCabe J, Peterson J, Foa EB (2007) D-cycloserine augmented exposure therapy for obsessive-compulsive disorder. Biol Psychiatry 62(8):835–838. https://doi.org/10.1016/j.biopsych.2006.12.020

Kvale G, Hansen B, Björgvinsson T, Børtveit T, Hagen K, Haseth S, Kristensen UB, Launes G, Ressler KJ, Solem S, Strand A, van den Heuvel OA, Öst L-G (2018) Successfully treating 90 patients with obsessive compulsive disorder in eight days: the Bergen 4-day treatment. BMC Psychiatry 18(1):323–323. https://doi.org/10.1186/s12888-018-1887-4

Lader MH, Mathews AM (1968) A physiological model of phobic anxiety and desensitization. Behav Res Ther 6(4):411–421. https://doi.org/10.1016/0005-7967(68)90021-1

Lang PJ (1971) The application of psychophysiological methods to the study of psychotherapy and behavior modification. In: Bergin A, Garfield S (eds) Handbook of psychotherapy and behavior change. Wiley, New York

Larsen DL, Attkisson CC, Hargreaves WA, Nguyen TD (1979) Assessment of client/patient satisfaction: development of a general scale. Eval Program Plann 2(3):197–207

Launes G, Hagen K, Sunde T et al (2019) A randomized controlled trial of concentrated ERP, self-help and waiting list for obsessive- compulsive disorder: the Bergen 4-day treatment. Front Psychol 10:2500. https://doi.org/10.3389/fpsyg.2019.02500

Ledgerwood L, Richardson R, Cranney J (2003) Effects of D-cycloserine on extinction of conditioned freezing. Behav Neurosci 117(2):341–349. https://doi.org/10.1037/0735-7044.117.2.341

Lenhard F, Andersson E, Mataix-Cols D, Ruck C, Vigerland S, Hogstrom J, Hillborg M, Brander G, Ljungstrom M, Ljotsson B, Serlachius E (2017) Therapist-guided, internet-delivered cognitive-behavioral therapy for adolescents with obsessive-compulsive disorder: a randomized controlled trial. J Am Acad Child Adolesc Psychiatry 56(1):10–19.e2. https://doi.org/10.1016/j.jaac.2016.09.515

Lindsay M, Crino R, Andrews G (1997) Controlled trial of exposure and response prevention in obsessive-compulsive disorder. Br J Psychiatry 171:135–139

Liu R-J, Lee FS, Li X-Y, Bambico F, Duman RS, Aghajanian GK (2012) Brain-derived neurotrophic factor Val66Met allele impairs basal and ketamine-stimulated synaptogenesis in prefrontal cortex. Biol Psychiatry 71(11):996–1005. https://doi.org/10.1016/j.biopsych.2011.09.030

Lovell K, Bower P, Gellatly J, Byford S, Bee P, McMillan D, Arundel C, Gilbody S, Gega L, Hardy G, Reynolds S, Barkham M, Mottram P, Lidbetter N, Pedley R, Molle J, Peckham E, Knopp-Hoffer J, Price O, Connell J, Heslin M, Foley C, Plummer F, Roberts C (2017) Clinical effectiveness, cost-effectiveness and acceptability of low-intensity interventions in the management of obsessive-compulsive disorder: the Obsessive-Compulsive Treatment Efficacy randomised controlled Trial (OCTET). Health Technol Assess 21(37):1–132. https://doi.org/10.3310/hta21370

Maher MJ, Huppert JD, Chen H, Duan N, Foa EB, Liebowitz MR, Simpson HB (2010) Moderators and predictors of response to cognitive-behavioral therapy augmentation of pharmacotherapy in obsessive-compulsive disorder. Psychol Med 40(12):2013–2023. https://doi.org/10.1017/S0033291710000620

Mancebo MC, Eisen JL, Pinto A, Greenberg BD, Dyck IR, Rasmussen SA (2006) The brown longitudinal obsessive-compulsive study: treatments received and patient impressions of improvement. J Clin Psychiatry 67(11):1713–1720. https://doi.org/10.4088/jcp.v67n1107

Mancebo MC, Eisen JL, Sibrava NJ, Dyck IR, Rasmussen SA (2011) Patient utilization of cognitive-behavioral therapy for OCD. Behav Ther 42(3):399–412. https://doi.org/10.1016/j.beth.2010.10.002

Marques L, LeBlanc NJ, Weingarden HM, Timpano KR, Jenike M, Wilhelm S (2010) Barriers to treatment and service utilization in an internet sample of individuals with obsessive-compulsive symptoms. Depress Anxiety 27(5):470–475. https://doi.org/10.1002/da.20694

Mataix-Cols D, Turner C, Monzani B, Isomura K, Murphy C, Krebs G, Heyman I (2014) Cognitive-behavioural therapy with post-session D-cycloserine augmentation for pediatric obsessive-compulsive disorder: pilot randomised controlled trial. Br J Psychiatry 204 (1):77–78. https://doi.org/10.1192/bjp.bp.113.126284

Mataix-Cols D, Fernández de la Cruz L, Nordsletten AE, Lenhard F, Isomura K, Simpson HB (2016) Towards an international expert consensus for defining treatment response, remission, recovery and relapse in obsessive-compulsive disorder. World Psychiatry 15(1):80–81. https://doi.org/10.1002/wps.20299

Mataix-Cols D, Fernandez de la Cruz L, Monzani B, Rosenfield D, Andersson E, Perez-Vigil A, Frumento P, de Kleine RA, Difede J, Dunlop BW, Farrell LJ, Geller D, Gerardi M, Guastella AJ, Hofmann SG, Hendriks G-J, Kushner MG, Lee FS, Lenze EJ et al (2017) D-cycloserine augmentation of exposure-based cognitive behavior therapy for anxiety, obsessive-compulsive, and posttraumatic stress disorders: a systematic review and meta-analysis of individual participant data. JAMA Psychiatry 74(5):501–510. https://doi.org/10.1001/jamapsychiatry.2016.3955

Matsumoto K, Sutoh C, Asano K, Seki Y, Urao Y, Yokoo M, Takanashi R, Yoshida T, Tanaka M, Noguchi R, Nagata S, Oshiro K, Numata N, Hirose M, Yoshimura K, Nagai K, Sato Y, Kishimoto T, Nakagawa A, Shimizu E (2018) Internet-based cognitive behavioral therapy with real-time clinician support via videoconference for patients with obsessive-compulsive disorder, panic disorder, and social anxiety disorder: pilot single-arm trial. J Med Internet Res 20 (12):e12091. https://doi.org/10.2196/12091

McGuire JF, Storch EA (2019) An inhibitory learning approach to cognitive-behavioral therapy for children and adolescents. Cogn Behav Pract 26(1):214–224. https://doi.org/10.1016/j.cbpra.2017.12.003

McTeague LM, Huemer J, Carreon DM, Jiang Y, Eickhoff SB, Etkin A (2017) Identification of common neural circuit disruptions in cognitive control across psychiatric disorders. Am J Psychiatr 174(7):676–685. https://doi.org/10.1176/appi.ajp.2017.16040400

Melin K, Skarphedinsson G, Thomsen PH, Weidle B, Torp NC, Valderhaug R, Hojgaard DRMA, Hybel KA, Nissen JB, Jensen S, Dahl K, Skarsater I, Haugland BS, Ivarsson T (2019) Treatment gains are sustainable in pediatric obsessive-compulsive disorder: three-year follow-up from the NordLOTS. J Am Acad Child Adolesc Psychiatry. https://doi.org/10.1016/j.jaac.2019.01.010

Milad MR, Quirk GJ (2012) Fear extinction as a model for translational neuroscience: ten years of progress. Annu Rev Psychol 63(1):129–151

Moritz S, Wittekind CE, Hauschildt M, Timpano KR (2011) Do it yourself? Self-help and online therapy for people with obsessive-compulsive disorder. Curr Opin Psychiatry 24(6):541–548. https://doi.org/10.1097/YCO.0b013e32834abb7f

National Institute for Health and Care Excellence (2005) Obsessive-compulsive disorder and body dysmorphic disorder: treatment. https://www.nice.org.uk/guidance/cg31

Niendam TA, Laird AR, Ray KL, Dean YM, Glahn DC, Carter CS (2012) Meta-analytic evidence for a superordinate cognitive control network sub serving diverse executive functions. Cogn Affect Behav Neurosci 12(2):241–268. https://doi.org/10.3758/s13415-011-0083-5

Olatunji BO, Davis ML, Powers MB, Smits JAJ (2013) Cognitive-behavioral therapy for obsessive-compulsive disorder: a meta-analysis of treatment outcome and moderators. J Psychiatr Res 47 (1):33–41. https://doi.org/10.1016/j.jpsychires.2012.08.020

Öst L-G, Havnen A, Hansen B, Kvale G (2015) Cognitive behavioral treatments of obsessive-compulsive disorder. A systematic review and meta-analysis of studies published 1993-2014. Clin Psychol Rev 40:156–169. https://doi.org/10.1016/j.cpr.2015.06.003

Patel SR, Galfavy H, Kimeldorf MB, Dixon LB, Simpson HB (2017) Patient preferences and acceptability of evidence-based and novel treatments for obsessive-compulsive disorder. Psychiatr Serv 68(3):250–257. https://doi.org/10.1176/appi.ps.201600092

Patel SR, Humensky JL, Olfson M, Simpson HB, Myers R, Dixon LB (2014) Treatment of obsessive-compulsive disorder in a nationwide survey of office-based physician practice. Psychiatr Serv 65(5):681–684. https://doi.org/10.1176/appi.ps.201300192

Patel SR, Wheaton MG, Andersson E, Rück C, Schmidt AB, La Lima CN, Galfavy H, Pascucci O, Myers RW, Dixon LB, Simpson HB (2018) Acceptability, feasibility, and effectiveness of internet-based cognitive-behavioral therapy for obsessive-compulsive disorder in New York. Behav Ther 49(4):631–641. https://doi.org/10.1016/j.beth.2017.09.003. Epub 2017 Sep 15. PMID: 29937263; PMCID: PMC6945297

Pediatric OCD Treatment Study [POTS] Team (2004) Cognitive-behavior therapy, sertraline, and their combination for children and adolescents with obsessive-compulsive disorder: the Pediatric OCD treatment study [POTS] randomized controlled trial. JAMA 292:1969–1976

Peris TS, Rozenman MS, Sugar CA, McCracken JT, Piacentini J (2017) Targeted family intervention for complex cases of pediatric obsessive-compulsive disorder: a randomized controlled trial. J Am Acad Child Adolesc Psychiatry 56(12):1034–1042.e1. https://doi.org/10.1016/j.jaac.2017.10.008

Piacentini J, Bergman RL, Chang S, Langley A, Peris T, Wood JJ, McCracken J (2011) Controlled comparison of family cognitive behavioral therapy and psychoeducation/relaxation training for child obsessive-compulsive disorder. J Am Acad Child Adolesc Psychiatry 50(11):1149–1161. https://doi.org/10.1016/j.jaac.2011.08.003

Pinto A, Liebowitz MR, Foa EB, Simpson HB (2011) Obsessive compulsive personality disorder as a predictor of exposure and ritual prevention outcome for obsessive compulsive disorder. Behav Res Ther 49(8):453–458. https://doi.org/10.1016/j.brat.2011.04.004

Ressler KJ, Rothbaum BO, Tannenbaum L, Anderson P, Graap K, Zimand E, Hodges L, Davis M (2004) Cognitive enhancers as adjuncts to psychotherapy: use of D-cycloserine in phobic individuals to facilitate extinction of fear. Arch Gen Psychiatry 61(11):1136–1144. https://doi.org/10.1001/archpsyc.61.11.1136

Robbins TW, Gillan CM, Smith DG, de Wit S, Ersche KD (2012) Neurocognitive endophenotypes of impulsivity and compulsivity: towards dimensional psychiatry. Trends Cogn Sci 16 (1):81–91. https://doi.org/10.1016/j.tics.2011.11.009

Rodriguez CI, Kegeles LS, Levinson A, Feng T, Marcus SM, Vermes D, Flood P, Simpson HB (2013) Randomized controlled crossover trial of ketamine in obsessive-compulsive disorder: proof-of-concept. Neuropsychopharmacology 38(12):2475–2483. https://doi.org/10.1038/npp.2013.150

Rück C, Lundstrom L, Flygare O, Enander J, Bottai M, Mataix-Cols D, Andersson E (2018) Study protocol for a single-blind, randomised controlled, non-inferiority trial of internet-based versus face-to-face cognitive behaviour therapy for obsessive-compulsive disorder. BMJ Open 8(9): e022254. https://doi.org/10.1136/bmjopen-2018-022254

Ruscio AM, Stein DJ, Chiu WT, Kessler RC (2010) The epidemiology of obsessive-compulsive disorder in the National Comorbidity Survey Replication. Mol Psychiatry 15(1):53–63. https://doi.org/10.1038/mp.2008.94

Scahill L, Riddle MA, McSwiggin-Hardin M et al (1997) Children's Yale-brown obsessive-compulsive scale: reliability and validity. J Am Acad Child Adolesc Psychiatry 36:844–8529183141

Schwartz C, Schlegl S, Kuelz AK, Voderholzer U (2013) Treatment-seeking in OCD community cases and psychological treatment actually provided to treatment-seeking patients: a systematic review. J Obsess-Compuls Rel Disord 2(4):448–456. https://doi.org/10.1016/j.jocrd.2013.10.006

Silverman WK, Albano AM (1996) The anxiety disorders interview schedule for children for DSM-IV: child and parent versions. Psychological Corporation, San Antonio

Simpson HB, Liebowitz MR, Foa EB, Kozak MJ, Schmidt AB, Rowan V, Petkova E, Kjernisted K, Huppert JD, Franklin ME, Davies SO, Campeas R (2004) Post-treatment effects of exposure therapy and clomipramine in obsessive-compulsive disorder. Depress Anxiety 19(4):225–233

Simpson HB, Huppert JD, Petkova E, Foa EB, Liebowitz MR (2006) Response versus remission in obsessive-compulsive disorder. J Clin Psychol 67(2):269–276

Simpson HB, Foa EB, Liebowitz MR, Ledley DR, Huppert JD, Cahill S, Vermes D, Schmidt AB, Hembree E, Franklin M, Campeas R, Hahn C-G, Petkova E (2008) A randomized, controlled trial of cognitive-behavioral therapy for augmenting pharmacotherapy in obsessive-compulsive disorder. Am J Psychiatry 165(5):621–630. https://doi.org/10.1176/appi.ajp.2007.07091440

Simpson HB, Maher M, Page JR, Gibbons CJ, Franklin ME, Foa EB (2010) Development of a patient adherence scale for exposure and response prevention therapy. Behav Ther 41(1):30–37. https://doi.org/10.1016/j.beth.2008.12.002

Simpson HB, Maher MJ, Wang Y, Bao Y, Foa EB, Franklin M (2011) Patient adherence predicts outcome from cognitive behavioral therapy in obsessive-compulsive disorder. J Consult Clin Psychol 79(2):247–252. https://doi.org/10.1037/a0022659

Simpson HB, Marcus SM, Zuckoff A, Franklin M, Foa EB (2012) Patient adherence to cognitive-behavioral therapy predicts long-term outcome in obsessive-compulsive disorder. J Clin Psychiatry 73(9):1265–1266. https://doi.org/10.4088/JCP.12l07879

Simpson HB, Foa EB, Liebowitz MR, Huppert JD, Cahill S, Maher MJ, McLean CP, Bender JJ, Marcus SM, Williams MT, Weaver J, Vermes D, Van Meter PE, Rodriguez CI, Powers M, Pinto A, Imms P, Hahn C-G, Campeas R (2013) Cognitive-behavioral therapy vs risperidone for augmenting serotonin reuptake inhibitors in obsessive-compulsive disorder: a randomized clinical trial. JAMA Psychiat 70(11):1190–1199. https://doi.org/10.1001/jamapsychiatry.2013.1932

Skoog G, Skoog I (1999) A 40-year follow-up of patients with obsessive-compulsive disorder. Arch Gen Psychiatry 56(2):121–127. https://doi.org/10.1001/archpsyc.56.2.121

Solem S, Håland AT, Vogel PA, Hansen B, Wells A (2009) Change in metacognitions predicts outcome in obsessive-compulsive disorder patients undergoing treatment with exposure and response prevention. Behav Res Ther 47(4):301–307. https://doi.org/10.1016/j.brat.2009.01.003

Steinman SA, Dunsmoor EJ, Gazman Z, Stovezky Y, Pascucci O, Pomerenke J, Phelps AE, Fyer A, Simpson HB (2020) Novelty does not facilitate extinction in individuals with anxiety disorders or obsessive-compulsive disorder. Behav Res Ther 109:68–74

Steketee G (1996) Treatment of obsessive-compulsive disorder (treatment manuals for practitioners). Guilford, New York

Steketee G, Siev J, Yovel I, Lit K, Wilhelm S (2019) Predictors and moderators of cognitive and behavioral therapy outcomes for OCD: a patient-level mega-analysis of eight sites. Behav Ther 50(1):165–176. https://doi.org/10.1016/j.beth.2018.04.004

Storch EA, Merlo LJ, Bengtson M, Murphy TK, Lewis MH, Yang MC, Jacob ML, Larson M, Hirsh A, Fernandez M, Geffken GR, Goodman WK (2007) D-cycloserine does not enhance exposure-response prevention therapy in obsessive-compulsive disorder. Int Clin Psychopharmacol 22(4):230–237. https://doi.org/10.1097/YIC.0b013e32819f8480

Storch EA, Murphy TK, Lack CW, Geffken GR, Jacob ML, Goodman WK (2008) Sleep-related problems in pediatric obsessive-compulsive disorder. J Anxiety Disord 22(5):877–885. https://doi.org/10.1016/j.janxdis.2007.09.003

Storch EA, Mariaskin A, Murphy TK (2009) Psychotherapy for obsessive-compulsive disorder. Curr Psychiatry Rep 11(4):296–301. https://doi.org/10.1007/s11920-009-0043-8

Storch EA, Caporino NE, Morgan JR, Lewin AB, Rojas A, Brauer L, Larson MJ, Murphy TK (2011) Preliminary investigation of web-camera delivered cognitive-behavioral therapy for youth with obsessive-compulsive disorder. Psychiatry Res 189(3):407–412. https://doi.org/10.1016/j.psychres.2011.05.047

Storch EA, Wilhelm S, Sprich S, Henin A, Micco J, Small BJ, McGuire J, Mutch PJ, Lewin AB, Murphy TK, Geller DA (2016) Efficacy of augmentation of cognitive behavior therapy with weight-adjusted d-cycloserine vs placebo in pediatric obsessive-compulsive disorder: a randomized clinical trial. JAMA Psychiatry 73(8):779–788. https://doi.org/10.1001/jamapsychiatry.2016.1128

Tenneij NH, van Megen HJ, Denys DA, Westenberg HG (2005) Behavior therapy augments response of patients with obsessive-compulsive disorder responding to drug treatment. J Clin Psychiatry 66(9):1169–1175. https://doi.org/10.4088/jcp.v66n0913

Thompson-Hollands J, Edson A, Tompson MC, Comer JS (2014a) Family involvement in the psychological treatment of obsessive-compulsive disorder: a meta-analysis. J Fam Psychol 28(3):287–298. https://doi.org/10.1037/a0036709

Thompson-Hollands J, Kerns CE, Pincus DB, Comer JS (2014b) Parental accommodation of child anxiety and related symptoms: range, impact, and correlates. J Anxiety Disord 28(8):765–773. https://doi.org/10.1016/j.janxdis.2014.09.007

Tolin DF, Hannan S, Maltby N, Diefenbach GJ, Worhunsky P, Brady RE (2007) A randomized controlled trial of self-directed versus therapist-directed cognitive-behavioral therapy for obsessive-compulsive disorder patients with prior medication trials. Behav Ther 38(2):179–191. https://doi.org/10.1016/j.beth.2006.07.001

Torous J, Roberts LW (2017) The ethical use of mobile health technology in clinical psychiatry. J Nerv Ment Dis 205(1):4–8. https://journals.lww.com/jonmd/Fulltext/2017/01000/The_Ethical_Use_of_Mobile_Health_Technology_in.2.aspx

Tryon WW (2005) Possible mechanisms for why desensitization and exposure therapy work. Clin Psychol Rev 25(1):67–95. https://doi.org/10.1016/j.cpr.2004.08.005

Uhre CF, Uhre VF, Lonfeldt NN, Pretzmann L, Vangkilde S, Plessen KJ, Gluud C, Jakobsen JC, Pagsberg AK (2020) Systematic review and meta-analysis: cognitive-behavioral therapy for obsessive-compulsive disorder in children and adolescents. J Am Acad Child Adolesc Psychiatry 59(1):64–77. https://doi.org/10.1016/j.jaac.2019.08.480

Van Ameringen M, Turna J, Khalesi Z, Pullia K, Patterson B (2017) There is an app for that! The current state of mobile applications (apps) for DSM-5 obsessive-compulsive disorder, posttraumatic stress disorder, anxiety and mood disorders. Depress Anxiety 34(6):526–539. https://doi.org/10.1002/da.22657

Vogel PA, Solem S, Hagen K, Moen EM, Launes G, Haland AT, Hansen B, Himle JA (2014) A pilot randomized controlled trial of videoconference-assisted treatment for obsessive-compulsive disorder. Behav Res Ther 63:162–168. https://doi.org/10.1016/j.brat.2014.10.007

Walker DL, Ressler KJ, Lu K-T, Davis M (2002) Facilitation of conditioned fear extinction by systemic administration or intra-amygdala infusions of D-cycloserine as assessed with fear-potentiated startle in rats. J Neurosci 22(6):2343–2351

Watts FN (1979) Habituation model of systematic desensitization. Psychol Bull 86(3):627–637

Wells A, Myers S, Simons M, Fisher P (2017) Metacognitive model and treatment of OCD. In: Abramowitz JS, McKay D, Storch EA (eds) The Wiley handbook of obsessive compulsive disorders

Wheaton MG, Rosenfield D, Foa EB, Simpson HB (2015) Augmenting serotonin reuptake inhibitors in obsessive-compulsive disorder: what moderates improvement? J Consult Clin Psychol 83(5):926–937. https://doi.org/10.1037/ccp0000025

Wheaton MG, Galfalvy H, Steinman SA, Wall MM, Foa EB, Simpson HB (2016) Patient adherence and treatment outcome with exposure and response prevention for OCD: which components of adherence matter and who becomes well? Behav Res Ther 85:6–12. https://doi.org/10.1016/j.brat.2016.07.010

Wilhelm S, Buhlmann U, Tolin DF, Meunier SA, Pearlson GD, Reese HE, Cannistraro P, Jenike MA, Rauch SL (2008) Augmentation of behavior therapy with D-cycloserine for obsessive-compulsive disorder. Am J Psychiatry 165(3):335–341.; quiz 409. https://doi.org/10.1176/appi.ajp.2007.07050776

Wootton BM, Dear BF, Johnston L, Terides MD, Titov N (2013) Remote treatment of obsessive-compulsive disorder: a randomized controlled trial. J Obsess-Compuls Rel Disord 2 (4):375–384. https://doi.org/10.1016/j.jocrd.2013.07.002

Zhou D-D, Wang W, Wang G-M, Li D-Q, Kuang L (2017) An updated meta-analysis: short-term therapeutic effects of repeated transcranial magnetic stimulation in treating obsessive-compulsive disorder. J Affect Disord 215:187–196

Pharmacotherapeutic Strategies and New Targets in OCD

Christopher Pittenger

Contents

C. Pittenger (✉)
Department of Psychiatry, Yale University School of Medicine, New Haven, CT, USA
e-mail: christopher.pittenger@yale.edu

Curr Topics Behav Neurosci (2021) 49: 331–384
https://doi.org/10.1007/7854_2020_204
Published Online: 10 March 2021

Abstract Effective pharmacological and psychotherapeutic treatments are well established for obsessive-compulsive disorder (OCD). Serotonin reuptake inhibitors (SRIs) are first-line treatment and are of benefit to about half of patients. Augmentation of SRI treatment with low-dose neuroleptics is an evidence-based second-line strategy. Specialty psychotherapy is also used as both first-line and second-line treatment and can benefit many. However, a substantial number of patients do not respond to these treatments. New alternatives are urgently needed. This review summarizes evidence for these established pharmacotherapeutic strategies, and for others that have been investigated in refractory disease but are not supported by the same level of evidence. We focus on three neurotransmitter systems in the brain: serotonin, dopamine, and glutamate. We summarize evidence from genetic, neuroimaging, animal model, and other lines of investigation that probe these three systems in patients with OCD. We also review recent work on predictors of response to current treatments. While many studies suggest abnormalities that may provide insight into the pathophysiology of the disorder, most studies have been small, and non-replication of reported findings has been common. Nevertheless, the gradual accrual of evidence for neurotransmitter dysregulation may in time lead the way to new pharmacological strategies.

Keywords Dopamine · Glutamate · Obsessive-compulsive disorder (OCD) · Pharmacotherapy · Selective serotonin reuptake inhibitors · Serotonin

1 Introduction

Obsessive-compulsive disorder (OCD) affects approximately one person in 40 over the course of a lifetime and causes significant morbidity, worldwide (Ruscio et al. 2010; Kessler et al. 2012; Hollander et al. 2016; Torres et al. 2017). Once considered rare (Rasmussen and Eisen 1990), OCD is now recognized as a significant source of both personal suffering and public health impact (Ruscio et al. 2010; Hollander et al. 2016; Boisseau et al. 2017; Torres et al. 2017) and is the focus of substantial research, as epitomized by the current volume. Pharmacological and psychotherapeutic treatments for OCD are well established and supported by a robust evidence base (Skapinakis et al. 2016). Unfortunately, a substantial minority of patients, perhaps as many as a third, get little benefit from available treatments. Remission of symptoms is rare in those with moderate to severe disease, even with the best treatments we currently have to offer. There is thus an urgent need for new insights into the causes and pathophysiology of the disorder, and new strategies for treatment, prevention, and cure.

This chapter provides a critical review of available pharmacological treatments for OCD. The goal is not to present an exhaustive summary of the current literature or a practical guide for clinicians; these have been recently provided by others (Fineberg et al. 2015; Skapinakis et al. 2016; Pittenger 2017a, b; Reid et al. 2017). The focus on psychopharmacology should not be construed as implying that medication treatment of OCD is all that is needed in most cases; specialized cognitive-behavioral therapy is a first-line treatment and is often a critical component of an optimal treatment program (Excellence, N. I. f. C. 2006; Koran et al. 2007). Here, however, the focus is on pharmacology, and more specifically on pharmacology in adults. We review current evidence-based treatments, and some other treatment strategies that are sometimes used clinically in refractory disease despite a weaker evidence base. We then consider what these strategies may be telling us about the underlying mechanisms of disease, and other lines of evidence (genetics, neuroimaging, pharmacological challenge studies, animal models) that may help inform our understanding of these mechanisms.

Importantly, the efficacy of a particular drug cannot by itself identify specific pathophysiological mechanisms. Pharmacological treatment may act by reversing a core disease-associated abnormality, as when levothyroxine supplementation replaces missing thyroid hormone in primary hypothyroidism, or L-DOPA treatment increases dopamine production to compensate for dopamine depletion in Parkinson's disease. In such a case, extrapolation from the mechanism of action of a drug to inferences about pathophysiology is straightforward. But a medication can also treat disease by indirectly counteracting pathophysiology through its downstream effects; for example, interferon gamma has historically been used to cure hepatitis C by stimulating the immune system to eliminate the virus, but this does not imply that hepatitis is caused by interferon deficiency. Finally, a medication can compensate for disease pathophysiology without reversing it. For instance, diuretics such as furosemide are a mainstay of treatment for congestive heart failure (CHF) but act on the kidney, not the heart; they do not reverse heart failure but rather mitigate some of its consequences by removing fluid, thus achieving a new, more functional equilibrium. It would be an error to infer that, because furosemide provides clinical benefit in patients, CHF must be, at its core, a disease of excessive fluid retention by the kidney.

Similarly, in the case of OCD, we cannot facilely conclude from the well-established efficacy of selective serotonin reuptake inhibitors (SSRIs) (Soomro et al. 2008; Issari et al. 2016; Skapinakis et al. 2016) that serotonin deficiency or dysregulation is central to the underlying pathophysiology of the condition. Indeed, as further discussed below, evidence from other sources (genetics, neuroimaging;

see chapters “The Future of Obsessive-Compulsive Spectrum Disorders: A Research Perspective”, “Genetics of OCD and Related Disorders; Searching for Shared Factors”, “Recent Developments in the Habit Hypothesis of OCD and Compulsive Disorders”, and “Invasive and Non-invasive Neurostimulation for OCD”) provides limited support for this theory. Nevertheless, thoughtful consideration of the pattern of medication response may provide some hints as to what is happening in the brain in patients, or at least facilitate the framing of new hypotheses.

It is widely appreciated, but nevertheless important to state, that OCD is an extremely heterogeneous disorder. This creates challenges both to studies of pathophysiology and to the development and validation of new treatments. It is clear that OCD does not have a unitary cause. Genetic factors, which determine 40–50% of the risk (Davis et al. 2013), are complex and not well understood (Fernandez et al. 2018), and developmental or environmental contributors are even more mysterious (see chapters “The Future of Obsessive-Compulsive Spectrum Disorders: A Research Perspective”, “On the Development of OCD”, “Genetics of OCD and Related Disorders; Searching for Shared Factors”, and “New Directions for Surgical Ablation Treatment of Obsessive Compulsive Disorder”). Given this causal complexity, and presumed heterogeneity, group-level studies may obscure important truths. For example, a recent genetic analysis identified 2 candidate rare genetic causes of OCD and suggested the existence of as many as 400 similar rare, large-effect mutations, each explaining a tiny fraction of cases (Cappi et al. 2019). Such rare causes may be determinative in individual cases but will never be seen in a population study (Stewart et al. 2013a, b; Mattheisen et al. 2015; International Obsessive Compulsive Disorder Foundation Genetics, C. and O. C. D. C. G. A. Studies 2018), as they explain too little of the population variance.

This heterogeneity in turn complicates the search for new treatments, pharmacological or otherwise. In the extreme case, a treatment might be curative in patients who carry a particular causal mutation but completely ineffective in those who do not; such a treatment will never show statistically significant benefit in the heterogeneous clinical populations enrolled in large controlled treatment studies. It may be necessary to stratify studies by underlying pathophysiology in order to reach the ultimate “precision medicine” goal of identifying the optimal treatment for each patient. Such stratification by individual characteristics may be a realistic near-term goal in the treatment of cancer (for example) (Friedman et al. 2015), but we are far from achieving this precision in the treatment of OCD, or any other psychiatric syndrome.

It is possible that clinical OCD represents a final common pathway that individuals can reach through a range of different trajectories (Pittenger et al. 2017b). Alternatively, it may not be a unitary condition at all but rather a multidimensional spectrum (Bloch et al. 2008a, b; Olatunji et al. 2017), or a collection of phenomenologically overlapping pathophysiological entities that share no common core. We must be modest in what we can expect of group-level pharmacological treatment studies; given the heterogeneity of OCD, it is unlikely that any treatment will produce dramatic benefit in all patients. Indeed, this prediction matches experience: SSRIs, the mainstay of the pharmacological treatment of OCD, are of significant benefit in 40–50% of patients, but of limited benefit in others, and none at all in a distressingly large minority.

With this context in mind – and these caveats – we turn to a review of current pharmacological treatments, and the brain systems that they implicate.

2 Serotonin Reuptake Inhibitors

2.1 Efficacy of Serotonin Reuptake Inhibitors (SRIs)

Clomipramine, a tricyclic antidepressant that potently blocks serotonin reuptake (Wishart et al. 2018), was first identified as a potential treatment for OCD in the 1960s (Fernandez Cordoba and Lopez-Ibor Alino 1967; Fineberg and Gale 2005). Randomized, placebo-controlled trials began to appear in the early 1980s (Marks et al. 1980; Insel et al. 1983), and two definitive multi-center trials were completed at the end of that decade, with robustly positive results (DeVeaugh-Geiss et al. 1989; Katz et al. 1990; Group, C. C. S 1991). Meta-analysis of multiple controlled studies unambiguously demonstrates benefit from clomipramine treatment (Skapinakis et al. 2016). However, clomipramine is not recommended as a first-line agent in current treatment guidelines, because of its problematic side effect profile (Baldwin et al. 2005; Excellence, N. I. f. C. 2006; Koran et al. 2007).

The selective serotonin reuptake inhibitors (SSRIs) – fluoxetine, fluvoxamine, sertraline, paroxetine, citalopram, and escitalopram – are the mainstay of the pharmacotherapy of OCD and the first-line pharmacological recommendation in all major treatment guidelines (Baldwin et al. 2005; Excellence, N. I. f. C. 2006; Koran et al. 2007; Fineberg et al. 2015). They were first shown to be efficacious in the late 1980s (Perse et al. 1987; Goodman et al. 1989a, b, c; Montgomery et al. 1993); efficacy has now been shown across multiple double-blind, placebo-controlled studies and is unambiguously supported by meta-analysis (Soomro et al. 2008; Issari et al. 2016; Skapinakis et al. 2016). The SSRIs appear to be slightly less efficacious than clomipramine, though this has never been demonstrated in a head-to-head comparison and may be an artifact of differential study design (Skapinakis et al. 2016). Nevertheless, their favorable side effect and safety profile makes them the first choice for pharmacological treatment in most or all cases (Excellence, N. I. f. C. 2006, Koran et al. 2007, Fineberg et al. 2015).

The six available SSRIs have all been shown to be efficacious in the treatment of OCD. Fluoxetine, fluvoxamine, sertraline, and paroxetine are approved by the United States Food and Drug Agency and by parallel regulatory agencies in other countries for this indication; escitalopram is also approved in Europe. Citalopram and escitalopram (in the United States) are often used off-label (Excellence, N. I. f. C. 2006; Koran et al. 2007; Fineberg et al. 2015; Reid et al. 2017). Head-to-head comparisons between these agents are sparse and have not shown differential efficacy; meta-analysis similarly shows no difference between them (Soomro et al. 2008, Issari et al. 2016, Skapinakis et al. 2016).

2.2 SSRI Dosage

Optimal SSRI response in OCD is achieved at high doses (Bloch et al. 2010), and doses in excess of the approved antidepressant dose range are often used. For example, treatment guidelines suggest doses of fluoxetine of up to 120 mg/dy, fluvoxamine up to 450 mg/dy, and sertraline up to 400 mg/dy (Koran et al. 2007; Reid et al. 2017). The clinical importance of this guidance is clear but should not be exaggerated; more typical SSRI doses are of benefit to many, and the added benefit of raising them beyond the usual range is modest, at the population level. That said, in the absence of problematic side effects and or other contraindications, it is good practice to push to high doses before declaring a trial a failure, lest a potentially beneficial treatment be abandoned prematurely. The highest doses are typically tried when more typical doses produce partial response without problematic side effects.

A caveat to this guidance arose about a decade ago with the observation that high doses of citalopram are sometimes associated with an increase in the QTc interval on the EKG (FDA 2012), which could in principle lead to dangerous arrhythmias. This led to a recommendation by the US Food and Drug Administration that citalopram not be dosed above 40 mg/dy, or above 20 mg/dy in those over 60 years old (FDA 2012). Similar concerns have been raised for escitalopram. Subsequent analyses have not revealed a clear association between citalopram use and negative cardiac outcomes (Zivin et al. 2013; Hasnain and Vieweg 2014; Hutton et al. 2017); evidence remains equivocal. Nevertheless, the FDA warning (and similar guidance in Canada and the United Kingdom (NHS 2011)) complicates the use of high-dose citalopram, and perhaps escitalopram, for OCD; at the very least, medicolegal considerations must be taken into account, and EKG monitoring is advisable. These factors have steered many prescribers away from the use of high-dose citalopram in OCD, despite the clear evidence for efficacy (Bloch et al. 2010; Issari et al. 2016).

2.3 Duration of SSRI Treatment

Full clinical response to SSRI pharmacotherapy in OCD often takes many weeks (Soomro et al. 2008). Treatment guidelines therefore recommend trials of 8–12 weeks before a particular medication can be declared ineffective (Baldwin et al. 2005; Koran et al. 2007). This can be a source of frustration for both prescribers and patients, as it may entail continuation of ineffective treatment for many weeks before moving on to an alternative strategy. There is evidence that SRI benefit begins early in treatment; in the largest, best-powered of the early clomipramine efficacy trials benefit was seen as early as 2 weeks (DeVeaugh-Geiss et al. 1989); and a recent meta-analysis including data from more than 3,200 patients concluded that benefit begins shortly after treatment initiation and indeed is greatest in the early weeks (Issari et al. 2016) (though it takes several weeks to accumulate to the point of being statistically detectable in clinical studies). Partial early response may therefore be a useful indicator of the likelihood of later benefit; indeed, one study examining this issue found that only 20% of patients who did not show partial response at 4 weeks went on to develop a full clinical response ($\geq$35% improvement) at 12 weeks (da Conceicao Costa et al. 2013). Early response may therefore be a clinically useful indicator of the likely ultimate benefit from a particular drug trial, though with some risk of premature termination in a fraction of patients who would ultimately respond to a full-length trial.

2.4 *Optimization of SSRI Treatment*

Unfortunately, a substantial fraction of patients – close to half – do not respond to a first SSRI trial. There are several reasonable strategies to optimize SRI response. While our focus in this review is on pharmacotherapy, evidence-based psychotherapy should always be considered in this context (Koran et al. 2007; Simpson et al. 2013). If response is partial, further escalation of SSRI dose or extension of a trial is a reasonable strategy. Higher doses are often more effective, as noted above (Bloch et al. 2010, Issari et al. 2016); and in some patients, full response becomes apparent only after many months (Rasmussen et al. 1997). Intravenous SRI treatment has been used in a few studies as a way to rapidly achieve high doses; there is some evidence of benefit in those who have not responded to standard dosing (Fallon et al. 1998; Pallanti et al. 2002; Karameh and Khani 2015), but this is not a widely available therapeutic strategy. Switching to a second SSRI may be of benefit in some patients, although the likelihood of response declines with subsequent trials (Reid et al. 2017). Switching from an SSRI to clomipramine is more likely to be of benefit, and clomipramine should therefore be considered as a second- or third-line intervention, when initial SSRI monotherapy is ineffective (Koran et al. 2007). Other strategies, such as combination of two different SSRIs, or of an SSRI and clomipramine, have been investigated in small trials (Reid et al. 2017), but the evidence is insufficient to support any of these as a clear next step after initial nonresponse. Such combination regimens increase the risk of drug–drug interactions and problematic side effects.

3 Other Serotonergic Agents

The efficacy of monotherapy with serotonin reuptake inhibitors (clomipramine and SSRIs) has motivated trials of other serotonergic strategies, especially in patients who do not respond to an initial SSRI trial. No other agents have been definitively shown to be effective; but most trials have been small, and the evidence for other serotonergic agents is thus best considered equivocal, rather than negative.

3.1 Serotonin-Norepinephrine Dual Reuptake Inhibitors

A few studies have investigated the dual reuptake inhibitor venlafaxine as monotherapy in OCD, with mixed results. An early study reported that patients who failed initial SSRI treatment were more likely to respond if they were switched to venlafaxine treatment than if they were switched to an alternative SSRI (Lochner et al. 2001); but a subsequent crossover study found the opposite result (Denys et al. 2004b). More recently, an observational study from India found a high rate of response to venlafaxine treatment in SSRI nonresponders (Balachander et al. 2019). There are no published trials of desvenlafaxine in OCD. Duloxetine treatment of OCD has been reported in a number of case reports, and in one controlled trial that found similar response to duloxetine versus a second SSRI after failure of response to a first SSRI trial (Mowla et al. 2016). It is difficult to draw firm conclusions from these small studies.

3.2 Serotonin Agonists and Antagonists

Buspirone is a partial agonist of the 5-HT1A receptor, with some affinity for dopamine D3 and D4 receptors. It is marketed for the treatment of generalized anxiety disorder. Controlled data in OCD are limited, and largely negative (Koran et al. 2007). High doses of buspirone are advocated by some experts as an augmentation strategy for SSRI-resistant OCD symptoms; however, this strategy has not been rigorously tested in controlled studies.

Pindolol is both a 5-HT1A and a beta-adrenergic antagonist. Small, early controlled studies reported some benefit from either monotherapy or augmentation of SSRI treatment (Koran et al. 2007). A recent meta-analysis of controlled augmentation studies showed a trend towards symptom improvement with pindolol augmentation, but it did not reach statistical significance (Sassano-Higgins and Pato 2015). More research is needed to establish the therapeutic role of this agent, if any.

Mirtazapine is an alpha-2 adrenergic agonist and an antagonist of the serotonin 2A, 2C, and 3 receptors. Two controlled studies have studied its effects in OCD and suggest that it may produce some benefit, or may accelerate the response to a concomitantly prescribed SSRI (Pallanti et al. 2004; Koran et al. 2005). Again, more work is needed before clear conclusions can be drawn.

Ondansetron is an agonist of the 5-HT3 serotonin receptor; it is indicated for the treatment of nausea. Five controlled studies examined its use in OCD, with some evidence of benefit even at very low doses (Serata et al. 2015). This motivated a large, multi-site study by Transcept Pharmaceuticals (NCT01275248). Unfortunately, this trial failed to meet its primary efficacy endpoint (BioSpace 2012) (though it remains unpublished). The appropriate role of ondansetron in the treatment of OCD therefore remains unclear.

4 Serotonin Dysregulation in OCD

The clear benefit of SRI treatment in OCD has motivated interest in serotonin dysregulation as a pathophysiological contributor (Graat et al. 2017; Szechtman et al. 2020). This has been one motivation for the many (largely inconclusive) trials of other serotoninergic agents summarized above. Importantly, the fact that SRIs often produce therapeutic benefit does not constitute proof that serotonin abnormalities are causally related to OCD symptoms, as noted earlier in this article; this remains an empirical question.

Genetic findings have suggested that alleles of genes involved in serotonergic neurotransmission, as a group, may contribute to OCD risk, though they have not yet unambiguously identified specific alleles or genes with adequate statistical confidence (Sinopoli et al. 2017).

4.1 The Serotonin Reuptake Transporter

Clomipramine and the SSRIs block the action of the serotonin reuptake transporter, also known as 5-HTT or SERT, which is encoded by the *SLC6A4* gene on chromosome 17. This is therefore an obvious candidate actor in the pathophysiology of OCD, and it has been intensively investigated.

Polymorphisms in the *SLC6A4* gene appear to make a small contribution to OCD risk. An intensively studied polymorphism in the promoter region of the gene influences expression levels and has been associated with neuroticism, depression, and the response to stress or trauma (Caspi et al. 2003; Risch et al. 2009; Clarke et al. 2010; Talati et al. 2017). Multiple candidate gene association studies have examined this polymorphism in OCD. Early results were mixed (Bloch et al. 2008b); but more recent meta-analysis suggests a positive association, with an odds ratio of approximately 1.25 (i.e., carriers of the risk allele have a 25% higher likelihood of developing OCD than those without) (Taylor 2013). It is important to note that candidate gene studies have a checkered history in neuropsychiatry, with most failing to replicate (Altshuler et al. 2008). In OCD, out of well over a hundred candidate gene studies, most have not replicated (Fernandez et al. 2018). Such associations must therefore be interpreted with caution. Genome-wide association studies (GWAS), which can largely avoid the pitfalls of candidate gene studies, have not to date identified either *SLC6A4* or any other serotonin-related genes as leading candidate risk loci in OCD (IOCDF-GC and OCGAS 2018), but much larger studies are needed to provide a comprehensive characterization of the genetic architecture of OCD risk.

Interestingly, the association of risk-associated polymorphisms in the promoter region with neuroticism appears to be quantitatively similar to the association with OCD (Talati et al. 2017), and the genetic correlation between neuroticism and major OCD symptom dimensions is substantial (Bergin et al. 2014). Therefore, while there does seem to be a statistically significant association between this polymorphism and OCD, it is unclear that this is a *specific* effect, rather than a more general association with neuroticism and related symptomatology (depression, anxiety, stress reactivity). Comparison with other psychiatric diagnoses does suggest some specificity to the association with OCD (Taylor 2016), but this analysis is limited by the specific discrete disorders examined; neuroticism was not included.

A parallel set of molecular brain imaging studies has examined 5-HTT protein in the brain. Importantly, while genetic associations provide prima facie evidence for a causal contribution of the implicated alleles in the development of disease, establishing the causal nature of an association of brain imaging findings with pathophysiology is a more complex undertaking: receptor abnormalities could be the cause of symptoms, but they could as easily be a consequence of either symptoms or treatment, a compensation, or an epiphenomenon that correlates with disease without contributing to it. Early studies with the positron emission tomography (PET) ligand ^{11}C-McN 5,652 and the single photon emission computerized tomography (SPECT) ligand ^{123}I-β-CIT produced equivocal results, suggesting either elevated (Pogarell et al. 2003) or unchanged 5-HTT levels (Simpson et al. 2003) relative to healthy controls (Nikolaus et al. 2016).

5-HTT is expressed at high levels in platelets, and studies examining receptor levels/binding here have been more consistent, with most showing reduced 5-HTT ligand binding in patients (Bastani et al. 1991; Marazziti et al. 1992, 1996, 1997, 1999; Weizman et al. 1992; Sallee et al. 1996; Delorme et al. 2004), and some showing correlation of this reduced binding with symptom severity (Marazziti et al. 1997) or relationships with treatment outcome (Marazziti et al. 1997; Delorme et al. 2004). The challenge here is, of course, to understand the relationship between these peripheral measures and what is happening in the brain. The status of brain 5-HTT in OCD remains unclear.

4.2 Serotonin 2A Receptor

The serotonin 2A receptor (5-HT2A), encoded by the *HTR2A* gene, is a primarily postsynaptic, $G_{\alpha}q$-coupled metabotropic receptor. Like 5-HTT it has been intensively studied in candidate gene association studies in OCD. Meta-analysis suggests an association with two linked alleles, rs6311 in the promoter region and rs6313 in the 5′ coding region (Taylor 2013, 2016). As noted above, candidate gene associations must be interpreted with caution; *HTR2A* has not emerged as a candidate from genome-wide association studies to date (Fernandez et al. 2018).

5-HT2A has also been studied using PET imaging. Of note, 5-HT2A levels are low in the basal ganglia, and between-group differences in this structure can thus be difficult to detect. An early study, using the tracer ^{18}F-altanserine, reported increased receptor availability in the caudate nucleus, bilaterally (Adams et al. 2005). In contrast, a smaller study with the tracer ^{11}C-MDL reported *reduced* receptor availability across multiple cortical regions; tracer binding in the caudate and putamen was not reported (Perani et al. 2008). Finally, a somewhat larger, methodologically rigorous 2011 study, also using ^{11}C-MDL, reported no differences between patients with OCD and controls in any brain region (Simpson et al. 2011). Exploratory analyses in the latter two studies reported correlations between tracer binding in the orbitofrontal cortex and different measures of disease (symptom severity in one case (Perani et al. 2008), earlier age of onset in the other (Simpson et al. 2011)). This heterogeneity of results, with one study reporting increased binding, one decreased binding, and one no change, leaves little clarity regarding the status of 5-HT2A receptors in adults with OCD.

Atypical antipsychotics are antagonists of the 5-HT2A (in addition to their blockade of dopamine D2 receptors), and it is possible that benefit from augmentation with agents such as aripiprazole relates to this binding. We return to the use of atypical antipsychotics in OCD treatment below, when discussing dopamine dysregulation in OCD. 5-HT2A is also targeted by psychedelic drugs such as LSD and psilocybin; they are potent agonists (Johnson et al. 2019), and this binding appears to be critical for their profound subjective effects (Kraehenmann et al. 2017; Preller et al. 2017). Interest in psychedelic agents as potential therapeutics has been rekindled in recent years, and research in this area is accelerating (Reiff et al. 2020). A single small uncontrolled study suggests that psilocybin can have lasting benefits in some patients with OCD (Moreno et al. 2006); controlled studies are underway (NCT03300947; NCT03356483).

4.3 Serotonin 1A Receptor

The serotonin 1A receptor (5-HT1A), encoded by the *HTR1A* gene, is a G_{α}i-coupled somatodendritic autoreceptor on both serotonergic neurons, where it mediates negative feedback in the raphe nuclei, and on other neurons throughout the brain. Desensitization of 5-HT1A receptors is thought to be an important event in the response to SSRIs (El Mansari and Blier 2005). 5-HT1A receptors have not been extensively studied in OCD. No genetic studies have implicated polymorphisms in *HTR1A* as increasing risk, and no PET studies have examined the expression of the receptor.

5-HT1A is the primary target of buspirone. As noted above, buspirone is sometimes used as a tertiary pharmacological agent in OCD, although controlled data supporting this strategy are lacking (Koran et al. 2007).

4.4 Serotonin 1B/1D Receptor

The 1B/1D receptor, encoded by the *HTR1B* gene, is a G_{α}i-coupled receptor that is primarily localized to axon terminals, both on serotonergic neurons themselves (where it can also contribute to negative feedback regulation of transmitter release) and on other neuronal types (Sari 2004). Some candidate gene association studies have implicated *HTR1B* in OCD risk, but these have not been consistently replicated and are not supported by meta-analysis (Taylor 2013).

An older literature examined the impact of 5-HT1B/1D agonists on OCD symptoms. Agonists such as sumatriptan have been reported to worsen either symptoms of OCD or neurochemical abnormalities in several studies (Koran et al. 2001; Gross-Isseroff et al. 2004; Zohar et al. 2004), though not in all (Pian et al. 1998; Boshuisen and den Boer 2000). In rodents, agonists of the 5-HT1B receptor impair prepulse inhibition (PPI) and increase anxiety and perseverative behavior, all of which have been argued to reflect aspects of an OCD-like response (Shanahan et al. 2009; Woehrle et al. 2013).

A single PET imaging study has examined 5-HT1B binding in unmedicated patients with OCD, using the tracer ^{11}C-P943. There was no difference in binding between 12 adults with OCD and 12 matched healthy controls (Pittenger et al. 2016). Motivated by the ability of 5-HT1B agonists to impair PPI in animals (Shanahan et al. 2009), the authors examined the relationship between ^{11}C-P943 binding and PPI in patients and controls, and an interesting pattern emerged: there was no correlation between PPI and ^{11}C-P943 binding in cortical regions in controls, but a strong positive correlation in individuals with OCD. In subcortical structures the pattern was quite different: a positive correlation between PPI and ^{11}C-P943 binding in controls was lost in OCD, and even reversed in some structures (pallidum, thalamus) (Pittenger et al. 2016). This result requires replication, and its import is unclear, but it suggests a subtle reorientation of the serotonergic system in OCD that may not be apparent in simple between-group analyses.

4.5 Serotonin-3 Receptors

The 5-HT3 receptor is unique among serotonin receptors in that it is a heteropentameric ligand-gated ion channel, rather than a metabotropic G-protein-coupled receptor. Interest in this receptor in OCD derives from treatment studies using 5-HT3 agonists such as ondansetron. As reviewed above, initial studies with these agents were promising, but a large double-blind, placebo-controlled study (NCT01275248) did not hit its primary efficacy endpoint (BioSpace 2012). There are five 5-HT3 genes, two of which (*HTR3A* and *HTR3B*) are expressed at significant levels in the brain. No genetic studies to date have implicated these genes as risk alleles in OCD.

Interestingly, ondansetron reduces activity in cortical circuitries involved in interoception and sensorimotor processing, suggesting potential utility in the treatment of OCD symptoms associated with disgust and/or sensory hypersensitivity (Stern et al. 2019). A controlled study to test this possibility is underway (NCT03239210).

4.6 Serotonin Levels in OCD?

Some of the findings reviewed above have been interpreted as suggesting that levels of serotonin itself are dysregulated in individuals with OCD. Unfortunately, serotonin in and around synapses is not accessible to experimental measurement in humans using currently available technologies, and so arguments for dysregulated serotonin levels are necessarily indirect.

The simplest explanation of the efficacy of SRI medications is that serotonin is low in OCD, and SRI treatment, by blocking serotonin reuptake, normalizes it. Consistent with this idea, the 5-HTT promoter allele associated with OCD risk – the L_a allele (Taylor 2013) – is more efficiently transcribed and thus can produce more 5-HTT protein in the brain (Heinz et al. 2000). This increased protein leads to more efficient reuptake of serotonin, at least in lymphoblasts (Lesch et al. 1996), suggesting that L_a allele carriers may have reduced basal serotonin levels.

Studies in animals with a 5-HT neurotoxin have shown that orbitofrontal serotonin depletion can produce cognitive inflexibility (Clarke et al. 2004). Brain serotonin can be acutely reduced in humans using tryptophan depletion, a dietary manipulation. If low brain serotonin contributes directly to symptomatology, then tryptophan depletion should exacerbate symptoms. The literature using this technique is small and mixed. Early studies found no effect on core OCD symptoms after tryptophan depletion in SRI-treated patients, though depression and/or subjective distress were worsened in most studies (Barr et al. 1994; Smeraldi et al. 1996; Berney et al. 2006; Kulz et al. 2007). In contrast, a recent small controlled study found tryptophan depletion to acutely worsen some OCD symptoms, increasing interfering thoughts and decreasing perceived control over them (Hood et al. 2017).

Brain 5-HT levels can be indirectly measured by quantifying the serotonin metabolite 5-hydroxyindoleacetic acid (5-HIAA) in cerebrospinal fluid (CSF). An old literature took this approach; again, results are mixed. One early study found reduced 5-HIAA in a group of obsessional patients (Insel et al. 1985). However, other studies have not identified any 5-HIAA abnormalities in patients, though there may be an association between pretreatment CSF 5-HIAA levels and response to SRI treatment (Baumgarten and Grozdanovic 1998).

Neurotransmitter levels can be inferred from PET imaging, with caveats. PET measures the specific binding of radiolabeled ligand to its molecular target. When a ligand binds to the same site on a neurotransmitter receptor as the transmitter itself, radioligand binding reflects a combination of the amount of receptor present, its affinity for the ligand (which can be modulated by molecular changes such as covalent posttranslational modifications or allosteric interactions), and competition with endogenous neurotransmitter. With appropriate experimental design it is possible to isolate the contribution of competition with endogenous neurotransmitter, and thus to make inferences about relative neurotransmitter levels between experimental groups. This strategy has been extensively used in measurements of endogenous dopamine, as further discussed below (Denys et al. 2013). However, comparable methods are not as well established to measure serotonin levels, and so, while some authors have made claims about endogenous serotonin concentrations from PET signal, any such inferences remain tentative. Furthermore, as reviewed above, the PET literature on both the serotonin transporter and serotonin receptors in OCD is quite mixed, with no clear consensus as to dysregulation of any individual serotonin-related protein.

In sum, the literature attempting to directly or inferentially measure serotonin levels in OCD is quite mixed, and firm conclusions are difficult to draw. The robust therapeutic response to SRI medications is the clearest finding in the treatment literature on OCD (Soomro et al. 2008; Issari et al. 2016), and characterization of endogenous serotonin levels and dynamics therefore remains a compelling goal. Unfortunately, such characterization is limited by available technology; brain serotonin cannot be directly assayed in humans, except in rare circumstances. This is an important area for future research.

5 Dopaminergic Agents in the Treatment of OCD

While SRI antidepressants are the mainstay and the universally endorsed first-line pharmacological treatment for OCD, nonresponse and limited response are distressingly common. There is thus a clear clinical need for second-line treatments and augmentation strategies. Pharmacologically, much attention has focused on modulators of dopaminergic neurotransmission. Clinically it is of course always important to consider expert cognitive behavioral therapy, either as an augmentation strategy or as a stand-alone treatment; but our focus here remains on pharmacological strategies.

5.1 D2 Antagonist Monotherapy

Neuroleptics – dopamine D2 antagonists – have been investigated for the treatment of OCD. While the total number of studies is not large, the consensus in the field is that neuroleptic monotherapy is ineffective (Koran et al. 2007). A study in the 1950s, before modern diagnostic criteria were developed, examined chlorpromazine in 75 patients with obsessive-compulsive symptoms and reported at least some benefit in nearly half (Trethowan and Scott 1955); but the lack of modern diagnostic criteria and of quantitative measures of symptom change complicates extrapolation from these early results to modern clinical practice. Case reports of typical and atypical neuroleptic monotherapy have been mixed (Hussain and Ahad 1970; O'Regan 1970; Rivers-Bulkeley and Hollender 1982; Chiou et al. 2015). An open-label study in the 1990s examined clozapine monotherapy and found significant side effects but no benefit (McDougle et al. 1995). A more recent open-label study examined aripiprazole monotherapy and found some patients to benefit; side effects were again a problem (Connor et al. 2005). Small sample size and open-label design limit the interpretability of these studies. More rigorous examination of neuroleptic monotherapy would be valuable; however, the limited evidence for benefit, both in the published literature and in the experience of expert clinicians, provides little motivation for such studies.

In fact, neuroleptic monotherapy can sometimes induce or exacerbate OCD symptoms (Grillault Laroche and Gaillard 2016; Poyurovsky 2017). This is particularly clear in the case of clozapine (Lykouras et al. 2003; Schirmbeck and Zink 2012; Grillault Laroche and Gaillard 2016) but may also be true of other atypical/second-generation neuroleptics, perhaps because of their antagonism of the 5-HT2A receptor (Schirmbeck and Zink 2013; Poyurovsky 2017). There is some evidence that specific genetic polymorphisms may contribute to this susceptibility (Cai et al. 2013), though this finding is based on a candidate gene approach, which is historically problematic, and has yet to be replicated (Schirmbeck et al. 2012).

5.2 *D2 Antagonist Augmentation*

In contrast, addition of neuroleptics to stable SRI treatment – that is, pharmacological augmentation – has been shown to be of benefit in a number of studies. Early work focused on the typical neuroleptics haloperidol and pimozide, as well as risperidone, with generally positive results (Bloch et al. 2006). Meta-analysis of this early work confirmed overall benefit, with one positive study for haloperidol augmentation and three concordant controlled studies for risperidone (Bloch et al. 2006; Skapinakis et al. 2007). More recent work has largely examined atypical/second-generation neuroleptics. A 2014 meta-analysis of controlled studies of atypical neuroleptic augmentation found evidence for benefit from aripiprazole and risperidone, but not from olanzapine or quetiapine (Veale et al. 2014). A 2015 meta-analysis including all studies to date, of both typical and atypical neuroleptics, reached similar conclusions, with good evidence for benefit from augmentation with haloperidol, risperidone, and aripiprazole (Dold et al. 2015). Based on this work, augmentation with these agents is included as a second-tier pharmacological intervention in all current treatment guidelines (Baldwin et al. 2005; Koran et al. 2007). The doses used are typically low; evidence for benefit from higher doses is quite limited. Restricting neuroleptic doses to the low end of the usual range can help mitigate the side effects of these agents, for which tolerability is often a limiting factor (Koran et al. 2007).

It is important to note that this literature is not uniform, and only a minority of patients respond to neuroleptic augmentation. Indeed, a recent controlled trial of risperidone versus CBT for adults with OCD who had an inadequate response to SRI monotherapy found no benefit from risperidone augmentation, in contrast to clear benefit from CBT (Simpson et al. 2013). It is clinically advisable to monitor closely over the course of a trial of limited duration (4-–6 weeks), and to discontinue the trial if there is no clear benefit of improvement.

What underlies this heterogeneity of response? As in all such cases, our understanding is limited. An early study suggested that OCD patients with current or historical tic comorbidity have a higher rate of response to augmentation with the typical/first-generation neuroleptic haloperidol (McDougle et al. 1994). This benefit was borne out by later meta-analysis (Bloch et al. 2006). In contrast, meta-analysis of more recent studies of atypical/second-generation neuroleptics found no evidence for particular benefit in patients with comorbid tics (although a majority of the studies reviewed did not stratify by tic comorbidity and thus this sub-analysis was of limited power) (Veale et al. 2014). One possible explanation is that the high D2 affinity of haloperidol makes it particularly efficacious in patients with tic comorbidity; low-dose haloperidol is a well-established monotherapy for tic disorders (Bloch 2008). It is also possible that the 5-HT2A antagonism that characterizes atypical/second-generation neuroleptics makes them less effective than the typical agents in the treatment of OCD patients with comorbid tics. A meta-regression analysis found benefit of antipsychotics to be related to their D2R/D3R affinity and not to their 5-HT2A affinity, supporting the former explanation (Ducasse et al. 2014). More study is needed to place these speculations on a firmer foundation; based on the current data, it may be appropriate to turn to the typical agents in patients with comorbid tics and to the atypical agents (specifically, aripiprazole) in those without this comorbidity.

Patients with OCD who have severe disease, limited insight, and/or bizarre obsessions can be both diagnostically and therapeutically quite challenging (Hamblin et al. 2017; Poyurovsky 2017). It is intuitive to turn more quickly to neuroleptic augmentation in this context, in an effort to simultaneously target OCD and psychotic symptomatology. While this approach is clinically reasonable, it is important to recognize that it rests on a rather thin empirical basis. Most controlled studies have not stratified patients by schizotypy or level of insight. One that did, an early study of risperidone augmentation, did not find a higher rate of response in patients with OCD and comorbid schizotypy (McDougle et al. 2000). One study found insight to predict poorer response to pharmacotherapy (Catapano et al. 2001), but another found no relationship (Eisen et al. 2001). Studies of body dysmorphic disorder (BDD), an OCD-related disorder that is similarly responsive to high-dose SSRI pharmacotherapy but in which impaired insight to the point of delusionality is much more common, provide a useful parallel. BDD patients with delusional levels of insight responded just as well as those with better insight to fluoxetine monotherapy in a placebo-controlled trial (Phillips et al. 2002). The utility of neuroleptic augmentation in OCD with poor insight remains unclear.

5.3 Other Dopaminergic Agents

Only small studies have examined other agents that target the brain's dopamine system, and in no case is the evidence base sufficient for such agents to be considered standard-of-care. One small controlled study found benefit from both dextroamphetamine and caffeine in OCD, suggesting a potential role for stimulants in some patients (Koran et al. 2009). A recent larger controlled trial of fluvoxamine plus either extended-release methylphenidate or placebo found a higher response rate in the methylphenidate group (Zheng et al. 2019). However, both methylphenidate (Kouris 1998; Serby 2003; Woolley and Heyman 2003; Jhanda et al. 2016) and methamphetamine (Iyo et al. 1999) have also been reported to worsen OCD symptoms, so caution is warranted.

An open-label study of bupropion, which blocks both dopamine and norepinephrine reuptake, found no net benefit in OCD; some patients improved over 8 weeks of treatment, but others worsened (Vulink et al. 2005).

Interestingly, SSRIs at high doses can increase brain dopamine (Koch et al. 2002). Given that high SSRI doses are more efficacious in the treatment of OCD than typical antidepressant doses (Soomro et al. 2008), it is plausible that dopamine reuptake blockade could contribute to therapeutic benefit (Graat et al. 2017). This speculation has yet to be rigorously tested.

6 Dopamine Dysregulation in OCD

As in the case of serotonin, the efficacy of dopaminergic drugs – specifically, pharmacological augmentation with D2 antagonists – has motivated interest in the dopamine system in the brain of individuals with OCD. The evidence for dopamine dysregulation is much clearer for Tourette syndrome (TS) and other tic disorders (Williams et al. 2013) which are often comorbid with OCD and in tic-related OCD (Pittenger 2017b). In OCD without tics, evidence for DA dysregulation is ambiguous.

6.1 Animal Studies

Much of the evidence for a relationship between dopamine dysregulation and OCD-like phenomenology comes from animal models. The construction, evaluation, and interpretation of animal models of potential relevance to understanding the pathophysiology of tics and OCD is a complicated and sometimes vexing topic (Pittenger 2014; Bortolato and Pittenger 2017; Pittenger et al. 2017a), which is treated in detail in chapter "Inflammation, Obsessive-Compulsive Disorder, and Related Disorders". With that caveat, a range of dopamine agonists, precursors, reuptake blockers, and releasers – amphetamine; L-DOPA; bromocriptine; apomorphine – can lead to repetitive, stereotypic behaviors, which have been argued to be of relevance to the symptoms of OCD (Szechtman et al. 1998, Denys et al. 2004, Graat et al. 2017, Szechtman et al. 2020). However, dopaminergic agonists can also ameliorate such behaviors in some contexts. For example, repetitive checking behaviors are induced by the dopamine D2/D3 agonist quinpirole (Szechtman et al. 1998); but dopamine agonists can ameliorate apparently similar behaviors that occur naturally in some species (Vandebroek and Odberg 1997; Korff et al. 2008). It is of course possible that some repetitive behaviors in animals recapitulate the mechanisms underlying OCD in humans, and others do not. Such interpretative ambiguities, combined with the underlying complexity of the dopamine system – which comprises multiple circuits, receptors, and functional roles – make it difficult to draw any clear conclusions from this animal literature as to what role dopamine dysregulation may have in the development or maintenance of obsessions and compulsions in humans.

6.2 Genetics

As noted elsewhere (see chapters "The Future of Obsessive-Compulsive Spectrum Disorders: A Research Perspective" and "Genetic of OCD and Related Disorders"), the genetics of OCD are not mature; early candidate gene association studies have largely failed to replicate, and GWAS studies are not yet of sufficient size to yield large convincing, statistically robust findings (Fernandez et al. 2018). However, two findings of dopamine-related genes from candidate gene studies have been sufficiently well replicated as to merit discussion. Interestingly, both show sexually dimorphic effects, with evidence for a genetic contribution to OCD risk in males but not in females.

Catechol O-methyltransferase (COMT) metabolizes dopamine and other catecholamines, especially in the prefrontal cortex. A well-characterized common polymorphism, rs4680 (Val158Met), has been studied extensively in OCD. The Met allele, which is associated with lower enzymatic activity and thus presumably with higher levels of dopamine, is over-represented in males with OCD (Taylor 2013). Monoamine oxidase A (MAO-A) is the primary enzyme that metabolizes dopamine in the basal ganglia. At a well-characterized polymorphism, EcoRV (also called c1460C > T), the T allele appears to be over-represented in males with OCD (Taylor 2013). This allele is associated with lower enzymatic activity; again, therefore, carriers are predicted to have higher basal dopamine levels.

Several other alleles of dopamine-related genes examined in the same meta-analysis of candidate gene association studies showed suggestive relationships with OCD that failed to reach statistical significance; these include polymorphisms in the dopamine transporter gene *DAT1* and the dopamine receptor gene *DRD3* (Taylor 2013). All of these associations – including the reported male-specific associations with COMT and MAOA polymorphisms – must be interpreted with great caution; non-replications of these findings continue to appear (Sampaio et al. 2015). The field awaits better-powered studies to provide a comprehensive picture of the genetic architecture of OCD (Fernandez et al. 2018).

6.3 Receptor Imaging Studies

Dopamine binds to 5 distinct receptors, which fall into two classes: D1-like receptors (D1R and D5R) and D2-like receptors (D2R, D3R, and D4R). SPECT and PET imaging have been used to visualize these receptors. As in the case of serotonin receptors, reviewed above, studies to date have been small and findings somewhat variable, and it is difficult to draw specific conclusions with confidence.

Two studies using SPECT imaging reported elevation of the dopamine transporter in the striatum in OCD (Kim et al. 2003; van der Wee et al. 2004); however, a third study reported reduced levels of the transporter (Hesse et al. 2005). The dopamine D1R receptor has been imaged using PET in two small studies of unmedicated patients with OCD, with evidence for reduced binding in the striatum and the anterior cingulate cortex (Olver et al. 2009, 2010). A single study examined dopamine metabolism using the PET tracer ^{18}F-DOPA, which labels dopaminergic vesicles, in five patients compared to six controls; increases in ^{18}F-DOPA uptake were seen in several cortical structures, though most were only at trend level (Hsieh et al. 2014).

More studies have examined the D2R family of receptors in OCD and in tic disorders. An early SPECT study using the tracer iodobenzamide indicated lower D2 receptors in the left caudate (Denys et al. 2004), although a subsequent study found this effect only in patients with comorbid social anxiety disorder (Schneier et al. 2008). A pair of PET imaging studies with the D2R/D3R antagonist ^{11}C-raclopride similarly found reduced D2/D3 binding in the caudate-putamen (Perani et al. 2008; Denys et al. 2013), making this the best replicated evidence for a dopaminergic abnormality in OCD.

6.4 Dopamine Levels

Several of the findings reviewed above could be consistent with alterations in basal dopamine levels. As with serotonin, the most direct way to assay dopamine levels is to examine dopamine and its metabolites in cerebrospinal fluid, serum, or urine of patients with OCD. Here the literature is sparse and inconclusive. An early study found homovanillic acid (HVA), a primary metabolite of dopamine, to be no different in the CSF of patients with OCD than in controls, though it was elevated in Tourette syndrome (Leckman et al. 1995). A study in patients with normal pressure hydrocephalus found both elevated HVA and elevated obsessive-compulsive symptoms, but the two were not correlated with one another (Markianos et al. 2009). In small studies, HVA was reduced after treatment of patients with OCD with clomipramine (Altemus et al. 1994) or risperidone (Yoshimura et al. 2006). It is difficult to draw any overall conclusions from this heterogeneous literature.

The alleles of COMT and MAO-A that are associated with increased OCD risk, in males, have lower metabolic activity and are therefore consistent with elevated dopamine levels. And the reduced D1R and D2R receptor binding seen in several

PET studies, summarized above, may suggest increased basal dopamine levels, and thus greater competition between dopamine and the PET tracers.

A classic way of measuring dopamine is through PET imaging before and after challenge with amphetamine. By releasing large amounts of dopamine, amphetamine floods the receptors with transmitter, displacing the radioligand and thus reducing PET signal. Modeling of this reduced PET signal in a within-subject design allows inferences about baseline dopamine levels. A single study has used this approach in patients with OCD, compared to a group with Tourette syndrome, using the D2/D3 antagonist tracer raclopride. Baseline raclopride binding in the caudate-putamen was reduced in both OCD and TS; it was unchanged after amphetamine challenge. TS symptoms worsened after amphetamine challenge and correlated with changes in raclopride signal (and thus presumably with dopamine release) in the ventral striatum; no such relationships were seen in OCD (Denys et al. 2013). This study thus provides clearer support for dopaminergic pathophysiology in TS than OCD. This is however a single, small study; further work is needed to permit clear conclusions.

In summary, the evidence for dysregulated dopamine, dopamine homeostasis, or dopamine receptors in the pathophysiology of OCD is equivocal. It may be that dopamine dysregulation contributes to disease in some patients but not in others. Indeed, given the clearer evidence for striatal dopamine abnormalities in tic disorders (Williams et al. 2013), it may be that OCD with comorbid tics is characterized by dopaminergic pathology, while OCD without tics is not. However, this remains speculative; better powered studies that are more able to discriminate plausible pathophysiological subtypes are needed.

7 Glutamate Modulators in the Treatment of Refractory OCD

The evidence-based algorithm for the pharmacological treatment of OCD is distressingly limited: only the SRIs (clomipramine and the selective serotonin reuptake inhibitors) and augmentation thereof with certain neuroleptics are supported by sufficiently robust evidence to be considered the clear standard of care. Symptoms that are refractory to these interventions are unfortunately common. Investigational use of other agents, either as monotherapy or (more often) as augmentation of SRI treatment, therefore represents an important area of research. Some of these studies have been summarized above in the discussions of serotonin and dopamine-centered mechanisms.

Much recent work has focused on the hypothesis that glutamate imbalance contributes to OCD, and that glutamate modulators may have therapeutic utility in treatment of otherwise refractory disease (Pittenger et al. 2006, 2011; Pittenger 2015; Marinova et al. 2017). It is important to note that these two claims are not equivalent: glutamate modulators may be of therapeutic benefit even if glutamate dysregulation is not central to pathophysiology; and glutamate abnormalities might contribute causally to the development of OCD and yet glutamate modulators not be therapeutically effective. In this section we briefly summarize recent work on glutamate modulators; this material has been recently reviewed in more detail, with more specific clinical guidance, elsewhere (Pittenger 2015; Marinova et al. 2017; Pittenger 2017c).

7.1 *Memantine*

The most studied glutamate modulator for the treatment of refractory OCD is the NMDA receptor antagonist memantine. Memantine is a low-affinity open-pore blocker with a rapid off-rate; these characteristics may explain why its effects differ from those of a higher-affinity blocker, ketamine, which is further discussed below.

Early uncontrolled studies in both adults and children suggested that some OCD sufferers benefit from memantine augmentation (Aboujaoude et al. 2009; Feusner et al. 2009; Hezel et al. 2009; Stewart et al. 2010). More recently, a series of placebo-controlled trials has suggested a large benefit, with a response rate of 81% in memantine-treated patients (Ghaleiha et al. 2013; Haghighi et al. 2013; Sahraian et al. 2017; Modarresi et al. 2018). These studies have recently been summarized in a systematic review and meta-analysis, suggesting substantial benefit from memantine treatment in adults with OCD (Modarresi et al. 2019). However, this conclusion has been questioned on methodological grounds (Andrade 2019). The four placebo-controlled trials of memantine are individually small and were performed in the same geographic area, in Iran. They produced effects larger than those seen in open-label studies performed elsewhere, with response rates in individual studies of up to 100%, raising questions about their generalizability. Given these concerns, memantine cannot be considered to be an established treatment for OCD at this point; larger studies, from a wider geographical range and with optimally rigorous methodology, are needed (Andrade 2019).

7.2 *Riluzole*

Riluzole is a glutamate modulator developed for the treatment of amyotrophic lateral sclerosis. It has several mechanisms; at physiologically realistic concentrations, the two most significant appear to be a reduction in synaptic glutamate release and an enhancement of glial reuptake of perisynaptic glutamate (Pittenger et al. 2008a). An open-label trial (Coric et al. 2005), a retrospective case series (Pittenger et al. 2008b), and two small controlled studies (Pittenger et al. 2015; Emamzadehfard et al. 2016) suggest some benefit in adults, though only one of the controlled studies reached its primary efficacy endpoint (Emamzadehfard et al. 2016), and both were small and had methodological limitations. A controlled study in pediatric OCD showed no evidence of benefit (Grant et al. 2014). As in the case of memantine, more and higher-quality studies are needed before any firm recommendations can be made as to the use of riluzole in OCD refractory to other pharmacological treatments.

Recently, a riluzole prodrug, troriluzole, has been developed and tested for several indications. A large, multi-site controlled study of troriluzole augmentation in SRI-refractory OCD (NCT03299166) recently showed promising evidence of benefit, though it did not hit its primary efficacy endpoint (Biohaven Pharmaceuticals 2020).

7.3 Ketamine

The dissociative anesthetic drug ketamine has garnered much recent interest due to the observation that it can produce a rapid-onset, lasting antidepressant effect, sometimes after a single dose (Berman et al. 2000; Zarate Jr. et al. 2006; Krystal et al. 2019). In OCD, the literature is sparse and the benefit is less clear. A first open-label trial in treatment-refractory patients found a clear improvement in depressive symptoms but no clinically significant change in OCD symptoms after a single dose of ketamine (Bloch et al. 2012). However, a controlled trial in less ill patients (no significant comorbidities, no other medication) found benefit in many, lasting over a week (Rodriguez et al. 2011, 2013). Ongoing work has consisted mostly of case studies, investigating the combination of ketamine with other interventions such as memantine (Rodriguez et al. 2016a) or CBT (Rodriguez et al. 2016b; Adams et al. 2017). Clinical experience suggests that repeated ketamine treatments may be helpful in a minority of treatment-refractory OCD patients (Sharma et al. 2020); but repeated dosing has not been systematically studied in OCD. Again, large controlled studies are needed to more clearly establish the role of ketamine treatment, if any, in a comprehensive OCD treatment algorithm.

7.4 Other Glutamatergic Agents

A number of other agents that can modulate brain glutamate have been investigated in the treatment of refractory OCD (Pittenger 2015, Marinova et al. 2017). These include the anti-epileptic drugs topiramate (Berlin et al. 2011; Afshar et al. 2014; Sahraian et al. 2014) and lamotrigine (Kumar and Khanna 2000; Uzun 2010; Bruno et al. 2012; Arrojo-Romero et al. 2013; Hussain et al. 2015; Khalkhali et al. 2016; Naguy et al. 2016), the nutraceuticals N-acetylcysteine (Lafleur et al. 2006; Afshar et al. 2012; Yazici and Percinel 2015; Paydary et al. 2016; Costa et al. 2017; Ghanizadeh et al. 2017) and sarcosine (Wu et al. 2011), and the amino acid glycine (Greenberg et al. 2009). Of these, N-acetylcysteine is perhaps the most promising; its efficacy is supported by a recent meta-analysis (Gadallah et al. 2020) (see chapter "Animal Models for OCD Research" for a description of the anti-inflammatory effects of N-acetylcysteine in relation to OCD). For the other agents there are some promising studies, but others that are equivocal, and there have been no large, placebo-controlled studies of sufficient quality to produce definitive treatment recommendations.

An interesting agent in this category is D-cycloserine (DCS), a partial agonist of the glycine site on the NMDA receptor. DCS has been found in animal studies to potentiate NMDA-dependent plasticity and NMDA-dependent extinction learning (Richardson et al. 2004; Kaplan and Moore 2011). This has motivated investigations of whether DCS pretreatment can potentiate the response to extinction-based psychotherapies in OCD, and in other disorders in which dysregulated anxiety is prominent. This is a complex literature, and beyond our scope to review in detail here (see chapter "Innovations in the Delivery of Exposure and Response Prevention for Obsessive-Compulsive Disorder" for a review of the use of DCS to augment CBT for OCD). Recent meta-analyses suggest no significant benefit (Burkner et al. 2017) or only subtle benefits (Mataix-Cols et al. 2017) from DCS augmentation of CBT in OCD. However, studies have differed enormously in the dose of DCS and, critically, in the temporal relationship of DCS dosage with CBT. In appropriately selected patients and with optimal dosing, DCS may be of some clinical benefit; but studies to date do not allow clear specification of which patients these might be, or of the optimal dosage and timing to produce clinical benefit. For these reasons, despite the theoretical attractiveness of the concept of pharmacologically augmenting effective behavioral treatments, DCS has not entered widespread clinical use.

8 Glutamate Dysregulation in the Pathophysiology of OCD

Several lines of evidence suggest that glutamate dysregulation may contribute to the pathophysiology of OCD. None of these is definitive – as in the case of serotonin and dopamine contributions to pathophysiology, reviewed above, studies are mixed and our understanding remains woefully incomplete. Nevertheless, a brief review of this literature helps to frame potentially fruitful future avenues for intervention development. More detailed descriptions of the glutamate system and how it may be perturbed in patients with OCD can be found elsewhere (Ting and Feng 2008; Pittenger et al. 2011; Karthik et al. 2020).

It is important to recognize in framing this discussion the ways that the function and organization of glutamatergic neurotransmission in the mammalian brain differs from that of the serotonin and dopamine systems. Serotonin and dopamine are modulatory neurotransmitters. In both cases, a relatively small number of neurons located in a specific brain area (the dorsal raphe nuclei for serotonin; the substantia nigra/ventral tegmental area for dopamine) project widely throughout the brain and modulate the function of other circuitries. It is reasonable to think of these modulatory transmitters as being involved in a small set of discrete processes. It is thus coherent, though doubtless a huge simplification, to think of excess or deficiency of these transmitters as a cause of disease or as an important aspect of pathophysiology.

Glutamate, in contrast, is the primary excitatory neurotransmitter in the mammalian brain. It is used as a neurotransmitter by over half of the neurons and in every structure and circuit in the central nervous system, and thus it is involved in virtually everything the brain does. A global excess or deficiency in glutamate would be likely to have dramatic, widespread effects. Conversely, pathological activity of virtually any circuit in the brain could be reflected in a local disruption of glutamate levels.

8.1 Glutamate-Related Genes in OCD

Much attention has focused on the possible role of polymorphisms in glutamate-related genes as determinants of OCD risk (see chapters "The Future of Obsessive-Compulsive Spectrum Disorders: A Research Perspective", "Genetics of OCD and Related Disorders; Searching for Shared Factors", and "Invasive and Non-invasive Neurostimulation for OCD" for a more detailed review). An early focus was the glutamate transporter EAAT3, encoded by the *SLC1A1* gene (Escobar et al. 2019). The *SLC1A1* region was initially identified as an OCD risk locus in two linkage studies (Hanna et al. 2002; Willour et al. 2004); reports of several different polymorphisms in the gene itself in patients OCD accumulated over the next several years (Escobar et al. 2019). The fact that *SLC1A1* arose as a risk gene from genome-wide linkage studies, rather than a specific a priori hypothesis about pathophysiology, makes these association studies somewhat more compelling than most studies of serotonergic and glutamatergic genes (Taylor 2013, 2016). OCD-associated alleles have been reported to modulate *SLC1A1* expression levels (Wendland et al. 2009). However, distinct alleles have been identified in different studies, weakening the statistical association of any one allele with OCD; perhaps as a result, the association is not supported by meta-analysis (Stewart et al. 2013a). Furthermore, *SLC1A1* has not emerged as a risk locus in more recent genome-wide association studies (Stewart et al. 2013a; IOCDF-GC and OCGAS 2018). The association is thus equivocal; *SLC1A1* alleles may contribute to risk in some cases and some contexts, but more work is needed to confirm the role of this gene and to clarify the details of its role.

As noted above, the two GWAS studies completed in OCD to date have not identified any disease-associated loci that reached genome-wide statistical significance (Stewart et al. 2013b; Mattheisen et al. 2015; IOCDF-GC and OCGAS 2018). However, examination of loci that approach significance allows some tentative conclusions about the genetic structure of OCD risk, even if individual risk alleles remain elusive. Glutamate-related genes are strikingly over-represented amongst these borderline-significant alleles. One of the strongest associations is with the *GRID2* gene, an orphan receptor with significant homology to ionotropic glutamatergic receptors. A slightly less strong association is with *GRIK2*, a subunit of the kainate receptor class of ionotropic glutamate receptors. Finally, the *DLGAP1* gene, which is involved in the structural organization of the postsynaptic glutamate receptors and associated proteins, may be associated with OCD risk (IOCDF-GC and OCGAS 2018).

Again, none of these alleles have reached genome-wide statistical significance, and thus more data are needed before any of these associations can be considered definitive. Nevertheless, the prominent representation of glutamate-related genes amongst probable risk alleles is striking. Since genetic contributors to pathophysiology are necessarily causes, at the level of a single organism, these associations – if replicated – constitute suggestive evidence that glutamate dysregulation may contribute to OCD pathophysiology, in at least some cases.

8.2 Glutamate Receptors and Related Proteins in OCD

No PET or SPECT imaging studies of glutamate receptors in OCD have been reported. This is in contrast to the literature on dopamine and serotonin reviewed above.

A recent post-mortem study, on the other hand, focused on glutamate receptors and other glutamate-related transcripts in the orbitofrontal cortex and basal ganglia (also see chapter "Animal Models for OCD Research") (Piantadosi et al. 2019). RNA for the gene *DLGAP1* and the related gene *DLGAP2*, also involved in the organization of the glutamatergic synapse, were both reduced in orbitofrontal cortex. So was *SLC1A1*. The total number of subjects in this study remains low (8 OCD, 8 controls), and more work is needed to confirm and extend these findings, but the preliminary work is supportive of a role for glutamate dysregulation in the pathophysiology of OCD, especially in the orbitofrontal cortex.

8.3 CSF and Brain Glutamate Levels

Two studies, from a single group, have examined glutamate levels in spinal fluid in unmedicated adults with OCD. Both found CSF glutamate to be elevated in a subset of patients (Chakrabarty et al. 2005; Bhattacharyya et al. 2009). This is a striking finding. Because glutamate is involved in so many circuits and processes, and because excess glutamate is toxic to neurons, glutamate homeostasis is tightly regulated. Spillover of enough glutamate to be detectable in lumbosacral CSF suggests a significant disruption of these homeostatic mechanisms. However, this elevation is seen in only a minority of OCD patients; most have CSF glutamate levels indistinguishable from healthy controls. It remains to be established whether those with elevated glutamate represent a subset with distinct pathophysiology.

Brain glutamate levels in patients with OCD have been measured using magnetic resonance spectroscopy (MRS), which takes advantage of the fact that the electromagnetic environment of protons in the glutamate molecule creates a unique magnetic resonance signature (Brennan et al. 2013). Most studies measure a composite of glutamate and the related amino acid glutamine, which are difficult to disambiguate in the MRS signal using standard techniques. Importantly, MRS measures total tissue glutamate, including intracellular and metabolic pools. This differs from other techniques, such as CSF sampling and inferences from PET imaging, which measure extracellular pools of transmitter.

Early MRS studies suggested that children with OCD have elevated glutamate/glutamine levels in the striatum (Rosenberg et al. 2000) and reduced glutamate/glutamine in the anterior cingulate cortex (Rosenberg et al. 2004) (for a review of MRS studies in OCD, please see chapter "Recent Developments in the Habit Hypothesis of OCD and Compulsive Disorders"). Caudate glutamate/glutamine levels normalized after pharmacotherapy with paroxetine (Rosenberg et al. 2000). These findings spurred great interest and a large number of follow-up MRS studies in both children and adults – indeed, these early studies were largely responsible for kindling interest in glutamate dysregulation as a potential pathophysiological mechanism in OCD (Rosenberg et al. 2001). However, subsequent findings have been equivocal, with many more negative studies than positive ones (Brennan et al. 2013). It remains unclear whether this is because glutamate dysregulation observable by MRS is unique to pediatric OCD (most subsequent studies have been in adults), is seen in only a subset of patients, is sensitively dependent on technical variables that differ between studies, or was a false positive in the early work.

One recent study has combined MRS measures of brain glutamate with prediction of treatment response, with intriguing early results (O'Neill et al. 2017). O'neill and colleagues found MRS-measured glutamate in the pregenual anterior cingulate cortex to decline after cognitive behavioral therapy in children with OCD. On the other hand, baseline glutamate in the ventral posterior cingulate cortex predicted treatment response (O'Neill et al. 2017). If this latter finding proves to be robust, such a baseline predictor of response may help stratify patients to individualized treatment protocols, moving us towards precision medicine in psychiatry – a topic to which we return below.

8.4 Animal Studies

A number of animal studies of the circuitry thought to be perturbed in OCD have implicated glutamatergic mechanisms. We review this work only briefly; it has been summarized in more detail elsewhere (see chapter "Inflammation, Obsessive-Compulsive Disorder, and Related Disorders") (Pittenger et al. 2011; Pittenger 2014; Ahmari and Dougherty 2015; Ahmari 2016; Szechtman et al. 2017). As noted above and reviewed in detail in the chapter "Animal Models for OCD Research," the use of animal models to shed light on the pathophysiology of OCD, or of any complex neuropsychiatric disease, raises complex interpretative issues (Pittenger 2014; Pittenger et al. 2017a; Bortolato and Pittenger 2017). Furthermore, since glutamate is involved in virtually every brain circuit, it is hardly surprising that circuit manipulations in animals can measurably perturb measures of glutamate function, and that alterations of glutamatergic neurotransmission may produce circuit abnormalities. Nevertheless, the convergence of several lines of animal work on the conclusion that glutamatergic pathophysiology within cortico-striatal circuits can lead to repetitive behaviors merits brief discussion in this chapter as well.

One of the best-studied animal models of repetitive behavior is the *Sapap3* knockout mouse, which has been interpreted as a mouse model of OCD (Welch et al. 2007) (see also chapter "Inflammation, Obsessive-Compulsive Disorder, and Related Disorders"). The mouse SAPAP proteins are orthologs of human DLGAPs and are important constituents of the postsynaptic protein network that coordinates and stabilizes glutamate receptors at synapse. As noted above, *DLGAP1* and *DLGAP2* have been tentatively implicated in initial genetic (IOCDF-GC and OCGAS 2018) and post-mortem (Piantadosi et al. 2019) studies of OCD, which renders a connection between SAPAP disruption in mice and OCD-relevant abnormalities intrinsically plausible. *Sapap3* knockout mice exhibit repetitive behavior and anxiety, both of which are responsive to treatment with fluoxetine, and abnormalities in glutamatergic synapses in the cortico-striatal circuitry (Welch et al. 2007).

Other targeted disruptions of glutamate-related genes have similarly produced repetitive behaviors in mice. For example, targeted disruption of the *Slc1a1* gene in mice reduced repetitive stereotypical behaviors after amphetamine in mice (Zike et al. 2017), while overexpression of *Slc1a1* leads to anxiety and behavioral stereotypy (Delgado-Acevedo et al. 2019), supporting a role for appropriate dosing of this gene in normal maintenance of behavioral flexibility. Mutation of the neural signaling molecule *Slitrk5*, which regulates the maturation of glutamatergic synapses (among other functions), produces an excessive grooming phenotype quite similar to that seen in the *Sapap3* knockout mice, reinforcing the conclusion that abnormal glutamatergic synapses can produce repetitive behaviors (Shmelkov et al. 2010).

In sum, multiple convergent lines of evidence suggest that glutamate dysregulation may contribute to the development of OCD and OCD-relevant brain circuit abnormalities. In conjunction with accumulating evidence that glutamate modulators can mitigate otherwise refractory OCD symptoms (Pittenger 2015; Marinova et al. 2017), this suggests that targeting glutamate neurotransmission may be a fruitful avenue forward for the development of new pharmacological treatment strategies.

9 Towards Personalized Medicine in the Treatment of OCD

In OCD, as in most of psychiatry, we have a collection of treatments – SRI medications; neuroleptic augmentation; specialized cognitive-behavioral therapy – that are helpful for many patients, but ineffective for many others. An important goal is the development of predictors and algorithms to determine, in advance, which patients are most likely to respond to which interventions, thus sparing them weeks or months spent in trials of treatments that ultimately prove to be ineffective. This goal of precision or personalized medicine has been identified by the U.S. National Institute of Mental Health as a key research priority (NIMH 2020, Goal 3.2).

We are far from achieving this goal in practice (Hazari et al. 2016). However, several studies have made initial steps. This final section reviews these efforts at prediction of individual treatment outcomes.

9.1 *Phenomenological, Symptomatic, and Cognitive Predictors of Treatment Response*

The most clinically useful predictor of treatment response would be one that is apparent on clinical evaluation, without the need for specialized testing.

Symptom content has been investigated as a predictor of treatment response in a number of studies. Several findings emerge from this literature. Hoarding symptoms, which were considered a part of OCD prior to the publication of DSM-5 in 2013 (American Psychiatric Association. and American Psychiatric Association. DSM-5 Task Force. 2013), predict poor response to both psychotherapy and pharmacotherapy in most studies, and in meta-analysis (Bloch et al. 2014). This poorer response may be mitigated by the deployment of more recently developed therapeutic strategies specifically targeting hoarding symptoms; this is not yet clear. Among the major dimensions of OCD symptomatology (Bloch et al. 2008a, b), somatic (Erzegovesi et al. 2001) and sexual/religious obsessions and compulsions (Ferrao et al. 2006) have been associated with poor response to pharmacotherapy. However, another study found good response to SSRI treatment among those with sexual and religious and, especially, harm-related obsessions and compulsions (Landeros-Weisenberger et al. 2010), while yet another found washing rituals to be predictive of poor outcome (Ravizza et al. 1995). This inconsistency is striking. Symptom subtype does not appear to be a strong predictor of response to pharmacotherapy in OCD.

One early study found that high scores on particular component scores on the Yale-Brown Obsessive Compulsive Scale (Y-BOCS) (Goodman et al. 1989b, c), the most widely used instrument for rating of OCD symptom severity, predicted good response to clomipramine. Specifically, clomipramine response was predicted by high scores on item #3 (distress caused by obsessions) and item #8 (distress caused if a patient is prevented from engaging in a compulsion) (Ackerman et al. 1996a, b). This suggests that pharmacotherapy, at least with clomipramine, is most effective in treating distress associated with OCD, and may be less effective in treating OCD symptoms that are more automatic or integrated into normal patterns of behavior. This speculation requires more investigation.

Some studies suggest that OCD with comorbid tics is less likely to respond to SSRI monotherapy (McDougle et al. 1993). As noted above, early studies of neuroleptic augmentation suggested that haloperidol is of greater benefit in patients with comorbid tics (McDougle et al. 1994). While this finding has not been robustly replicated, it remains significant in meta-analysis (Bloch et al. 2006). This may guide a prescriber towards haloperidol augmentation in patients with significant comorbid tics. More robust studies are needed to confirm this practice and to quantify what the differential benefit in patients with tics may be.

Other baseline clinical characteristics have been reported to predict poor SRI response. These include:

- early age of onset (Ravizza et al. 1995; Ackerman et al. 1996a, b; Erzegovesi et al. 2001);
- duration and chronicity of illness (DeVeaugh-Geiss et al. 1990; Ravizza et al. 1995; Ferrao et al. 2006; Stein et al. 2007);
- severity of symptoms (Mataix-Cols et al. 2002; Denys et al. 2003; Ferrao et al. 2006; Storch et al. 2006; Stein et al. 2007);
- comorbid depression (Ackerman et al. 1996a, b) or OCPD (Cavedini et al. 1997);
- poor insight (Catapano et al. 2001; Erzegovesi et al. 2001) [though see (Eisen et al. 2001) for a contrary finding]; and,
- previous failure to respond to psychopharmacology (Ackerman et al. 1998).

It should be noted, however, that most of these reported effects are small. Furthermore, the ability to identify patients who are less likely to respond to treatment is not of great practical use, in the absence of guidance as to what they *are* likely to respond to.

SSRI trials in OCD are typically lengthy. Characteristics of early response that predict long-term benefit could be of great utility if they permitted clinical decisions (e.g., to continue the trial, increase the dose, or switch strategies) to be made more quickly. As discussed above, there is increasing evidence that early clinical response predicts ultimate benefit with reasonable accuracy (da Conceicao Costa et al. 2013; Issari et al. 2016). There is also evidence, from trials of both clomipramine (Ackerman et al. 1996a, b) and fluoxetine (Ackerman et al. 1998) that side effects early in a drug trial may predict ultimate clinical benefit. This may be because some side effects reflect individual sensitivity to the serotonergic effects of these drugs, which ultimately contribute to therapeutic response.

Cognitive measures have been investigated as predictors of response to both CBT and SSRIs. In a study of 38 adults randomized to receive either CBT or SSRI treatment, verbal IQ, verbal memory, and performance on the Stroop color-word conflict task predicted better response to both psychotherapy and pharmacotherapy (D'Alcante et al. 2012). These may be nonspecific markers of better overall neuropsychological status. Cognitive flexibility is impaired in OCD (Gruner and Pittenger 2017) (see also chapter "Neurocognitive Endophenotypes of OCD"); preserved cognitive flexibility (operationalized as perseveration on the California Verbal Learning Test) predicted better response to CBT, but worse response to SSRIs (D'Alcante et al. 2012). Another small study produced a consistent finding: impaired performance on the Wisconsin Card Sort task, a different measure of cognitive flexibility, predicted good response to pharmacotherapy (Fontenelle et al. 2001). A more recent study found an association between executive function scores and response to CBT in children (Hybel et al. 2017). These studies suggest that measures of cognitive flexibility might be able to provide the sort of guidance a clinician needs: preserved cognitive flexibility may indicate that CBT is appropriate, while impaired cognitive flexibility may steer treatment towards pharmacotherapy. This guidance is not yet ready for broad clinical use, as the studies are small, measures of cognitive flexibility vary and are not readily deployed in clinical practice, and the effect sizes and predictive value of measures of cognitive inflexibility are not established. Nevertheless, these studies suggest that sort of approach that might inform treatment algorithms in the future.

9.2 Genetic Predictors of Treatment Response and Side Effects

With rapid advances in clinical genetics of the past two decades, a persistent hope is that genetic testing will assist with treatment decisions in complex disorders such as OCD (see also chapter "Invasive and Non-invasive Neurostimulation for OCD"). Indeed, a number of companies will provide patients with genetic testing, and use it to derive specific treatment recommendations. Unfortunately, these recommendations rest on a thin evidence base and are rarely useful to guide treatment decisions, at this point. The evidence-based integration of genetic information into OCD treatment algorithms awaits further refinement of our understanding of the genetic architecture of the disorder.

Nevertheless, a few studies have examined genetic predictors of treatment response in OCD, and these merit brief mention here. As in the discussion of the genetics of serotonin, dopamine, and glutamate reviewed above, it is important to note that early candidate gene studies have been plagued by non-replication; GWAS and other studies that interrogate the whole genome are needed to identify genetic predictors of treatment response with confidence.

Candidate gene studies have focused on the CYP enzymes of the liver, which are responsible for the first-pass metabolism of many therapeutic drugs, and on some of the genes in the serotonin, dopamine, and glutamate systems (Brandl et al. 2012; Zai et al. 2014). Suggestive results have been reported for the serotonin transporter (*SLC6A4*), the 5-HT1B and 5-HT2A receptors, and the EAAT3 glutamate receptor encoded by *SLC1A1* (Brandl et al. 2012). Targeted testing of CYP enzymes suggests an association between CYP2D6 metabolizer status and the likelihood of nonresponse to SSRIs, as well as the incidence of side effects after venlafaxine treatment (Brandl et al. 2014). However, replication is required before any of these reports can be interpreted with confidence or deployed in clinical practice.

A single study suggests that alternate alleles of *SLC6A4*, the serotonin transporter gene, differentially predict response to paroxetine versus venlafaxine (Denys et al. 2007). This result, if replicated, is the sort of finding that may have clinical utility, as it could steer clinical decision making (i.e., choice of drug). However, it should be noted that the evidence supporting venlafaxine as a first-line agent in the treatment of OCD is equivocal, as reviewed above. A genetic predictor of response to venlafaxine would have to be quite strong to justify using it in preference to an SSRI, when the latter are so much better supported by the treatment literature.

A single genome-wide association study has examined genetic predictors of self-reported response to pharmacotherapy in adult OCD (Qin et al. 2016). Top hits in this analysis included SNPs near the genes *DISP1* and *PCDH10*, neither of which leads immediately to a mechanistic model. Pathway analysis of all possible and probable genetic loci suggested involvement of the glutamatergic and serotonergic systems. Much more work of this type is needed if we are to move towards actionable genetic predictors of treatment response.

9.3 Structural and Functional Neuroimaging Predictors of Treatment Response

A number of studies have examined structural and functional brain imaging measures as predictors of treatment response. While this work, too, has not converged on clinically actionable insights, it may provide initial clues as to what such imaging-informed treatment algorithms might look like in the future.

Structural neuroimaging has been examined as a predictor of treatment response in a few small studies. Attention has generally been focused on the cortico-striatal loops implicated in OCD by functional neuroimaging studies (Brennan and Rauch 2017). One comparative study found gray matter density (measured using voxel based morphometry, or VBM) in the orbitofrontal cortex (OFC) to predict response to pharmacotherapy, while VBM in the medial prefrontal cortex predicted response to psychotherapy (Hoexter et al. 2013). A follow-up analysis from the same group found thickness of the orbitofrontal cortex to predict treatment response in two small cohorts of patients (Hoexter et al. 2015). A more recent study examined cortical morphology across the whole brain using structural covariance analysis and identified a network of structures in which cortical surface area and/or thickness predicted response to pharmacotherapy with 89% accuracy (Yun et al. 2015). This supports the potential of brain structural analyses to predict treatment response; but much more work is needed before such approaches can be used clinically.

More studies have examined measures of brain function as predictors of treatment response. Again, the focus has largely been on the frontal cortico-striatal circuitry; early PET studies of brain metabolism have converged on the importance of the OFC. The first such study found lower right OFC metabolism to predict response to clomipramine at the end of 2 months of treatment (Swedo et al. 1989), though not at 1 year (Swedo et al. 1992). In a subsequent study, higher metabolic activity in the left OFC was found to predict better response to behavioral therapy, but worse response to fluoxetine (Brody et al. 1998). A later study from the same group found concordant results: lower baseline OFC metabolism (bilaterally) was associated with better response to paroxetine (Saxena et al. 1999). Finally, in yet another study, lower baseline metabolism in the OFC and higher metabolism in the posterior cingulate cortex (PCC) predicted response to paroxetine (Rauch et al. 2002). The association of low relative metabolism in the OFC as a predictor of response to pharmacotherapy appears robust across these studies.

A pair of studies have found similar relationships with activity of the caudate nucleus, another node within the cortico-striatal circuitry canonically implicated in OCD (Brennan and Rauch 2017). In one, using PET imaging at baseline, higher pretreatment metabolism in the right caudate was found to predict response to paroxetine in patients with OCD; there was no such relationship in patients with major depressive disorder, suggesting some specificity to the effect (Saxena et al. 2003). A study using SPECT imaging found similar results: perfusion in the right caudate, as well as in the dorsal anterior cingulate, predicted response to sertraline in OCD (Hendler et al. 2003). In contrast, another study using SPECT found baseline perfusion in the right cerebellum, as well as in the brain as a whole, to be predictive of response to fluvoxamine (Ho Pian et al. 2005).

Interestingly, the two nodes most robustly identified as predictors of response to treatment using metabolic imaging – the OFC and the caudate – are also the brain regions whose activity has most consistently been shown to be modulated by treatment (van der Straten et al. 2017). This concordance increases confidence that these relationships are meaningful, despite the fact that most individual studies have been small.

Compared to PET, fMRI is more widely available, cheaper, and less invasive; it thus has greater theoretical potential for broad clinical use. On the other hand, fMRI typically has a lower signal-to-noise than PET and generally produces relative measures, rather than absolute ones, with sensitive dependence on the behavioral task used and other details of acquisition that may complicate clinical deployment. A single study has examined brain activation during symptom provocation as a predictor of response to 12 weeks of fluvoxamine treatment (Sanematsu et al. 2010). Activation in the right cerebellum and left superior temporal gyrus – not regions that figure prominently in the larger OCD neuroimaging literature – predicted response.

fMRI can also be used to make inferences about brain functional connectivity and network organization. This analysis can be performed on data collected while subjects are at rest, rather than engaged in a task, which reduces the complexity of data collection and may make such approaches more appropriate for clinical deployment in the future. Several recent studies have applied this approach to baseline characterization in OCD (e.g., Harrison et al. 2009, 2013; Anticevic et al. 2014; Shin et al. 2014). Others have used this approach to investigate predictors of treatment response. fMRI-derived functional connectivity at rest between the dorsal raphe nuclei and the left medial temporal gyrus has been reported to be elevated in OCD, and the degree of elevation predicts nonresponse to SSRI treatment (Kim et al. 2019).

Several studies have examined baseline resting-state predictors of response to CBT. Low functional connectivity between the basolateral amygdala and the ventromedial prefrontal cortex, both structures known to be involved in OCD and in emotion regulation more generally, was found to predict response to CBT in a large sample of adults with OCD (Fullana et al. 2017). Using a brain-wide machine learning-based analysis of functional connectivity, another study found connectivity patterns within the brain's default mode network to predict CBT response (Reggente et al. 2018). Functional connectivity analyses have also been used to investigate predictors of response to surgical treatment of refractory OCD (Yin et al. 2018) and to an experimental neurofeedback approach (Scheinost et al. 2014).

9.4 Neurochemical Predictors of Treatment Response

Finally, a number of studies have examined neurochemical measures as potential predictors of response. The focus has again been on serotonin, dopamine, and glutamate.

Measurement of plasma or platelet serotonin provides one accessible assay of serotonergic function, though its relationship with brain serotonin levels and homeostasis is not fully understood. As reviewed above, a number of studies have found platelet expression of the serotonin transporter, 5-HTT, to be reduced in patients with OCD. In one of these, higher whole blood serotonin concentration was found to predict response to SRI treatment (Delorme et al. 2004). Plasma serotonin levels decline with treatment; interestingly, another study found that rapid decline in plasma serotonin was associated with poor response to clomipramine or paroxetine, while a more gradual decline predicted clinical improvement (Humble et al. 2001).

Finally, MRS measurement of glutamate and related compounds in the posterior cingulate cortex has been found in a single study to predict response to CBT in children, as reviewed above (O'Neill et al. 2017). MRS is an attractive modality for such studies, as it is non-invasive and can be targeted to specific brain areas, potentially providing more relevant measures than (for example) peripheral blood. More studies of this sort are needed.

In sum, many different approaches have been used to identify potential predictors of both pharmacotherapy and psychotherapy in OCD, and there are a number of promising leads. The most robustly replicated predictors of poor response appear to be the presence of comorbid hoarding symptoms (Bloch et al. 2014) and the baseline metabolic activity of the orbitofrontal cortex; most other measures reviewed are more equivocal, or are supported by only a few studies. More work is needed to develop these predictors to the point that they can be deployed in widespread clinical practice. Importantly, prediction of which patients are likely to respond poorly is of limited utility; ultimately, we need measures that will help the clinician decide between available treatments. Algorithms based on such measures could bring relief more quickly than the trial-and-error method that is typically used today and would represent a significant step towards the long-term goal of personalized medicine in psychiatry.

10 Conclusion

Pharmacotherapeutic options for the treatment of OCD are distressingly limited. Clear evidence supports the use of serotonin reuptake inhibitors – the selective serotonin reuptake inhibitors (SSRIs) and the older drug clomipramine – as monotherapy (Koran et al. 2007; Soomro et al. 2008; Issari et al. 2016; Skapinakis et al. 2016). Somewhat weaker but still convincing evidence supports the use of neuroleptics, especially haloperidol, risperidone, and aripiprazole, as augmentation of stable SRI treatment when symptom response is incomplete (Bloch et al. 2006; Skapinakis et al. 2007; Veale et al. 2014). But these strategies, together with expert cognitive-behavioral psychotherapy, leave many patients – perhaps as many as a third – with significant ongoing symptoms and attendant suffering. The need for new treatment options is clear.

In this chapter we have reviewed the evidence behind both standard-of-care and less well-established pharmacological treatment strategies. We have focused on three neurotransmitter systems in the brain: serotonin, dopamine, and glutamate. This focus is unavoidably myopic. These neurotransmitter systems do not operate in isolation, either in the course of normal brain function or during the response to pharmacotherapy. Rather, they interact with one another, both developmentally and in the adult, to modulate complex brain circuitries whose function and dysfunction underpin both normal behavior and the symptoms and course of neuropsychiatric disease. Furthermore, even keeping this heuristic focus on specific neurochemical systems, OCD surely involves other aspects of brain function: GABA (Simpson et al. 2012); histamine (Castellan Baldan et al. 2014; Pittenger 2017a); neuropeptides (McDougle et al. 1999; Marroni et al. 2007; Dietrich et al. 2017); and others. However, more treatment and neurobiological research has focused on serotonin, dopamine, and glutamate than on any other neurochemical systems. They are thus an appropriate place to start a consideration of the current state of pharmacological treatment of OCD, and possible ways forward.

Several conclusions arise from this review.

First, OCD is an extremely complex and heterogeneous disease entity. This complicates the formulation of any general model of how it develops, manifests, and is treated (Pittenger et al. 2017b; Szechtman et al. 2020). As our understanding matures, we are likely to appreciate that two patients with a diagnosis of OCD may have quite different underlying pathophysiology. Indeed, some hints of possible subdivisions of the diagnosis have begun to emerge. Tic-related OCD may be genetically distinct (Davis et al. 2013) and may be characterized by more dopaminergic pathophysiology than non-tic associated OCD (Pittenger 2017b), which may motivate earlier use of augmentation with high-affinity D2 antagonist such as haloperidol. Glutamate dysregulation may characterize only a subset of patients (Chakrabarty et al. 2005; Bhattacharyya et al. 2009) and may prove to predict treatment response (O'Neill et al. 2017). Future study is likely to refine these distinctions and identify new ones, complexifying our conception of OCD as a diagnosis but also placing it on a firmer foundation. As we better understand the heterogeneity of OCD and how it may predict treatment response, algorithms for personalized medicine will become all the more important.

Second, much larger and more careful studies of novel agents for the treatment of refractory disease are needed. In many cases initial work suggests the possibility of therapeutic benefit, but without better-powered investigations treatment guidelines cannot be formulated with confidence. The case of ondansetron is an informative cautionary tale: several small controlled studies suggested that it led to improvement in patients with OCD (Serata et al. 2015), but a better-powered, more definitive study did not show benefit (NCT01275248). It remains to be seen whether initial promising studies with such agents as memantine (Andrade 2019; Modarresi et al. 2019) and riluzole (Pittenger et al. 2008a, b, 2015; Emamzadehfard et al. 2016) will replicate in well-powered studies, which are desperately needed.

Third, our understanding of the neurobiological underpinnings of OCD remains woefully incomplete. Throughout this review we have been forced over and over again to repeat the mantra "more work is needed." Often small studies suggest a neurobiological insight, but subsequent work fails to replicate it, or at least suggests greater complexity. For example, PET studies of the 5-HT2A receptor have shown it to be increased (Adams et al. 2005), reduced (Perani et al. 2008), or unchanged (Simpson et al. 2011) in patients with OCD. Early MRS studies suggested an important contribution of dysregulated brain glutamate in patients (Rosenberg et al. 2000, 2004), but follow-up studies have been largely negative (Brennan et al. 2013). Early genetic findings have hinted at potentially important new insights, such as the implication of mutations in *Slc1a1* in OCD (Hanna et al. 2002, Willour et al. 2004), but follow-up studies have been heterogeneous and more equivocal (Stewart et al. 2013a, b).

Lack of ready replication of the simple conclusions drawn by early studies may mean that the early work was incorrect, or that technical differences between studies are muddying the waters. But this heterogeneity of results may simply be a reflection of the complexity of the underlying condition. For example, if glutamate dysregulation characterizes only a fraction of patients with OCD, as suggested by CSF measures (Chakrabarty et al. 2005, Bhattacharyya et al. 2009), then it is to be expected that measures of glutamate function will be variable depending on the chance composition of the specific population being examined, especially in small studies. Much larger, rigorous studies can simultaneously yield reproducible results in which we can have confidence – most obviously in the case of genetics, but also in imaging and other neurobiological modalities. Insights from such well-powered descriptive studies may pave the way for the identification of more homogeneous groups of patients for future investigations of treatment, making it easier to detect meaningful signal in modestly sized therapeutic trials.

Fourth, pathophysiology and treatment may not map cleanly onto one another. SRI medications are the best pharmacological treatment we have for OCD, and the evidence for the efficacy is unequivocal (Soomro et al. 2008, Issari et al. 2016, Skapinakis et al. 2016). But neuroimaging (Pogarell et al. 2003, Simpson et al. 2003) and genetic evidence (Taylor 2013, 2016) for dysregulation of the serotonin transporter, the target of SRI medications, as a key contributor to OCD pathophysiology is decidedly weak. Given the inadequacy of our current understanding of the pathophysiology of OCD, it is perhaps fortunate that the development of new treatments need not wait for more complete knowledge and rational design, which may be long in coming.

From where are the next advances in the treatment of OCD likely to come? It remains unclear, and any predictions are fraught with uncertainty. The development of more robust predictors of response to currently available interventions may help to optimize treatment algorithms. It is important that these studies focus on predictors that inform treatment decisions, rather than simply predict higher or lower likelihood of response to a single intervention, and that they move beyond the fairly small proof of concept studies to better powered, more generalizable investigations. As to new agents, the glutamate modulators memantine and riluzole (and troriluzole; NCT03299166) are perhaps the most promising agents currently under investigation. Dopamine dysregulation in OCD is not particularly well understood, and only the typical and atypical neuroleptics have been investigated with any rigor; there may be room here for exploration of other agents. Brain stimulation techniques such as transcranial magnetic stimulation (Carmi et al. 2019) show promise, though these lie beyond the primary focus of the current review. Finally, the use of pharmacological or other somatic manipulations to enhance the efficacy of CBT holds great theoretical promise, as it may allow us to take advantage of both the potency of neurobiological interventions and the individualized specificity of psychotherapy. The first attempt at such a mechanistic synergy, using D-cycloserine, has not proven to be of robust enough benefit to enter widespread use (Burkner et al. 2017, Mataix-Cols et al. 2017), but it is a valuable proof of concept that may light the way for future work along the same conceptual line.

In sum, there is reason for both optimism and caution. We have a few robustly proven treatments for OCD, and these can benefit many patients and mitigate much suffering. But more is needed to help the severely affected and those whose symptoms are refractory to standard care. It is to be hoped that in the coming decade, developing insights such as those summarized in this volume will light the way towards the development of qualitatively new interventions, to the benefit of many.

References

Aboujaoude E, Barry JJ, Gamel N (2009) Memantine augmentation in treatment-resistant obsessive-compulsive disorder: an open-label trial. J Clin Psychopharmacol 29(1):51–55

Ackerman DL, Greenland S, Bystritsky A (1996a) Use of receiver-operator characteristic (ROC) curve analysis to evaluate predictors of response to clomipramine therapy. Psychopharmacol Bull 32(1):157–165

Ackerman DL, Greenland S, Bystritsky A, Katz RJ (1996b) Relationship between early side effects and therapeutic effects of clomipramine therapy in obsessive-compulsive disorder. J Clin Psychopharmacol 16(4):324–328

Ackerman DL, Greenland S, Bystritsky A (1998) Clinical characteristics of response to fluoxetine treatment of obsessive-compulsive disorder. J Clin Psychopharmacol 18(3):185–192

Adams KH, Hansen ES, Pinborg LH, Hasselbalch SG, Svarer C, Holm S, Bolwig TG, Knudsen GM (2005) Patients with obsessive-compulsive disorder have increased 5-HT2A receptor binding in the caudate nuclei. Int J Neuropsychopharmacol 8(3):391–401

Adams TG, Bloch MH, Pittenger C (2017) Intranasal ketamine and cognitive-behavioral therapy for treatment-refractory obsessive-compulsive disorder. J Clin Psychopharmacol 37(2):269–271

Afshar H, Roohafza H, Mohammad-Beigi H, Haghighi M, Jahangard L, Shokouh P, Sadeghi M, Hafezian H (2012) N-acetylcysteine add-on treatment in refractory obsessive-compulsive disorder: a randomized, double-blind, placebo-controlled trial. J Clin Psychopharmacol 32 (6):797–803

Afshar H, Akuchekian S, Mahaky B, Zarean E (2014) Topiramate augmentation in refractory obsessive-compulsive disorder: a randomized, double-blind, placebo-controlled trial. J Res Med Sci 19(10):976–981

Ahmari SE (2016) Using mice to model obsessive compulsive disorder: from genes to circuits. Neuroscience 321:121–137

Ahmari SE, Dougherty DD (2015) Dissecting OCD circuits: from animal models to targeted treatments. Depress Anxiety 32(8):550–562

Altemus M, Swedo SE, Leonard HL, Richter D, Rubinow DR, Potter WZ, Rapoport JL (1994) Changes in cerebrospinal fluid neurochemistry during treatment of obsessive-compulsive disorder with clomipramine. Arch Gen Psychiatry 51(10):794–803

Altshuler D, Daly MJ, Lander ES (2008) Genetic mapping in human disease. Science 322 (5903):881–888

American Psychiatric Association. and American Psychiatric Association. DSM-5 Task Force (2013) Diagnostic and statistical manual of mental disorders: DSM-5. American Psychiatric Association, Washington

Andrade C (2019) Augmentation with memantine in obsessive-compulsive disorder. J Clin Psychiatry 80(6)

Anticevic A, Hu S, Zhang S, Savic A, Billingslea E, Wasylink S, Repovs G, Cole MW, Bednarski S, Krystal JH, Bloch MH, Li CS, Pittenger C (2014) Global resting-state functional magnetic resonance imaging analysis identifies frontal cortex, striatal, and cerebellar dysconnectivity in obsessive-compulsive disorder. Biol Psychiatry 75(8):595–605

Arrojo-Romero M, Tajes Alonso M, de Leon J (2013) Lamotrigine augmentation of serotonin reuptake inhibitors in severe and long-term treatment-resistant obsessive-compulsive disorder. Case Rep Psychiatry 2013:612459

Balachander S, Kodancha PG, Arumugham SS, Sekharan JT, Narayanaswamy JC, Reddy YCJ (2019) Effectiveness of venlafaxine in selective serotonin reuptake inhibitor-resistant obsessive-compulsive disorder: experience from a specialty Clinic in India. J Clin Psychopharmacol 39 (1):82–85

Baldwin DS, Anderson IM, Nutt DJ, Bandelow B, Bond A, Davidson JR, den Boer JA, Fineberg NA, Knapp M, Scott J, Wittchen HU, P. British Association for (2005) Evidence-based guidelines for the pharmacological treatment of anxiety disorders: recommendations from the British Association for Psychopharmacology. J Psychopharmacol 19(6):567–596

Barr LC, Goodman WK, McDougle CJ, Delgado PL, Heninger GR, Charney DS, Price LH (1994) Tryptophan depletion in patients with obsessive-compulsive disorder who respond to serotonin reuptake inhibitors. Arch Gen Psychiatry 51(4):309–317

Bastani B, Arora RC, Meltzer HY (1991) Serotonin uptake and imipramine binding in the blood platelets of obsessive-compulsive disorder patients. Biol Psychiatry 30(2):131–139

Baumgarten HG, Grozdanovic Z (1998) Role of serotonin in obsessive-compulsive disorder. Br J Psychiatry 35(Suppl):13–20

Bergin J, Verhulst B, Aggen SH, Neale MC, Kendler KS, Bienvenu OJ, Hettema JM (2014) Obsessive compulsive symptom dimensions and neuroticism: an examination of shared genetic and environmental risk. Am J Med Genet B Neuropsychiatr Genet 165B(8):647–653

Berlin HA, Koran LM, Jenike MA, Shapira NA, Chaplin W, Pallanti S, Hollander E (2011) Double-blind, placebo-controlled trial of topiramate augmentation in treatment-resistant obsessive-compulsive disorder. J Clin Psychiatry 72(5):716–721

Berman RM, Cappiello A, Anand A, Oren DA, Heninger GR, Charney DS, Krystal JH (2000) Antidepressant effects of ketamine in depressed patients. Biol Psychiatry 47(4):351–354

Berney A, Sookman D, Leyton M, Young SN, Benkelfat C (2006) Lack of effects on core obsessive-compulsive symptoms of tryptophan depletion during symptom provocation in remitted obsessive-compulsive disorder patients. Biol Psychiatry 59(9):853–857

Bhattacharyya S, Khanna S, Chakrabarty K, Mahadevan A, Christopher R, Shankar SK (2009) Anti-brain autoantibodies and altered excitatory neurotransmitters in obsessive-compulsive disorder. Neuropsychopharmacology 34(12):2489–2496

Biohaven Pharmaceuticals (2020). https://www.biohavenpharma.com/investors/news-events/press-releases/06-24-2020

BioSpace (2012) Transcept Pharmaceuticals, Inc.'s drug for obsessive compulsive Disroder fails trial. https://www.biospace.com/article/releases/transcept-pharmaceuticals-inc-s-drug-for-obsessive-compulsive-disorder-fails-trial–/. Accessed 22 May 2020

Bloch MH (2008) Emerging treatments for Tourette's disorder. Curr Psychiatry Rep 10(4):323–330

Bloch MH, Landeros-Weisenberger A, Kelmendi B, Coric V, Bracken MB, Leckman JF (2006) A systematic review: antipsychotic augmentation with treatment refractory obsessive-compulsive disorder. Mol Psychiatry 11(7):622–632

Bloch MH, Landeros-Weisenberger A, Rosario MC, Pittenger C, Leckman JF (2008a) Meta-analysis of the symptom structure of obsessive-compulsive disorder. Am J Psychiatry 165 (12):1532–1542

Bloch MH, Landeros-Weisenberger A, Sen S, Dombrowski P, Kelmendi B, Coric V, Pittenger C, Leckman JF (2008b) Association of the serotonin transporter polymorphism and obsessive-compulsive disorder: systematic review. Am J Med Genet B Neuropsychiatr Genet 147B (6):850–858

Bloch MH, McGuire J, Landeros-Weisenberger A, Leckman JF, Pittenger C (2010) Meta-analysis of the dose-response relationship of SSRI in obsessive-compulsive disorder. Mol Psychiatry 15 (8):850–855

Bloch MH, Wasylink S, Landeros-Weisenberger A, Panza KE, Billingslea E, Leckman JF, Krystal JH, Bhagwagar Z, Sanacora G, Pittenger C (2012) Effects of ketamine in treatment-refractory obsessive-compulsive disorder. Biol Psychiatry 72(11):964–970

Bloch MH, Bartley CA, Zipperer L, Jakubovski E, Landeros-Weisenberger A, Pittenger C, Leckman JF (2014) Meta-analysis: hoarding symptoms associated with poor treatment outcome in obsessive-compulsive disorder. Mol Psychiatry 19(9):1025–1030

Boisseau CL, Schhwartzman CM, Rasmussen SA (2017) Quality of life and psychosocial functioning in OCD. In: Pittenger C (ed) Obsessive-compulsive disorder: phenomenology, pathophysiology, and treatment. Oxford University Press, New York

Bortolato M, Pittenger C (2017) Modeling tics in rodents: conceptual challenges and paths forward. J Neurosci Methods 292:12–19

Boshuisen ML, den Boer JA (2000) Zolmitriptan (a 5-HT1B/1D receptor agonist with central action) does not increase symptoms in obsessive compulsive disorder. Psychopharmacology 152(1):74–79

Brandl EJ, Muller DJ, Richter MA (2012) Pharmacogenetics of obsessive-compulsive disorders. Pharmacogenomics 13(1):71–81

Brandl EJ, Tiwari AK, Zhou X, Deluce J, Kennedy JL, Muller DJ, Richter MA (2014) Influence of CYP2D6 and CYP2C19 gene variants on antidepressant response in obsessive-compulsive disorder. Pharmacogenomics J 14(2):176–181

Brennan BP, Rauch SL (2017) A review and synthesis of functional neuroimaging studies of obsessive-compulsive disorder. In: Pittenger C (ed) Obsessive-copulsive disorder: phenomenology, pathophysiology, and treatment. Oxford Univeristy Press, New York

Brennan BP, Rauch SL, Jensen JE, Pope HG Jr (2013) A critical review of magnetic resonance spectroscopy studies of obsessive-compulsive disorder. Biol Psychiatry 73(1):24–31
Brody AL, Saxena S, Schwartz JM, Stoessel PW, Maidment K, Phelps ME, Baxter LR Jr (1998) FDG-PET predictors of response to behavioral therapy and pharmacotherapy in obsessive compulsive disorder. Psychiatry Res 84(1):1–6
Bruno A, Mico U, Pandolfo G, Mallamace D, Abenavoli E, Di Nardo F, D'Arrigo C, Spina E, Zoccali RA, Muscatello MR (2012) Lamotrigine augmentation of serotonin reuptake inhibitors in treatment-resistant obsessive-compulsive disorder: a double-blind, placebo-controlled study. J Psychopharmacol 26(11):1456–1462
Burkner PC, Bittner N, Holling H, Buhlmann U (2017) D-cycloserine augmentation of behavior therapy for anxiety and obsessive-compulsive disorders: a meta-analysis. PLoS One 12(3): e0173660
Cai J, Zhang W, Yi Z, Lu W, Wu Z, Chen J, Yu S, Fang Y, Zhang C (2013) Influence of polymorphisms in genes SLC1A1, GRIN2B, and GRIK2 on clozapine-induced obsessive-compulsive symptoms. Psychopharmacology 230(1):49–55
Cappi C, Oliphant ME, Peter Z, Zai G, do Rosario MC, Sullivan CAW, Gupta AR, Hoffman EJ, Virdee M, Olfson E, Abdallah SB, Willsey AJ, Shavitt RG, Miguel EC, Kennedy JL, Richter MA, Fernandez TV (2019) De novo damaging DNA coding mutations are associated with obsessive-compulsive disorder and overlap with Tourette's disorder and autism. Biol Psychiatry
Carmi L, Tendler A, Bystritsky A, Hollander E, Blumberger DM, Daskalakis J, Ward H, Lapidus K, Goodman W, Casuto L, Feifel D, Barnea-Ygael N, Roth Y, Zangen A, Zohar J (2019) Efficacy and safety of deep transcranial magnetic stimulation for obsessive-compulsive disorder: a prospective multicenter randomized double-blind placebo-controlled trial. Am J Psychiatry 176(11):931–938
Caspi A, Sugden K, Moffitt TE, Taylor A, Craig IW, Harrington H, McClay J, Mill J, Martin J, Braithwaite A, Poulton R (2003) Influence of life stress on depression: moderation by a polymorphism in the 5-HTT gene. Science 301(5631):386–389
Castellan Baldan L, Williams KA, Gallezot JD, Pogorelov V, Rapanelli M, Crowley M, Anderson GM, Loring E, Gorczyca R, Billingslea E, Wasylink S, Panza KE, Ercan-Sencicek AG, Krusong K, Leventhal BL, Ohtsu H, Bloch MH, Hughes ZA, Krystal JH, Mayes L, de Araujo I, Ding YS, State MW, Pittenger C (2014) Histidine decarboxylase deficiency causes tourette syndrome: parallel findings in humans and mice. Neuron 81(1):77–90
Catapano F, Sperandeo R, Perris F, Lanzaro M, Maj M (2001) Insight and resistance in patients with obsessive-compulsive disorder. Psychopathology 34(2):62–68
Cavedini P, Erzegovesi S, Ronchi P, Bellodi L (1997) Predictive value of obsessive-compulsive personality disorder in antiobsessional pharmacological treatment. Eur Neuropsychopharmacol 7(1):45–49
Chakrabarty K, Bhattacharyya S, Christopher R, Khanna S (2005) Glutamatergic dysfunction in OCD. Neuropsychopharmacology 30(9):1735–1740
Chiou YJ, Lin PY, Lee Y (2015) Risperidone as monotherapy for a patient with obsessive compulsive disorder comorbid with schizoaffective disorder: a case report. Clin Neuropharmacol 38(3):114–116
Clarke HF, Dalley JW, Crofts HS, Robbins TW, Roberts AC (2004) Cognitive inflexibility after prefrontal serotonin depletion. Science 304(5672):878–880
Clarke H, Flint J, Attwood AS, Munafo MR (2010) Association of the 5- HTTLPR genotype and unipolar depression: a meta-analysis. Psychol Med 40(11):1767–1778
Connor KM, Payne VM, Gadde KM, Zhang W, Davidson JR (2005) The use of aripiprazole in obsessive-compulsive disorder: preliminary observations in 8 patients. J Clin Psychiatry 66 (1):49–51
Coric V, Taskiran S, Pittenger C, Wasylink S, Mathalon DH, Valentine G, Saksa J, Wu YT, Gueorguieva R, Sanacora G, Malison RT, Krystal JH (2005) Riluzole augmentation in treatment-resistant obsessive-compulsive disorder: an open-label trial. Biol Psychiatry 58 (5):424–428

Costa DLC, Diniz JB, Requena G, Joaquim MA, Pittenger C, Bloch MH, Miguel EC, Shavitt RG (2017) Randomized, double-blind, placebo-controlled trial of N-Acetylcysteine augmentation for treatment-resistant obsessive-compulsive disorder. J Clin Psychiatry 78(7):e766–e773

D'Alcante CC, Diniz JB, Fossaluza V, Batistuzzo MC, Lopes AC, Shavitt RG, Deckersbach T, Malloy-Diniz L, Miguel EC, Hoexter MQ (2012) Neuropsychological predictors of response to randomized treatment in obsessive-compulsive disorder. Prog Neuro-Psychopharmacol Biol Psychiatry 39(2):310–317

da Conceicao Costa DL, Shavitt RG, Castro Cesar RC, Joaquim MA, Borcato S, Valerio C, Miguel EC, Diniz JB (2013) Can early improvement be an indicator of treatment response in obsessive-compulsive disorder? Implications for early-treatment decision-making. J Psychiatr Res 47 (11):1700–1707

Davis LK, Yu D, Keenan CL, Gamazon ER, Konkashbaev AI, Derks EM, Neale BM, Yang J, Lee SH, Evans P, Barr CL, Bellodi L, Benarroch F, Berrio GB, Bienvenu OJ, Bloch MH, Blom RM, Bruun RD, Budman CL, Camarena B, Campbell D, Cappi C, Cardona Silgado JC, Cath DC, Cavallini MC, Chavira DA, Chouinard S, Conti DV, Cook EH, Coric V, Cullen BA, Deforce D, Delorme R, Dion Y, Edlund CK, Egberts K, Falkai P, Fernandez TV, Gallagher PJ, Garrido H, Geller D, Girard SL, Grabe HJ, Grados MA, Greenberg BD, Gross-Tsur V, Haddad S, Heiman GA, Hemmings SM, Hounie AG, Illmann C, Jankovic J, Jenike MA, Kennedy JL, King RA, Kremeyer B, Kurlan R, Lanzagorta N, Leboyer M, Leckman JF, Lennertz L, Liu C, Lochner C, Lowe TL, Macciardi F, McCracken JT, McGrath LM, Mesa Restrepo SC, Moessner R, Morgan J, Muller H, Murphy DL, Naarden AL, Ochoa WC, Ophoff RA, Osiecki L, Pakstis AJ, Pato MT, Pato CN, Piacentini J, Pittenger C, Pollak Y, Rauch SL, Renner TJ, Reus VI, Richter MA, Riddle MA, Robertson MM, Romero R, Rosario MC, Rosenberg D, Rouleau GA, Ruhrmann S, Ruiz-Linares A, Sampaio AS, Samuels J, Sandor P, Sheppard B, Singer HS, Smit JH, Stein DJ, Strengman E, Tischfield JA, Valencia Duarte AV, Vallada H, Van Nieuwerburgh F, Veenstra-Vanderweele J, Walitza S, Wang Y, Wendland JR, Westenberg HG, Shugart YY, Miguel EC, McMahon W, Wagner M, Nicolini H, Posthuma D, Hanna GL, Heutink P, Denys D, Arnold PD, Oostra BA, Nestadt G, Freimer NB, Pauls DL, Wray NR, Stewart SE, Mathews CA, Knowles JA, Cox NJ, Scharf JM (2013) Partitioning the heritability of Tourette syndrome and obsessive compulsive disorder reveals differences in genetic architecture. PLoS Genet 9(10):e1003864

Delgado-Acevedo C, Estay SF, Radke AK, Sengupta A, Escobar AP, Henriquez-Belmar F, Reyes CA, Haro-Acuna V, Utreras E, Sotomayor-Zarate R, Cho A, Wendland JR, Kulkarni AB, Holmes A, Murphy DL, Chavez AE, Moya PR (2019) Behavioral and synaptic alterations relevant to obsessive-compulsive disorder in mice with increased EAAT3 expression. Neuropsychopharmacology 44(6):1163–1173

Delorme R, Chabane N, Callebert J, Falissard B, Mouren-Simeoni MC, Rouillon F, Launay JM, Leboyer M (2004) Platelet serotonergic predictors of clinical improvement in obsessive compulsive disorder. J Clin Psychopharmacol 24(1):18–23

Denys D, Burger H, van Megen H, de Geus F, Westenberg H (2003) A score for predicting response to pharmacotherapy in obsessive-compulsive disorder. Int Clin Psychopharmacol 18 (6):315–322

Denys D, van der Wee N, Janssen J, De Geus F, Westenberg HG (2004a) Low level of dopaminergic D2 receptor binding in obsessive-compulsive disorder. Biol Psychiatry 55 (10):1041–1045

Denys D, van Megen HJ, van der Wee N, Westenberg HG (2004b) A double-blind switch study of paroxetine and venlafaxine in obsessive-compulsive disorder. J Clin Psychiatry 65(1):37–43

Denys D, Zohar J, Westenberg HG (2004c) The role of dopamine in obsessive-compulsive disorder: preclinical and clinical evidence. J Clin Psychiatry 65(Suppl 14):11–17

Denys D, Van Nieuwerburgh F, Deforce D, Westenberg HG (2007) Prediction of response to paroxetine and venlafaxine by serotonin-related genes in obsessive-compulsive disorder in a randomized, double-blind trial. J Clin Psychiatry 68(5):747–753

Denys D, de Vries F, Cath D, Figee M, Vulink N, Veltman DJ, van der Doef TF, Boellaard R, Westenberg H, van Balkom A, Lammertsma AA, van Berckel BN (2013) Dopaminergic activity in Tourette syndrome and obsessive-compulsive disorder. Eur Neuropsychopharmacol 23 (11):1423–1431

DeVeaugh-Geiss J, Landau P, Katz R (1989) Preliminary results from a multicenter trial of clomipramine in obsessive-compulsive disorder. Psychopharmacol Bull 25(1):36–40

DeVeaugh-Geiss J, Katz R, Landau P, Goodman W, Rasmussen S (1990) Clinical predictors of treatment response in obsessive compulsive disorder: exploratory analyses from multicenter trials of clomipramine. Psychopharmacol Bull 26(1):54–59

Dietrich MO, Zimmer MR, Bober J, Horvath TL (2017) Hypothalamic Agrp neurons drive stereotypic behaviors beyond feeding. Cell 169(3):559

Dold M, Aigner M, Lanzenberger R, Kasper S (2015) Antipsychotic augmentation of serotonin reuptake inhibitors in treatment-resistant obsessive-compulsive disorder: an update meta-analysis of double-blind, randomized, placebo-controlled trials. Int J Neuropsychopharmacol 18(9)

Ducasse D, Boyer L, Michel P, Loundou A, Macgregor A, Micoulaud-Franchi JA, Courtet P, Abbar M, Leboyer M, Fond G (2014) D2 and D3 dopamine receptor affinity predicts effectiveness of antipsychotic drugs in obsessive-compulsive disorders: a metaregression analysis. Psychopharmacology 231(18):3765–3770

Eisen JL, Rasmussen SA, Phillips KA, Price LH, Davidson J, Lydiard RB, Ninan P, Piggott T (2001) Insight and treatment outcome in obsessive-compulsive disorder. Compr Psychiatry 42 (6):494–497

El Mansari M, Blier P (2005) Responsiveness of 5-HT(1A) and 5-HT2 receptors in the rat orbitofrontal cortex after long-term serotonin reuptake inhibition. J Psychiatry Neurosci 30 (4):268–274

Emamzadehfard S, Kamaloo A, Paydary K, Ahmadipour A, Zeinoddini A, Ghaleiha A, Mohammadinejad P, Zeinoddini A, Akhondzadeh S (2016) Riluzole in augmentation of fluvoxamine for moderate to severe obsessive-compulsive disorder: randomized, double-blind, placebo-controlled study. Psychiatry Clin Neurosci 70(8):332–341

Erzegovesi S, Cavallini MC, Cavedini P, Diaferia G, Locatelli M, Bellodi L (2001) Clinical predictors of drug response in obsessive-compulsive disorder. J Clin Psychopharmacol 21 (5):488–492

Escobar AP, Wendland JR, Chavez AE, Moya PR (2019) The neuronal glutamate transporter EAAT3 in obsessive-compulsive disorder. Front Pharmacol 10:1362

Excellence, N. I. f. C (2006) Obsessive-compulsive disorder: core interventions in the treatment of obsessive-compulsive disorder and body dysmorphic disorder. https://www.nice.org.uk/guidance/cg31. Accessed 15 Apr 2020

Fallon BA, Liebowitz MR, Campeas R, Schneier FR, Marshall R, Davies S, Goetz D, Klein DF (1998) Intravenous clomipramine for obsessive-compulsive disorder refractory to oral clomipramine: a placebo-controlled study. Arch Gen Psychiatry 55(10):918–924

FDA (2012) FDA drug safety communications: revised recommendations for celexa (citalopram hydrobromide) related to a potential risk of abnormal heart rhythms with high doses. http://www.fda.gov/Drugs/DrugSafety/ucm297391.htm. Accessed 20 May 2020

Fernandez Cordoba E, Lopez-Ibor Alino J (1967) Use of monochlorimipramine in psychiatric patients who are resistant to other therapy. Actas Luso Esp Neurol Psiquiatr 26(2):119–147

Fernandez TV, Leckman JF, Pittenger C (2018) Genetic susceptibility in obsessive-compulsive disorder. Handb Clin Neurol 148:767–781

Ferrao YA, Shavitt RG, Bedin NR, de Mathis ME, Carlos Lopes A, Fontenelle LF, Torres AR, Miguel EC (2006) Clinical features associated to refractory obsessive-compulsive disorder. J Affect Disord 94(1–3):199–209

Feusner JD, Kerwin L, Saxena S, Bystritsky A (2009) Differential efficacy of memantine for obsessive-compulsive disorder vs. generalized anxiety disorder: an open-label trial. Psychopharmacol Bull 42(1):81–93

Fineberg NA, Gale T (2005) Evidence-based pharmacological treatments for obsessive-compulsive disorder. Int J Neuropsychopharmacol 8:107–129

Fineberg NA, Reghunandanan S, Simpson HB, Phillips KA, Richter MA, Matthews K, Stein DJ, Sareen J, Brown A, Sookman D, D. Accreditation Task Force of The Canadian Institute for Obsessive Compulsive (2015) Obsessive-compulsive disorder (OCD): practical strategies for pharmacological and somatic treatment in adults. Psychiatry Res 227(1):114–125

Fontenelle L, Marques C, Engelhardt E, Versiani M (2001) Impaired set-shifting ability and therapeutic response in obsessive-compulsive disorder. J Neuropsychiatry Clin Neurosci 13 (4):508–510

Friedman AA, Letai A, Fisher DE, Flaherty KT (2015) Precision medicine for cancer with next-generation functional diagnostics. Nat Rev Cancer 15(12):747–756

Fullana MA, Zhu X, Alonso P, Cardoner N, Real E, Lopez-Sola C, Segalas C, Subira M, Galfalvy H, Menchon JM, Simpson HB, Marsh R, Soriano-Mas C (2017) Basolateral amygdala-ventromedial prefrontal cortex connectivity predicts cognitive behavioural therapy outcome in adults with obsessive-compulsive disorder. J Psychiatry Neurosci 42(6):378–385

Gadallah AHA, Ebada MA, Gadallah A, Ahmed H, Rashad W, Eid KA, Bahbah E, Alkanj S (2020) Efficacy and safety of N-acetylcysteine as add-on therapy in the treatment of obsessive-compulsive disorder: a systematic review and meta-analysis. J OCD Relat Disord 25:100529

Ghaleiha A, Entezari N, Modabbernia A, Najand B, Askari N, Tabrizi M, Ashrafi M, Hajiaghaee R, Akhondzadeh S (2013) Memantine add-on in moderate to severe obsessive-compulsive disorder: randomized double-blind placebo-controlled study. J Psychiatr Res 47(2):175–180

Ghanizadeh A, Mohammadi MR, Bahraini S, Keshavarzi Z, Firoozabadi A, Alavi Shoshtari A (2017) Efficacy of N-Acetylcysteine augmentation on obsessive compulsive disorder: a multicenter randomized double blind placebo controlled clinical trial. Iran J Psychiatry 12 (2):134–141

Goodman WK, Price LH, Rasmussen SA, Delgado PL, Heninger GR, Charney DS (1989a) Efficacy of fluvoxamine in obsessive-compulsive disorder. A double-blind comparison with placebo. Arch Gen Psychiatry 46(1):36–44

Goodman WK, Price LH, Rasmussen SA, Mazure C, Delgado P, Heninger GR, Charney DS (1989b) The Yale-Brown obsessive compulsive scale. II. Validity. Arch Gen Psychiatry 46 (11):1012–1016

Goodman WK, Price LH, Rasmussen SA, Mazure C, Fleischmann RL, Hill CL, Heninger GR, Charney DS (1989c) The Yale-Brown obsessive compulsive scale. I. Development, use, and reliability. Arch Gen Psychiatry 46(11):1006–1011

Graat I, Figee M, Denys D (2017) Neurotransmitter dysregulation in OCD. In: Pittenger C (ed) Obsessive-compulsive disorder: phenomenology, pathophysiology, and treatment. Oxford University Press, New York

Grant PJ, Joseph LA, Farmer CA, Luckenbaugh DA, Lougee LC, Zarate CA Jr, Swedo SE (2014) 12-week, placebo-controlled trial of add-on riluzole in the treatment of childhood-onset obsessive-compulsive disorder. Neuropsychopharmacology 39(6):1453–1459

Greenberg WM, Benedict MM, Doerfer J, Perrin M, Panek L, Cleveland WL, Javitt DC (2009) Adjunctive glycine in the treatment of obsessive-compulsive disorder in adults. J Psychiatr Res 43(6):664–670

Grillault Laroche D, Gaillard A (2016) Induced obsessive compulsive symptoms (OCS) in schizophrenia patients under atypical 2 antipsychotics (AAPs): review and hypotheses. Psychiatry Res 246:119–128

Gross-Isseroff R, Cohen R, Sasson Y, Voet H, Zohar J (2004) Serotonergic dissection of obsessive compulsive symptoms: a challenge study with m-chlorophenylpiperazine and sumatriptan. Neuropsychobiology 50(3):200–205

Group, C. C. S (1991) Clomipramine in the treatment of patients with obsessive-compulsive disorder. The Clomipramine Collaborative Study Group. Arch Gen Psychiatry 48(8):730–738

Gruner P, Pittenger C (2017) Cognitive inflexibility in obsessive-compulsive disorder. Neuroscience 345:243–255

Haghighi M, Jahangard L, Mohammad-Beigi H, Bajoghli H, Hafezian H, Rahimi A, Afshar H, Holsboer-Trachsler E, Brand S (2013) In a double-blind, randomized and placebo-controlled trial, adjuvant memantine improved symptoms in inpatients suffering from refractory obsessive-compulsive disorders (OCD). Psychopharmacology 228(4):633–640

Hamblin RJ, Park JM, Wu MS, Storch EA (2017) Variable insight in OCD. In: Pittenger C (ed) Obsessive-compulsive disorder: phenomenology, pathophysiology, and treatment. Oxford University Press, New York

Hanna GL, Veenstra-VanderWeele J, Cox NJ, Boehnke M, Himle JA, Curtis GC, Leventhal BL, Cook EH Jr (2002) Genome-wide linkage analysis of families with obsessive-compulsive disorder ascertained through pediatric probands. Am J Med Genet 114(5):541–552

Harrison BJ, Soriano-Mas C, Pujol J, Ortiz H, Lopez-Sola M, Hernandez-Ribas R, Deus J, Alonso P, Yucel M, Pantelis C, Menchon JM, Cardoner N (2009) Altered corticostriatal functional connectivity in obsessive-compulsive disorder. Arch Gen Psychiatry 66 (11):1189–1200

Harrison BJ, Pujol J, Cardoner N, Deus J, Alonso P, Lopez-Sola M, Contreras-Rodriguez O, Real E, Segalas C, Blanco-Hinojo L, Menchon JM, Soriano-Mas C (2013) Brain corticostriatal systems and the major clinical symptom dimensions of obsessive-compulsive disorder. Biol Psychiatry 73(4):321–328

Hasnain M, Vieweg WV (2014) QTc interval prolongation and torsade de pointes associated with second-generation antipsychotics and antidepressants: a comprehensive review. CNS Drugs 28 (10):887–920

Hazari N, Narayanaswamy JC, Arumugham SS (2016) Predictors of response to serotonin reuptake inhibitors in obsessive-compulsive disorder. Expert Rev Neurother 16(10):1175–1191

Heinz A, Jones DW, Mazzanti C, Goldman D, Ragan P, Hommer D, Linnoila M, Weinberger DR (2000) A relationship between serotonin transporter genotype and in vivo protein expression and alcohol neurotoxicity. Biol Psychiatry 47(7):643–649

Hendler T, Goshen E, Tzila Zwas S, Sasson Y, Gal G, Zohar J (2003) Brain reactivity to specific symptom provocation indicates prospective therapeutic outcome in OCD. Psychiatry Res 124 (2):87–103

Hesse S, Muller U, Lincke T, Barthel H, Villmann T, Angermeyer MC, Sabri O, Stengler-Wenzke K (2005) Serotonin and dopamine transporter imaging in patients with obsessive-compulsive disorder. Psychiatry Res 140(1):63–72

Hezel DM, Beattie K, Stewart SE (2009) Memantine as an augmenting agent for severe pediatric OCD. Am J Psychiatry 166(2):237

Ho Pian KL, van Megen HJ, Ramsey NF, Mandl R, van Rijk PP, Wynne HJ, Westenberg HG (2005) Decreased thalamic blood flow in obsessive-compulsive disorder patients responding to fluvoxamine. Psychiatry Res 138(2):89–97

Hoexter MQ, Dougherty DD, Shavitt RG, D'Alcante CC, Duran FL, Lopes AC, Diniz JB, Batistuzzo MC, Evans KC, Bressan RA, Busatto GF, Miguel EC (2013) Differential prefrontal gray matter correlates of treatment response to fluoxetine or cognitive-behavioral therapy in obsessive-compulsive disorder. Eur Neuropsychopharmacol 23(7):569–580

Hoexter MQ, Diniz JB, Lopes AC, Batistuzzo MC, Shavitt RG, Dougherty DD, Duran FL, Bressan RA, Busatto GF, Miguel EC, Sato JR (2015) Orbitofrontal thickness as a measure for treatment response prediction in obsessive-compulsive disorder. Depress Anxiety 32(12):900–908

Hollander E, Doernberg E, Shavitt R, Waterman RJ, Soreni N, Veltman DJ, Sahakian BJ, Fineberg NA (2016) The cost and impact of compulsivity: a research perspective. Eur Neuropsychopharmacol 26(5):800–809

Hood SD, Broyd A, Robinson H, Lee J, Hudaib AR, Hince DA (2017) Effects of tryptophan depletion on selective serotonin reuptake inhibitor-remitted patients with obsessive compulsive disorder. J Psychopharmacol 31(12):1615–1623

Hsieh HJ, Lue KH, Tsai HC, Lee CC, Chen SY, Kao PF (2014) L-3,4-Dihydroxy-6-[F-18] fluorophenylalanine positron emission tomography demonstrating dopaminergic system

abnormality in the brains of obsessive-compulsive disorder patients. Psychiatry Clin Neurosci 68(4):292–298
Humble M, Bejerot S, Bergqvist PB, Bengtsson F (2001) Reactivity of serotonin in whole blood: relationship with drug response in obsessive-compulsive disorder. Biol Psychiatry 49 (4):360–368
Hussain MZ, Ahad A (1970) Treatment of obsessive-compulsive neurosis. Can Med Assoc J 103 (6):648
Hussain A, Dar MA, Wani RA, Shah MS, Jan MM, Malik YA, Chandel RK, Margoob MA (2015) Role of lamotrigine augmentation in treatment-resistant obsessive compulsive disorder: a retrospective case review from South Asia. Indian J Psychol Med 37(2):154–158
Hutton LMJ, Cave AJ, St-Jean R, Banh HL (2017) Should we be worried about QTc prolongation using citalopram? A review. J Pharm Pract 30(3):353–358
Hybel KA, Mortensen EL, Lambek R, Hojgaard D, Thomsen PH (2017) Executive function predicts cognitive-behavioral therapy response in childhood obsessive-compulsive disorder. Behav Res Ther 99:11–18
Insel TR, Murphy DL, Cohen RM, Alterman I, Kilts C, Linnoila M (1983) Obsessive-compulsive disorder. A double-blind trial of clomipramine and clorgyline. Arch Gen Psychiatry 40 (6):605–612
Insel TR, Mueller EA, Alterman I, Linnoila M, Murphy DL (1985) Obsessive-compulsive disorder and serotonin: is there a connection? Biol Psychiatry 20(11):1174–1188
International Obsessive Compulsive Disorder Foundation Genetics Consortium (IOCDF-GC), C. and OCD Collaborative Genetic Association (OCGAS) (2018) Revealing the complex genetic architecture of obsessive-compulsive disorder using meta-analysis. Mol Psychiatry 23 (5):1181–1188
Issari Y, Jakubovski E, Bartley CA, Pittenger C, Bloch MH (2016) Early onset of response with selective serotonin reuptake inhibitors in obsessive-compulsive disorder: a meta-analysis. J Clin Psychiatry 77(5):e605–e611
Iyo M, Sekine Y, Matsunaga T, Tsukamoto T, Takei N, Mori N (1999) Methamphetamine-associated obsessional symptoms and effective risperidone treatment: a case report. J Clin Psychiatry 60(5):337–338
Jhanda S, Singla N, Grover S (2016) Methylphenidate-induced obsessive-compulsive symptoms: a case report and review of literature. J Pediatr Neurosci 11(4):316–318
Johnson MW, Hendricks PS, Barrett FS, Griffiths RR (2019) Classic psychedelics: an integrative review of epidemiology, therapeutics, mystical experience, and brain network function. Pharmacol Ther 197:83–102
Kaplan GB, Moore KA (2011) The use of cognitive enhancers in animal models of fear extinction. Pharmacol Biochem Behav 99(2):217–228
Karameh WK, Khani M (2015) Intravenous clomipramine for treatment-resistant obsessive-compulsive disorder. Int J Neuropsychopharmacol 19(2)
Karthik S, Sharma LP, Narayanaswamy JC (2020) Investigating the role of glutamate in obsessive-compulsive disorder: current perspectives. Neuropsychiatr Dis Treat 16:1003–1013
Katz RJ, DeVeaugh-Geiss J, Landau P (1990) Clomipramine in obsessive-compulsive disorder. Biol Psychiatry 28(5):401–414
Kessler RC, Petukhova M, Sampson NA, Zaslavsky AM, Wittchen HU (2012) Twelve-month and lifetime prevalence and lifetime morbid risk of anxiety and mood disorders in the United States. Int J Methods Psychiatr Res 21(3):169–184
Khalkhali M, Aram S, Zarrabi H, Kafie M, Heidarzadeh A (2016) Lamotrigine augmentation versus placebo in serotonin reuptake inhibitors-resistant obsessive-compulsive disorder: a randomized controlled trial. Iran J Psychiatry 11(2):104–114
Kim CH, Koo MS, Cheon KA, Ryu YH, Lee JD, Lee HS (2003) Dopamine transporter density of basal ganglia assessed with [123I]IPT SPET in obsessive-compulsive disorder. Eur J Nucl Med Mol Imaging 30(12):1637–1643

Kim M, Kwak S, Yoon YB, Kwak YB, Kim T, Cho KIK, Lee TY, Kwon JS (2019) Functional connectivity of the raphe nucleus as a predictor of the response to selective serotonin reuptake inhibitors in obsessive-compulsive disorder. Neuropsychopharmacology 44(12):2073–2081

Koch S, Perry KW, Nelson DL, Conway RG, Threlkeld PG, Bymaster FP (2002) R-fluoxetine increases extracellular DA, NE, as well as 5-HT in rat prefrontal cortex and hypothalamus: an in vivo microdialysis and receptor binding study. Neuropsychopharmacology 27(6):949–959

Koran LM, Pallanti S, Quercioli L (2001) Sumatriptan, 5-HT(1D) receptors and obsessive-compulsive disorder. Eur Neuropsychopharmacol 11(2):169–172

Koran LM, Gamel NN, Choung HW, Smith EH, Aboujaoude EN (2005) Mirtazapine for obsessive-compulsive disorder: an open trial followed by double-blind discontinuation. J Clin Psychiatry 66(4):515–520

Koran LM, Hanna GL, Hollander E, Nestadt G, Simpson HB, A. American Psychiatric (2007) Practice guideline for the treatment of patients with obsessive-compulsive disorder. Am J Psychiatry 164(7 Suppl):5–53

Koran LM, Aboujaoude E, Gamel NN (2009) Double-blind study of dextroamphetamine versus caffeine augmentation for treatment-resistant obsessive-compulsive disorder. J Clin Psychiatry 70(11):1530–1535

Korff S, Stein DJ, Harvey BH (2008) Stereotypic behaviour in the deer mouse: pharmacological validation and relevance for obsessive compulsive disorder. Prog Neuro-Psychopharmacol Biol Psychiatry 32(2):348–355

Kouris S (1998) Methylphenidate-induced obsessive-compulsiveness. J Am Acad Child Adolesc Psychiatry 37(2):135

Kraehenmann R, Pokorny D, Vollenweider L, Preller KH, Pokorny T, Seifritz E, Vollenweider FX (2017) Dreamlike effects of LSD on waking imagery in humans depend on serotonin 2A receptor activation. Psychopharmacology 234(13):2031–2046

Krystal JH, Abdallah CG, Sanacora G, Charney DS, Duman RS (2019) Ketamine: a paradigm shift for depression research and treatment. Neuron 101(5):774–778

Kulz AK, Meinzer S, Kopasz M, Voderholzer U (2007) Effects of tryptophan depletion on cognitive functioning, obsessive-compulsive symptoms and mood in obsessive-compulsive disorder: preliminary results. Neuropsychobiology 56(2–3):127–131

Kumar TC, Khanna S (2000) Lamotrigine augmentation of serotonin re-uptake inhibitors in obsessive-compulsive disorder. Aust N Z J Psychiatry 34(3):527–528

Lafleur DL, Pittenger C, Kelmendi B, Gardner T, Wasylink S, Malison RT, Sanacora G, Krystal JH, Coric V (2006) N-acetylcysteine augmentation in serotonin reuptake inhibitor refractory obsessive-compulsive disorder. Psychopharmacology 184(2):254–256

Landeros-Weisenberger A, Bloch MH, Kelmendi B, Wegner R, Nudel J, Dombrowski P, Pittenger C, Krystal JH, Goodman WK, Leckman JF, Coric V (2010) Dimensional predictors of response to SRI pharmacotherapy in obsessive-compulsive disorder. J Affect Disord 121 (1–2):175–179

Leckman JF, Goodman WK, Anderson GM, Riddle MA, Chappell PB, McSwiggan-Hardin MT, McDougle CJ, Scahill LD, Ort SI, Pauls DL et al (1995) Cerebrospinal fluid biogenic amines in obsessive compulsive disorder, Tourette's syndrome, and healthy controls. Neuropsychopharmacology 12(1):73–86

Lesch KP, Bengel D, Heils A, Sabol SZ, Greenberg BD, Petri S, Benjamin J, Muller CR, Hamer DH, Murphy DL (1996) Association of anxiety-related traits with a polymorphism in the serotonin transporter gene regulatory region. Science 274(5292):1527–1531

Lochner C et al (2001) Venlafaxine in obsessive compulsive disorder. In: Fifth Internatonal obsessive compulsive disorder conference, Sardinia, Italy

Lykouras L, Alevizos B, Michalopoulou P, Rabavilas A (2003) Obsessive-compulsive symptoms induced by atypical antipsychotics. A review of the reported cases. Prog Neuro-Psychopharmacol Biol Psychiatry 27(3):333–346

Marazziti D, Hollander E, Lensi P, Ravagli S, Cassano GB (1992) Peripheral markers of serotonin and dopamine function in obsessive-compulsive disorder. Psychiatry Res 42(1):41–51

Marazziti D, Rossi A, Gemignani A, Giannaccini G, Pfanner C, Milanfranchi A, Presta S, Lucacchini A, Cassano GB (1996) Decreased platelet 3H-paroxetine binding in obsessive-compulsive patients. Neuropsychobiology 34(4):184–187
Marazziti D, Pfanner C, Palego L, Gemignani A, Milanfranchi A, Ravagli S, Lensi P, Presta S, Cassano GB (1997) Changes in platelet markers of obsessive-compulsive patients during a double-blind trial of fluvoxamine versus clomipramine. Pharmacopsychiatry 30(6):245–249
Marazziti D, Dell'Osso L, Presta S, Pfanner C, Rossi A, Masala I, Baroni S, Giannaccini G, Lucacchini A, Cassano GB (1999) Platelet [3H]paroxetine binding in patients with OCD-related disorders. Psychiatry Res 89(3):223–228
Marinova Z, Chuang DM, Fineberg N (2017) Glutamate-modulating drugs as a potential therapeutic strategy in obsessive-compulsive disorder. Curr Neuropharmacol 15(7):977–995
Markianos M, Lafazanos S, Koutsis G, Sfagos C, Seretis A (2009) CSF neurotransmitter metabolites and neuropsychiatric symptomatology in patients with normal pressure hydrocephalus. Clin Neurol Neurosurg 111(3):231–234
Marks IM, Stern RS, Mawson D, Cobb J, McDonald R (1980) Clomipramine and exposure for obsessive-compulsive rituals: I. Br J Psychiatry 136:1–25
Marroni SS, Nakano FN, Gati CD, Oliveira JA, Antunes-Rodrigues J, Garcia-Cairasco N (2007) Neuroanatomical and cellular substrates of hypergrooming induced by microinjection of oxytocin in central nucleus of amygdala, an experimental model of compulsive behavior. Mol Psychiatry 12(12):1103–1117
Mataix-Cols D, Marks IM, Greist JH, Kobak KA, Baer L (2002) Obsessive-compulsive symptom dimensions as predictors of compliance with and response to behaviour therapy: results from a controlled trial. Psychother Psychosom 71(5):255–262
Mataix-Cols D, Fernandez de la Cruz L, Monzani B, Rosenfield D, Andersson E, Perez-Vigil A, Frumento P, de Kleine RA, Difede J, Dunlop BW, Farrell LJ, Geller D, Gerardi M, Guastella AJ, Hofmann SG, Hendriks GJ, Kushner MG, Lee FS, Lenze EJ, Levinson CA, McConnell H, Otto MW, Plag J, Pollack MH, Ressler KJ, Rodebaugh TL, Rothbaum BO, Scheeringa MS, Siewert-Siegmund A, Smits JAJ, Storch EA, Strohle A, Tart CD, Tolin DF, van Minnen A, Waters AM, Weems CF, Wilhelm S, Wyka K, Davis M, Ruck C, D. C. S. A. C. the, Altemus M, Anderson P, Cukor J, Finck C, Geffken GR, Golfels F, Goodman WK, Gutner C, Heyman I, Jovanovic T, Lewin AB, McNamara JP, Murphy TK, Norrholm S, Thuras P (2017) D-Cycloserine augmentation of exposure-based cognitive behavior therapy for anxiety, obsessive-compulsive, and posttraumatic stress disorders: a systematic review and meta-analysis of individual participant data. JAMA Psychiat 74(5):501–510
Mattheisen M, Samuels JF, Wang Y, Greenberg BD, Fyer AJ, McCracken JT, Geller DA, Murphy DL, Knowles JA, Grados MA, Riddle MA, Rasmussen SA, McLaughlin NC, Nurmi EL, Askland KD, Qin HD, Cullen BA, Piacentini J, Pauls DL, Bienvenu OJ, Stewart SE, Liang KY, Goes FS, Maher B, Pulver AE, Shugart YY, Valle D, Lange C, Nestadt G (2015) Genome-wide association study in obsessive-compulsive disorder: results from the OCGAS. Mol Psychiatry 20(3):337–344
McDougle CJ, Goodman WK, Leckman JF, Barr LC, Heninger GR, Price LH (1993) The efficacy of fluvoxamine in obsessive-compulsive disorder: effects of comorbid chronic tic disorder. J Clin Psychopharmacol 13(5):354–358
McDougle CJ, Goodman WK, Leckman JF, Lee NC, Heninger GR, Price LH (1994) Haloperidol addition in fluvoxamine-refractory obsessive-compulsive disorder. A double-blind, placebo-controlled study in patients with and without tics. Arch Gen Psychiatry 51(4):302–308
McDougle CJ, Barr LC, Goodman WK, Pelton GH, Aronson SC, Anand A, Price LH (1995) Lack of efficacy of clozapine monotherapy in refractory obsessive-compulsive disorder. Am J Psychiatry 152(12):1812–1814
McDougle CJ, Barr LC, Goodman WK, Price LH (1999) Possible role of neuropeptides in obsessive compulsive disorder. Psychoneuroendocrinology 24(1):1–24

McDougle CJ, Epperson CN, Pelton GH, Wasylink S, Price LH (2000) A double-blind, placebo-controlled study of risperidone addition in serotonin reuptake inhibitor-refractory obsessive-compulsive disorder. Arch Gen Psychiatry 57(8):794–801

Modarresi A, Sayyah M, Razooghi S, Eslami K, Javadi M, Kouti L (2018) Memantine augmentation improves symptoms in serotonin reuptake inhibitor-refractory obsessive-compulsive disorder: a randomized controlled trial. Pharmacopsychiatry 51(6):263–269

Modarresi A, Chaibakhsh S, Koulaeinejad N, Koupaei SR (2019) A systematic review and meta-analysis: Memantine augmentation in moderate to severe obsessive-compulsive disorder. Psychiatry Res 282:112602

Montgomery SA, McIntyre A, Osterheider M, Sarteschi P, Zitterl W, Zohar J, Birkett M, Wood AJ (1993) A double-blind, placebo-controlled study of fluoxetine in patients with DSM-III-R obsessive-compulsive disorder. The Lilly European OCD Study Group. Eur Neuropsychopharmacol 3(2):143–152

Moreno FA, Wiegand CB, Taitano EK, Delgado PL (2006) Safety, tolerability, and efficacy of psilocybin in 9 patients with obsessive-compulsive disorder. J Clin Psychiatry 67 (11):1735–1740

Mowla A, Boostani S, Dastgheib SA (2016) Duloxetine augmentation in resistant obsessive-compulsive disorder: a double-blind controlled clinical trial. J Clin Psychopharmacol 36 (6):720–723

Naguy A, Alamiri B, Al-Khadhari S, Francis K (2016) Lamotrigine augmentation in treatment-resistant pediatric obsessive-compulsive disorder with a 16 month follow-up. J Child Adolesc Psychopharmacol 26(8):769–772

NHS (2011) Citalopram and escitalopram: QT interval prolongation. Drug Saf Update 5(5):A1

Nikolaus S, Muller HW, Hautzel H (2016) Different patterns of 5-HT receptor and transporter dysfunction in neuropsychiatric disorders--a comparative analysis of in vivo imaging findings. Rev Neurosci 27(1):27–59

NIMH (2020). https://www.nimh.nih.gov/about/strategic-planning-reports/goal-3-strive-for-prevention-and-cures.shtml. Accessed 1 July 2020

Olatunji BO, Ebesutani C, Abramowitz JS (2017) Examination of a Bifactor model of obsessive-compulsive symptom dimensions. Assessment 24(1):45–59

Olver JS, O'Keefe G, Jones GR, Burrows GD, Tochon-Danguy HJ, Ackermann U, Scott A, Norman TR (2009) Dopamine D1 receptor binding in the striatum of patients with obsessive-compulsive disorder. J Affect Disord 114(1–3):321–326

Olver JS, O'Keefe G, Jones GR, Burrows GD, Tochon-Danguy HJ, Ackermann U, Scott AM, Norman TR (2010) Dopamine D(1) receptor binding in the anterior cingulate cortex of patients with obsessive-compulsive disorder. Psychiatry Res 183(1):85–88

O'Neill J, Piacentini J, Chang S, Ly R, Lai TM, Armstrong CC, Bergman L, Rozenman M, Peris T, Vreeland A, Mudgway R, Levitt JG, Salamon N, Posse S, Hellemann GS, Alger JR, McCracken JT, Nurmi EL (2017) Glutamate in pediatric obsessive-compulsive disorder and response to cognitive-behavioral therapy: randomized clinical trial. Neuropsychopharmacology 42 (12):2414–2422

O'Regan JB (1970) Treatment of obsessive--compulsive neurosis with haloperidol. Can Med Assoc J 103(2):167–168

Pallanti S, Quercioli L, Koran LM (2002) Citalopram intravenous infusion in resistant obsessive-compulsive disorder: an open trial. J Clin Psychiatry 63(9):796–801

Pallanti S, Quercioli L, Bruscoli M (2004) Response acceleration with mirtazapine augmentation of citalopram in obsessive-compulsive disorder patients without comorbid depression: a pilot study. J Clin Psychiatry 65(10):1394–1399

Paydary K, Akamaloo A, Ahmadipour A, Pishgar F, Emamzadehfard S, Akhondzadeh S (2016) N-acetylcysteine augmentation therapy for moderate-to-severe obsessive-compulsive disorder: randomized, double-blind, placebo-controlled trial. J Clin Pharm Ther 41(2):214–219

Perani D, Garibotto V, Gorini A, Moresco RM, Henin M, Panzacchi A, Matarrese M, Carpinelli A, Bellodi L, Fazio F (2008) In vivo PET study of 5HT(2A) serotonin and D(2) dopamine dysfunction in drug-naive obsessive-compulsive disorder. NeuroImage 42(1):306–314

Perse TL, Greist JH, Jefferson JW, Rosenfeld R, Dar R (1987) Fluvoxamine treatment of obsessive-compulsive disorder. Am J Psychiatry 144(12):1543–1548

Phillips KA, Albertini RS, Rasmussen SA (2002) A randomized placebo-controlled trial of fluoxetine in body dysmorphic disorder. Arch Gen Psychiatry 59(4):381–388

Pian KL, Westenberg HG, van Megen HJ, den Boer JA (1998) Sumatriptan (5-HT1D receptor agonist) does not exacerbate symptoms in obsessive compulsive disorder. Psychopharmacology 140(3):365–370

Piantadosi SC, Chamberlain BL, Glausier JR, Lewis DA, Ahmari SE (2019) Lower excitatory synaptic gene expression in orbitofrontal cortex and striatum in an initial study of subjects with obsessive compulsive disorder. Mol Psychiatry

Pittenger C (2014) Animal models of Tourette syndrome and obsessive-compulsive disorder. In: LeDoux M (ed) Animal models of movement disorders. Elsevier, Academic Press, San Deigo, pp 748–766

Pittenger C (2015) Glutamate modulators in the treatment of obsessive-compulsive disorder. Psychiatr Ann 45(6):308–315

Pittenger C (2017a) Histidine decarboxylase knockout mice as a model of the pathophysiology of Tourette syndrome and related conditions. Handb Exp Pharmacol 241:189–215

Pittenger C (2017b) The neurobiology of tic disorders and obsessive-compulsive disorder: human and animal studies. In: Nestler EJ, Buxbaum JD, Sklar P, Charney DS (eds) Nestler and Charney's neurobiology of mental illness. Oxford University Press, New York, pp 879–890

Pittenger C (2017c) The pharmacological treatment of refractory OCD. In: Pittenger C (ed) Obsessive-compulsive disorder: phenomenology, pathophysiology, and treatment. Oxford University Press, New York

Pittenger C, Krystal JH, Coric V (2006) Glutamate-modulating drugs as novel pharmacotherapeutic agents in the treatment of obsessive-compulsive disorder. NeuroRx 3(1):69–81

Pittenger C, Coric V, Banasr M, Bloch M, Krystal JH, Sanacora G (2008a) Riluzole in the treatment of mood and anxiety disorders. CNS Drugs 22(9):761–786

Pittenger C, Kelmendi B, Wasylink S, Bloch MH, Coric V (2008b) Riluzole augmentation in treatment-refractory obsessive-compulsive disorder: a series of 13 cases, with long-term follow-up. J Clin Psychopharmacol 28(3):363–367

Pittenger C, Bloch MH, Williams K (2011) Glutamate abnormalities in obsessive compulsive disorder: neurobiology, pathophysiology, and treatment. Pharmacol Ther 132(3):314–332

Pittenger C, Bloch MH, Wasylink S, Billingslea E, Simpson R, Jakubovski E, Kelmendi B, Sanacora G, Coric V (2015) Riluzole augmentation in treatment-refractory obsessive-compulsive disorder: a pilot randomized placebo-controlled trial. J Clin Psychiatry 76(8):1075–1084

Pittenger C, Adams TG Jr, Gallezot JD, Crowley MJ, Nabulsi N, James R, Gao H, Kichuk SA, Simpson R, Billingslea E, Hannestad J, Bloch M, Mayes L, Bhagwagar Z, Carson RE (2016) OCD is associated with an altered association between sensorimotor gating and cortical and subcortical 5-HT1b receptor binding. J Affect Disord 196:87–96

Pittenger C, Dulawa S, Thompson SL (2017a) Animal models of OCD: a conceptual framework. In: Pittenger C (ed) Obsessive-compulsive disorder: phenomenology, pathophysiology, and treatment. Oxford University Press, New York

Pittenger C, Gruner P, Adams TA, Kelmendi B (2017b) The dynamics of obsessive-compulsive disorder: a heuristic framework. In: Pittenger C (ed) Obsessive-compulsive disorder: phenomenology, pathophysiology, and treatment. Oxford University Press, New York

Pogarell O, Hamann C, Popperl G, Juckel G, Chouker M, Zaudig M, Riedel M, Moller HJ, Hegerl U, Tatsch K (2003) Elevated brain serotonin transporter availability in patients with obsessive-compulsive disorder. Biol Psychiatry 54(12):1406–1413

Poyurovsky M (2017) Obsessive-compulsive symptoms in schizophrenia. In: Pittenger C (ed) Obsessive-compulsive disorder: phenomenology, pathophysiology, and treatment. Oxford University Press, New York

Preller KH, Herdener M, Pokorny T, Planzer A, Kraehenmann R, Stampfli P, Liechti ME, Seifritz E, Vollenweider FX (2017) The fabric of meaning and subjective effects in LSD-induced states depend on serotonin 2A receptor activation. Curr Biol 27(3):451–457

Qin H, Samuels JF, Wang Y, Zhu Y, Grados MA, Riddle MA, Greenberg BD, Knowles JA, Fyer AJ, McCracken JT, Murphy DL, Rasmussen SA, Cullen BA, Piacentini J, Geller D, Stewart SE, Pauls D, Bienvenu OJ, Goes FS, Maher B, Pulver AE, Valle D, Lange C, Mattheisen M, McLaughlin NC, Liang KY, Nurmi EL, Askland KD, Nestadt G, Shugart YY (2016) Whole-genome association analysis of treatment response in obsessive-compulsive disorder. Mol Psychiatry 21(2):270–276

Rasmussen SA, Eisen JL (1990) Epidemiology of obsessive compulsive disorder. J Clin Psychiatry 51(Suppl):10–13; discussion 14

Rasmussen S, Hackett E, DuBoff E, Greist J, Halaris A, Koran LM, Liebowitz M, Lydiard RB, McElroy S, Mendels J, O'Connor K (1997) A 2-year study of sertraline in the treatment of obsessive-compulsive disorder. Int Clin Psychopharmacol 12(6):309–316

Rauch SL, Shin LM, Dougherty DD, Alpert NM, Fischman AJ, Jenike MA (2002) Predictors of fluvoxamine response in contamination-related obsessive compulsive disorder: a PET symptom provocation study. Neuropsychopharmacology 27(5):782–791

Ravizza L, Barzega G, Bellino S, Bogetto F, Maina G (1995) Predictors of drug treatment response in obsessive-compulsive disorder. J Clin Psychiatry 56(8):368–373

Reggente N, Moody TD, Morfini F, Sheen C, Rissman J, O'Neill J, Feusner JD (2018) Multivariate resting-state functional connectivity predicts response to cognitive behavioral therapy in obsessive-compulsive disorder. Proc Natl Acad Sci U S A 115(9):2222–2227

Reid JE, Reghunandanan S, Roberts A, Fineberg NA (2017) Standard evidence-based pharmacological treatment for OCD. In: Pittenger C (ed) Obsessive-compulsive disorder: phenomenology, pathophysiology, and treatment. Oxford University Press, New York

Reiff CM, Richman EE, Nemeroff CB, Carpenter LL, Widge AS, Rodriguez CI, Kalin NH, McDonald WM, B. and the Work Group on and a. D. o. t. A. P. A. C. o. R. Novel Treatments (2020) Psychedelics and psychedelic-assisted psychotherapy. Am J Psychiatry 177(5):391–410

Richardson R, Ledgerwood L, Cranney J (2004) Facilitation of fear extinction by D-cycloserine: theoretical and clinical implications. Learn Mem 11(5):510–516

Risch N, Herrell R, Lehner T, Liang KY, Eaves L, Hoh J, Griem A, Kovacs M, Ott J, Merikangas KR (2009) Interaction between the serotonin transporter gene (5-HTTLPR), stressful life events, and risk of depression: a meta-analysis. JAMA 301(23):2462–2471

Rivers-Bulkeley N, Hollender MH (1982) Successful treatment of obsessive-compulsive disorder with loxapine. Am J Psychiatry 139(10):1345–1346

Rodriguez CI, Kegeles LS, Flood P, Simpson HB (2011) Rapid resolution of obsessions after an infusion of intravenous ketamine in a patient with treatment-resistant obsessive-compulsive disorder. J Clin Psychiatry 72(4):567–569

Rodriguez CI, Kegeles LS, Levinson A, Feng T, Marcus SM, Vermes D, Flood P, Simpson HB (2013) Randomized controlled crossover trial of ketamine in obsessive-compulsive disorder: proof-of-concept. Neuropsychopharmacology 38(12):2475–2483

Rodriguez CI, Levinson A, Zwerling J, Vermes D, Simpson HB (2016a) Open-label trial on the effects of memantine in adults with obsessive-compulsive disorder after a single ketamine infusion. J Clin Psychiatry 77(5):688–689

Rodriguez CI, Wheaton M, Zwerling J, Steinman SA, Sonnenfeld D, Galfalvy H, Simpson HB (2016b) Can exposure-based CBT extend the effects of intravenous ketamine in obsessive-compulsive disorder? An open-label trial. J Clin Psychiatry 77(3):408–409

Rosenberg DR, MacMaster FP, Keshavan MS, Fitzgerald KD, Stewart CM, Moore GJ (2000) Decrease in caudate glutamatergic concentrations in pediatric obsessive-compulsive disorder patients taking paroxetine. J Am Acad Child Adolesc Psychiatry 39(9):1096–1103

Rosenberg DR, MacMillan SN, Moore GJ (2001) Brain anatomy and chemistry may predict treatment response in paediatric obsessive--compulsive disorder. Int J Neuropsychopharmacol 4(2):179–190

Rosenberg DR, Mirza Y, Russell A, Tang J, Smith JM, Banerjee SP, Bhandari R, Rose M, Ivey J, Boyd C, Moore GJ (2004) Reduced anterior cingulate glutamatergic concentrations in childhood OCD and major depression versus healthy controls. J Am Acad Child Adolesc Psychiatry 43(9):1146–1153

Ruscio AM, Stein DJ, Chiu WT, Kessler RC (2010) The epidemiology of obsessive-compulsive disorder in the National Comorbidity Survey Replication. Mol Psychiatry 15(1):53–63

Sahraian A, Bigdeli M, Ghanizadeh A, Akhondzadeh S (2014) Topiramate as an adjuvant treatment for obsessive compulsive symptoms in patients with bipolar disorder: a randomized double blind placebo controlled clinical trial. J Affect Disord 166:201–205

Sahraian A, Jahromi LR, Ghanizadeh A, Mowla A (2017) Memantine as an adjuvant treatment for obsessive compulsive symptoms in manic phase of bipolar disorder: a randomized, double-blind, placebo-controlled clinical trial. J Clin Psychopharmacol 37(2):246–249

Sallee FR, Richman H, Beach K, Sethuraman G, Nesbitt L (1996) Platelet serotonin transporter in children and adolescents with obsessive-compulsive disorder or Tourette's syndrome. J Am Acad Child Adolesc Psychiatry 35(12):1647–1656

Sampaio AS, Hounie AG, Petribu K, Cappi C, Morais I, Vallada H, do Rosario MC, Stewart SE, Fargeness J, Mathews C, Arnold P, Hanna GL, Richter M, Kennedy J, Fontenelle L, de Braganca Pereira CA, Pauls DL, Miguel EC (2015) COMT and MAO-A polymorphisms and obsessive-compulsive disorder: a family-based association study. PLoS One 10(3):e0119592

Sanematsu H, Nakao T, Yoshiura T, Nabeyama M, Togao O, Tomita M, Masuda Y, Nakatani E, Nakagawa A, Kanba S (2010) Predictors of treatment response to fluvoxamine in obsessive-compulsive disorder: an fMRI study. J Psychiatr Res 44(4):193–200

Sari Y (2004) Serotonin1B receptors: from protein to physiological function and behavior. Neurosci Biobehav Rev 28(6):565–582

Sassano-Higgins SA, Pato MT (2015) Pindolol augmentation of selective serotonin reuptake inhibitors and clomipramine for the treatment of obsessive-compulsive disorder: a meta-analysis. J Pharmacol Pharmacother 6(1):36–38

Saxena S, Brody AL, Maidment KM, Dunkin JJ, Colgan M, Alborzian S, Phelps ME, Baxter LR Jr (1999) Localized orbitofrontal and subcortical metabolic changes and predictors of response to paroxetine treatment in obsessive-compulsive disorder. Neuropsychopharmacology 21 (6):683–693

Saxena S, Brody AL, Ho ML, Zohrabi N, Maidment KM, Baxter LR Jr (2003) Differential brain metabolic predictors of response to paroxetine in obsessive-compulsive disorder versus major depression. Am J Psychiatry 160(3):522–532

Scheinost D, Stoica T, Wasylink S, Gruner P, Saksa J, Pittenger C, Hampson M (2014) Resting state functional connectivity predicts neurofeedback response. Front Behav Neurosci 8:338

Schirmbeck F, Zink M (2012) Clozapine-induced obsessive-compulsive symptoms in schizophrenia: a critical review. Curr Neuropharmacol 10(1):88–95

Schirmbeck F, Zink M (2013) Comorbid obsessive-compulsive symptoms in schizophrenia: contributions of pharmacological and genetic factors. Front Pharmacol 4:99

Schirmbeck F, Nieratschker V, Frank J, Englisch S, Rausch F, Meyer-Lindenberg A, Rietschel M, Zink M (2012) Polymorphisms in the glutamate transporter gene SLC1A1 and obsessive-compulsive symptoms induced by second-generation antipsychotic agents. Psychiatr Genet 22 (5):245–252

Schneier FR, Martinez D, Abi-Dargham A, Zea-Ponce Y, Simpson HB, Liebowitz MR, Laruelle M (2008) Striatal dopamine D(2) receptor availability in OCD with and without comorbid social anxiety disorder: preliminary findings. Depress Anxiety 25(1):1–7

Serata D, Kotzalidis GD, Rapinesi C, Janiri D, Di Pietro S, Callovini G, Piacentino D, Gasperoni C, Brugnoli R, Ferri VR, Girardi N, Tatarelli R, Ferracuti S, Angeletti G, Girardi P, Del Casale A

(2015) Are 5-HT3 antagonists effective in obsessive-compulsive disorder? A systematic review of literature. Hum Psychopharmacol 30(2):70–84
Serby M (2003) Methylphenidate-induced obsessive-compulsive symptoms in an elderly man. CNS Spectr 8(8):612–613
Shanahan NA, Holick Pierz KA, Masten VL, Waeber C, Ansorge M, Gingrich JA, Geyer MA, Hen R, Dulawa SC (2009) Chronic reductions in serotonin transporter function prevent 5-HT1B-induced behavioral effects in mice. Biol Psychiatry 65(5):401–408
Sharma LP, Thamby A, Balachander S, Janardhanan CN, Jaisoorya TS, Arumugham SS, Reddy YCJ (2020) Clinical utility of repeated intravenous ketamine treatment for resistant obsessive-compulsive disorder. Asian J Psychiatr 52:102183
Shin DJ, Jung WH, He Y, Wang J, Shim G, Byun MS, Jang JH, Kim SN, Lee TY, Park HY, Kwon JS (2014) The effects of pharmacological treatment on functional brain connectome in obsessive-compulsive disorder. Biol Psychiatry 75(8):606–614
Shmelkov SV, Hormigo A, Jing D, Proenca CC, Bath KG, Milde T, Shmelkov E, Kushner JS, Baljevic M, Dincheva I, Murphy AJ, Valenzuela DM, Gale NW, Yancopoulos GD, Ninan I, Lee FS, Rafii S (2010) Slitrk5 deficiency impairs corticostriatal circuitry and leads to obsessive-compulsive-like behaviors in mice. Nat Med 16(5):598–602, 591p following 602
Simpson HB, Lombardo I, Slifstein M, Huang HY, Hwang DR, Abi-Dargham A, Liebowitz MR, Laruelle M (2003) Serotonin transporters in obsessive-compulsive disorder: a positron emission tomography study with [(11)C]McN 5652. Biol Psychiatry 54(12):1414–1421
Simpson HB, Slifstein M, Bender J Jr, Xu X, Hackett E, Maher MJ, Abi-Dargham A (2011) Serotonin 2A receptors in obsessive-compulsive disorder: a positron emission tomography study with [11C]MDL 100907. Biol Psychiatry 70(9):897–904
Simpson HB, Shungu DC, Bender J Jr, Mao X, Xu X, Slifstein M, Kegeles LS (2012) Investigation of cortical glutamate-glutamine and gamma-aminobutyric acid in obsessive-compulsive disorder by proton magnetic resonance spectroscopy. Neuropsychopharmacology 37(12):2684–2692
Simpson HB, Foa EB, Liebowitz MR, Huppert JD, Cahill S, Maher MJ, McLean CP, Bender J Jr, Marcus SM, Williams MT, Weaver J, Vermes D, Van Meter PE, Rodriguez CI, Powers M, Pinto A, Imms P, Hahn CG, Campeas R (2013) Cognitive-behavioral therapy vs risperidone for augmenting serotonin reuptake inhibitors in obsessive-compulsive disorder: a randomized clinical trial. JAMA Psychiat 70(11):1190–1199
Sinopoli VM, Burton CL, Kronenberg S, Arnold PD (2017) A review of the role of serotonin system genes in obsessive-compulsive disorder. Neurosci Biobehav Rev 80:372–381
Skapinakis P, Papatheodorou T, Mavreas V (2007) Antipsychotic augmentation of serotonergic antidepressants in treatment-resistant obsessive-compulsive disorder: a meta-analysis of the randomized controlled trials. Eur Neuropsychopharmacol 17(2):79–93
Skapinakis P, Caldwell DM, Hollingworth W, Bryden P, Fineberg NA, Salkovskis P, Welton NJ, Baxter H, Kessler D, Churchill R, Lewis G (2016) Pharmacological and psychotherapeutic interventions for management of obsessive-compulsive disorder in adults: a systematic review and network meta-analysis. Lancet Psychiatry 3(8):730–739
Smeraldi E, Diaferia G, Erzegovesi S, Lucca A, Bellodi L, Moja EA (1996) Tryptophan depletion in obsessive-compulsive patients. Biol Psychiatry 40(5):398–402
Soomro GM, Altman D, Rajagopal S, Oakley-Browne M (2008) Selective serotonin re-uptake inhibitors (SSRIs) versus placebo for obsessive compulsive disorder (OCD). Cochrane Database Syst Rev 1:CD001765
Stein DJ, Andersen EW, Overo KF (2007) Response of symptom dimensions in obsessive-compulsive disorder to treatment with citalopram or placebo. Braz J Psychiatry 29(4):303–307
Stern ER, Shahab R, Grimaldi SJ, Leibu E, Murrough JW, Fleysher L, Parides MK, Coffey BJ, Burdick KE, Goodman WK (2019) High-dose ondansetron reduces activation of interoceptive and sensorimotor brain regions. Neuropsychopharmacology 44(2):390–398
Stewart SE, Jenike EA, Hezel DM, Stack DE, Dodman NH, Shuster L, Jenike MA (2010) A single-blinded case-control study of memantine in severe obsessive-compulsive disorder. J Clin Psychopharmacol 30(1):34–39

Stewart SE, Mayerfeld C, Arnold PD, Crane JR, O'Dushlaine C, Fagerness JA, Yu D, Scharf JM, Chan E, Kassam F, Moya PR, Wendland JR, Delorme R, Richter MA, Kennedy JL, Veenstra-VanderWeele J, Samuels J, Greenberg BD, McCracken JT, Knowles JA, Fyer AJ, Rauch SL, Riddle MA, Grados MA, Bienvenu OJ, Cullen B, Wang Y, Shugart YY, Piacentini J, Rasmussen S, Nestadt G, Murphy DL, Jenike MA, Cook EH, Pauls DL, Hanna GL, Mathews CA (2013a) Meta-analysis of association between obsessive-compulsive disorder and the 3′ region of neuronal glutamate transporter gene SLC1A1. Am J Med Genet B Neuropsychiatr Genet 162B(4):367–379

Stewart SE, Yu D, Scharf JM, Neale BM, Fagerness JA, Mathews CA, Arnold PD, Evans PD, Gamazon ER, Davis LK, Osiecki L, McGrath L, Haddad S, Crane J, Hezel D, Illman C, Mayerfeld C, Konkashbaev A, Liu C, Pluzhnikov A, Tikhomirov A, Edlund CK, Rauch SL, Moessner R, Falkai P, Maier W, Ruhrmann S, Grabe HJ, Lennertz L, Wagner M, Bellodi L, Cavallini MC, Richter MA, Cook EH Jr, Kennedy JL, Rosenberg D, Stein DJ, Hemmings SM, Lochner C, Azzam A, Chavira DA, Fournier E, Garrido H, Sheppard B, Umana P, Murphy DL, Wendland JR, Veenstra-VanderWeele J, Denys D, Blom R, Deforce D, Van Nieuwerburgh F, Westenberg HG, Walitza S, Egberts K, Renner T, Miguel EC, Cappi C, Hounie AG, do Rosario MC, Sampaio AS, Vallada H, Nicolini H, Lanzagorta N, Camarena B, Delorme R, Leboyer M, Pato CN, Pato MT, Voyiaziakis E, Heutink P, Cath DC, Posthuma D, Smit JH, Samuels J, Bienvenu OJ, Cullen B, Fyer AJ, Grados MA, Greenberg BD, McCracken JT, Riddle MA, Wang Y, Coric V, Leckman JF, Bloch M, Pittenger C, Eapen V, Black DW, Ophoff RA, Strengman E, Cusi D, Turiel M, Frau F, Macciardi F, Gibbs JR, Cookson MR, Singleton A, C. North American Brain Expression, Hardy J, U. K. B. E. Database, Crenshaw AT, Parkin MA, Mirel DB, Conti DV, Purcell S, Nestadt G, Hanna GL, Jenike MA, Knowles JA, Cox N, Pauls DL (2013b) Genome-wide association study of obsessive-compulsive disorder. Mol Psychiatry 18(7):788–798

Storch EA, Larson MJ, Shapira NA, Ward HE, Murphy TK, Geffken GR, Valerio H, Goodman WK (2006) Clinical predictors of early fluoxetine treatment response in obsessive-compulsive disorder. Depress Anxiety 23(7):429–433

Swedo SE, Schapiro MB, Grady CL, Cheslow DL, Leonard HL, Kumar A, Friedland R, Rapoport SI, Rapoport JL (1989) Cerebral glucose metabolism in childhood-onset obsessive-compulsive disorder. Arch Gen Psychiatry 46(6):518–523

Swedo SE, Pietrini P, Leonard HL, Schapiro MB, Rettew DC, Goldberger EL, Rapoport SI, Rapoport JL, Grady CL (1992) Cerebral glucose metabolism in childhood-onset obsessive-compulsive disorder. Revisualization during pharmacotherapy. Arch Gen Psychiatry 49 (9):690–694

Szechtman H, Sulis W, Eilam D (1998) Quinpirole induces compulsive checking behavior in rats: a potential animal model of obsessive-compulsive disorder (OCD). Behav Neurosci 112 (6):1475–1485

Szechtman H, Ahmari SE, Beninger RJ, Eilam D, Harvey BH, Edemann-Callesen H, Winter C (2017) Obsessive-compulsive disorder: Insights from animal models. Neurosci Biobehav Rev 76(Pt B):254–279

Szechtman H, Harvey BH, Woody EZ, Hoffman KL (2020) The psychopharmacology of obsessive-compulsive disorder: a preclinical roadmap. Pharmacol Rev 72(1):80–151

Talati A, Odgerel Z, Wickramaratne PJ, Norcini-Pala A, Skipper JL, Gingrich JA, Weissman MM (2017) Associations between serotonin transporter and behavioral traits and diagnoses related to anxiety. Psychiatry Res 253:211–219

Taylor S (2013) Molecular genetics of obsessive-compulsive disorder: a comprehensive meta-analysis of genetic association studies. Mol Psychiatry 18(7):799–805

Taylor S (2016) Disorder-specific genetic factors in obsessive-compulsive disorder: a comprehensive meta-analysis. Am J Med Genet B Neuropsychiatr Genet 171B(3):325–332

Ting JT, Feng G (2008) Glutamatergic synaptic dysfunction and obsessive-compulsive disorder. Curr Chem Genom 2:62–75

Torres AR, Fontenelle LH, Shhavitt RG, Hoexter MQ, Pittenger C, Miguel EC (2017) Epidemiology, comorbidity, and burden of OCD. In: Pittenger C (ed) Obsessive-compulsive disorder: phenomenology, pathophysiology, and treatment. Oxford University Press, New York
Trethowan WH, Scott PA (1955) Chlorpromazine in obsessive-compulsive and allied disorders. Lancet 268(6868):781–785
Uzun O (2010) Lamotrigine as an augmentation agent in treatment-resistant obsessive-compulsive disorder: a case report. J Psychopharmacol 24(3):425–427
van der Straten AL, Denys D, van Wingen GA (2017) Impact of treatment on resting cerebral blood flow and metabolism in obsessive compulsive disorder: a meta-analysis. Sci Rep 7(1):17464
van der Wee NJ, Stevens H, Hardeman JA, Mandl RC, Denys DA, van Megen HJ, Kahn RS, Westenberg HM (2004) Enhanced dopamine transporter density in psychotropic-naive patients with obsessive-compulsive disorder shown by [123I]{beta}-CIT SPECT. Am J Psychiatry 161 (12):2201–2206
Vandebroek I, Odberg FO (1997) Effect of apomorphine on the conflict-induced jumping stereotypy in bank voles. Pharmacol Biochem Behav 57(4):863–868
Veale D, Miles S, Smallcombe N, Ghezai H, Goldacre B, Hodsoll J (2014) Atypical antipsychotic augmentation in SSRI treatment refractory obsessive-compulsive disorder: a systematic review and meta-analysis. BMC Psychiatry 14:317
Vulink NC, Denys D, Westenberg HG (2005) Bupropion for patients with obsessive-compulsive disorder: an open-label, fixed-dose study. J Clin Psychiatry 66(2):228–230
Weizman A, Mandel A, Barber Y, Weitz R, Cohen A, Mester R, Rehavi M (1992) Decreased platelet imipramine binding in Tourette syndrome children with obsessive-compulsive disorder. Biol Psychiatry 31(7):705–711
Welch JM, Lu J, Rodriguiz RM, Trotta NC, Peca J, Ding JD, Feliciano C, Chen M, Adams JP, Luo J, Dudek SM, Weinberg RJ, Calakos N, Wetsel WC, Feng G (2007) Cortico-striatal synaptic defects and OCD-like behaviours in Sapap3-mutant mice. Nature 448(7156):894–900
Wendland JR, Moya PR, Timpano KR, Anavitarte AP, Kruse MR, Wheaton MG, Ren-Patterson RF, Murphy DL (2009) A haplotype containing quantitative trait loci for SLC1A1 gene expression and its association with obsessive-compulsive disorder. Arch Gen Psychiatry 66 (4):408–416
Williams K, Bloch MH, State MW, Pittenger C (2013) In: Buxbaum JD, Sklar P, Nestler EJ et al (eds) Tourette syndrome. Neurobiology of mental illness, 4th edn. Oxford University Press, New York
Willour VL, Yao Shugart Y, Samuels J, Grados M, Cullen B, Bienvenu OJ 3rd, Wang Y, Liang KY, Valle D, Hoehn-Saric R, Riddle M, Nestadt G (2004) Replication study supports evidence for linkage to 9p24 in obsessive-compulsive disorder. Am J Hum Genet 75(3):508–513
Wishart DS, Feunang YD, Guo AC, Lo EJ, Marcu A, Grant JR, Sajed T, Johnson D, Li C, Sayeeda Z, Assempour N, Iynkkaran I, Liu Y, Maciejewski A, Gale N, Wilson A, Chin L, Cummings R, Le D, Pon A, Knox C, Wilson M (2018) DrugBank 5.0: a major update to the DrugBank database for 2018. Nucleic Acids Res 46(D1):D1074–D1082
Woehrle NS, Klenotich SJ, Jamnia N, Ho EV, Dulawa SC (2013) Effects of chronic fluoxetine treatment on serotonin 1B receptor-induced deficits in delayed alternation. Psychopharmacology 227(3):545–551
Woolley JB, Heyman I (2003) Dexamphetamine for obsessive-compulsive disorder. Am J Psychiatry 160(1):183
Wu PL, Tang HS, Lane HY, Tsai CA, Tsai GE (2011) Sarcosine therapy for obsessive compulsive disorder: a prospective, open-label study. J Clin Psychopharmacol 31(3):369–374
Yazici KU, Percinel I (2015) N-Acetylcysteine augmentation in children and adolescents diagnosed with treatment-resistant obsessive-compulsive disorder: case series. J Clin Psychopharmacol 35 (4):486–489
Yin D, Zhang C, Lv Q, Chen X, Zeljic K, Gong H, Zhan S, Jin H, Wang Z, Sun B (2018) Dissociable frontostriatal connectivity: mechanism and predictor of the clinical efficacy of capsulotomy in obsessive-compulsive disorder. Biol Psychiatry 84(12):926–936

Yoshimura R, Kaneko S, Shinkai K, Nakamura J (2006) Successful treatment for obsessive-compulsive disorder with addition of low-dose risperidone to fluvoxamine: implications for plasma levels of catecholamine metabolites and serum brain-derived neurotrophic factor levels. Psychiatry Clin Neurosci 60(3):389–393

Yun JY, Jang JH, Kim SN, Jung WH, Kwon JS (2015) Neural correlates of response to pharmacotherapy in obsessive-compulsive disorder: individualized cortical morphology-based structural covariance. Prog Neuro-Psychopharmacol Biol Psychiatry 63:126–133

Zai G, Brandl EJ, Muller DJ, Richter MA, Kennedy JL (2014) Pharmacogenetics of antidepressant treatment in obsessive-compulsive disorder: an update and implications for clinicians. Pharmacogenomics 15(8):1147–1157

Zarate CA Jr, Singh JB, Carlson PJ, Brutsche NE, Ameli R, Luckenbaugh DA, Charney DS, Manji HK (2006) A randomized trial of an N-methyl-D-aspartate antagonist in treatment-resistant major depression. Arch Gen Psychiatry 63(8):856–864

Zheng H, Jia F, Han H, Wang S, Guo G, Quan D, Li G, Huang H (2019) Combined fluvoxamine and extended-release methylphenidate improved treatment response compared to fluvoxamine alone in patients with treatment-refractory obsessive-compulsive disorder: a randomized double-blind, placebo-controlled study. Eur Neuropsychopharmacol 29(3):397–404

Zike ID, Chohan MO, Kopelman JM, Krasnow EN, Flicker D, Nautiyal KM, Bubser M, Kellendonk C, Jones CK, Stanwood G, Tanaka KF, Moore H, Ahmari SE, Veenstra-VanderWeele J (2017) OCD candidate gene SLC1A1/EAAT3 impacts basal ganglia-mediated activity and stereotypic behavior. Proc Natl Acad Sci U S A 114(22):5719–5724

Zivin K, Pfeiffer PN, Bohnert AS, Ganoczy D, Blow FC, Nallamothu BK, Kales HC (2013) Evaluation of the FDA warning against prescribing citalopram at doses exceeding 40 mg. Am J Psychiatry 170(6):642–650

Zohar J, Kennedy JL, Hollander E, Koran LM (2004) Serotonin-1D hypothesis of obsessive-compulsive disorder: an update. J Clin Psychiatry 65(Suppl 14):18–21

Pharmacogenetics of Obsessive-Compulsive Disorder: An Evidence-Update

Gwyneth Zai

Contents

Abstract Pharmacogenetics has become increasingly important in the treatment of psychiatric disorders because approximately 50% of individuals who take psychotropic medications do not typically respond to them. Obsessive-compulsive disorder (OCD) is one such chronic and often debilitating mental illness with significant non-response to even the first-line medication, serotonin reuptake inhibitors. Precision medicine utilizing genetic testing panels has received significant attention based on the evidence that the variability of antidepressant response and adverse effects is partly due to the variability in an individual's genome. Thus, extensive research has investigated the role of genetic factors on antidepressant response in major depressive disorder (MDD) and their utility for guiding antidepressant treatment to

G. Zai (✉)
Neurogenetics Section, Molecular Brain Science Department, Campbell Family Mental Health Research Institute, and General Adult Psychiatry and Health Systems Division, Centre for Addiction and Mental Health, Toronto, ON, Canada

Department of Psychiatry, University of Toronto, Toronto, ON, Canada

Institute of Medical Science, University of Toronto, Toronto, ON, Canada

Behavioural and Clinical Neuroscience Institute, Department of Psychiatry, University of Cambridge, Cambridge, UK
e-mail: gwyneth.zai@camh.ca

Curr Topics Behav Neurosci (2021) 49: 385–398
https://doi.org/10.1007/7854_2020_205
Published Online: 7 February 2021

significantly improving outcomes in individuals with MDD. However, limited progress remains in the pharmacogenetics of OCD. This chapter will provide an overview of the recent findings in the pharmacogenetics of OCD. Promising results with limited replications have been reported for the cytochrome P450 liver metabolism genes in addition to several serotonergic and glutamatergic system genes, which may play an important role in antidepressant response in the treatment of OCD.

Keywords Antidepressant/drug/treatment response · Cytochrome P450 drug metabolism/system · Genetics · Obsessive-compulsive disorder (OCD) · Pharmacogenetics

1 Introduction

The increasing importance of the role of pharmacogenetics has emerged in the pharmacological treatment of mental illness. Evidence suggests that inter-individual genetic variations partly determine drug response and adverse effects. Extensive pharmacogenetic research in major depressive disorder (MDD) and schizophrenia has provided fundamental knowledge and support to utilize genetic data in the pharmacological treatment of these disorders. Genetic factors influencing antidepressant response have been well-established in MDD but have not been extensively studied in obsessive-compulsive disorder (OCD) (Zai et al. 2014).

Selective serotonin reuptake inhibitors (SSRIs) represent the first-line pharmacological treatment of OCD (Fineberg et al. 2015; Katzman et al. 2014; Koran et al. 2007). The general consensus and evidence from dose-finding-studies indicate that individuals with OCD tend to respond best at higher doses of SSRI. A clinically relevant improvement can be observed within 10–12 weeks at an adequate dose of SSRI treatment (Fineberg et al. 2015; Koran et al. 2007). The tricyclic antidepressant, clomipramine, is considered to be a second-line medication for the treatment of OCD due to its significant side-effect profile. It is also thought to exert its effect in OCD via serotonin reuptake inhibition (Katzman et al. 2014; Koran et al. 2007). However, typically, more than 50% of patients do not respond to the initial treatment. Dopamine antagonists, especially the atypical antipsychotics such as risperidone and aripiprazole (partial dopamine agonist), are useful as a first-line adjunctive medication for the treatment of OCD in patients with limited or partial response to a serotonin reuptake inhibitor (SRI) (Fineberg et al. 2015; Katzman et al. 2014; Koran et al. 2007).

Pharmacological treatment decisions for psychiatric disorders are traditionally made using guidance that has been derived from study data representative of the "average" patient and are not usually tailored to the individual patient. Pharmacogenetics may significantly contribute to better prediction of drug response and side effect tolerance by personalizing the choice of the best antidepressant for a particular patient, taking into account the genetic information that influences drug metabolism, drug transport, and drug mechanism. This should ultimately decrease the duration of patient suffering, reduce the number of medication trials, lessen adverse effects, lower health care costs, and improve patients' overall quality of life.

Our group has previously reviewed the pharmacogenetics of OCD (Brandl et al. 2012; Zai et al. 2014). At the time, there was a limited number of studies that reported promising findings related to several serotonergic and glutamatergic system genes as well as the cytochrome CYP450 liver enzyme genes, which appear to play a role in SRI-response in OCD. In this chapter, we have focused on the most robust recent studies that have emerged over the past 5 years.

2 Methods

We reviewed all studies on the pharmacogenetics of OCD that were published in PubMed, Medline, and Google Scholar after February 2014 (until December 2019). A summary and review of previous studies on the pharmacogenetics of OCD can be found in the review article by Zai et al. (2014). The specific search terms used included: genetics, pharmacogenetics, obsessive-compulsive gene response, treatment response, drug response, antidepressant response, antidepressant, OCD, obsessive-compulsive disorder, side effects, and adverse effects.

3 Results

Four original studies and two case reports were included in this review chapter. Given the scarcity of the published papers post-2014, we have also included a brief summary of several articles that may be important in OCD. Pharmacogenetics can be subdivided into two separate categories, pharmacokinetic factors including the cytochrome P450 liver enzyme system, and pharmacodynamic factors including the central nervous system (CNS). Table 1 presents all pharmacogenetic studies of OCD to date including studies prior to 2014.

3.1 Pharmacokinetic Factors

Evidence has supported the effects of cytochrome P450 liver enzymes in drug metabolism (Caley 2011; Flockhart 2007; Hicks et al. 2015, 2017). In particular, the cytochrome P450 2D6 liver enzyme (CYP2D6) metabolizes most antidepressants including all tricyclic antidepressants (TCAs) including clomipramine, the serotonin–norepinephrine reuptake inhibitor (SNRI) venlafaxine, and most selective SSRIs, predominantly fluoxetine and paroxetine. Another important cytochrome

Table 1 Pharmacogenetic studies of OCD to date ("+" indicates positive findings; "–" indicates negative findings; **bold** indicates recent studies)

Gene	Location	Results	References
CYP2D6	22q13.2	+	Brandl et al. (2014), **Brown et al.** (2017) **(case report), and** Müller et al. (2012)
		–	van Nieuverburge et al. (2009)
CYP2C19	10q23.33	+	Brandl et al. (2014)
DISP1	1q41-q42	–	**Qin et al.** (2016) **(GWAS) and Lisoway et al.** (2018)
HTR1B	6q14.1	–	Denys et al. (2007) and Miguita et al. (2011)
SLC1A1	9p24.2	+	**Abdolhosseinzadeh et al.** (2019)
BDNF	11p14.1	+	Real et al. (2009)
5HTR2A	13q14.2	+	Denys et al. (2007) and Zhang et al. (2004)
		–	Miguita et al. (2011) and Tot et al. (2003)
SLC6A4	17q11.2	+	**Ahmadipour et al.** (2018)**,** Billett et al. (1997), Denys et al.
		–	(2007), Di Bella et al. (2002), McDougle et al. (1998), Miguita et al. (2011) and Zhang et al. (2004)
COMT	22p11.21	+	Vulink et al. (2012)
		–	Miguita et al. (2011) and Zhang et al. (2004)
MAO-A	Xp11.3	–	Zhang et al. (2004)

P450 liver enzyme is CYP2C19, which metabolizes several TCAs (including clomipramine) and SSRIs including escitalopram and citalopram. Furthermore, certain genetic variations in *CYP2D6* and *CYP2C19* may alter the metabolism of their substrates, thereby affecting drug efficiency and tolerability. Although other cytochrome P450 liver enzymes such as CYP1A2, CYP2B6, CYP2C9, and CYP3A4 are also important for the metabolism of some classes of antidepressant including TCA, norepinephrine reuptake inhibitor (NDRI), bupropion, serotonin receptor antagonist and reuptake inhibitor (SARI), trazodone, and some SSRIs (to a variable degree), only limited pharmacogenetic studies of these enzymes and their effect on antidepressant response have been published in the literature (Altar et al. 2013) and none has so far been investigated in OCD. Thus, we have included only studies investigating *CYP2D6* and *CYP2C19* in this chapter.

The genotypes of the cytochrome P450 liver enzymes *CYP2D6* and *CYP2C19* provide an accurate prediction of their enzymatic activity levels in the absence of confounding factors such as co-medication. Enzymatic activity levels based on pharmacogenetic testing is generally grouped into four different metabolism classifications including: "poor" with no or very little enzymatic activity, "intermediate" with metabolism between poor and extensive, "extensive" with normal activity, and "ultra-rapid" with significant increase in metabolism (Jann and Cohen 2000). Having a poor metabolism status indicates that the individual will metabolize the substrate (i.e., SSRIs) at a very low/slow rate. This can in turn potentially increase side effects given the higher cumulative drug concentration that remains in the individual's body. Conversely, an individual with an ultra-rapid metabolism status will break down the substrate (i.e., SSRIs) at a higher/faster rate, resulting in low cumulative serum concentration of the medication and possibly a sub-therapeutic level. However, the association between plasma level and clinical outcome has not been clearly established thus far for psychotropic medications. Ethnicity also plays an important factor in the cytochrome P450 liver metabolism variability (Neafsey et al. 2009). On average, 10% of the general population are poor metabolizers, 35% are intermediate metabolizers, and 7% are ultra-rapid metabolizers. More specifically, for *CYP2D6*, the prevalence of poor metabolizers ranges from approximately 2 to 10% in Caucasians, 2 to 7% in Hispanics, and 1 to 2% in Asians, Africans, and African Americans (Bernard et al. 2006). Regarding *CYP2C19*, similar inter-ethnic differences have been observed, with Asians having a much higher prevalence (ranging from 12 to 23%) of poor metabolizers when compared to Caucasians (1 to 6%) and Black Africans (1 to 7.5%) (Desta et al. 2002).

Prescribing physicians can potentially adjust the dose of the medication according to the genetically predetermined liver metabolizing status by either lowering the dose in cases with poor metabolizing status or increasing the dose in cases with ultra-rapid metabolizing status. The Clinical Pharmacogenetics Implementation Consortium (CPIC) has published clinical guidelines for the recommended use of pharmacogenetic testing of *CYP2D6* and *CYP2C19* prior to prescribing TCAs (Hicks et al. 2017) and SSRIs (Hicks et al. 2015). The overall recommendation from CPIC is that pharmacogenetic tests are one of the several pieces of clinical information that should be considered before initiating drug therapy, though the relationship between the pharmacogenomic profile and the efficacy and safety of medications requires further investigation in controlled studies (FDA).

A recent case report from the USA (Brown et al. 2017) and two previous studies from our group (Brandl et al. 2013; Müller et al. 2012) have reported an influence of genetic variations in *CYP2D6* and *CYP2C19* on the medication response of individuals with OCD. One case report presented an individual with depression and OCD who had trialed and failed multiple psychotropic medications, many of which are subject to CYP2D6 and CYP2C19 metabolism (Brown et al. 2017). This patient had been taking clomipramine and pharmacogenetic testing revealed an intermediate metabolism status for both CYP2D6 and CYP2C19, which resulted in considerably elevated serum trough concentrations of clomipramine and its active metabolite, desmethylclomipramine. In this case, the prescribing physician made use of the genetic information (intermediate metabolizing status) as a guide to prescribing clomipramine at a lower than the standard dose. The author concluded that pharmacogenetic test results can provide guidance for adjusting medication dosage in OCD.

The first of the two previously published retrospective studies by Müller et al. (2012) included 39 patients with OCD who were assessed for SSRI (including fluoxetine, sertraline, fluvoxamine, paroxetine, and citalopram) response based on the Clinical Global Impression-Improvement (CGI-I) scale. The second study by Brandl et al. (2013) examined 184 individuals with OCD who had data related to treatment response, side-effect profile, in addition to CYP2D6 and CYP2C19 metabolism status. Although both Müller et al. (2012) and Brandl et al. (2013) did not detect a significant interaction between CYP2D6 or CYP2C19 metabolizer status and treatment response, Müller et al. (2012) reported that CYP2D6 poor metabolism status was nominally associated with lack of response to medication (i.e., close to the cut-off *P*-value for significance and unlikely to survive correction for multiple testing), whereas Brandl et al. (2013) observed CYP2D6 non-extensive metabolism (poor, intermediate, and ultra-rapid metabolism status) as having a significantly increased number of failed medication trials when compared to extensive metabolizers (48% with $\geq$4 trials versus 22% with $\geq$4 trials; $P = 0.007$). The extent to which the historical treatment-failures resulted from efficacy or tolerability issues was not fully clear. However, the use of venlafaxine [a treatment not found to be efficacious in OCD in randomized controlled trials but nevertheless recommended by some guidelines as a second-line treatment for OCD (Katzman et al. 2014)] in individuals with CYP2D6 non-extensive metabolism was found to be significantly associated with a higher side-effect burden (Brandl et al. 2013).

As yet there are no pharmacogenetic data examining a possible link between metabolism status and response to dopamine antagonists in SRI-resistant OCD.

To summarize, two cytochrome P450 liver enzyme genes, *CYP2D6* and *CYP2C19*, have been studied in relation to medication response in OCD. The genetic status of *CYP2D6* and *CYP2C19* has been implicated in adversely influencing either the clinical response to or tolerability of antidepressants including serotonin reuptake inhibitors and venlafaxine in OCD patients. However, the literature in relation to these pharmacokinetic genomics in OCD comprises retrospective studies of small sample size, and, since 2014, only case reports have been added. Thus, further validation of these findings is necessary using prospective designs, before genetic analysis for these CYP variants should be introduced routinely in OCD treatment.

3.2 Pharmacodynamic Factors

Pharmacodynamic factors affect drug actions at a molecular level within our central nervous system (CNS). Psychotropic medications act on various neurotransmitter systems, which hypothetically lead to a change in neurochemical balance. The most well-known and most studied systems are the serotonergic, glutamatergic, and dopaminergic systems, which interact with one another and are associated with drug response and adverse effects of medications in a broad range of disorders including OCD (Gvozdic et al. 2012). In this section, we present a brief overview of these neurotransmitter systems. However, there has been limited research into the specific genes across these systems and their effect on medication response in OCD. We identified one genome-wide association study (GWAS) (Qin et al. 2016) and one negative replication study (Lisoway et al. 2018) of the top finding from GWAS as well as two original articles (Abdolhosseinzadeh et al. 2019; Ahmadipour et al. 2018) for this section of the review.

3.3 Serotonergic System Genes

The serotonergic system has remained the primary focus of pharmacogenetic studies in OCD because of the efficacy of SSRIs, representing the first-line pharmacological treatment (APA 2007; CPA 2006; Fineberg et al. 2015; Katzman et al. 2014; Koran et al. 2007; NICE 2006). SSRIs increase the level of serotonin in the synaptic cleft by inhibiting its reuptake into the presynaptic cells, thereby allowing more serotonin to bind to the postsynaptic receptors and interact in a complex manner with other neurotransmitter systems. This is thought to be the mechanism by which these medications reduce OCD symptoms. Genes that are involved in serotonin metabolism, transport, storage, and the signaling cascade have been studied extensively in OCD (Pauls et al. 2014; Pauls 2010; Sinopoli et al. 2017) and those implicated in the pharmacogenetics of OCD include the serotonin transporter gene (*SLC6A4*) and its promoter (*HTTLPR*) and the serotonin 2A receptor gene (*HTR2A*).

The first evidence supporting the role of genes affecting the serotonergic system in the treatment response came from pharmacogenetic studies of MDD, with the most robust findings implicating *HTR2A* single nucleotide polymorphism (SNPs) (Fabbri et al. 2013; Singh et al. 2014). In addition to a number of unpublished studies prior to 2014 (Zai et al. 2014), one study examined the *SLC6A4* gene in 279 Iranian patients with OCD who were treated with 150–300 mg of fluvoxamine for 12 weeks (Ahmadipour et al. 2018) (Table 1). This study found an association between the single nucleotide polymorphism (SNP) rs25533 and OCD risk, but neither of the studied SNPs, rs25533 nor rs8914832 (I425V), were associated with fluvoxamine response.

To summarize, there are limited data informing the genetics of medication response across serotonergic system genes in OCD. Although one recent negative pharmacogenetic study in OCD was published (Ahmadipour et al. 2018), it did not investigate the major SNPs already implicated in the medication response in disorders such as depression and OCD. Thus, additional research effort will be required to further delineate the role of the serotonergic system genes in predicting medication response in individuals with OCD.

3.4 Glutamatergic System Genes

The glutamatergic system has been one of the most significantly implicated neurotransmitter systems in the etiology of OCD (Kariuki-Nyuthe et al. 2014; Wu et al. 2012). Support for the involvement of this system comes from research linking the complex neurotransmitter network, the cortico-striatal-thalamo-cortical (CSTC) circuit, in the underlying neuro-anatomical basis of OCD (Robbins et al. 2019; Pauls et al. 2014). Altered glutamatergic function has been hypothesized as one of the major drivers of observed overactivity in this brain circuit of individuals with OCD (Menzies et al. 2008; Milad and Rauch 2012). Glutamate has also been implicated in this disorder based on evidence from magnetic resonance spectroscopy (MRS) studies, knock-out mouse models (i.e., SAP90/PSD95-associated proteins 3 [SAPAP3] knock-out mice with OCD-like compulsive grooming behavior, SLIT and NTRK like family member 5 [SLITRK5] knock-out mice with excessive grooming), clinical studies of glutamatergic agents, and candidate gene association studies (i.e., solute carrier family 1 member 1 [*SLC1A1*], glutamate ionotropic receptor *N*-methyl-D-aspartate [NMDA] type subunit 2B [GRIN2B]) (Pauls et al. 2014; Wu et al. 2012). MRS studies showed that there is a possible tonic-phasic dysregulation of glutamate within the cortico-striatal circuitry in OCD with reduced tonic glutamate in anterior cingulate cortex combined with phasic overactivity in the striatum and orbitofrontal cortex (Wu et al. 2012).

Glutamatergic medications such as phenobarbital, lamotrigine, topiramate, riluzole, *N*-acetylcysteine (NAC), memantine, ketamine, and D-cycloserine have each been used as an augmenting agent in the management of OCD (Fineberg et al. 2015; Grados et al. 2013; Marinova et al. 2017; Kariuki-Nyuthe et al. 2014). One recent study reported an association between the activity of the glutamate transporter gene (*SLC1A1*) with fluvoxamine response in OCD (Abdolhosseinzadeh et al. 2019). Two hundred and forty-three individuals with OCD were treated with 150–300 mg of fluvoxamine over a 12-week period and symptom severity was recorded using the Yale-Brown Obsessive-Compulsive Scale (Y-BOCS). The authors detected significant associations between two of the three SNPs examined, rs2228622 (intronic synonymous SNP with no known functional implication) and rs3789413 (intronic SNP with unknown functional significance) with fluvoxamine response.

The first and only GWAS of medication response was performed by Qin et al. (2016) with 804 OCD cases. However, drug response status was determined based on retrospective recall using the Clinical Global Impression-Improvement scale, which is subject to confounding influences. The SNP most implicated was rs17162912, which is near the dispatched RND transporter family member 1 gene (*DISP1*) mapped onto chromosome 1q41–q42. However, when our group conducted a replication study (hypothesis-driven candidate gene study), investigating the *DISP1* rs17162912 with medication response in 112 Caucasian individuals with OCD, the result was negative (Lisoway et al. 2018).

To summarize, so far we have only small and inconsistent findings in relation to the role of glutamatergic genes and OCD treatment. However, in the light of evidence from the use of glutamatergic agents in the treatment of OCD, and genetic and neurobiological studies implicating the role of glutamate in the etiology of OCD, further investigation in this area is warranted.

3.5 Dopaminergic System Genes

Clinical trials of the use of antipsychotic medications, preclinical, neuroimaging, and neurochemical studies have provided evidence for dopamine dysfunction in the complex etiology of OCD (Koo et al. 2010; Pauls et al. 2014; Pauls 2010). Further evidence of dopamine influence is provided by neuroimaging, neurochemical, and candidate gene studies (Koo et al. 2010). Dopamine antagonist and partial agonist medications have established utility as augmentors of antidepressants, specifically risperidone and aripiprazole, and are first-line adjunctive medications for the treatment of treatment resistant OCD (APA 2007; Dold et al. 2013; Fineberg et al. 2015; Katzman et al. 2014).

Dopaminergic system genes have thus received some interest in genetic association studies of OCD (Zai et al. 2019). However, only limited attention has been focused on the pharmacogenetics of dopamine-related genes in OCD and no new studies have been published in the past 5 years. Two pharmacogenetic studies, examining the dopaminergic genes, dopamine D2 receptor (*DRD2*) and dopamine D4 receptor (*DRD4*), in OCD (Viswanath et al. 2013; Vulink et al. 2012) were previously reported. Only the study by Vulink et al. (2012) reported a functional SNP, Val158Met (rs4680; Val with higher COMT enzymatic activity and Met with lower COMT enzymatic activity), within the catechol-O-methyltransferase gene (*COMT*) to be associated with the citalopram response. Specifically, 48% of the responders to citalopram had COMT Met/Met genotype as compared to 0% of the non-responders ($\chi^2 = 10.06$, $P = 0.007$), implicating the Met/Met genotype with a better response to citalopram.

To summarize, limited pharmacogenetics effort has been concentrated on the dopaminergic system genes in OCD. Additional investigations are needed to clarify the role of dopamine-related genes in predicting response to SSRIs, dopamine antagonists, or other medications in OCD.

4 Conclusion and Future Perspective

Considering its high prevalence and morbidity, there has been a significant increase in interest in the pharmacogenetics of OCD over the past decade in the search to refine and improve treatment outcomes. However, although a number of international research groups have examined candidate genetic factors that might potentially determine medication response in OCD, there remains a paucity of reliable data upon which to base conclusions. Specifically, there are no definitive data supporting a single genetic variation or gene that determines clinical response to medication in OCD. However, many SNPs across specific genes within the cytochrome P450 (i.e., CYP2D6, CYP2C19), glutamatergic, and serotonergic systems appear interesting and potentially involved (Table 1).

Currently, recommendations have been provided by the FDA (https://www.fda.gov/drugs/science-and-research-drugs/table-pharmacogenomic-biomarkers-drug-labeling; https://www.fda.gov/medical-devices/precision-medicine/table-pharmacogenetic-associations) and CPIC (Hicks et al. 2015; https://cpicpgx.org/guidelines/guideline-for-selective-serotonin-reuptake-inhibitors-and-cyp2d6-and-cyp2c19/) that relate to pharmacogenetic testing of *CYP2D6* and *CYP2C19* when considering the use of SSRIs and TCAs such as clomipramine. These recommendations are not related to psychiatric diagnosis but instead give guidance related to the medications being prescribed (https://ispg.net/genetic-testing-statement/). Ethical, legal, and social implications must be considered with the use of genetic profiles in maximizing benefits and minimizing harms to individuals and society (Lazaro-Munoz et al. 2019). According to the recently updated expert statement on Genetic Testing and Psychiatric Disorders from the International Society of Psychiatric Genetics (https://ispg.net/genetic-testing-statement/), whereas evidence to support widespread use of pharmacogenetic tests at this time is still inconclusive, pharmacogenetic testing may nevertheless be viewed as a decision-support tool to assist implementation of good clinical care. If pharmacogenetic testing results are already available, clinicians are encouraged to integrate this information into medication selection and dosing decisions.

Taking account of the fact that allelic variation in the genes encoding the CYP2D6 and CYP2C19 enzymes leads to differences in the metabolism of medications relevant to OCD, and considering that individuals with extreme drug metabolism phenotypes have been shown to be at increased risk for adverse events, and some evidence suggests such individuals are less likely to respond well to treatment (Milosavljevic et al. 2020), genetic information for CYP2C19 and CYP2D6 would probably be most beneficial for those patients with OCD who have experienced an inadequate response or adverse reaction to a previous treatment trial, or those for whom pharmacokinetic interactions with co-prescribed medication involving these CYP enzymes could be predicted. Indeed, in patients with OCD known to carry CYP2D6 alleles associated with unusually rapid or slow metabolism, it may be advisable to use an SSRI that is not predominantly metabolized by CYP2D6 such as escitalopram, citalopram, or sertraline. In such cases, if clomipramine is required, it would be advisable to check the serum clomipramine level at a sub-therapeutic dose in order to reduce the risk of adverse effects. Alternatively, in the case of carriers of CYP2C19 alleles associated with ultra-rapid metabolism, it may be advisable to avoid citalopram and escitalopram due to the likelihood that "supra-formulary" doses of escitalopram or citalopram would be needed to produce a clinical effect. Such doses are currently contraindicated by many regulatory authorities owing to the risk of QTc prolongation. In contrast, carriers of CYP2C19 alleles associated with poor drug metabolism may benefit from receiving a reduction in starting dose with the maximum recommended dose of citalopram at 20 mg per day (https://www.fda.gov/drugs/drug-safety-and-availability/fda-drug-safety-communication-revised-recommendations-celexa-citalopram-hydrobromide-related).

As multiple genetic factors are likely to be involved in influencing any given patient's medication response and side effects, the development of prediction algorithms that include SNPs across multiple genes and at remote locations that regulate gene activity will no doubt be an important next step to refine the current pharmacogenetic testing panel. Future research should focus on increasing sample sizes and fine mapping these chromosomal regions by investigating them in more detail and examining genetic markers located closer together, in order to replicate these findings and to refine our understanding of the functionality of these genetic variations and their relation to the pathophysiology of OCD and antidepressant response. Computational and bioinformatics advances to address the complex facets (i.e., clinical, pharmacokinetic, pharmacodynamic, genetics, and environmental factors) of each patient will be required. Identification of genetic variations that predict drug response and tolerability will likely lead to precision medicine in the relatively near future, with attendant improvement in clinical outcome and reduction in morbidity and mortality.

References

Abdolhosseinzadeh S, Sina M, Ahmadiani A, Asadi S, Shams J (2019) Genetic and pharmacogenetic study of glutamate transporter (SLC1A1) in Iranian patients with obsessive-compulsive disorder. J Clin Pharm Ther 44(1):39–48

Ahmadipour E, Rashidi FS, Ahmadiani A, Asadi S, Shams J (2018) SLC6A4 polymorphisms rs25533 and I425V: association with obsessive-compulsive disorder and its treatment response in Iranian patients. Personalized Med Psychiatry 11-12:23–29

Altar CA, Hornberger J, Shewade A, Cruz V, Garrison J, Mrazek D (2013) Clinical validity of cytochrome P450 metabolism and serotonin gene variants in psychiatric pharmacotherapy. Int Rev Psychiatry 25(5):509–533

American Psychiatric Association (APA) (2007) Practice guideline for the treatment of patients with obsessive-compulsive disorder. Arlington, VA, USA. http://www.psych.org/psych_pract/treatg/pg/prac_guide.cfm

Bernard S, Neville KA, Nguyen AT, Flockhart DA (2006) Interethnic differences in genetic polymorphisms of CYP2D6 in the U.S. population: clinical implications. Oncologist 11 (2):126–135

Billett EA, Richter MA, King N, Heils A, Lesch KP, Kennedy JL (1997) Obsessive compulsive disorder, response to serotonin reuptake inhibitors and the serotonin transporter gene. Mol Psychiatry 2(5):403–406

Brandl EJ, Müller DJ, Richter MA (2012) Pharmacogenetics of obsessive-compulsive disorders. Pharmacogenomics 13(1):71–81

Brandl EJ, Tiwari AK, Zhou X, Deluce J, Kennedy JL, Müller DJ, Richter MA (2013) Influence of CYP2D6 and CYP2C19 gene variants on antidepressant response in obsessive-compulsive disorder. Pharmacogenomics J 14(2):176–181

Brandl EJ, Tiwari AK, Zhou X, Deluce J, Kennedy JL, Muller DJ, Richter MA (2014) Influence of CYP2D6 and CYP2C19 gene variants on antidepressant response in obsessive-compulsive disorder. Pharm J 14(2):176–181

Brown JT, Schneiderhan M, Eum S, Bishop JR (2017) Serum clomipramine and desmethylclomipramine levels in a CYP2C19 and CYP2D6 intermediate metabolizer. Pharmacogenomics 18(7):601–605

Caley CF (2011) Interpreting and applying CYP450 genomic test results to psychotropic medications. J Pharm Pract 24(5):439–446

Canadian Psychiatric Association (CPA) (2006) Clinical practice guidelines: management of anxiety disorders. Can J Psychiatr 51(2):43S–49S

Denys D, Van Nieuwerburge F, Deforce D, Westenberg HG (2007) Prediction of response to paroxetine and venlafaxine by serotonin-related genes in obsessive-compulsive disorder in a randomized, double-blind trial. J Clin Psychiatry 68(5):747–753

Desta Z, Zhao X, Shin JG, Flockhart DA (2002) Clinical significance of the cytochrome P450 2C19 genetic polymorphism. Clin Pharmacokinet 41(12):913–958

Di Bella D, Erzegovesi S, Cavallini MC, Bellodi L (2002) Obsessive-compulsive disorder, 5-HTTLPR polymorphism and treatment response. Pharm J 2(3):176–181

Dold M, Aigner M, Lanzenberger R, Kasper S (2013) Antipsychotic augmentation of serotonin reuptake inhibitors in treatment-resistant obsessive-compulsive disorder: a meta-analysis of double-blind, randomized, placebo-controlled trials. Int J Neuropsychopharmacol 16 (3):557–574

Fabbri C, Di Girolamo G, Serretti A (2013) Pharmacogenetics of antidepressant drugs: an update after almost 20 years of research. Am J Med Genet B Neuropsychiatr Genet 162B(6):487–520

Fineberg NA, Reghunandanan S, Simpson HB, Phillips KA, Richter MA, Matthews K, Stein DJ, Sareen J, Brown A, Sookman D (2015) Accreditation Task Force of The Canadian Institute for Obsessive Compulsive Disorders. Obsessive-compulsive disorder (OCD): practical strategies for pharmacological and somatic treatment in adults. Psychiatry Res 227(1):114–125

Flockhart DA (2007) Drug interactions: cytochrome P450 drug interaction table. Indiana University School of Medicine. http://medicine.iupui.edu/clinpharm/ddis/clinical-table/. Accessed 10 Jan 2014

Grados MA, Specht MW, Sung HM, Fortune D (2013) Glutamate drugs and pharmacogenetics of OCD: a pathway-based exploratory approach. Expert Opin Drug Discov 8(12):1515–1527

Gvozdic K, Brandl EJ, Taylor DL, Müller DJ (2012) Genetics and personalized medicine in antidepressant treatment. Curr Pharm Des 18(36):5853–5878

Hicks JK, Bishop JR, Sangkuhl K, Müller DJ, Li U, Leckband SG, Leeder JS, Graham RL, Chiulli DL, LLerena A, Skaar TC, Scott SA, Stingl JC, Klein TE, Caudle KE, Gaedigk A (2015) Clinical Pharmacogenetics Implementation Consortium (CPIC) guideline for CYP2D6 and CYP2C19 genotypes and dosing of selective serotonin reuptake inhibitors. Clin Pharmacol Ther 98(2):127–134

Hicks JK, Sangkuhl K, Swen JJ, Ellingrod VL, Müller DJ, Shimoda K, Bishop JR, Kharasch ED, Skaar TC, Gaedigk A, Dunnenberger HM, Klein TE, Caudle KE, Stingl JC (2017) Clinical Pharmacogenetics Implementation Consortium guideline (CPIC) for CYP2D6 and CYP2C19 genotypes and dosing of tricyclic antidepressants: 2016 update. Clin Pharmacol Ther 102 (1):37–44

Jann MW, Cohen LJ (2000) The influence of ethnicity and antidepressant pharmacogenetics in the treatment of depression. Drug Metabol Drug Interact 16(1):39–67

Kariuki-Nyuthe C, Gomez-Mancilla B, Stein DJ (2014) Obsessive-compulsive disorder and the glutamatergic system. Curr Opin Psychiatry 27(1):32–37

Katzman MA, Bleau P, Blier P, Chokka P, Kjernisted K, Van Ameringen M, Canadian Anxiety Guidelines Initiative Group on behalf of the Anxiety Disorders Association of Canada/Association Canadienne des troubles anxieux and McGill University (2014) Canadian clinical practice guidelines for the management of anxiety, posttraumatic stress and obsessive-compulsive disorders. BMC Psychiatry 14(Suppl 1):S1

Koo MS, Kim EJ, Roh D, Kim CH (2010) Role of dopamine in the pathophysiology and treatment of obsessive-compulsive disorder. Expert Rev Neurother 10(2):275–290

Koran LM, Hanna GL, Hollander E, Nestadt G, Simpson HB, American Psychiatric Association (2007) Practice guideline for the treatment of patients with obsessive-compulsive disorder. Am J Psychiatry 164(7 Suppl):5–53

Lazaro-Munoz G, Sabatello M, Huckins L, Peay H, Degenhardt F, Meiser B et al (2019) International society of psychiatric genetics ethics committee: issues facing us. Am J Med Genet B Neuropsychiatr Genet 180(8):543–554

Lisoway AJ, Zai G, Tiwari AK, Zai CC, Wigg K, Goncalves V, Zhang D, Freeman N, Müller DJ, Kennedy JL, Richter MA (2018) Pharmacogenetic evaluation of a DISP1 gene variant in antidepressant treatment of obsessive-compulsive disorder. Hum Psychopharmacol 33(4):e2659

Marinova Z, Chuang DM, Fineberg N (2017) Glutamate-modulating drugs as a potential therapeutic strategy in obsessive-compulsive disorder. Curr Neuropharmacol 15(7):977–995

McDougle CJ, Epperson CN, Price LH, Gelernter J (1998) Evidence for linkage disequilibrium between serotonin transporter protein gene (*SLC6A4*) and obsessive compulsive disorder. Mol Psychiatry 3(3):270–273

Menzies L, Chamberlain SR, Laird AR, Thelen SM, Sahakian BJ, Bullmore ET (2008) Integrating evidence from neuroimaging and neuropsychological studies of obsessive-compulsive disorder: the orbitofronto-striatal model revisited. Neurosci Biobehav Rev 32(3):525–549

Miguita K, Cordeiro Q, Shavitt RG, Miguel EC, Vallada H (2011) Association study between genetic monoaminergic polymorphisms and OCD response to clomipramine treatment. Arq Neuropsiquiatr 69(2B):283–287

Milad MR, Rauch SL (2012) Obsessive-compulsive disorder: beyond segregated cortico-striatal pathways. Trends Cogn Sci 16(1):43–51

Milosavljevic F, Bukvic N, Pavlovic Z, Miljevic C, Pesic V, Molden E, Ingelman-Sundberg M, Leucht S, Jukic MM (2020) Association of CYP2C19 and CYP2D6 poor and intermediate metabolizer status with antidepressant and antipsychotic exposure: a systematic review and meta-analysis. JAMA Psychiatry:e203643. http://doi.org/10.1001/jamapsychiatry.2020.3643

Müller DJ, Brandl EJ, Hwang R, Tiwari AK, Sturgess JE, Zai CC, Lieberman JA, Kennedy JL, Richter MA (2012) The AmpliChip® CYP450 test and response to treatment in schizophrenia and obsessive-compulsive disorder: a pilot study and focus on cases with abnormal CYP2D6 drug metabolism. Genet Test Mol Biomarkers 16(8):897–903

National Collaborating Centre for Mental Health (UK) (2006) Obsessive-compulsive disorder: core interventions in the treatment of obsessive-compulsive disorder and body dysmorphic disorder. British Psychological Society, Leicester (NICE Clinical Guidelines, No. 31.) https://www.ncbi.nlm.nih.gov/books/NBK56458/

Neafsey P, Ginsberg G, Hattis D, Sonawane B (2009) Genetic polymorphism in cytochrome P450 2D6 (CYP2D6): population distribution of CYP2D6 activity. J Toxicol Environ Health B Crit Rev 12(5-6):334–361

Pauls DL (2010) The genetics of obsessive-compulsive disorder: a review. Dialogues Clin Neurosci 12(2):149–163

Pauls DL, Abramovitch A, Rauch SL, Geller DA (2014) Obsessive-compulsive disorder: an integrative genetic and neurobiological perspective. Nat Rev Neurosci 15(6):410–424

Real E, Gratacòs M, Soria V, Escaramís G et al (2009) A brain-derived neurotrophic factor haplotype is associated with therapeutic response in obsessive-compulsive disorder. Biol Psychiatry 66(7):674–680

Robbins TW, Vaghi MM, Banca P (2019) Obsessive-compulsive disorder: puzzles and prospects. Neuron 102(1):27–47

Qin H, Samuels JF, Wang Y, Zhu Y, Grados MA, Riddle MA, Greenberg BD, Knowles JA, Fyer AJ, McCracken JT, Murphy DL, Rasmussen SA, Cullen BA, Piacentini J, Geller D, Stewart SE, Pauls D, Bienvenu OJ, Goes FS, Maher B, Pulver AE, Valle D, Lange C, Mattheisen M, McLaughlin NC, Liang KY, Nurmi EL, Askland KD, Nestadt G, Shugart YY (2016) Whole-genome association analysis of treatment response in obsessive-compulsive disorder. Mol Psychiatry 21(2):270–276

Singh AB, Bousman CA, Ng C, Berk M (2014) Antidepressant pharmacogenetics. Curr Opin Psychiatry 27:43–51

Sinopoli VM, Burton CL, Kronenberg S, Arnold PD (2017) A review of the role of serotonin system genes in obsessive-compulsive disorder. Neurosci Biobehav Rev 80:372–381

Tot S, Erdal ME, Yazici K, Yazici AE, Metin O (2003) T102C and -1438 G/A polymorphisms of the *5-HT2A* receptor gene in Turkish patients with obsessive-compulsive disorder. Eur Psychiatry 18(5):249–254

Van Nieuverburge FC, Denys DA, Westenberg HG, Deforce DL (2009) Response to serotonin reuptake inhibitors in OCD is not influenced by common CYP2D6 polymorphisms. Int J Psychiatry Clin Pract 13(1):345–348

Viswanath B, Taj MJRJ, Purushottam M, Kandavel T, Shetty PH, Reddy YC, Jain S (2013) No association between DRD4 gene and SRI treatment response in obsessive-compulsive disorder: need for a novel approach. Asian J Psychiatr 6(4):347–348

Vulink NC, Westenberg HG, van Nieuwerburgh F, Deforce D, Fluitman SB, Meinardi JS, Denys D (2012) Catechol-O-methyltransferase gene expression is associated with response to citalopram in obsessive-compulsive disorder. Int J Psychiatry Clin Pract 16(4):277–283

Wu K, Hanna GL, Rosenberg DR, Arnold PD (2012) The role of glutamate signaling in the pathogenesis and treatment of obsessive-compulsive disorder. Pharmacol Biochem Behav 100 (4):726–735

Zai G, Brandl EJ, Müller DJ, Richter MA, Kennedy JL (2014) Pharmacogenetics of antidepressant treatment in obsessive-compulsive disorder: an update and implications for clinicians. Pharmacogenomics 15(8):1147–1157

Zai G, Barta C, Cath D, Eapen V, Geller D, Grünblatt E (2019) New insights and perspectives on the genetics of obsessive-compulsive disorder. Psychiatr Genet 29(5):142–151

Zhang L, Liu X, Li T, Yang Y, Hu X, Collier D (2004) Molecular pharmacogenetic studies of drug responses to obsessive-compulsive disorder and six functional genes. Zhonghua Yi Xue Yi Chuan Xue Za Zhi 21(5):479–481

Invasive and Non-invasive Neurostimulation for OCD

Isidoor O. Bergfeld, Eva Dijkstra, Ilse Graat, Pelle de Koning, Bastijn J. G. van den Boom, Tara Arbab, Nienke Vulink, Damiaan Denys, Ingo Willuhn, and Roel J. T. Mocking

Contents

I. O. Bergfeld
Department of Psychiatry, Amsterdam UMC, University of Amsterdam, Amsterdam, The Netherlands

Amsterdam Brain and Cognition, Amsterdam, The Netherlands

E. Dijkstra, I. Graat, P. de Koning, N. Vulink, and R. J. T. Mocking (✉)
Department of Psychiatry, Amsterdam UMC, University of Amsterdam, Amsterdam, The Netherlands
e-mail: R.J.Mocking@amsterdamumc.nl

B. J. G. van den Boom and T. Arbab
Department of Psychiatry, Amsterdam UMC, University of Amsterdam, Amsterdam, The Netherlands

Netherlands Institute for Neuroscience, Royal Netherlands Academy of Arts and Sciences, Amsterdam, The Netherlands

D. Denys and I. Willuhn
Department of Psychiatry, Amsterdam UMC, University of Amsterdam, Amsterdam, The Netherlands

Amsterdam Brain and Cognition, Amsterdam, The Netherlands

Netherlands Institute for Neuroscience, Royal Netherlands Academy of Arts and Sciences, Amsterdam, The Netherlands

Curr Topics Behav Neurosci (2021) 49: 399–436
https://doi.org/10.1007/7854_2020_206
Published Online: 7 February 2021

Abstract It becomes increasingly clear that (non-)invasive neurostimulation is an effective treatment for obsessive-compulsive disorder (OCD). In this chapter we review the available evidence on techniques and targets, clinical results including a meta-analysis, mechanisms of action, and animal research. We focus on deep brain stimulation (DBS), but also cover non-invasive neurostimulation including transcranial magnetic stimulation (TMS). Data shows that most DBS studies target the ventral capsule/ventral striatum (VC/VS), with an overall 76% response rate in treatment-refractory OCD. Also TMS holds clinical promise. Increased insight in the normalizing effects of neurostimulation on cortico-striatal-thalamic-cortical (CSTC) loops – through neuroimaging and animal research – provides novel opportunities to further optimize treatment strategies. Advancing clinical implementation of neurostimulation techniques is essential to ameliorate the lives of the many treatment-refractory OCD patients.

Keywords Deep brain stimulation · Neuromodulation · Transcranial magnetic stimulation · Treatment outcome · Working mechanisms

1 Introduction

Obsessive-compulsive disorder (OCD) is a disabling psychiatric disorder that, if left untreated, generally follows a chronic course. Conventionally, clinical management of OCD comprises cognitive behavior therapy (CBT) and pharmacotherapy (Abramowitz 2006; Denys 2006; Foa et al. 2005; March et al. 1997; O'Connor et al. 2006). Both treatments have their limitations: drugs have significant side effects and the exposure and response prevention in CBT often provoke intense anxiety resulting in a 25% drop out of patients (Franklin et al. 2000). Moreover, even when optimal treatment is provided, more than an estimated 10% of patients remain treatment-refractory, with enduring disabling consequences for patients, their families, and society (Denys 2006).

In the past decades, neurostimulation has emerged as a new treatment option for patients with treatment-refractory OCD. Neurostimulation is a rapidly evolving field of therapy, altering neural activity through delivery of electrical stimulation to specific targeted brain areas. The modulation of neural activity is effected by means of invasive microelectrodes such as deep brain stimulation (DBS) or non-invasive stimulation, such as transcranial magnetic stimulation (TMS), transcranial direct current stimulation (tDCS), or transcranial alternating current stimulation (tACS). We focus our review on the main applications for treating therapy-refractory OCD patients: DBS and TMS.

DBS is a treatment that involves the implantation of electrodes that send electrical impulses to specific locations in the brain, selected according to the type of symptoms to be addressed. DBS permits focal, adjustable, and reversible neuromodulation of the dysfunctional neural network that eventually leads to improvement of severe neurological or psychiatric symptoms in difficult-to-treat patient populations.

TMS is a form of stimulation in which an alternating magnetic field is used to cause electric current through electromagnetic induction. TMS is non-invasive, and does not require surgery or electrode implantation. During the procedure, a magnetic coil that is connected to an electric pulse generator is positioned at the head of the person receiving the treatment. The electric pulse generator produces a changing electric current within the coil, which induces a magnetic field that passes through the scalp. Directing the pulse at a targeted area in the cortex causes a localized electrical current, which can either depolarize or hyperpolarize neurons at that site. Essentially, "single-pulse" TMS is used to investigate brain functioning, whereas repetitive TMS (rTMS) is used to modulate brain activity that lasts beyond the stimulation period. During rTMS, variations in stimulation parameters, such as target, frequency, and duration might be used for the treatment of various neurological and psychiatric disorders, including OCD.

In the following chapter, we will review techniques and targets of DBS, clinical results, mechanism of action in humans and animals, and rTMS.

2 Deep Brain Stimulation

2.1 Techniques and Targets

The DBS system comprises three parts: an implanted pulse generator (IPG), the lead (a coated wire with four electrodes at the tip), and the extension cables that connect the lead to the IPG.

The electrical impulses are generated by the IPG, which includes a battery, and is surgically implanted under the skin in the chest, below the collarbone or in the abdomen. In bilateral DBS (which is usually performed in OCD), two insulated extension cables run from the IPG under the skin of the neck and scalp, through burr holes in the skull, and connect to the leads, which are bilaterally surgically implanted in the brain. The precise anatomical position of implantation of the leads is calculated beforehand, commonly on the basis of magnetic resonance scans. Recent studies show that diffusion weighted imaging (DWI) could be of additional value in specifying the target (Liebrand et al. 2019).

As IPGs will eventually become depleted, regular checks are carried out to monitor the IPG's power status. Anticipating on forthcoming depletion, the IPG is replaced during a short, usually outpatient procedure. The recent introduction of rechargeable (r)IPGs increased device longevity from 3 years to approximately 10–15 years.

Clinicians adjust the IPG activity of the electrodes using a portable appliance communicating with the IPG through telemetry. Adjustments can be made to the current (voltage), frequency, pulse width, and by activating separate electrodes on the lead. Typical stimulation parameters for DBS in OCD mostly vary within the following ranges: current between 2 and 6 V, frequency between 100 and 185 Hz (high-frequency stimulation), and pulse widths between 60 and 150 ms. The specific location of the stimulation can be adjusted by activating one or more of the electrodes on the lead.

Following implantation, parameters are adjusted to maximize symptom reduction and minimize side effects. Typically, the adjustment phase takes three to 6 months, but may last up to 12 months or more until the best outcome is achieved. Parameters are adjusted every 2 weeks on the basis of an individual patient's response and side effects (Denys et al. 2020). In some patients, symptoms decrease within days or weeks, but in most patients it can take months before an optimal effect is reached. When adjustment appears to result in a clinically relevant response, the stimulation is maintained continuously. In the event of absolute non-response and/or occurrence of side effects, the device can be switched off and may eventually be removed. As a final step, some centers add CBT when a plateau of effectiveness is reached after parameter optimization (Mantione et al. 2014).

2.1.1 Targets

Most centers use a target in striatal and surrounding capsular areas, which are specified in different papers as bed nucleus stria terminalis, nucleus accumbens, (ventral) anterior limb of the internal capsule, ventral caudate or ventral capsule/ ventral striatum (VC/VS). In this section, we will refer to all of these targets as VC/VS, although exact positioning might differ between studies based on target coordinates, angle of the implantation, and/or electrode type used. Other targets used for OCD are the subthalamic nucleus (STN) and the inferior thalamic peduncle (ITP).

2.2 *Clinical Results*

Since the seminal report of Nuttin and colleagues in 1999 (Nuttin et al. 1999), clinical results of more than 40 papers on approximately 240 unique patients have been published. All clinical trials define a clinical response as $\geq$35% reduction of the Yale-Brown Obsessive-Compulsive Scale (Y-BOCS) score. Most papers also classify patients with a Y-BOCS score reduction between 25 and 35% as partial responders. To estimate the efficacy of DBS for OCD, we have meta-analyzed the clinical results of all papers with four or more included patients with a follow-up of 12 months or more to give a general overview of the clinical results. We wish to clarify that the meta-analyses and forest plots are shown as a service to readers, to demonstrate the sample sizes and the confidence intervals associated with the published results. We did not perform a quality assessment of the trials. An extensive meta-analysis including sensitivity analyses, meta-regressions, and systematic review with risk-of-bias and GRADE assessments is beyond the scope of this general overview.

We found 14 papers reporting response rates and Y-BOCS scores, of which 13 reported open-label results at 12 months of follow-up. These included 10 papers on VC/VS DBS (Denys et al. 2020; Fayad et al. 2016; Greenberg et al. 2006; Huff et al. 2010; Huys et al. 2019; Islam et al. 2015; Luyten et al. 2016; Menchon et al. 2019; Roh et al. 2012; Tsai et al. 2012), which is by far the most extensively investigated target. In addition, we included two papers on STN DBS (Chabardes et al. 2013; Mallet et al. 2019) and two papers on ITP DBS (Jimenez et al. 2013; Lee et al. 2019). Typically, patients are considered eligible for DBS if they are treatment-resistant to all available therapies, including selective serotonin reuptake inhibitors, clomipramine, and psychotherapy. The symptoms of these patients are usually very severe (Y-BOCS > 25) and persistent (duration ≥ 5 years). In practice, however, patients have much longer disease durations than that. For instance, the average Y-BOCS score ranges between 32 and 36, and the average disease duration between 8 and 31 years (see Table 1 for a detailed overview of the studies).

In the studies reporting 12 month data, an estimated 76% of the patients show a response, with 57% full responders (95% confidence interval [CI]: 44–70%; $k = 13$; $N = 208$, see Fig. 1) and 19% partial responders (95% CI: 14–26%; $k = 11$; $N = 153$). Of these 13 studies, 9 also gave information of the average Y-BOCS scores before and 12 months following DBS. The mean Y-BOCS reduction was 13.3 points (95% CI: 11.1–15.6; $k = 9$; $N = 149$, see Fig. 2), which corresponds to a very large effect size (Hedges g: 1.9, 95% CI: 1.6–2.3). Given the open-label nature of the included studies, this effect size may not only reflect DBS effects. Other possible factors contributing to this effect could be additional treatments (see also section on cognitive-behavioral therapy below), changes in social environment (e.g. employment status) or placebo effects, that may also interact with DBS effects.

Given the limited number of papers on STN and ITP DBS, it is currently impossible to estimate systematic differences between targets.

On a descriptive level, VC/VS and STN DBS seem to have similar full response rates: 95 of 182 patients (52%) and 8 of 14 patients (57%), respectively. All patients that underwent ITP DBS showed a full response, but this is based on two small samples of 5 and 6 patients, respectively (Jimenez et al. 2013; Lee et al. 2019).

The latest follow-up point was, on average, 21.3 (range 6–100) months after implantation, at which 83% of the patients showed a response. The estimated full response rate was 64% (95% CI: 49–78%; $k = 14$, $N = 197$); the estimated additional partial response rate was 19% (95% CI: 14–26%; $k = 13$; $N = 167$). The mean Y-BOCS reduction was 16.2 points (95% CI: 13.4–19.1; $k = 11$; $N = 167$), which corresponds to a very large effect size (Hedges g: 2.5, 95% CI: 1.9–3.0). The results of only 28 patients following 5 years of DBS or longer have been published (Fayad et al. 2016; Islam et al. 2015; Luyten et al. 2016). Of these, 21 patients showed a full response (75%), 5 a partial response (17.9%) and 2 did not respond (7.1%). These rates and symptom reductions at long-term follow-up show that the beneficial effects of DBS are maintained over the passage of time and might even increase. Note, however, that the results at longer follow-up durations of only a

Table 1 Descriptive of the studies in the meta-analyses

Study	Target	Detailed target	*N* (total)	*N* (male)	*N* (female)	Age (*M*)	Age (SD)	Last follow-up (*M*, months)	Age of onset (*M*, years)	Duration OCD (*M*, years)	Y-BOCS (*M*)	Y-BOCS (SD)
Open-label studies												
Jimenez et al. (2013)	ITP	ITP	6	3	3	34.7	16.2	24	N/R	N/R	16.2	10
Lee et al. (2019)	ITP	ITP	5	2	3	32.4	9.5	50	16.2	6.7	16.2	13.1
Chabardes et al. (2013)	STN	STN	4	2	2	38.3	3.6	6	17.8	4.6	20.5	7.8
Mallet et al. (2019)	STN	STN	14	8	6	43.6	7.9	41	12.6	4.9	31.1	8
Denys et al. (2020)	VC/VS	vALIC	70	22	48	41.7	11.2	12	16.8	8.7	25	11
Fayad et al. (2016)	VC/VS	VC/VS	6	2	4	44.5	8.5	100	N/R	N/R	N/R	N/R
Greenberg et al. (2006)	VC/VS	ALIC	10	6	4	35.3	11.2	32.1	12.8	4.3	22.5	8.1
Huff et al. (2010)[a]	VC/VS	RUL NAc	10	6	4	36.3	6.4	12	14.1	6.3	22.2	8.7
Huys et al. (2019)	VC/VS	NAc/ALIC	20	10	10	43.2	13.8	12	14.2	5.6	26.1	12
Islam et al. (2015)	VC/VS	BNST	8	7	1	43.6	9.7	34	14.4	9.3	29.3	8.4
Menchon et al. (2019)	VC/VS	ALIC	30	15	15	41	9.9	12	16.5	7.9	24.5	9
Raymaekers et al. (2017)	VC/VS	BNST	24	12	12	40.5	10.6	76.5	N/R	N/R	N/R	N/R

Roh et al. (2012)	VC/VS	VC/VS	4	2	2	33.8	12.2	24	17	10.4	16.8	5.9
Tsai et al. (2012)	VC/VS	VC/VS	4	4	0	25.5	5.2	18	13	2.9	8.3	2.5
Randomized, controlled studies												
Mallet et al. (2008)	STN	STN	16	9	7	43.8	7.6	3	13.4	5.8	30.9	7.1
Abelson et al. (2005)	VC/VS	ALIC	4	2	2	40.3	11.7	13	17.8	10.7	22.5	15.9
Denys et al. (2010)	VC/VS	vALIC	16	9	7	42.6	11.4	21	14.2	7.4	28.4	11.6
Goodman et al. (2010)	VC/VS	VC/VS	6	2	4	36.2	N/R	12	12.2	N/R	24	N/R
Luyten et al. (2016)	VC/VS	BNST	24	12	12	40.5	10.6	77	N/R	N/R	N/R	N/R

Overview of studies included in the meta-analysis
(v)ALIC ventral anterior limb of the internal capsule, *BNST* bed nucleus of the stria terminalis, *ITP* inferior thalamic peduncle, *M* mean, *NAc* nucleus accumbens, *OCD* obsessive-compulsive disorder, *SD* standard deviation, *STN* subthalamic nucleus, *Y-BOCS* Young-Brown Obsessive-Compulsive Scale (before DBS surgery), N/R not reported
[a]Huff et al. (2010) contributed data to the meta-analysis of open-label studies, and also to the meta-analysis of the randomized, controlled studies

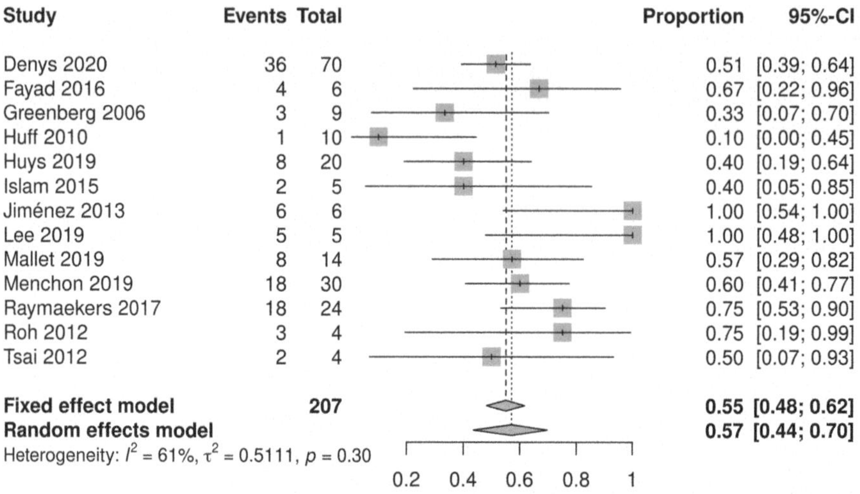

Fig. 1 Forest plot of full response rates after 12 months of deep brain stimulation. Note: Forest plot shows the result of a fixed and random intercept logistic regression meta-analysis with logit transformation. Outcome concerns the full response rates after 12 months of deep brain stimulation in patients with obsessive-compulsive disorder. The analysis was done with package "meta" in *R*, version 3.6 (Balduzzi et al. 2019)

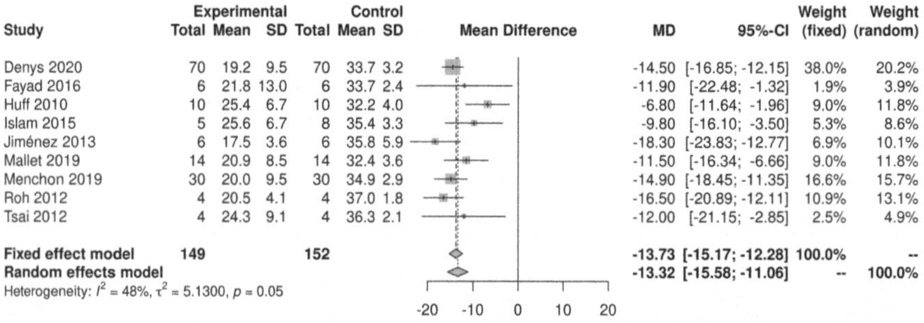

Fig. 2 Forest plot of difference in Y-BOCS scores after 12 months of DBS. Note: Forest plot shows the result of a fixed and random-effects meta-analysis with inverse variance weighing. Outcome measure is the Y-BOCS scores after 12 months of DBS (referred to as Experimental here) and before DBS (referred to as Control here). Four studies included in the meta-analysis of response rates (see Fig. 1) did not give sufficient information to be included in this analysis (Greenberg et al. 2006; Huys et al. 2019; Lee et al. 2019; Raymaekers et al. 2017). The analysis was done with package "meta" in *R*, version 3.6 (Balduzzi et al. 2019). *DBS* deep brain stimulation, *Y-BOCS* Yale-Brown Obsessive-Compulsive Scale

fraction of patients have been published about, so these might not be representative of the entire population. For example, the results at longer follow-ups might be biased because non-responders might have chosen to stop the treatment.

In the 239 patients reported in the above-mentioned 14 studies, 2 hemorrhages (0.8%) and 8 infections (3.2%) have been reported after surgery. During DBS optimization and follow-up, one completed suicide (0.4%) and 10 suicide attempts (4.0%) have been reported. Since most adverse events have a low incidence and are only recorded in a small total sample, it is difficult to establish which are caused by DBS. However, most frequently reported adverse events associated with stimulation are increased impulsivity, temporary hypomanic symptoms, or temporary sleep disturbances. These are all estimated to occur in 10–15% of the patients during the course of parameter optimization. Increased impulsivity usually is mild (e.g. increased talkativeness) or moderate (increased urge to shop), but sometimes can be classified as hypomania necessitating a parameter change or in rare cases, an inpatient admittance to a clinic.

2.2.1 Results of Randomized, Controlled Trials

Six studies with a total of 64 patients compared active DBS with sham DBS, including five studies on VC/VS DBS (Abelson et al. 2005; Denys et al. 2010; Goodman et al. 2010; Huff et al. 2010; Luyten et al. 2016) and one on STN DBS (Mallet et al. 2008). In Table 1 descriptive of the included samples can be found. Some of these studies included sub-samples of the open-label studies discussed before (Denys et al. 2010 is a sub-sample of Denys et al. 2020; Goodman et al. 2010 of Fayad et al. 2016; Luyten et al. 2016 of Raymaekers et al. 2017; Mallet et al. 2008 of Mallet et al. 2019), and one study contributed data to both the open-label as well as the randomized controlled meta-analysis (Huff et al. 2010). We compared the Y-BOCS scores following active with sham DBS in a separate meta-analysis. On average, patients scored 7.1 points (95% CI: 4.4–9.9) lower on the Y-BOCS following active compared to sham DBS (see Fig. 3). This corresponds to a large effect size (Hedges *g*: 0.9, 95% CI: 0.5–1.3), although this is smaller than the Y-BOCS reduction found after 12 months in open-label studies. As mentioned above, this may suggest that other factors than the direct effect of stimulation play

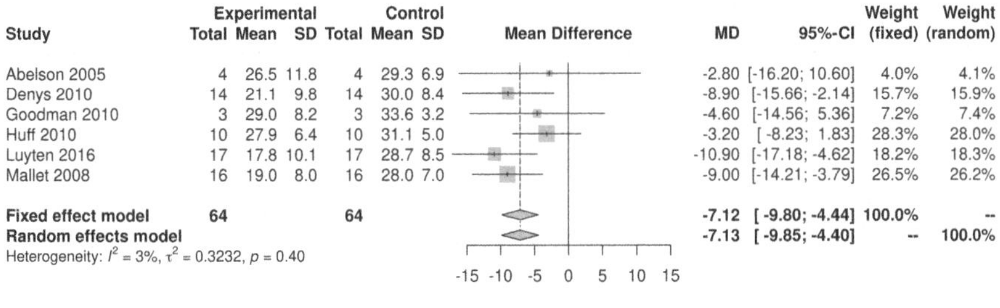

Fig. 3 Forest plot of difference in Y-BOCS scores between active and sham DBS. Note: Forest plot shows the result of a fixed and random-effects meta-analysis with inverse variance weighing. Outcome measure is the Y-BOCS scores after active (referred to as Experimental here) and sham (referred to as Control here) DBS. The analysis was done with package "meta" in *R*, version 3.6 (Balduzzi et al. 2019). *DBS* deep brain stimulation, *Y-BOCS* Yale-Brown Obsessive-Compulsive Scale

a role in the overall effect, or interact with the effect of direct stimulation to modulate effects over time.

2.2.2 Addition of Cognitive-Behavioral Therapy

One factor interacting with DBS could be the addition of CBT. Two uncontrolled studies examined the additional effects of CBT in patients with DBS for OCD. One study found that CBT augmented the effects of VC/VS DBS in 16 patients (Mantione et al. 2014). Following, on average, 14 weeks of DBS, a standardized 24-week CBT treatment program was added. Mean Y-BOCS scores decreased with an additional 22% of total YBOCS score. However, a study including 6 patients receiving both VC/VS and STN DBS found no statistically significant beneficial effects of CBT (Tyagi et al. 2019). A possible explanation for the negative results in the second study was that 4 out of 6 patients were already experiencing only mild OCD symptoms, potentially reflecting a floor effect induced by DBS. The addition of CBT, therefore, might be only indicated in those who experience only a moderate reduction of symptoms after DBS. In addition, the uncontrolled nature of these studies precludes establishing a causative effect of CBT.

2.2.3 Effects on Other Symptoms and Quality of Life

DBS reduces depression and anxiety in the majority of patients with OCD as measured with the Hamilton Depression Scale and the Hamilton Anxiety Scale (Abelson et al. 2005; Denys et al. 2020, 2010; Goodman et al. 2010). This effect has also been noted in patients with OCD after ITP DBS (Lee et al. 2019), as well as one patient with depression (Jimenez et al. 2013). However, the effect on affective symptoms seems to be restricted to DBS of the VC/VS or ITP, since stimulation of the STN does not cause an antidepressant effect (Mallet et al. 2008). Interestingly, VC/VS DBS first reduces symptoms of depression and anxiety, and later on OCD symptoms (Denys et al. 2020).

Initial reports show that symptom improvement is reflected in improved quality of life. Not surprisingly, psychological well-being improves, but measures of physical well-being and satisfaction in environmental factors also improve, as measured by the WHO Quality of Life scale (Ooms et al. 2014b). Quality of life improves even further on the long term (3–5 years after surgery) (Ooms et al. 2014b).

When costs and improvements in quality of life are jointly analyzed, DBS turned out to be cost-effective over a 4-year period. Costs were estimated on the basis of health care utilization (DBS as well as all other medical costs including those for diagnoses different from OCD), productivity (or the loss thereof), and travel costs. Quality of life was measured using the 5-dimensional EuroQol (EQ-5D), which can be used to estimate quality adjusted life years (QALY), a measure that quantifies the gain in overall health following a specific treatment. Taking a time horizon of 4 years, DBS costed around 70,000 euro per gained QALY compared to treatment as usual, which is generally considered cost-effective. The models show that cost-effectiveness could be improved and would even be potentially cost saving by using rechargeable batteries (Ooms et al. 2017).

2.2.4 Outcome Predictors

Some studies suggest that certain baseline variables might be associated with response to DBS. Patients with sexual or religious obsessions, a later age of OCD onset or good insight in illness might respond better to DBS (Alonso et al. 2015; Denys et al. 2010). In contrast, patients with more depressive symptoms at baseline, and the need for symmetry or perfectionism might respond worse to DBS (Denys et al. 2010; Raymaekers et al. 2017). However, the number of patients included in these studies was too small (maximum $N = 24$) to reliably implement these predictors in clinical practice at this moment.

2.2.5 Conclusion

In conclusion, around 75% of patients experience at least a partial improvement over the first year of DBS, and this response rate seems to be maintained at a longer follow-up. DBS results in large to very large reductions in OCD symptoms, and also reduces symptoms of depression and anxiety. The results of the comparison between active and sham DBS indicate that this is not merely a placebo effect. Although the available studies have relatively small sample sizes and also include uncontrolled designs, effect sizes of DBS are among the largest known in medicine in general and psychiatry in particular and persist over several years of follow-up in the small samples that have been reported on. Furthermore, severe side effects of the surgery and stimulation are rare, and the frequently observed increases in impulsivity and hypomania are usually temporary or disappear after parameter adjustment. These results are particularly remarkable given that the target population had an insufficient response to various other forms of treatment.

2.3 Mechanisms of Action: Neuroimaging and Circuitry

Development of neurostimulation techniques goes hand in hand with emerging insights in the pathophysiology of brain disorders. On the one hand, the decision where, when, and how to stimulate the brain requires a hypothesis on the neurobiological substrate of the targeted symptoms. On the other hand, the clinical and neurobiological effects of various forms of neurostimulation have fundamentally enhanced our understanding of the brain and its disorders. In this section we review the ongoing shared accomplishments of neuroscientists and psychiatrists in unraveling the neurobiological effects of neurostimulation on the brain, with a focus on neuroimaging the effects of DBS on brain circuitry in OCD.

2.3.1 Cortico-Striatal-Thalamic-Cortical Loops in OCD as a Target for DBS

Of all psychiatric disorders, the neurobiological basis of OCD is relatively well understood. Groundbreaking work in the 1980s used various neuroimaging techniques to show that OCD is accompanied by alterations in structure, connectedness or function in cortico-striatal-thalamic-cortical (CSTC) loops (Robbins et al. 2019). In brief, some of these loops are thought to become hyperactive and/or hyperconnected in OCD, resulting in self-exciting positive feedback (Dougherty et al. 2018). This provided some explanation for the efficacy of psychosurgical treatments such as ventral capsulotomy and cingulotomy that directly disrupted these loops.

Although the basic hypothesis of hyperactive CSTC-loops is largely maintained, several nuances have been made since. More recent evidence suggests a more extended network, also including limbic, insular, temporal, parietal, and cerebellar alterations, which may correspond with OCD heterogeneity (Robbins et al. 2019). In addition, it seems relevant to distinguish lateral and medial orbitofrontal (OFC) compartments. The lateral OFC – involved in reversal learning – and the medial OFC – involved in reward learning – may have distinct activity patterns depending on the context (Chamberlain et al. 2008; Fettes et al. 2017). Moreover, instead of distinct CSTC-loops, these networks may be better conceptualized as interacting spirals (Milad and Rauch 2012).

Notwithstanding these advances in neurobiological understanding, hyperactive CSTC-loops have remained the primary target of modern neurostimulation techniques like DBS in OCD (Lee et al. 2019; Mulders et al. 2016). DBS allows continuous stimulation of brain structures, e.g. those involved in or relating to the CSTC-loops. Traditionally, high-frequency stimulation (>100 Hz) has been thought to exert its effects by locally disrupting pathological hyperactivity, like ablative surgical interventions. More recently, it is increasingly acknowledged that enabling restoration of normal functioning and connectivity in brain networks by DBS contributes to its beneficial clinical effects in OCD.

2.3.2 Difficulties in Studying the Neurobiological Effects of DBS

Before delving more into the mechanisms of action, a disclaimer is in place. Studying the neurobiological effects of DBS is complicated by several factors (Greenberg et al. 2010). First, the small sample sizes in DBS studies increase the risk of type I and II errors, especially with complex outcomes like neuroimaging data. Second, DBS has profound clinical effects, not only on obsessions and compulsions, but also primarily on anxiety and depression. It may be hard to delineate the direct effects of DBS from indirect neurobiological changes due to improvement in associated symptoms. Third, the precise local neurophysiological effects of DBS at the cellular and molecular level remain heterogeneous, complex, and incompletely understood (Jakobs et al. 2019; Montgomery Jr. and Gale 2008). Effects range from stimulation, modulation, normalization, or reduction of neuronal firing rates, depending on, e.g. stimulation settings and positioning of neuronal fibers relative to the electrode. In addition, there is a large methodological heterogeneity between studies, e.g. in the localization of stimulation, stimulation settings, and study design including acute vs. chronic effects (Voon 2019). Finally, in addition to electrical effects, histological effects around the electrode potentially including edema, microhemorrhages, vascular changes, and long-term gliosis may contribute to the observed effects (Jakobs et al. 2019). Nevertheless, a growing number of studies described below applied increasingly advanced designs that provided more insight in the effects of DBS on brain circuits in OCD.

2.3.3 Effects of DBS on Brain Activity

Using fMRI and EEG in an on/off design, we previously showed that VC/VS DBS targeted at the nucleus accumbens (NAc) normalized NAc hypoactivity and frontostriatal NAc-prefrontal (PFC) hyperconnectivity in OCD, which was associated with symptom reduction (Figee et al. 2013; Smolders et al. 2013). This fits with the idea that DBS is able to reset the neural output of a stimulated nucleus and overrides disruptive brain network oscillations in CSTC circuits.

Several other studies using positron emission tomography in on/off designs support the view that VC/VS DBS can influence the activity of CSTC structures. VC/VS DBS has been shown to increase activation of the orbitofrontal cortex (OFC), anterior cingulate cortex (ACC), striatum, globus pallidus, and thalamus (Rauch et al. 2006). Another study showed that stimulation of the most ventral contact increased activity of the dorsal ACC, whereas the most dorsal contact increased activity in the thalamus, striatum, and globus pallidus (Dougherty et al. 2016). Moreover, VC/VS DBS has been shown to decrease PFC activity (Park et al. 2019; Suetens et al. 2014; Van Laere et al. 2006). However, increases in PFC brain metabolism have been reported as well (Baldermann et al. 2019).

Altogether, VC/VS DBS is thought to interrupt pathological CSTC circuitry, normalizing activity, and connectivity. This allows patients to shift from habitual excessive processing of disease related stimuli towards restoration of goal-directed behavior. However, dynamic causation modeling – i.e. Bayesian estimation of changes in coupling between brain regions – suggested that additional effects of VC/VS DBS on fronto-limbic interaction were important for rapid improvements in mood and anxiety (Fridgeirsson et al. 2019).

In addition to the VC/VS, several other targets have been investigated in OCD. STN DBS has been shown to decrease activity in the ACC and frontal medial gyrus (Le Jeune et al. 2010; Senova et al. 2019). DBS targeted at the ITP has been suggested to result in more widespread changes in brain activity (Lee et al. 2019).

2.3.4 Brain Networks Targeted by DBS

Several investigators have tried to apply knowledge about the effects of DBS on brain activity to better understand the underlying networks and thereby enhance clinical outcomes in OCD. One way to understand these networks is by comparing the networks that are stimulated in DBS-responders with the networks that are stimulated in non-responders. Hartmann showed that VC/VS DBS had a better outcome in patients where white matter was stimulated that was connected to the right anterior middle frontal gyrus (Hartmann et al. 2015). We previously showed that VC/VS DBS was more effective when the superolateral branch of the medial forebrain bundle (slMFB) was stimulated (Liebrand et al. 2019).

Other investigators modeled the connection pattern of the DBS stimulation site using connectivity profiles derived from a large database of healthy subjects (Makris et al. 2016). Using resting state functional connectivity from healthy subjects, Voon et al. hypothesized that STN DBS in OCD works by activating a network that is involved in decisional impulsivity and consists of the dorsolateral PFC and the VS (Voon et al. 2017). Likewise, Coenen et al. used the human connectome project data to show that commonly used DBS stimulation sites involve affect and reward networks through the anterior thalamic radiation and slMFB, respectively (Coenen et al. 2020).

A recent RCT compared VC/VS with STN DBS in OCD and found evidence for interesting dissociable effects, despite comparable overall clinical efficacy (Tyagi et al. 2019). VC/VS DBS had profound beneficial effects on mood, whereas STN DBS improved cognitive flexibility. This may be explained by the networks that were targeted by both stimulation locations: the VC/VS is connected to the medial OFC, while the STN is connected to the lateral OFC, dorsal ACC, and dorsolateral PFC. There was little additional gain from combined stimulation (Tyagi et al. 2019).

2.3.5 Using Brain Networks to Optimize DBS Targeting

A next step is to use brain network information to improve DBS targeting. Usually, DBS is targeted at specified stereotactic coordinates, lacking personal specificity with regard to targeted networks. Baldermann et al. showed that connectivity analyses could predict 40% of the variance in clinical improvement for individual OCD patients after VC/VS DBS (Baldermann et al. 2019). In detail, response could be predicted by DBS placement in a CSTC fronto-thalamic white matter tract channeling through the ventral anterior limb of the internal capsule, connecting the PFC and the thalamus, dorsally traversing to the NAc. Likewise, Widge et al. found that increased task-related power of PFC theta oscillations evoked by VC/VS DBS may serve as a biomarker of DBS response, through an association with cognitive control (Widge et al. 2019). In addition, Barcia et al. performed an RCT and showed that DBS was most effective when targeted in specific striatal sites that had CSTC projections to prefrontal areas that became activated during symptom provocation in each individual patient. On the basis of this effortful study (due to the relatively long-term testing of many different stimulation sites), they suggest that this personalized strategy could potentially result in a 86% response rate as seen in their RCT (Barcia et al. 2019). However, these three studies investigated the stimulated networks still a posteriori. Coenen et al. describe that they use a priori machine learning tractography on DTI neuroimaging to facilitate optimal placement of DBS in the slMFB, clinical results will follow (Coenen et al. 2019). Dougherty et al. are working on a trial that tests the longitudinal correlation between task-induced theta oscillations and clinical response by simultaneously implanting bilateral stimulating/recording electrodes at both the VC/VS and the dorsolateral PFC (NCT03184454).

2.3.6 Conclusion

In conclusion, improved neuroscientific insight in the pathophysiology of OCD facilitated more guided DBS treatment, while the effects of DBS enhanced knowledge about the neuronal networks relevant for OCD. Several studies show that DBS is capable of normalizing brain activity in main CSTC brain circuits, likely by targeting CSTC white matter tracts connecting frontal and thalamic and/or striatal regions. This potentially disturbs aberrant signaling and restores the capacity for physiological CSTC communication. Importantly, evidence suggests that different electrode localizations target distinguishable networks with correspondingly differential effects on clinical symptoms, e.g. VC/VS DBS with more affective effects vs. STN DBS with more cognitive effects.

2.3.7 Future Perspectives

Future studies could build upon this knowledge by further improving spatial and temporal precision. Spatially, a priori definition of personalized networks that should be targeted in each individual patient could enhance clinical effectiveness. This could be accomplished using advanced connectomics, combining different high resolution imaging modalities including DTI and functional neuroimaging with, e.g. symptom provocation, to inform the neurosurgeon where to implant and the psychiatrist where to stimulate. This may entail differential targeting according to symptom dimensions, e.g. VC/VS targeting for predominant affective symptoms and STN targeting for primarily cognitive inflexibility. Steering electrodes will facilitate this process. Using adaptive trial designs, clinical outcomes could inform and enhance future targeting. Temporally, the electrical stimulation could be timed more precisely. If the electrodes only stimulate when needed – i.e. during aberrant brain activity causing symptoms – the effect may be enhanced, the battery life prolonged and side effects reduced. To this end, more insight is needed in the aberrant neurophysiological signals that form the substrate of OCD symptoms in each individual patient, i.e. recording neuronal signatures of "OCD attacks" (Trial NL7486) and e.g. model them using computational models (Fradkin et al. 2020).

Technological innovations of novel DBS systems that can stimulate and record simultaneously will facilitate these developments towards closed-loop stimulation that adjust stimulation automatically and real-time based on recorded brain biomarkers, helped by animal research reviewed below. Ultimately, the goal is to further increase response rates and reduce side effects of this already powerful intervention, while at the same time improving our understanding of the brain and its disorders.

2.4 Animal Research and DBS

Animal studies can provide essential insight into both the neurobiological underpinnings of OCD and the DBS mechanism of action in which many of the difficulties faced in studying human patients can be addressed (Ahmari and Dougherty 2015; Feenstra and Denys 2012; Haynes and Mallet 2010). However, animal studies come with their own limitations. The first, most obvious limitation is neuroanatomical inter-species differences. Although DBS target structures are generally homologous between humans, non-human primates, and rodents, the VC/VS differs in its relative arrangement. Specifically, the anterior limb of the internal capsule (vALIC), a dense white matter tract that separates caudate from putamen in primates, is absent in rodents. Instead, the rodent caudoputamen is an unseparated structure penetrated by numerous, isolated white matter fascicules. Consequently, targeting the VC/VS necessitates a more caudal placement of DBS electrodes in rodents, where the internal capsule is a more dense fiber tract and carries similar projection fibers as stimulated in patients (Coizet et al. 2017). Anatomical differences are of less concern for STN and ITP. Technical constraints are the second limitation: DBS is rarely applied continuously across weeks or months, but solutions are becoming available (Alpaugh et al. 2019; de Haas et al. 2012; Paralikar et al. 2015). Moreover, many different types of electrodes (e.g., differing in material, thickness, etc.) have been used in rodents, which hampers comparability. Third, no animal model features all clinical symptoms of a psychiatric disorder. Thus, deconstructing clinical conditions into more basic, broad-dimensional domains such as endophenotypes (Chamberlain and Menzies 2009; Gottesman and Gould 2003; Gould and Gottesman 2006) is generally better suited for translational endeavors.

The first invaluable advantage is that in animals, DBS effects on healthy and pathological brains can be compared, which improves our knowledge on how DBS exerts its therapeutic effects and allows evaluation of side effects and safety measures. Second, neuroscience using rodents offers an arsenal of invasive neuromeasurement methods that can probe brain activity with temporal precision and brain-cell, projection, and neurotransmitter specificity on both the microscopic and macroscopic scale. Such measurements are impossible in humans and provide incomparable insight into all aspects of DBS. Third, animal studies allow testing novel applications of DBS, including stimulation parameters, brain targets, and electrode material and type, and thereby are likely to advance efficacy in clinical settings. To attempt this in humans would be unethical, too costly due to the wide range of changeable parameters, and too time-consuming. Below, we review animal studies investigating DBS and compulsivity mechanisms of action, illustrating how preclinical models may help refine DBS treatment.

2.4.1 DBS in Animal Models Exhibiting Compulsive Behavior

The validity of DBS in animal models for compulsive behavior is underlined by the fact that results are overall consistent with DBS effects in humans (Baunez 2011; Feenstra and Denys 2012; Hamani and Temel 2012; van Kuyck et al. 2007; Winter et al. 2010). High-frequency DBS diminishes compulsive behaviors in a number of animal models, including 8-OH-DPAT-induced perseveration, schedule-induced polydipsia (SIP), quinpirole-induced checking, compulsive lever-pressing after signal attenuation, bicuculline-induced tics, and genetic models such as the SAPAP3 mutant mouse (see chapter "Animal Models for OCD Research" of this book). For example, perseverative and compulsive behavior provoked by 8-OH-DPAT or quinpirole administration is diminished by high-frequency DBS of the entopeduncular nucleus (EPN), STN, or VS (but not globus pallidus (GP)) (Andrade et al. 2010; Djodari-Irani et al. 2011; Mundt et al. 2009; Winter et al. 2008). Such behavior is also diminished by low-frequency DBS (but not high-frequency DBS) of the reticular thalamic nucleus, but is increased by low-frequency DBS of the VC/VS (van Kuyck et al. 2003). Similarly, compulsive actions induced behaviorally via signal attenuation and schedule-induced polydipsia can be alleviated by high-frequency, but not low-frequency, DBS in the STN, GP, EPN, VS, BNST, and mediodorsal thalamus. Consistently, neuronal activity-suppressing muscimol injections into STN and EPN have similar effects as DBS in the above-mentioned brain structures (Djodari-Irani et al. 2011; Klavir et al. 2009; Winter et al. 2008), suggesting that DBS inactivates or disrupts function in the vicinity of the electrode. However, VS lesions increase compulsive checking behavior, instead of decreasing it (van Kuyck et al. 2003; Winter et al. 2008), thus complicating the picture and demanding closer investigation. Another cardinal symptom of OCD, avoidance behavior, is decreased by DBS in VS (Rodriguez-Romaguera et al. 2016).

In non-human primates, stereotypies can be induced by intracranial micro-injection of bicuculline. Such tic-like behavior induced by intra-GPe bicuculline infusion in non-human primates can be remedied by high-frequency DBS of limbic aspects of the STN (Baup et al. 2008). Furthermore, bicuculline-provoked neuronal activity in both external and internal segments of the GP, related to such motor tics, can be remedied by high-frequency DBS in the internal GP via temporal locking of cell firing with stimulus pulses (McCairn et al. 2012, 2013). This last finding further complicates the idea of DBS acting via neuronal inhibition. In summary, these DBS studies in animal models of compulsivity indicate the efficacy of different DBS parameters and brain targets. Although brain targets are variable, they all have in common that they likely affect CSTC networks, making a case for the CSTC hypothesis of OCD, but raising questions regarding the DBS mechanism of action.

Notably, the studies discussed above mainly focused on altering compulsive behavior itself, which is an obvious and important goal. However, this focus sets aside other prominent OCD symptoms such as anxiety and avoidance behavior (Figee et al. 2016; Hauser et al. 2016; Milad and Rauch 2012; Robbins et al. 2012). In contrast, some genetic mouse models for OCD exhibit multiple symptoms of OCD. SAPAP3 mutant mice (SAPAP3$^{-/-}$) are arguably the best genetic model for OCD (Welch et al. 2007). Both OCD patients (Chakrabarty et al. 2005; Milad and Rauch 2012; Pittenger et al. 2011; Ting and Feng 2011) and SAPAP3$^{-/-}$ (Wan et al. 2014; Welch et al. 2007) exhibit aberrant brain activity in cortico-striatal projections. SAPAP3$^{-/-}$ display compulsive-like grooming (Ehmer et al. 2020a, b; Manning et al. 2019; van den Boom et al. 2019, 2017; Welch et al. 2007; Wood et al. 2018), similar to one of the biggest subtypes of OCD patients. Other similarities with OCD comprise increased anxiety-like behavior (Pinhal et al. 2018; van den Boom et al. 2019; Welch et al. 2007), compromised behavioral flexibility (Manning et al. 2019; van den Boom et al. 2019), and altered habit formation (Ehmer et al. 2020b; Hadjas et al. 2019). Optogenetic stimulation of cortico-striatal projections and viral rescue of the striatal SAPAP3 protein can restore normal grooming (Burguiere et al. 2013; Welch et al. 2007). Similarly, administration of selective serotonin reuptake inhibitors (SSRIs), the primary pharmacotherapy for OCD, normalizes self-grooming and anxiety-like behavior in SAPAP3$^{-/-}$ (Welch et al. 2007). Thus, these mice enable the study of a range of emotional and cognitive functions in relation to (genetically determined) spontaneous compulsive behavior, a state that does not have to be induced by conditioning or pharmacology. Previously, we implanted DBS electrodes in VC and VS (Pinhal et al. 2018), targets known to improve conditioned anxiety (van Dijk et al. 2013) and conditioned fear (Rodriguez-Romaguera et al. 2012). High-frequency DBS in both areas decreased excessive grooming and recruited prefrontal cortical neurons, whereby VS-DBS was more effective (Pinhal et al. 2018). Taken together, SAPAP3$^{-/-}$ studies support conclusions from previous DBS studies in animals and humans, are consistent with contemporary theories of OCD brain malfunction, and provide a powerful research tool in the quest for a mechanism of action of DBS and brain signatures of OCD-like behavior.

2.4.2 Unraveling a DBS Mechanism of Action and a Brain Signature of Compulsive Behavior

To identify a mechanism of action, neuronal activity needs to be recorded before, during, and after DBS. However, electrophysiological recordings during DBS are complicated by (1) electrical interference by DBS itself and (2) their difficulty to identify brain cells by type and projection target (McIntyre and Anderson 2016). Recent developments in optogenetic technology make it possible to circumvent these limitations (Warden et al. 2014) by optically measuring neuronal activity using genetically targeted and encoded fluorescence-based calcium indicators (GCaMPs) (Resendez et al. 2016).

A key step is to identify neuronal responses near the tip of the DBS electrode (McIntyre and Anderson 2016). Recent studies attempted this by using 2-photon calcium imaging in the cortex of awake, head-fixed mice. Stimulation induced a wide range of responses, potentially by affecting axons. Electrical stimulation activated a sparse and distributed network of neurons in an intensity-dependent way (Histed et al. 2009). At lower frequencies (<90 Hz), stimulation affected both local and distant neurons. At higher frequencies only local neurons were affected (Michelson et al. 2019), which may explain previously reported frequency-dependent effects on the behavioral level. In addition, local axonal responses (connected to distant somas) differed from local somatic responses – instead of sustained effects, activity decreased over time in a frequency-dependent manner, demonstrating differential DBS effects on soma and axon. However, to increase the translational value of such studies, DBS should be performed, not in cortex, but in deeper brain structures that are targeted in psychiatric patients. Although previously inaccessible, the latest advances in calcium imaging employ specialized lenses that enable imaging deep brain structures in rodents (Resendez et al. 2016).

Of equal importance is the evaluation of long-range effects of DBS on CSTC circuits and other brain-wide networks. Marked differences between DBS and pharmacological inactivation of basal-ganglia neurons (Djodari-Irani et al. 2011; Rodriguez-Romaguera et al. 2012) suggest that DBS effects are, at least partially, mediated by processes that extend beyond the DBS target region. Indeed, clinical DBS often targets white-fiber tracts (vALIC, VS/VC) that connect different brain regions. We recently mapped DBS network effects in SAPAP3$^{-/-}$ after DBS in VS and VC, by measuring brain-wide molecular changes using the neuronal activity marker c-Fos, and found that DBS in both locations recruited PFC regions, albeit activity in the lateral orbitofrontal cortex (lOFC) was increased significantly exclusively after VC-DBS (Pinhal et al. 2018). These data are consistent with fMRI results in OCD patients that show DBS normalization of abnormal PFC activity (Figee et al. 2013; Le Jeune et al. 2010) and reduced electrophysiological activity in OFC in anaesthetized rats during DBS (McCracken and Grace 2007). Moreover, optogenetic activation of lOFC projections to the striatum normalizes grooming in SAPAP3$^{-/-}$ and optogenetic activation of the medial OFC projection to the striatum induces excessive grooming in "healthy" mice, pointing to a causal involvement of the PFC in therapeutic effects of DBS (Ahmari and Dougherty 2015; Burguiere et al. 2013). More recently, it was found that optogenetic stimulation of basolateral-amygdala inputs to PFC induces compulsive behavior (Sun et al. 2019). Together, these findings highlight long-range brain-network effects of DBS and provide causal evidence for the role of CSTC circuits, and in particular PFC, in compulsivity.

Although multiple studies implicate the lOFC in compulsivity, thus far, no animal study has monitored lOFC activity in real-time during ongoing compulsive behavior. To elucidate the specifics of lOFC involvement in the future, neuronal activity in lOFC during compulsive and non-compulsive behavior should be compared using methods like calcium imaging, ideally under varying DBS parameters. A potentially resulting "compulsivity signature" could inspire changes in clinical DBS parameters such as frequency, amplitude, and pulse width, but also DBS targets.

2.4.3 Identification of an Electrophysiological Compulsivity Signature for (Closed-Loop) DBS

Two main challenges are posed by current treatment of compulsive behavior with DBS (referred to as "open-loop DBS"): (1) open-loop DBS is applied continuously, which strains battery life and (2) stimulation parameters are programmed manually, which is time-consuming and unrelated to the underlying neurophysiology. A closed-loop (CL) device would monitor neuronal signals characteristic of psychiatric symptoms continuously and apply DBS only as an immediate, targeted response-on-demand to a specific neuronal signature of compulsivity. CL-DBS may be effective in symptom reduction while requiring less energy and maintaining adaptability: the algorithm can dynamically adjust DBS parameters based on feedback (Parastarfeizabadi and Kouzani 2017). To enable CL-DBS, it is imperative to find a reliable biomarker for compulsivity, ideally one that reflects its causal mechanisms, and preferably one that can be measured with the DBS electrode itself. The range of potential electrophysiological biomarkers is wide, from activity of single neurons, to field potentials of local or distal networks, and any combination thereof. Animal models of compulsivity have a unique advantage of testing the applicability of (combinations of) these biomarkers for CL-DBS. Their efficacy can readily be evaluated, providing that the model exhibits neural activity characteristic of compulsivity homologous to that of human patients.

The schedule-induced polydipsia (SIP) model of compulsive behavior yields both compulsive and non-compulsive (resistant) rats within the same cohort, and is therefore ideally suited to identify biomarkers specific to compulsivity. In SIP, altered single neuronal firing (Welkenhuysen et al. 2013) and local field potentials (LFPs) (Wu et al. 2016) in the BNST are related to compulsivity. Conversely, high-frequency DBS in the BNST effectively reduces compulsive behavior in SIP (van Kuyck et al. 2008), and its efficacy correlates with changes in the observed compulsivity-related LFPs. Accordingly, VC/VS DBS targeted at the BNST-region has proven effective in human OCD patients (Nuttin et al. 1999; Winter et al. 2018). Therefore, compulsivity-related neuronal activity discovered in the SIP rodent model might translate to a CL-DBS biomarker in humans.

Another example of pathological neuronal activity-patterns characteristic of compulsive behavior in humans that has been explored in greater detail in rodents is altered OFC activity (Fettes et al. 2017). One such study in the SAPAP3 mutant mouse model describes single-neuron OFC activity associated with compulsive behavior, which can be remedied by striatal pathway-specific stimulation (Burguiere et al. 2013). Translated to their human equivalents, these findings show potential for online detection and characterization of aberrant neuronal activity underlying compulsive behavior, to serve as biomarkers for CL-DBS.

Overall, the described progress in this section on DBS in animal studies supports an optimistic outlook on future research endeavors aimed at characterizing DBS and compulsivity mechanisms of action. Although animal studies have not yet informed human DBS therapy significantly, recent technical developments will potentially enable steps towards a more refined CSTC hypothesis of OCD and provide novel and actionable insights into the DBS mechanism of action.

3 Transcranial Magnetic Stimulation

Repetitive transcranial magnetic stimulation (rTMS), a *non*-invasive neuromodulation technique, is increasingly being investigated in treatment-resistant OCD. rTMS has been proposed for patients with an inadequate response to pharmacologic or behavioral therapy, potentially preceding DBS in the treatment algorithm (Rapinesi et al. 2019). During rTMS, a rapidly changing electromagnetic field continuously generates short electrical pulses, modulating the targeted brain region beneath the head coil. Depending on the parameters of stimulation, rTMS is assumed to excite ($\geq$5 Hz) or inhibit ($\leq$1 Hz) the underlying cortical region (Fox et al. 2014). The most often reported side effects include transient headaches and localized scalp pain (Rapinesi et al. 2019). The most serious TMS-related adverse effect, the occurrence of seizures, is extremely rare and it is mostly associated with stimulation that exceeds rTMS safety limits (Rossi et al. 2009).

Variations to rTMS are deep TMS (dTMS), in which a so-called H-coil is used to modulate deeper cortical regions (Carmi et al. 2019), and accelerated intermittent or continuous theta-burst stimulation (aiTBS or cTBS), in which excitatory (aiTBS) and inhibitory (cTBS) stimulation is delivered in pulses of 50 Hz in order to shorten treatment sessions and potentiate neuroplasticity (Di Lazzaro et al. 2005).

3.1 Effects of TMS on Brain Networks

rTMS modulates deregulated brain networks rather than just affecting the local stimulation target (To et al. 2018). OCD is associated with dysfunction of the extended CSTC circuitry (Milad and Rauch 2012; Rapinesi et al. 2019), which includes the dorsolateral PFC (DLPFC), the medial PFC (mPFC), OFC, ACC, the supplementary motor area (SMA), and the basal ganglia. Interestingly, even though different types of brain stimulation (DBS, rTMS) for OCD are targeted at different brain areas, these targets are often nodes of the same brain network (Fox et al. 2014). Whereas DBS focuses on subcortical areas, rTMS focuses on the cortical targets, including the DLPFC, mPFC-ACC, OFC, and SMA (Lusicic et al. 2018).

3.2 Clinical Effects of TMS

rTMS stimulation of each of these targets has been investigated in terms of potential efficacy for OCD as monotherapy or as adjunctive therapy (Table 2). Neuroimaging studies show that the DLPFC, SMA, and OFC are hyperactive in patients with OCD (Del Casale et al. 2011). Therefore, it is hypothesized that low-frequency inhibitory rTMS over these areas might reduce this hyperactivity (Rapinesi et al. 2019). However, confirmatory neuroimaging studies are needed to prove this hypothesis. A recent study showed that low-frequency (1 Hz) stimulation over the right DLPFC significantly reduced symptoms in 45 OCD patients compared to 10 Hz and sham stimulation (Elbeh et al. 2016). The Y-BOCS score of 15 patients who received 1 Hz stimulation dropped on average from 25.3 (SD 6.5) at baseline to 14.7 (SD 5.9, Cohen's $d = 1.71$) after rTMS and the mean Y-BOCS score was still 15.7 3 months after treatment (SD 7.9, $d = 1.33$). Another study also found that 1 Hz stimulation was superior to sham stimulation in 27 OCD patients (Seo et al. 2016). A recent review concluded that low-frequency ($\leq$5 Hz) stimulation over the right DLPFC is possibly effective for the treatment of OCD, based on the therapeutic efficacy of real versus sham rTMS replicated in a sufficient number of independent studies (Lefaucheur et al. 2020). Thirteen RCTs investigated the DLPFC as rTMS-target in OCD patients (Table 2).

Even though other cortical areas seem promising targets as well, they have not reached that level of clinical relevance yet due to an insufficient number of studies. Respectively, 7, 2, and 1 RCT('s) targeted the (pre-)SMA, OFC, and mPFC/ACC (Table 2). A meta-analysis with data obtained from 18 randomized controlled trials up till 2016 implied that low-frequency rTMS over the SMA might offer the greatest effectiveness in the treatment of chronic and resistant OCD ($g = 1.68$, very large effect) (Rehn et al. 2018). In line with this finding, a systematic review of 17 papers on brain stimulation in OCD in 2018 concluded that low-frequency rTMS stimulation of the SMA and OFC has the best treatment potential (Rapinesi et al. 2019). Effectiveness of SMA-rTMS remained at 3-month follow-up and subgroup analyses suggested that low-frequency rTMS was more effective than high-frequency rTMS (Rehn et al. 2018).

Finally, high-frequency (20 Hz) dTMS over the mPFC, including the ACC, resulted in a significant larger decrease of OCD symptoms in 42 patients that received active stimulation compared to 45 patients that received sham stimulation (effect size = 0.69) (Carmi et al. 2019). This has led to FDA approval of dTMS as an adjunctive treatment option for severe treatment-resistant OCD (Lefaucheur et al. 2020).

It should be acknowledged that findings in these clinical studies should be interpreted carefully. Many of the RCTs were heterogeneous as to clinical variables and rTMS stimulation parameters and the quality of sham rTMS differs between studies. Moreover, most included patients had treatment-resistant OCD and were maintained on pharmacological treatments. This limits the possibility to determine rTMS effectiveness for patients who are naïve to medication. In addition, even

Table 2 Overview of randomized controlled trials on transcranial magnetic stimulation for obsessive-compulsive disorder

Study	Design	*N*	Cortical target	Frequency (Hz)	% RMT	Sessions	Total pulses	Duration (weeks)	Follow-up (weeks)	Outcome
Alonso et al. (2001)	TMS vs sham	18	R-DLPFC	1	110	18	21,600	6	N/A	No significant difference
Prasko et al. (2006)	TMS vs sham	30	L-DLPFC	1	110	10	18,000	5	2	No significant difference
Sachdev et al. (2007)	TMS vs sham	10	L-DLPFC	10	110	10	15,000	2	N/A	No significant difference
Kang et al. (2009)	TMS vs sham	20	R-DLPFC + pre-SMA	1	110	10	12,000	2	2	No significant difference
Ruffini et al. (2009)	TMS vs sham	23	L-OFC	1	80	15	9,000	3	12	Significant but time-limited improvement compared of active TMS to sham
Badawy et al. (2010)	TMS monotherapy vs add-on	40	L-DLPFC	20	?	15	12,000	3	N/A	Monotherapy not effective, add-on therapy effective
Mantovani et al. (2010)	TMS vs sham	18	Pre-SMA	1	100	20	24,000	4	N/A	Active TMS significantly better than sham
Sarkhel et al. (2010) and Morishita et al. (2014)	Add-on TMS vs sham	42	R-DLPFC	10	110	10	8,000	2	2	No significant difference
Mansur et al. (2011)	TMS vs sham	27	R-DLPFC	10	110	30	60,000	6	2	No significant difference
Gomes et al. (2012)	TMS vs sham	22	Pre-SMA	1	100	10	12,000	2	12	Active TMS significantly better than sham

(continued)

Table 2 (continued)

Study	Design	*N*	Cortical target	Frequency (Hz)	% RMT	Sessions	Total pulses	Duration (weeks)	Follow-up (weeks)	Outcome
Ma et al. (2014)	Alpha electroencephalogram-guided TMS vs sham	46	B-DLPFC	8–12	80	10	6,480–8,720	2	1	Active TMS significantly better than sham
Nauczyciel et al. (2014)	TMS vs sham cross-over design	19	R-OFC	1	120	10	12,000	1	N/A	Active TMS better than sham ($p = 0.07$), 4 Y-BOCS points difference
Haghighi et al. (2015)	TMS vs sham cross-over design	22	B-DLPFC	20	100	20	7,500	5	N/A	Active TMS significantly better than sham
Elbeh et al. (2016)	TMS vs sham	45	R-DLPFC	10/1	100	10	20,000	2	12	No significant difference sham vs 10 Hz TMS, 1 Hz TMS significantly better than sham
Hawken et al. (2016)	TMS vs sham	22	SMA	1	110	25	?	6	N/A	Active TMS significantly better than sham
Jahangard et al. (2016)	TMS vs sham cross-over design	10	B-DLPFC	20	100	10	7,500	2	N/A	Active TMS significantly better than sham
Pelissolo et al. (2016)	TMS vs sham	16	Pre-SMA	1	100	20	30,000	4	N/A	No significant difference
Seo et al. (2016)	Adjunctive TMS vs sham	13	R-DLPFC	1	100	15	18,000	3	N/A	Adjunctive TMS significantly better than sham
Shayganfard et al. (2016)	Adjunctive TMS vs sham cross-over design	10	L-DLPFC	20	100	10	750	2	N/A	Adjunctive TMS significantly better than sham

Arumugham et al. (2018)	Adjunctive TMS vs sham cross-over design	40	Pre-SMA	1	100	18	1,200	3	N/A	No significant difference
Carmi et al. (2019)	Multicenter trial, deep TMS vs sham	99	mPFC and ACC	20	100	29	2,000	6	4	Deep TMS significantly better than sham
Harika-Germaneau et al. (2019)	Theta-burst TMS vs sham	30	Pre-SMA	5	70	30	600	6	12	No significant difference

L left, *R* right, *B* bilateral, *DLPFC* dorsolateral prefrontal cortex, *OFC* orbitofrontal cortex, *SMA* supplementary motor area, *mPFC* medial prefrontal cortex, *ACC* anterior cingulate cortex though studies have investigated both high- and low-frequency rTMS, it remains unclear whether high- or low-frequency rTMS is better for OCD (Rapinesi et al. 2019).

3.3 *tDCS and OCD*

rTMS should be distinguished from other non-invasive neuromodulation techniques, such as transcranial direct current stimulation (tDCS), which are also recently being investigated in OCD (Brunelin et al. 2018). During tDCS, a weak direct electrical current (1–2 mA) is applied between two electrodes that are placed on the scalp. Anodal tDCS stimulates the underlying brain region, whereas cathodal tDCS inhibits it (Shivakumar et al. 2019). Only a few ($N = 11$) studies have investigated the effects of tDCS for OCD, but they have showed promising results (Brunelin et al. 2018; Shivakumar et al. 2019). The studies reported a significant benefit on OCD symptoms, suggesting OFC and pre-SMA/SMA as the ideal targets. Stimulation of the DLPFC mostly contributed to improvement of comorbid affective symptoms, rather than OC symptoms. These results are preliminary and further sham-controlled studies are needed to define the role of tDCS in the treatment of OCD.

3.4 *Conclusion*

Overall, (adjunctive) rTMS seems to represent a promising option for OCD patients who have responded poorly to medication and behavioral therapy. Low-frequency stimulation of the right DLPFC, OFC, and SMA, and high-frequency dTMS of the mPFC-ACC seem to be the most effective approaches. However, although these results are encouraging, future large-scale studies are needed to establish long-term efficacy and to improve the rTMS parameters for site, frequency, and treatment duration in both treatment-resistant but also non-resistant OCD patients. Moreover, future studies should incorporate neuroimaging to determine the neurobiological mechanism underlying clinical response to rTMS. Finally, further developments with the application of newer forms of rTMS, including dTMS and TBS, could represent a step forward in the direction of non-invasive brain stimulation as an add-on treatment approach for OCD.

4 Conclusion and Discussion

DBS and rTMS, as invasive and non-invasive neurostimulation methods, are effective for treatment of OCD. Meta-analyses show both methods resulting in large symptom reductions and relatively few side effects. Where rTMS has a role as adjunctive therapy and appears to be effective in patients with an inadequate response to pharmacological therapy, the more invasive DBS has been primarily investigated in extremely severe treatment-refractory patients. Both methods are thought to target the same extended CSTC brain circuits, albeit from a different node in the network. Likewise, both methods have been targeted at a wide array of nodes in these networks, with comparable efficacy. Innovations in neuroscientific insight and technology can further improve targeting and stimulation parameters. This will facilitate further embedding of these powerful interventions in clinical practice.

Up till now, the number of OCD patients treated with DBS and rTMS is still limited. For DBS, the difference is especially notable compared to the number of Parkinson's disease patients being treated. In the past 30 years, an estimated 300–400 patients have been treated with DBS for OCD compared to an estimated 300,000 patients with Parkinson's disease, despite a comparable technique and response rate, which cannot be explained by differences in prevalence. This could mean that, unlike neurology, psychiatry is not ready yet to fully embrace invasive neuromodulation. This may be because clinicians feel uncomfortable with DBS, e.g. due to unfamiliarity with the side effects or management of DBS parameters. This would imply that better education can improve this translational gap. The non-invasive rTMS is investigated to a similar extent in neurological and psychiatric disorders. However, level A evidence (definite efficacy) has only been established for neuropathic pain, depression, and motor recovery after stroke (Lefaucheur et al. 2020). Due to an insufficient number of studies and unsystematic use of treatment parameters (target, frequency, duration), it is challenging to prove efficacy in other disorders, such as OCD. Another important limitation is the lack of long-term studies, thus long-term efficacy of rTMS is not yet clear.

This stresses the urgent need for larger clinical trials in which the efficacy of DBS and rTMS is investigated. These trials would help to improve neuromodulation therapy and to convince psychiatrists of the importance and impact of neuromodulation for complex psychiatric disorders. In order to acquire more data, it seems necessary to conduct more multicenter trials and to analyze and interpret the data on a global scale in view of the small research centers with their small samples. Important in this regard is the large variability in the execution of DBS and rTMS studies. There are different targets, different neurosurgical techniques or stimulation protocols, different diagnostic methods, and different interpretations of improvement, especially if one abandons the standardized Diagnostic and Statistical Manual of Mental Disorders (DSM) assessment, which is often necessary in DBS studies. Nevertheless, a global data collection of DBS and rTMS in OCD patients would be a huge step forward. Ideally, this would make it possible to make use of the global variability in the execution of neurostimulation studies, i.e. by directly comparing different treatment strategies.

In spite of the small samples, in recent years we have already come a long way. We can locate the so-called hotspot better and better. Whereas originally, the internal capsule was targeted with extremely high voltages (9–11 V), the target moved to the more selective area around the NAc with lower voltages (3.5–5.5 V), and now to specific white matter bundles such as the slMFB (1.5–3.5 V). In addition, studies on rTMS might eventually result in a shift towards more efficacious targets and stimulation protocols. The SMA appears to be the target of-choice in the treatment of OCD, and this is attributed to the possible normalization of hyperactive orbitofronto-striatal circuits, induced by low-frequency rTMS (Rehn et al. 2018). Furthermore, other promising forms of rTMS, including dTMS and TBS, have the potential to advance the current treatment possibilities. The progress illustrates the importance of neuroimaging studies in both DBS and rTMS.

Besides these important refinements of both treatments, future studies should also focus on outcome prediction. The clinical, personality, and cognitive profile, as well as neuroimaging parameters all might have predictive value, but are understudied to implement in clinical practice as of yet. Furthermore, specific patient profiles might be associated with better clinical response following DBS of a specific target or specific stimulation parameters. Large-scale comparative studies could provide an answer for target distinction in both DBS and rTMS.

Several other questions are still open-ended. For instance, patients may show anxious and depressive rebound effects after de-activating DBS (Ooms et al. 2014a), and we need to establish if, and after what time interval these withdrawal effects disappear. In the same vein, we need to establish the relapse rates and after what interval we can expect relapse after rTMS. Moreover, the clinical relation between rTMS and DBS remains understudied. Can patients that did not benefit from rTMS neuromodulation still respond to DBS? Should these two forms of neuromodulation be used sequentially in a staged approach, or could we predict beforehand in which patients rTMS suffices and which patients need DBS? Finally, it remains a challenge to offer a perspective to the small but important group of patients who have been treated with all available therapies, including rTMS and DBS, but still do not experience an effect. There are several options including experimental pharmacotherapy (e.g. ketamine), new forms of psychotherapy (e.g. acceptance and commitment therapy), or irreversible ablative neurosurgery (Stein et al. 2019).

A number of fundamental questions can be addressed using animal models. There are numerous rodent models exhibiting compulsivity in which state-of-the-art neuronal recording and manipulation techniques can be employed to evaluate the efficacy of DBS in a homogeneous experimental setting, circumventing sample size, and variability problems that plague human studies. Animal studies improve our understanding of the neuromodulation affected by DBS on immediate and long-term time scales, with a level of detail unparalleled by human work. Through work in rodents we may identify the most effective implant targets and stimulation parameters, based fundamentally on compulsivity and DBS mechanisms of action. Looking ahead, integrating these results into findings from human studies may lay the groundwork for closed-loop DBS systems that target specifically the neuronal circuitry underlying compulsivity, improving patient outcomes.

In conclusion, DBS and rTMS show great promise in the treatment of OCD. Both treatment modalities show large proportions of responding patients. Symptom reductions are strikingly large, especially given that most of the treated patients have a long track record of failure to respond to conventional treatments. These already promising results could be further improved by refining the neuromodulation targets and establishing predictors for clinical outcome. The greatest challenge for the future lies in setting up large, international multicenter trials to increase the level of evidence to convince mental health clinicians of the efficacy of these treatments.

References

Abelson JL, Curtis GC, Sagher O, Albucher RC, Harrigan M, Taylor SF et al (2005) Deep brain stimulation for refractory obsessive-compulsive disorder. Biol Psychiatry 57(5):510–516. https://doi.org/10.1016/j.biopsych.2004.11.042

Abramowitz JS (2006) The psychological treatment of obsessive-compulsive disorder. Can J Psychiatr 51(7):407–416. https://doi.org/10.1177/070674370605100702

Ahmari SE, Dougherty DD (2015) Dissecting OCD circuits: from animal models to targeted treatments. Depress Anxiety 32(8):550–562. https://doi.org/10.1002/da.22367

Alonso P, Pujol J, Cardoner N, Benlloch L, Deus J, Menchón JM et al (2001) Right prefrontal repetitive transcranial magnetic stimulation in obsessive-compulsive disorder: a double-blind, placebo-controlled study. Am J Psychiatr 158(7):1143–1145. https://doi.org/10.1176/appi.ajp.158.7.1143

Alonso P, Cuadras D, Gabriels L, Denys D, Goodman W, Greenberg BD et al (2015) Deep brain stimulation for obsessive-compulsive disorder: a Meta-analysis of treatment outcome and predictors of response. PLoS One 10(7):e0133591. https://doi.org/10.1371/journal.pone.0133591

Alpaugh M, Saint-Pierre M, Dubois M, Aube B, Arsenault D, Kriz J et al (2019) A novel wireless brain stimulation device for long-term use in freely moving mice. Sci Rep 9(1):6444. https://doi.org/10.1038/s41598-019-42910-7

Andrade P, Carrillo-Ruiz JD, Ramirez Y, Jimenez F (2010) Effects of thalamic reticular nucleus electrical stimulation in rats in a T-maze perseverative behavior model induced by 8-OH-DPAT. Neuromodulation 13(1):2–9. https://doi.org/10.1111/j.1525-1403.2009.00242.x

Arumugham SS, Vs S, Hn M, B V, Ravi M, Sharma E et al (2018) Augmentation effect of low-frequency repetitive transcranial magnetic stimulation over presupplementary motor area in obsessive-compulsive disorder: a randomized controlled trial. J ECT 34(4):253–257. https://doi.org/10.1097/YCT.0000000000000509

Badawy AA, El Sawy H, El Hay M (2010) Efficacy of repetitive transcranial magnetic stimulation in the management of obsessive compulsive disorder. Egyptian J Neurol Psychiatry Neurosurg 47:393–397

Baldermann JC, Bohn KP, Hammes J, Schuller CB, Visser-Vandewalle V, Drzezga A, Kuhn J (2019) Local and global changes in brain metabolism during deep brain stimulation for obsessive-compulsive disorder. Brain Sci 9(9):220. https://doi.org/10.3390/brainsci9090220

Balduzzi S, Rucker G, Schwarzer G (2019) How to perform a meta-analysis with R: a practical tutorial. Evid Based Ment Health 22(4):153–160. https://doi.org/10.1136/ebmental-2019-300117

Barcia JA, Avecillas-Chasin JM, Nombela C, Arza R, Garcia-Albea J, Pineda-Pardo JA et al (2019) Personalized striatal targets for deep brain stimulation in obsessive-compulsive disorder. Brain Stimul 12(3):724–734. https://doi.org/10.1016/j.brs.2018.12.226

Baunez C (2011) A few examples of the contribution of animal research in rodents for clinical application of deep brain stimulation. In: Progress in brain research, vol 194. Elsevier, pp 105–116

Baup N, Grabli D, Karachi C, Mounayar S, Francois C, Yelnik J et al (2008) High-frequency stimulation of the anterior subthalamic nucleus reduces stereotyped behaviors in primates. J Neurosci 28(35):8785–8788. https://doi.org/10.1523/JNEUROSCI.2384-08.2008

Brunelin J, Mondino M, Bation R, Palm U, Saoud M, Poulet E (2018) Transcranial direct current stimulation for obsessive-compulsive disorder: a systematic review. Brain Sci 8(2). https://doi.org/10.3390/brainsci8020037

Burguiere E, Monteiro P, Feng G, Graybiel AM (2013) Optogenetic stimulation of lateral orbitofronto-striatal pathway suppresses compulsive behaviors. Science 340 (6137):1243–1246. https://doi.org/10.1126/science.1232380

Carmi L, Tendler A, Bystritsky A, Hollander E, Blumberger DM, Daskalakis J et al (2019) Efficacy and safety of deep transcranial magnetic stimulation for obsessive-compulsive disorder: a prospective multicenter randomized double-blind placebo-controlled trial. Am J Psychiatry 176(11):931–938. https://doi.org/10.1176/appi.ajp.2019.18101180

Chabardes S, Polosan M, Krack P, Bastin J, Krainik A, David O et al (2013) Deep brain stimulation for obsessive-compulsive disorder: subthalamic nucleus target. World Neurosurg 80(3–4):S31 e31–38. https://doi.org/10.1016/j.wneu.2012.03.010

Chakrabarty K, Bhattacharyya S, Christopher R, Khanna S (2005) Glutamatergic dysfunction in OCD. Neuropsychopharmacology 30(9):1735–1740. https://doi.org/10.1038/sj.npp.1300733

Chamberlain SR, Menzies L (2009) Endophenotypes of obsessive-compulsive disorder: rationale, evidence and future potential. Expert Rev Neurother 9(8):1133–1146. https://doi.org/10.1586/ern.09.36

Chamberlain SR, Menzies L, Hampshire A, Suckling J, Fineberg NA, del Campo N et al (2008) Orbitofrontal dysfunction in patients with obsessive-compulsive disorder and their unaffected relatives. Science 321(5887):421–422. https://doi.org/10.1126/science.1154433

Coenen VA, Schlaepfer TE, Reinacher PC, Mast H, Urbach H, Reisert M (2019) Machine learning-aided personalized DTI tractographic planning for deep brain stimulation of the superolateral medial forebrain bundle using HAMLET. Acta Neurochir 161(8):1559–1569. https://doi.org/10.1007/s00701-019-03947-9

Coenen VA, Schlaepfer TE, Sajonz B, Dobrossy M, Kaller CP, Urbach H, Reisert M (2020) Tractographic description of major subcortical projection pathways passing the anterior limb of the internal capsule. Corticopetal organization of networks relevant for psychiatric disorders. Neuroimage Clin 25:102165. https://doi.org/10.1016/j.nicl.2020.102165

Coizet V, Heilbronner SR, Carcenac C, Mailly P, Lehman JF, Savasta M et al (2017) Organization of the anterior limb of the internal capsule in the rat. J Neurosci 37(10):2539–2554. https://doi.org/10.1523/JNEUROSCI.3304-16.2017

de Haas R, Struikmans R, van der Plasse G, van Kerkhof L, Brakkee JH, Kas MJ, Westenberg HG (2012) Wireless implantable micro-stimulation device for high frequency bilateral deep brain stimulation in freely moving mice. J Neurosci Methods 209(1):113–119. https://doi.org/10.1016/j.jneumeth.2012.05.028

Del Casale A, Kotzalidis GD, Rapinesi C, Serata D, Ambrosi E, Simonetti A et al (2011) Functional neuroimaging in obsessive-compulsive disorder. Neuropsychobiology 64(2):61–85. https://doi.org/10.1159/000325223

Denys D (2006) Pharmacotherapy of obsessive-compulsive disorder and obsessive-compulsive spectrum disorders. Psychiatr Clin North Am 29(2):553–584., xi. https://doi.org/10.1016/j.psc.2006.02.013

Denys D, Mantione M, Figee M, van den Munckhof P, Koerselman F, Westenberg H et al (2010) Deep brain stimulation of the nucleus accumbens for treatment-refractory obsessive-compulsive disorder. Arch Gen Psychiatry 67(10):1061–1068. https://doi.org/10.1001/archgenpsychiatry.2010.122

Denys D, Graat I, Mocking R, de Koning P, Vulink N, Figee M et al (2020) Efficacy of deep brain stimulation of the ventral anterior limb of the internal capsule for refractory obsessive-compulsive disorder: a clinical cohort of 70 patients. Am J Psychiatry 177(3):265–271. https://doi.org/10.1176/appi.ajp.2019.19060656

Di Lazzaro V, Pilato F, Saturno E, Oliviero A, Dileone M, Mazzone P et al (2005) Theta-burst repetitive transcranial magnetic stimulation suppresses specific excitatory circuits in the human motor cortex. J Physiol 565(Pt 3):945–950. https://doi.org/10.1113/jphysiol.2005.087288

Djodari-Irani A, Klein J, Banzhaf J, Joel D, Heinz A, Harnack D et al (2011) Activity modulation of the globus pallidus and the nucleus entopeduncularis affects compulsive checking in rats. Behav Brain Res 219(1):149–158. https://doi.org/10.1016/j.bbr.2010.12.036

Dougherty DD, Chou T, Corse AK, Arulpragasam AR, Widge AS, Cusin C et al (2016) Acute deep brain stimulation changes in regional cerebral blood flow in obsessive-compulsive disorder. J Neurosurg 125(5):1087–1093. https://doi.org/10.3171/2015.9.JNS151387

Dougherty DD, Brennan BP, Stewart SE, Wilhelm S, Widge AS, Rauch SL (2018) Neuroscientifically informed formulation and treatment planning for patients with obsessive-compulsive disorder: a review. JAMA Psychiat 75(10):1081–1087. https://doi.org/10.1001/jamapsychiatry.2018.0930

Ehmer I, Crown L, van Leeuwen W, Feenstra M, Willuhn I, Denys D (2020a) Evidence for distinct forms of compulsivity in the SAPAP3 mutant-mouse model for obsessive-compulsive disorder. eNeuro:0245–0219.2020. https://doi.org/10.1523/ENEURO.0245-19.2020

Ehmer I, Feenstra M, Willuhn I, Denys D (2020b) Instrumental learning in a mouse model for obsessive-compulsive disorder: impaired habit formation in Sapap3 mutants. Neurobiol Learn Mem 168:107162. https://doi.org/10.1016/j.nlm.2020.107162

Elbeh KAM, Elserogy YMB, Khalifa HE, Ahmed MA, Hafez MH, Khedr EM (2016) Repetitive transcranial magnetic stimulation in the treatment of obsessive-compulsive disorders: double blind randomized clinical trial. Psychiatry Res 238:264–269. https://doi.org/10.1016/j.psychres.2016.02.031

Fayad SM, Guzick AG, Reid AM, Mason DM, Bertone A, Foote KD et al (2016) Six-nine year follow-up of deep brain stimulation for obsessive-compulsive disorder. PLoS One 11(12): e0167875. https://doi.org/10.1371/journal.pone.0167875

Feenstra MGP, Denys D (2012) Animal studies in deep brain stimulation research. In: Denys D, Feenstra MGP, Schuurman R (eds) Deep brain stimulation: a new frontier in psychiatry. Springer, Heidelberg, pp 217–224

Fettes P, Schulze L, Downar J (2017) Cortico-striatal-thalamic loop circuits of the orbitofrontal cortex: promising therapeutic targets in psychiatric illness. Front Syst Neurosci 11:25. https://doi.org/10.3389/fnsys.2017.00025

Figee M, Luigjes J, Smolders R, Valencia-Alfonso CE, van Wingen G, de Kwaasteniet B et al (2013) Deep brain stimulation restores frontostriatal network activity in obsessive-compulsive disorder. Nat Neurosci 16(4):386–387. https://doi.org/10.1038/nn.3344

Figee M, Pattij T, Willuhn I, Luigjes J, van den Brink W, Goudriaan A et al (2016) Compulsivity in obsessive-compulsive disorder and addictions. Eur Neuropsychopharmacol 26(5):856–868. https://doi.org/10.1016/j.euroneuro.2015.12.003

Foa EB, Liebowitz MR, Kozak MJ, Davies S, Campeas R, Franklin ME et al (2005) Randomized, placebo-controlled trial of exposure and ritual prevention, clomipramine, and their combination in the treatment of obsessive-compulsive disorder. Am J Psychiatry 162(1):151–161. https://doi.org/10.1176/appi.ajp.162.1.151

Fox MD, Buckner RL, Liu H, Chakravarty MM, Lozano AM, Pascual-Leone A (2014) Resting-state networks link invasive and noninvasive brain stimulation across diverse psychiatric and neurological diseases. Proc Natl Acad Sci U S A 111(41):E4367–E4375. https://doi.org/10.1073/pnas.1405003111

Fradkin I, Adams RA, Parr T, Roiser JP, Huppert JD (2020) Searching for an anchor in an unpredictable world: a computational model of obsessive compulsive disorder. Psychol Rev. https://doi.org/10.1037/rev0000188

Franklin ME, Abramowitz JS, Kozak MJ, Levitt JT, Foa EB (2000) Effectiveness of exposure and ritual prevention for obsessive-compulsive disorder: randomized compared with nonrandomized samples. J Consult Clin Psychol 68(4):594–602. Retrieved from https://www.ncbi.nlm.nih.gov/pubmed/10965635

Fridgeirsson EA, Figee M, Luigjes J, van den Munckhof P, Schuurman PR, van Wingen G, Denys D (2019) Deep brain stimulation modulates directional frontolimbic connectivity in obsessive-compulsive disorder. bioRxiv:809269. https://doi.org/10.1101/809269

Gomes PV, Brasil-Neto JP, Allam N, Rodrigues de Souza E (2012) A randomized, double-blind trial of repetitive transcranial magnetic stimulation in obsessive-compulsive disorder with three-month follow-up. J Neuropsychiatry Clin Neurosci 24(4):437–443. https://doi.org/10.1176/appi.neuropsych.11100242

Goodman WK, Foote KD, Greenberg BD, Ricciuti N, Bauer R, Ward H et al (2010) Deep brain stimulation for intractable obsessive compulsive disorder: pilot study using a blinded, staggered-onset design. Biol Psychiatry 67(6):535–542. https://doi.org/10.1016/j.biopsych.2009.11.028

Gottesman II, Gould TD (2003) The endophenotype concept in psychiatry: etymology and strategic intentions. Am J Psychiatry 160(4):636–645. https://doi.org/10.1176/appi.ajp.160.4.636

Gould TD, Gottesman II (2006) Psychiatric endophenotypes and the development of valid animal models. Genes Brain Behav 5(2):113–119. https://doi.org/10.1111/j.1601-183X.2005.00186.x

Greenberg BD, Malone DA, Friehs GM, Rezai AR, Kubu CS, Malloy PF et al (2006) Three-year outcomes in deep brain stimulation for highly resistant obsessive-compulsive disorder. Neuropsychopharmacology 31(11):2384–2393. https://doi.org/10.1038/sj.npp.1301165

Greenberg BD, Rauch SL, Haber SN (2010) Invasive circuitry-based neurotherapeutics: stereotactic ablation and deep brain stimulation for OCD. Neuropsychopharmacology 35(1):317–336. https://doi.org/10.1038/npp.2009.128

Hadjas LC, Luscher C, Simmler LD (2019) Aberrant habit formation in the Sapap3-knockout mouse model of obsessive-compulsive disorder. Sci Rep 9(1):12061. https://doi.org/10.1038/s41598-019-48637-9

Haghighi M, Shayganfard M, Jahangard L, Ahmadpanah M, Bajoghli H, Pirdehghan A et al (2015) Repetitive transcranial magnetic stimulation (rTMS) improves symptoms and reduces clinical illness in patients suffering from OCD--results from a single-blind, randomized clinical trial with sham cross-over condition. J Psychiatr Res 68:238–244. https://doi.org/10.1016/j.jpsychires.2015.06.020

Hamani C, Temel Y (2012) Deep brain stimulation for psychiatric disease: contributions and validity of animal models. Sci Transl Med 4(142):142rv148. https://doi.org/10.1126/scitranslmed.3003722

Harika-Germaneau G, Rachid F, Chatard A, Lafay-Chebassier C, Solinas M, Thirioux B et al (2019) Continuous theta burst stimulation over the supplementary motor area in refractory obsessive-compulsive disorder treatment: a randomized sham-controlled trial. Brain Stimul 12 (6):1565–1571. https://doi.org/10.1016/j.brs.2019.07.019

Hartmann CJ, Lujan JL, Chaturvedi A, Goodman WK, Okun MS, McIntyre CC, Haq IU (2015) Tractography activation patterns in dorsolateral prefrontal cortex suggest better clinical responses in OCD DBS. Front Neurosci 9:519. https://doi.org/10.3389/fnins.2015.00519

Hauser TU, Eldar E, Dolan RJ (2016) Neural mechanisms of harm-avoidance learning: a model for obsessive-compulsive disorder? JAMA Psychiat 73(11):1196–1197. https://doi.org/10.1001/jamapsychiatry.2016.1938

Hawken ER, Dilkov D, Kaludiev E, Simek S, Zhang F, Milev R (2016) Transcranial magnetic stimulation of the supplementary motor area in the treatment of obsessive-compulsive disorder: a multi-site study. Int J Mol Sci 17(3):420. https://doi.org/10.3390/ijms17030420

Haynes WI, Mallet L (2010) High-frequency stimulation of deep brain structures in obsessive-compulsive disorder: the search for a valid circuit. Eur J Neurosci 32(7):1118–1127. https://doi.org/10.1111/j.1460-9568.2010.07418.x

Histed MH, Bonin V, Reid RC (2009) Direct activation of sparse, distributed populations of cortical neurons by electrical microstimulation. Neuron 63(4):508–522. https://doi.org/10.1016/j.neuron.2009.07.016

Huff W, Lenartz D, Schormann M, Lee SH, Kuhn J, Koulousakis A et al (2010) Unilateral deep brain stimulation of the nucleus accumbens in patients with treatment-resistant obsessive-compulsive disorder: outcomes after one year. Clin Neurol Neurosurg 112(2):137–143. https://doi.org/10.1016/j.clineuro.2009.11.006

Huys D, Kohl S, Baldermann JC, Timmermann L, Sturm V, Visser-Vandewalle V, Kuhn J (2019) Open-label trial of anterior limb of internal capsule-nucleus accumbens deep brain stimulation

for obsessive-compulsive disorder: insights gained. J Neurol Neurosurg Psychiatry 90 (7):805–812. https://doi.org/10.1136/jnnp-2018-318996
Islam L, Franzini A, Messina G, Scarone S, Gambini O (2015) Deep brain stimulation of the nucleus accumbens and bed nucleus of stria terminalis for obsessive-compulsive disorder: a case series. World Neurosurg 83(4):657–663. https://doi.org/10.1016/j.wneu.2014.12.024
Jahangard L, Haghighi M, Shyayganfard M, Ahmadpanah M, Sadeghi Bahmani D, Bajoghli H et al (2016) Repetitive transcranial magnetic stimulation improved symptoms of obsessive-compulsive disorder, but also cognitive performance: results from a randomized clinical trial with a cross-over Design and sham condition. Neuropsychobiology 73(4):224–232. https://doi.org/10.1159/000446287
Jakobs M, Fomenko A, Lozano AM, Kiening KL (2019) Cellular, molecular, and clinical mechanisms of action of deep brain stimulation-a systematic review on established indications and outlook on future developments. EMBO Mol Med 11(4):e9575. https://doi.org/10.15252/emmm.201809575
Jimenez F, Nicolini H, Lozano AM, Piedimonte F, Salin R, Velasco F (2013) Electrical stimulation of the inferior thalamic peduncle in the treatment of major depression and obsessive compulsive disorders. World Neurosurg 80(3–4):S30 e17–25. https://doi.org/10.1016/j.wneu.2012.07.010
Kang JI, Kim CH, Namkoong K, Lee CI, Kim SJ (2009) A randomized controlled study of sequentially applied repetitive transcranial magnetic stimulation in obsessive-compulsive disorder. J Clin Psychiatry 70(12):1645–1651. https://doi.org/10.4088/JCP.08m04500
Klavir O, Flash S, Winter C, Joel D (2009) High frequency stimulation and pharmacological inactivation of the subthalamic nucleus reduces 'compulsive' lever-pressing in rats. Exp Neurol 215(1):101–109. https://doi.org/10.1016/j.expneurol.2008.09.017
Le Jeune F, Verin M, N'Diaye K, Drapier D, Leray E, Du Montcel ST et al (2010) Decrease of prefrontal metabolism after subthalamic stimulation in obsessive-compulsive disorder: a positron emission tomography study. Biol Psychiatry 68(11):1016–1022. https://doi.org/10.1016/j.biopsych.2010.06.033
Lee DJ, Dallapiazza RF, De Vloo P, Elias GJB, Fomenko A, Boutet A et al (2019) Inferior thalamic peduncle deep brain stimulation for treatment-refractory obsessive-compulsive disorder: a phase 1 pilot trial. Brain Stimul 12(2):344–352. https://doi.org/10.1016/j.brs.2018.11.012
Lefaucheur JP, Aleman A, Baeken C, Benninger DH, Brunelin J, Di Lazzaro V et al (2020) Evidence-based guidelines on the therapeutic use of repetitive transcranial magnetic stimulation (rTMS): an update (2014-2018). Clin Neurophysiol 131(2):474–528. https://doi.org/10.1016/j.clinph.2019.11.002
Liebrand LC, Caan MWA, Schuurman PR, van den Munckhof P, Figee M, Denys D, van Wingen GA (2019) Individual white matter bundle trajectories are associated with deep brain stimulation response in obsessive-compulsive disorder. Brain Stimul 12(2):353–360. https://doi.org/10.1016/j.brs.2018.11.014
Lusicic A, Schruers KR, Pallanti S, Castle DJ (2018) Transcranial magnetic stimulation in the treatment of obsessive-compulsive disorder: current perspectives. Neuropsychiatr Dis Treat 14:1721–1736. https://doi.org/10.2147/NDT.S121140
Luyten L, Hendrickx S, Raymaekers S, Gabriels L, Nuttin B (2016) Electrical stimulation in the bed nucleus of the stria terminalis alleviates severe obsessive-compulsive disorder. Mol Psychiatry 21(9):1272–1280. https://doi.org/10.1038/mp.2015.124
Ma X, Huang Y, Liao L, Jin Y (2014) A randomized double-blinded sham-controlled trial of alpha electroencephalogram-guided transcranial magnetic stimulation for obsessive-compulsive disorder. Chin Med J 127(4):601–606. Retrieved from https://www.ncbi.nlm.nih.gov/pubmed/24534207
Makris N, Rathi Y, Mouradian P, Bonmassar G, Papadimitriou G, Ing WI et al (2016) Variability and anatomical specificity of the orbitofrontothalamic fibers of passage in the ventral capsule/ventral striatum (VC/VS): precision care for patient-specific tractography-guided targeting of deep brain stimulation (DBS) in obsessive compulsive disorder (OCD). Brain Imaging Behav 10(4):1054–1067. https://doi.org/10.1007/s11682-015-9462-9

Mallet L, Polosan M, Jaafari N, Baup N, Welter ML, Fontaine D et al (2008) Subthalamic nucleus stimulation in severe obsessive-compulsive disorder. N Engl J Med 359(20):2121–2134. https://doi.org/10.1056/NEJMoa0708514

Mallet L, Du Montcel ST, Clair AH, Arbus C, Bardinet E, Baup N et al (2019) Long-term effects of subthalamic stimulation in obsessive-compulsive disorder: follow-up of a randomized controlled trial. Brain Stimul 12(4):1080–1082. https://doi.org/10.1016/j.brs.2019.04.004

Manning EE, Dombrovski AY, Torregrossa MM, Ahmari SE (2019) Impaired instrumental reversal learning is associated with increased medial prefrontal cortex activity in Sapap3 knockout mouse model of compulsive behavior. Neuropsychopharmacology 44(8):1494–1504. https://doi.org/10.1038/s41386-018-0307-2

Mansur CG, Myczkowki ML, de Barros Cabral S, Sartorelli Mdo C, Bellini BB, Dias AM et al (2011) Placebo effect after prefrontal magnetic stimulation in the treatment of resistant obsessive-compulsive disorder: a randomized controlled trial. Int J Neuropsychopharmacol 14 (10):1389–1397. https://doi.org/10.1017/S1461145711000575

Mantione M, Nieman DH, Figee M, Denys D (2014) Cognitive-behavioural therapy augments the effects of deep brain stimulation in obsessive-compulsive disorder. Psychol Med 44 (16):3515–3522. https://doi.org/10.1017/S0033291714000956

Mantovani A, Simpson HB, Fallon BA, Rossi S, Lisanby SH (2010) Randomized sham-controlled trial of repetitive transcranial magnetic stimulation in treatment-resistant obsessive-compulsive disorder. Int J Neuropsychopharmacol 13(2):217–227. https://doi.org/10.1017/S1461145709990435

March JS, Frances A, Carpenter D, Kahn DA (1997) The expert consensus guideline series: treatment of obsessive–compulsive disorder. J Clin Psychiatry 58:5–71

McCairn KW, Iriki A, Isoda M (2012) High-frequency pallidal stimulation eliminates tic-related neuronal activity in a nonhuman primate model of Tourette syndrome. Neuroreport 23 (4):206–210. https://doi.org/10.1097/WNR.0b013e32834fab9d

McCairn KW, Iriki A, Isoda M (2013) Deep brain stimulation reduces Tic-related neural activity via temporal locking with stimulus pulses. J Neurosci 33(15):6581–6593. https://doi.org/10.1523/JNEUROSCI.4874-12.2013

McCracken CB, Grace AA (2007) High-frequency deep brain stimulation of the nucleus accumbens region suppresses neuronal activity and selectively modulates afferent drive in rat orbitofrontal cortex in vivo. J Neurosci 27(46):12601–12610. https://doi.org/10.1523/JNEUROSCI.3750-07.2007

McIntyre CC, Anderson RW (2016) Deep brain stimulation mechanisms: the control of network activity via neurochemistry modulation. J Neurochem 139(Suppl 1):338–345. https://doi.org/10.1111/jnc.13649

Menchon JM, Real E, Alonso P, Aparicio MA, Segalas C, Plans G et al (2019) A prospective international multi-center study on safety and efficacy of deep brain stimulation for resistant obsessive-compulsive disorder. Mol Psychiatry. https://doi.org/10.1038/s41380-019-0562-6

Michelson NJ, Eles JR, Vazquez AL, Ludwig KA, Kozai TDY (2019) Calcium activation of cortical neurons by continuous electrical stimulation: frequency dependence, temporal fidelity, and activation density. J Neurosci Res 97(5):620–638. https://doi.org/10.1002/jnr.24370

Milad MR, Rauch SL (2012) Obsessive-compulsive disorder: beyond segregated cortico-striatal pathways. Trends Cogn Sci 16(1):43–51. https://doi.org/10.1016/j.tics.2011.11.003

Montgomery EB Jr, Gale JT (2008) Mechanisms of action of deep brain stimulation(DBS). Neurosci Biobehav Rev 32(3):388–407. https://doi.org/10.1016/j.neubiorev.2007.06.003

Morishita T, Fayad SM, Higuchi MA, Nestor KA, Foote KD (2014) Deep brain stimulation for treatment-resistant depression: systematic review of clinical outcomes. Neurotherapeutics 11 (3):475–484. https://doi.org/10.1007/s13311-014-0282-1

Mulders AEP, Plantinga BR, Schruers K, Duits A, Janssen MLF, Ackermans L et al (2016) Deep brain stimulation of the subthalamic nucleus in obsessive-compulsive disorder: neuroanatomical and pathophysiological considerations. Eur Neuropsychopharmacol 26(12):1909–1919. https://doi.org/10.1016/j.euroneuro.2016.10.011

Mundt A, Klein J, Joel D, Heinz A, Djodari-Irani A, Harnack D et al (2009) High-frequency stimulation of the nucleus accumbens core and shell reduces quinpirole-induced compulsive checking in rats. Eur J Neurosci 29(12):2401–2412. https://doi.org/10.1111/j.1460-9568.2009.06777.x

Nauczyciel C, Le Jeune F, Naudet F, Douabin S, Esquevin A, Verin M et al (2014) Repetitive transcranial magnetic stimulation over the orbitofrontal cortex for obsessive-compulsive disorder: a double-blind, crossover study. Transl Psychiatry 4:e436. https://doi.org/10.1038/tp.2014.62

Nuttin B, Cosyns P, Demeulemeester H, Gybels J, Meyerson B (1999) Electrical stimulation in anterior limbs of internal capsules in patients with obsessive-compulsive disorder. Lancet 354 (9189):1526. https://doi.org/10.1016/S0140-6736(99)02376-4

O'Connor KP, Aardema F, Robillard S, Guay S, Pelissier MC, Todorov C et al (2006) Cognitive behaviour therapy and medication in the treatment of obsessive-compulsive disorder. Acta Psychiatr Scand 113(5):408–419. https://doi.org/10.1111/j.1600-0447.2006.00767.x

Ooms P, Blankers M, Figee M, Mantione M, van den Munckhof P, Schuurman PR, Denys D (2014a) Rebound of affective symptoms following acute cessation of deep brain stimulation in obsessive-compulsive disorder. Brain Stimul 7(5):727–731. https://doi.org/10.1016/j.brs.2014.06.009

Ooms P, Mantione M, Figee M, Schuurman PR, van den Munckhof P, Denys D (2014b) Deep brain stimulation for obsessive-compulsive disorders: long-term analysis of quality of life. J Neurol Neurosurg Psychiatry 85(2):153–158. https://doi.org/10.1136/jnnp-2012-302550

Ooms P, Blankers M, Figee M, Bergfeld IO, van den Munckhof P, Schuurman PR, Denys D (2017) Cost-effectiveness of deep brain stimulation versus treatment as usual for obsessive-compulsive disorder. Brain Stimul 10(4):836–842. https://doi.org/10.1016/j.brs.2017.04.120

Paralikar K, Santa W, Iyer R, Thom A, Su X, Hovland E et al (2015) A fully implantable and rechargeable neurostimulation system for animal research. Paper presented at the 2015 7th International IEEE/EMBS Conference on Neural Engineering (NER)

Parastarfeizabadi M, Kouzani AZ (2017) Advances in closed-loop deep brain stimulation devices. J Neuroeng Rehabil 14(1):79. https://doi.org/10.1186/s12984-017-0295-1

Park HR, Kim IH, Kang H, McCairn KW, Lee DS, Kim BN et al (2019) Electrophysiological and imaging evidence of sustained inhibition in limbic and frontal networks following deep brain stimulation for treatment refractory obsessive compulsive disorder. PLoS One 14(7):e0219578. https://doi.org/10.1371/journal.pone.0219578

Pelissolo A, Harika-Germaneau G, Rachid F, Gaudeau-Bosma C, Tanguy ML, BenAdhira R et al (2016) Repetitive transcranial magnetic stimulation to supplementary motor area in refractory obsessive-compulsive disorder treatment: a sham-controlled trial. Int J Neuropsychopharmacol 19(8). https://doi.org/10.1093/ijnp/pyw025

Pinhal CM, van den Boom BJG, Santana-Kragelund F, Fellinger L, Bech P, Hamelink R et al (2018) Differential effects of deep brain stimulation of the internal capsule and the striatum on excessive grooming in Sapap3 mutant mice. Biol Psychiatry 84(12):917–925. https://doi.org/10.1016/j.biopsych.2018.05.011

Pittenger C, Bloch MH, Williams K (2011) Glutamate abnormalities in obsessive compulsive disorder: neurobiology, pathophysiology, and treatment. Pharmacol Ther 132(3):314–332. https://doi.org/10.1016/j.pharmthera.2011.09.006

Prasko J, Paskova B, Zalesky R, Novak T, Kopecek M, Bares M, Horacek J (2006) The effect of repetitive transcranial magnetic stimulation (rTMS) on symptoms in obsessive compulsive disorder. A randomized, double blind, sham controlled study. Neuro Endocrinol Lett 27 (3):327–332. Retrieved from https://www.ncbi.nlm.nih.gov/pubmed/16816829

Rapinesi C, Kotzalidis GD, Ferracuti S, Sani G, Girardi P, Del Casale A (2019) Brain stimulation in obsessive-compulsive disorder (OCD): a systematic review. Curr Neuropharmacol 17 (8):787–807. https://doi.org/10.2174/1570159X17666190409142555

Rauch SL, Dougherty DD, Malone D, Rezai A, Friehs G, Fischman AJ et al (2006) A functional neuroimaging investigation of deep brain stimulation in patients with obsessive-compulsive disorder. J Neurosurg 104(4):558–565. https://doi.org/10.3171/jns.2006.104.4.558

Raymaekers S, Vansteelandt K, Luyten L, Bervoets C, Demyttenaere K, Gabriels L, Nuttin B (2017) Long-term electrical stimulation of bed nucleus of stria terminalis for obsessive-compulsive disorder. Mol Psychiatry 22(6):931–934. https://doi.org/10.1038/mp.2016.124

Rehn S, Eslick GD, Brakoulias V (2018) A meta-analysis of the effectiveness of different cortical targets used in repetitive transcranial magnetic stimulation (rTMS) for the treatment of obsessive-compulsive disorder (OCD). Psychiatry Q 89(3):645–665. https://doi.org/10.1007/s11126-018-9566-7

Resendez SL, Jennings JH, Ung RL, Namboodiri VM, Zhou ZC, Otis JM et al (2016) Visualization of cortical, subcortical and deep brain neural circuit dynamics during naturalistic mammalian behavior with head-mounted microscopes and chronically implanted lenses. Nat Protoc 11 (3):566–597. https://doi.org/10.1038/nprot.2016.021

Robbins TW, Gillan CM, Smith DG, de Wit S, Ersche KD (2012) Neurocognitive endophenotypes of impulsivity and compulsivity: towards dimensional psychiatry. Trends Cogn Sci 16 (1):81–91. https://doi.org/10.1016/j.tics.2011.11.009

Robbins TW, Vaghi MM, Banca P (2019) Obsessive-compulsive disorder: puzzles and prospects. Neuron 102(1):27–47. https://doi.org/10.1016/j.neuron.2019.01.046

Rodriguez-Romaguera J, Do Monte FH, Quirk GJ (2012) Deep brain stimulation of the ventral striatum enhances extinction of conditioned fear. Proc Natl Acad Sci U S A 109(22):8764–8769. https://doi.org/10.1073/pnas.1200782109

Rodriguez-Romaguera J, Greenberg BD, Rasmussen SA, Quirk GJ (2016) An avoidance-based rodent model of exposure with response prevention therapy for obsessive-compulsive disorder. Biol Psychiatry 80(7):534–540. https://doi.org/10.1016/j.biopsych.2016.02.012

Roh D, Chang WS, Chang JW, Kim CH (2012) Long-term follow-up of deep brain stimulation for refractory obsessive-compulsive disorder. Psychiatry Res 200(2–3):1067–1070. https://doi.org/10.1016/j.psychres.2012.06.018

Rossi S, Hallett M, Rossini PM, Pascual-Leone A, Safety of TMS Consensus Group (2009) Safety, ethical considerations, and application guidelines for the use of transcranial magnetic stimulation in clinical practice and research. Clin Neurophysiol 120(12):2008–2039. https://doi.org/10.1016/j.clinph.2009.08.016

Ruffini C, Locatelli M, Lucca A, Benedetti F, Insacco C, Smeraldi E (2009) Augmentation effect of repetitive transcranial magnetic stimulation over the orbitofrontal cortex in drug-resistant obsessive-compulsive disorder patients: a controlled investigation. Prim Care Companion J Clin Psychiatry 11(5):226–230. https://doi.org/10.4088/PCC.08m00663

Sachdev PS, Loo CK, Mitchell PB, McFarquhar TF, Malhi GS (2007) Repetitive transcranial magnetic stimulation for the treatment of obsessive compulsive disorder: a double-blind controlled investigation. Psychol Med 37(11):1645–1649. https://doi.org/10.1017/S0033291707001092

Sarkhel S, Sinha VK, Praharaj SK (2010) Adjunctive high-frequency right prefrontal repetitive transcranial magnetic stimulation (rTMS) was not effective in obsessive-compulsive disorder but improved secondary depression. J Anxiety Disord 24(5):535–539. https://doi.org/10.1016/j.janxdis.2010.03.011

Senova S, Clair AH, Palfi S, Yelnik J, Domenech P, Mallet L (2019) Deep brain stimulation for refractory obsessive-compulsive disorder: towards an individualized approach. Front Psych 10 (905):905. https://doi.org/10.3389/fpsyt.2019.00905

Seo HJ, Jung YE, Lim HK, Um YH, Lee CU, Chae JH (2016) Adjunctive low-frequency repetitive transcranial magnetic stimulation over the right dorsolateral prefrontal cortex in patients with treatment-resistant obsessive-compulsive disorder: a randomized controlled trial. Clin Psychopharmacol Neurosci 14(2):153–160. https://doi.org/10.9758/cpn.2016.14.2.153

Shayganfard M, Jahangard L, Nazaribadie M, Haghighi M, Ahmadpanah M, Sadeghi Bahmani D et al (2016) Repetitive transcranial magnetic stimulation improved symptoms of obsessive-compulsive disorders but not executive functions: results from a randomized clinical trial with crossover design and sham condition. Neuropsychobiology 74(2):115–124. https://doi.org/10.1159/000457128

Shivakumar V, Dinakaran D, Narayanaswamy JC, Venkatasubramanian G (2019) Noninvasive brain stimulation in obsessive-compulsive disorder. Indian J Psychiatry 61(Suppl 1):S66–S76. https://doi.org/10.4103/psychiatry.IndianJPsychiatry_522_18

Smolders R, Mazaheri A, van Wingen G, Figee M, de Koning PP, Denys D (2013) Deep brain stimulation targeted at the nucleus accumbens decreases the potential for pathologic network communication. Biol Psychiatry 74(10):e27–e28. https://doi.org/10.1016/j.biopsych.2013.03.012

Stein DJ, Costa DLC, Lochner C, Miguel EC, Reddy YCJ, Shavitt RG et al (2019) Obsessive-compulsive disorder. Nat Rev Dis Primers 5(1):52–52. https://doi.org/10.1038/s41572-019-0102-3

Suetens K, Nuttin B, Gabriels L, Van Laere K (2014) Differences in metabolic network modulation between capsulotomy and deep-brain stimulation for refractory obsessive-compulsive disorder. J Nucl Med 55(6):951–959. https://doi.org/10.2967/jnumed.113.126409

Sun T, Song Z, Tian Y, Tian W, Zhu C, Ji G et al (2019) Basolateral amygdala input to the medial prefrontal cortex controls obsessive-compulsive disorder-like checking behavior. Proc Natl Acad Sci U S A 116(9):3799–3804. https://doi.org/10.1073/pnas.1814292116

Ting JT, Feng G (2011) Neurobiology of obsessive-compulsive disorder: insights into neural circuitry dysfunction through mouse genetics. Curr Opin Neurobiol 21(6):842–848. https://doi.org/10.1016/j.conb.2011.04.010

To WT, De Ridder D, Hart J Jr, Vanneste S (2018) Changing brain networks through non-invasive neuromodulation. Front Hum Neurosci 12:128. https://doi.org/10.3389/fnhum.2018.00128

Tsai HC, Chang CH, Pan JI, Hsieh HJ, Tsai ST, Hung HY, Chen SY (2012) Pilot study of deep brain stimulation in refractory obsessive-compulsive disorder ethnic Chinese patients. Psychiatry Clin Neurosci 66(4):303–312. https://doi.org/10.1111/j.1440-1819.2012.02352.x

Tyagi H, Apergis-Schoute AM, Akram H, Foltynie T, Limousin P, Drummond LM et al (2019) A randomized trial directly comparing ventral capsule and anteromedial subthalamic nucleus stimulation in obsessive-compulsive disorder: clinical and imaging evidence for dissociable effects. Biol Psychiatry 85(9):726–734. https://doi.org/10.1016/j.biopsych.2019.01.017

van den Boom BJG, Pavlidi P, Wolf CJH, Mooij AH, Willuhn I (2017) Automated classification of self-grooming in mice using open-source software. J Neurosci Methods 289:48–56. https://doi.org/10.1016/j.jneumeth.2017.05.026

van den Boom BJG, Mooij AH, Miseviciute I, Denys D, Willuhn I (2019) Behavioral flexibility in a mouse model for obsessive-compulsive disorder: impaired Pavlovian reversal learning in SAPAP3 mutants. Genes Brain Behav 18(4):e12557. https://doi.org/10.1111/gbb.12557

van Dijk A, Klanker M, van Oorschot N, Post R, Hamelink R, Feenstra MG, Denys D (2013) Deep brain stimulation affects conditioned and unconditioned anxiety in different brain areas. Transl Psychiatry 3(7):e289. https://doi.org/10.1038/tp.2013.56

van Kuyck K, Demeulemeester H, Feys H, De Weerdt W, Dewil M, Tousseyn T et al (2003) Effects of electrical stimulation or lesion in nucleus accumbens on the behaviour of rats in a T-maze after administration of 8-OH-DPAT or vehicle. Behav Brain Res 140(1–2):165–173. https://doi.org/10.1016/s0166-4328(02)00295-4

van Kuyck K, Gabriels L, Cosyns P, Arckens L, Sturm V, Rasmussen S, Nuttin B (2007) Behavioural and physiological effects of electrical stimulation in the nucleus accumbens: a review. Acta Neurochir Suppl 97(Pt 2):375–391. https://doi.org/10.1007/978-3-211-33081-4_43

van Kuyck K, Brak K, Das J, Rizopoulos D, Nuttin B (2008) Comparative study of the effects of electrical stimulation in the nucleus accumbens, the mediodorsal thalamic nucleus and the bed nucleus of the stria terminalis in rats with schedule-induced polydipsia. Brain Res 1201:93–99. https://doi.org/10.1016/j.brainres.2008.01.043

Van Laere K, Nuttin B, Gabriels L, Dupont P, Rasmussen S, Greenberg BD, Cosyns P (2006) Metabolic imaging of anterior capsular stimulation in refractory obsessive-compulsive disorder: a key role for the subgenual anterior cingulate and ventral striatum. J Nucl Med 47(5):740–747. Retrieved from https://www.ncbi.nlm.nih.gov/pubmed/16644742

Voon V (2019) Toward precision medicine: prediction of deep brain stimulation targets of the ventral internal capsule for obsessive-compulsive disorder. Biol Psychiatry 85(9):708–710. https://doi.org/10.1016/j.biopsych.2019.03.969

Voon V, Droux F, Morris L, Chabardes S, Bougerol T, David O et al (2017) Decisional impulsivity and the associative-limbic subthalamic nucleus in obsessive-compulsive disorder: stimulation and connectivity. Brain 140(2):442–456. https://doi.org/10.1093/brain/aww309

Wan Y, Ade KK, Caffall Z, Ilcim Ozlu M, Eroglu C, Feng G, Calakos N (2014) Circuit-selective striatal synaptic dysfunction in the Sapap3 knockout mouse model of obsessive-compulsive disorder. Biol Psychiatry 75(8):623–630. https://doi.org/10.1016/j.biopsych.2013.01.008

Warden MR, Cardin JA, Deisseroth K (2014) Optical neural interfaces. Annu Rev Biomed Eng 16 (1):103–129. https://doi.org/10.1146/annurev-bioeng-071813-104733

Welch JM, Lu J, Rodriguiz RM, Trotta NC, Peca J, Ding JD et al (2007) Cortico-striatal synaptic defects and OCD-like behaviours in Sapap3-mutant mice. Nature 448(7156):894–900. https://doi.org/10.1038/nature06104

Welkenhuysen M, Gligorijevic I, Ameye L, Prodanov D, Van Huffel S, Nuttin B (2013) Neuronal activity in the bed nucleus of the stria terminalis in a rat model for obsessive-compulsive disorder. Behav Brain Res 240:52–59. https://doi.org/10.1016/j.bbr.2012.11.019

Widge AS, Zorowitz S, Basu I, Paulk AC, Cash SS, Eskandar EN et al (2019) Deep brain stimulation of the internal capsule enhances human cognitive control and prefrontal cortex function. Nat Commun 10(1):1536. https://doi.org/10.1038/s41467-019-09557-4

Winter C, Mundt A, Jalali R, Joel D, Harnack D, Morgenstern R et al (2008) High frequency stimulation and temporary inactivation of the subthalamic nucleus reduce quinpirole-induced compulsive checking behavior in rats. Exp Neurol 210(1):217–228. https://doi.org/10.1016/j.expneurol.2007.10.020

Winter C, Harnack D, Kupsch A (2010) Deep brain stimulation for neurological and psychiatric diseases: animal experiments on effect and mechanisms. Nervenarzt 81(6):711–718. https://doi.org/10.1007/s00115-010-2936-5

Winter L, Heitland I, Saryyeva A, Lutjens G, Schwabe K, Heissler HE et al (2018) Acute effects of electrical stimulation of the bed nucleus of the stria terminalis/internal capsule in obsessive-compulsive disorder. World Neurosurg 111:e471–e477. https://doi.org/10.1016/j.wneu.2017.12.084

Wood J, LaPalombara Z, Ahmari SE (2018) Monoamine abnormalities in the SAPAP3 knockout model of obsessive-compulsive disorder-related behaviour. Philos Trans R Soc Lond Ser B Biol Sci 373(1742). https://doi.org/10.1098/rstb.2017.0023

Wu H, Tambuyzer T, Nica I, Deprez M, van Kuyck K, Aerts JM et al (2016) Field potential oscillations in the bed nucleus of the stria terminalis correlate with compulsion in a rat model of obsessive-compulsive disorder. J Neurosci 36(39):10050–10059. https://doi.org/10.1523/JNEUROSCI.1872-15.2016

New Directions for Surgical Ablation Treatment of Obsessive Compulsive Disorder

Lori-An Etherington, Keith Matthews, and Harith Akram

Contents

Abstract Although there are effective treatments available for many, probably most, patients with OCD, a significant number do not respond, or fail to experience a sustained beneficial response. For patients with such chronic, disabling and 'treatment-refractory' OCD, neurosurgical treatments may be considered. The best-established neurosurgical treatments are so-called ablative procedures, where targeted lesions are created with the intention of interrupting and modifying specific

L.-A. Etherington and K. Matthews (✉)
Division of Molecular and Clinical Medicine, Ninewells Hospital and Medical School, University of Dundee, Dundee, UK
e-mail: k.matthews@dundee.ac.uk

H. Akram
Unit of Functional Neurosurgery, UCL Queen Square Institute of Neurology and The National Hospital for Neurology and Neurosurgery (UCLH), London, UK

Curr Topics Behav Neurosci (2021) 49: 437–460
https://doi.org/10.1007/7854_2020_207
Published Online: 10 February 2021

circuitry functions. There is a lengthy history of such procedures and a substantial literature although this is largely of an observational nature. However, both stereotactic radiosurgery (gamma knife) and MR-guided high intensity focused ultrasound are methods of lesion generation that lend themselves to the conduct of blinded randomised trial designs and these are beginning to be utilised. In this chapter, we present a narrative review of the key recent literature that describes the evidence for the safety and efficacy of lesion procedures for OCD. For context, we also consider the strength and quality of evidence relating to intensive residential treatment for OCD (sometimes proposed as an alternative to neurosurgery), furthermore, we also present some comparative data for lesion surgery and deep brain stimulation (DBS).

Keywords Ablative neurosurgery · Anterior capsulotomy · Anterior cingulotomy · Lesion · Magnetic resonance-guided focused ultrasound (MRgFUS) · Radiosurgery

1 Introduction

Neurosurgical procedures are considered 'established' or 'standard of care' treatment approaches for patients with severe treatment refractory mental disorders in many healthcare systems around the world. These procedures are offered by numerous centres internationally, with greatly increased activity in recent decades (Nuttin et al. 2014). Much of this activity is described in a burgeoning peer-reviewed clinical and scientific literature. There is, perhaps surprisingly, substantial international consensus around procedure selection, patient selection and the ethical and scientific standards for the conduct of such procedures (Nuttin et al. 2014). Although much of the resurgent interest in psychiatric neurosurgery in recent years has been driven by the exploration of the utility of deep brain stimulation (DBS) techniques, it should be recognised that stereotactically guided, focal lesion surgery, or 'ablative neurosurgery', is more commonly considered to represent 'established' practice in most countries and also represents the more important group of interventions in resource-poor healthcare environments. Ablative procedures have even been used as 'rescue' treatments when trials of DBS have failed (Luyten et al. 2016).

2 What Is Ablative Neurosurgery?

The term 'ablative' refers to neurosurgical procedures that create precise, targeted, circumscribed lesions within either the grey matter or nuclei of the brain, or more typically, the white matter tracts that carry fibres of connection. Despite the Latin etymology, contemporary ablation does not involve removal of any brain tissue, rather the in situ interruption, or irreversible termination of function of an area of tissue. This is typically achieved by one of three techniques: the direct application of radiofrequency (RF) generated heat through a brain electrode or probe; by focused irradiation; or by focused ultrasound. The last two methods do not involve the insertion of a probe into brain tissue. Whilst high intensity focused ultrasound ablation is a promising novel tool for making brain lesions, radiofrequency and Gamma Knife ablation, on the other hand, are long established techniques within the field of functional neurosurgery, where they have been widely used for several decades as a highly effective and safe method for preserving and improving function in patients with, for example, disabling movement disorders such as Parkinson's Disease and dystonia (e.g. see Cif and Hariz 2017). Crucially, the targeting of lesions is achieved to millimetric accuracy using advanced stereotactic neuronavigation techniques that are guided by high resolution structural brain imaging, typically magnetic resonance imaging.

3 Why Is Ablative Neurosurgery Considered for Some Patients with OCD?

Standard and 'first-line' therapies for patients with OCD comprise a combination of psychological therapies (usually variants of cognitive behavioural therapy with a strong emphasis on graded exposure with response prevention (McKay et al. 2015) although other promising therapies are in development – e.g. Inference Based Therapy (e.g. see Aardema et al. 2005) and there are numerous effective pharmacotherapies. There is reasonable quality evidence available to suggest that these therapies are effective for a substantial proportion of OCD patients (Skapinakis et al. 2016). However, the most compelling evidence relates to those with relatively mild, 'pure' OCD with shorter duration of illness. The literature on the management of patients with chronic, complex, severe and more 'treatment refractory' forms of OCD is sparse and much less compelling. It is within this context that the literature on ablative neurosurgery should be considered.

The true prevalence of these more treatment refractory forms of OCD is relatively poorly described. However, it is generally considered that around 30–50% of patients do not respond satisfactorily to 'standard' treatments in an enduring manner (Eisen et al. 2013). Certainly, due to the relatively high general population prevalence of OCD and despite the widespread availability of established pharmacotherapy and psychological therapy in many healthcare systems, there remain many patients with demonstrably 'refractory' forms of OCD (Mancebo et al. 2006; Eisen et al. 2013). In most circumstances, a marked improvement in clinical status is relatively rare after 3 years of continuous illness. In part, this relative ignorance of the epidemiology and the clinical course of more severe and 'refractory' presentations of OCD are because the conceptual and practical operationalisation of 'refractoriness' or 'treatment resistance' is a relatively recent endeavour and has been driven by either the instantiation of novel therapies (e.g. DBS) where patients are required to be demonstrably non-responsive to standard therapies, or by the emergent need to standardise recruitment to studies testing any 'non-standard' interventions that may be compared with 'treatment as usual'.

For OCD, as with other mental disorder diagnoses, there are many limitations to current definitions and proposed taxonomic schemes, including for 'treatment resistance'. These limitations include, for example, an absence of validating studies, false (untested) assumptions about treatment hierarchies, a tendency to dichotomisation of response and non-response, arbitrary definitions of response thresholds and limited elucidation of any underpinning neurobiology. There is also a general tendency to ignore the important contribution and relevance of comorbid diagnoses that occur alongside OCD. OCD is known to be associated with a high number of comorbidities, which contribute to heightened distress and disability as well as having deleterious effects on family relationships, work and clinical outcomes (Pallanti et al. 2011).

For the purposes of neurosurgical interventions for OCD (neurostimulation and lesion surgery), 'refractoriness to treatment' generally infers a biological substrate that is relatively (or absolutely) unresponsive to the pharmacological effects of 'classic' SRI drug therapies, antipsychotic drugs and the evidence-based psychological therapies. Furthermore, this infers that the underlying biology of the patient is qualitatively different from 'standard' patients with OCD who DO respond to therapies. These assumptions remain untested and unproven, but represent opportunities for future studies and procedural refinement.

4 Therapeutic Approaches to Treatment Refractory OCD

4.1 Intensive Treatment Programmes

There are numerous specialist OCD treatment centres across the world that provide multimodal intensive treatment programmes for patients with 'treatment refractory' OCD. Failure to have experienced a sustained, beneficial response to standard treatments is the general criterion for admission to such programmes. As a background context for the consideration and evaluation of the quality and magnitude of the literature relating to the outcomes for psychiatric neurosurgery for OCD, it is worth briefly reviewing the equivalent literature for such multimodal intensive treatment programmes. Key elements of these studies are summarised in Table 1.

There have been two key papers on this topic published by Veale and colleagues in recent years; the first a large observational case series describing residential treatment outcomes (2016a), the other a formal systematic review and meta-analysis of intensive treatment programme outcomes (2016b).

In a large open case series, Veale et al. (2016a) summarised 11 years of experience (2000–2012) of the provision of intensive therapy for OCD in a residential setting from a single centre. There were two separate interventions included in the report. A 'standard' programme of 12 weeks with CBT as the core therapeutic ingredient (sessions scheduled four times per week) and a more intensive 2-week

Table 1 Summary of recent evidence relating to intensive residential therapy for patients with OCD

	Study design	Entry *n*	Entry OCD severity	Duration of intervention	Nature of therapy[a]	Change in Y-BOCS during treatment	Follow-up	Comments
Veale et al. (2016a)	Retrospective observational case series	418 'standard' 52 'intensive' (2001–12)	85/418 "severe & treatment refractory" Y-BOCS 30.75 (+5.95)	'Standard' 12 weeks 'intensive' 2 weeks	Multimodal medication optimised CBT/ERP 4 sessions per week	Exitscore = 18.55 (±7.84) Average decline in Y-BOCS = 12.2 (16.2 for 'intensive' subgroup) 50% reached 25% Y-BOCS reduction	1–3 m FU score for $n = 75$ 19.60 (±9.39)	84/418 (20.1%) 'early or self-discharge' 33% missing data at discharge, 57% at 1–3 m follow-up
Veale et al. (2016b)	Systematic review of 19 studies	2,306	Mean Y-BOCS range 24.1–34.7	Mean duration between 5 and 19 weeks	Multimodal medication optimised CBT/ERP	Mean group change in Y-BOCS range = 8.0–12.9	6 of 19 studies only. Typically 3–4 point increase on Y-BOCS by 3 m	Extensive missing data
Siwiec et al. (2019)	Retrospective observational case series	379 (2012–17)	Self-report Y-BOCS 27.5 ± 5.4 for 'remitters' Self-report Y-BOCS 24.2 ± 5.1 for 'non-remitters'	Open ended mean duration of stay 58.8. days	Multimodal 27 h 'therapy' per week CBT/ERP	Self-report Y-BOCS 8.3 ± 3.3 for 'remitters' –33% of population Self-report Y-BOCS 20.3 ± 5.2 for 'non-remitters'	N/A	Admission Y-BOCS threshold of 14, substantive comorbidities excluded all data presented according to discharge 'remission' status defined as Y-BOCS ≤12

(continued)

Table 1 (continued)

	Study design	Entry n	Entry OCD severity	Duration of intervention	Nature of therapy[a]	Change in Y-BOCS during treatment	Follow-up	Comments
Taube-Schiffe et al. (2020)	Prospective observational case series	19	Self-report Y-BOCS 29.2 ± 5.9	12 weeks	Multimodal medication optimised CBT/ERP 3 h per day	Exitscore = 17.00 (±6.94) "42% mean reduction in Y-BOCS"	N/A	"<5 non-completers" qualitative interview data summarised
Balachander et al. (2020)	Retrospective observational case series	420 of 451 admissions	Y-BOCS 29.9 ± 4.5	Open ended mean duration of stay 42.7 (± 25.3) days	Multimodal medication optimised CBT/ERP 4–5 h per week	Mean Y-BOCS decline = 11.8 (±7.2) n = 38 'remitters' n = 211 'responders' (includes 'remitters') n = 86 'partial responders'	n = 262 @6 m n = 232 @12 m n = 169 @24 m Latent class analysis no Y-BOCS scores	Family member (first degree relative or spouse) also resident during stay around 50% 'minimal' or non response 14.5% longer term 'remission'

[a]'Nature of Therapy' during each programme is generally described in greater detail within each report, but commonalities included the emphasis on cognitive behavioural therapy (CBT) /exposure and response prevention (ERP) to varying degrees, psychoeducation, social skills and communication training, optimisation of medication and support with activities of daily living

programme with 20 h of CBT. Of 472 admitted patients, 418 entered the 'standard programme', with 52 directed to the 'intensive programme'. However, it should be noted that only 85 (20%) entering the 'standard programme' were specifically for the treatment of severe refractory OCD and thus these data mostly relate to a different population. Medication was withdrawn, stabilised or optimised during admission. Understandably, the results from this 'real world' case series were complex, with considerable missing data and considerable variability in the patient entrant characteristics and individual treatment experiences. Indeed, 57% of data were missing at 1–3 month follow-up and 65% missing at 6–12 m follow-up. Also, 20% of the entrants to the standard programme were discharged early, or self-discharged for a variety of reasons, but this included deterioration in their health and/or inability to engage with treatment. Overall, for those who did remain within the 'standard programme' (a completer analysis) there were substantial drops in scores on the Y-BOCS during the acute treatment period, but even in this limited 'completer' sample, symptom score rises were observed at longer-term follow-up. For those designated as 'severe and treatment refractory' a similar pattern was observed. From this series (the largest published), there were certainly suggestions that intensive treatment could be beneficial with clearly positive clinical change, at least in the short term, during the period of admission and treatment, but the evidence for the key desired outcome of *sustained improvement* was, regrettably, very weak. Indeed, as Veale and colleagues themselves noted '. . .the follow-up data should be treated with great caution because of the amount of missing data'.

Taking a broader perspective of the relevant literature, Veale et al. (2016b) conducted a systematic review and meta-analysis of published outcomes from intensive treatment programmes for OCD in adults. They identified 19 studies with 2,306 individual patients, published over a period of 22 years, thereby encompassing the pre-SSRI period as well as contemporary treatment. Remarkably consistent with the outcomes reported from their own (Veale et al. 2016a) observational series, the mean overall reduction in clinician rated Y-BOCS scoring from treatment initiation to termination was 10.7. The majority of studies, unfortunately, reported no follow-up data. Veale et al. further noted that being married, or being in a cohabiting relationship (both possible markers of less impactful and prolonged illness), predicted superior outcomes, with alcohol misuse, or hoarding symptoms, predicting poorer outcomes. However, perhaps because of the very substantial study and patient heterogeneity, there was no evidence that any particular programme details (e.g. numbers of hours of therapy) exerted meaningful and reliable influence on outcomes.

Three further observational series have been published since the systematic review by Veale et al. (2016b). Siwiec et al. (2019) reported on the intensive residential treatment outcomes for 379 patients treated with CBT/ERP between 2012 and 2017. A threshold of self-rated symptom severity of only 14 on the Y-BOCS was required for admission to the programme. Patients with substantive comorbidities (e.g. psychotic symptoms, substance misuse, suicidality) were not enrolled. Again, this programme was multi-faceted, with medication optimisation, non-OCD group activities and 27 h of targeted individual therapy each week. By the end of the programme (mean length of stay 58.8 days), 126 (33%) had achieved 'remission', a self-report Y-BOCS score of $\leq$12. In an exploration of predictors of clinical symptom improvement, the authors reported that '. . .higher admission OCD severity, comorbid PTSD and related disorders diagnosis, intolerance of uncertainty' all predicted non-remission.

Taube-Schiffe et al. (2020) reported on the outcomes for 19 patients who entered their intensive treatment programme between July 2017 and June 2018. This – typical of such residential programmes – was a complex 12 week intervention with residential or 'day hospital' support, optimisation of medication, group and individual CBT focused upon OCD, in addition to other therapeutic and supportive group activities. Individual CBT was delivered for 3–4 h per day under the direction of an expert OCD therapist. An array of self-report outcome measures was described and the change in status from time of admission to time of discharge is reported. There were also additional qualitative data presented. The quantitative data were presented as changes in group mean scores analysed by t-test. Over the period of inclusion in the treatment programme, the group self-report Y-BOCS scores fell from 29.2 (5.9) to 17 (6.9). There were accompanying positive changes in depression symptom scores and quality of life measurements. It should be noted, however, that there was a relatively high proportion of programme 'non-completers' ($n < 5$) and it is not clear how these were accounted for in the analysis. Also, these intervention outcome data only described improvement by the end of the programme, with no longer-term follow-up data.

In the most recent observational study of intensive treatment for OCD, Balachander et al. (2020) described characteristics and outcomes for a retrospective case series of 420 patients who entered a residential intensive treatment programme. This programme, like the others, was 'multimodal' with optimisation of medication and a variety of therapeutic activities in addition to 4–5 h of CBT/ERP weekly. Interestingly, as a distinction from the other programmes, but consistent with a model of care often used in hospitals in India, a family member (spouse or first degree relative) was also 'admitted' into the programme and supported to act as a co-therapist. The impact of residential intensive therapy resulted in a mean Y-BOCS decline of 11.8 ($\pm$7.2) in this group of patients, with 38 (9%) achieving criteria for 'remission', 173 (41%) were 'responders' and 86 (20%) 'partial responders'. A major strength of this report was the reporting of follow-up data collected at 6, 12 and 24 m ($n = 262$, 232 & 169, respectively). These data were presented as a latent class growth analysis to model response trajectories. Using this approach, a best-fit four 'class' solution was identified. There were 14.5% deemed 'remitters', with 14.6% designated as 'non-responders' with a flat trajectory across the study, 34.7% 'minimal responders and 36.3% as 'responders'. Shorter duration of illness, 'better insight' and a lower burden of contamination symptoms were predictive of improved outcomes.

A reasonable conclusion based on these reports (see Table 1.) might, therefore, be that intensive residential therapy, with optimisation of medication, *may* reduce OCD symptom scores by a mean of 10–13 points on the Y-BOCS, by the time of discharge, in those who are able to adhere to the programme. However, where data does exist, the evidence for sustained remission suggests that this is uncommon, even within less severe and refractory populations, and probably does not exceed around 15% of treated patients.

For the purposes of contextualising the relative strength and value of the literature that describes the experiences with ablative neurosurgery, these intensive treatment programme data are highly relevant and useful. The summary by Veale et al. (2016b) concludes – 'Inpatient and residential treatment is an encouraging option for those with severe or treatment refractory OCD. Such clients can make significant improvements with intensive residential or in-patient therapy, but little is known either about its long-term benefits or cost effectiveness compared with an alternative. However, this study highlights the diversity in the types of treatments offered (length of stay, treatment program content etc.). In addition, the study highlights that data available on predictors of drop out or treatment outcome is inconsistent, with studies concluding an array of different results'.

It is probably also reasonable to conclude that we know very little about even the medium term efficacy (3–6 m), let alone the longer-term efficacy (>12 m) of these (relatively) expensive treatment approaches that are systematically selecting for patients with the greatest likelihood of responding. This selection is, arguably, entirely appropriate for any intervention, but it does mean that the evidence for intensive treatment for patients with the most chronic, severe, comorbid and refractory forms of OCD is somewhere between extremely weak and non-existent. It is against this background of low quality, non-randomised evidence, with substantial loss to follow-up and scant information on long-term benefits that we now consider the quality of evidence to support ablative neurosurgery for OCD. It should also be noted that it is usual for patients proceeding to ablative neurosurgery to have previously tried and failed to benefit from intensive residential therapy.

4.2 Neurosurgical Procedure Evolution

Neurosurgical approaches to the treatment of mental disorder have a fraught history that has, understandably, resulted in polarised and often negative perceptions within both clinical and lay populations (Cabrera et al. 2020). The emergence and subsequent retraction of neurosurgery in the field of psychiatry in the early-mid twentieth century has been extensively debated and considered elsewhere (see, for example, Mahoney and Green 2020). Briefly, it is worth noting that this began with relatively untargeted, 'freehand' operations in small cohorts of psychiatric patients with poorly categorised and diverse difficulties. Contemporary ablative neurosurgery is very different in almost every respect. Ablative procedures today are performed using stereotactic methods, targeted using neuroimaging guidance in carefully selected patients who are demonstrably unresponsive to non-surgical therapies and where full informed consent is provided.

4.3 Lesion Surgery for OCD

Whilst the pathophysiology of OCD remains to be fully elucidated at the present time, altered metabolism, connectivity and function within cortico-basal ganglia-thalamocortical (CBTC) circuits have long been believed to have an important role. For OCD, the frontal–striatal–thalamic limbic circuit consisting of both the cingulate and orbitofrontal cortices, the ventral striatum and the mediodorsal nucleus of the thalamus seems to be particularly strongly implicated. The underpinning rationale for all OCD neurosurgical procedures – both neuromodulation and lesion surgery – is that of deliberate modulation of function within either the fibres of communication linking these regions, or by targeting key nuclei themselves. Historically, there have been four key neurosurgical ablation targets; anterior capsulotomy, anterior cingulotomy, subcaudate tractotomy and the combination of the latter two procedures in what is known as 'limbic leucotomy'. These procedures have been offered to patients with both treatment refractory depressive disorders and OCD. Internationally the most commonly performed procedures, by far, are the anterior cingulotomy and anterior capsulotomy. In this chapter, we will, therefore, focus upon these two.

As a general point, it is worth noting that the literature is broadly vulnerable to exactly the same categories of quality issues that affect the 'intensive treatment' literature. That is, almost all data are drawn from observational settings, with meagre descriptions of illness characteristics and previous treatments, vagueness in description of detail of interventions delivered (typically the targets for surgery are described but not necessarily the confirmed lesion characteristics), missing data and non-negligible loss to follow-up over variable timescales. Here we present a brief narrative review of the most recent and substantive literature relating to these two procedures, with particular emphasis on recently published systematic reviews with synthesis and meta-analysis and large, high quality observational series.

5 Anterior Cingulotomy

An anterior cingulotomy is a targeted lesion of the dorsal anterior cingulate cortex and the cingulum bundle bilaterally (see Figs. 1 and 2b). It is believed to interrupt communication between the cingulate and the orbitofrontal cortex, the ventral striatum and other limbic structures. It has been performed for several decades (see Foltz and White 1962) as a therapeutic intervention for pain, depression and OCD. There are no published randomised trials of anterior cingulotomy, by any lesion generation method, but several large observational studies have been reported over the decades. In a recent systematic review of ablative procedures for OCD, Brown et al. (2016) brought together data from two radiofrequency thermablation cingulotomy case series (summarised in Table 2) that met their rigorous inclusion criteria, including 51 individual patients followed up over 24–69 months. The

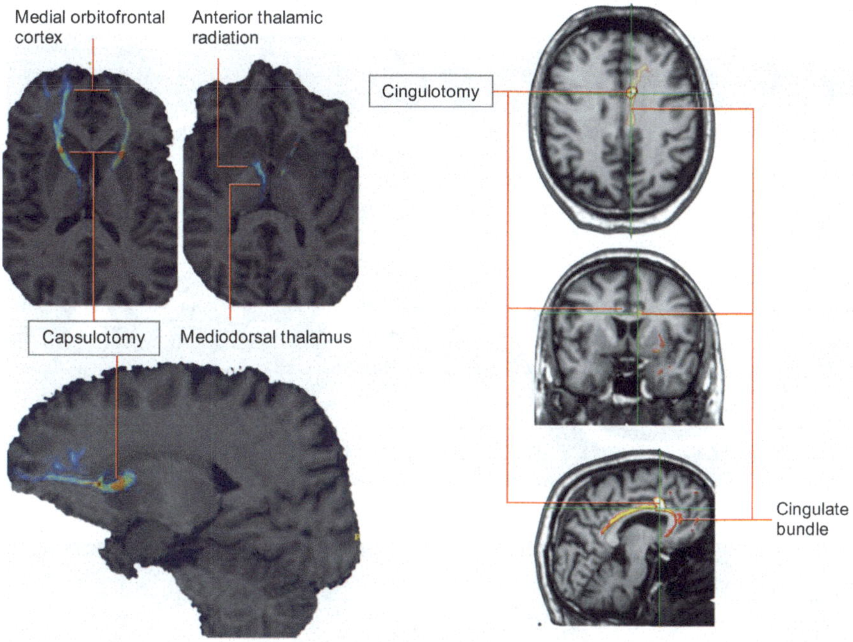

Fig. 1 The left panel shows an effective capsulotomy in axial (left top) and sagittal (left bottom) shown in orange fused with preoperative tractography showing the lesion interrupting fibres connecting the medial orbitofrontal cortex and the mediodorsal thalamus. The right panel shows an effective cingulotomy (bright red) in axial, coronal and sagittal planes with the lesion disconnecting the cingulate bundle as identified on pre-operative tractography

change in Y-BOCS scoring at 12 months post-surgery was −36% for one series ($n = 17$) and −37% for the other ($n = 34$). At long-term follow-up, 47% were classified as having achieved a 'full response' in one series and 38% 'full response' with an additional 25% 'partial response' in the other. For both series, the overall rates of adverse events were low and mostly of short duration (e.g. post-operative subjective memory difficulties and urinary incontinence) although there were two eventual suicides in the larger of the series (from a total of 99 total procedures within a cohort of 64) and one intracranial infection requiring further surgery.

6 Anterior Capsulotomy

The anterior capsulotomy procedure – whether achieved using radiofrequency (thermal) lesions, radiosurgery (focused gamma irradiation) or magnetic resonance imaging focused ultrasound (MRgFUS), targets the anterior limb of the internal capsule (ALIC) between the head of the caudate nucleus and the lenticular nucleus where white matter fibres converge from the orbitofrontal and cingulate cortices,

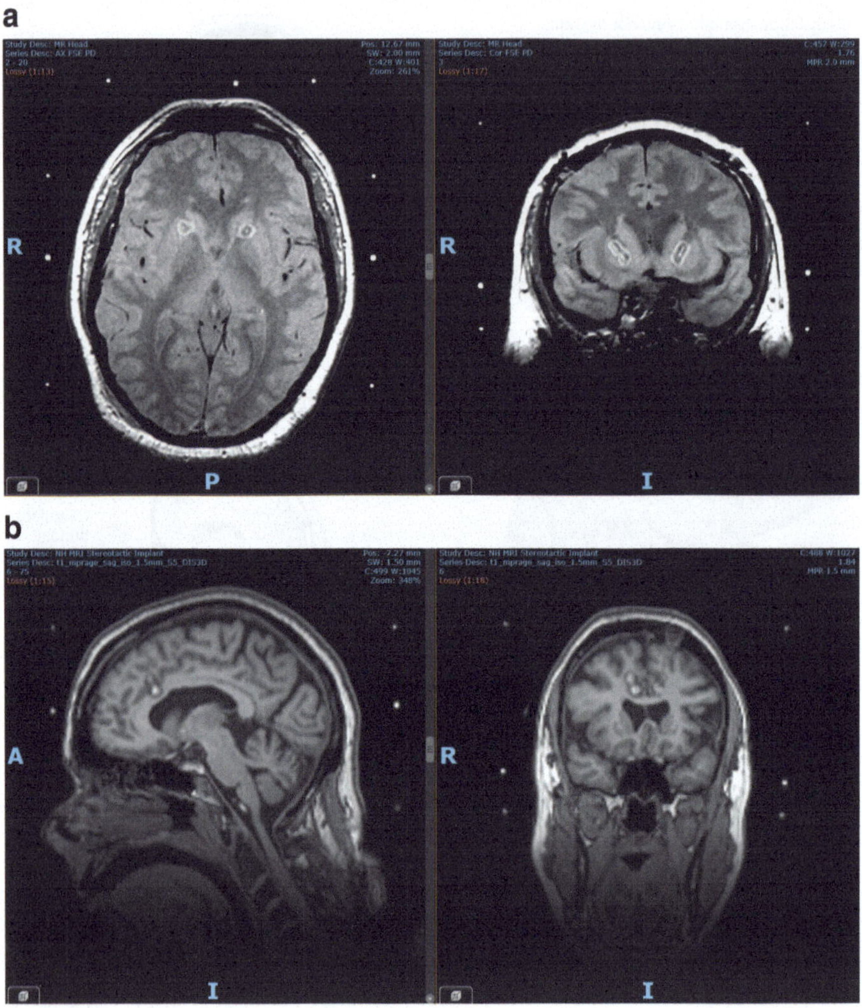

Fig. 2 (**a**) Anterior Capsulotomy. Axial (L) and coronal (R) proton-density weighted 1.5T stereotactic MRI immediately after bilateral radiofrequency thermal anterior capsulotomy. (image supplied by Professor Ludvic Zrinzo, Unit of Functional Neurosurgery, UCL Queen Square Institute of Neurology and The National Hospital for Neurology and Neurosurgery (UCLH), London UK). (**b**) Anterior Cingulotomy. Sagittal (l) and coronal (R) stereotactic T1-weighted 1.5T MRI of bilateral thermal anterior cingulotomy

projecting onwards to ventral striatum and the mediodorsal thalamus (see Figs. 1 and 2a). The first bilateral anterior capsulotomy was performed in 1949, using radiofrequency thermablation, by Jean Talairach. The target rationale had been derived from the hypothesis proposed by Papez, that the internal capsule carried connections with the limbic system circuit that included the cingulum bundle, hippocampus, thalamus, and hypothalamus and was believed to mediate emotions.

Table 2 Summary of studies of anterior cingulotomy included in systematic review and meta-analysis by Brown et al. (2016)

Authors	Study design	*n*	Baseline Y-BOCS	12 m follow-up Y-BOCS	24 m follow-up Y-BOCS	% 'responders' (Y-BOCS decline of ≥ 35%)	Lesion procedure	Adverse events
Jung et al. (2006)	Prospective case series, confirmed chronic and refractory OCD	17 7 females	Mean = 35 (±3.9)	Mean = 22.4 (±6.5) mean % change −36	Mean = 18.2 (±4.3) mean % change −36	At 12 m and 24 m 8/17 47%	Bilateral radiofrequency thermal lesions	Prospective surveillance no surgical adverse effects no adverse neuropsychological effects detected
Sheth et al. (2013)	Prospective case series, confirmed chronic and refractory OCD	64 22 females	Mean = 31 (±7.6)	Mean = 19.5 (±10.4) mean % change −37		At last follow-up (>12 m) 38%	Bilateral radiofrequency thermal lesions	Prospective surveillance 6 individual events deemed 'serious' in total of 99 procedures[a] 13 individual events deemed 'transient' in total of 99 procedures[a]

The inclusion criteria for this systematic review followed the PRISMA guidelines (Moher D, Liberati A, Tetzlaff J, Altman DG: *Preferred reporting items for systematic reviews and meta-analyses: the PRISMA statement*. BMJ 339:b2535, 2009) and AHRQ recommendations for comparative effectiveness (Viswanathan M, Ansari M, Berkman N, Hartling L, McPheeters M, Santaguida PL, et al.: *Assessing the risk of bias of individual studies in systematic reviews of health care interventions*, in Methods Guide for Comparative Effectiveness Reviews. Rockville, MD: Agency for Healthcare Research and Quality, 2012) Eligibility criteria required that the individual reports were for adults (>18) with OCD and no prior history of neurosurgery and that procedures were conducted under MRI guidance with OCD symptom severity quantified using the Y-BOCS

[a]Reporting of surgical case series is often complicated by the inclusion of further procedures. In this report, the adverse effect estimates relate to the performance of 99 separate procedures in a patient series of 64

Four years after the first description of capsulotomy, in 1953, the Swedish neurosurgeon Lars Leksell performed the first radiosurgical capsulotomy using focused ionised radiation with the lesion targeted in the anterior part of the internal capsule. Worldwide, anterior capsulotomy has been favoured as the procedure of choice for the treatment of OCD in most centres and most of the existing literature reflects this.

7 Observational Evidence

As with the literature for anterior cingulotomy, there are numerous reviews of the use of anterior capsulotomy, but we will here summarise the most recent and informative (see Table 3).

The first stop is to revisit the systematic review by Brown et al. (2016) considering the ablation literature specific to OCD. They identified 8 studies of anterior capsulotomy meeting their inclusion criteria, including 112 patients. Six were prospective observational studies, one retrospective cohort study and one small prospective, controlled, cohort study. Notably, this review highlights the variability in reporting of key clinical data within these case series and the accompanying risk of bias, rendering the task of synthesis very challenging. However, they reported a 36–75% drop in Y-BOCS scoring from baseline to 12 m following capsulotomy in these studies, with longer-term changes of 32–79% in Y-BOCS scoring and overall 'responder rates' of 37–80%.

Adverse event rates in the prospective case series were categorised as common, but mostly transient (e.g. urinary incontinence 8–40% for a period of days, confusion/memory impairment 30–50% for a period of up to 10 days), with more persistent adverse effects occurring at a much lower frequency, but potentially of greater concern, e.g. 'personality or behaviour change' at around 5% and perioperative haemorrhage with resultant paresis (1 event). The latter, needless to say, is not a 'target-specific' complication but rather a rare risk of intracranial surgery.

The highest reported rates of adverse effects were derived from a largely retrospective series (Ruck et al. 2008) where there had been very limited and inconsistent baseline data collection and attribution of 'events' or putative changes to surgery was highly problematic.

As with the 'intensive treatment' literature *and* the cingulotomy literature, inconsistent and incomplete reporting presents a major challenge to synthesising and summarising this literature. There are also separate challenges that arise when trying to bring together data over a period of many years when clinical practice (neurosurgical techniques, imaging) has changed dramatically. Nonetheless, the overall impression from the observational literature is of relatively high rates of clinical benefit in terms of OCD symptom relief, with low rates of adverse events although some of the rarest can be significant and potentially life changing.

Lai et al. (2020) recently reported their own updated systematic review and meta-analysis of the ablative literature for OCD and identified 23 studies with 487 individual patients, with 21 studies and 459 patients in their meta-analysis. Again, the

Table 3 Summary of key reviews and individual studies of Anterior Capsulotomy

Authors	Study design	*n*	Baseline Y-BOCS	12 m follow-up Y-BOCS	24 m follow-up Y-BOCS	% 'responders' (Y-BOCS decline of ≥ 35%)	Lesion procedure	Adverse events
Brown et al. (2016)	[a]systematic review of 8 studies	112	Range = 21.2 (±4) to 38.2 (+1.8)	Decline range −12.0 to −18.6% fall range −36.0 to −75.0	N/A	Range 37–80%	Bilateral radiofrequency thermal lesions – 4 studies bilateral gamma knife radiosurgery – 3 studies bilateral leucotome – 1 study	24 individual events deemed 'serious' in total of 112 procedures[b] 63 individual events deemed 'transient' in total of 99 procedures[b]
Rasmussen et al. (2018)	Prospective case series [c] confirmed chronic and refractory OCD	55	Mean = 33.3 (±4.8) mean = 34.2 (±3.2)	Mean = 30.7 (±7.6) mean = 20.3 (±7.3)	Mean = 24.1 (±9.3) mean = 17.8 (±7.6)	46% at 36 m 75% at 36 m	Bilateral gamma knife radiosurgery	Prospective surveillance structured with SAFTEE[d] 3 radiation-induced cysts "no adverse effects of GVC on personality or neuropsychological function at follow-up" – with exception of 1 case of "apathy" and 1 of radionecrosis
Lopes et al. (2014)	Double blinded randomised controlled trial confirmed chronic and refractory OCD	16	'Sham' mean = 34.8 (±4.0) 'active' mean = 32.5 (+2.7)	'Sham' mean = 31.9 (±4.1) 'active' mean = 20.9 (+11.0)	[e] 'now active' mean = 16.3 (+19.7) $n = 4$ 'active'	No "sham" responders at 12 m 4 "active" responders at 12 m by this	Bilateral gamma knife radiosurgery or "sham" procedure	Prospective surveillance structured with SAFTEE "no permanent deleterious neuropsychologic or personality changes after 1–5 years of

(continued)

Table 3 (continued)

Authors	Study design	*n*	Baseline Y-BOCS	12 m follow-up Y-BOCS	24 m follow-up Y-BOCS	% 'responders' (Y-BOCS decline of $\geq$ 35%)	Lesion procedure	Adverse events
					mean = 17.8 (+10.0)	single criterion		follow-up However, the risk for delayed brain cyst development (n = 1) is a concern. Other adverse effects, including the manic episodes observed here (n = 2), also require clinical vigilance"
Davidson et al. (2020)	Prospective case series confirmed chronic and refractory OCD	6 of 8	Mean = 33 range 21–40	6 m FU data decline range −2% to −58%	N/A	4/6 met criteria for "response"	MRgFUS	Primarily a safety study No significant A/E's 6 m follow-up lesion generation failed in 2 of 8 in OCD series

Eligibility criteria required that the individual reports were for adults (>18) with OCD and no prior history of neurosurgery and that procedures were conducted under MRI guidance with OCD symptom severity quantified using the Y-BOCS

[a]The inclusion criteria for this systematic review followed the PRISMA guidelines (Moher D, Liberati A, Tetzlaff J, Altman DG: *Preferred reporting items for systematic reviews and meta-analyses: the PRISMA statement*. BMJ 339:b2535, 2009) and AHRQ recommendations for comparative effectiveness (Viswanathan M, Ansari M, Berkman N, Hartling L, McPheeters M, Santaguida PL, et al.: *Assessing the risk of bias of individual studies in systematic reviews of health care interventions*, in Methods Guide for Comparative Effectiveness Reviews. Rockville, MD: Agency for Healthcare Research and Quality, 2012)

[b]Reporting of surgical case series is often complicated by the inclusion of further procedures and by the assignment of more than one adverse effect to individual patients. In this report, the adverse effect estimates are a pooled estimate for 112 patients who were studied and it is not possible to calculate meaningful rates

[c]Report includes two subgroups treated by 'single' and 'double shot' capsulotomy by gamma knife

[d]Levine J, Schooler NR. *SAFTEE: a technique for the systematic assessment of side effects in clinical trials*. Psychopharmacol Bull. 1986;22(2):343–381

[e]With the RCT study design, 4 of the 'sham' treated participants went on to have gamma ventral capsulotomy (unblinded). Two met criteria for response at 12 m. Five of the original 'active' treatment group met criteria for response by 54 m

overall conclusion offered was that ablative surgery was associated with a response rate at last follow-up of 59% (95% CI 50–61) for anterior capsulotomy procedures and 36% (95% CI 23–50) for anterior cingulotomy procedures. Lai and colleagues also noted that adverse effects were not reported systematically and, indeed, some studies made no mention of such. With these caveats, focusing on those studies with more systematic and prospective surveillance for adverse effects, a similar pattern emerged to that reported in previous reviews, with the most frequent adverse effects being transient and mild (median rate of 1.7%), with low rates of more severe and potentially irreversible changes – for example, 0.3% for anterior capsulotomy.

There has recently been a further informative review, using a different search strategy and inclusion method, by Pepper et al. (2019). In this report, a systematic search strategy was followed to identify all literature relating specifically to anterior capsulotomy for OCD. In this search, unlike other reviews, they did not apply an exclusion criterion relating to the use of the Y-BOCS scale, thus enabling capture of data from older literature (pre-1989) that has been excluded from the other recent systematic reviews. They identified 25 separate publications between 1961 and 2018, with 512 individual patients. In those series reporting Y-BOCS scores (a mixture of radiofrequency thermal, gamma knife and focused ultrasound lesions), their consolidated estimate was that 73% of patients achieved 'response' ($\geq$35% drop in Y-BOCS score), with 24% 'remission' (Y-BOCS score < 8). Within the older literature (mostly radiofrequency thermal lesions with a few gamma knife surgeries), using a bespoke grading scheme, 90% were rated as having 'responded', with 39% 'symptom free'. In terms of adverse effects, there were no surgical deaths, 2% had an intracranial haemorrhage (most either asymptomatic or with no sustained neurological deficit) and 2 patients developed seizures (from 512). The commonest adverse effect was weight gain in 69 patients (13%). The literature on adverse effects such as fatigue, cognitive and neuropsychological functioning was, of course, noted to be more variable and inconsistently reported, but with no obvious signal of concern.

8 Comparisons of Ablative Surgery with DBS for OCD

With the enthusiastic uptake internationally of deep brain stimulation (DBS) as a potential intervention for patients with OCD although there have been no direct, controlled comparisons, there have been three interesting comparative reviews.

Kumar et al. (2019) compiled a systematic review and meta-analysis of impact on symptoms and also quality of life measures in 56 studies including 367 patients with OCD treated by lesion surgery and 314 by DBS. They focused on deriving a composite measure of 'utility' – quantification of change in Quality of Life as a function of change in Y-BOCS scoring – in order to directly compare the two groups of procedures. Interestingly, they found that the overall utility, as described above, was greater for lesion surgery than for DBS, with lower rates of adverse effects. There are many caveats, of course, with acknowledgement that the same quality limitations in the literature that were described earlier were captured within this study also.

Adopting a different approach, Pepper et al. (2015) reviewed the published literature on anterior capsulotomy (by any surgical method provided Y-BOCS scores were available) and ventral capsule/ventral striatum or nucleus accumbens DBS. They included 20 studies with 170 individual patients, with 10 reports on 108 lesion surgery patients and 10 studies of 62 DBS patients. Amongst the interesting exploratory analyses that they performed was to look at outcomes following categorisation of patients by severity of symptoms according to the baseline Y-BOCS scores. Anterior capsulotomy patients with more severe OCD symptoms ('severe' or 'extreme' according to Y-BOCS) were much more likely to achieve response, or remission, than were the DBS patients. In terms of adverse effects, the only clear difference between the two procedures was with respect to post-operative weight gain, which was more prevalent and of greater magnitude in the capsulotomy patients.

Of course, there are many limitations and vulnerabilities when attempting to make the types of comparisons that are described above, whether this is by lesion versus DBS, type of lesion, or target of lesion. Non-randomised comparisons are all vulnerable to multiple sources of uncontrolled bias. However, there are some comparative data that are a little more interesting because they are the (non-randomised) outcomes from surgery performed by the same group, from within the same population of patients and using similar clinical protocols. Suetens et al. (2014) reported (primarily) on metabolic imaging data in two groups of patients who were prospectively followed up over the same time period after receiving either VC/VS DBS ($n = 16$) or thermal anterior capsulotomy for OCD ($n = 13$). The inclusion criteria for DBS were somewhat stricter than for capsulotomy, requiring that they be able to follow a demanding evaluation and programming protocol. Interestingly, the mean percentage change in Y-BOCS scores was higher (50.6% ±26.3) in the capsulotomy series than in the DBS series (43.0% ±26.4). As described by the authors, the design of this comparison, if anything, favoured the DBS outcomes '...the stricter criteria for DBS may have led to some selection bias, as patients selected for DBS needed to be more functionally competent to ascertain a complete follow-up'.

9 Radiosurgical Capsulotomy

Published *after* the systematic review by Brown et al. (2016) was a further substantial radiosurgical series. Rasmussen et al. (2018) reported on an extended series of patients with OCD (n-55) treated by gamma knife lesions (focused irradiation) over a period of 20 years, with 3 years of follow-up for each patient. Thirty-one of the 55 patients (56%) showed substantial improvement (>35%) on the primary outcome measure, the Y-BOCS score. Three patients, however, developed radiation-induced cysts during the follow-up period (3–5 years), one of whom experienced neurological sequelae. Interpretation of the data presented in this series is complicated by the evolution of the surgical procedure and the equipment used over the prolonged study period (switch from 'single shot' to 'dual shot' capsulotomy with apparently greater efficacy with the latter procedure) and the apparent association between one particular model of gamma knife and cyst formation. Major strengths of this study, however, were the rigorous inclusion criteria and the structured, systematic, prospective surveillance for adverse effects with a full 36 m follow-up after each procedure. Personality and neuropsychological functioning did not appear to be adversely effected (Table 3).

10 Randomised Evidence

Unlike radiofrequency-generated thermal lesion surgery, both irradiation using the gamma knife and MRI guided focused ultrasound (MgFUS) as lesion creation methods do not require craniotomy. It is the requirement for craniotomy, with the attendant risks of intracranial haemorrhage and infection that have resulted in a general consensus that randomised lesion studies with 'sham surgery' would breach contemporary ethical standards, presenting unacceptable risk to participants, without likely benefit. However, this means that both of these other lesion creation techniques can be used to conduct clinical trials with double-blinding and true 'sham' lesions. In 2014, Lopes and colleagues reported a landmark double-blind, placebo-controlled, randomised clinical trial of gamma ventral capsulotomy for treatment refractory OCD. This study included 16 patients with demonstrably refractory and severe chronic OCD, eight of whom received active treatment with focused gamma irradiation targeting the ALIC bilaterally and 8 randomised to 'sham' treatment with no exposure to radiation. Two of the eight 'active' treatment patients met robust pre-defined criteria for response at 12 months. None of the 'sham' treated patients did. The median Y-BOCS scores decreased during the period of blinding (12 m) by 28.6% in the 'active' treatment group and 5.8% in the 'sham' group. At 54 months, 3 *additional* patients in the 'active' treatment group met criteria for response. Four of the 'sham' treatment group subsequently underwent gamma capsulotomy (*unblinded this time*). Two patients responded within the 12 m FU period. There was single significant adverse effect from the 16 patients who participated, an asymptomatic radiation-induced cyst. The remaining patients who did not receive active treatment remained unchanged over the duration of follow-up.

11 Magnetic Resonance-Guided Focused Ultrasound (MRgFUS)

Focused ultrasound was considered as a possible method for the creation of targeted brain lesions in combination with the stereotactic frame in the 1950s and 1960s, but technical challenges favoured the earlier development of radiosurgery instead. To develop a transcranial method that did not require a craniotomy, the twin challenges of excessive heating of the skull and uneven distribution/scattering of signal needed to be overcome. In combination with magnetic resonance imaging techniques that allowed measurement of heat signatures within tissue (MR thermometry), early in the 2000s, systems were engineered that were capable of creating lesions, without craniotomy, in patients with brain tumours. Eventually, technical developments allowed MR-guided focused ultrasound (MRgFUS) to create deep structure brain lesions without craniotomy – and using randomised trial designs – most notably thalamic lesions to treat essential tremor patients (Elias et al. 2016). The use of these technologies has been strongly influenced by the status of lesion surgeries for various clinical indications using other methods and, therefore, MRgFUS has recently been tested as a method to generate anterior capsulotomy lesions for patients with OCD (Jung et al. 2015; Kim et al. 2018; Davidson et al. 2020).

The first small 'proof of concept' series of four OCD patients was reported by Jung et al. in 2015, but with a substantive follow-up of an extended series subsequently reported by Kim and colleagues in Korea (2018). In this report, they described the outcomes at 2 years for 11 patients with OCD treated by MRgFUS. At 12 m follow-up, 6 patients met criteria for 'response' with 3 'partial responders'. At 24 m follow-up, 6 were 'responders, 2 were 'partial responders' and 1 was fully 'in remission'. Importantly, the procedures were well tolerated, with apparently minimal and transient adverse effects with '. . .no significant physical adverse events (such as fatigue, urinary incontinence or seizure) or behavioural changes (such as hypomania, personality changes, emotional blunting, indifference or carelessness), which were reported in previous capsulotomy studies. . . We observed no significant changes in weight during the 24-month follow-up period'.

Most recently, there is a detailed report of the first 7 OCD patients treated by MRgFUS in Toronto (Davidson et al., 2020). The patient variability in suitability for treatment is a major issue for MRgFUS, with the series containing 8 patients with OCD and 8 with Major Depression but there were 4 patients (2 with OCD) for whom it was simply impossible to create the required target tissue heating. Although the primary focus of the study was safety, of those who had reached 6 m follow-up post-surgery ($n = 12$), 6 had achieved criteria for 'response' (see Table 3). Four of these were patients with OCD diagnoses (4/6). The adverse event rates were reported to be low and mild, with no serious adverse events at the time of publication. To quote the authors 'Testing revealed no negative effects of MRgFUS capsulotomy on patients' cognitive or behavioral functioning. There were some significant, but mild improvements on select cognitive tests, however it is unclear whether these cognitive changes represent true improvements or practice effects. In terms of behavioral changes, modest, but significant improvements were observed on self-report measures of apathy and executive function following treatment'.

12 'Horizon Scanning'

Although lesion surgery is relatively firmly 'established' as part of the therapeutic armamentarium for patients with severe and refractory OCD, there remains much that we do not know and there is considerable scope – particularly with the newer methods of lesion generation – to advance our understanding not only of the efficacy and safety of surgery, but also of the biology of OCD itself. The feasibility of blinded randomised trials of lesions, particularly anterior capsulotomy, has now been established for radiosurgery (Lopes et al. 2014) and is clearly possible for MRgFUS (see, for example, Elias et al. 2016). Although blinded, randomised trials of neurosurgical procedures are still relatively uncommon within neurosurgical culture and practice, the pressure for high grade evidence within interventional psychiatry will hopefully motivate further trials. Currently, the choice of which lesion procedure is offered for the neurosurgical treatment of OCD is primarily based on historical institutional practice and/or surgeon preference. An 'evidence-based' approach to the deployment of neurosurgical treatments for psychiatric disorders, including OCD, has been hindered by the scarcity of studies, with the literature dominated by small scale, open case series with limited detail of key patient characteristics, including previous treatment histories and adequacy of treatment. These criticisms, of course, apply to intensive treatment studies, and DBS studies as well as lesion studies. As a consequence, much clinical consensus and opinion, irrespective of the rigour of the review processes undertaken, is framed around interpretation of diverse observational studies with incomplete data, incomplete follow-up and uncontrolled sources of bias. For both DBS and lesion surgery, a fundamental challenge yet to be overcome is the asymmetry between the precision and sophistication of the surgical procedures and the comparatively crude and simplistic approaches to patient selection (diagnosis, comorbidities, evaluation of previous treatments, capturing and quantifying severity of core symptoms and measuring change reliably).

13 Procedure-Specific 'New Directions'

With the advent of MRgFUS and the existing capabilities for radiosurgery, there is a clear opportunity for robust, blinded, RCT studies that can unequivocally resolve residual issues around procedure efficacy. Thus far, the weight of evidence does suggest that surgery is effective in a relatively high proportion of patients, but repeated careful audit of case series has been insufficient to install firmly lesion surgery procedures within interventional psychiatry. Essentially, if psychiatry (more broadly defined) is to be convinced of the utility and safety of such procedures, the level of evidence will require further elevation. MRgFUS and radiosurgery can do this.

Similarly, the conduct of rigorous blinded studies can also address the unresolved questions around optimal lesion placement and inform our understanding of the relationship between lesion characteristics, fibres of connection and impact upon symptoms. Just as neuroimaging techniques (including, for example, tractography) can generate novel information and hypotheses about circuitry function and dysfunction in OCD, neuromodulation and lesion studies can address such hypotheses and advance and refine the armamentarium of therapeutics.

There are no data at this stage to inform whether any one method of lesion generation is safer, more efficacious or more cost effective (see Kumar et al. 2019). Direct comparative studies would be challenging to perform, but are clearly possible with increasing availability of two or more lesion methods in some large neurosurgical centres. There is also scope for increased synergy between OCD DBS and lesion surgery programmes. With DBS for OCD likely generating a 'reversible' lesion effect, usually within the anterior limb of the internal capsule, it would be possible to offer radiofrequency lesions targeting the most effective DBS stimulation site following a trial period with full exploration of individual stimulation parameters and carefully relating these to the clinical effects, thus 'personalising' the lesion procedure.

14 Clinical Guidance

The use of neurosurgical procedures in the treatment of mental disorders remains a sensitive and controversial topic and it is appropriate that the clinical, ethical and regulatory frameworks for its conduct are robust. Developing clinical guidance in the absence of extensive Level 1 evidence, derived from multiple large randomised studies, requires that cautious interpretations are made of data derived other study designs. Such guidance appears in numerous publications by different professional bodies and societies. Following review of the landscape of clinical evidence in 2017, the UK Royal College of Psychiatrists produced a Position Statement that encompassed the use of both DBS and lesion surgeries for OCD (RCPsych 2017).

First, it was held that '...for carefully selected patients, with difficulties in specific symptom domains – specifically those with Depressive Disorders and Obsessive Compulsive Disorders – neurosurgical therapies may reasonably be considered. In each individual case, consideration of the appropriateness of offering any form of NMD must balance the risks and benefits of surgery with the risks and benefits of continuing with 'treatment as usual' and should also acknowledge patient preference'. Further, that '...the delivery of safe and effective neurosurgical interventions represents an important element of the ethical and optimised management of patients with chronic, otherwise treatment refractory mental disorder – specifically mood disorders (Major Depression and Bipolar Disorder) and Obsessive Compulsive Disorder (OCD)'. Importantly, several core principles were specified.

These included the requirement that all neurosurgical provisions for the treatment of psychiatric disorders should be subject to independent ethical and clinical governance oversight, with special emphasis upon the assessment of capacity and the nature of informed consent. Furthermore, procedures should only be offered by closely collaborating, experienced, clinical teams where there is significant expertise within the team membership in the diagnosis, treatment and long-term management of OCD and also in the use of functional stereotactic neurosurgical techniques. Neurosurgical procedures should not be provided in the absence of robust pre- and post-operative evaluation and audit and an excellent track record of safety. Detailed post-operative care plans must be developed and agreed in advance of surgery.

Such core principles are, of course, not new and have been articulated by field leaders since at least 1975 (Laitinen 1975). However, their importance in ensuring the safe, ethical and scientifically robust exploration of the opportunities presented by increasingly refined and newer technologies, such as the gamma knife and MRgFUS, is fundamental.

References

Aardema F, Emmelkamp P, O'Connor K (2005) Inferential confusion, cognitive change and treatment outcome in obsessive-compulsive disorder. Clin Psychol Psychother 12:337–345

Balachander S, Bajaj A, Hazari N et al (2020) Long-term outcomes of intensive inpatient care for severe, resistant obsessive-compulsive disorder. Can J Psychiatr 65(11):779–789

Brown L, Mikell C, Youngerman B, Zhang Y, McKhann G, Seth S (2016) Dorsal anterior cingulotomy and anterior capsulotomy for severe, refractory obsessive-compulsive disorder: a systematic review of observational studies. J Neurosurg 124:77–89

Cabrera L, Courchesne C, Bittlinger M et al (2020) Authentic self and last resort: international perceptions of psychiatric neurosurgery. Cult Med Psychiatry. https://doi.org/10.1007/s11013-020-09679-1. Online ahead of print

Cif L, Hariz M (2017) Seventy years of pallidotomy for movement disorders. Movement Disord 32:972–982

Davidson B, Hamani C, Huang Y et al (2020) Magnetic resonance-guided focused ultrasound capsulotomy for refractory obsessive compulsive disorder and major depressive disorder: clinical and imaging results from two phase I trials. Mol Psychiatry 25:1946–1957

Eisen J, Sibrava N, Boissea C et al (2013) Five-year course of obsessive-compulsive disorder: predictors of remission and relapse. J Clin Psychiat 74(3):233–239

Elias W, Lipsman N, Ondo W et al (2016) A randomized trial of focused ultrasound thalamotomy for essential tremor. N Engl J Med 375:730–739

Foltz EL, White LE Jr (1962) Pain "relief" by frontal cingulumotomy. J Neurosurg 19:89–100

Jung H, Kim C-H, Chang J, Park Y, Chung S, Chang J (2006) Bilateral anterior cingulotomy for refractory obsessive-compulsive disorder: long-term follow-up results. Stereotact Funct Neurosurg 84:184–189

Jung H, Kim S, Chang J et al (2015) Bilateral thermal capsulotomy with MR-guided focused ultrasound for patients with treatment-refractory obsessive-compulsive disorder: a proof-of-concept study. Mol Psychiatry 20:1205–1211

Kim S, Roh D, Jung H, Chang S, Kim CH, Chang J (2018) A study of novel bilateral thermal capsulotomy with focused ultrasound for treatment-refractory obsessive–compulsive disorder: 2-year follow-up. J Psychiatry Neurosci 43(5):327–337

Kumar K, Appelboom G, Lamsam L et al (2019) Comparative effectiveness of neuroablation and deep brain stimulation for treatment-resistant obsessive compulsive disorder: a meta-analytic study. J Neurol Neurosurg Psychiatry 90:469–473

Lai Y, Wang T, Zhang C et al (2020) Effectiveness and safety of neuroablation for severe and treatment-resistant obsessive–compulsive disorder: a systematic review and meta-analysis. J Psychiatry Neurosci 45(5):356–369

Laitinen L (1975) Psychosurgery on trial (letter). Lancet 19(2(7925)):131–132

Lopes A, Greenberg B, Canteras M et al (2014) Gamma ventral capsulotomy for obsessive-compulsive disorder. A randomized clinical trial. JAMA Psychiat 71(9):1066–1076

Luyten L, Hendrickx S, Raymaekers S, Gabriëls L, Nuttin B (2016) Electrical stimulation in the bed nucleus of the stria terminalis alleviates severe obsessive-compulsive disorder. Mol Psychiatry 21:1272–1280

Mahoney D, Green A (2020) Psychosurgery: history of the neurosurgical management of psychiatric disorders. World Neurosurg 137:327–334

Mancebo M, Eisen J, Pinto A, Greenberg B, Dyck I, Rasmussen S (2006) The Brown longitudinal obsessive compulsive study: treatments received and patient impressions of improvement. J Clin Psychiatry 67(11):1713–1720

McKay D, Sookman D, Neziroglu F et al (2015) Efficacy of cognitive-behavioral therapy for obsessive–compulsive disorder. Psychiatry Res 225:236–246

Nuttin B, Wu H, Mayberg H et al (2014) Consensus on guidelines for stereotactic neurosurgery for psychiatric disorders. J Neurol Neurosurg Psychiatry 85:1003–1008

Pallanti S, Grassi G, Sarrechia E, Cantisani A, Pellegrini M (2011) Obsessive-compulsive disorder comorbidity: clinical assessment and therapeutic implications. Front Psych 2:70–80

Pepper J, Hariz M, Zrinzo L (2015) Deep brain stimulation versus anterior capsulotomy for obsessive-compulsive disorder: a review of the literature. J Neurosurg 122:1028–1037

Pepper J, Zrinzo L, Hariz M (2019) Anterior capsulotomy for obsessive-compulsive disorder: a review of old and new literature. J Neurosurg 11:1–10. Online ahead of print

Rasmussen S, Noren G, Greenberg B et al (2018) Gamma ventral capsulotomy in intractable obsessive-compulsive disorder. Biol Psychiatry 84:355–364

Royal College of Psychiatrists, Committee on ECT and Related Treatments (2017) Statement on neurosurgery for Mental Disorder (NMD), also known as psychiatric neurosurgery. Position statement CERT 05/17 2017. Royal College of Psychiatrists, London

Ruck C, Karlsson A, Steele J et al (2008) Capsulotomy for obsessive-compulsive disorder: long-term follow-up of 25 patients. Arch Gen Psychiatry 65(8):914–921

Sheth S, Neal J, Tangherlini F et al (2013) Limbic system surgery for treatment-refractory obsessive-compulsive disorder: a prospective long-term follow-up of 64 patients. J Neurosurg 118:491–497

Siwiec S, Riemann B, Lee HJ (2019) Predictors of acute outcomes for intensive residential treatment of obsessive–compulsive disorder. Clin Psychol Psychother 26:661–672

Skapinakis P, Caldwell D, Hollingworth W et al (2016) Pharmacological and psychotherapeutic interventions for management of obsessive-compulsive disorder in adults: a systematic review and network meta-analysis. Lancet Psychiatry 3(8):730–739

Suetens K, Nuttin B, Gabriels L, Van Laere K (2014) Differences in metabolic network modulation between capsulotomy and deep-brain stimulation for refractory obsessive-compulsive disorder. J Nucl Med 55:951–959

Taube-Schiffe M, Rector N, Larkin P, Mehak A, Richter M (2020) Effectiveness of intensive treatment services for obsessive compulsive disorder: outcomes from the first Canadian residential treatment program. Int J Psychiatry Clin Pract 24(1):59–67

Veale D, Naismith I, Childs G, Ball J, Miles S, Darnley S (2016a) Outcome of intensive cognitive behavior therapy in a residential setting for people with severe obsessive compulsive disorder: a large open case series. Behav Cogn Psychother 44(3):331–346

Veale D, Naismith I, Miles S, Gledhill L, Stewart G, Hodsoll J (2016b) Outcomes for residential or inpatient intensive treatment of obsessive–compulsive disorder: a systematic review and meta-analysis. J Obs Compuls Relat Disord 8:38–49

The Future of Obsessive-Compulsive Spectrum Disorders: A Research Perspective

T. Vats, N. A. Fineberg, and E. Hollander

Contents

Abstract Obsessive-compulsive disorder (OCD) sits at the epicenter of a spectrum of related conditions (often referred to as obsessive-compulsive related disorders (OCRD) or obsessive-compulsive spectrum disorders (OCSD)) that can be as disabling as they are varied in presentation. Research in the field now encompasses diverse disciplines ranging from inflammatory mechanisms to computational psychiatry, to neurocognitive endophenotypes to functional imaging to pharmacogenomics to brain stimulation approaches. As these disorders become more clearly elucidated, there is a need to continually re-evaluate the implications of research

T. Vats and E. Hollander (✉)
Autism and Obsessive-Compulsive Spectrum Disorders Program, Psychiatric Research Institute of Montefiore-Einstein, Albert Einstein College of Medicine and Montefiore Medical Center, Bronx, NY, USA
e-mail: eholland@montefiore.org

N. A. Fineberg
Center for Clinical and Health Research Services, School of Life and Medical Sciences, University of Hertfordshire, Hatfield, UK

Hertfordshire Partnership University NHS Foundation Trust, Rosanne House, Welwyn Garden City, Hertfordshire, UK

University of Cambridge School of Clinical Medicine, Cambridge, UK

Curr Topics Behav Neurosci (2021) 49: 461–477
https://doi.org/10.1007/7854_2020_208
Published Online: 7 February 2021

findings and to incorporate these findings into new treatment approaches that benefit both patients and clinicians. Even the Diagnostic and Statistical Manual of Mental Disorders 5th Edition (DSM-5) is intended to be flexible and to incorporate validated and reliable biomarkers and neuroscience findings as they become available. This concluding chapter highlights just a few areas of study that promise to influence our understanding of the pathophysiology and clinical practice of OCRD. These include patient-centered outcomes research, the study of developmental brain trajectories in spectrum conditions, robot models of OCRDs, goal-directed versus habit-based behaviors, pharmacogenomics, problematic use of the Internet, and digital interventions. For example, digital medicine may become increasingly useful by identifying patients early on in the course of their illness; providing biomarkers to subtype patients; predicting treatment response; serving as a more proximal outcome measure of treatment response; or providing easily accessible and less costly forms of care. In order to address unmet clinical needs in OCRD, it is helpful to take an interdisciplinary perspective, and the work described in this collection of articles is likely to be invaluable in shaping the future of the field.

Keywords Obsessive-compulsive disorder (OCD) · Patient-centered outcomes research · Robot model · Pharmacogenomics · Problematic use of the Internet (PUI) · Digital interventions

1 Introduction

Once a neglected illness, obsessive-compulsive disorder (OCD) is now recognized as a common, highly disabling and treatable brain-based disorder. Changes in the DSM-5 (American Psychiatric Association 2013) and more recently ICD-11 (World Health Organization 2016) have set OCD as the exemplar of a new family of obsessive-compulsive and related (OCRD) or obsessive-compulsive spectrum disorders (OCSD). Clinical and translational research in OCD has grown over the years with substantial advances in our understanding of the phenomenology, brain-based biology and treatment response of these related conditions, leading to innovations in our nosological conceptualization, therapeutic interventions and services (Fineberg et al. 2020). This concluding chapter highlights a few selected topics which either have the potential for influencing our understanding and especially treatment of OCSD or have not been fully addressed in the preceding chapters, including patient-centered outcomes research, assessing developmental trajectories in OCSD, robot models of OCRDs, practical implications of pharmacogenomics, problematic use of the Internet, and digital interventions.

1.1 Patient-Centered Outcomes Research

One important area for future study is patient-centered outcomes research to test the "real world" clinical effectiveness of treatments whose individual efficacy has already been established by reference to placebo with confidence. Stakeholder feedback, meta-analytic studies, US and UK Practice Guidelines, and our own feasibility trial (Fineberg et al. 2018c) have identified important gaps in our knowledge on the treatment of Obsessive-Compulsive Disorder (OCD), especially with regard to what treatment to start first, what treatment to try if patients have not received full remission with first-line treatment, and what is the most effective treatment over the long term. These aims might investigate the comparative clinical effectiveness of selective serotonin reuptake inhibitors (SSRIs) and cognitive behavior therapy (CBT) both as monotherapy treatments and as part of a sequential treatment regimen.

While meta analyses and systematic reviews have demonstrated SSRIs (chapter "Pharmacotherapeutic Strategies and New Targets in OCD") and CBT, including CBT with exposure and response prevention (ERP) (chapter "Innovations in the Delivery of Exposure and Response Prevention for Obsessive-Compulsive Disorder"), to be more effective than placebo, in fact most psychotherapy trials should be considered variants of combination trials since most patients in these studies were first stabilized on SSRI treatment (Skapinakis et al. 2016). In addition, OCD is recognized to follow a chronic or relapsing course, at least in treatment seeking patients (Skoog and Skoog 1999) and relapse is associated with significant loss of health-related quality of life (Hollander et al. 2010). Yet, few randomized controlled trials (RCT) have compared the relative benefit of SSRI, specific modalities of CBT or their combination over an extended treatment phase to allow a comparative evaluation of longer-term outcomes that patients greatly care about, as determined by our stakeholder engagement exercises (using Orchard (https://www.orchardocd.org/) and OCD Action (https://www.ocdaction.org.uk/); by soliciting feedback from the broader OCD community via interviews about the evidence gaps and design issues to address these gaps in online media (Hollander 2019a); as well as blogs in ADAA (Hollander et al. 2018) and Psychology Today (Hollander 2019b). A recent pilot study demonstrated the feasibility of investigating these treatments in adults with OCD over a 52-week period and found significant clinical and cost-effectiveness benefits for first-line treatment with SSRI compared to CBT with ERP, signaling the need for more robustly powered longer-term patient-centered studies to address the gaps in knowledge experienced by both clinicians and patients (Fineberg et al. 2018b).

Most studies parse outcomes in terms of those who achieve a clinical response–measured conservatively as a percentage improvement ranging from 25 to 35% in OCD symptom-scores (e.g., Yale-Brown Obsessive-Compulsive Scale). Yet patients achieving such a response status tend to remain significantly symptomatic and functionally impaired. Consultation with patients highlights the importance for them of achieving full remission (Hollander 2019a, b; Hollander et al. 2018). Furthermore, naturalistic follow-up of patients under treatment shows that those who achieve a more robust level of clinical improvement under acute treatment are less likely to relapse (Eisen et al. 2013).

We therefore propose that future studies of SSRIs and CBT include more clinically relevant designs, including patient-centered outcomes. Study designs might include sequential phases, such that the initial treatment phase could compare remission rates over an extended period of time, as well as measures of health-related quality of life, in order to determine, with greater confidence than hitherto, the optimal first-line treatment, and taking patients' well-being more fully into account. Similarly, non-remitters could be randomized to different treatment strategies, e.g. combine (coadministration of SSRI+CBT) or switch (to the alternative treatment modality) for an additional sustained period to determine the best next step treatments, and in the case of further non-response additional stages of intervention may be sequentially tested, e.g. augment SSRI with antipsychotic (chapter "Pharmacotherapeutic Strategies and New Targets in OCD") or a form of non-invasive neurostimulation such as repetitive transcranial magnetic stimulation (chapter "Invasive and Non-invasive Neurostimulation for OCD").

Findings from patient-centered studies such as these could be expected to address gaps in our knowledge that would inform these key clinical decisional dilemmas and help patients, clinicians, and stakeholders to decide which initial and subsequent treatments are optimal for patients.

1.2 Assessing the Developmental Trajectory of Brain Development During Adolescence

Adolescence may be viewed as the pinnacle of brain development and may be the last point for any permanent change before a sensitive period closes. Studies have shown how information from the maturational stages that precede or occur during adolescence is likely to hold the key toward optimizing development to produce an adult who is resilient and well adapted to their environment (Andersen 2016). Conversely, as the unique cognitive, academic, and social developmental tasks of childhood and adolescence are concentrated into a relatively few number of years, if these developmental milestones are disrupted, it may have lasting consequences well beyond childhood, with an altered educational, professional or social trajectory (Fineberg et al. 2019). Interestingly, no studies so far have tracked these developmental trajectories in OCSD.

OCSD may be considered developmental disorders, as they generally show an early age of onset, particularly around puberty and early adolescence, and the trajectory of these conditions are influenced not only by genetic vulnerabilities but also shaped by learning and environmental factors (chapter "On the Development of OCD"). Better understanding of the developmental trajectories of these conditions may lead to earlier detection–and thereby more efficacious intervention (Fineberg et al. 2019) and ultimately better long-term outcome. Measures taken at a single time point may not be nearly as informative as viewing these measures over a period of time, which may detect critical windows in brain development. This is especially true for impulsivity and compulsivity, neuropsychological domains of great relevance for conferring vulnerability to OCSD (Fineberg et al. 2018a), as myelin formation and the maturation of prefrontal control mechanisms play a critical role in regulating impulse control (Casey et al. 2017). Critical periods are integral for the foundation of building a brain. Indeed, periods occur when the brain is uniquely sensitive to environmental impacts. Such increased sensitivity to environmental events is especially heightened during adolescence (Andersen 2016), and may be relevant to the formation of habit behavior (Casey et al. 2017; Gillan et al. 2017) which is also found to be integral to OCSD (chapter "Recent Developments in the Habit Hypothesis of OCD and Compulsive Disorders").

In order to identify the developmental trajectories and the congenital and environmental factors that predispose to the development of mental disorders as opposed to healthy development, the National Institute of Health (NIH) funded its largest long-term study of brain development and child health in the USA, called the Adolescent Brain Cognitive Development (ABCD) Study. The ABCD study (https://medicine.umich.edu/dept/psychiatry/adolescent-brain-cognitive-development-abcd-study) will recruit 10,000 healthy children, ages 9–10 across the USA, and follow them with repeated measures of health and well-being into early adulthood, collecting alongside a plethora of bio-specimens for assessment of substance abuse, hormones, genetic/epigenetic markers, and developmental exposures. Sequential measures of neurocognition and brain imaging will also be collected. Given the complex convergence of key developmental factors that shape an individual's development, this cohort is well suited to undergo various analyses which may have relevance to OCSD (Kristina et al. 2018). For example, studies could look at the impact of digital screen time and social media use on shaping developmental brain trajectories and associated symptom profiles (Paulus et al. 2019). Alternatively, studies might examine the impact of substance use disorders (SUD) in the developing brain on neurocognition and brain structure in subjects with OCRD over time (Mancebo et al. 2009).

We would suggest that future studies might utilize the ABCD resource via different methods to answer unique questions regarding OCSD. For example, by being particularly vigilant for OCSD developing in the cohort, we could look back to identify prodromal changes and related factors. Alternatively, we could follow-up cases prospectively to see how development diverges in the symptomatic group. These approaches could increase our understanding of the environmental, social, genetic, and other biological factors that specifically affect brain and cognitive development in OCRD, and help determine those various factors that could enhance or disrupt a young person's life trajectory. Thus, studies comparing subjects with OCSD to matched control populations collected in the ABCD would help determine when and how relevant brain circuits go awry, and long-term follow-up studies of the healthy adolescent cohort could potentially identify biomarkers for those at high risk of developing OCSD, as a first step toward preventative intervention.

1.3 Robot Models for OCRD

Robot models represent a new and not uncontroversial complement to existing research techniques, e.g. those using non-human animals or computational modeling, as a method for investigating mental disorders such as OCRD–i.e. disorders characterized by distinctive behavioral abnormalities that can be readily simulated by a robot. In the case of OCRD (OCD, body dysmorphic disorder, hair pulling disorder, skin picking disorder), many of these behaviors are manifested as stereotyped compulsive rituals–typically excessive cleaning, grooming, checking, etc. Robot models have certain advantages as they allow precisely controlled and reproducible manipulations (i.e., highly reliable) for simulating repetitive behavior that might not be possible in live subjects for ethical, methodological, or practical reasons. And by being "embodied," robots incorporate interaction with the real environment–so models of the disorder can be tested more realistically than with a computer program alone. In addition to the behavioral "compulsion" aspect accessed by animal models, robot models of motivation also allow us to consider the internal "obsession" aspect of OCD, since elements of the model can be viewed as "thoughts" even when they do not result in action.

However, as with other modeling techniques, there are a number of limitations of this approach. For example, whereas robots may be useful for modeling motivational components of disorders like OCRDs, they do not capture the distressing mental and emotional experiences. Furthermore, as they exist in a non-biological system, robot models are far removed from behavior that occurs in a biological system and the same sorts of biological interventions used for treating OCRD, such as psychopharmacology, cannot be applied.

Robot models could nevertheless be helpful for refining our understanding of aspects of the brain processes that control particular features of OCRDs, which could then lead on to identifying new treatment targets. From conventional research, we already have a number of theoretical models of OCD that indicate changes in communication within brain systems involved in fine-tuning instrumental behavior, but important gaps in our understanding remain. Using robots, we can attempt to simulate some of these changes, e.g. by altering the different values of a certain response threshold or by adding "noise" to disrupt communication within a certain motivational system that governs the robot's behavior. By reproducing behavior resembling particular aspects of OCD using these techniques, we can theoretically start to build a framework for better understanding the complex brain basis of the disorder as whole.

We are still at the very earliest stages, but the first prototypical robot model of aspects of compulsive behavior has recently been developed (Lewis et al. 2019). This model is based on embodied motivational control architecture for decision-making and takes inspiration from both the "cybernetic" computational model of Pitman (1987) and the theory-based "signal attenuation" animal model of Joel (2006). According to these models, instrumental behavior normally results from the attempt to correct "perceptual errors" indicating a mismatch between an "actual" perceived signal and an internal reference, ideal or "target" value that the system aims to reach. Thus, Pitman proposed that compulsive behaviors result from "the persistence of high error signals, or mismatch that cannot be reduced to zero through behavioural output" (Pitman 1987, p. 336). This model was further developed by Joel (2006), who proposed that a deficit in the feedback associated with the normal performance of goal-directed responses is responsible for generating the mismatch and leads to compulsive repetition (Joel 2006). Alongside compulsions, this model has also been used to explain other features of OCRD such as perfectionism, indecision, need for control, over-specification, and obsessive thoughts, with the presence of the error signal itself potentially being subjectively experienced as a pathological sense of incompleteness and doubt.

In the prototype model, an autonomous robot was programed to follow various decision-making (motivated behavior selection) strategies governing the decision to eat (based on energy level) or groom (based on integrity of integument) or avoid damage to the integument caused by bumping into things, measured in terms of contribution to maintenance of homeostasis (Lewis and Canamero 2016). From previous work, it was known that a robot will typically stop attending to one need (e.g., to groom) before the related physiological variable (integument integrity) reaches its ideal value, due to competition from other needs (e.g., to eat). The point when the value of the active motivation is overtaken by another motivation, which becomes the new active motivation, can either be thought of as the "stop signal" for that behavior, or as an "attend to another need and switch behavior" signal.

Through controlled manipulation of the robot's perception of the internal errors linked to its physiological variables, i.e. the robot's perceived ideal ("target") value for the integument, alterations were then made to the decision-making (action selection) process related to the calculation of the "error." To model "OCD pathology," values for the perceived ideal value of the status of the integument were programed to include those that were not actually achievable. Since the unrealistic values are not valid, having such a target value can be viewed as a "perceptual error" by the robot (Lewis et al. 2019).

As expected, under the highly unrealistic perceptual target, excessive persistence in behavioral execution (grooming) was observed, that bore a resemblance to decision-making problems seen in OCRDs and other conditions such as addictions. Indeed, the self-grooming behavior resulting from this perceptual manipulation was executed for long periods, (related directly to hair pulling disorder, skin picking disorder) and continued beyond the point where the "normal" robot would have stopped. This phenomenon was thought possibly related to perfectionism, a characteristic of several OCRDS (OCD, hair pulling disorder, skin picking disorder, body dysmorphic disorder, obsessive-compulsive personality disorder) (Fineberg et al. 2007; Fineberg et al. 2015). Thus, by distorting perception, the robot model had interfered with the normal dynamics of decision-making, and in particular with the reasons to stop grooming. The model achieved face validity with compulsions and obsessions, as well as a degree of construct validity, as it provided experimental support that Pitman's theoretical model can generate behavior akin to OCD.

Alongside behavioral assessment, metrics related to the regulation of physiology (well-being, physiological balance, maintenance of physiological variables away from dangerously low values, maintenance of integument), showed that the highly unrealistic perceptual target was disadvantageous, resulting in poorer well-being and physiological values than either the realistic target value or a mildly unrealistic target value. However, contrary to expectation, a mildly unrealistic target value was not disadvantageous–on the contrary, it showed a non-significant trend toward greater well-being than did the realistic target value. The potential for a mildly unrealistic motivational target to confer adaptive advantage is an interesting finding and indicates an avenue for future research. Such potential advantages also contribute to the discourse around the possible evolutionary origins of OCRDs and compulsive behavioral in general (Stein et al. 2016).

Predictive validity indicates that the behavior of a model can be used to make reliable predictions about outcomes in the human condition, such as which interventions will, or will not, work. This aspect of validity is highly important for clinical research purposes but to date, this model has not been investigated under any form of "treatment," such as exposure and response prevention. Indeed, the authors suggest that at this early stage in the development of the model, only limited predictive validity is expected (Lewis et al. 2019). However, this is clearly an important objective for future work, and it is to be seen whether, as this and other robot models are refined, these predictions could also be refined.

Engagement with the public, especially those with the lived experience of OCRD, is another important future ingredient of robotic research. Using robots to demonstrate that relatively simple processing errors can generate compulsive behaviors may possibly help people with OCRD who feel stigmatized to challenge fears that they are somehow "morally wrong" and contribute to breaking down the humiliation and shame associated with these illnesses.

1.4 Goal-Directed Versus Habit Behaviors

Other converging evidence, from computational studies (Voon et al. 2014) and translational neuroscience (Gillan et al. 2016), posits compulsivity (i.e., a tendency to behave in a compulsive way) as an imbalance between goal-direction and habit learning; more specifically, stimulus-response habit learning is framed against deficient goal-directed control over response inhibition (Gillan et al. 2016) (chapter "Recent Developments in the Habit Hypothesis of OCD and Compulsive Disorders"). Patients with OCD demonstrate habit maintenance even upon withdrawal of desired outcome (Gillan et al. 2011). Insight with regard to compulsions remains intact, but OCD patients have a reduced sense of control over external events (Gillan et al. 2014).

From a clinical perspective, a decrease in goal-directed and structured activities often results in an increase in compulsive behavior, whereas an increase in goal-directed behavior is associated with improvement. Conversely, stress and anxiety symptoms may also drive the transition from goal-directed actions to habit formation; neuroanatomical mechanisms for this transition have been proposed elsewhere (Adams et al. 2018). The combination of these findings in the patient with OCSD leads to the development of habits that are reinforced, to dysfunctional responses to outcome withdrawal, and to specific compulsion-defined syndromes such as trichotillomania, body dysmorphic disorder, dermatillomania, and gambling disorder.

This psychopathology challenges current practice. Current thinking posits obsessions as being the primary symptoms and compulsions, which develop in response to the obsessions and are aimed at neutralizing them, as being secondary. However, patients with OCD show persistent dissociation of habits from initial avoidance-focused stimuli relative to healthy controls. Compulsions may instead be primary, and obsessions a secondary elaboration (chapter "Recent Developments in the Habit Hypothesis of OCD and Compulsive Disorders"). Taking this avenue of research forward into the clinical domain, if goal and habit-oriented behaviors, as well as dissociation between goal-directed planning and action, can be reliably identified and associated with neuroanatomical (e.g., Gillan et al. 2015) and genetic correlates (see below), these biomarkers can then be studied as quantitative measures predisposing to hereditary risk, disease severity, and risk of relapse, and may help differentiate OCD from comorbid conditions that may present like OCD (such as ruminations in depression).

1.5 Genetic and Pharmacogenomic Factors

Identification of biological factors that may relate to treatment response may be valuable in tailoring clinical management. Genetic factors have long been associated with incidence and severity of many psychiatric disorders, including OCSD, and may reflect variable outcomes in pharmacological management (Brandl et al. 2012). Many SNPs across candidate genes within the CYP450, glutamatergic and serotonergic systems have been identified. Specific gene variants may modulate outcome to drugs used in the treatment of OCD. Recent advances in technologies such as GWAS, Open Array, DNA and RNA sequencing, and fine mapping have supported more powerful investigations of genetic variations across the entire genome. The recent ENCODE project has explored large regions between genes that contain functional elements and regulators of a large number of known genes. This new framework examines DNA sequences outside of recognized coding regions, and elucidates the importance of a majority of these stretches of non-expressed DNA sequences, previously known as "junk" DNA (Zai et al. 2014).

Typically, 40–60% of OCRD patients are deemed non-responders to SSRI medications and clinical factors have only been modestly correlated with treatment response (Brandl et al. 2012). Despite numerous clinical and demographic factors believed to influence treatment response, no satisfactory prediction algorithm has yet been developed to predict pharmacotherapeutic efficacy. Although there is burgeoning evidence regarding the etiological significance of genetic factors in

OCD, exploration of pharmacogenetic and pharmacodynamic/pharmacokinetic factors in this illness is just beginning.

Clinical pharmacogenomics is the practice of using patient genetics, and resultant biochemical implications, to inform and tailor the use of pharmacotherapy. A classic clinical use is in evaluating risk of drug–drug interactions, and how these risks may be modified by patient polymorphisms in the cytochrome P450 system (chapter "Pharmacogenetics of Obsessive-Compulsive Disorder: An Evidence-Update"). The use of pharmacogenomics has begun to appear in the clinical setting, but the clinical utility of such testing remains unknown. Currently, major candidates of pharmacokinetic research are variants of the CYP450 family, particularly CYP1A2, CYP2C19, CYP2D6, and CYP3A4, and some of these pharmacokinetic measures are used by clinicians to help determine estimated plasma levels of various drug treatments in OCD.

It is understood that a strong genetic component in the pathogenesis of OCD exists (Zai et al. 2014), but testing is currently of limited value. When considering the etiological, neurobiological, and therapeutic complexity of OCD, it is unsurprising that the identification of specific genetic risk factors has thus far been relatively unproductive. Findings have implicated impairment of the cortico-striatal function and the glutamatergic neurotransmitter system, leading to preliminary data supporting the use of glutamatergic agents such as riluzole, memantine, and N-acetylcysteine, in the management of OCD (Hirschtritt et al. 2017). However, these treatments tend to have only modest effects (Marinova et al. 2017). There has been some clinical utility in identifying common gene variants associated with vulnerability to major comorbidities such as bipolar disorder, attention deficit hyperactivity disorder, and depression (Nezgovorova et al. 2018). This utility has yet to be fully realized in the case of OCD. As the cost of genetic sequencing continues to fall, attention needs to shift from technical questions of feasibility to clinical significance. Associating endophenotypes with trends in treatment outcomes offers observable correlates to genetic analysis. Large-scale, algorithmic analyses are likely to be of value in answering complex and polygenic questions. Such work will also rely on longitudinal databases that may suggest relationships between genotypes and clinical milestones or outcomes.

1.6 Problematic Usage of the Internet (PUI)

OCSD include compulsive and impulsive conditions that reflect both genetic vulnerabilities and environmental exposures. Changes in our environment may be seen in evolving patterns in OCSD. PUI is an umbrella term for a range of repetitive functionally impairing compulsive behaviors including excessive online gambling, gaming, sexual behavior, shopping, video-streaming or social media use that have emerged in the past decade or so, and are particularly relevant during the COVID-19 pandemic when general use of digital technology is high (Fineberg et al. 2020). There is considerable heterogeneity within PUI, which is thought to comprise a cluster of separate disorders with different degrees of compulsivity and impulsivity as underlying mechanisms. Some types of PUI (e.g., gambling, pornography viewing, shopping) show phenotypic similarities with impulse-control disorders or substance addictions, suggesting impulsive mechanisms. Other PUI behaviors share more similarities with compulsive disorders such as OCD (e.g., repeatedly checking emails or social media, digital hoarding) or social anxiety disorder (e.g., excessive use of social media as an avoidance of face-to-face social contact), suggesting they are linked more closely to compulsive or anxiety mechanisms, respectively. Some forms of PUI are markers of impulsive traits or symptoms within OCD patients. Other forms of PUI that are more phenotypically compulsive, such as gaming disorder or cyberchondria (Vismara et al. 2020) or anxiety related social media use, occur in individuals with higher levels of trait or state compulsivity or anxiety and show increased rates of co-occurrence with OCSDs, such as OCD or obsessive-compulsive personality disorder or anxiety disorders (Ioannidis et al. 2016).

While advances have been made in defining diagnostic criteria and developing rating scales for some forms of PUI (e.g., King et al. 2020; Starcevic et al. 2020), a considerable amount of research is needed to understand better the broad range of PUI phenomena; to translate the known behavioral phenotypes into valid and reliable diagnostic criteria and assessment tools; to facilitate the systematic investigation of etiological factors and brain-based mechanisms; and to serve as a platform for the development of preventative and therapeutic interventions (Fineberg et al. 2018a). Various instruments have been developed to assess specific types of problem behavior, but there nevertheless remains room for the development of a standardized contemporary measure focusing on PUI severity that could be broadly applied across all–or almost all clinically relevant forms of PUI. Consideration of PUI from different but non-mutually exclusive and complementary perspectives, as promoted by the recently established COST Action 16207 European Network for Problematic Usage of the Internet (www.internetandme.eu), including those related to addiction, impulsivity, and compulsivity, may in the future mean that more of its features can be captured, understood, and addressed.

Other key objectives for research in this area include the need to more rigorously assess the impact of comorbidities (e.g., OCD, ADHD, impulse-control disorders) on functional impairment and treatment outcomes, further work on the longitudinal profiles of different types of PUI, and the extent to which assessment instruments can be shown to be sensitive to the effects of treatment. There is a need for the validation of instruments besides factor validity using external measures confirming concurrent and predictive validity. In general, there is a need to harmonize assessment tools to assure comparability of findings (Fineberg et al. 2018b).

It would therefore be of interest to investigate PUI in an OCD specialized service; to explore how OCD expresses itself in digital compulsive and impulsive behaviors (digital hoarding, cyberchondria, Internet gaming, etc.), and to assess dimensional compulsive and impulsive traits in a PUI population. More work is also needed to determine the impact of PUI on long-term outcomes in OCSD, as well as the long-term physical health consequences, including obesity and metabolic disorder. However, studies have also shown that patients with a broad range of serious mental disorder including even schizophrenia, may be at increased risk of PUI. Studies are therefore needed to determine the rates of different forms of PUI in OCSD with more confidence, and to compare these rates to subjects with non-OCSD mental disorders to determine if there is true specificity in the relationships.

Some authors (Vismara et al. 2020) underline the clinical importance of cyberchondria (CYB) as a potentially novel form of compulsive digital behavior. The compulsions of cyberchondria resemble those of hypochondriasis or OCD. Furthermore, the need for certainty, which drives some forms of CYB, may reflect an underlying obsessive-compulsive personality disorder. Importantly, CYB also manifests characteristics that can be conceptualized within a framework of behavioral addiction and for which the distressing loss of control over online activity, resulting in time-consuming, compulsive behavior, represents a major source of interference with functioning. A few studies have attempted to characterize CYB, but most of the existing data is cross-sectional in nature and derived from self-report instruments in community samples. The extent to which these findings can be generalized to the clinical setting remains unclear. Given the increasingly widespread use of the Internet and the potential negative effects of online health searches, CYB is likely to represent an increasing public health burden. Further investigations are needed to understand the longitudinal course and the impact of this phenomenon at an individual and societal level, but certain preliminary conclusions can be drawn as a basis for further research.

1.7 Novel Digital Interventions

The digital era and the technology accompanying it offer new opportunities for monitoring and interventions. Accordingly, active online interventions have become readily available for OCRD (Whiteside et al. 2013). Such interventions may include social media groups, comprising patient and staff members, in which the patient may report in real time their difficulties, daily achievement and progress. Such digital groups enable continuing communication, real-time reports and enable prompt responses and rapid intervention when needed. In addition, the digital intervention may serve as a platform for continuous monitoring of tasks delivered in face-to-face meetings. Another example of existing digital interventions is the proactive use of webcams and smartphone cameras (Fineberg et al. 2020).

As the digital platform bridges the elapsed time between therapeutic sessions, it can enhance involvement and adherence with the treatment. Moreover, this may overcome geographical distances and enables therapeutic practice in the patient's natural environment, where symptoms are manifested daily (rather than in the neutral clinic). In practice, this approach breaks down the traditional terminology of "outpatient," "in-patient," and "day hospitalization," by allowing real time, objective, and continuous monitoring. It also allowed communication with clinical staff, including (if needed) prompted intervention. This kind of digital monitoring and communication could even be considered as "virtualized hospitalization," as it offers more comprehensive and intensive treatment. This aspect is important as key aspects of the treatment can take place while the patient is in their natural environment, where the OCRD usually occurs, and not within the artificial setting of the clinic.

The Internet also opens new avenues for different types of biomarkers for PUI. Developments in imaging and biomarkers, along with medical informatics, may well allow for better assessments and interventions in the future. Recent advances in the concept of the "digital phenotype," which involves using computerized measurement tools to capture the characteristics of a given psychiatric disorder, is one paradigmatic example. The extensive use of smartphones and vast amounts of information they contain has positioned their use as a proxy for behavior and social interactions (WHO 2016). Harnessing smartphone technology along with smart wearable (e.g., smart watches) is expected to be a valuable source of continuous, objective, and reliable data for clinical characterization, behavioral monitoring, and treatment support (Marzano et al. 2015). This is true for several disorders but especially true for PUI as the digital media that is directly linked to the disorder is the same one that can accurately monitor the behavior (Fineberg et al. 2020).

Accordingly, careful continuous monitoring of behaviors of individuals via digital tools with informed consent along with big data analyses may help to characterize the "digital phenotype" of PUI and to identify the "finger print" of individuals at risk (e.g., by monitoring the online Internet usage in comparison with changes in diurnal variation, lack of human contact, lack of geographical movement, restricted circles of friends, etc.). Considering the digital form of the disorder, digital tools may be used as an objective and continuous monitor of the pathological behavior. Once developed and validated in clinical trials, these Internet-based tools may serve both as monitoring and as interventional utensils (Fineberg et al. 2018c). One potential research step using digital monitoring could be to alert the individual whenever a "PUI pattern" has emerged, help him/her to adjust their behavior accordingly, and to monitor and feedback progress (Fineberg et al. 2020).

Digital tools may also eventually be helpful as early onset biomarkers to identify patients early in the course of their illness. For example, by detecting PUI early on via digital monitoring, this might alert clinicians to subjects at high risk for developing binge alcohol and other compulsive, impulsive, and addictive problems. Extending "biomarkers" research to young individuals at theoretical risk of developing various forms of addiction (such as the unaffected relatives of people with PUI) or prodromal phases of PUI, may permit early interventions that might alter the trajectory of the disorder toward a better long-term outcome. For example, if various forms of PUI could be reliably predicted from recognized traits and symptomatology, such as tendencies toward disordered compulsive (urge-driven) behaviors, this could potentially have important implications for early identification of "at risk" individuals and timely intervention before problems take hold (Fineberg et al. 2018a).

Digital measures may also be studied to help to subtype OCRD populations; to predict response to specific treatment modalities like mobile CBT apps (Roncero et al. 2019); and to be useful for relapse prevention. Digital measures deserve study to determine if they are sensitive as outcome biomarkers in therapeutic trials, perhaps by being more proximal to the underlying illness, and by being less subject to rater, caregiver, and subject expectation bias. If so, this could be helpful in driving down placebo response rates and maximizing active treatment vs. placebo separation in clinical trials.

2 Conclusion

A greater understanding of the psychobiological substrate of OCSD ranging from genes to observable behaviors may eventually give us new insights into more personalized and targeted treatments for these complex conditions. There is also an unmet need to detect early on alterations in the developmental brain trajectory in OCSD, since this might lead to better early intervention strategies. There is certainly a need for better patient-centered outcomes and measures of well-being in our clinical trials of OCSD. Finally, looking ahead to the future, new avenues such as the use of advanced robotic models and digital tools to diagnose, monitor, and treat OCRD are needed, and this may be a promising approach.

Acknowledgements This chapter is based upon work from COST Action CA16207 "European Network for Problematic Usage of the Internet", supported by COST (European Cooperation in Science and Technology) – www.cost.eu.

Dr. Hollander has received grants from Department of Defense (DOD), Orphan Products Division of the Food and Drug Administration (OPD-FDA), Roche, and GW Pharma in the past 3 years.

Dr. Fineberg has, in the past 3 years, held research or networking grants from the ECNP, UK NIHR, EU H2020, MRC, University of Hertfordshire, accepted travel and/or hospitality expenses from the BAP, ECNP, RCPsych, CINP, International Forum of Mood and Anxiety Disorders, World Psychiatric Association, Indian Association for Biological Psychiatry, Sun, received payment from Taylor and Francis and Elsevier for editorial duties, and accepted a paid speaking engagement in a webinar sponsored by Abbott. Previously, she has accepted paid speaking engagements in various industry supported symposia and has recruited patients for various industry sponsored studies in the field of OCD treatment. She leads an NHS treatment service for OCD and holds Board membership for various registered charities linked to OCD. She has given expert advice on psychopharmacology to the UK MHRA and NICE.

Contributors All authors agreed on the title, scope, and content of the manuscript, contributed to the literature searches, and approved the final manuscript.

References

Adams TG, Kelmendi B, Brake CA, Gruner P, Badour CL, Pittenger C (2018) The role of stress in the pathogenesis and maintenance of obsessive-compulsive disorder. Chronic Stress (Thousand Oaks), 2. https://doi.org/10.1177/2470547018758043. Epub 2018 Mar 4

American Psychiatric Association (2013) Diagnostic and statistical manual of mental disorders (DSM-5). American Psychiatric Pub

Andersen SL (2016) Commentary on the special issue on the adolescent brain: adolescence, trajectories, and the importance of prevention. Neurosci Biobehav 70:329–333

Brandl EJ, Müller DJ, Richter MA (2012) Pharmacogenetics of obsessive–compulsive disorders. Pharmacogenomics 13(1):71–81. https://doi.org/10.2217/pgs.11.133

Casey BJ, Heller AS, Gee DG, Cohen AO (2017) Development of the emotional brain. Neurosci Lett. S0304-3940(17):30964–30963

Eisen JL, Sibrava NJ, Boisseau CL, Mancebo MC, Stout RL, Pinto A, Rasmussen SA (2013) Five-year course of obsessive-compulsive disorder: predictors of remission and relapse. J Clin Psychiatry 74(3):233–239

Fineberg NA, Sharma P, Sivakumaran T, Sahakian B, Chamberlain S (2007) Does obsessive compulsive personality disorder belong within the obsessive-compulsive spectrum? CNS Spectrums 12(6):467–482

Fineberg NA, Day GA, de Koenigswarter N, Reghunandanan S, Kolli S, Jefferies-Sewell K, Hranov G, Laws KR (2015) The neuropsychology of obsessive-compulsive personality disorder: a new analysis. CNS Spectr:1–10. [Epub ahead of print]

Fineberg NA, Apergis-Schoute AM, Vaghi MM, Banca P, Gillan CM, Voon V, Chamberlain SR, Cinosi E, Reid J, Shahper S, Bullmore ET, Sahakian BJ, Robbins TW (2018a) Mapping compulsivity in the DSM-5 obsessive compulsive and related disorders: cognitive domains, neural circuitry, and treatment. Int J Neuropsychopharmacol 21(1):42–58. https://doi.org/10.1093/ijnp/pyx088

Fineberg NA, Baldwin DS, Drummond LM, Wyatt S, Hanson J, Gopi S, Kaur S, Reid J, Marwah V, Sachdev RA, Pampaloni I, Shahper S, Varlakova Y, Mpavaenda D, Manson C, O'Leary C, Irvine K, Monji-Patel D, Shodunke A, Dyer T, Dymond A, Barton G, Wellsted D (2018b) Optimal treatment for obsessive compulsive disorder: a randomized controlled feasibility study of the clinical-effectiveness and cost-effectiveness of cognitive-behavioural therapy, selective serotonin reuptake inhibitors and their combination in the management of obsessive compulsive disorder. Int Clin Psychopharmacol 33(6):334–348

Fineberg NA, Demetrovics Z, Stein DJ, Loannidis K, Poetnza MN, Grunblatt E, Brand M, Billeieux J, Carmi L, King DL, Grant JE, Hollander E, Goudriaan A, Pallanti S (2018c) Manifesto for a European research network into problematic usage of the internet. Eur Neuropharmacol 28:1232–1246

Fineberg NA, Dell'Osso B, Albert A, Maina G, Geller D, Carmi L, Sireau N, Walitza S, Grassi G, Pallanti S, Hollander E, Brakoulias V, Menchon JM, Marazziti D, Ioannidis K, Apergis-Schoute A, Stein DJ, Cath DC, Veltman DJ, Van Ameringen M, Fontenelle LF, Shavitt RG, Costa D, Diniz JB, Zohar J (2019) Early intervention for obsessive compulsive disorder: an expert consensus statement. Eur Neuropsychopharmacol 29(4):549–565. https://doi.org/10.1016/j.euroneuro.2019.02.002. Epub 2019 Feb 14. PMID:30773387

Fineberg NA, Hollander E, Pallanti S, Walitza S, Grünblatt E, Dell'Osso BM, Albert U, Geller DA, Brakoulias V, Janardhan Reddy YC, Arumugham SS, Shavitt RG, Drummond L, Grancini B, De Carlo V, Cinosi E, Chamberlain SR, Ioannidis K, Rodriguez CI, Garg K, Castle D, Van Ameringen M, Stein DJ, Carmi L, Zohar J, Menchon JM (2020) Clinical advances in obsessive-compulsive disorder: a position statement by the international college of obsessive-compulsive spectrum disorders. Int Clin Psychopharmacol 35(4):173–193. https://doi.org/10.1097/YIC.0000000000000314

Gillan CM, Papmeyer M, Morein-Zamir S, Sahakian BJ, Fineberg NA, Robbins TW, de Wit S (2011) Disruption in the balance between goal-directed behavior and habit learning in obsessive-compulsive disorder. Am J Psychiatry 168(7):718–726

Gillan CM, Morein-Zamir S, Urcelay GP, Sule A, Voon V, Apergis-Schoute AM, Fineberg NA, Sahakian BJ, Robbins TW (2014) Enhanced avoidance habits in obsessive-compulsive disorder. Biol Psychiatry 75(8):631–638. https://doi.org/10.1016/j.biopsych.2013.02.002. Epub 2013 Mar 16

Gillan CM, Apergis-Schoute AM, Morein-Zamir S, Urcelay GP, Sule A, Fineberg NA, Sahakian BJ, Robbins TW (2015) Functional neuroimaging of avoidance habits in obsessive-compulsive disorder. Am J Psychiatry 172(3):284–293

Gillan CM, Robbins TW, Sahakian BJ, Van den Heuvel OA, Van Wingen G (2016) The role of habit in compulsivity. Eur Neuropsychopharmacol 26(5):828–840

Gillan CM, Fineberg NA, Robbins TW (2017) A trans-diagnostic perspective on obsessive-compulsive disorder. Psychol Med 47(9):1528–1548. https://doi.org/10.1017/S0033291716002786. Epub 2017 Mar 27. PMID: 28343453; PMCID: PMC5964477

Hirschtritt ME, Bloch MH, Mathews CA (2017) Obsessive-compulsive disorder: advances in diagnosis and treatment. JAMA 317(13):1358–1367

Hollander E (2019a) Why my OCD went undiagnosed for 25 years. https://elemental.medium.com/c81b26168a12

Hollander E (2019b) Overcoming obsessions and shifting focus 5 ways- break the OCD stigma. https://www.psychologytoday.com/ca/blog/201904

Hollander E, Stein D, Fineberg NA, Legault M (2010) Quality of life outcomes in patients with obsessive-compulsive disorder: relationship to treatment response and symptom relapse. J Clin Psychiatry 71(6):784–792

Hollander E, Nezgovorova V, Citron K (2018) Consumer patient centered study digital treatment. https://adaa.org/learn-from-us/from-the-experts/blogposts/

Ioannidis K, Chamberlain SR, Treder MS, Kiraly F, Leppink EW, Redden SA, Stein DJ, Lochner C, Grant JE (2016) Problematic internet use (PIU): associations with the impulsive-compulsive spectrum. An application of machine learning in psychiatry. J Psychaitr Res 83:94–102. https://doi.org/10.1016/j.jpsychires.2016.08.010. Epub 2016 Aug 15

Joel D (2006) Current animal models of obsessive compulsive disorder: a critical review. Prog Neuropsychopharmacol Biol Psychiatry 30(3):374–388. https://doi.org/10.1016/j.pnpbp.2005.11.006

King DL, Chamberlain SR, Carragher N et al (2020) Screening and assessment tools for gaming disorder: a comprehensive systematic review. Clin Psychol Rev. https://doi.org/10.1016/j.cpr.2020.101831

Kristina A. Uban, Megan K. Horton, Joanna Jacobus, Charles Heyser, Wesley K. Thompson, Susan F. Tapert, Pamela A.F. Madden, Elizabeth R. Sowell. Adolescent Brain Cognitive Development Study. Biospecimens and the ABCD study: Rationale, methods of collection, measurement and early data. Dev Cogn Neurosci. 2018 Aug;32:97–106. https://doi.org/10.1016/j.dcn.2018.03.005. Epub 2018 Mar 16.

Lewis M, Canamero L (2016) Hedonic quality or reward? A study of basic pleasure in homeostasis and decision making of a motivated autonomous robot. Adaptive Behav 24:267–291. https://doi.org/10.1177/1059712316666331

Lewis M, Fineberg NA, Canamero L (2019) A robot model of OC spectrum disorders: design framework, implementation and first experiments. Comput Psychiatry. Accepted for publication, April 2019

Mancebo MC, Grant JE, Pinto A, Eisen JL, Rasmussen SA (2009) Substance use disorders in an obsessive compulsive disorder clinical sample. J Anxiety Disord 23(4):429–435

Marinova Z, Chuang DM, Fineberg N (2017) Glutamate-modulating drugs as a potential therapeutic strategy in obsessive-compulsive disorder. Curr Neuropharmacol. 15(7):977–995. https://doi.org/10.2174/1570159X15666170320104237

Marzano L, Bardill A, Fields B, Herd K, Veale D, Grey N, Moran P (2015) The application of mHealth to mental health: opportunities and challenges. Lancet Psychiatry 2:942–948

Nezgovorova V, Rigby N, Battles J, Krause D, Fineberg NA, Van Ameringen M, Hollander E (2018) Genetic variants of OCD phenotypes and comorbid conditions, ACNP Poster, 2018. The American College of Neuropsychopharmacology, Hollywood, Florida.

Paulus MP, Squeglia LM, Bagot K, Jacobus J, Kuplicki R, Breslin FJ, Bodurka J, Morris AS, Thompson WK, Bartsch H, Tapert SF (2019) Neuroimage. Author manuscript; available in PMC 2019 Apr 29Screen Media Activity and Brain Structure in Youth: Evidence for Diverse

Structural Correlation Networks from the ABCD Study. Published in final edited form as: Neuroimage 185:140–153. Published online 2018 Oct 16. https://doi.org/10.1016/j.neuroimage.2018.10.040

Pitman RK (1987) Acybernetic model of obsessive-compulsive psychopathology. Comprehensive Psychiatry 28(4):334–343. https://doi.org/10.1016/0010-440X(87)90070-8

Roncero M, Belloch A, Doron G (2019) Can brief, daily training using a mobile app help change maladaptive beliefs? Crossover randomized controlled trial. JMIR Mhealth Uhealth 7(2): e11443

Skapinakis P, Caldwell DM, Hollingworth W, Bryden P, Fineberg NA, Salkovskis P, Welton N, Baxter H, Kessler D, Churchill R (2016) Pharmacological and psychotherapeutic interventions for management of obsessive-compulsive disorder in adults: a systematic review and network meta-analysis. Lancet Psychiatry 3:730–739

Skoog G, Skoog I (1999) A 40 year follow up of patients with obsessive compulsive disorder. Arch Gen Psychiatry 56:121–127

Starcevic V, Berle D, Arnáez S, Vismara M, Fineberg NA (2020) The assessment of cyberchondria: instruments for assessing problematic online health-related research. Curr Addict Rep. Published online 30th April 2020. https://doi.org/10.1007/s40429-020-00308-w

Stein DJ, Hermesh H, Eilam D, Segalas C, Zohar J, Menchon J, Nesse RM (2016) Human compulsivity: a perspective from evolutionary medicine. Eur Neuropsychopharmacol. 26 (5):869–876. https://doi.org/10.1016/j.euroneuro.2015.12.004. Epub 2015 Dec 10

Vismara M, Caricasole V, Starcevic V, Cinosi E, Dell'Osso B, Martinotti G, Fineberg NA (2020) Is cyberchondria a new transdiagnostic digital compulsive syndrome? A systematic review of the evidence. Comprehensive Psychiatry 99:152167. https://doi.org/10.1016/j.comppsych.2020.152167. [Epub ahead of print]

Voon V, Derbyshire K, Rück C, Irvine MA, Worbe Y, Enander J, Schreiber LR, Gillan C, Fineberg NA, Sahakian BJ, Robbins TW, Harrison NA, Wood J, Daw ND, Dayan P, Grant JE, Bullmore ET (2014) Disorders of compulsivity: a common bias towards learning habits. Mol Psychiatry. https://doi.org/10.1038/mp.2014.44. [Epub ahead of print]

Whiteside S, Ale C, Vickers K, Tiede M, Dammann J (2013) Case examples of enhancing pediatric OCD treatment with a smartphone application. Clin Case Stud 13:80–94

World Health Organization (2016) Monitoring and evaluating digital health interventions: a practical guide to conducting research and assessment. WHO

Zai G, Brandl EJ, Müller DJ, Richter MA, Kennedy JL (2014) Pharmacogenetics of antidepressant treatment in obsessive-compulsive disorder: an update and implications for clinicians. Pharmacogenomics 15(8):1147–1157

The manufacturer's authorised representative in the EU is Springer Nature Customer Service Centre GmbH, Europaplatz 3, 69115 Heidelberg, Germany. If you have any concerns regarding our products, please contact ProductSafety@springernature.com

Printed and bound by CPI Group (UK) Ltd, Croydon, CR0 4YY
30/09/2026
02225687-0011